T0135177

Lecture Notes in Computer Science   13610

More information about this subseries at https://link.springer.com/bookseries/7151

James F. Peters · Andrzej Skowron ·
Rabi Nanda Bhaumik · Sheela Ramanna (Eds.)

Transactions on
Rough Sets XXIII

 Springer

Editors-in-Chief
James F. Peters ⓘ
University of Manitoba
Winnipeg, MB, Canada

Andrzej Skowron ⓘ
Systems Research Institute, Polish Academy
of Sciences
Warsaw, Poland

Guest Editors
Rabi Nanda Bhaumik
Tripura University
Agartala, India

Sheela Ramanna ⓘ
University of Winnipeg
Winnipeg, MB, Canada

ISSN 0302-9743   ISSN 1611-3349 (electronic)
Lecture Notes in Computer Science
ISSN 1861-2059   ISSN 1861-2067 (electronic)
Transactions on Rough Sets
ISBN 978-3-662-66543-5   ISBN 978-3-662-66544-2 (eBook)
https://doi.org/10.1007/978-3-662-66544-2

This Springer imprint is published by the registered company Springer-Verlag GmbH, DE,
part of Springer Nature
The registered company address is: Heidelberger Platz 3, 14197 Berlin, Germany

Preface

Volume XXIII of Transactions on Rough Sets (TRS) commemorates the 95th birthday of Zdzisław Pawlak, and part of this volume creates a special issue dedicated to the International e-conference on Rough Sets organized by Rabi Nanda Bhaumik, President of the Fuzzy and Rough Sets Association, and held during November 10–13, 2021, to celebrate the seminal work of Zdzisław Pawlak[1].

The Fuzzy and Rough Sets Association (FRSA), with an International Advisory Board, was inaugurated in January 2009 at the Department of Mathematics, Tripura University, India, and held its first International Conference on Rough Sets, Fuzzy Sets and Soft Computing during November 5–11, 2009. Several international conferences and workshops were organized under the aegis of FRSA in the subsequent years of 2011, 2012, 2013, and 2015 to encourage young researchers in the field of fuzzy and rough sets through deliberations on recent developments by well-known scientists. FRSA meetings are typically held in November to observe the birthday of Zdzisław Pawlak.

The 13 papers presented at the FRSA 2021 e-conference were submitted to this issue. After an extensive peer review process, with at least two reviewers per paper including one reviewer from TRS Editorial Board, five papers were accepted (38% acceptance rate).

The paper co-authored by Mohua Banerjee and Mihir K. Chakraborty traces their journey with rough sets since their first meeting with Pawlak. Purbita Jana investigates fuzzy α-cut, fuzzy strict α-cut, and their significance in probabilistic rough set theory and studies the properties of Gödel-like arrow. The paper co-authored by Seeratpal Jaura and Sheela Ramanna presents an application of a semi-supervised learning algorithm (TPL) based on tolerance rough sets to the task of biomedical named entity recognition (BioNER) in respect of scientific articles related to COVID-19. A. Mani discusses graded rough sets in the context of rational approximation from a granular perspective, and examines algebraic semantic aspects of granular graded rough sets. The paper coauthored by Surapati Pramanik, Suman Das, Rakhal Das, and Binod Chandra Tripathy presents two multi-attribute decision-making (MADM) strategies based on a rough-bipolar neutrosophic arithmetic mean operator and a rough-bipolar neutrosophic geometric mean operator. The paper coauthored by Binod Chandra Tripathy, Suman Das, and Rakhal Das introduces the notion of single-valued neutrosophic rough continuous mapping, single-valued neutrosophic rough compactness via single-valued neutrosophic rough topological spaces.

---

[1] See, *e.g.*, Pawlak, Z. Rough sets. International Journal of Computer and Information Sciences 11, 341–356 (1982). https://doi.org/10.1007/BF01001956. See also, *e.g.*, Pawlak, Z., A Treatise on Rough Sets, Transactions on Rough Sets IV, (2006), 1–17. See, also, Pawlak, Z., Skowron, A.: Rudiments of rough sets, Information Sciences 177 (2007) 3–27; Pawlak, Z., Skowron, A.: Rough sets: Some extensions, Information Sciences 177 (2007) 28–40; Pawlak, Z., Skowron, A.: Rough sets and Boolean reasoning, Information Sciences 177 (2007) 41–73.

Zbigniew Suraj and P. Grochowalski present an overview of a new version of the Rough Set Database System (RSDS) for creating bibliographies on rough sets and related fields, as well as sharing and analysis. The new version of the RSDS includes a number of modifications, extensions, and functional improvements. The last two papers of the volume are thoroughly reviewed PhD theses dedicated to the rough set approach. The paper by Marek Grzegorowski describes interactive feature extraction from raw sensor readings, and proposes several innovative approaches to automating feature creation and selection processes including the rough set-based feature reduction method. The paper by Anuj Kumar More presents a study of algebraic structures and logics based on categories of rough sets.

The editors of the special issue (which is a part of this volume) wish to express their gratitude to the Editors-in-Chief for agreeing to publish this special issue and for their guidance during the its preparation.

All editors would like to express gratitude to the authors of all submitted papers. Special thanks are due to the following reviewers: K. Anita, Piotr Artiemjew, Jaydip Bhattacharya, Mihir Chakraborty, Purbita Jana, Anjan Mukherjee, A. Mani, Surapati Pramanik, Dominik Slezak, Tanmay Som, Binod Chandra Tripathy, Marcin Wolski, and Wei-Zhi Wu.

All editors and authors of this volume extend their gratitude to the LNCS team at Springer for their support in making this volume of TRS possible.

August 2022

Rabi Nanda Bhaumik
Sheela Ramanna
James F. Peters
Andrzej Skowron

LNCS Transactions on Rough Sets

The Transactions on Rough Sets series has as its principal aim the fostering of professional exchanges between scientists and practitioners who are interested in the foundations and applications of rough sets. Topics include foundations and applications of rough sets as well as foundations and applications of hybrid methods combining rough sets with other approaches important for the development of intelligent systems. The series includes high-quality research articles accepted for publication on the basis of thorough peer reviews. Dissertations and monographs up to 250 pages that include new research results can also be considered as regular papers. Extended and revised versions of selected papers from conferences can also be included in regular or special issues of TRS.

Editors-in-Chief

James F. Peters and Andrzej Skowron

Managing Editor

Sheela Ramanna

Technical Editor

Marcin Szczuka

Editorial Board

Contents

FRSA 2021 Conference Papers

Zdzisław Pawlak and Our Journey with Rough Sets

Mohua Banerjee[1]([⊠])([iD]) and Mihir K. Chakraborty[2]

[1] Department of Mathematics and Statistics, IIT Kanpur, Kanpur, India
mohua@iitk.ac.in
[2] School of Cognitive Science, Jadavpur University, Kolkata, India

Abstract. The authors trace their journey with rough sets since their first interactions with Z. Pawlak. The article is a narration of how work on rough sets was initiated in India, and how it continues to thrive in research groups connected to the authors and others in the country.

Keyword: Rough sets

1 Introduction of Rough Sets to India

In late 1989, Banerjee got the junior research fellowship of the Council of Scientific and Industrial Research (CSIR), Government of India, to do a doctoral dissertation. She had just completed her M.Phil. dissertation at the University of Calcutta under Chakraborty's supervision, and decided to approach the latter to be her Ph.D. advisor as well. It so happened that Chakraborty had met Pawlak in the beginning of 1989 and was excited about rough set theory. They had spent one wonderful evening in Pawlak's living room having animated discussions. At the end of the meeting, Chakraborty promised to work in this field if he got a good young researcher to work with him. Fortunately, Banerjee was willing and it was thus that work in India on rough set theory was initiated, with the Ph.D. thesis of Banerjee.

Enthused by the fact that work on rough set theory had begun in India, Pawlak invited Banerjee to his institute in Warsaw, Poland, for a research visit in 1992. Banerjee and Chakraborty started with a proposal on categories of rough sets – the first of its kind. It was presented in the first conference on rough sets held at Kiekrz near Poznań, Poland, in 1992 (Fig. 1). Pawlak was not so keen on category theory himself, and as Banerjee recalls: "the day before we were travelling to Poznań, Prof. Pawlak and I were having lunch at the University cafeteria. In the middle of a discussion, he suddenly asked – are you sure you want to present work on category theory in the conference?!" The conference was graced by several important researchers working on different aspects of rough set theory. Amongst them was Cecylia Rauszer, whose untimely demise two years later left the community shocked. Close interactions with her and particularly,

J. F. Peters et al. (Eds.): TRS XXIII, LNCS 13610, pp. 3–11, 2022.
https://doi.org/10.1007/978-3-662-66544-2_1

Fig. 1. The first conference on rough sets held at Kiekrz near Poznań, Poland, 1992.

discussions at her beautiful apartment overlooking the river Vistula, are etched in Banerjee's mind.

After these initial interactions, there was no looking back. Groups in Poland started taking great interest in Chakraborty and Banerjee's work. Boosted by tremendous support from Pawlak and Andrzej Skowron, the authors were regularly invited for visits to Poland. Apart from Pawlak, Skowron and Rauszer, among the important researchers Banerjee in particular had the privilege to interact with in Warsaw, were M. Krynicki, A. Obtułowicz, E. Orłowska, L. Szcerba, A. Wasilewska and A. Wiweger. It was an enriching experience for Banerjee, when she was invited to give talks at institutes in Warsaw, Gdańsk and Kraków. Along with Orłowska, Rauszer was instrumental in inspiring the authors to work on logics for rough sets. Chakraborty subsequently collaborated with Orłowska as well. E-mail or access to internet was not so common in those days in India, and access to books and journals was not easy either. These visits to and extensive interactions in Poland proved to be immensely beneficial for the work of the authors. The first few papers by them were published in the Bulletin of the Polish Academy of Sciences. Several new notions of algebras and logics related to rough sets were proposed. A breakthrough proposal was in the form of a new algebraic structure called the *Topological quasi-Boolean algebra* (tqBa). For work on this structure, Banerjee was awarded the Young Scientist Medal of the Indian National Science Academy (INSA) in 1995. Moreover, the notion and work was commended by a pioneer in the area of non-classical logic, Helena Rasiowa. The authors owe an immense lot to the Polish researchers on rough sets, in particular to Pawlak and Skowron. A photograph of the two stalwarts in Fig. 2:

Fig. 2. Pawlak and Skowron.

2 Propagation of Rough Sets in India

Rough set theory, in the meantime, had gained much ground in different areas of applications. After the completion of Banerjee's Ph.D., Chakraborty scouted whether applications of rough sets could be studied in India as well. One of the important research groups of the country working on fuzzy set theory, was centred at the Indian Statistical Institute (ISI), Kolkata, led by S.K. Pal. Banerjee joined the group and started a collaboration with S. Mitra and S.K. Pal. In fact, it was through Banerjee that rough sets were introduced at ISI, Kolkata. Some significant and highly cited work on integration of rough sets with fuzzy sets, neural networks and genetic algorithms came about as a result of this collaboration. While at ISI, Banerjee also gained enormously from interactions with another stalwart in applications of fuzzy sets, D. Dutta Majumdar.

Collaboration between the research groups of Kolkata and Warsaw entered a new phase with the approval of an Indo-Polish project (1996–1998) under the title "Reasoning under uncertainty about complex objects". The Indian team consisted of Chakraborty (PI), T.K. Ghosal, S.K. Pal and Banerjee. On the Polish side were Skowron (PI), Orłowska, Pawlak and L. Polkowski. This project resulted in several exchange visits, seminars, meetings and publication of books and papers.

In 1997, Banerjee joined the Department of Mathematics and Statistics, Indian Institute of Technology (IIT) Kanpur, as a faculty member. She had continued her collaboration on rough sets with Chakraborty during her fellowship at ISI, and the work continued after she joined IIT Kanpur as well. But new developments on rough sets took place at IIT. Work on tolerance intervals and rough sets was one such area, where Banerjee collaborated with an important name in Artificial Intelligence, A. Mukerjee. She worked on applications with P. Mitra. Moreover, she started supervising doctoral, post-doctoral and Masters students, all on various facets of rough sets. Rough set theory was introduced as part of a course in the curriculum of the department.

The year 2010 saw the formation of the Indian Society of Rough Sets (ISRS). The chief architect of ISRS was C.R. Rao of the University of Hyderabad. S.K. Pal was the founder President, C.R. Rao the Secretary, and Banerjee the joint Secretary. Chakraborty was in the advisory board. The society has been religiously observing the birthday of Pawlak every year, through seminars and meetings. At present, the concept of rough sets is not novel anymore in India. Many people have joined in the work on and propagation of rough set theory. Tripura University has established a centre for fuzzy and rough sets under the leadership of R.N. Bhaumik, and events are organized regularly. So, Pawlak's ideas have established firm roots in the country. The International Rough Set Society (IRSS) has given full support to all activities on rough sets in India, with particular interest taken by Dominik Ślezak. It is worth mentioning in this context that Chakraborty and Banerjee have been made Fellows of the IRSS, and from among their collaborators, the awardees of the IRSS are S.K. Pal (Fellow), Md. Aquil Khan (Senior Member), A. Mani (Senior Member) and P. Mitra (Senior Member). Both the authors are on the editorial board of Transactions on Rough Sets, the special Springer journal on the subject.

3 Facets of Work on Rough Sets by the Authors

One of the prime intentions of the authors has been to show the mainstream community in Mathematical Logic and Algebra, that some beautiful mathematics can emerge from rough set theory. There was a conscious choice of journals (such as Studia Logica, Journal of Philosophical Logic, International Journal of Approximate Reasoning) in communicating the work done. Rough sets were introduced as a focus area at conferences on logic wherever the authors were involved – e.g. at the Indian Conference of Logic and its Applications (ICLA) and Indian School of Logic and its Applications (ISLA), the two flagship events of the Association for Logic in India (ALI). Further, research visits of the authors and their students also helped propagate the subject and establish new collaborations. For instance, Banerjee designed and taught a course on rough set theory while on a research visit at the Institute of Logic and Cognition (ILC), China.

As contributions on rough sets by the authors, one of the significant developments was in showing that the well-known "discussive" logic of Jaśkowski is just a logic with rough truth semantics. With M. Bunder, different forms of rough modus ponens were explored. Work with Md. Aquil Khan defined new formal dynamic logic frameworks that accommodate updates in information systems, temporal logic frameworks that capture information systems evolving with time, and logics for approximation spaces from multiple sources. With the group at ILC, China, Banerjee formulated an application of rough sets to the problem of open world models. In the meantime, collaboration also took place between D. Dubois and Y. Yao, during Banerjee's visits to their institutes (Université Paul Sabatier, France and University of Regina, Canada respectively). Arun Kumar and Banerjee investigated generalised rough sets in covering spaces, and formulated new algebras and corresponding logics. Anuj K. More and Banerjee followed up on the pioneering work on categories of rough sets done by the authors. Again new algebras and logics emerged. With P. Howlader, Banerjee started investigating the connections of formal concept analysis and rough set theory, with the work still continuing. With G. Panicker, Banerjee established relationships of McCarthy's algebras of conditional logic and rough set algebras. It is gratifying to note that work on rough sets has now percolated into the next generation of researchers. V.S. Patel just completed her Ph.D. under the supervision of Md. Aquil Khan. Her dissertation includes investigations of connections between epistemic logic and approximation operators in rough set theory.

Meanwhile, researchers had completed their doctoral dissertations under the guidance of Chakraborty as well. Jayanta Sen worked partially on tqBa, Pulak Samanta on covering based rough sets, Anirban Saha on algebraic structures in the vicinity of tqBa and pre-rough algebra. A. Mani formally registered under Chakraborty, but worked independently on algebraic and logical aspects of rough sets. Rough set theory had also become highly popular in China. It may be remarked in the context that, it is through Chakraborty that Minghui Ma (Sun Yat Sen University, Guangzhou) and Lin Zhe (South West University, Chongqing) got interested in the algebra and logic of rough sets and have written some joint papers with him.

4   Treasured Memories

Pawlak visited Kolkata in February 2002 to attend AFSS (an international conference on Fuzzy Sets and Systems). In this conference, many luminaries including Bezdek, Sugeno and Dubois were present. Pawlak's talk entitled "The Rough Set View of Bayes' Theorem" was greatly admired. Both the authors possess nice memories of those few days of togetherness with Pawlak in their own city Kolkata. Here are two photographs (Figs. 3 and 4), one of a chat session during that conference and another of an evening at Banerjee's home with her parents (who are both no more).

Fig. 3. Pawlak at AFSS Kolkata, 2002. L to R in front: Bezdek, Dubois, Bhaumik, Pawlak, Chakraborty.

Fig. 4. Pawlak at Banerjee's residence in Kolkata with her parents and Chakraborty, 2002.

Chakraborty recalls the wonderful time he spent with Pawlak walking by the river Seine in Paris in 2005, when both had gone to attend two different conferences. Pawlak took several photographs of an embarrassed Chakraborty by the river, but unfortunately none is available. For Banerjee, there are just too many memories with Pawlak to be reminisced here. Pawlak affectionately called her his "granddaughter". The numerous long walks, concerts of Western classical music, visits to different beautiful places in and around Warsaw, dinners at his home with the charming Mrs. Pawlak, picnics in the forest outside Warsaw accompanied by Skowron, a trip to the stunning Tatra mountains at Zakopane during a conference, the time in Banff and a visit to the Niagara falls when in Canada to attend the 2nd conference on rough sets in 1993 – all are now treasured memories. Pawlak's erudition, down-to-earth nature devoid of any airs of the stalwart that he was, great sense of humour, kind and supportive personality – all were a source of tremendous inspiration and learning experience for Banerjee.

Chakraborty attended the memorial meeting of Pawlak held in Warsaw on his first death anniversary in 2007. Zadeh was also present in the meeting. Chakraborty, with Skowron, paid an emotional visit to Pawlak's grave at the time. Both of them also went to Cecylia Rauszer's remains and observed a little silence. An emotional visit to the graves of the two beautiful people was also a ritual with Banerjee and Skowron, whenever Banerjee visited Warsaw subsequently.

Namita Chaudhuri, the Bengali poet accompanying Chakraborty to the memorial meeting in Warsaw, composed a poem for the occasion. She had never met Pawlak, but had known about him and his personality from others. This poem was read (in Bengali) by her in the meeting and was rendered a liberal prose translation by Chakraborty.

We would like to mention the condolence message from us sent to Skowron after the sudden demise of Pawlak (Fig. 5). It was translated into Polish and read in the memorial meeting.

Pawlak was an artist. He loved to paint landscapes with rough strokes of the brush. On his visit to Kolkata, Pawlak gifted one such painting to Banerjee (Fig. 6) – it hangs on the wall of Banerjee's residence in Kanpur as a precious

Fig. 5. Condolence message by authors.

possession. We consider Pawlak's work on rough sets as an everlasting bouquet of beautiful landscapes too.

Fig. 6. An original painting by Pawlak gifted by him to Banerjee in 2002.

A list of some publications on rough sets by us and our group is given in the References [1–32].

References

1. Banerjee, M., Chakraborty, M.K.: A category for rough sets. Found. Comput. Decis. Sci. **18**(3–4), 167–180 (1993)
2. Chakraborty, M.K., Banerjee, M.: Rough logic with rough quantifiers. Bull. Pol. Acad. Sci. (Math.) **41**, 305–315 (1993)
3. Banerjee, M.: A categorial approach to the algebra and logic of the indiscernible. Ph.D. dissertation, University of Calcutta, India (1995)
4. Banerjee, M., Chakraborty, M.K.: Rough sets through algebraic logic. Fund. Inform. **28**(3–4), 211–221 (1996)

5. Banerjee, M., Pal, S.K.: Roughness of a fuzzy set. Inf. Sci. (Inform. Comput. Sci.) **93**(3–4), 235–246 (1996)
6. Banerjee, M.: Rough sets and 3-valued Łukasiewicz logic. Fund. Inform. **32**(1), 213–220 (1997)
7. Chakraborty, M.K., Orłowska, E.: Substitutivity principles in some theories of uncertainty. Fund. Inform. **32**, 107–120 (1997)
8. Sen, J., Chakraborty, M.K.: A study of interconnections between rough, 3-valued and linear logics. Fund. Inform. **51**, 311–324 (2002)
9. Banerjee, M., Chakraborty, M.K.: Algebras from rough sets. In: Pal, S.K., Polkowski, L., Skowron, A. (eds.) Rough-Neuro Computing: Techniques for Computing with Words. COGTECH, pp. 157–184. Springer, Heidelberg (2004). https://doi.org/10.1007/978-3-642-18859-6_7
10. Banerjee, M.: Logic for rough truth. Fund. Inform. **71**(2–3), 139–151 (2006)
11. Chakraborty, M.K.: Pawlak's landscaping with rough sets. In: Peters, J.F., Skowron, A., Düntsch, I., Grzymała-Busse, J., Orłowska, E., Polkowski, L. (eds.) Transactions on Rough Sets VI. LNCS, vol. 4374, pp. 51–63. Springer, Heidelberg (2007). https://doi.org/10.1007/978-3-540-71200-8_4
12. Banerjee, M., Yao, Y.: A categorial basis for granular computing. In: An, A., Stefanowski, J., Ramanna, S., Butz, C.J., Pedrycz, W., Wang, G. (eds.) RSFDGrC 2007. LNCS (LNAI), vol. 4482, pp. 427–434. Springer, Heidelberg (2007). https://doi.org/10.1007/978-3-540-72530-5_51
13. Banerjee, M., Mitra, S., Banka, H.: Evolutionary-rough feature selection in gene expression data. IEEE Trans. Syst. Man Cybern. Part C Appl. Rev. **37**(4), 622–631 (2007)
14. Bunder, M.W., Banerjee, M., Chakraborty, M.K.: Some rough consequence logics and their interrelations. In: Peters, J.F., Skowron, A. (eds.) Transactions on Rough Sets VIII. LNCS, vol. 5084, pp. 1–20. Springer, Heidelberg (2008). https://doi.org/10.1007/978-3-540-85064-9_1
15. Pagliani, P., Chakraborty, M.K.: A Geometry of Approximation. Rough Set Theory: Logic, Algebra and Topology of Conceptual Patterns. Trends in Logic, Springer, Dordrecht (2008). https://doi.org/10.1007/978-1-4020-8622-9
16. Khan, M.A., Banerjee, M.: A logic for multiple-source approximation systems with distributed knowledge base. J. Philos. Log. **40**(5), 663–692 (2011). https://doi.org/10.1007/s10992-010-9163-1
17. Samanta, P., Chakraborty, M.K.: Generalized rough sets and implication lattices. In: Peters, J.F., et al. (eds.) Transactions on Rough Sets XIV. LNCS, vol. 6600, pp. 183–201. Springer, Heidelberg (2011). https://doi.org/10.1007/978-3-642-21563-6_10
18. Chakraborty, M.K., Banerjee, M.: Rough sets: some foundational issues. Fund. Inform. **127**(1–4), 1–15 (2013)
19. Khan, M.A., Banerjee, M.: Algebras for information systems. In: Skowron, A., Suraj, Z. (eds.) Rough Sets and Intelligent Systems - Professor Zdzisław Pawlak in Memoriam. ISRL, vol. 42, pp. 381–407. Springer, Heidelberg (2013). https://doi.org/10.1007/978-3-642-30344-9_14
20. Chakraborty, M.K.: Membership function based rough set. Int. J. Approximate Reasoning **55**(1), 402–411 (2014)
21. Banerjee, M., Dubois, D.: A simple logic for reasoning about incomplete knowledge. Int. J. Approximate Reasoning **55**(2), 639–653 (2014)
22. Khan, M.A., Banerjee, M.: Logics for some dynamic spaces - parts I and II. J. Log. Comput. **25**(3), 827–878 (2015)

23. Banerjee, M., Ju, S., Khan, M.A., Tang, L.: Open world models: a view from rough set theory. In: Chakraborty, M.K., Skowron, A., Maiti, M., Kar, S. (eds.) Facets of Uncertainties and Applications. SPMS, vol. 125, pp. 77–86. Springer, New Delhi (2015). https://doi.org/10.1007/978-81-322-2301-6_6

24. Chakraborty, M.K.: On some issues in the foundation of rough sets: the problem of definition. Fund. Inform. 148(1–2), 123–132 (2016)

25. Saha, A., Sen, J., Chakraborty, M.K.: Algebraic structures in the vicinity of pre-rough algebra and their logics II. Inf. Sci. 333, 44–60 (2016)

26. Ma, M., Chakraborty, M.K.: Covering-based rough sets and modal logics, part I. Int. J. Approximate Reasoning 77, 55–65 (2016)

27. More, A.K., Banerjee, M.: Categories and algebras of rough sets: new facets. Fund. Inform. 148(1–2), 173–190 (2016)

28. Kumar, A., Banerjee, M.: Kleene algebras and logic: Boolean and rough set representations, 3-valued, rough set and perp semantics. Stud. Logica. 105(3), 439–469 (2017). https://doi.org/10.1007/s11225-016-9696-6

29. More, A.K., Banerjee, M.: Transformation semigroups for rough sets. In: Nguyen, H.S., Ha, Q.-T., Li, T., Przybyła-Kasperek, M. (eds.) IJCRS 2018. LNCS (LNAI), vol. 11103, pp. 584–598. Springer, Cham (2018). https://doi.org/10.1007/978-3-319-99368-3_46

30. Panicker, G., Banerjee, M.: Rough sets and the algebra of conditional logic. In: Mihálydeák, T., et al. (eds.) IJCRS 2019. LNCS (LNAI), vol. 11499, pp. 28–39. Springer, Cham (2019). https://doi.org/10.1007/978-3-030-22815-6_3

31. Lin, Z., Chakraborty, M.K., Ma, M.: Residuated algebraic structures in the vicinity of pre-rough algebra and decidability. Fund. Inform. 179(3), 239–274 (2021)

32. Patel, V.S., Khan, M.A., Chakraborty, M.K.: Modal systems for covering semantics and boundary operator. Int. J. Approximate Reasoning 135, 110–126 (2021)

Fuzzy α-Cut in Rough Sets and Its Application

Purbita Jana[✉][iD]

Indian Institute of Technology, Kanpur (IITK), Kanpur, India
purbita_presi@yahoo.co.in

Abstract. This work indicates application of the notion of fuzzy α-cut in rough set theory and studies the properties of Gödel-like arrow in details.

Keywords: L-fuzzy set · α-cut · Fuzzy α-cut · Fuzzy strict α-cut · Gödel arrow · Gödel-like arrow · Frame · Semilinear · Rough sets

1 Introduction

This work is based on the work done in [3]. Properties of Gödel-like arrow are studied in detail in this article. Especially the notion of semilinearity with respect to Gödel-like arrow and its connection with the notion of so-called prelinearity is one of the main focus points. The application of fuzzy α-cut in the area of rough set theory is also taken into account. To make this paper self contained the necessary concepts will be recalled and mentioned. The notion of fuzzy α-cut was introduced by us in [4].

Definition 1 (Fuzzy α-cut of a fuzzy set) [4]. *Let $(X, \tilde{A})$ be an L-fuzzy set, where L is a complete lattice. Then for $\alpha \in L$, the **fuzzy α-cut** of $(X, \tilde{A})$ is the fuzzy subset $(X, \tilde{A}_\alpha)$ where $\tilde{A}_\alpha$ is defined by:*

$$\tilde{A}_\alpha(x) = \begin{cases} \tilde{A}(x) & \text{if } \tilde{A}(x) \geq \alpha \\ 0_L & \text{otherwise.} \end{cases}$$

We will denote fuzzy α-cut of an L-fuzzy set $(X, \tilde{A})$ simply by $\tilde{A}_\alpha$ if the base set X is understood.

It is to be noted that the fuzzy α-cut of a fuzzy set while considering $L = [0, 1]$ is also known as level fuzzy set [7]. In the paper [7] the author defined the algebraic operations viz. Intersection, union, complementation of level sets as is done in fuzzy set theory by min, max and $1 - (\cdot)$, in the value set $[0, 1]$ and established certain elementary properties. The set of level sets is closed with respect to these operations because of the specific characteristics of the value set $[0, 1]$. Since instead of $[0, 1]$ we are taking a complete lattice, defining algebraic operations in the case of fuzzy α-cut is not that straight forward. Throughout this article L will stand for complete lattice unless it will be specified.

© Springer-Verlag GmbH Germany, part of Springer Nature 2022
J. F. Peters et al. (Eds.): TRS XXIII, LNCS 13610, pp. 12–22, 2022.
https://doi.org/10.1007/978-3-662-66544-2_2

Definition 2 (Fuzzy strict α-cut of a fuzzy set). *Let $(X, \tilde{A})$ be an L-fuzzy set. Then for $\alpha \in L$, the* **fuzzy strict α-cut** *of $(X, \tilde{A})$ is the fuzzy subset $(X, \tilde{A}_{\alpha+})$ where $\tilde{A}_{\alpha+}$ is defined by:*

$$\tilde{A}_{\alpha+}(x) = \begin{cases} \tilde{A}(x) & \text{if } \tilde{A}(x) > \alpha \\ 0_L & \text{otherwise.} \end{cases}$$

We will denote fuzzy strict α-cut of an L-fuzzy set $(X, \tilde{A})$ simply by $\tilde{A}_{\alpha+}$ if the base set X is understood.

Definition 3 (L-implication). *Let L be a complete lattice. Then a binary operation $\rightarrow: L \times L \rightarrow L$ is said to be an L-implication if for all $l, l_1, l_2 \in L$, it satisfies the following conditions: (i) $l_1 \leq l_2$ implies $(l_1 \rightarrow l) \geq (l_2 \rightarrow l)$; (ii) $l_1 \leq l_2$ implies $(l \rightarrow l_1) \leq (l \rightarrow l_2)$; (iii) $1_L \rightarrow l = l$; (iv) $l_1 \rightarrow (l_2 \rightarrow l) = l_2 \rightarrow (l_1 \rightarrow l)$; (v) $l_1 \leq l_2 = 1_L$ iff $l_1 \leq l_2$.*

Definition 4 (Gödel-like arrow) [3]. *Let L be any complete lattice. Then the Gödel-like arrow is a binary operation in L defined as follows:*

$$l_1 \rightarrow l_2 = \begin{cases} 1_L & \text{if } l_1 \leq l_2 \\ l_2 & \text{otherwise.} \end{cases}$$

for all $l_1, l_2 \in L$ where 1_L is the top element of L.

There is another kind of implication in L called residuated implication defined as below.

Definition 5 (Residuated implication) [2]. *Let L be any complete lattice. Then the residuated implication is defined by $a \rightarrow b = sup\{c \in L \mid c \wedge a \leq b\}$ for all $a, b \in L$.*

Relationship between these two types of implications is discussed in Sect. 3.

Definition 6 (Prelinearity) [2]. *A complete lattice L together with an L-implication, $\rightarrow$, is prelinear if $(l_1 \rightarrow l_2) \vee (l_2 \rightarrow l_1) = 1_L$, for all $l_1, l_2 \in L$.*

Let $(X, \tilde{A})$, $(X, \tilde{B})$ be two L-fuzzy sets. Then $\tilde{A} \subseteq \tilde{B}$ if and only if $\tilde{A}(x) \leq \tilde{B}(x)$, for any $x \in X$. One can easily check that the following properties hold: (i) $\tilde{A}_{\alpha+} \subseteq \tilde{A}_{\alpha} \subseteq \chi_{\tilde{A}_{\alpha}}$; (ii) $\alpha_1 \leq \alpha_2$ implies $\tilde{A}_{\alpha_1} \supseteq \tilde{A}_{\alpha_2}$; (iii) $(\tilde{A} \cap \tilde{B})_{\alpha} = \tilde{A}_{\alpha} \cap \tilde{B}_{\alpha}$; (iv) $(\tilde{A} \cup \tilde{B})_{\alpha} = \tilde{A}_{\alpha} \cup \tilde{B}_{\alpha}$; (v) $\alpha_1 \leq \alpha_2$ implies $\tilde{A}_{\alpha_1+} \supseteq \tilde{A}_{\alpha_2+}$; (vi) $(\tilde{A} \cap \tilde{B})_{\alpha+} = \tilde{A}_{\alpha+} \cap \tilde{B}_{\alpha+}$; (vii) $(\tilde{A} \cup \tilde{B})_{\alpha+} = \tilde{A}_{\alpha+} \cup \tilde{B}_{\alpha+}$.

Proposition 1 [3]. *Let L be linear and $\tilde{A}$ be a fixed L-fuzzy set. Then (i) $\tilde{A}_{\alpha} \cap \tilde{A}_{\beta} = \tilde{A}_{\alpha \vee \beta}$; (ii) $\tilde{A}_{\alpha} \cup \tilde{A}_{\beta} = \tilde{A}_{\alpha \wedge \beta}$; (iii) $\tilde{A}_{\alpha+} \cap \tilde{A}_{\beta+} = \tilde{A}_{(\alpha \vee \beta)+}$; (iv) $\tilde{A}_{\alpha+} \cup \tilde{A}_{\beta+} = \tilde{A}_{(\alpha \wedge \beta)+}$.*

Theorem 1 [3]. *Let L be linear and $\tilde{A}$ be a fixed L-fuzzy set. Then $(\{\tilde{A}_{\alpha} \mid \alpha \in L\}, \cap, \cup)$ and $(\{\tilde{A}_{\alpha+} \mid \alpha \in L\}, \cap, \cup)$ are lattices.*

If we consider non-linear L then the result will not be that straightforward that was already done in [3] and we will observe explicitly in the sequel.

Let us discuss the properties of the frame together with Gödel-like arrow. It is well known that frame (c.f. Definition 7) together with residuated implication is complete Heyting algebra.

2 Prelinear and Semilinear Frame

In this section we will consider prelinear and semilinear frames in some detail along with examples.

Definition 7 (Frame). *A **frame** is a complete lattice such that,*

$$x \wedge \bigvee Y = \bigvee \{x \wedge y : y \in Y\}.$$

i.e., the binary meet distributes over arbitrary join.

Definition 8 (Prelinear Frame) [1]. *A prelinear frame $L = (L, \wedge, \bigvee, \to)$ is a frame $(L, \wedge, \bigvee)$ together with an L-implication, $\to$, such that for all $l_1, l_2 \in L$,*

$$(l_1 \to l_2) \vee (l_2 \to l_1) = 1_L,$$

where 1_L is the top element of L.

For our purpose we will take $\to$ as Gödel-like arrow. In this section henceforth all the arrows are Gödel-like arrows.

Definition 9 (Semilinear Frame). *A semilinear frame $L = (L, \wedge, \bigvee, \to)$ is a frame $(L, \wedge, \bigvee)$ together with a binary operation $\to$ such that for all distinct $l_1, l_2, l_3 \in L$, $(l_1 \to l_2) \wedge (l_1 \to l_3) = (l_1 \to l_2 \wedge l_3)$.*

It can be verified by considering all possible cases that any frame with up to 4-elements is always preilinear. We shall prove in Sect. 3 that with respect to Gödel-like arrow preliniearity implies semilinierity. We give below an example of a 5-element lattice which is the smallest semilinear but not prelinear frame.

Example 1. The following frame is not prelinear but semilinear.

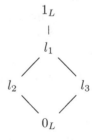

For this frame $(l_2 \to l_3) \lor (l_3 \to l_2) = l_3 \lor l_2 = l_1 \neq 1_L$. Hence it is not prelinear.

The following is an example of a frame with six elements which is not semilinear. It is to be noted that the following frame is the smallest non semilinear frame. In other words a non-semilinear distributive lattice contains at least six elements.

Example 2. The following frame is not semilinear.

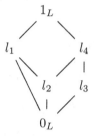

For this frame $(l_2 \to l_1) \land (l_2 \to l_3) = 1_L \land l_3 = l_3$, whereas $l_2 \to (l_1 \land l_3) = l_2 \to 0_L = 0_L$. So, Corollary 1 fails.

Example 3. The following frame is the smallest Boolean algebra which is not semilinear.

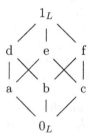

For this frame $(a \to c) \land (a \to d) = c \land 1_L = c$, whereas $a \to (c \land d) = a \to 0_L = 0_L$. So, Corollary 1 fails.

One can see that the concepts of prelinearity and semilinearity are based on the underlying lattice of the frame and the arrow. While prelinearity is an well known concept, semilinearity is not so and which is a more general concept [c.f. Property 5].

Let us enlist below some properties of Gödel-like arrow.

3   Properties of Gödel-Like Arrow

In this section some properties of Gödel-like arrow are listed along with verification of some of them.

Property 1. $l \to l = 1_L$, for any $l \in L$.

Property 2. $(l_1 \to l_2) \wedge (l_2 \to l_3) \leq (l_1 \to l_3)$, for any $l_1, l_2, l_3 \in L$.

Proof. It may be observed that the value of $l_1 \to l_2$ is either 1_L or l_2. Similarly for $l_2 \to l_3$ the values are either 1_L or l_3 and for $l_1 \to l_3$ values are either 1_L or l_3. Now the following cases may arise:

Case 1: $l_1 \to l_2 = 1_L$ and $l_2 \to l_3 = 1_L$.
Here $l_1 \leq l_2$ and $l_2 \leq l_3$ and hence as L is transitive, $l_1 \leq l_3$. Consequently $l_1 \to l_3 = 1_L$. Therefore $(l_1 \to l_2) \wedge (l_2 \to l_3) = 1_L \wedge 1_L = 1_L = (l_1 \to l_3)$.

Case 2: $l_1 \to l_2 = 1_L$ and $l_2 \to l_3 = l_3$.
We have $(l_1 \to l_2) \wedge (l_2 \to l_3) = 1_L \wedge l_3 = l_3 \leq l_3 \; (or \; 1_L) = (l_1 \to l_3)$.

Case 3: $l_1 \to l_2 = l_2$ and $l_2 \to l_3 = 1_L$.
As $l_2 \leq l_3$, $(l_1 \to l_2) \wedge (l_2 \to l_3) = l_2 \wedge 1_L = l_2 \leq l_3 \; (or \; 1_L) = (l_1 \to l_3)$.

Case 4: $l_1 \to l_2 = l_2$ and $l_2 \to l_3 = l_3$.
In this case $(l_1 \to l_2) \wedge (l_2 \to l_3) = l_2 \wedge l_3 = l_3 \leq l_3 \; (or \; 1_L) = (l_1 \to l_3)$.

Property 3. $l_1 \leq l_2$ implies $(l_1 \to l) \geq (l_2 \to l)$, for any $l_1, l_2, l \in L$.

Property 4. $l_1 \leq l_2$ implies $(l \to l_1) \leq (l \to l_2)$, for any $l_1, l_2, l \in L$.

Property 5. If L is prelinear frame with atleast three elements then it is semilinear.

Proof. Let us consider any three distinct elements l_1, l_2 and l_3 of L. Since L is prelinear, $(l_1 \to l_2) \vee (l_2 \to l_1) = 1_L$, for any $l_1, l_2 \in L$. If possible let $(l_1 \to l_2) \wedge (l_1 \to l_3) \neq l_1 \to (l_2 \wedge l_3)$. Then two cases may arise. Case 1: $l_1 < l_2$, (l_1, l_3) and (l_2, l_3) are incomparable [where (l_1, l_2) represents the pair of points from L]. Case 2: $l_1 < l_3$, (l_1, l_2) and (l_2, l_3) are incomparable.

Case 1: In this case notice that $l_1 \vee l_3 = 1_L$ as $(l_3 \to l_1) \vee (l_1 \to l_3) = 1_L$ and l_1, l_3 are incomparable. Hence $l_2 \wedge (l_3 \vee l_1) = l_2 \wedge 1_L = l_2$. Now $(l_2 \wedge l_3) \vee (l_2 \wedge l_1) = (l_2 \wedge l_3) \vee l_1$. The following three cases may arise under this situation. Either $l_2 \wedge l_3 \leq l_1$ or $l_2 \wedge l_3 > l_1$ or the pair $(l_2 \wedge l_3, l_1)$ is incomparable. If $l_2 \wedge l_3 \leq l_1$, then $(l_2 \wedge l_3) \vee l_1 = l_1 \neq l_2$. When $l_2 \wedge l_3 > l_1$ then $(l_2 \wedge l_3) \vee l_1 = l_2 \wedge l_3 \neq l_2$ as if $l_2 \wedge l_3 = l_2$ then $l_2 \leq l_3$, a contradiction. As L is prelinear $(l_1 \to (l_2 \wedge l_3)) \vee ((l_2 \wedge l_3) \to l_1) = 1_L$. When the pair $(l_2 \wedge l_3, l_1)$ is incomparable then $(l_1 \to (l_2 \wedge l_3)) \vee ((l_2 \wedge l_3) \to l_1) = (l_2 \wedge l_3) \vee l_1 = 1_L \neq l_2$ (as l_2 and l_3 are incomparable).

Hence for either cases $l_2 \wedge (l_3 \vee l_1) = l_2 \neq (l_2 \wedge l_3) \vee (l_2 \wedge l_1)$, but L is distributive.

Similarly for Case 2 also we get a contradiction.

Corollary 1. *If L is totally ordered frame then* $(l_1 \to l_2) \wedge (l_1 \to l_3) = l_1 \to (l_2 \wedge l_3)$, *for any* $l_1, l_2, l_3 \in L$.

Property 6. $\inf_i\{(l_i \to l)\} = \sup_i\{l_i\} \to l$, for any $l_i, l \in L$.

Proof. $sup_i\{l_i\} \rightarrow l = \begin{cases} 1_L & \text{if } sup_i\{l_i\} \leq l \\ l & \text{otherwise.} \end{cases}$

Now for $sup_i\{l_i\} \leq l$ we have $l_i \leq sup_i\{l_i\} \leq l$.

Hence for this case $(l_i \rightarrow l) = 1_L$, for each i and consequently $inf_i\{l_i \rightarrow l\} = 1_L$. If $sup_i\{l_i\} > l$ then there exist atleast one l_i such that $l_i > l$ and rest will be either bellow l or equal to l. Now for the case $l_i > l$, $l_i \rightarrow l = l$ and for all other cases $l_i \rightarrow l = 1_L$. As $l \leq 1_L$, $inf_i\{l_i \rightarrow l\} = l$. If $sup\{l_i\}$ and l are incomparable then atleast one of the l_i's, say l_j is incomparable to l and consequently $l_j \rightarrow l = l$. Hence $inf_i\{l_i \rightarrow l\} = l$.

Property 7. $l_1 \leq l_2$ iff $l_1 \rightarrow l_2 = 1_L$.

Property 8. $l_1 \wedge (l_1 \rightarrow l_2) \leq l_2$.

Proposition 2 [3]. *Let L be a complete lattice which is prelinear (with respect to Gödel-like arrow) and distributive. Then L contains at most two incomparable elements and their join is 1_L.*

Consideration of Gödel-like arrow and its properties enables us to get Proposition 2. That is, prelinearity and distributivity give us either linear lattice or a special lattice structure which may be called kite-like lattice (as the Hasse diagram of the lattice looks like a kite with a tail). Kite-like lattice is almost linear except the upper most part containing only two incomparable elements with top as their join. This small change in the value set destroys the results of Proposition 1 and we will get Theorem 2. It is to be noted that to get such structures Gödel-like arrow plays a crucial role. It is not observed with respect to known arrows in fuzzy set theory. We have already studied the known arrows listed in [5]. None of them are able to give such structures. The Gödel-like arrow proposed by us gives such a nice study. Although at this stage of study we are unable to conclude that it is the only arrow to enjoy such properties to produce such structure.

Comparison Between Gödel-Like Arrow and Residuated Implication

Now we will deal with two kinds of fuzzy implications (arrows) defined in Definition 4 and Definition 5. First of all it may be noted that Gödel-like arrow and residuated implication are the generalisations of Gödel arrow while generalising the value set from $[0, 1]$ to a frame L. Gödel like arrow does not enjoy adjointness condition (i.e., $c \leq a \rightarrow b$ iff $c \wedge a \leq b$, for any $a, b, c \in L$) whereas residuated implication satisfies adjointness condition. If $\rightarrow$ is a Gödel-like arrow then it can be verified that if $c \leq a \rightarrow b$ then $c \wedge a \leq b$ holds for any $a, b, c \in L$ but the converse is not true in general. For example if we consider Example 3 then we have $f \wedge a = 0_L \leq c$ but $f \nleq a \rightarrow c = c$. Hence the frame with respect to residuated implication forms a complete Heyting algebra but with respect to Gödel-like arrow is not a complete Heyting algebra in general. Frame together with Gödel-like arrow is neither semilinear nor prelinear in general whereas, frame together with residuated implication is always semilinear but not prelinear in general. All the properties discussed in Sect. 3 are satisfied by residuated implication. It may be noted that considering residuated implication all through this article will give all the results in a very obvious way. Considering Gödel-like arrow in this article not only produces the results but gave some very interesting properties of frame (c.f. Sect. 2 and Sect. 3) which was not present in the literature.

Theorem 2 [3]. $(\{\tilde{A}_\alpha\}_{\alpha \in L} \cup \{\tilde{A}_{\alpha+}\}_{\alpha \in L}, \subseteq)$ *is a lattice, where L is prelinear (with respect to Gödel-like arrow) and distributive lattice.*

Let us consider the following example to demonstrate Theorem 2.

Example 4. Let us consider the following distributive lattice which is prelinear with respect to Gödel-like arrow:

Now consider the fuzzy set

$$(X, \tilde{A}) = \{(x_1, 0_L), (x_2, l_1), (x_3, l_2), (x_4, l_3), (x_5, l_4), (x_6, 1_L), (x_7, l_4), (x_8, l_2)\},$$

where $X = \{x_1, x_2, x_3, x_4, x_5, x_6, x_7, x_8\}$ and $\tilde{A} : X \to L$. we have denoted $\tilde{A}(x) = l$ by (x, l). We have

$$\tilde{A}_{l_1} = \{(x_1, 0_L), (x_2, l_1), (x_3, 0_L), (x_4, 0_L), (x_5, 0_L), (x_6, 1_L), (x_7, 0_L), (x_8, 0_L)\}$$

$$\tilde{A}_{l_2} = \{(x_1, 0_L), (x_2, 0_L), (x_3, l_2), (x_4, 0_L), (x_5, 0_L), (x_6, 1_L), (x_7, 0_L), (x_8, l_2)\}$$

$$\tilde{A}_{l_3} = \{(x_1, 0_L), (x_2, l_1), (x_3, l_2), (x_4, l_3), (x_5, 0_L), (x_6, 1_L), (x_7, 0_L), (x_8, l_2)\}$$

$$\tilde{A}_{l_4} = \{(x_1, 0_L), (x_2, l_1), (x_3, l_2), (x_4, l_3), (x_5, l_4), (x_6, 1_L), (x_7, l_4), (x_8, l_2)\}$$

$$\tilde{A}_{l_1+} = \{(x_1, 0_L), (x_2, 0_L), (x_3, 0_L), (x_4, 0_L), (x_5, 0_L), (x_6, 1_L), (x_7, 0_L), (x_8, 0_L)\}$$

$$\tilde{A}_{l_2+} = \{(x_1, 0_L), (x_2, 0_L), (x_3, 0_L), (x_4, 0_L), (x_5, 0_L), (x_6, 1_L), (x_7, 0_L), (x_8, 0_L)\}$$

$$\tilde{A}_{l_3+} = \{(x_1, 0_L), (x_2, l_1), (x_3, l_2), (x_4, 0_L), (x_5, 0_L), (x_6, 1_L), (x_7, 0_L), (x_8, l_2)\}$$

$$\tilde{A}_{l_4+} = \{(x_1, 0_L), (x_2, l_1), (x_3, l_2), (x_4, l_3), (x_5, 0_L), (x_6, 1_L), (x_7, 0_L), (x_8, l_2)\}$$

Then,

$- \tilde{A}_{l_1} \cap \tilde{A}_{l_2} = \tilde{A}_{1_L} = \tilde{A}_{l_1 \vee l_2}$ and $\tilde{A}_{l_1} \cup \tilde{A}_{l_2} = \tilde{A}_{l_3+} = \tilde{A}_{(l_1 \wedge l_2)+}.$

- $\tilde{A}_{l_1} \cap \tilde{A}_{l_2} = \tilde{A}_{1_L} = \tilde{A}_{l_1 \vee l_2}$
- $\tilde{A}_{l_3} \cap \tilde{A}_{l_4} = \tilde{A}_{l_3} = \tilde{A}_{l_3 \vee l_4}$ and $\tilde{A}_{l_3} \cup \tilde{A}_{l_4} = \tilde{A}_{l_4} = \tilde{A}_{l_3 \wedge l_4}$.
- $\tilde{A}_{l_3} \cap \tilde{A}_{l_4} = \tilde{A}_{l_3} = \tilde{A}_{l_3 \vee l_4}$
- $\tilde{A}_{l_1+} \cap \tilde{A}_{l_2+} = \tilde{A}_{1_L} = \tilde{A}_{l_1 \vee l_2}$ and $\tilde{A}_{l_1+} \cup \tilde{A}_{l_2+} = \tilde{A}_{1_L} = \tilde{A}_{l_1 \vee l_2}$.
- $\tilde{A}_{l_3+} \cap \tilde{A}_{l_4+} = \tilde{A}_{l_3+} = \tilde{A}_{(l_3 \vee l_4)+}$ and $\tilde{A}_{l_3+} \cup \tilde{A}_{l_4+} = \tilde{A}_{l_4+} = \tilde{A}_{(l_3 \wedge l_4)+}$.
- $\tilde{A}_{l_4+} \cap \tilde{A}_{l_1} = \tilde{A}_{l_1}$, $\tilde{A}_{l_1+} \cap \tilde{A}_{l_2} = \tilde{A}_{l_1+}$ and $\tilde{A}_{l_1+} \cap \tilde{A}_{l_4} = \tilde{A}_{l_1+}$.
- $\tilde{A}_{l_4+} \cup \tilde{A}_{l_1} = \tilde{A}_{l_4+}$, $\tilde{A}_{l_1+} \cup \tilde{A}_{l_2} = \tilde{A}_{l_2}$ and $\tilde{A}_{l_1+} \cup \tilde{A}_{l_4} = \tilde{A}_{l_4}$.

The following is an example in which the lattice L is distributive and prelinear with respect to the residuated arrow but fails to give a lattice structure for above mentioned families. If instead of Gödel-like arrow residuated arrow is taken, Theorem 2 fails also. Thus, Gödel-like arrow becomes significant in this context.

Example 5. Let us consider the following lattice L:

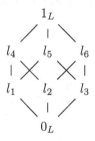

Now consider the fuzzy set

$$(X, \tilde{A}) = \{(x_1, 0_L), (x_2, l_1), (x_3, l_2), (x_4, l_3), (x_5, l_4), (x_6, 1_L), (x_7, l_5), (x_8, l_6)\},$$

where $X = \{x_1, x_2, x_3, x_4, x_5, x_6, x_7, x_8\}$ and $\tilde{A} : X \to L$. we have denoted $\tilde{A}(x) = l$ by (x, l). We have

$$\tilde{A}_{l_3} = \{(x_1, 0_L), (x_2, 0_L), (x_3, 0_L), (x_4, l_3), (x_5, 0_L), (x_6, 1_L), (x_7, l_5), (x_8, l_6)\}$$

$$\tilde{A}_{l_4} = \{(x_1, 0_L), (x_2, 0_L), (x_3, 0_L), (x_4, 0_L), (x_5, l_4), (x_6, 1_L), (x_7, 0_L), (x_8, 0_L)\}$$

$$\tilde{A}_{l_3} \cup \tilde{A}_{l_4} = \{(x_1, 0_L), (x_2, 0_L), (x_3, 0_L), (x_4, l_3), (x_5, l_4), (x_6, 1_L), (x_7, l_5), (x_8, l_6)\}$$

Notice that $\tilde{A}_{l_3} \cup \tilde{A}_{l_4}$ is neither a fuzzy α cut nor a fuzzy strict α-cut of the fuzzy set $\tilde{A}$.

Similarly it is possible to observe from this example that operation on fuzzy strict α-cut is also not closed.

Example 6. Let us consider the following lattice L:

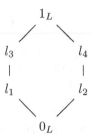

The lattice is complete, prelinear with respect to Gödel-like arrow but not distributive. Now consider the fuzzy set

$$(X, \tilde{A}) = \{(x_1, 0_L), (x_2, l_1), (x_3, l_2), (x_4, l_3), (x_5, l_4), (x_6, 1_L), (x_7, l_4), (x_8, l_1)\},$$

where $X = \{x_1, x_2, x_3, x_4, x_5, x_6, x_7, x_8\}$ and $\tilde{A} : X \to L$. we have denoted $\tilde{A}(x) = l$ by (x, l). We have

$$\tilde{A}_{l_1+} = \tilde{A}_{l_3} = \{(x_1, 0_L), (x_2, 0_L), (x_3, 0_L), (x_4, l_3), (x_5, 0_L), (x_6, 1_L), (x_7, 0_L), (x_8, 0_L)\}$$

$$\tilde{A}_{l_2+} = \tilde{A}_{l_4} = \{(x_1, 0_L), (x_2, 0_L), (x_3, 0_L), (x_4, 0_L), (x_5, l_4), (x_6, 1_L), (x_7, l_4), (x_8, 0_L)\}$$

$$\begin{aligned}
\tilde{A}_{l_1+} \cup \tilde{A}_{l_2+} &= \tilde{A}_{l_3} \cup \tilde{A}_{l_4} \\
&= \{(x_1, 0_L), (x_2, 0_L), (x_3, 0_L), (x_4, l_3), (x_5, l_4), (x_6, 1_L), (x_7, l_4), (x_8, 0_L)\}
\end{aligned}$$

Notice that $\tilde{A}_{l_1+} \cup \tilde{A}_{l_2+} = \tilde{A}_{l_3} \cup \tilde{A}_{l_4}$ is neither a fuzzy α cut nor a fuzzy strict α-cut of the fuzzy set $\tilde{A}$.

Notice that the examples that we dealt with need to be complete lattice and distributive as well. As we confined ourselves with finite cases for the examples, distributive lattice will be enough to consider but in case of general settings frame needs to be considered. One of the prominent usages of frame was reflected in [3] while considering the application of fuzzy α-cut in topological structure. As it is a study in continuation of [3] we would like to stick with the frame here as well.

Now we shall show some usage of the notion of fuzzy α-cut in the area of probabilistic rough set theory to some extent. For the application of this concept in the areas of algebra and topology are mentioned in detail in [3].

4   Probabilistic Rough Set Theory

In [3] we had mentioned how the notion of fuzzy α-cut applicable in the context of rough set theory [6,8] to define lower and upper approximations as fuzzy sets. Now we will list some of the properties of the proposed lower and upper approximations. Let us recall the necessary notions to describe the lower and upper approximations as a fuzzy set with the help of the notion of fuzzy α-cut. It is well known that an approximation space is a tuple (X, R), consisting of a set of objects X and an equivalence relation R, known as indiscernibility relation on X. For any $A \subseteq X$, the lower and upper approximations of A in the approximation space (X, R) are denoted by $\underline{A}$ and $\overline{A}$ respectively and defined as follows.

$$\underline{A} = \bigcup \{[x] \mid [x] \subseteq A\}; \quad \overline{A} = \bigcup \{[x] \mid A \cap [x] \neq \emptyset\}.$$

A rough membership function of A, denoted by μ_A, is a function from X to $[0, 1]$ such that $\mu_A(x) = \frac{|[x] \cap A|}{|[x]|} \leq 1$, where $\mid S \mid$ stands for the cardinality of the set S and $[x]$ stands for the equivalence class of $x \in X$. In this definition X is taken to be a finite set.

In [8], while defining lower and upper approximations of a set A, α-cuts and strict β-cuts are used with $0 \leq \beta < \alpha \leq 1$ in the following way.

$$\underline{A_\alpha} = \{x \in X \mid \mu_A(x) \geq \alpha\}; \quad \overline{A_\beta} = \{x \in X \mid \mu_A(x) > \beta\}.$$

These are crisp sets. Hence the grade disappears in the final approximations.

But in [3], the lower and upper approximations became fuzzy sets and defined as follows.

$$\underline{A_\alpha} : X \longrightarrow [0, 1] \text{ s.t. } \underline{A_\alpha}(x) = \begin{cases} \mu_A(x) & \text{if } \mu_A(x) \geq \alpha \\ 0 & \text{otherwise.} \end{cases} = (\mu_A)_\alpha$$

$$\overline{A_\beta} : X \longrightarrow [0, 1] \text{ s.t. } \overline{A_\beta}(x) = \begin{cases} \mu_A(x) & \text{if } \mu_A(x) \geq \beta \\ 0 & \text{otherwise.} \end{cases} = (\mu_A)_\beta$$

Now we are enlisting some of the properties of these approximations which can be shown using the properties of fuzzy α-cut: for $\alpha, \alpha' \in (0, 1]$ and $\beta, \beta' \in [0, 1)$,

(P1) $\underline{A_\alpha} \subseteq (\underline{A_\alpha^c})^c$;

(P2) $\overline{A_\beta} \subseteq (\overline{A_\beta^c})^c$;

(P3) $\underline{A_\alpha} \cap \underline{B_\alpha} = \underline{(A \cap B)_\alpha}$;

(P4) $\underline{A_\alpha} \cup \underline{B_\alpha} = \underline{(A \cup B)_\alpha}$;

(P5) $\overline{A_\beta} \cap \overline{B_\beta} = \overline{(A \cap B)_\beta}$;

(P6) $\overline{A_\beta} \cup \overline{B_\beta} = \overline{(A \cup B)_\beta}$;

(P7) If $A \subseteq B$ then $\underline{A_\alpha} \subseteq \underline{B_\alpha}$;

(P8) If $A \subseteq B$ then $\overline{A_\beta} \subseteq \overline{B_\beta}$;

(P9) If $\alpha \geq \alpha'$ then $\underline{A_\alpha} \subseteq \underline{A_{\alpha'}}$;

(P10) If $\beta \geq \beta'$ then $\overline{A_\beta} \subseteq \overline{A_{\beta'}}$.

Notice that, here instead of two different types of cuts viz. α-cuts and strict β-cuts one type of cut has been used uniformly in determining lower and upper approximations. It is quite expected that one may apply this approximation in the decision-theoretic rough set model following the work in [8].

5   Concluding Remarks

In this article we have dealt with the notion of fuzzy α-cut, fuzzy strict α-cut and their significance in probabilistic rough set theory. However further investigation in these aspects is required. Study of the family of fuzzy α-cuts and fuzzy strict α-cuts is done in detail. In this work we generalise the notion of frame one step further and call it 'semilinear frame'. It is to emphasise that the notion semilinearity introduced in this work is more general than 'prelinearity'; while the latter notion has been widely discussed in literature, the former notion is not. A detailed study of Gödel-like arrow provides a nice result about the relation between prelinearity and semilinearity property. The algebraic notion of a semilinear frame needs to be studied in more detail. Taking a general fuzzy arrow instead of Gödel-like arrow may also be considered as an interesting future project.

Acknowledgement. The research work of the author was supported by the Department of Science & Technology, Government of India under Women Scientist Scheme (reference no. SR/WOS-A/PM-52/2018). The author is thankful to Prof. Mihir K. Chakraborty for several discussions during the preparation of this article. I am grateful to the anonymous reviewer whose comments/suggestions helped to improve and clarify the final version of this manuscript.

References

1. Bělohlávek, R.: Fuzzy Relational Systems: Foundations and Principles. Kluwer Academic Publishers, New York (2002)
2. Hájek, P.: Metamathematics of Fuzzy Logic. Kluwer Academic Publishers, New York (1998)
3. Jana, P., Chakraborty, M.K.: Fuzzy α-cut and related mathematical structures. Soft. Comput. **25**(1), 207–213 (2021). https://doi.org/10.1007/s00500-020-05131-z
4. Jana, P., Chakraborty, M.K.: Categorical relationships of fuzzy topological systems with fuzzy topological spaces and underlying algebras-II. Ann. Fuzzy Math. Inform. **10**(1), 123–137 (2015)
5. Klir, G., Yuan, B.: Fuzzy Sets and Fuzzy Logic: Theory and Applications. Prentice Hall PTR, Upper Saddle River (1995)
6. Pawlak, Z.: Rough sets. Int. J. Comput. Inf. Sci. **11**, 341–356 (1982). https://doi.org/10.1007/BF01001956
7. Radecki, T.: Level fuzzy sets. J. Cybern. **7**, 189–198 (1977)
8. Yao, Y.: Probabilistic rough set approximations. Int. J. Approximate Reasoning **49**, 255–271 (2007)

Named Entity Recognition on CORD-19 Bio-Medical Dataset with Tolerance Rough Sets

Seeratpal Jaura[1] and Sheela Ramanna[2]([✉]) [iD]

[1] Department of Computing Science, University of Alberta, Edmonton, AB, Canada
[2] University of Winnipeg, Winnipeg, MB R3B 2E9, Canada
s.ramanna@uwinnipeg.ca

Abstract. Biomedical named entity recognition is becoming increasingly important to biomedical research due to a proliferation of articles and also due to the current pandemic disease. This paper addresses the task of automatically finding and recognizing biomedical entity types related to COVID (e.g., virus, cell, therapeutic) with tolerance rough sets. The task includes i) extracting nouns and their co-occurring contextual patterns from a large BioNER dataset related to COVID-19 and, ii) annotating unlabelled data with a semi-supervised learning algorithm using co-occurence statistics. 465,250 noun phrases and 6,222,196 contextual patterns were extracted from 29,500 articles using natural language text processing methods. Three categories were successfully classified at this time: virus, cell and therapeutic. Early *precision@N* results demonstrate that our proposed tolerant pattern learner (TPL) is able to constrain concept drift in all 3 categories during the iterative learning process.

Keywords: Named entity recognition · Semi-supervised learning · Text mining · Tolerance rough sets · Unstructured text categorization

1 Introduction

Named-entity recognition (NER), is a subtask of information extraction that seeks to discover and categorize specific entities such as nouns or relation in unstructured text [8]. Biomedical named entity recognition (BioNER) and BioRD (relation detection) are becoming increasingly important for biomedical research due to tremendous growth in the volume of publications for instance, PubMEd and MEDLINE[1] [15] databases. Most early BioNER annotation methods relied on dictionary or rule-based methods. The success of deep-learning methods for natural language processing has led to its application in BioNER [2].

For natural language processing applications, classical rough set theory [14] based on equivalence relations is considered too restrictive. In contrast, a tolerance form of rough set theory based [5,13,16] on *tolerance relations* is more

---

[1] https://www.nlm.nih.gov/bsd/pmresources.html.

S. Ramanna—This work is dedicated to Prof. Z. Pawlak on his 95th birthday.

J. F. Peters et al. (Eds.): TRS XXIII, LNCS 13610, pp. 23–32, 2022.
https://doi.org/10.1007/978-3-662-66544-2_3

appropriate. Mathematically, tolerance relations are reflexive and symmetric but are not necessarily transitive, so the classes induced by such relations may have overlapping members. An in-depth survey of application of tolerance rough sets in text categorization (structured and unstructured) can be found in [17].

Document clustering using tolerance rough sets model (TRSM) was introduced by several authors [3,4,9,11,24]. In [25] document clustering using a lexicon-based document representation based on the TRSM model. A framework for retrieval of indonesian language text based on TRSM was proposed in [26]. TRSM model was applied to semantic document indexing in [10,12,23].

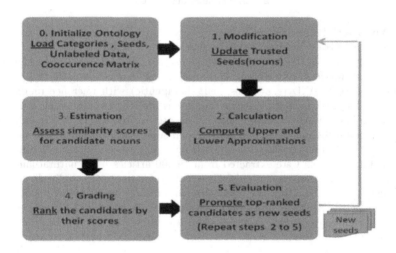

Fig. 1. TPL process flow retrieved from [7]

In this paper, we apply semi-supervised learning algorithm (TPL) based on tolerance rough sets for annotating entities (nouns) from CORD-19 dataset consisting of scientific articles related to COVID-19 [27] (overall process flow shown in Fig. 1). The tolerance rough set-based learner (TPL) was successfully applied to categorize noun and relations derived from two large web corpora [7,17,19,20]. The web corpora consisted of: ClueWeb12 [1] (a collection on 33,019,372 English web pages) and ClueWeb09[2]. More details regarding pre-processing of these corpora can be found in [6,18].

The motivation for applying TPL to a biomedical dataset, is to address the challenge of categorizing common biomedical entity types such as genes, chemicals and diseases as well several new entities related to severe acute respiratory syndrome virus (fine grained entity types) from a large repository. 465,250 noun phrases and 6,222,196 contextual patterns were extracted from 29,500 articles. Three categories were successfully classified at this time: virus, cell and therapeutic. All categories showed improvement during the iterative learning process. The category *virus* performed best with a precision of 83% on the 6th iteration.

---

[2] http://lemurproject.org/clueweb09/.

This paper is organized as follows. In Sect. 2, we recall the definitions underlying the tolerance rough set model to represent linguistic entities. In Sect. 3, we describe the data preparation process. In Sect. 4, we give the TPL algorithm and discuss our annotation results.

2   Preliminaries

In this section, we briefly recall the formal model that was introduced in [19] and based on the tolerance approximation space model described in [21]. Let $\mathcal{N} = \{n_1, n_2, .., n_M\}$ be the set of noun phrases and $\mathcal{C} = \{c_1, c_2, ..., c_P\}$ be the set of contextual patterns in our corpus. Furthermore, let $C : \mathcal{N} \to \mathbb{P}(\mathcal{C})$ denote the co-occurring contexts $\forall n_i \in \mathcal{N}$ and let $N : \mathcal{C} \to \mathbb{P}(\mathcal{N})$ denote the co-occurring nouns $\forall c_i \in \mathcal{C}$. We define a named entity tolerance space $A = (U, I, \tau, \nu, P)$ where universe $U = \mathcal{C}$ and the tolerance classes are determined by the uncertainty function I_θ using the contextual overlap function:

$$I_\theta(c_i) = \{c_j : \omega(N(c_i), N(c_j)) \geq \theta\}. \tag{1}$$

Here, θ is the tolerance threshold parameter and ω is the overlap index function which is the Sørensen-Dice index [22]:

$$\omega(A, B) = \frac{2|A \cap B|}{|A| + |B|}, \tag{2}$$

where ω is reflexive and symmetric. We further define $\nu(X, Y) = \frac{|X \cap Y|}{|X|}$ and $P(I(c_i)) = 1$. The resulting tolerance classes are then used to approximate the noun phrases n_i as the target concept:

$$U_A(n_i) = \{c_j \in C : \nu(I_\theta(c_j), C(n_i)) > 0\}, \tag{3}$$

$$L_A(n_i) = \{c_j \in C : \nu(I_\theta(c_j), C(n_i)) = 1\}. \tag{4}$$

The semi-supervised algorithm (described in Sect. 4) promotes a candidate instance n_i based on the following three sets: $C(n_i)$, $U_A(n_i)$ and $L_A(n_i)$. These descriptors are then used to calculate a micro-score for a candidate instance n_j, by the trusted instance n_i using the overlap index ω:

$$micro(n_i, n_j) = \omega(C(n_i), C(n_j))\alpha + \omega(U_A(n_i), C(n_j))\beta + \omega(L_A(n_i), C(n_j))\gamma, \tag{5}$$

where α, β, γ are weights that contribute to the trustworthiness of a candidate noun.

3   Data Preparation

The original data source is the CORD-19: The COVID-19 Open Research [27]. It is a collection of scientific papers related COVID-19 coronavirus research. Since it's release this collection has been growing; however, for this paper, we have worked with 29,500 articles. A snapshot of the contents of a typical annotated article is shown in Fig. 2. The annotation marked in orange shows entity types such as virus, cell and therapeutic categories to be recognized.

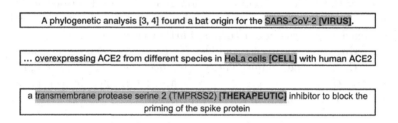

Fig. 2. Annotated corpus (Color figure online)

The overall process of preparing the files can be summarized in 3 steps:

Step 1. Preparing the files for preprocessing which starts with extracting each document as a separate pdf file.

Step 2. Extraction of 465,250 noun phrases using a pre-trained Spacy model for POS tagging and a custom tokenizer with multithreading.

Step 3. Extraction of 6,222,196 contextual patterns using regular expressions in Python to make chunking patterns as described in [7].

Figure 3 shows the steps for data preparation phase used in this paper. A multi-threaded programming approach was implemented on Unix system with 36 cores and 256 GB of RAM. The preprocessing steps include converting each document into a separate pdf document. Each pdf document consists of the title, abstract and body of the document. Then the documents were cleaned by removing punctuations, followed by tokenization and lemmatization. The first step was to select the appropriate attributes from the files for preprocessing. Each file contained the following information: id, source, doi, pmcid, pubmed_id, publish_time, journal, author, title, abstract, body and entities (shown in Fig. 4). The text preprocessing steps include tokenization, performing lemmatization and removing punctuation (shown in Fig. 5). After data cleaning, the next step is to extract the noun phrases (shown in Fig. 6). The extracted noun phrases were then compared with the nouns (entity types) that were provided in the benchmark dataset. The nouns common to both methods were retained along with the unique set of nouns extracted by our extraction method.

After extracting contextual patterns for each noun phrase, the co-occurrence frequency for the nouns and their corresponding contextual patterns were

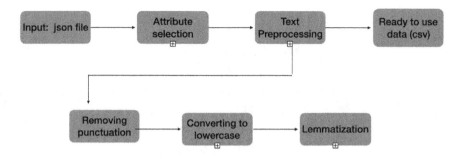

Fig. 3. Data preparation phase

id	doi	title	abstract	body
0	" 10.1007/s00134-020-05985-9"	"Angiotensin-converting enzyme 2 (ACE2) as a SARS-CoV-2 receptor: molecular mechanisms and potential therapeutic target"	" "	"'SARS-CoV-2 has been sequenced [3] . A phylogenetic analysis [3, 4] found a bat origin for the SARS-CoV-2. ... "

Fig. 4. Attribute selection step

	title	body
Before text cleaning	"Angiotensin-converting enzyme 2 (ACE2) as a SARS-CoV-2 receptor: molecular mechanisms and potential therapeutic target"	"'SARS-CoV-2 has been sequenced [3] . A phylogenetic analysis [3, 4] found a bat origin for the SARS-CoV-2. ... "
After text cleaning	"angiotensin-converting enzyme 2 ace2 as a sars-cov-2 receptor molecular mechanisms and potential therapeutic target"	"sars-cov-2 has been sequenced 3 a phylogenetic analysis 3 4 found a bat origin for the sars-cov-2 ..."

Fig. 5. Text cleaning illustration

recorded in a co-occurrence matrix (CCM) as shown in Fig. 7. All frequencies of less than two occurrences (contextual patterns and nouns) were removed. The size of the CCM was 2.76 GB.

4   Annotation Results

The TPL algorithm that was first introduced in [19] is given in Algorithm 1.

For each entity category, for the first iteration, 5–8 seed instances (trusted entities) were used. With each iteration, the model finds the candidate noun phrases for each category c_i. Using the score-based ranking described in Sect. 2, each candidate noun phrase is ranked in its own category. The total score for each candidate noun np_j in category c_i is calculated as follows:

$$TS_{c_i}(np_j) = \sum_{\forall np_i \in TN_{c_i}} score(np_i, np_j),$$

```
79724  fprs
79725  b4-wt
79726  upload
79727  stat5b
79728  t30like
79729  potika
79730  fv-3-like
79731  hypothesizing
79732  pct-h24
79733  anti-sarsr-cov
79734  hochberg
79735  vagueness
```

Fig. 6. Sample nouns

Context/Noun	coronavirus	influenza	sars-cov	vaccine	antibody	lung
respiratory syndrome-associated _	145	16	63	8	3	4
binding of_	32	39	25	21	445	12
revealed that_	59	93	28	67	166	110
virus replication in_	5	47	5	5	5	84

Fig. 7. Sample co-occurence matrix

where $TS_{c_i}(np_j)$ is the total score for the jth noun phrase np_j for the ith category c_i. After calculating the total score of each noun phrase for the ith category c_i, candidate nouns are ranked and the top candidate nouns are added to the seed set for next round.

Table 1 gives a list of initial seed nouns for each category based on the benchmark dataset.

However, some of the categories such as substrate, wildlife, material did not have sufficient examples for promotion and ended up being misclassified. Hence, we decided to make the classes more general. Therefore, our final classification categories were: virus, cell and therapeutic. The virus category broadly classifies the virus and more specifically virus related to COVID-19. The therapeutic category refers to the therapeutic procedures and substances such as chemicals and drugs that could act as a therapeutic. The cell category includes cell, gene and protein instances. Each category was initialized with 8 seed instances. The

---

Algorithm 1: Tolerant Pattern Learner for Categories [19]

Input : An ontology O defining categories and a small set of seed examples; a large corpus U

Output: Trusted instances (entities) for each category

1   **for** $r = 1 \rightarrow \infty$ **do**
2   **for** each category cat **do**
3   **for** each new trusted noun phrase n_i of cat **do**
4   Calculate the approximations $U_A(n_i)$ and $L_A(n_i)$;
5   **for** each candidate noun phrase n_j **do**
6   Calculate $micro(n_i, n_j)$;

7   **for** each candidate noun phrase n_j **do**
8      $macro_{cat}(n_j) = \sum\limits_{\forall n_i \in cat} micro(n_i, n_j)$;

9   Rank instances by $macro_{cat}/|cat|$;
10   Promote top instances as trusted;

Table 1. Seed nouns for each entity type.

Entity	Initial seed nouns
Virus	Sars, cov, mers, covid-19, sars-cov-2
Livestock	Calve, chicken, chick, pig, poultry
Evolution	Mutation, phylogenetics, evolution, recombination, substitution
Material	Air, plastic, fluid, copper, silica
Sign_or_symptom	Cough, respiratory, diarrhea, vomiting, wheezing
Substrate	Blood, urine, sputum, saliva, fecal
Therapeutic	Detection, inoculation, isolation, stimulation, vaccination
Diagnostics	Imaging, immunohistochemistry, scanning, necropsy, biopsy
Wildlife	Bat, wild-birds, animal, wild-mosquitoes, Pteropus

results over six iterations are reported using the parameters values $\alpha = 0.5$, $\beta = 0.25$, $\gamma = 0.25$. In each iteration, 1–5 new trusted nouns were promoted to the seed set. Since the dataset is unlabelled, the precision was manually calculated using *precision@N* measure, which is a ratio of number of correct examples to the N top ranked examples. Table 2 shows the preliminary results of our annotation.

As the dataset was not labelled the precision of the results was calculated manually. The results clearly demonstrate that the TPL model precision improves in all 3 categories at the sixth iteration. Overall, the virus category tends to perform better with a precision of 83% on 6th iteration.

Table 2. Precision@30 of TPL per category.

Categories	Iteration 1	Iteration 3	Iteration 6
Virus	47%	75%	83%
Therapeutic	20%	40%	76%
Cell	13%	43%	66%

5　Conclusion

In this paper, we presented results based on extracted nouns and their co-occurring contextual patterns from a large BioNER dataset related to COVID-19. This is a fundamental step in text mining to facilitate biomedical knowledge extraction. We applied a semi-supervised annotation method with tolerance rough sets to label three categories of well-known biomedical entities. Early precision results clearly demonstrate that our proposed tolerant pattern learner (TPL) model improves in all 3 categories during the iterative learning process. However, this study was limited due to sheer size of the CCM and the limited number of examples in some categories. The repository of articles has now grown to more than 50,000 scholarly works. Future work will involve extraction of more examples as well as examining of the CCM for inclusion of less frequently appearing patterns.

Acknowledgments. Seeratpal' s work was supported by University of Winnipeg 2020 and 2021 NSERC Undergraduate Research Award (USRA). Sheela Ramanna' s was supported by NSERC Discovery Grant # 194376. The authors wish to acknowledge the help of Rajesh Jaiswal for preprocessing the dataset and Christopher Henry for providing the GPU computing platform.

References

1. Callan, J.: The lemur project and its clueweb12 dataset. In: Invited Talk at the SIGIR 2012 Workshop on Open-Source Information Retrieval (2012)
2. Cho, H., Lee, H.: Biomedical named entity recognition using deep neural networks with contextual information. BMC Bioinform. **20**, 1–11 (2019). Article number: 735. https://doi.org/10.1186/s12859-019-3321-4
3. Ho, T.B., Nguyen, N.B.: Nonhierarchical document clustering based on a tolerance rough set model. Int. J. Intell. Syst. **17**, 199–212 (2002)
4. Kawasaki, S., Binh, N., Bao, T.: Hierarchical document clustering based on tolerance rough set model. In: Zighed, D.A., Komorowski, J., Żytkow, J. (eds.) PKDD 2000. LNCS (LNAI), vol. 1910, pp. 458–463. Springer, Heidelberg (2000). https://doi.org/10.1007/3-540-45372-5_51
5. Marcus, S.: Tolerance rough sets, Cech topologies, learning processes. Bull. Pol. Acad. Sci. Tech. Sci. **42**(3), 471–487 (1994)
6. Moghaddam, H.: Exploring scalability and concept drift issues in learning categorical facts with tolerance rough sets. Master's thesis, University of Winnipeg (2019). Supervisor: S. Ramanna

7. Moghaddam, H., Ramanna, S.: Harvesting patterns from textual web sources with tolerance rough sets. Patterns **1**(4), 100053 (2020)
8. Nadeau, D., Sekine, S.: A survey of named entity recognition and classification. Linguisticae Investigationes **30**(1), 3–26 (2007). http://www.ingentaconnect.com/content/jbp/li/2007/00000030/00000001/art00002
9. Ngo, C.L.: A tolerance rough set approach to clustering web search results. Master's thesis, Warsaw University (2003)
10. Nguyen, H.S.: Applications of tolerance rough set model semantic text analysis. In: Ropiak, K., Polkowski, L., Artiemjew, P. (eds.) Proceedings of the 28th International Workshop on Concurrency, Specification and Programming. CEUR Workshop Proceedings, Olsztyn, Poland, 24–26 September 2019, vol. 2571. CEUR-WS.org (2019). http://ceur-ws.org/Vol-2571/CSP2019_paper_18.pdf
11. Nguyen, H.S., Ho, T.B.: Rough document clustering and the internet. In: Handbook of Granular Computing, pp. 987–1003 (2008)
12. Nguyen, S.H., Nguyen, H.S.: An approach to semantic indexing based on tolerance rough set model. In: Nguyen, N.T., van Do, T., le Thi, H.A. (eds.) Advanced Computational Methods for Knowledge Engineering. SCI, vol. 479, pp. 343–354. Springer International Publishing, Heidelberg (2013). https://doi.org/10.1007/978-3-319-00293-4_26
13. Nieminen, J.: Rough tolerance equality and tolerance black boxes. Fund. Inform. **11**, 289–296 (1988)
14. Pawlak, Z.: Rough sets. Int. J. Comput. Inf. Sci. **11**(5), 341–356 (1982). https://doi.org/10.1007/BF01001956
15. Perera, N., Dehmer, M., Emmert-Streib, F.: Named entity recognition and relation detection for biomedical information extraction. Front. Cell Dev. Biol. **8**, 673 (2020). https://www.frontiersin.org/article/10.3389/fcell.2020.00673
16. Polkowski, L., Skowron, A., Zytkow, J.: Tolerance based rough sets. In: Lin, T.Y., Wildberger, M. (eds.) Soft Computing: Rough Sets, Fuzzy Logic, Neural Networks, Uncertainty Management, Knowledge Discovery, pp. 55–58. Simulation Councils Inc., San Diego (1994)
17. Ramanna, S., Peters, J.F., Sengoz, C.: Application of tolerance rough sets in structured and unstructured text categorization: a survey. In: Wang, G., Skowron, A., Yao, Y., Ślęzak, D., Polkowski, L. (eds.) Thriving Rough Sets. SCI, vol. 708, pp. 119–138. Springer, Cham (2017). https://doi.org/10.1007/978-3-319-54966-8_6
18. Sengoz, C.: A granular-based approach for semi-supervised web information labeling. Master's thesis, University of Winnipeg (2014). Supervisor: S. Ramanna
19. Sengoz, C., Ramanna, S.: A semi-supervised learning algorithm for web information extraction with tolerance rough sets. In: Ślęzak, D., Schaefer, G., Vuong, S.T., Kim, Y.-S. (eds.) AMT 2014. LNCS, vol. 8610, pp. 1–10. Springer, Cham (2014). https://doi.org/10.1007/978-3-319-09912-5_1
20. Sengoz, C., Ramanna, S.: Learning relational facts from the web: a tolerance rough set approach. Pattern Recogn. Lett. **67**(P2), 130–137 (2015)
21. Skowron, A., Stepaniuk, J.: Tolerance approximation spaces. Fund. Inform. **27**(2–3), 245–253 (1996)
22. Sørensen, T.: A method of establishing groups of equal amplitude in plant sociology based on similarity of species content and its application to analyses of the vegetation on Danish commons. Biologiske skrifter, I kommission hos E. Munksgaard (1948). http://books.google.co.in/books?id=rpS8GAAACAAJ
23. Swieboda, W., Krasuski, A., Nguyen, H.S., Janusz, A.: Interactive method for semantic document indexing based on explicit semantic analysis. Fund. Inform. **132**(3), 423–438 (2014). https://doi.org/10.3233/FI-2014-1052

24. Świeboda, W., Meina, M., Nguyen, H.S.: Weight learning for document tolerance rough set model. In: Lingras, P., Wolski, M., Cornelis, C., Mitra, S., Wasilewski, P. (eds.) RSKT 2013. LNCS (LNAI), vol. 8171, pp. 385–396. Springer, Heidelberg (2013). https://doi.org/10.1007/978-3-642-41299-8_37

25. Virginia, G., Nguyen, H.S.: Lexicon-based document representation. Fundamenta Informatica **124**(1–2), 27–46 (2013)

26. Virginia, G., Nguyen, H.S.: A semantic text retrieval for Indonesian using tolerance rough sets models. In: Peters, J.F., Skowron, A., Ślęzak, D., Nguyen, H.S., Bazan, J.G. (eds.) Transactions on Rough Sets XIX. LNCS, vol. 8988, pp. 138–224. Springer, Heidelberg (2015). https://doi.org/10.1007/978-3-662-47815-8_9

27. Wang, X., Song, X., Li, B., Guan, Y., Han, J.: Comprehensive named entity recognition on CORD-19 with distant or weak supervision. arXiv preprint arXiv:2003.12218 (2020)

Granularity and Rational Approximation: Rethinking Graded Rough Sets

A. Mani$^{(\boxtimes)}$ⓘ

Machine Intelligence Unit, Indian Statistical Institute, 203, B. T. Road, Kolkata
(Calcutta) 700108, India
a.mani.cms@gmail.com,amani.rough@isical.ac.in

Abstract. The concept of rational discourse is typically determined by
subjective, normative, and rule based constraints in the context under
consideration. It is typically determined by related ontologies, and coher-
ence between associated concepts employed in the discourse. Classical rough approximations, and variants of variable precision rough sets
(VPRS) including graded rough sets embody at least some aspects of
potentially useful concepts of rational approximation, but can be very
lacking in application contexts, and rough set theoretical frameworks for
cluster validation. While the literature on knowledge from general rough
perspectives is rich and diverse, not much work has been done from
the perspective of rationality in explicit terms. In this research, the gap
is addressed by the present author in variants of high granular partial
algebras. Specifically, the nature of optimal concepts of rational approximations is examined, and formalized by her in such frameworks. Graded
rough sets are generalized from a granular perspective, and the compatibility of the introduced concepts are studied over it. Further aspects of
algebraic semantics of granular graded rough sets are examined. Some
incorrect results in graded rough sets in the literature are also corrected.

Keywords: General rough sets · Rational approximation · Graded
rough sets · Mereology · Contamination · High granular partial
algebras · Ingredient grading problem

1 Introduction

A number of theories of rationality, rational inference, belief, and judgment are
known in the literature on philosophy, epistemology, social sciences, and artificial
intelligence (see [14]. In all these, concepts of rationality refer to much more
than the act of drawing inferences from a given set of premises or constructing
facets of reasoning from a given collection of conclusions or state of affairs.
*Often it is about drawing inferences and conclusions or actions that satisfy a
specific beneficial or normative pattern.* When the possible choices are bounded,
then rational actions may be those that ensure beneficial choices – this may be
called the idea of *bounded general rationality.* Ideas of rational beliefs are, for

© Springer-Verlag GmbH Germany, part of Springer Nature 2022
J. F. Peters et al. (Eds.): TRS XXIII, LNCS 13610, pp. 33–59, 2022.
https://doi.org/10.1007/978-3-662-66544-2_4

example, defined by conditions that involve belief operators subject to additional restrictive assumptions. Reasonable and rational approximations in the context of general rough approximations are essential for applications to contexts that require high quality approximations, or predictions (especially when robustness cannot be expected). This is very relevant in applications to human learning, automated evaluation frameworks in education [27], general rough analysis of cluster validation [29] and other areas. Apart from these a number of potential application areas like intelligent robust image segmentation with unlearning, epidemiology (where feature selection fails badly) and medical diagnostics can be indicated.

Specifically, in the application contexts mentioned, a rich collection of mereological relations such as *is an essential part of, is an integral part of, is an apparent part of, is a substantial part of, is a functional part of* and related expression may be found. For example, the concept of solving systems of linear equations has a very large number of parts of the types mentioned. How does one proceed to identify the appropriate ones especially when many nonstandard solution strategies are possible? How does one accommodate alternative conceptions (instead of dismissing them as mistakes) [27,39]? Cladistics of various kinds including biology [3,17] also involve more complex parthood relations.

In all this it should be noted that frameworks such as those of variable precision rough sets [13,51], generalizations thereof [19,37,41,50], graded rough sets [45], dominance based rough sets, and soft set approaches that seek to manipulate attributes from numeric valuation-based considerations have too many shortcomings for the proposed application domains where such valuations or fuzzy relations (as in [37]) are not usually reasonable. Modal connections of VPRS and graded rough sets are known (this is outlined in the following subsection), but are not granular in the axiomatic sense. *In this context, it should be mentioned that graded rough sets are relatively more realistic than VPRS despite them being closely related.* For these reasons, in this research,

- aspects of graded rough sets are reviewed and corrected,
- graded rough sets are generalized from a granular perspective,
- a new minimal framework for rationality of approximations is introduced,
- semantic aspects are considered, and
- rationality of approximations in granular graded rough sets are explored.

This research is organized as follows: in the next subsection the background required for the framework is mentioned in brief. In the following section, graded rough sets are reviewed and generalized from a granular perspective by the present author. A recurring series of abstract examples, and an application to a practical problem are introduced in the section. Concepts of substantial parthood are also defined by her. New frameworks for rough rationality are introduced in the third section. Extended examples involving bited and graded approximations are considered in the following section. In the fifth section, connections of the framework with graded rough sets are explored. The sixth section features a few open problems.

1.1   Background

In theoretical understandings of granularity, the term *granules* refer to parts or building blocks of the computational process and *granulations* to collections of such granules in the context. According to the present author, the three main ways of doing granular computing are according to the *primitive granular computing paradigm* (PGCP), *classical granular computing paradigm* (CGCP) [15,16,40,44,46,47] and the *axiomatic granular computing paradigm* (AGCP) [25]. In the axiomatic frameworks (AGCP) [21,25] that does not refer to numeric precision for defining granules, the problem of defining or rather extracting concepts that qualify require much work in the specification of semantic domains and process abstraction.

The reader is expected to be familiar with the axiomatic approach to granularity in [21,23,25]. Related history of granular computing can be found in [25]. The mereology [3,24,25] assumed in the approach is a minimalist and non-transitive one without atomisticity (objects are sums of atoms) or atomicity (every object has an atom as its part), and scope for adding additional conditions on the part-of relation – so correctly a minimalist mereology is assumed. The differences with the numeric function dependent rough mereology of [33,35] is explained in [21]. A more detailed explanation will appear separately. Some essential notions are repeated for convenience.

In a *high general granular operator space* (GGS), defined below, aggregation and co-aggregation operations ($\vee, \wedge$) are conceptually separated from the binary parthood (**P**), and a basic partial order relation ($\leqslant$). Parthood is assumed to be reflexive and antisymmetric. It may satisfy additional generalized transitivity conditions in many contexts. Real-life information processing often involves many non-evaluated instances of aggregations (fusions), commonalities (conjunctions) and implications because of laziness or supporting meta data or for other reasons – this justifies the use of partial operations. Specific versions of a GGS and granular operator spaces have been studied in [23]. Partial operations in GGS permit easier handling of adaptive granules [40] through morphisms. The universe $\underline{\mathbb{S}}$ may be a set of collections of attributes, labeled or unlabeled objects among other things. The granularity constraint cannot be usually realized in pointwise approximation [22] based rough sets. So most generalized approaches to rough sets and many hybrid variants thereof fit into the frameworks. The rough mereological approach of [34,36] are distinct, and have a strong dependence on numeric membership functions.

Notation: Quantifiers are uniformly enclosed in braces for easier reading. Thus, $\forall a \exists b\, \Phi(a,b)$ is the same as $(\forall a)(\exists b)\, \Phi(a,b)$.

Definition 1. *A* High General Granular Operator Space *(GGS)* $\mathbb{S}$ *is a partial algebraic system of the form* $\mathbb{S} = \langle \underline{\mathbb{S}}, \gamma, l, u, \mathbf{P}, \leqslant, \vee, \wedge, \perp, \top \rangle$ *with* $\underline{\mathbb{S}}$ *being a set,* γ *being a unary predicate that determines* $\mathcal{G}$ *(by the condition* γx *if and only if* $x \in \mathcal{G}$) *an admissible granulation(defined below) for* $\mathbb{S}$ *and* l, u *being operators* $: \underline{\mathbb{S}} \longmapsto \underline{\mathbb{S}}$ *satisfying the following (*$\underline{\mathbb{S}}$ *is replaced with* $\mathbb{S}$ *if clear from the context.* $\vee$ *and* $\wedge$ *are idempotent partial operations and* $\mathbf{P}$ *is a binary predicate. Further*

γx *will be replaced by* $x \in \mathcal{G}$ *for convenience.):*

$$(\forall x)\mathbf{P}xx \qquad \text{(PT1)}$$

$$(\forall x, b)(\mathbf{P}xb \,\&\, \mathbf{P}bx \longrightarrow x = b) \qquad \text{(PT2)}$$

$$(\forall a, b)a \vee b \overset{\omega}{=} b \vee a \,;\, (\forall a, b)a \wedge b \overset{\omega}{=} b \wedge a \qquad \text{(G1)}$$

$$(\forall a, b)(a \vee b) \wedge a \overset{\omega}{=} a \,;\, (\forall a, b)(a \wedge b) \vee a \overset{\omega}{=} a \qquad \text{(G2)}$$

$$(\forall a, b, c)(a \wedge b) \vee c \overset{\omega}{=} (a \vee c) \wedge (b \vee c) \qquad \text{(G3)}$$

$$(\forall a, b, c)(a \vee b) \wedge c \overset{\omega}{=} (a \wedge c) \vee (b \wedge c) \qquad \text{(G4)}$$

$$(\forall a, b)(a \leqslant b \leftrightarrow a \vee b = b \leftrightarrow a \wedge b = a) \qquad \text{(G5)}$$

$$(\forall a \in \mathbb{S})\,\mathbf{P}a^l a \,\&\, a^{ll} = a^l \,\&\, \mathbf{P}a^u a^{uu} \qquad \text{(UL1)}$$

$$(\forall a, b \in \mathbb{S})(\mathbf{P}ab \longrightarrow \mathbf{P}a^l b^l \,\&\, \mathbf{P}a^u b^u) \qquad \text{(UL2)}$$

$$\bot^l = \bot \,\&\, \bot^u = \bot \,\&\, \mathbf{P}\top^l \top \,\&\, \mathbf{P}\top^u \top \qquad \text{(UL3)}$$

$$(\forall a \in \mathbb{S})\,\mathbf{P}\bot a \,\&\, \mathbf{P}a\top \qquad \text{(TB)}$$

Let $\mathbb{P}$ *stand for proper parthood, defined via* $\mathbb{P}ab$ *if and only if* $\mathbf{P}ab \,\&\, \neg\mathbf{P}ba$). *A granulation is said to be admissible if there exists a term operation* t *formed from the weak lattice operations such that the following three conditions hold:*

$$(\forall x \exists x_1, \ldots x_r \in \mathcal{G})\, t(x_1, x_2, \ldots x_r) = x^l$$

$$\text{and } (\forall x)\,(\exists x_1, \ldots x_r \in \mathcal{G})\, t(x_1, x_2, \ldots x_r) = x^u, \qquad \text{(Weak RA, WRA)}$$

$$(\forall a \in \mathcal{G})(\forall x \in \mathbb{S}))\,(\mathbf{P}ax \longrightarrow \mathbf{P}ax^l), \qquad \text{(Lower Stability, LS)}$$

$$(\forall x, a \in \mathcal{G})(\exists z \in \mathbb{S}))\,\mathbb{P}xz, \,\&\, \mathbb{P}az \,\&\, z^l = z^u = z, \qquad \text{(Full Underlap, FU)}$$

Definition 2.
- *In the above definition, if the antisymmetry condition PT2 is dropped, then the resulting system will be referred to as a* Pre-GGS. *If the restriction* $\mathbf{P}a^l a$ *is removed from UL1, then it will be referred to as a* Pre*-GGS.
- *In a GGS, if the parthood is defined by* $\mathbf{P}ab$ *if and only if* $a \leqslant b$ *then the GGS is said to be a high granular operator space GS.*
- *A higher granular operator space (HGOS)* $\mathbb{S}$ *is a GS in which the lattice operations are total.*
- *In a higher granular operator space, if the lattice operations are set theoretic union and intersection, then the HGOS will be said to be a set HGOS.*

For basics of partial algebras, the reader is referred to [4, 18].

Definition 3. *A partial algebra* P *is a tuple of the form*

$$\langle \underline{P}, f_1, f_2, \ldots, f_n, (r_1, \ldots, r_n) \rangle$$

with $\underline{P}$ *being a set,* f_i *'s being partial function symbols of arity* r_i. *The interpretation of* f_i *on the set* $\underline{P}$ *should be denoted by* $f_i^{\underline{P}}$, *but the superscript will be dropped in this paper as the application contexts are simple enough. If predicate symbols enter into the signature, then* P *is termed a* partial algebraic system.

In this paragraph the terms are not interpreted. For two terms s, t, s $\overset{\omega}{=}$ t shall mean, if both sides are defined then the two terms are equal (the quantification is implicit). $\overset{\omega}{=}$ is the same as the existence equality (also written as $\overset{e}{=}$) in the present paper. s $\overset{\omega^*}{=}$ t shall mean if either side is defined, then the other is and the two sides are equal (the quantification is implicit). Note that the latter equality can be defined in terms of the former as

$$(s \overset{\omega}{=} s \longrightarrow s \overset{\omega}{=} t) \,\&\, (t \overset{\omega}{=} t \longrightarrow s \overset{\omega}{=} t)$$

In [26], it is shown that the binary predicates in a GGS and variants can be replaced by partial two-place operations and γ is replaceable by a total unary operation. This results in a semantically equivalent partial algebra called a *high granular operator partial algebra* (GGSp).

1.2   Modal Connections

A number of logical approaches to reasoning in rough sets have been proposed since the early days of the subject. Those based on modal logic are [2, 7, 30–32, 42] and these are based on pointwise approximations of a simpler kind (relative to the definition in [22] by the present author).

A graded modal logic [5, 6, 10, 11] may be viewed as an extension of propositional logic with graded modalities that count the number of successors of a given state. This perspective is not in perfect sync with the graded rough set interpretation offered in [45] that is closer to the idea of degrees of discernibility relative to special approximations (interpretible as necessities and possibilities). The epistemic interpretation in [12] corresponds to *something is provable in the sense of the logic if it is up to* k *exceptions.*

Studies on graded modal logics have been in relation to completeness, decidability [8], finite model property (FMP) [8, 38], and expressibility in particular through multiple approaches like filtration and bisimulation. These results naturally help in restricting considerations to finite models when the graded rough set model admits relevant correspondences. Parts of modal connections of VPRS and graded rough sets are reviewed in [1], but these are not granular in the axiomatic sense.

2   Semantics of Graded Rough Sets

In graded rough sets, approximations are constructed relative to integral grades that are related to the cardinality of sets. The granulation is assumed to be the set of arbitrary neighborhoods generated by points in [45] and the approximations are defined pointwise. Granular approximations generated by equivalence relations are studied in [49] (though the definition used is different from [45], they coincide). Here this will be generalized to arbitrary granulations and explored. While these can be related to variable precision rough sets, they can be suggestive of the use of mutually inconsistent procedures in their construction. A number of incorrect claims in [49] are also rectified.

The version explored in [45] has the following form:

$$A^{u^p_k} = \{z : n(z) \in \mathcal{G} \,\&\, \#(n(z) \cap A) > k\} \tag{k-upper1}$$

$$A^{l^p_k} = \{z : n(z) \in \mathcal{G} \,\&\, \#(n(z)\backslash A) \leqslant k\} \tag{k-lower1}$$

$$Pos^p_k(A) = A^{u^p_k} \cap A^{l^p_k} \tag{k-positive region1}$$

$$Neg^p_k(A) = H\backslash(A^{l^p_k} \cup A^{u^p_k}) \tag{k-negative region1}$$

$$Bnd_{u^p_k}(A) = A^{u^p_k}\backslash A^{l^p_k} \tag{upper k-boundary1}$$

$$Bnd_{l^p_k}(A) = A^{l^p_k}\backslash A^{u^p_k} \tag{lower k-boundary1}$$

It is obvious that even if $\mathcal{G}$ is a partition, it can happen that $A^{l^p_k} \not\subseteq A$ and $A \not\subseteq A^{u^p_k}$. For this reason alone, it is not a very rational kind of approximation. Further the construction presupposes complete knowledge of the context or domain (with no scope for vagueness). Semantic duality results have limited scope in the situation. The approximation operators are graded modal operators though, and satisfy the following properties [1, 45]:

Proposition 1. *In the above context,*

$$A^{l^p} = A^{l^p_0} \tag{GL0}$$

$$A^{l^p_k} = \neg((\neg A)^{u^p_k}) \tag{GL1}$$

$$H^{l^p_k} = H \tag{GL2}$$

$$(A \cap B)^{l^p_k} \subseteq A^{l^p_k} \cap B^{l^p_k} \tag{GL3}$$

$$A^{l^p_k} \cup B^{l^p_k} \subseteq (A \cup B)^{l^p_k} \tag{GL4}$$

$$A \subseteq B \longrightarrow A^{l^p_k} \subseteq B^{l^p_k} \tag{GL5}$$

$$k \leqslant t \longrightarrow A^{l^p_k} \subseteq A^{l^p_t} \tag{GL6}$$

$$A^{u^p_0} = A^{u^p} \tag{GU0}$$

$$A^{u^p_k} = \neg((\neg A)^{l^p_k}) \tag{GU1}$$

$$\emptyset^{u^p_k} = \emptyset \tag{GU2}$$

$$A^{u^p_k} \cup B^{u^p_k} \subseteq (A \cup B)^{u^p_k} \tag{GU3}$$

$$(A \cap B)^{u^p_k} \subseteq A^{u^p_k} \cap B^{u^p_k} \tag{GU4}$$

$$A \subseteq B \longrightarrow A^{u^p_k} \subseteq B^{u^p_k} \tag{GU5}$$

$$k \leqslant t \longrightarrow A^{u^p_t} \cap A^{u^p_k} \tag{GU6}$$

This makes it a graded modal logic of a form. Semantics of graded modal logics (see [9, 38] and related references) and their limitations are applicable to this non-granular approach. When R is an equivalence, the additional modal axioms satisfied are (there are a few typos in [45]):

$$A^{l^p_0} \subseteq A^{u^p_0} \tag{GD}$$

$$A^{l^p_0} \subseteq A \tag{GT}$$

$$A \subseteq (A^{u^p_0})^{l^p_0} \tag{GB}$$

$$A^{l^p_k} \subseteq (A^{l^p_k})^{l^p_0} \tag{G4}$$

$$A^{u^p_k} \subseteq (A^{u^p_k})^{l^p_0} \tag{G5}$$

Examples for pointwise graded approximations can be found in [45] and related papers.

2.1   Granular Generalization

The motivation for granular generalizations is grounded in the need to construct approximations in a granular way within the same perspective. While, the dependence on k look beyond the number of granules included or intersected with a given set, the generality is of limited (but specific) value in the broader discourse on rationality. Practical examples are easy to construct.

A *grade* k (as before) can be any fixed positive integer, and is to be related to the cardinality of granules or sets used in the context. Let S be a collection of sets (that are subsets of a H), and $\mathcal{G}$ a subset of S. The basic part of relation will be taken to set inclusion. If k is a fixed positive integer and x is a set in the collection S, then the k-*lower* and k-*upper* approximations, and related regions will be

$$x^{u_k} = \bigcup\{h : h \in \mathcal{G} \,\&\, \#(h \cap x) > k\} \tag{k-upper}$$

$$x^{l_k} = \bigcup\{h : h \in \mathcal{G} \,\&\, \#(h) - \#(h \cap x) \leqslant k\} \tag{k-lower}$$

$$Pos_k(x) = x^{u_k} \cap x^{l_k} \tag{k-positive region}$$

$$Neg_k(x) = H\backslash(x^{l_k} \cup x^{u_k}) \tag{k-negative region}$$

$$Bnd^u_k(x) = x^{u_k}\backslash x^{l_k} \tag{upper k-boundary}$$

$$Bnd^l_k(x) = x^{l_k}\backslash x^{u_k} \tag{lower k-boundary}$$

The lower approximation can result in strange values, and the following regularized version is of natural interest:

$$x^{l^r_k} = \bigcup\{h : h \in \mathcal{G} \,\&\, h \subseteq x \,\&\, \#(h) - \#(h \cap x) \leqslant k\} \tag{k-reg.lower}$$

Proposition 2. *The range of the operations* u_k *and* l_k *on* $\wp(H)$ *need not be equal even when* $\mathcal{G}$ *is a partition of* H, *and* $S = \wp(S)$.

Proof. The range of definition of the operations u_k and l_k are subsets of the set of all unions of neighborhood granules. Suppose these are H_{uk} and H_{lk}. Counterexamples for showing the inequality of the two can be based on the following general considerations:

If $[z]$ is an equivalence class (with $z \in H$) with less than or equal to k elements, then it's k-lower approximation is $[z]$, but its k-upper approximation would be the empty set. In fact, $[z] \notin H_{uk}$ will hold (and neither will a superset of $[z]$ be in H_{uk}). If $[z]$ is an equivalence class with more than k elements, then it will be in both H_{lk} and H_{uk}. So it can happen that $H_{lk} \neq H_{uk}$.   □

2.2   Abstract Example

This abstract toy example can be used to illustrate various aspects of the theoretical framework developed. Let $H, \mathbb{S}, \mathcal{G}$ and the predicates $\mathbf{P}_\alpha$ and $\mathbf{P}_\beta$ be defined as follows (strings of letters like bef are used to denote the set $\{b, e, f\}$ for convenience)

$$H = \{a, b, c, e, f\}$$
$$\mathbb{S} = \wp(H) \backslash \{e, b, bf, bef, abce\}$$
$$\mathcal{G} = \{ab, bce, ef, a, b, f\}$$

In the computation of the pointwise approximations below, it is assumed that

$$n(a) = ab, n(b) = bce, n(c) = bce, n(e) = ef, n(f) = f$$

Relative to $\subseteq$ as the part of relation, if $A = aef$, then it can be checked that

$$A^{l_0} = aef; \; A^{u_0} = abcef \qquad \text{(0-lower/upper)}$$
$$A^{l_1} = aef; \; A^{u_1} = abcef \qquad \text{(1-lower/upper)}$$
$$A^{l_2} = abef; \; A^{u_2} = ef \qquad \text{(2-lower/upper)}$$
$$Pos_1(A) = c \qquad \text{(1-positive)}$$
$$A^{l_1^p} = aef; \; A^{u_1^p} = e \text{ over } \wp(H) \qquad \text{(1-pointwise)}$$

Note that $A^{u_1^p} \subset A^{l_1^p}$ in this example. Even if the requirement $n(x) \subseteq A$ is imposed in the definition of the lower approximation, this would yield $\{x : n(x) \subseteq A \,\&\, \#(n(x)) - \#(n(x) \cap A) \leqslant 1\} = ef$ and this would properly contain $A^{u_1^p}$.

2.3   Example: Dynamic Classification

This example illustrates the construction of granules without any explicit specification of relations in particular.

For ensuring uniform quality of processed food products that also need to meet specific standards, dynamic intelligent classification of the ingredients required need to be used. Computer vision based methods do not follow universal standards, often reinvent the wheel, are opaque (source code are rarely available) and have other limitations. The situation for food grains is surveyed in [43]. Some manufacturing units resort to simplified automated color based sorting of grains and fruits procured from different farms, but this is known to lead to wide variations in quality of the end products. The computational power required to collect more data from *flowing produce* is marginal, and it is possible to use graded rough sets for classification and approximation.

Suppose that n different attributes of the produce are measured, and let their associated value sets be $\{V_a : a \in \{1, 2, \ldots n\}\}$. An f-valued classification (with $f \geqslant 3$) into the disjoint categories: reject, perfect, sub-perfect-(1) ... sub-perfect-$(f-2)$ can be assumed to be the result of the procedure. Similar sorting patterns can be assumed for other types of ingredients. A subsequent transfer of the things falling under the category reject to the reject bin, perfect to the perfect bin and sub-perfect-i to the sub-perfect-i bin ensures an initial classification. This stage can also involve graded rough or other soft methods. But it is the use in subsequent stages that will be highlighted in this example (Fig. 1).

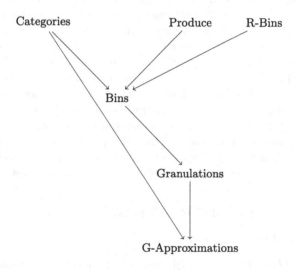

Fig. 1. Industrial application

After a time t of sorting the produce into f number of bins, the problem would be of combining them to form blocks of admissible grade that satisfy the quality criteria. It is assumed that some bins have residual material from previous runs. The problem may also be read as a number of optimization problems (as it is not required to exhaust bins in each cycle), that may or may not have solutions – but such an approach would require human intervention and decision-making.

- A granule h is a standardized set of objects determined by a set $V(h)$ of n-tuples of values that are computed to be admissible (from the mass of ingredients and associated classification) according to an external criteria. For example, an ingredient of type-1 and another ingredient of type-2 in the ratio 1:2 may be known to satisfy the criteria. But there is no need to explicitly mention the ratio

- Lower graded approximations are combinations from the bins that satisfy the criteria
- Upper graded approximations on the other hand can be used to satisfy a lower quality criteria.
- The objective is to produce admissible combinations of ingredients to approximate the quality criteria.
- The various mereological features associated are made concrete below.

2.4   Example: Concrete Bins

Suppose the granules are collections of ingredients of the form $x_i j$ with i indicating the ingredient and j it's type. ij can be assumed to be the bin identity indicator as well. It is assumed that bin ij contains many approximate copies of x_{ij}. Note that if the sorting is imperfect, then this would mean that *most of the content of bin* ij *is of the* j*th type of ingredient* i. For this example, also assume that there are three ingredients each of five types. A sum formula like $x_{11} \oplus x_{21} \oplus x_{33}$ is intended to mean a combination of the granules in some ratio. In the simplest case this can be a union of a number of granules.

Now, a rule like there should be *at least two ingredients of perfect type, and at least three ingredients for an acceptable combination* is equivalent to constructing a lower graded approximation. For a purely set theoretic version the granules can be numbered in addition (relative to mass of ingredients).

While a combination need not exactly correspond to a quality, the goal is to approximate it. Thus a lower graded approximation can assure a certain quality, while a upper approximation can possibly assure a certain quality.

2.5   Partitions and Graded Rough Sets

It is proved in [49] that

Theorem 1. *When* $H = \bigcup S = \bigcup \mathcal{G}$, *and* $\mathcal{G}$ *is a collection of disjoint sets (that is a partition), then*

$$Bnd_k^l x = \emptyset \leftrightarrow x^{l_k} \subseteq x^{u_k} \tag{1}$$

$$Bnd_k^u x = \emptyset \leftrightarrow x^{u_k} \subseteq x^{l_k} \tag{2}$$

$$x^{u_k u_k} = x^{u_k} \subseteq x^{u_k l_k} \tag{3}$$

$$x^{u_k l_k} = x^{u_k} \cup \{\cup\{h : h \in \mathcal{G} \,\&\, \#(h) \leqslant k\}\} \tag{4}$$

$$x^{l_k u_k} \subseteq x^{l_k} = x^{l_k l_k} \tag{5}$$

$$x^{l_k} = (x^{l_k})^{u_k} \cup \{\cup\{h : h \in \mathcal{G} \,\&\, \#(h) \leqslant k\} \tag{6}$$

Concepts of outer and inner boundaries are defined as follows:

$$\text{Bnd}_k^{ou}(a,b) = \cup\{h : h \subseteq a^{uk}, \& h \subseteq b^{uk} \& h \not\subseteq (a \cup b)^{uk}\} \quad \text{(o-u k-boundary)}$$

$$\text{Bnd}_k^{ol}(a,b) = \cup\{h : h \subseteq a^{lk}, \& h \subseteq b^{lk} \& h \not\subseteq (a \cup b)^{lk}\} \quad \text{(o-l k-boundary)}$$

$$\text{Bnd}_k^{iu}(a,b) = \cup\{h : h \not\subseteq a^{uk}, \& h \not\subseteq b^{uk} \& h \subseteq (a \cup b)^{uk}\} \quad \text{(i-u k-boundary)}$$

$$\text{Bnd}_k^{il}(a,b) = \cup\{h : h \not\subseteq a^{lk}, \& h \not\subseteq b^{lk} \& h \subseteq (a \cup b)^{lk}\} \quad \text{(i-l k-boundary)}$$

Erroneous claims (Definition 4 and beyond) are made in [49] in relation to definable operations on the power set ($\wp(H)$). The authors claim that $\sqcup, \overline{\cap}$, and $\sim^*$ are total operations and that the graded approximations are epimorphisms, but in their formulation these are wrong and ill-defined. *For example, the following definition is improper as z and w can be written as k-upper approximations in multiple ways to yield different results as demonstrated in the counterexample below:* For any $z, w \in \wp(H)$, if $z = a^{uk} \& w = b^{uk}$, then $z \sqcup w := a^{uk} \cup b^{uk} \cup \text{Bnd}_k^{iu}(a,b)$. These are pointed out in a forthcoming note (also see [28]). The errors are carried through Proposition 5 of [49], and the remarks following that. The same pattern emerges in Definition 7 and Proposition 6 of that paper.

Example 1 (Counterexample).

$$\text{Let } U = \{a, b, c, e, f, g, h, i, j\} \quad \text{(Universe)}$$
$$U|R = \{\{a\}, \{b\}, \{c, e, f\}, \{g, h, i, j\}\} \quad \text{(Classes)}$$
$$A = \{a, c, e, g\} \quad \text{(A)}$$
$$B = \{b, h,\} \quad \text{(B)}$$
$$E = \{e, f, h\} \quad \text{(E)}$$
$$F = \{b\} \quad \text{(F)}$$

It can be checked that

1. $A^{u_1} = \{c, e, f\} = E^{u_1}$
2. $B^{u_1} = \emptyset = F^{u_1}$
3. $\text{Bnd}_1^{iu}(A, B) = \{g, h, i, j\}$
4. $\text{Bnd}_1^{iu}(E, F) = \emptyset$
5. It follows that

$$A^{uk} \cup B^{uk} \cup \text{Bnd}_k^{iu}(A, B) \neq E^{uk} \cup F^{uk} \cup \text{Bnd}_k^{iu}(E, F)$$

The mistakes are not resolvable from [49]. So the following new definitions are introduced by the present author:

Definition 4. *For any* $a, b \in \wp(H)$

$$a \sqcup_u b := a^{uk} \cup b^{uk} \cup Bnd^{iu}_k(a, b) \qquad (sqcup_u)$$

$$a \sqcup_l b := a^{lk} \cup b^{lk} \cup Bnd^{il}_k(a, b) \qquad (sqcup_l)$$

$$a \sqcap_u b := (a^{uk} \cap b^{uk}) \backslash Bnd^{ou}_k(a, b) \qquad (sqcap_u)$$

$$a \sqcap_l b := (a^{lk} \cap b^{lk}) \backslash Bnd^{ol}_k(a, b) \qquad (sqcap_l)$$

$$\neg_u a = (a^{uk})^c \qquad (\neg_u)$$

$$\neg_l a = (a^{lk})^c \qquad (\neg_l)$$

Further, if c *denotes complementation in* $\wp(H)$, *then for any* $z \in \wp(H)$

$$\neg z := \begin{cases} (a^{uk})^c & \text{if } z = a^{uk} \\ (a^{lk})^c & \text{if } z = a^{lk} \\ \text{undefined otherwise} \end{cases} \qquad (\neg)$$

Theorem 2. *The partial and total operations of Definition 4 are well-defined on* $\wp(H)$.

Proof. It is obvious that $\sqcup_u$, $\sqcup_l$, $\sqcap_u$, $\sqcap_l$, $\neg_l$ and $\neg_u$ are all well-defined as their evaluation can be done in only one way. Also if $z = a^{uk} = b^{lk}$ for some $a, b, z \in \wp(H)$, then the evaluation of $\neg z$ remains unique.   □

The defined operations satisfy the following properties

Theorem 3. *1.* $\sqcup_l, \sqcup_u, \sqcap_l$ *and* $\sqcap_u$ *are all commutative, idempotent operations.*
2. $\sqcup_l, \sqcup_u, \sqcap_l$ *and* $\sqcap_u$ *are not necessarily associative.*
3. For any $a, b \in \wp(H)$, $a \sqcup_u b = (a \cup b)^{uk}$
4. For any $a, b \in \wp(H)$, $a \sqcap_l b = (a \cap b)^{lk}$
5. For any $a, b \in \wp(H)$, $a \sqcup_l b = (a \cup b)^{lk}$
6. For any $a, b \in \wp(H)$, $a \sqcap_u b = (a \cap b)^{uk}$

Proof. 1. Commutativity and idempotence of the four operations can be directly verified.
2. The last four hold because of the definition of outer and inner boundaries, that accounts for the different types of elements in $(a \cup b)^{uk}$, $(a \cap b)^{uk}$, $(a \cup b)^{lk}$, and $(a \cap b)^{lk}$.
3. The failure of associativity can be verified by simple counterexamples or by a contradiction argument.
   (a) If $\sqcup_l$ is always associative, then for any a, b, c the expanded form of the equation for associativity must hold.
   (b) This would mean, that $a^{lk} \cup b^{lk} \cup c^{lk} \cup Bnd^{il}_k(a, b) \cup Bnd^{il}_k(a \sqcup_l b, c)$ must be equal to $a^{lk} \cup b^{lk} \cup c^{lk} \cup Bnd^{il}_k(b, b) \cup Bnd^{il}_k(a, b \sqcup_l c)$
   (c) Note that if $z = b^{lk}$, then for any c, $Bnd^{il}_k(z, c) = \emptyset$ because z is a union of granules. $a \sqcup_l b$ and $b \sqcup_l c$ are unions of granules.
   (d) So this would mean that $Bnd^{il}_k(a, b) = Bnd^{il}_k(b, c)$ or that these are subsets of $a^{lk} \cup b^{lk} \cup c^{lk}$ always.
   (e) This is obviously false in general.
                                                                    □

Other Cases. The definitions of relative boundaries can be extended to granular graded rough contexts as follows ($\mathcal{G}$ being the set of granules used):

Definition 5. *For any* $a, b \in S$,

$$\mathrm{Bnd}_k^{ou}(a, b) = \cup\{h : h \in \mathcal{G} \,\&\, h \subseteq a^{uk}, \,\&\, h \subseteq b^{uk} \,\&\, h \nsubseteq (a \cup b)^{uk}\}$$
$$\text{(o-u k-boundary)}$$

$$\mathrm{Bnd}_k^{ol}(a, b) = \cup\{h : h \in \mathcal{G} \,\&\, h \subseteq a^{lk}, \,\&\, h \subseteq b^{lk} \,\&\, h \nsubseteq (a \cup b)^{lk}\}$$
$$\text{(o-l k-boundary)}$$

$$\mathrm{Bnd}_k^{iu}(a, b) = \cup\{h : h \in \mathcal{G} \,\&\, h \nsubseteq a^{uk}, \,\&\, h \nsubseteq b^{uk} \,\&\, h \subseteq (a \cup b)^{uk}\}$$
$$\text{(i-u k-boundary)}$$

$$\mathrm{Bnd}_k^{il}(a, b) = \cup\{h : h \in \mathcal{G} \,\&\, h \nsubseteq a^{lk}, \,\&\, h \nsubseteq b^{lk} \,\&\, h \subseteq (a \cup b)^{lk}\}$$
$$\text{(i-l k-boundary)}$$

The meaning of such boundaries is analogous to the case considered earlier, but depends on the properties of the granulation. This will be investigated separately.

2.6   Substantial Parthood

In the context of granular graded rough sets, predicates possibly interpretible as forms of substantial parthood can be defined as follows:

For any $a, b \in S$, and a fixed positive integer k

$$\mathbf{P}_s^1 ab \text{ if and only if } a^{lk} \subseteq b^{lk} \tag{s1}$$

$$\mathbf{P}_s^2 ab \text{ if and only if } a \subseteq b \,\&\, \#(a) > k \tag{s2}$$

$$\mathbf{P}_s^* ab \text{ if and only if } \#(a \cap b) > k \,\&\, b \nsubseteq a \tag{s*}$$

$$\mathbf{P}_s^3 ab \text{ if and only if } \#(a \cap b) > k \,\&\, a \subseteq b \tag{s3}$$

Some examples that correspond to the insight in the properties are mentioned in the subsection below.

The basic properties of these relations are stated in the next four propositions.

Proposition 3. $\mathbf{P}_s^1$ *satisfies the following:*

1. $(\forall a)\, \mathbf{P}_s^1 aa$
2. $(\forall a, b, c)\, (\mathbf{P}_s^1 ab \,\&\, \mathbf{P}_s^1 bc \longrightarrow \mathbf{P}_s^1 ac)$
3. $\mathbf{P}_s^1$ *is not antisymmetric or symmetric in general.*

Proof. 1. For any $a \in S$, $a^{lk} = a^{lk}$. So $\mathbf{P}_s^1 aa$ holds.
2. Transitivity holds because $a^{lk} \subseteq b^{lk} \subseteq c^{lk}$ implies $a^{lk} \subseteq c^{lk}$.
3. Follows from the basic properties of l_k

$\square$

Proposition 4. $\mathbf{P}_s^2$ *satisfies the following:*

1. $\mathbf{P}_s^2 aa$ *if and only if* $\#(a) > k$

2. $(\forall a, b, c) (\mathbf{P}_s^2 ab \, \& \, \mathbf{P}_s^2 bc \longrightarrow \mathbf{P}_s^2 ac)$
3. $\mathbf{P}_s^2$ is antisymmetric but is not symmetric in general.

Proof. The proof is by direct verification and obvious counterexamples.   □

Proposition 5. $\mathbf{P}_s^*$ *satisfies the following:*

1. $\mathbf{P}_s^* aa$ *if and only if* $\#(a) > k$,
2. $\mathbf{P}_s^*$ *is not transitive in general,*
3. $\mathbf{P}_s^*$ *is neither antisymmetric nor symmetric in general.*

Proof. 1. For any a, b, $\mathbf{P}_s^* ab$ if and only if $\#(a \cap b) > k \, \& \, b \not\subset a$ by definition. So if $\#(a) > k$, $a = a$ and $\mathbf{P}_s^* aa$ follows.
2. Let $k = 4$, $a = \{1, 2, 3, 4, 5, 6, 7, 8, 9\}$,
   $b = \{20, 15, 1, 2, 3, 4, 5\}$, and $c = \{20, 12, 1, 2, 3, 30$. Then it can be checked that $\mathbf{P}_s^* ab$, and $\mathbf{P}_s^* bc$ hold, but $\mathbf{P}_s^* ac$ does not hold. So transitivity fails in general.
3. In the context of the same example, if $h = \{1, 2, 3, 4, 5\}$, $\mathbf{P}_s^* ah$ holds, but $\mathbf{P}_s^* ha$ fails. The failure of antisymmetry can be checked with an easy counterexample.

     □

Proposition 6. $\mathbf{P}_s^3$ *satisfies the following:*

1. $\mathbf{P}_s^3 aa$ *holds if and only if* $\#(a) > k$,
2. $\mathbf{P}_s^3$ *is transitive,*
3. $\mathbf{P}_s^3$ *is antisymmetric and not symmetric in general.*

Proof. The proof is by direct verification and obvious counterexamples.   □

The predicates defined above will be checked for possible use in defining concepts of *rational approximation* in the contexts after introducing the framework in Sect. 3.

2.7   Abstract Example-2

In the context of Sect. 2.2, the substantial part-of predicates defined on $\mathbb{S}$ are as follows:

For $k = 1$, it can be checked that 1-lower approximations of some of the sets are as in Table 1.

Table 1. Some 1-lower approximations

Subset	a	b	c	e	f	ab	ac	ae	af	aef
l_1		a	b	∅	∅	f	ab	∅	∅	∅	aef

From this, it can be checked that $\mathbf{P}_s^1$ is a superset of

$$\{(c, a), (ac, a), (ae, b), (af, b), (a, aef), (a, a), (f, f),$$
$$(b, b), (b, ac), (ac, b), (aef, aef)\}$$

Some elements of $\mathbf{P}_s^2$ for $k = 1$ are

$$\{(ab, abc), (ac, abc), (bc, abc)\} \subset \mathbf{P}_s^2$$

For $k = 3$,

$$\mathbf{P}_s^* = \{(abce, abcef), (abcf, abcef), (bcef, abcef), (abef, abcef), (acef, abcef)\}$$

For $k = 3$, $\mathbf{P}_s^3$ is the set

$$\{(abce, abcef), (abcf, abcef), (bcef, abcef), (abef, abcef), (acef, abcef),$$
$$(abce, abce), (abcf, abcf), (bcef, bcef), (abef, abef), (acef, acef)\}$$

3   Framework for Rational Approximations

A simple set theoretic realization of rationality or of being a substantial approximation is not possible without an external valuation or an additional predicate in most situations. If external valuations are simply based on cardinalities of sets or probabilistic measures, then often it is at the cost of accuracy and meaning. Associated performance in different contexts can be predicted.

In the context of general rough sets, it is possible to distinguish between the approximation operator being rational and rationality being a property of concrete approximations. VPRS-style definitions try to ensure a form of the latter (and that may work under very nice uncommon conditions). Graded rough sets and the granular variants introduced also try to define a rationality of the latter type based on the size of granules.

Essentially definitions of instances of the following schema are considered: l is rational for a general rough object x if and only if x^l is a *rational lower approximation* of x if and only if $\Re x^l x$. This requires another predicate to be feasible. Possible candidates are the closely related predicates *is an essential part of* $\mathbf{P}_e$, *is an inessential part of* $\mathbf{P}_n$, *is a relevant part of* $\mathbf{P}_r$, *is a substantial defining part of* $\mathbf{P}_s$ and variations thereof. As these are not always definable in an easy way in a GGSP relative to observations in a real-life situation (especially when numeric valuations are difficult), it is preferable to add one of them to the model. In this research $\mathbf{P}_s$ will be preferred because it does not commit to a specific ontology. The predicate is very closely related to being *a substantial part of* (that is somewhat known in the mereology literature).

Definition 6. *A Pre-General Pre-Substantial Granular Space (pGpsGS)* $\mathbb{S}$ *will be a partial algebraic system of the form*

$$\mathbb{S} = \langle \underline{\mathbb{S}}, \gamma, l, u, \mathbf{P}, \mathbf{P}_s, \leqslant, \vee, \wedge, \perp, \top \rangle$$

with $\langle \underline{\mathbb{S}}, \gamma, l, u, \mathbf{P}, \leqslant, \vee, \wedge, \bot, \top \rangle$ *being a* **Pre*-GGS** *and satisfies* **sub3-sub4** *($\approx$ is intended as any definable rough equality):*

$$(\forall a, b)\,(\mathbf{P}_s ab \,\&\, \mathbf{P}_s ba \longrightarrow a \approx b) \tag{sub3}$$

$$(\forall a, b, e)\,(\mathbf{P}_s ae \,\&\, \mathbf{P}_s ab \longrightarrow \mathbf{P}_s a(b \vee e)) \tag{sub4}$$

If in addition, **sub1, sub2, sub5, UL1** *and* **sub6** *are satisfied, then* $\mathbb{S}$ *will be said to be a* Pre-General Substantial Granular Space *(pGsGS).*

$$(\forall a)\,\mathbf{P}_s aa \tag{sub1}$$

$$(\forall a, b)\,(\mathbf{P}_s ab \longrightarrow \mathbf{P}ab) \tag{sub2}$$

$$(\forall a, b, e)\,(\mathbf{P}_s ba \,\&\, \mathbf{P}_s be \,\&\, \mathbf{P}ae \longrightarrow \mathbf{P}_s ae) \tag{sub5}$$

$$(\forall a, b, e)\,(\mathbf{P}_s ab \,\&\, \mathbf{P}_s eb \,\&\, \mathbf{P}ae \longrightarrow \mathbf{P}_s ae) \tag{sub6}$$

If a pGsGS satisfies the condition **PT1,** *then it will be referred to as a* General Substantial Granular Space.

The defining conditions **sub1-sub2** say that anything is a d-substantial part of itself, if something is a d-substantial part of something else, then it is also a part of that something else. The condition **sub3** ensures that if something is a d-substantial part of something else and conversely, then they have something in common. **sub4** says that if one thing is a substantial part of two other things, then it is a substantial part of their generalized join (possibly a generalized s-norm when $\mathbf{P}$ and $\mathbf{P}_s$ are transitive and antisymmetric). It may be noted that a special case is when $\mathbf{P}$ and $\mathbf{P}_s$ are partial orders and $\vee$ is a generalized s-norm (see [48]). It might appear that **sub1, sub5** and **sub6** are too strong in some cases, and that is the motivation for introducing at least two levels of the concept. These are illustrated in Example 4.1 and the recurring abstract examples.

A condition that can fail to hold in many contexts because of conflict with parthood is this (it was considered as a possible defining condition in an earlier draft of this research paper):

$$(\forall a, b, e)\,(\mathbf{P}_s ae \,\&\, \mathbf{P}_s be \longrightarrow \mathbf{P}_s(a \vee b)e) \tag{Asy}$$

Note that a binary relation R is said to be *r-euclidean* if and only if the condition $(\forall a, b, c)(Rab \,\&\, Rac \longrightarrow Rbc)$. R is *l-euclidean* if and only if $(\forall a, b, c)(Rba \,\&\, Rca \longrightarrow Rbc)$. The conditions **sub5** and **sub6** are respectively generalizations of these through two predicates.

Definition 7. *In a pGpsGS* $\mathbb{S}$, *rational lower approximation* (l_s) *of a* $a \in \mathbb{S}$ *will be a lower approximation of some* b *that satisfies*

- $\mathbf{P}ba$
- $(\forall e)(e^l = e \,\&\, \mathbf{P}eb \longrightarrow \mathbf{P}_s ea)$

Definition 8. *In the context of the above definition, a* rational upper approximation (u_s) *of* $a \in \mathbb{S}$ *will be some* b *that satisfies*

- $(\exists z)z^u = b$
- $\mathbf{P}ab \,\&\, \mathbf{P}ba^u$
- $(\forall x)(x^l = x \,\&\, \mathbf{P}xb \longrightarrow \mathbf{P}_sxb)$

Clearly there is a big difference between upper and rational upper approximations. It can happen that rational upper approximations do not exist even in the context of classical rough sets (when it is enhanced with restrictions on *substantial inclusion*). So it is necessarily a partial operation on a pGpsGS. By contrast, rational lower approximations exist always.

Proposition 7. *In the context of Definition 8, the following hold in a pGsGS:*

$$(\forall x)x^{l_s l_s} = x^{l_s} \qquad\qquad \text{(Idempotence)}$$

$$(\forall x)\mathbf{P}x^{l_s}x^l \qquad\qquad \text{(Low-comp1)}$$

$$(\forall a, b)(\mathbf{P}ab \longrightarrow \mathbf{P}a^{l_s}b^{l_s}) \qquad\qquad \text{(s-Monotony)}$$

$$(\forall x)\mathbf{P}x^{u_s}x^u \qquad\qquad \text{(Up-comp1)}$$

Proof. 1. Idempotence is inherited from l in the context of the definition.
2. This again follows from the definition
3. s-Monotony is induced from that of l again

$\square$

An important compatibility problem in a pGpsGS is about finding conditions that ensure the following:

$$(\forall x)\mathbf{P}_sx^{l_s}x^l \qquad\qquad \text{(Low-comp2)}$$

$$(\forall x)\mathbf{P}_sx^{u_s}x^u \qquad\qquad \text{(Up-comp2)}$$

3.1   Abstract Example-3

The context of the Example 2.2, can be directly enhanced to form pGsGS, by taking the same base set, granules, l_1, and u_1 as the approximations, $\subseteq$ as the part of relation $\mathbf{P}$, and a suitable substantial parthood from Example 2.7.

$\subseteq$ can be replaced by a non-transitive parthood defined by

$$\mathbf{P}_\alpha = \{(a, b),\, (b, c),\, (e, f),\, (a, ab),\, (ab, ac),\, (ab, abc)\} \cup \{(x, x) : x \in H\}$$

The more detailed Example 4.1 illustrates a lot more.

3.2   Philosophy of Rational Intents

This subsection is intended to contextualize the approach adopted in this research.

An approximation operator has rational intent if that which is being approximated has much in common with those that are involved in the approximation process (as specified by the operator). This idea involves multiple semantic domains as listed below:

- An approximation operator has rational intent: this can be realized in the classical domain or a domain that is a part of the classical domain
- The object being approximated: can be assumed to be in the classical domain by default.
- The approximation can be in distinct rough domains $\mathfrak{R}_1 \ldots \mathfrak{R}_n$ because a single operator may be associated with multiple domains. For simplicity or a relatively reductionist approach all these can be taken as one.

A number of perspectives for defining ideas of *rational intent* are possible within the contexts of general rough sets. While the best may require external frameworks for their validation, it is always better to internalize most aspects while modeling.

Ri1. An approximation operator has rational intent if its range of errors is small.

Ri2. An approximation operator has rational intent if its errors are uniformly distributed.

Ri3. An approximation operator has rational intent if it is consistent.

Ri4. An approximation operator has rational intent if its realization is coherent with an ideal rough context.

Ri5. An approximation operator has rational intent if its realization is part of an ideal rough context.

Ri6. An approximation operator has rational intent if it is a substantial part of the thing being approximated or the converse is true.

While Ri4 and Ri5 explicitly refer to an external context, Ri1–Ri3 may also involve relatively external context for a definition of errors and consistency. In principle, if all aspects of the context are properly done, then Ri6 can be the best approach.

4   Specific Cases, Examples

In the bited approach, on Let $H = \langle \underline{H}, R \rangle$ be a general approximation space with R being a binary relation on the set $\underline{H}$. For a granulation $\mathcal{G} \subset \wp(H)$ (understood as a proper cover), the lower, upper and bited upper approximations of a $a \subseteq H$ are defined as follows:

$$a^l = \bigcup \{x : x \subseteq a, \, a \in \mathcal{G}\} \qquad \text{(Lower)}$$

$$a^u = \bigcup \{x : a \cap x \neq \emptyset, \, x \in \mathcal{G}\} \qquad \text{(Upper)}$$

$$\text{POS}_{\mathcal{G}}(a) = a^l \qquad \text{(Positive Region)}$$

$$\text{NEG}_{\mathcal{G}}(a) = a^{cl} \qquad \text{(Negative Region)}$$

$$a^{ub} = a^u \backslash a^{cl} \qquad \text{(Bited Upper)}$$

Consider the following relations on the powerset $\wp(H)$

bsub1. $\mathbf{P}_s ab$ if and only if $a^{ub} \subseteq b^{ub}$.
bsub2. $\mathbf{P}_s ab$ if and only if $(\exists e)e^l \subseteq a^u \subseteq b^u \& \emptyset \subset e^l$.
bsub3. $\mathbf{P}_s ab$ if and only if $(\exists e)e^{ub} \subseteq a \& e^{ub} \subseteq b \& \emptyset \subset e^{ub} \& b \not\subseteq a$.
bsub4. $\mathbf{P}_s ab$ if and only if $(\exists e)e^{ub} \subseteq a \& e^{ub} \subseteq b \& \emptyset \subset e^{ub}$
bsub5. $\mathbf{P}_s ab$ if and only if $a^u \backslash a^l \subseteq b^u \backslash b^l \& a^l \subseteq b^l$
bsub6. $\mathbf{P}_s ab$ if and only if $a^u \backslash a^l \subseteq b^u \backslash b^l \& a^{ub} \subseteq b^{ub}$

All six are of interest for potential use as substantial parthoods in the context of bited approximations in general approximation spaces (and specifically in the context of tolerance approximation spaces). It may be necessary to weaken parthood for the condition sub2 to hold.

Theorem 4. *In the case of tolerance approximation spaces, with lower, upper and bited upper approximations as defined above,*

1. ***bsub1*** *generally defines a reflexive, transitive and generally non-antisymmetric and non symmetric relation.*
2. ***bsub2*** *generally defines a partially reflexive, transitive, non-antisymmetric and non symmetric relation.*
3. ***bsub3*** *generally defines a partially reflexive, antisymmetric, non-transitive relation that is not right Euclidean or left Euclidean or symmetric.*
4. ***bsub4*** *generally defines a partially reflexive, symmetric relation that is not right Euclidean or left Euclidean or antisymmetric.*
5. ***bsub5*** *generally defines a reflexive, transitive relation that is not right Euclidean or left Euclidean or antisymmetric or symmetric.*
6. ***bsub6*** *generally defines a reflexive, transitive relation that is not right Euclidean or left Euclidean or antisymmetric or symmetric.*

Proof. 1. **bsub1** obviously defines a reflexive and transitive relation. If $\mathbf{P}_s ac$ and $\mathbf{P}_s ca$, then it follows that $a^{ub} = c^{ub}$, which does not generally imply $a = c$.

2. **bsub2** is partially reflexive. If an element does not contain a granule (a block), then it cannot be a substantial part of itself. It is transitive because $\subseteq$ is transitive. However antisymmetry can fail as the defining condition only guarantees $a^u = b^u$ whenever both $\mathbf{P}_s ab$ and $\mathbf{P}_s ba$ hold. Counterexamples can be easily constructed for the context, and are well-known.

3. Rest of the proof is mostly direct, and is left to the reader.

$\square$

4.1 Example

Part of this example was originally constructed for bited approximations in [20] by the present author and additional details considered by her in [23]. Let $H = \{x_1, x_2, x_3, x_4\}$ and T be a tolerance T on it generated by

$$\{(x_1, x_2), (x_2, x_3)\}.$$

Denoting the statement that the granule generated by x_1 is (x_1, x_2) by $(x_1 : x_2)$, let the granules be the set of predecessor neighborhoods:

$$\mathcal{G} = \{(x_1 : x_2), (x_2 : x_1, x_3), (x_3 : x_2), (x_4 :)\}$$

The different approximations (lower (l), upper (u) and bited upper (u_b)) are then as in Table 2 below (for the actual definitions of the approximations see [23] – the symbols are changed here). The 1-grade granular approximations (l_1 and u_1) are also included in the table. It is of interest to look at possible substantial part of relations that make the graded, standard, and bited approximations rational or are otherwise interesting.

$\sim$ is an equivalence on $\wp(H) = \underline{S}$ defined by

$$a \sim c \text{ if and only if } a^l = c^l \,\&\, a^{u_b} = c^{u_b}$$

The first column in the table is for keeping track of the elements in the quotient $\wp(H)|\sim$. The braces on sets have been omitted.

Table 2. Bited+ 1-grade approximations

$\wp(S)|\sim$	Subset	a	a^l	a^u	a^{cl}	a^{u_b}	a^{l_1}	a^{u_1}
B_1	A_1	x_1	$\emptyset$	x_2, x_1	x_2, x_3, x_4	x_1	$\emptyset$	x_1, x_2
B_2	A_2	x_2	$\emptyset$	x_1, x_2, x_3	x_4	x_1, x_2, x_3	$\emptyset$	x_1, x_2, x_3
B_3	A_3	x_3	$\emptyset$	x_1, x_2, x_3	x_1, x_2, x_4	x_3	$\emptyset$	x_1, x_2, x_3
B_4	A_4	x_4	x_4	x_4	x_1, x_2, x_3	x_4	$\emptyset$	$\emptyset$
B_5	A_5	x_1, x_2	x_1, x_2	x_1, x_2, x_3	x_4	x_1, x_2, x_3	x_1, x_2	x_1, x_2, x_3
B_2	A_6	x_1, x_3	$\emptyset$	x_1, x_2, x_3	x_4	x_1, x_2, x_3	$\emptyset$	x_1, x_2, x_3
B_6	A_7	x_1, x_4	x_4	H	x_2, x_3	x_1, x_4	$\emptyset$	x_1, x_2, x_3
B_7	A_8	x_2, x_3	x_2, x_3	x_1, x_2, x_3	x_4	x_1, x_2, x_3	x_2, x_3	x_1, x_2, x_3
B_8	A_9	x_2, x_4	x_4	H	$\emptyset$	H	$\emptyset$	x_1, x_2, x_3
B_9	A_{10}	x_3, x_4	x_4	H	x_1, x_2	x_3, x_4	$\emptyset$	x_1, x_2, x_3
B_{10}	A_{11}	x_1, x_2, x_3	x_1, x_2, x_3	x_1, x_2, x_3	x_4	x_1, x_2, x_3	x_1, x_2, x_3	x_1, x_2, x_3
B_{11}	A_{12}	x_1, x_2, x_4	x_1, x_2, x_4	H	$\emptyset$	H	x_1, x_2	x_1, x_2, x_3
B_{12}	A_{13}	x_2, x_3, x_4	x_2, x_3, x_4	H	$\emptyset$	H	x_2, x_3	x_1, x_2, x_3
B_8	A_{14}	x_1, x_3, x_4	x_4	H	$\emptyset$	H	$\emptyset$	x_1, x_2, x_3
B_{14}	A_{15}	H	H	H	$\emptyset$	H	x_1, x_2, x_3	x_1, x_2, x_3
B_{13}	A_{16}	$\emptyset$	$\emptyset$	$\emptyset$	H	$\emptyset$	$\emptyset$	$\emptyset$

In the context of this example, relative to the substantial parthood defined by **bsub2**, that is $\mathbf{P}_s ab$ if and only if $(\exists e)e^l \subseteq a^u \subseteq b^u \,\&\, \emptyset \subset e^l$, it can be seen that the computation is cumbersome, but can be simplified by a few if then case-based rules such as

1. e should be such that e^l is a member of

$$\{\{x_4\}, \{x_1, x_2\}, \{x_3, x_2\}, \{x_1, x_2, x_3\}, \{x_1, x_2\,x_4\}, \{x_2, x_3, x_4\}, H\}$$

2. If $e^l = H$, then $e = a = H = b$.
3. If $e^l = \{x_4\}$, then e can be any of $\{\{x_4\}, \{x_1, x_4\}, \{x_2, x_4\}, \{x_3, x_4\}, \{x_1, x_3, x_4\}\}$. Relative to this, the first component of the substantial parthood instance should be such that its upper approximation is $\{x_4\}$ or H.

To see that the relation is distinct from $\subseteq$, it may be noted that

$$\mathbf{P}_s\{x_2, x_3\}\{x_1, x_3, x_4\}$$

Graded Sub-case. By contrast, the computation of the substantial parthoods (except for $\mathbf{P}_s^1$) in Subsect. 2.6 are easy. $\mathbf{P}_s^3$ is defined as follows

$$\mathbf{P}_s^3 ab \text{ if and only if } \#(a \cap b) > k \,\& \, a \subseteq b \tag{s3}$$

In the context of this example, for $k = 1$, it can be seen that

$$\mathbf{P}_s^3 = \{(A_5, A_5), (A_5, A_{11}), (A_5, A_{12}), (A_5, A_{15}), (A_6, A_6), (A_6, A_{11}), (A_6, A_{14}),$$
$$(A_6, A_{15}), (A_7, A_7), (A_7, A_{12}), (A_7, A_{14}), (A_7, A_{15}), (A_8, A_8), (A_8, A_{11}),$$
$$(A_8, A_{13}), (A_8, A_{15}), (A_9, A_9), (A_9, A_{12}), (A_9, A_{13}), (A_9, A_{15}), (A_{10}, A_{10}),$$
$$(A_{10}, A_{13}), (A_{10}, A_{14}), (A_{10}, A_{15}), (A_{11}, A_{11}), (A_{11}, A_{15}), (A_{12}, A_{12}), (A_{12}, A_{15}),$$
$$(A_{13}, A_{13}), (A_{13}, A_{15}), (A_{14}, A_{14}), (A_{14}, A_{15}), (A_{15}, A_{15})\}$$

Interpretation of the context of this example in the pGpsGS framework with the above substantial parthood is straightforward.

- Now consider the set $\{x_1, x_2, x_4\}$. From the table, it is clear that

$$\{x_1, x_2, x_4\}^{l_1} = \{x_1, x_2\} = A_5$$

- It can be checked that this is an instance of a substantial lower approximation from the definition (relative to $\mathbf{P}_s^3$).
- All other instances of l_1 can be similarly checked.
- So l_1 is an example of a rational lower approximation (the full proof of this is in the theorem in the next section)
- By contrast l is not a rational lower approximation relative to $\mathbf{P}_s^3$.

Remark 1. This example also shows that definitions of substantial parthoods that involve approximations may have additional computational load to deal with.

4.2   Representation of Concept Maturation

A practical example concerning evaluation contexts (see [27, 39]) is constructed to demonstrate aspects of the framework in this section. Suppose that labeled or partially labeled data about the understanding of concepts in a class is available in the form (Student_Id, Concept). This can be used to generate a pGpsGS $\mathbb{S}$

under the following interpretation $\mathbb{S}$: a set of sets of crisp and vague concepts, $\vee$: union, $\wedge$: intersection, $\top$: a universal set of collections of concepts, $\perp$: the set containing the empty set of concepts or whatever that should be regarded as empty, l: a greatest set of crisp concepts that are part of the given element, u: a smallest set of crisp concepts of which the given element is part of, and the granulation predicate γ is determined by a suitable subset of sets of concepts. A parthood relation $\mathbf{P}$ can be defined on the set of concepts in a perspective as follows:

$$\mathbf{P}ab \leftrightarrow b \text{ is an accessible super concept from } a$$

Transitivity of $\mathbf{P}$ can fail possibly because the agency of accessibility may require more. In many cases it may be transitive. Antisymmetry cannot be expected to be true without an additional layer of equivalencing because different concepts that mutually explain each other are always possible. For example, straight lines can be modeled by different types of algebraic equations.

Proposition 8. *$\mathbf{P}$ is a reflexive relation that is not antisymmetric on the set of concepts, that in turn induces a reflexive non-antisymmetric relation on the granulation $\mathcal{G}$.*

Irrespective of whether $\mathbf{P}$ is a quasi-order or not, a *filter* of $\mathbf{P}$ is any subset $F \subseteq \mathbb{S}$ that satisfies

$$(\forall x \in F)(\forall z \in \mathbb{S})(\mathbf{P}xz \Rightarrow z \in F)$$

Such a filter can be used to define a *substantial part of* relation as follows: $\mathbf{P}_s ab$ if and only if $a \in F$ and $\mathbf{P}ab$. Concrete definitions of rational lower approximation relative to the substantial parthood can then be formulated as:

$$(\forall x)x^{l_s} = \vee\{a : a^l = a \,\&\, \mathbf{P}_s ax\}$$

Rational upper approximations on the other hand need to computed by the Definition 8.

5   Graded Rough Sets and the Framework

In Subsect. 2.6, a number of candidates for substantial parthood were introduced. The compatibility of these with defining conditions in the pGpsGS framework introduced are considered here.

Theorem 5. *In the context of Subsect. 2.6, taking $\vee = $ cup and $\mathbf{P}$ as set inclusion ($\subseteq$),*

s1 $\mathbf{P}_s^1$ *satisfies sub3, sub4, sub1, sub5 and sub6. In fact, it satisfies the much stronger condition sub9.*

$$(\forall a, b)\,(\mathbf{P}ab \longrightarrow \mathbf{P}_s ab) \tag{sub9}$$

s2 $\mathbf{P}_s^2$ *satisfies sub3, sub4, sub2, and sub5. A stronger version* (**Asy** *antisymmetry*) *of sub3 is satisfied by it.*

$$(\forall a, b)\,(\mathbf{P}_s a b \,\&\, \mathbf{P}_s b a \longrightarrow a = b) \tag{Asy}$$

s* $\mathbf{P}_s^*$ *satisfies sub4, sub5 and sub6.*
s3 $\mathbf{P}_s^3$ *satisfies Asy, sub3, sub4, sub2 and sub5*

Proof. **s1**

1. sub9 follows from the basic properties of l_k, and so sub6 and sub5 follow from that.
2. If $a^{lk} \subseteq b^{lk}$ and $b^{lk} \subseteq a^{lk}$ for any a and b then $a^{lk} = b^{lk}$. So sub3 holds when $\approx$ is interpreted as the k-lower rough equality for a fixed integer k.
3. sub4 follows from the monotonicity of l_k.
4. For any a, from $a^{lk} = a^{lk}$ it follows that $\mathbf{P}_s^1 a a$. So sub1 holds.

s2

1. For any a and b, if $\mathbf{P}_s^2 a b$ and $\mathbf{P}_s^2 b a$ holds then it is necessary that $a = b$. So both Asy and sub3 follow from this. Note that $\approx$ can be defined in a few different ways for this.
2. If for any a, b and e, $\mathbf{P}_s^2 a e \,\&\, \mathbf{P}_s^2 a b$, then $\#(a) > k$, $a \subseteq b$, and $a \subseteq e$. From the latter, it follows that $a \subseteq (b \cup e)$. So $\mathbf{P}_s^2 a (b \cup e)$ must hold. This proves sub4.
3. sub2 holds because $\mathbf{P} a b$ (or $a \subseteq b$) is part of the defining conditions of $\mathbf{P}_s^2$.
4. If for any a, b and e, $\mathbf{P}_s^2 b a \,\&\, \mathbf{P}_s^2 b e$, then $\#(a) > k$, as $b \subseteq a$. In conjunction with $a \subseteq e$ this yield $\mathbf{P}_s^2 a e$. So sub5 is proved.

s*

1. If for any a, b and e, $\mathbf{P}_s^2 a e \,\&\, \mathbf{P}_s^2 a b$, then $\#(a \cap e) > k$, $\#(a \cap b) > k$. So $\#(a \cap (b \cup e)) > k$. $(b \cup e) \not\subseteq a$ follows from $b \not\subseteq a$ and $e \not\subseteq a$. Therefore sub4 holds.
2. sub5 holds because if for any a, b and e, $\mathbf{P}_s^2 b a \,\&\, \mathbf{P}_s^2 b e$ and $a \subseteq e$, then $\#(a \cap b) > k$, $\#(e \cap b) > k$ ensures that $\#(a \cap e) > k$ and $e \not\subseteq a$. It may happen that $a = e$.
3. sub6 holds because if for any a, b and e, $\mathbf{P}_s^2 a b \,\&\, \mathbf{P}_s^2 e b$ and $a \subseteq e$, then $\#(a \cap b) > k$, $\#(e \cap b) > k$ ensures that $\#(a \cap e) > k$ and $e \not\subseteq a$. It may happen that $a = e$.
4. Under the conditions, sub3, Asy, sub9, sub1, and sub2 can all be shown to fail in general through simple examples even in the classical rough set context that is a special case of k-grade rough sets.

s3

1. Since $\subseteq$ is antisymmetric, therefore $\mathbf{P}_s^3$ satisfies Asy and sub3.
2. If for any a, b and e, $\mathbf{P}_s^2 a e \,\&\, \mathbf{P}_s^2 a b$, then $\#(a \cap e) > k$, $\#(a \cap b) > k$, $a \subseteq e$, and $a \subseteq b$. So $a \subseteq (e \cup b)$. This proves sub4.
3. sub2 holds because $\subseteq$ is part of the defining condition of $\mathbf{P}_s^3$.
4. sub5 holds because if for any a, b and e, $\mathbf{P}_s^2 b a \,\&\, \mathbf{P}_s^2 b e$ and $a \subseteq e$, then $\#(a \cap b) > k$, $\#(e \cap b) > k$ ensures that $\#(a \cap e) > k$.

5. sub6 holds because if for any a, b and e, $\mathbf{P}_s^2 ab \,\&\, \mathbf{P}_s^2 eb$ and $a \subseteq e$, then $\#(a \cap b) > k$, $\#(e \cap b) > k$ ensures that $\#(a \cap e) > k$.

$\square$

For convenience, the above theorem and related properties are tabulated below (∗ indicates the property is satisfied, and − that it is not necessarily satisfied) (Table 3):

Table 3. Summary k-grade rough sets

Rel	sub1	Tr	Sy	Asy	sub2	sub3	sub4	sub5	sub6	sub9
$\mathbf{P}_s^1$	∗	∗	−	−	−	∗	∗	∗	∗	∗
$\mathbf{P}_s^2$	−	−	−	∗	∗	∗	∗	∗	−	−
$\mathbf{P}_s^*$	−	−	−	−	−	−	∗	∗	∗	−
$\mathbf{P}_s^3$	−	∗	−	∗	∗	∗	∗	∗	−	−

Remark 2. From the above, it is clear that the introduced framework is quite close to handling substantial parthood in the context of granular k-grade rough sets. It also highlights the need for additional variants defined by a modified set of axioms.

6   Discussion

In this research,

- graded rough sets are generalized from a granular perspective,
- the literature is corrected and algebraic semantic aspects are proposed,
- the idea of substantial parthood implicit in its meta theory is explicitly reworked,
- a new granular framework for handling substantial part of relations, and concepts of rational approximations are introduced, related results are proved,
- detailed examples illustrating many aspects of the above are constructed, industrial applications are proposed, and
- it is shown that rationality of approximations is closely related to that of substantiality of part of relations used in the context.

Of these, the main achievements of this paper are in the introduction of the framework, recognition of the need for its fluidity, and the practical application to the industrial problem of sorting and grading (that can be developed further). Many aspects of the proposed framework need to be investigated, especially for the granular rough framework for rethinking cluster validity due to the present author [29]. This is one reason for highlighting the focus on graded rough sets by her. The latter is also related to variable precision rough sets and generalizations thereof. This will be considered separately.

The granular framework does not have nice modal logic properties. So the question of relating it to ideas of rationality in modal perspectives is an area for further exploration.

Acknowledgment. This research is supported by woman scientist grant No. WOS-A/PM-22/2019 of the Department of Science and Technology, India.

References

1. Akama, S., Murai, T., Kudo, Y.: Reasoning with Rough Sets. Springer, Cham (2018). https://doi.org/10.1007/978-3-319-72691-5
2. Banerjee, M., Chakraborty, M.K.: Rough sets through algebraic logic. Fund. Inform. **28**, 211–221 (1996)
3. Burkhardt, H., Seibt, J., Imaguire, G., Gerogiorgakis, S. (eds.): Handbook of Mereology. Philosophia Verlag, Munich (2017)
4. Burmeister, P.: A Model-Theoretic Oriented Approach to Partial Algebras. Akademie-Verlag (1986, 2002)
5. de Caro, F.: Graded modalities, II (canonical models). Stud. Logica. **47**, 1–10 (1988). https://doi.org/10.1007/BF00374047
6. de Caro, F., Fattorsi-Barnaba, M.: Graded modalities. III. Stud. Logica. **47**, 99–110 (1988). https://doi.org/10.1007/BF00370285
7. Cattaneo, G., Ciucci, D., Dubois, D.: Algebraic models of deviant modal operators based on de Morgan and Kleene lattices. Inf. Sci. **181**, 4075–4100 (2011). https://doi.org/10.1016/J.INS.2011.05.008
8. Cerrato, C.: Decidability by filtrations for graded normal logics (graded modalities V). Stud. Logica. **53**(1), 61–74 (1994)
9. Chen, J., van Ditmarsch, H., Greco, G., Tzimoulis, A.: Neighbourhood semantics for graded modal logic, pp. 1–12. ArXiv:2105.09202 (2021)
10. Fattorsi-Barnaba, M., de Caro, F.: Graded modalities. I. Stud. Logica. **44**, 197–221 (1985). https://doi.org/10.1007/BF00379767
11. Goble, L.F.: Grades of modality. Logique et Anal. (N.S.) **13**, 323–334 (1970)
12. van der Hoek, W., Meyer, J.-J.C.: Graded modalities in epistemic logic. In: Nerode, A., Taitslin, M. (eds.) LFCS 1992. LNCS, vol. 620, pp. 503–514. Springer, Heidelberg (1992). https://doi.org/10.1007/BFb0023902
13. Katzberg, T., Ziarko, W.: Variable precision extension of rough sets. Fund. Inform. **27**(2–3), 155–168 (1996)
14. Kolodny, N., Brunero, J.: Instrumental rationality. In: Zalta, E.N. (ed.) The Stanford Encyclopedia of Philosophy (Spring 2020). Stanford University (2020). https://plato.stanford.edu/archives/spr2020/entries/rationality-instrumental/
15. Lin, T.Y.: Granular computing-1: the concept of granulation and its formal model. Int. J. Granular Comput. Rough Sets Intell. Syst. **1**(1), 21–42 (2009)
16. Lin, T., Liu, Q.: Rough approximate operators: axiomatic rough set theory. In: Ziarko, W.P. (ed.) Rough Sets, Fuzzy Sets and Knowledge Discovery. WORKSHOPS COMP., pp. 256–260. Springer, London (1994). https://doi.org/10.1007/978-1-4471-3238-7_31
17. Lipscomb, D.: Basics of Cladistic Analysis. George Washington University, Washington D.C. (1998). https://www.researchgate.net/publication/221959140

18. Ljapin, E.S.: Partial Algebras and Their Applications. Kluwer Academic (1996)
19. Mani, A.: Esoteric rough set theory: algebraic semantics of a generalized VPRS and VPFRS. In: Peters, J.F., Skowron, A. (eds.) Transactions on Rough Sets VIII. LNCS, vol. 5084, pp. 175–223. Springer, Heidelberg (2008). https://doi.org/10.1007/978-3-540-85064-9_9
20. Mani, A.: Algebraic semantics of similarity-based bitten rough set theory. Fund. Inform. **97**(1–2), 177–197 (2009)
21. Mani, A.: Dialectics of counting and the mathematics of vagueness. In: Peters, J.F., Skowron, A. (eds.) Transactions on Rough Sets XV. LNCS, vol. 7255, pp. 122–180. Springer, Heidelberg (2012). https://doi.org/10.1007/978-3-642-31903-7_4
22. Mani, A.: Generalized ideals and co-granular rough sets. In: Polkowski, L., et al. (eds.) IJCRS 2017. LNCS (LNAI), vol. 10314, pp. 23–42. Springer, Cham (2017). https://doi.org/10.1007/978-3-319-60840-2_2
23. Mani, A.: Algebraic methods for granular rough sets. In: Mani, A., Cattaneo, G., Düntsch, I. (eds.) Algebraic Methods in General Rough Sets. TM, pp. 157–335. Springer, Cham (2018). https://doi.org/10.1007/978-3-030-01162-8_3
24. Mani, A.: Dialectical rough sets, parthood and figures of opposition-I. In: Peters, J.F., Skowron, A. (eds.) Transactions on Rough Sets XXI. LNCS, vol. 10810, pp. 96–141. Springer, Heidelberg (2019). https://doi.org/10.1007/978-3-662-58768-3_4
25. Mani, A.: Comparative approaches to granularity in general rough sets. In: Bello, R., Miao, D., Falcon, R., Nakata, M., Rosete, A., Ciucci, D. (eds.) IJCRS 2020. LNCS (LNAI), vol. 12179, pp. 500–517. Springer, Cham (2020). https://doi.org/10.1007/978-3-030-52705-1_37
26. Mani, A.: Functional extensions of knowledge representation in general rough sets. In: Bello, R., Miao, D., Falcon, R., Nakata, M., Rosete, A., Ciucci, D. (eds.) IJCRS 2020. LNCS (LNAI), vol. 12179, pp. 19–34. Springer, Cham (2020). https://doi.org/10.1007/978-3-030-52705-1_2
27. Mani, A.: Towards student centric rough concept inventories. In: Bello, R., Miao, D., Falcon, R., Nakata, M., Rosete, A., Ciucci, D. (eds.) IJCRS 2020. LNCS (LNAI), vol. 12179, pp. 251–266. Springer, Cham (2020). https://doi.org/10.1007/978-3-030-52705-1_19
28. Mani, A.: A note on graded granular rough sets. In: FRSA International Conference on Z Pawlak's 95th Birth Anniversary, pp. 1–9. Preprint, November 2021. https://doi.org/10.5281/zenodo.5623951
29. Mani, A.: General rough modeling of cluster analysis. In: Ramanna, S., Cornelis, C., Ciucci, D. (eds.) IJCRS 2021. LNCS (LNAI), vol. 12872, pp. 75–82. Springer, Cham (2021). https://doi.org/10.1007/978-3-030-87334-9_6
30. Orlowska, E.: Logic of indiscernible relations. Bull. Pol. Acad. Sci. (Math.) **33**, 475–485 (1987)
31. Pagliani, P.: Three lessons on the topological and algebraic hidden core of rough set theory. In: Mani, A., Cattaneo, G., Düntsch, I. (eds.) Algebraic Methods in General Rough Sets. TM, pp. 337–415. Springer, Cham (2018). https://doi.org/10.1007/978-3-030-01162-8_4
32. Pagliani, P., Chakraborty, M.: A Geometry of Approximation: Rough Set Theory: Logic, Algebra and Topology of Conceptual Patterns. Springer, Berlin (2008). https://doi.org/10.1007/978-1-4020-8622-9
33. Polkowski, L.: A rough-neural computation model based on rough mereology. In: Pal, S.K., et al. (eds.) Rough-Neural Computing. COGTECH, pp. 85–108. Springer, Heidelberg (2004). https://doi.org/10.1007/978-3-642-18859-6_4

34. Polkowski, L.: Approximate Reasoning by Parts. Springer, Heidelberg (2011). https://doi.org/10.1007/978-3-642-22279-5
35. Polkowski, L., Semeniuk–Polkowska, M.: Reasoning about concepts by rough mereological logics. In: Wang, G., Li, T., Grzymala-Busse, J.W., Miao, D., Skowron, A., Yao, Y. (eds.) RSKT 2008. LNCS (LNAI), vol. 5009, pp. 205–212. Springer, Heidelberg (2008). https://doi.org/10.1007/978-3-540-79721-0_31
36. Polkowski, L., Skowron, A.: Rough mereology: a new paradigm for approximate reasoning. Int. J. Approximate Reasoning 15(4), 333–365 (1996)
37. Qiao, J., Hu, B.Q.: Granular variable precision L-fuzzy rough sets based on residuated lattices. Fuzzy Sets Syst. 336, 148–166 (2018)
38. de Rijke, M.: A note on graded modal logic. Stud. Logica. 64(2), 271–283 (2000). https://doi.org/10.1023/A:1005245900406
39. Sands, D., Parker, M., Hedgeland, H., Jordan, S., Galloway, R.: Using concept inventories to measure understanding. High. Educ. Pedagogies 3(1), 173–182 (2018)
40. Skowron, A., Jankowski, A., Dutta, S.: Interactive granular computing. Granular Comput. 1(2), 95–113 (2016). https://doi.org/10.1007/s41066-015-0002-1
41. Syau, Y.R., Lin, E.B., Liau, C.J.: Neighborhood systems and variable precision generalized rough sets. Fund. Inform. 153, 271–290 (2017)
42. Vakarelov, D.: A modal logic for similarity relations in Pawlak knowledge representation systems. Fund. Inform. 15, 61–79 (1991)
43. Velesaca, H.O., Suárez, P.L., Mira, R., Sappa, A.D.: Computer vision based food grain classification: a comprehensive survey. Comput. Electron. Agric. 187, 106287:1–106287:13 (2021)
44. Yao, Y.: The art of granular computing. In: Kryszkiewicz, M., Peters, J.F., Rybinski, H., Skowron, A. (eds.) RSEISP 2007. LNCS (LNAI), vol. 4585, pp. 101–112. Springer, Heidelberg (2007). https://doi.org/10.1007/978-3-540-73451-2_12
45. Yao, Y.Y., Lin, T.Y.: Generalizing rough sets using modal logics. Intell. Autom. Soft Comput. 2(2), 103–120 (1996)
46. Zadeh, L.A.: Fuzzy sets and information granularity. In: Gupta, N., et al. (eds.) Advances in Fuzzy Set Theory and Applications, pp. 3–18. North Holland, Amsterdam (1979)
47. Zadeh, L.A.: Toward a theory of fuzzy information granulation and its centrality in human reasoning and fuzzy logic. Fuzzy Sets Syst. 90(2), 111–127 (1997)
48. Zhang, D.: Triangular norms on partially ordered sets. Fuzzy Sets Syst. 153, 195–209 (2005)
49. Zhang, X., Mo, Z., Xiong, F., Cheng, W.: Comparative study of variable precision rough set model and graded rough set model. Int. J. Approximate Reasoning 53, 104–116 (2012)
50. Zhao, S., Tsang, E.C., Chen, D.: The model of fuzzy variable precision rough sets. IEEE Trans. Fuzzy Syst. 17(2), 451–467 (2009). https://doi.org/10.1109/TFUZZ.2009.2013204
51. Ziarko, W.: Variable precision rough set model. J. Comput. Syst. Sci. 46, 39–59 (1993)

MADM Strategies Based on Arithmetic and Geometric Mean Operator Under Rough-Bipolar Neutrosophic Set Environment

Surapati Pramanik[1]([✉]) [ID], Suman Das[2] [ID], Rakhal Das[2] [ID],
and Binod Chandra Tripathy[2] [ID]

[1] Department of Mathematics, Nandalal Ghosh B.T. College, Narayanpur 743126,
West Bengal, India
sura_pati@yahoo.co.in, surapati.math@gmail.com
[2] Department of Mathematics, Tripura University, Agartala 799022,
West Tripura, India

Abstract. The main focus of this paper is to introduce some aggregation operators namely, Rough-Bipolar Neutrosophic Arithmetic Mean (RBNAM) operator and Rough-Bipolar Neutrosophic Geometric Mean (RBNGM) operator under Rough-Bipolar Neutrosophic Set (RBNS) environment. Besides, we present the concept of score and accuracy functions under the RBNS environment. Further, we propose two multi-attribute decision-making (MADM) strategies based on RBNAM operator and RBNGM operator respectively under the RBNS environment. Finally, we provide a real-life numerical example to validate the proposed MADM strategy.

Keywords: Rough set · RBNS · MADM · RBNAM · RBNGM

1 Introduction

Broumi et al. [1] introduced the Rough Neutrosophic Set (RNS) by combining rough set [2] and neutrosophic set [3] in 2014. Thereafter, Deli et al. [4] presented the bipolar neutrosophic set by combining Bipolar Fuzzy Set (BFS) [5] and neutrosophic set [3]. RNS [6–24] and BNS [25–29] are emerging as powerful mathematical tools for dealing with the indeterminate, incomplete and inconsistent information. Pramanik [30] presented an overview of RNS in 2020. In 2016, Pramanik and Mondal [31] introduced the notion of a new hybrid set structure, namely, Rough Bipolar Neutrosophic Set (RBNS) by combing the RNS and BFS. In our daily life, decision making plays an important role for selecting the best one from a set of alternatives based on certain conflicting criteria. For decision

Supported by organization FRS.

© Springer-Verlag GmbH Germany, part of Springer Nature 2022
J. F. Peters et al. (Eds.): TRS XXIII, LNCS 13610, pp. 60–76, 2022.
https://doi.org/10.1007/978-3-662-66544-2_5

making, operators play an important role. In RNS environment, Mondal and Pramanik [32] proposed two aggregation operators, namely, a Rough Neutrosophic Arithmetic Mean Operator (RNAMO) and a Rough Neutrosophic Geometric Mean Operator (RNGMO) and established their properties. In the same study Mondal and Pramanik [32] developed four decision making strategies.

We illustrate the following example to explain how the membership are determined.

Assume that a factory wants to fill a position in the office with work experience, technical knowledge, and age person. There are three candidates for the post. The selection committee uses the neutrosophic bipolar set. Suppose 30 questions are asked from the expert community and out of 30, 10 questions have to be answered correctly. The memberships for truth-ness (known), unknown and the falseness for bipolar can be determined for the candidate after the interview is over are as follows:

Then, $U = \{(u, 0.3, 0.5, 0.2, -0.5, -0.5, -0.3), \quad (v, 0.9, 0.5, 0.7, -0.6, -0.3, -0.4), (w, 0.6, 0.8, 0.8, -0.7, -0.4, -0.4)\}$.

Here in $(u, 0.3, 0.5, 0.2, -0.5, -0.5, -0.3)$, 0.3 is the positive truth membership value of the candidate u, 0.5 means 5 questions could not conclude, and 0.2 means 2 questions answered wrongly and corresponding negative membership represents the degree of confessionness on candidates face recognize by the experts of true, indeterminacy and false membership.

Remark: We are calculating bipolar neutrosophic values since, the confidence of answer is to be determined by the expert committee.

There is no study in the literature relating to MADM strategy under the RBNS environment. To explore the unexplored MADM strategy in RBNS environment, we extend the concept of arithmetic and geometric mean for RNSs to RBNSs and develop two MADM strategies under RBNS-environment Rough-Bipolar Neutrosophic Arithmetic Mean (RBNAM) operator and Rough-Bipolar Neutrosophic Geometric Mean (RBNGM) operator.

The article has been divided into different sections as follows:

Section 2 is on preliminaries and definitions. In this section, we present some relevant definitions and preliminary results, which are useful for the development of the main results of this article. In Sect. 3, we introduce the notion of rough-bipolar neutrosophic arithmetic mean operator and rough-bipolar neutrosophic geometric mean operator under the RBNS environment. In Sect. 4, we define score and accuracy functions under the RBNS environment, and studied the ranking system of Rough-Bipolar Neutrosophic Numbers (RBNNs). Section 5 presents two MADM strategies based on RBNAM operator and RBNGM operator under the RBNS environment. Section 6 validates the proposed MADM strategies, and provides a realistic numerical example to show the applicability and effectiveness of the proposed MADM strategies. In Sect. 7, we present comparison of the results from the developed strategies. Finally, in Sect. 8, we conclude the paper by stating some future scope of research.

2 Some Relevant Results

In this section, we provide some relevant definitions and results those are very useful for the preparation of the main results of this article.

Definition 1. *[3] A Neutrosophic Set (NS) U over a non-empty set W is defined by: $U = \{(u, T_U(u), I_U(u), F_U(u)) : u \in W,$ and $T_U(u), I_U(u), F_U(u) \in [0,1]\}$, and so $0 \leq T_U(u) + I_U(u) + F_U(u) \leq 3$, for all $u \in W$.*

Example 1. Assume that $W = \{u, v\}$ is a non-empty set. Then, $U = \{(u, 0.3, 0.5, 0.2), (v, 0.9, 0.5, 0.7)\}$ is an NS over W.

Definition 2. *[4] A Bipolar Neutrosophic Set U over a non-empty set W is defined by: $U = \{(u, T_U^+(u), I_U^+(u), F_U^+(u), T_U^-(u), I_U^-(u), F_U^-(u)): u \in W\}$, where $T_U^+(u), I_U^+(u), F_U^+(u) \in [0,1]$, and $T_U^-(u), I_U^-(u), F_U^-(u) \in [-1,0]$.*

Example 2. Suppose that $W = \{u, v\}$ is a fixed set. Then, $U = \{(u, 0.3, 0.5, 0.2, -0.5, -0.5, -0.3), (v, 0.9, 0.5, 0.7, -0.6, -0.3, -0.4)\}$ is a BNS over W.

Definition 3. *[4] Suppose that $U = \{(u, T_U^+(u), I_U^+(u), F_U^+(u), T_U^-(u), I_U^-(u), F_U^-(u)): u \in W\}$ is any BNS. Then, for each $u \in W$, $[T_U^+(u), I_U^+(u), F_U^+(u), T_U^-(u), I_U^-(u), F_U^-(u)]$ is called a Bipolar Neutrosophic Number.*

Definition 4. *[4] Assume that $u = [T_U^+(u), I_U^+(u), F_U^+(u), T_U^-(u), I_U^-(u), F_U^-(u)]$ and $v = [T_U^+(v), I_U^+(v), F_U^+(v), T_U^-(v), I_U^-(v), F_U^-(v)]$ are two BNNs. Then,*

1. $k.u = [1 - (1 - T_U^+(u))^k, (I_U^+(u))^k, (F_U^+(u))^k, -(-T_U^-(u))^k, -(1 - (1 - (-I_U^-(u)))^k, -(1 - (1 - (-F_U^-(u)))^k]$, where $k > 0$;
2. $u^k = [(T_U^+(u))^k, 1 - (1 - I_U^+(u))^k, 1 - (1 - F_U^+(u))^k, -(1 - (1 - (-T_U^-(u)))^k), -(-I_U^-(u))^k, -(-F_U^-(u))^k]$, where $k > 0$;
3. $u + v = [T_U^+(u) + T_U^+(v) - T_U^+(u).T_U^+(v), I_U^+(u).I_U^+(v), F_U^+(u).F_U^+(v), -T_U^-(u).T_U^-(u), (-I_U^-(u) - I_U^-(v) - I_U^-(u).I_U^-(v)), (-F_U^-(u) - F_U^-(v) - F_U^-(u).F_U^-(v))]$;
4. $u.v = [T_U^+(u).T_U^+(v), I_U^+(u) + I_U^+(v) - I_U^+(u).I_U^+(v), F_U^+(u) + F_U^+(v) - F_U^+(u).F_U^+(v), (-T_U^-(u) - T_U^-(v) - T_U^-(u).T_U^-(v)), -I_U^-(u).I_U^-(v), -F_U^-(u).F_U^-(v)]$.

Definition 5. *[1] Suppose that R is an equivalence relation on W, and $U = \{(u, T_U(u), I_U(u), F_U(u)) : u \in W\}$ is an NS over W. Then, the lower approximation $(\underline{N}(U))$ and the upper approximation $(\overline{N}(U))$ of U in the approximation space (W, R) are defined by:*

$$\underline{N}(U) = \{(u, T_{\underline{N}(U)}(u), I_{\underline{N}(U)}(u), F_{\underline{N}(U)}(u)): p \in [u]_R, u \in W\};$$
$$\overline{N}(U) = \{(u, T_{\overline{N}(U)}(u), I_{\overline{N}(U)}(u), F_{\overline{N}(U)}(u)): p \in [u]_R, u \in W\};$$

where $T_{\underline{N}(U)} = \bigwedge_{p \in [u]_R} T_U(u), I_{\underline{N}(U)} = \bigwedge_{p \in [u]_R} I_U(u), F_{\underline{N}(U)} = \bigwedge_{p \in [u]_R} F_U(u), T_{\overline{N}(U)} = \bigvee_{p \in [u]_R} T_U(u), I_{\overline{N}(U)} = bigvee_{p \in [u]_R} I_U(u)$ and $F_{\overline{N}(U)} = \bigvee_{p \in [u]_R} F_U(u)$. The pair $(\underline{N}(U), \overline{N}(U))$ is called a rough neutrosophic set in (W, R).

Example 3. Assume that $W = \{a, b, c, d, e\}$ is a fixed set, and R is an equivalence relation such that the partition of W is given by $W/R = \{(a, c), (b, e), (d)\}$. Suppose that $U = \{(a, 0.5, 0.3, 0.6), (b, 0.7, 0.6, 0.4), (c, 0.5, 0.5, 0.4), (d, 0.6, 0.5, 0.7), (e, 0.6, 0.3, 0.8)\}$ be an N-Set over W.

Then, $\underline{N}(U) = \{(a, 0.5, 0.3, 0.4), (b, 0.6, 0.3, 0.4), (c, 0.5, 0.3, 0.4), (d, 0.6, 0.5, 0.7), (e, 0.6, 0.3, 0.4)\}$ and $\overline{N}(U) = \{(a, 0.5, 0.5, 0.6), (b, 0.7, 0.6, 0.8), (c, 0.5, 0.5, 0.6), (d, 0.6, 0.5, 0.7), (e, 0.7, 0.6, 0.8)\}$.

Therefore, the pair $(\underline{N}(U), \overline{N}(U)) = (\{(a, 0.5, 0.3, 0.4), (b, 0.6, 0.3, 0.4), (c, 0.5, 0.3, 0.4), (d, 0.6, 0.5, 0.7), (e, 0.6, 0.3, 0.4)\}, \{(a, 0.5, 0.5, 0.6), (b, 0.7, 0.6, 0.8), (c, 0.5, 0.5, 0.6), (d, 0.6, 0.5, 0.7), (e, 0.7, 0.6, 0.8)\})$ is a *RNS* over W.

Definition 6. *[31] Suppose that R is an equivalence relation on a fixed set W, and $U = \{(u, T_U^+(u), I_U^+(u), F_U^+(u), T_U^-(u), I_U^-(u), F_U^-(u)) : u \in W\}$ is a BNS over W. Then, the lower approximation($\underline{N}(U)$) and the upper approximation ($\overline{N}(U)$) of U in (W, R) are defined by:*

$$\underline{N}(Q) = \{(u, T_{\underline{N}(Q)}^+(u), I_{\underline{N}(Q)}^+(u), F_{\underline{N}(Q)}^+(u), T_{\underline{N}(Q)}^-(u), I_{\underline{N}(Q)}^-(u), F_{\underline{N}(Q)}^-(u) :$$
$$\Lambda_p \in [u]_R, u \in W\};$$
$$\overline{N}(Q) = \{(u, T_{\overline{N}(Q)}^+(u), I_{\overline{N}(Q)}^+(u), F_{\overline{N}(Q)}^+(u), T_{\overline{N}(Q)}^-(u), I_{\overline{N}(Q)}^-(u), F_{\overline{N}(Q)}^-(u) :$$
$$\Lambda_p \in [u]_R, u \in W\};$$
$$T_{\underline{N}(Q)}^+(u) = \Lambda_p \in [u]_R, T_U^+(u), \quad I_{\underline{N}(Q)}^+(u) = \Lambda_p \in [u]_R, I_U^+(u), \quad F_{\underline{N}(Q)}^+(u) =$$
$$\Lambda_p \in [u]_R, F_U^+(u), \quad T_{\overline{N}(Q)}^+(u) = \Lambda_p \in [u]_R, T_U^+(u), \quad I_{\overline{N}(Q)}^+(u) = \Lambda_p \in$$
$$[u]_R, I_U^+(u), \quad F_{\overline{N}(Q)}^+(u) = \Lambda_p \in [u]_R, F_U^+(u), \text{ for all } u \in w;$$
$$T_{\overline{N}(Q)}^+(u) = \Lambda_p \in [u]_R, T_U^+(u), \quad I_{\overline{N}(Q)}^+(u) = \Lambda_p \in [u]_R, I_U^+(u), \quad F_{\overline{N}(Q)}^+(u) =$$
$$\Lambda_p \in [u]_R, F_U^+(u), \quad T_{\underline{N}(Q)}^+(u) = \Lambda_p \in [u]_R, T_U^+(u), \quad I_{\underline{N}(Q)}^+(u) = \Lambda_p \in$$
$$[u]_R, I_U^+(u), \quad F_{\underline{N}(Q)}^+(u) = \Lambda_p \in [u]_R, F_U^+(u), \text{ for all } u \in w;$$

Then, the pair $(\underline{N}(U), \overline{N}(U))$ is called a RBNS in (W, R).

Example 4. Suppose that $W = \{a, b, c, d, e\}$ is a fixed set, and R be an equivalence relation, where its partition of W is given by $W/R = \{(a, c), (b, e), (d)\}$. Suppose that $U = \{(a, 0.5, 0.3, 0.6, -0.3, -0.5, -0.3), (b, 0.7, 0.6, 0.4, -0.8, -0.5, -0.6), (c, 0.5, 0.5, 0.4, -0.4, -0.6, -0.1), (d, 0.6, 0.5, 0.7, -0.1, -0.8, -0.5), (e, 0.6, 0.3, 0.8, 0.0, -0.6, -0.7)\}$ be a *BN*-Set over W.

Then, $\underline{N}(U) = \{(a, 0.5, 0.3, 0.4, -0.4, -0.6, -0.3), (b, 0.6, 0.3, 0.4, -0.8, -0.6, -0.7), (c, 0.5, 0.3, 0.4, -0.4, -0.6, -0.3), (d, 0.6, 0.5, 0.7, -0.1, -0.8, -0.5), (e, 0.6, 0.3, 0.4, -0.8, -0.6, -0.7)\}$ and $\overline{N}(U) = \{(a, 0.5, 0.5, 0.6, -0.3, -0.5, -0.1), (b, 0.7, 0.6, 0.8, 0.0, -0.5, -0.6), (c, 0.5, 0.5, 0.6, -0.3, -0.5, -0.1), (d, 0.6, 0.5, 0.7, -0.1, -0.8, -0.5), (e, 0.7, 0.6, 0.8, 0.0, -0.5, -0.6)\}$.

Therefore, $(\underline{N}(U), \overline{N}(U)) = (\{(a, 0.5, 0.3, 0.4, -0.4, -0.6, -0.3), (b, 0.6, 0.3, 0.4, -0.8, -0.6, -0.7), (c, 0.5, 0.3, 0.4, -0.4, -0.6, -0.3), (d, 0.6, 0.5, 0.7, -0.1, -0.8, -0.5), (e, 0.6, 0.3, 0.4, -0.8, -0.6, -0.7)\}, \{(a, 0.5, 0.5, 0.6, -0.3, -0.5, -0.1), (b, 0.7, 0.6, 0.8, 0.0, -0.5, -0.6), (c, 0.5, 0.5, 0.6, -0.3, -0.5, -0.1), (d, 0.6, 0.5, 0.7, -0.1, -0.8, -0.5), (e, 0.7, 0.6, 0.8, 0.0, -0.5, -0.6)\})$ is a *RBNS* over W.

Definition 7. *[31] Suppose that $U = (\underline{N}(U), \overline{N}(U))$ is a RBNS over a fixed set W. Then, the complement of $U = (\underline{N}(U), \overline{N}(U))$ is defined as follows: $U^c = (\underline{N}(U)^c, \overline{N}(U)^c)$*

$$\underline{N}(Q)^c = \{(u, 1 - T^+_{\underline{N}(Q)}(u), 1 - I^+_{\underline{N}(Q)}(u), 1 - F^+_{\underline{N}(Q)}(u), -1 - T^-_{\underline{N}(Q)}(u), -1 -$$
$$I^-_{\underline{N}(Q)}(u), -1 - F^-_{\underline{N}(Q)}(u) : \Lambda_p \in [u]_R, u \in W\};$$

$$\overline{N}(Q)^c = \{(u, 1 - T^+_{\overline{N}(Q)}(u), 1 - I^+_{\overline{N}(Q)}(u), 1 - F^+_{\overline{N}(Q)}(u), 1 - T^-_{\overline{N}(Q)}(u), -1 -$$
$$I^-_{\overline{N}(Q)}(u), -1 - F^-_{\overline{N}(Q)}(u) : \Lambda_p \in [u]_R, u \in W\};$$

Definition 8. *[31] Assume that $U = (\underline{N}(U), \overline{N}(U))$ and $V = (\underline{N}(V), \overline{N}(V))$ are any two RBNSs over a fixed set W. Then, $U \subseteq V$ if and only if $T^+_{\underline{N}(Q)}(u) \le T^+_{\overline{N}(Q)}(u)$ $I^+_{\underline{N}(Q)}(u) \le I^+_{\overline{N}(Q)}(u)$ $F^+_{\underline{N}(Q)}(u) \le F^+_{\overline{N}(Q)}(u)$ $T^-_{\underline{N}(Q)}(u) \ge T^-_{\overline{N}(Q)}(u)$ $I^-_{\underline{N}(Q)}(u) \ge I^-_{\overline{N}(Q)}(u)$ $F^-_{\underline{N}(Q)}(u) \ge F^-_{\overline{N}(Q)}(u)$ for all $u \in W$.*

Definition 9. *[31] Suppose that $U = (\underline{N}(U), \overline{N}(U))$ and $V = (\underline{N}(V), \overline{N}(V))$ are any two RBNSs over a fixed set W. Then, their intersection i.e., $U \cap V$ is defined as follows:*
$$\underline{N}(U \cap V) = (\underline{N}(U \cap V), \overline{N}(U \cap V)), \text{ where}$$
$$\underline{N}(U \cap V) = \{u, min\{T^+_{\underline{N}(u)}(u), T^+_{\overline{N}(v)}(u)\}, \frac{I^+_{\underline{N}(u)}(u), I^+_{\overline{N}(v)}(u)}{2}, max\{F^+_{\underline{N}(u)}(u),$$
$$F^+_{\overline{N}(v)}(u), max\{T^-_{\underline{N}(u)}(u), T^-_{\overline{N}(v)}(v)\}, \frac{I^-_{\underline{N}(u)}(u), I^-_{\overline{N}(v)}(v)}{2},$$
$$max\{F^-_{\underline{N}(u)}(u), F^-_{\overline{N}(v)}(v) : p \in [u]_R, u \in W\}$$
and
$$\overline{N}(U \cap V) = \{u, min\{T^+_{\overline{N}(u)}(u), T^+_{\underline{N}(v)}(v)\}, \frac{I^+_{\overline{N}(u)}(u), I^+_{\underline{N}(v)}(v)}{2},$$
$$max\{F^+_{\overline{N}(u)}(u), F^+_{\underline{N}(v)}(u), min\{T^-_{\overline{N}(u)}(u), T^-_{\underline{N}(v)}(u)\}, \frac{I^-_{\overline{N}(u)}(u), I^-_{\underline{N}(v)}(u)}{2},$$
$$min\{F^-_{\overline{N}(u)}(u), F^-_{\underline{N}(v)}(u) : p \in [u]_R, u \in W\}$$

Definition 10. *[31] Suppose that $U = (\underline{N}(U), \overline{N}(U))$ and $V = (\underline{N}(V), \overline{N}(V))$ are any two RBNSs over a fixed set W. Then, their union i.e., $U \cup V$ is defined as follows:*
$$\underline{N}(U \cup V) = (\underline{N}(U \cup V), \overline{N}(U \cap V)), \text{ where } \underline{N}(U \cap V) = \{u, min\{T^+_{\underline{N}(u)}(u), T^+_{\overline{N}(v)}$$
$$(u)\}, \frac{I^+_{\underline{N}(u)}(u), I^+_{\overline{N}(v)}(u)}{2}, min\{F^+_{\underline{N}(u)}(u), F^+_{\overline{N}(v)}(u), min\{T^-_{\underline{N}(u)}(u), T^-_{\overline{N}(v)}(v)\},$$
$$\frac{I^-_{\underline{N}(u)}(u), I^-_{\overline{N}(v)}(u)}{2}, max\{F^-_{\underline{N}(u)}(u), F^-_{\overline{N}(u)}(u) : p \in [u]_R, u \in W\} \text{ and } \overline{N}(U \cap V) =$$
$$\{u, min\{T^+_{\overline{N}(u)}(u), T^+_{\underline{N}(v)}(v)\}, \frac{I^+_{\overline{N}(u)}(u), I^+_{\underline{N}(v)}(v)}{2}, max\{F^+_{\overline{N}(u)}(u), F^+_{\underline{N}(v)}(u),$$
$$min\{T^-_{\overline{N}(u)}(u), T^-_{\underline{N}(v)}(u)\}, \frac{I^-_{\overline{N}(u)}(u), I^-_{\underline{N}(v)}(u)}{2}, min\{F^-_{\overline{N}(u)}(u), F^-_{\underline{N}(v)}(u) : p \in [u]_R, u \in W\}$$

3   Arithmetic and Geometric Mean Operator Under RBNS Environment

In this section, we introduce the RBNAM operator and RBNGM operator under the rough-bipolar neutrosophic environment.

3.1   Rough-Bipolar Neutrosophic Arithmetic Mean Operator

Definition 11. *Assume that $U = (\underline{N}(U), \overline{N}(U))$ is a RBNS in (W, R), where $\underline{N}(U) = \{(u, T^+_{\underline{N}(U)}(u), I^+_{\underline{N}(U)}(u), F^+_{\underline{N}(U)}(u), T^-_{\underline{N}(U)}(u), I^-_{\underline{N}(U)}(u), F^-_{\underline{N}(U)}(u))\colon p \in [u]_R, u \in W\}$ and $\overline{N}(U) = \{(u, T^+_{\overline{N}(U)}(u), I^+_{\overline{N}(U)}(u), F^+_{\overline{N}(U)}(u), T^-_{\overline{N}(U)}(u), I^-_{\overline{N}(U)}(u), F^-_{\overline{N}(U)}(u))\colon p \in [u]_R, u \in W\}$ are the lower approximation and upper approximation of U respectively in (W, R). Then, for each $u \in W$, $[(T^+_{\underline{N}(U)}(u), I^+_{\underline{N}(U)}(u), F^+_{\underline{N}(U)}(u), T^-_{\underline{N}(U)}(u), I^-_{\underline{N}(U)}(u), F^-_{\underline{N}(U)}(u)), (\overline{N}(U) = \{(u, T^+_{\overline{N}(U)}(u), I^+_{\overline{N}(U)}(u), F^+_{\overline{N}(U)}(u), T^-_{\overline{N}(U)}(u), I^-_{\overline{N}(U)}(u), F^-_{\overline{N}(U)}(u))]$ is called a Rough Bipolar Neutrosophic Number (RBNN).*

Definition 12. *Suppose that $u_i = [\underline{N}(u_i), \overline{N}(u_i)]$, $i = 1, 2, 3, \ldots, n$ be a family of RBN-Numbers in (W, R). Then, the rough-bipolar neutrosophic arithmetic mean (RBNAM) operator is defined by*

$$RBNAM\ (u_1, u_2, \ldots \ldots, u_n) = [\frac{1}{n}\sum_{i=1}^{n}\underline{N}(u_i), \frac{1}{n}\sum_{i=1}^{n}\overline{N}(u_i)]. \qquad (1)$$

Example 5. Let $u = [(0.5, 0.3, 0.6, -0.3, -0.5, -0.3), (0.6, 0.2, 0.5, -0.1, -0.8, -0.5)]$, $v = [(0.3, 0.5, 0.4, -0.5, -0.3, -0.7), (0.6, 0.4, 0.2, -0.3, -0.5, -0.8)]$ be two *RBN*-Numbers. Then, *RBNAM* $(u, v) = [0.5((0.5, 0.3, 0.6, -0.3, -0.5, -0.3) + (0.3, 0.5, 0.4, -0.5, -0.3, -0.7)), 0.5((0.6, 0.2, 0.5, -0.1, -0.8, -0.5) + (0.6, 0.4, 0.2, -0.3, -0.5, -0.8))] = [(0.41, 0.39, 0.49, -0.39, -0.81, -0.89), (0.6, 0.28, 0.32, -0.17, -0.95, -0.95)]$. It is an *RBN*-Number.

3.2   Rough-Bipolar Neutrosophic Geometric Mean Operator

Definition 13. *Assume that $u_i = [\underline{N}(u_i), \overline{N}(u_i)]$, $i = 1, 2, 3, \ldots, n$ is a family of RBN-Numbers in (W, R). Then, the rough-bipolar neutrosophic geometric mean operator is defined by*

$$RBNGM(u_1, u_2, \ldots \ldots, u_n) = [\prod_{i=1}^{n}(\underline{N}(u_i))^{\frac{1}{n}}, \prod_{i=1}^{n}(\overline{N}(u_i))^{\frac{1}{n}}]. \qquad (2)$$

Theorem 3.1. Let $u_i = [\underline{N}(u_i), \overline{N}(u_i)]$, $i = 1, 2, 3, \ldots, n$ be a family of RBN-Numbers in (W, R). Then, the aggregated value RBNGM $(u_1, u_2, \ldots, u_n)$ of $u_i = [\underline{N}(u_i), \overline{N}(u_i)]$, $i = 1, 2, 3, \ldots, n$ is also an RBN-Number.

Proof. Suppose that $u_i = [\underline{N}(u_i), \overline{N}(u_i)]$, $i = 1, 2, 3, \ldots, n$ is a family of RBN-Numbers in (W, R). Therefore, for $i = 1, 2, 3, \ldots, n$, each of $[\underline{N}(u_i), \overline{N}(u_i)]$, are bipolar neutrosophic numbers. Then, from the Definition 2.4., it is clear that, $[\prod_{i=1}^{n}(\underline{N}(u_i))^{\frac{1}{n}}$ and $\prod_{i=1}^{n}(\overline{N}(u_i))^{\frac{1}{n}}]$. are the bipolar neutrosophic numbers. Hence, RBNGM $(u_1, u_2, \ldots, u_n) = [\prod_{i=1}^{n}(\underline{N}(u_i))^{\frac{1}{n}}, \prod_{i=1}^{n}(\overline{N}(u_i))^{\frac{1}{n}}]$ is an RBN-Number.

Example 6. Suppose that $u = [(0.5, 0.3, 0.6, -0.3, -0.5, -0.3), (0.6, 0.2, 0.5, -0.1, -0.8, -0.5)]$, $v = [(0.3, 0.5, 0.4, -0.5, -0.3, -0.7), (0.6, 0.4, 0.2, -0.3, -0.5, -0.8)]$ be two RBN-Numbers. Then, $RBNGM(u, v) = [((0.5, 0.3, 0.6, -0.3, -0.5, -0.3). (0.3, 0.5, 0.4, -0.5, -0.3, -0.7))^{0.5}, ((0.6, 0.2, 0.5, -0.1, -0.8, -0.5). (0.6, 0.4, 0.2, -0.3, -0.5, -0.8))^{0.5}] = [(0.15, 0.65, 0.76, -0.65, -0.15, -0.21)^{0.5}, (0.36, 0.52, 0.6, -0.37, -0.4, -0.4)^{0.5}] = [(0.39, 0.41, 0.51, -0.41, -0.39, -0.46), (0.6, 0.31, 0.37, -0.21, -0.63, -0.63)]$. It is an RBN-Number.

4 Score and Accuracy Function Under RBNS Environment

Definition 14. *Let* $u = [\underline{N}(u), \overline{N}(u)]$ *be an RBN-Number in the approximation space* (W, R). *Then, the score and accuracy functions of* u *are defined as follows:*

$$S(u) = \tfrac{1}{12} [T^+_{\underline{N}(U)}(u) + 1 - I^+_{\underline{N}(U)}(u) + 1 - F^+_{\underline{N}(U)}(u) + 1 + T^-_{\underline{N}(U)}(u) - I^-_{\underline{N}(U)}(u) - F^-_{\underline{N}(U)}(u) + T^+_{\overline{N}(U)}(u) + 1 - I^+_{\overline{N}(U)}(u) + 1 - F^+_{\overline{N}(U)}(u) + 1 + T^-_{\overline{N}(U)}(u) - I^-_{\overline{N}(U)}(u) - F^-_{\overline{N}(U)}(u)]$$

$$A(u) = \frac{1}{4}[T^+_{\underline{N}(U)}(u) - F^+_{\underline{N}(U)}(u) + T^-_{\underline{N}(U)}(u) - F^-_{\underline{N}(U)}(u) + T^+_{\overline{N}(U)}(u) - F^+_{\overline{N}(U)}(u) + T^-_{\overline{N}(U)}(u) - F^-_{\overline{N}(U)}(u)] \tag{3}$$

Theorem 4.1. If $u = [\underline{N}(u), \overline{N}(u)]$ be an RBNN in the approximation space (W, R), then $0 \leq S(u) \leq 1$ and $0 \leq A(u) \leq 1$.

Proof. Let $u = [\underline{N}(u), \overline{N}(u)]$ be an RBNN in the approximation space (W, R). So, both $\underline{N}(u), \overline{N}(u)$ are bipolar neutrosophic numbers.

Therefore, $0 \leq T^+_{\underline{N}(u)}(u), I^+_{\underline{N}(u)}(u), F^+_{\underline{N}(u)}(u), T^+_{\overline{N}(u)}(u), I^+_{\overline{N}(u)}(u), F^+_{\overline{N}(u)}(u) \leq 1$ and $-1 \leq T^-_{\underline{N}(u)}(u), I^-_{\underline{N}(u)}(u), F^-_{\underline{N}(u)}(u), T^-_{\overline{N}(u)}(u), I^-_{\overline{N}(u)}(u), F^-_{\overline{N}(u)}(u) \leq 0$. This implies, $0 \leq T^+_{\underline{N}(u)}(u) + 1 - I^+_{\underline{N}(u)}(u) + 1 - F^+_{\underline{N}(u)}(u) + 1 + T^-_{\underline{N}(u)}(u) - I^-_{\underline{N}(u)}(u) - F^-_{\underline{N}(u)}(u) + T^+_{\overline{N}(u)}(u) + 1 - I^+_{\overline{N}(u)}(u) + 1 - F^+_{\overline{N}(u)}(u) + 1 + T^-_{\overline{N}(u)}(u) - I^-_{\overline{N}(u)}(u) - F^-_{\overline{N}(u)}(u) \leq 12$

and $0 \leq T^+_{\underline{N}(u)}(u) - F^+_{\underline{N}(u)}(u) + T^-_{\underline{N}(u)}(u) - F^-_{\underline{N}(u)}(u) + T^+_{\overline{N}(u)}(u) - F^+_{\overline{N}(u)}(u) + T^-_{\overline{N}(u)}(u) - F^-_{\overline{N}(u)}(u) \leq 4$

This implies, $0 \leq \frac{1}{12}[T^+_{\underline{N}(U)}(u) + 1 - I^+_{\underline{N}(U)}(u) + 1 - F^+_{\underline{N}(U)}(u) + 1 + T^-_{\underline{N}(U)}(u) - I^-_{\underline{N}(U)}(u) - F^-_{\underline{N}(U)}(u) + T^+_{\overline{N}(U)}(u) + 1 - I^+_{\overline{N}(U)}(u) + 1 - F^+_{\overline{N}(U)}(u) + 1 + T^-_{\overline{N}(U)}(u) - I^-_{\overline{N}(U)}(u) - F^-_{\overline{N}(U)}(u)] \leq 1$

and $0 \leq \frac{1}{4}[T^+_{\underline{N}(U)}(u) - F^+_{\underline{N}(U)}(u) + T^-_{\underline{N}(U)}(u) - F^-_{\underline{N}(U)}(u) + T^+_{\overline{N}(U)}(u) - F^+_{\overline{N}(U)}(u) + T^-_{\overline{N}(U)}(u) - F^-_{\overline{N}(U)}(u)] \leq 1$

This implies, $0 \leq S(u) \leq 1$ and $0 \leq A(u) \leq 1$.

Theorem 4.2. Let $u = [\underline{N}(u), \overline{N}(u)]$ be an RBNN in the approximation space (W, R). Then,

(i) $S(u) = 0$ iff $T^+_{\underline{N}(U)}(u) = T^+_{\overline{N}(U)}(u) = 0$, $T^-_{\underline{N}(U)}(u) = T^-_{\overline{N}(U)}(u) = -1$, $I^+_{\underline{N}(U)}(u) = I^+_{\overline{N}(U)}(u) = F^+_{\underline{N}(U)}(u) = F^+_{\overline{N}(U)}(u) = 1$, $I^-_{\underline{N}(U)}(u) = I^-_{\overline{N}(U)}(u) = F^-_{\underline{N}(U)}(u) = F^-_{\overline{N}(U)}(u) = 0$.

(ii) $S(u) = 1$ iff $T^+_{\underline{N}(U)}(u) = T^+_{\overline{N}(U)}(u) = 1$, $T^-_{\underline{N}(U)}(u) = T^-_{\overline{N}(U)}(u) = 0$, $I^+_{\underline{N}(U)}(u) = I^+_{\overline{N}(U)}(u) = F^+_{\underline{N}(U)}(u) = F^+_{\overline{N}(U)}(u) = 0$, $I^-_{\underline{N}(U)}(u) = I^-_{\overline{N}(U)}(u) = F^-_{\underline{N}(U)}(u) = F^-_{\overline{N}(U)}(u) = -1$.

Proof. (i) Let $u = [\underline{N}(u), \overline{N}(u)]$ be an RBNN in the approximation space (W, R) such that $T^+_{\underline{N}(U)}(u) = T^+_{\overline{N}(U)}(u) = 0$, $T^-_{\underline{N}(U)}(u) = T^-_{\overline{N}(U)}(u) = -1$, $I^+_{\underline{N}(U)}(u) = I^+_{\overline{N}(U)}(u) = F^+_{\underline{N}(U)}(u) = F^+_{\overline{N}(U)}(u) = 1$, $I^-_{\underline{N}(U)}(u) = I^-_{\overline{N}(U)}(u) = F^-_{\underline{N}(U)}(u) = F^-_{\overline{N}(U)}(u) = 0$.

Thus, $T^+_{\underline{N}(U)}(u) + 1 - I^+_{\underline{N}(U)}(u) + 1 - F^+_{\underline{N}(U)}(u) + 1 + T^-_{\underline{N}(U)}(u) - I^-_{\underline{N}(U)}(u) - F^-_{\underline{N}(U)}(u) + T^+_{\overline{N}(U)}(u) + 1 - I^+_{\overline{N}(U)}(u) + 1 - F^+_{\overline{N}(U)}(u) + 1 + T^-_{\overline{N}(U)}(u) - I^-_{\overline{N}(U)}(u) - F^-_{\overline{N}(U)}(u) = 0$.

This implies, $\frac{1}{12}[T^+_{\underline{N}(U)}(u) + 1 - I^+_{\underline{N}(U)}(u) + 1 - F^+_{\underline{N}(U)}(u) + 1 + T^-_{\underline{N}(U)}(u) - I^-_{\underline{N}(U)}(u) - F^-_{\underline{N}(U)}(u) + T^+_{\overline{N}(U)}(u) + 1 - I^+_{\overline{N}(U)}(u) + 1 - F^+_{\overline{N}(U)}(u) + 1 + T^-_{\overline{N}(U)}(u) - I^-_{\overline{N}(U)}(u) - F^-_{\overline{N}(U)}(u)] = 0$

This implies, $S(u) = 0$.

Conversely, let $u = [\underline{N}(u), \overline{N}(u)]$ be an RBNN in the approximation space (W, R) such that $S(u) = 0$.

Therefore, $\frac{1}{12}[T^+_{\underline{N}(U)}(u) + 1 - I^+_{\underline{N}(U)}(u) + 1 - F^+_{\underline{N}(U)}(u) + 1 + T^-_{\underline{N}(U)}(u) - I^-_{\underline{N}(U)}(u) - F^-_{\underline{N}(U)}(u) + T^+_{\overline{N}(U)}(u) + 1 - I^+_{\overline{N}(U)}(u) + 1 - F^+_{\overline{N}(U)}(u) + 1 + T^-_{\overline{N}(U)}(u) - I^-_{\overline{N}(U)}(u) - F^-_{\overline{N}(U)}(u)] = 0$

Hence, $T^+_{\underline{N}(U)}(u) + 1 - I^+_{\underline{N}(U)}(u) + 1 - F^+_{\underline{N}(U)}(u) + 1 + T^-_{\underline{N}(U)}(u) - I^-_{\underline{N}(U)}(u) - F^-_{\underline{N}(U)}(u) + T^+_{\overline{N}(U)}(u) + 1 - I^+_{\overline{N}(U)}(u) + 1 - F^+_{\overline{N}(U)}(u) + 1 + T^-_{\overline{N}(U)}(u) - I^-_{\overline{N}(U)}(u) -

$F^-_{\overline{N}(U)}(u) = 0$.

This implies, $T^+_{\underline{N}(U)}(u) = T^+_{\overline{N}(U)}(u) = 0$, $T^-_{\underline{N}(U)}(u) = T^-_{\overline{N}(U)}(u) = -1$, $I^+_{\underline{N}(U)}(u) = I^+_{\overline{N}(U)}(u) = F^+_{\underline{N}(U)}(u) = F^+_{\overline{N}(U)}(u) = 1$, $I^-_{\underline{N}(U)}(u) = I^-_{\overline{N}(U)}(u) = F^-_{\underline{N}(U)}(u) = F^-_{\overline{N}(U)}(u) = 0$.

Hence, $S(u) = 0$ iff $T^+_{\underline{N}(U)}(u) = T^+_{\overline{N}(U)}(u) = 0$, $T^-_{\underline{N}(U)}(u) = T^-_{\overline{N}(U)}(u) = -1$, $I^+_{\underline{N}(U)}(u) = I^+_{\overline{N}(U)}(u) = F^+_{\underline{N}(U)}(u) = F^+_{\overline{N}(U)}(u) = 1$, $I^-_{\underline{N}(U)}(u) = I^-_{\overline{N}(U)}(u) = F^-_{\underline{N}(U)}(u) = F^-_{\overline{N}(U)}(u) = 0$.

(ii) Let $u = [\underline{N}(u), \overline{N}(u)]$ be an RBNN in the approximation space (W, R) such that $T^+_{\underline{N}(U)}(u) = T^+_{\overline{N}(U)}(u) = 1$, $T^-_{\underline{N}(U)}(u) = T^-_{\overline{N}(U)}(u) = 0$, $I^+_{\underline{N}(U)}(u) = I^+_{\overline{N}(U)}(u) = F^+_{\underline{N}(U)}(u) = F^+_{\overline{N}(U)}(u) = 0$, $I^-_{\underline{N}(U)}(u) = I^-_{\overline{N}(U)}(u) = F^-_{\underline{N}(U)}(u) = F^-_{\overline{N}(U)}(u) = -1$.

Thus, $T^+_{\underline{N}(U)}(u)+1-I^+_{\underline{N}(U)}(u)+1-F^+_{\underline{N}(U)}(u)+1+T^-_{\underline{N}(U)}(u)-I^-_{\underline{N}(U)}(u)-F^-_{\underline{N}(U)}(u)+ T^+_{\overline{N}(U)}(u)+1-I^+_{\overline{N}(U)}(u)+1-F^+_{\overline{N}(U)}(u)+1+T^-_{\overline{N}(U)}(u)-I^-_{\overline{N}(U)}(u)-F^-_{\overline{N}(U)}(u) = 12$.

This implies, $\frac{1}{12}[T^+_{\underline{N}(U)}(u) + 1 - I^+_{\underline{N}(U)}(u) + 1 - F^+_{\underline{N}(U)}(u) + 1 + T^-_{\underline{N}(U)}(u) - I^-_{\underline{N}(U)}(u) - F^-_{\underline{N}(U)}(u) + T^+_{\overline{N}(U)}(u) + 1 - I^+_{\overline{N}(U)}(u) + 1 - F^+_{\overline{N}(U)}(u) + 1 + T^-_{\overline{N}(U)}(u) - I^-_{\overline{N}(U)}(u) - F^-_{\overline{N}(U)}(u)] = 1$

This implies, $S(u) = 1$.

Conversely, let $u = [\underline{N}(u), \overline{N}(u)]$ be an RBNN in the approximation space (W, R) such that $S(u) = 1$.

Therefore, $\frac{1}{12}[T^+_{\underline{N}(U)}(u) + 1 - I^+_{\underline{N}(U)}(u) + 1 - F^+_{\underline{N}(U)}(u) + 1 + T^-_{\underline{N}(U)}(u) - I^-_{\underline{N}(U)}(u) - F^-_{\underline{N}(U)}(u) + T^+_{\overline{N}(U)}(u) + 1 - I^+_{\overline{N}(U)}(u) + 1 - F^+_{\overline{N}(U)}(u) + 1 + T^-_{\overline{N}(U)}(u) - I^-_{\overline{N}(U)}(u) - F^-_{\overline{N}(U)}(u)] = 1$

Hence, $T^+_{\underline{N}(U)}(u)+1-I^+_{\underline{N}(U)}(u)+1-F^+_{\underline{N}(U)}(u)+1+T^-_{\underline{N}(U)}(u)-I^-_{\underline{N}(U)}(u)- F^-_{\underline{N}(U)}(u)+T^+_{\overline{N}(U)}(u)+1-I^+_{\overline{N}(U)}(u)+1-F^+_{\overline{N}(U)}(u)+1+T^-_{\overline{N}(U)}(u)-I^-_{\overline{N}(U)}(u)- F^-_{\overline{N}(U)}(u) = 12$.

This implies, $T^+_{\underline{N}(U)}(u) = T^+_{\overline{N}(U)}(u) = 1$, $T^-_{\underline{N}(U)}(u) = T^-_{\overline{N}(U)}(u) = 0$, $I^+_{\underline{N}(U)}(u) = I^+_{\overline{N}(U)}(u) = F^+_{\underline{N}(U)}(u) = F^+_{\overline{N}(U)}(u) = 0$, $I^-_{\underline{N}(U)}(u) = I^-_{\overline{N}(U)}(u) = F^-_{\underline{N}(U)}(u) = F^-_{\overline{N}(U)}(u) = -1$.

Hence, $S(u) = 1$ iff $T^+_{\underline{N}(U)}(u) = T^+_{\overline{N}(U)}(u) = 1$, $T^-_{\underline{N}(U)}(u) = T^-_{\overline{N}(U)}(u) = 0$, $I^+_{\underline{N}(U)}(u) = I^+_{\overline{N}(U)}(u) = F^+_{\underline{N}(U)}(u) = F^+_{\overline{N}(U)}(u) = 0$, $I^-_{\underline{N}(U)}(u) = I^-_{\overline{N}(U)}(u) = F^-_{\underline{N}(U)}(u) = F^-_{\overline{N}(U)}(u) = -1$.

Theorem 4.3. Let $u = [\underline{N}(u), \overline{N}(u)]$ be an RBNN in the approximation space (W, R). Then,

(i) $A(u) = 0$ iff $T^+_{\underline{N}(U)}(u) = T^+_{\overline{N}(U)}(u) = 0$, $T^-_{\underline{N}(U)}(u) = T^-_{\overline{N}(U)}(u) = -1$, $F^+_{\underline{N}(U)}(u) = F^+_{\overline{N}(U)}(u) = 1$, $F^-_{\underline{N}(U)}(u) = F^-_{\overline{N}(U)}(u) = 0$.

(ii) $A(u) = 1$ iff $T^+_{\underline{N}(U)}(u) = T^+_{\overline{N}(U)}(u) = 1$, $T^-_{\underline{N}(U)}(u) = T^-_{\overline{N}(U)}(u) = 0$, $F^+_{\underline{N}(U)}(u) = F^+_{\overline{N}(U)}(u) = 0$, $F^-_{\underline{N}(U)}(u) = F^-_{\overline{N}(U)}(u) = -1$.

Proof. (i) Let $u = [\underline{N}(u), \overline{N}(u)]$ be an RBNN in the approximation space (W, R) such that $T^+_{\underline{N}(U)}(u) = T^+_{\overline{N}(U)}(u) = 0$, $T^-_{\underline{N}(U)}(u) = T^-_{\overline{N}(U)}(u) = -1$, $F^+_{\underline{N}(U)}(u) = F^+_{\overline{N}(U)}(u) = 1$, $F^-_{\underline{N}(U)}(u) = F^-_{\overline{N}(U)}(u) = 0$.

Therefore, $T^+_{\underline{N}(U)}(u) + 1 - F^+_{\underline{N}(U)}(u) + 1 + T^-_{\underline{N}(U)}(u) - F^-_{\underline{N}(U)}(u) + T^+_{\overline{N}(U)}(u) + 1 - F^+_{\overline{N}(U)}(u) + 1 + T^-_{\overline{N}(U)}(u) - F^-_{\overline{N}(U)}(u) = 0$.

This implies, $\frac{1}{4}[T^+_{\underline{N}(U)}(u) + 1 - F^+_{\underline{N}(U)}(u) + 1 + T^-_{\underline{N}(U)}(u) - F^-_{\underline{N}(U)}(u) + T^+_{\overline{N}(U)}(u) + 1 - F^+_{\overline{N}(U)}(u) + 1 + T^-_{\overline{N}(U)}(u) - F^-_{\overline{N}(U)}(u)] = 0$

This implies, $A(u) = 0$.

Conversely, let $u = [\underline{N}(u), \overline{N}(u)]$ be an RBNN in the approximation space (W, R) such that $A(u) = 0$.

Therefore, $\frac{1}{4}[T^+_{\underline{N}(U)}(u) + 1 - F^+_{\underline{N}(U)}(u) + 1 + T^-_{\underline{N}(U)}(u) - F^-_{\underline{N}(U)}(u) + T^+_{\overline{N}(U)}(u) + 1 - F^+_{\overline{N}(U)}(u) + 1 + T^-_{\overline{N}(U)}(u) - F^-_{\overline{N}(U)}(u)] = 0$

This implies, $T^+_{\underline{N}(U)}(u) + 1 - F^+_{\underline{N}(U)}(u) + 1 + T^-_{\underline{N}(U)}(u) - F^-_{\underline{N}(U)}(u) + T^+_{\overline{N}(U)}(u) + 1 - F^+_{\overline{N}(U)}(u) + 1 + T^-_{\overline{N}(U)}(u) - F^-_{\overline{N}(U)}(u) = 0$.

This implies, $T^+_{\underline{N}(U)}(u) = T^+_{\overline{N}(U)}(u) = 0$, $T^-_{\underline{N}(U)}(u) = T^-_{\overline{N}(U)}(u) = -1$, $F^+_{\underline{N}(U)}(u) = F^+_{\overline{N}(U)}(u) = 1$, $F^-_{\underline{N}(U)}(u) = F^-_{\overline{N}(U)}(u) = 0$.

Hence, $A(u) = 0$ iff $T^+_{\underline{N}(U)}(u) = T^+_{\overline{N}(U)}(u) = 0$, $T^-_{\underline{N}(U)}(u) = T^-_{\overline{N}(U)}(u) = -1$, $F^+_{\underline{N}(U)}(u) = F^+_{\overline{N}(U)}(u) = 1$, $F^-_{\underline{N}(U)}(u) = F^-_{\overline{N}(U)}(u) = 0$.

(ii) Let $u = [\underline{N}(u), \overline{N}(u)]$ be an RBNN in the approximation space (W, R) such that $T^+_{\underline{N}(U)}(u) = T^+_{\overline{N}(U)}(u) = 1$, $T^-_{\underline{N}(U)}(u) = T^-_{\overline{N}(U)}(u) = 0$, $F^+_{\underline{N}(U)}(u) = F^+_{\overline{N}(U)}(u) = 0$, $F^-_{\underline{N}(U)}(u) = F^-_{\overline{N}(U)}(u) = -1$.

Therefore, $T^+_{\underline{N}(U)}(u) + 1 - F^+_{\underline{N}(U)}(u) + 1 + T^-_{\underline{N}(U)}(u) - F^-_{\underline{N}(U)}(u) + T^+_{\overline{N}(U)}(u) + 1 - F^+_{\overline{N}(U)}(u) + 1 + T^-_{\overline{N}(U)}(u) - F^-_{\overline{N}(U)}(u) = 4$.

This implies, $\frac{1}{4}[T^+_{\underline{N}(U)}(u) + 1 - F^+_{\underline{N}(U)}(u) + 1 + T^-_{\underline{N}(U)}(u) - F^-_{\underline{N}(U)}(u) + T^+_{\overline{N}(U)}(u) + 1 - F^+_{\overline{N}(U)}(u) + 1 + T^-_{\overline{N}(U)}(u) - F^-_{\overline{N}(U)}(u)] = 1$

This implies, $A(u) = 1$.

Conversely, let $u = [\underline{N}(u), \overline{N}(u)]$ be an RBNN in the approximation space (W, R) such that $A(u) = 1$.

Therefore, $\frac{1}{4}[T^+_{\underline{N}(U)}(u) + 1 - F^+_{\underline{N}(U)}(u) + 1 + T^-_{\underline{N}(U)}(u) - F^-_{\underline{N}(U)}(u) + T^+_{\overline{N}(U)}(u) + 1 - F^+_{\overline{N}(U)}(u) + 1 + T^-_{\overline{N}(U)}(u) - F^-_{\overline{N}(U)}(u)] = 1$

This implies, $T^+_{\underline{N}(U)}(u) + 1 - F^+_{\underline{N}(U)}(u) + 1 + T^-_{\underline{N}(U)}(u) - F^-_{\underline{N}(U)}(u) + T^+_{\overline{N}(U)}(u) + 1 - F^+_{\overline{N}(U)}(u) + 1 + T^-_{\overline{N}(U)}(u) - F^-_{\overline{N}(U)}(u) = 4$.

This implies, $T^+_{\underline{N}(U)}(u) = T^+_{\overline{N}(U)}(u) = 1$, $T^-_{\underline{N}(U)}(u) = T^-_{\overline{N}(U)}(u) = 0$, $F^+_{\underline{N}(U)}(u) = F^+_{\overline{N}(U)}(u) = 0$, $F^-_{\underline{N}(U)}(u) = F^-_{\overline{N}(U)}(u) = -1$.

Hence, $A(u) = 1$ iff $T^+_{\underline{N}(U)}(u) = T^+_{\overline{N}(U)}(u) = 1$, $T^-_{\underline{N}(U)}(u) = T^-_{\overline{N}(U)}(u) = 0$, $F^+_{\underline{N}(U)}(u) = F^+_{\overline{N}(U)}(u) = 0$, $F^-_{\underline{N}(U)}(u) = F^-_{\overline{N}(U)}(u) = -1$.

Theorem 4.4. If $u = [\underline{N}(u), \overline{N}(u)]$ and $v = [\underline{N}(v), \overline{N}(v)]$ be two RBNNs in the approximation space (W, R) such that $u \subseteq v$, then $S(u) \leq S(v)$.

Proof. Suppose that $u = [\underline{N}(u), \overline{N}(u)]$ and $v = [\underline{N}(v), \overline{N}(v)]$ be two RBNNs in the approximation space (W, R) such that $u \subseteq v$.

Therefore, $T^+_{\underline{N}(U)}(u) \leq T^+_{\overline{N}(U)}(v), I^+_{\underline{N}(U)}(u) \geq I^+_{\overline{N}(U)}(v), F^+_{\underline{N}(U)}(u) \geq F^+_{\overline{N}(U)}(v), T^-_{\underline{N}(U)}(u) \geq T^-_{\overline{N}(U)}(v), I^-_{\underline{N}(U)}(u) \leq I^-_{\overline{N}(U)}(v)$ and $F^-_{\underline{N}(U)}(u) \leq F^-_{\overline{N}(U)}(v)$.

Now, $S(u) = \frac{1}{12}[T^+_{\underline{N}(U)}(u) + 1 - I^+_{\underline{N}(U)}(u) + 1 - F^+_{\underline{N}(U)}(u) + 1 + T^-_{\underline{N}(U)}(u) - I^-_{\underline{N}(U)}(u) - F^-_{\underline{N}(U)}(u) + T^+_{\overline{N}(U)}(u) + 1 - I^+_{\overline{N}(U)}(u) + 1 - F^+_{\overline{N}(U)}(u) + 1 + T^-_{\overline{N}(U)}(u) - I^-_{\overline{N}(U)}(u) - F^-_{\overline{N}(U)}(u)] \leq \frac{1}{12}[T^+_{\underline{N}(U)}(v) + 1 - I^+_{\underline{N}(U)}(v) + 1 - F^+_{\underline{N}(U)}(v) + 1 + T^-_{\underline{N}(U)}(v) - I^-_{\underline{N}(U)}(v) - F^-_{\underline{N}(U)}(v) + T^+_{\overline{N}(U)}(v) + 1 - I^+_{\overline{N}(U)}(v) + 1 - F^+_{\overline{N}(U)}(v) + 1 + T^-_{\overline{N}(U)}(v) - I^-_{\overline{N}(U)}(v) - F^-_{\overline{N}(U)}(v)] = S(v)$

Hence, $u \subseteq v$ implies $S(u) \leq S(v)$.

Theorem 4.5. If $u = [\underline{N}(u), \overline{N}(u)]$ and $v = [\underline{N}(v), \overline{N}(v)]$ be two RBNNs in the approximation space (W, R) such that $u \subseteq v$, then $A(u) \leq A(v)$.

Proof. Suppose that $u = [\underline{N}(u), \overline{N}(u)]$ and $v = [\underline{N}(v), \overline{N}(v)]$ be two RBNNs in the approximation space (W, R) such that $u \subseteq v$. Therefore, $T^+_{\underline{N}(U)}(u) \leq T^+_{\overline{N}(U)}(v), F^+_{\underline{N}(U)}(u) \geq F^+_{\overline{N}(U)}(v), T^-_{\underline{N}(U)}(u) \geq T^-_{\overline{N}(U)}(v)$ and $F^-_{\underline{N}(U)}(u) \leq F^-_{\overline{N}(U)}(v)$.

Now, $A(u) = \frac{1}{4}[T^+_{\underline{N}(U)}(u) + 1 - F^+_{\underline{N}(U)}(u) + 1 + T^-_{\underline{N}(U)}(u) - F^-_{\underline{N}(U)}(u) + T^+_{\overline{N}(U)}(u) + 1 - F^+_{\overline{N}(U)}(u) + 1 + T^-_{\overline{N}(U)}(u) - F^-_{\overline{N}(U)}(u)] \leq \frac{1}{4}[T^+_{\underline{N}(U)}(v) + 1 - F^+_{\underline{N}(U)}(v) + 1 + T^-_{\underline{N}(U)}(v) - F^-_{\underline{N}(U)}(v) + T^+_{\overline{N}(U)}(v) + 1 - F^+_{\overline{N}(U)}(v) + 1 + T^-_{\overline{N}(U)}(v) - F^-_{\overline{N}(U)}(v)] = A(v)$

Hence, $u \subseteq v$ implies $A(u) \leq A(v)$.

Definition 15. *Suppose that $u = [\underline{N}(u), \overline{N}(u)]$ and $v = [\underline{N}(v), \overline{N}(v)]$ are any two RBN-Numbers in (W, R). Then,*

1. *if $S(u) > S(v)$, then $u > v$;*
2. *if $S(u) = S(v)$, $A(u) > A(v)$, then $u > v$;*
3. *if $S(u) = S(v)$, $A(u) = A(v)$, $T^+_{\underline{N}(U)}(u) > T^+_{\underline{N}(U)}(v)$, $T^-_{\underline{N}(U)}(u) > T^-_{\underline{N}(U)}(v)$, $T^+_{\overline{N}(U)}(u) > T^+_{\overline{N}(U)}(v)$, $T^-_{\overline{N}(U)}(u) > T^-_{\overline{N}(U)}(v)$, then $u > v$.*

Example 7. Suppose that $(\underline{N}(U), \overline{N}(U))$ is an *RBNS* in (W, R) as shown in Example 2.4.

It
is clear that $a = [(0.5, 0.3, 0.4, -0.4, -0.6, -0.3), (0.5, 0.5, 0.6, -0.3, -0.5, -0.1)]$

and $b = [(0.6, 0.3, 0.4, -0.8, -0.6, -0.7), (0.7, 0.6, 0.8, 0.0, -0.5, -0.6)]$ are two RBN-Numbers in (W, R) for a and $b(\in W)$ respectively. Then, $S(a) = 0.5$ and $S(b) = 0.5667$. Here, $S(b) > S(a)$. Therefore, $b > a$.

5   *MADM* Strategy Under the *RBNS* Environment

In this section, we propose two $MADM$ strategies by using the $RBNAM$ operator and $RBNGM$ operator respectively under the $RBNS$-environment.

Consider a family of possible alternatives $\{L_1, L_2, L_3, \ldots \ldots, L_m\}$, and let $\{A_1, A_2, \ldots \ldots, A_n\}$ be a collection of all attributes. Now, the decision maker provides his/her evaluation information for each alternative based on the attributes in terms of RBN-Numbers.

5.1   RBNAM Operator Based *MADM* Strategy

The MADM strategy is formulated using the following steps.

Step-1: The decision maker provides the evaluation information for each alternative against the attributes by using the RBN-Number. Here, $L_{A_{ij}}) = [\underline{N}(L_{A_{ij}}), \overline{N}(L_{A_i})]$, $i = 1, 2, \ldots, m$; $j = 1, 2, 3, \ldots, n$ is the RBN-Number, which is the evaluation information of the alternative L_i against the attribute A_j. Using this evaluation information, we construct a decision matrix D_M.

Step-2: In this step, we determine the value of $RBNAM$ $(A_1, A_2, A_3, \ldots, A_n)$ corresponding to each $L_i(i = 1, 2, \ldots, m)$ i.e., the aggregated value $(L_i | A_1, A_2, A_3, \ldots, A_n)$ of all the attributes for each alternative by utilizing Eq. (1). After the determination of aggregated values, we construct an aggregated decision matrix D_M.

Step-3: In this step, we determine the score values and accuracy values of each aggregated value for all alternatives by employing the Eq. (2) & Eq. (3).

Step-4: In this step, we rank the alternatives by employing the score values as defined in Definition 4.2.

Step-5: End.

6   Validation of the Proposed *MADM* Model

In this section we present a realistic example to validate our developed $MADM$ model.

Example 6.1. "Franchisee Selection of an E-commerce Company for Courier Delivery". In the market there are so many courier companies namely E-kart, DTDC, Xpress-bees, etc. These companies provide their franchisee to the common people for the growth of their business. Further, a person can earn 700$ or above by taking franchisee of such companies.

Suppose an entrepreneur wants to start an entrepreneurship by taking a franchisee of a courier delivery company for a particular pin code or a particular area. Suppose there are four possible alternatives (or courier delivery companies) $L_i(i = 1, 2, 3, 4)$. Then, the entrepreneur needs to choose the franchisee of a most suitable courier company based on the attribute, by which the entrepreneur will earn more money. For this the entrepreneur must select some attributes by the experts for choosing the suitable franchisee. Hence, the selection of "franchisee of an e-commerce company for courier delivery under the RBNS environment" can be considered as an $MADM$ problem. Assume that, the decision maker/entrepreneur choose three major attributes namely A_1: no of products sell in that pin code/area, A_2: ratings of the delivery company, A_3: Fastest delivery, for their further evaluation. By using the decision makers' evaluation information, we obtain the decision matrix (D_M) as follows (Table 1).

Table 1. Decision matrix (D_M)

D_M	A_1
L_1	[(0.3,0.5,0.2,−0.5,−0.6,−0.3), (0.5.0.3.0.1,−0.2,−0.7,−0.4)]
L_2	[(0.3,0.4,0.2,−0.7,−0.5,−0.3), (0.4,0.2.0.2,−0.5,−0.7,−0.6)]
L_3	[(0.3,0.6,0.8,−0.5,−0.7,−0.8), (0.6,0.5,0.6,−0.4,−0.7,−0.9)]
L_4	[(0.7,0.5,0.6,−0.5,−0.8,−0.8), (0.6,0.4,0.6,−0.2,−0.8,−0.9)]

A_2
[(0.4,0.5,0.7,−0.7,−0.6,−0.6), (0.6,0.3.0.4,−0.5,−0.6,−0.7)]
[(0.5,0.5,0.5,−0.6,−0.6,−0.4), (0.6,0.4,0.4,−0.5,−0.7,−0.4)]
[(0.5,0.6,0.4,−0.5,−0.7,−0.4), (0.7,0.3,0.2,−0.3,−0.8,−0.5)]
[(0.6,0.8,0.4,−0.5,−0.6,−0.3), (0.7,0.5,0.3,−0.3,−0.7,−0.5)]

A_3
[(0.7,0.5,0.5,−0.6,−0.4,−0.5), (0.8,0.4,0.4,−0.3,−0.8,−0.5)]
[(0.6,0.5,0.2,−0.5,−0.7,−0.4), (0.6,0.2,0.1,−0.4,−0.8,−0.5)]
[(0.9,0.4,0.6,−0.5,−0.8,−0.4), (0.9,0.1,0.1,−0.3,−0.9,−0.4)]
[(0.3,0.5,0.2,−0.5,−0.6,−0.3), (0.4,0.3,0.2,−0.3,−0.8,−0.5)]

Now, by using the $RBNAM$ operator as defined in Eq. (1), we find the aggregation values $(L_i|A_1, A_2, A_3)$ of all the attributes for each alternative $L_i(i = 1, 2, 3, 4)$ as follows (Tables 2):

By using Eq. (3), we have $S(L_1) = 0.5867$; $S(L_2) = 0.5808$; $S(L_3) = 0.6467$; $S(L_4) = 0.6225$. Therefore, $S(L_2) < S(L_1) < S(L_4) < S(L_3)$. Hence, L_3 is the best choice among the alternatives.

Now, by using the $RBNGM$ operator as defined in Eq. (2), we find the aggregation values $(L_i|A_1, A_2, A_3)$ of all the attributes for each alternative $L_i(i =$

Table 2. Aggregation values table

Aggregate-D_M	$(L_i\|A_1, A_2, A_3)$
L_1	$[(0.49,0.49,0.41,-0.59,-0.54,-0.48), (0.66,0.34,0.27,-0.31,-0.73,-0.55)]$
L_2	$[(0.48,0.46,0.27,-0.59,-0.61,-0.37), (0.54,0.27,0.22,-0.46,-0.73,-0.51)]$
L_3	$[(0.66,0.52,0.57,-0.49,-0.73,-0.59), (0.73,0.27,0.22,-0.34,-0.78,-0.68)]$
L_4	$[(0.57,0.58,0.37,-0.49,-0.69,-0.54), (0.59,0.39,0.33,-0.27,-0.78,-0.73)]$

$1, 2, 3, 4)$ as follows: By using Eq. (3), we have S(L1) = 0.555; S(L2) = 0.560833; S(L3) = 0.591667; S(L4) = 0.5725. Therefore, S(L1) < S(L2) < S(L4) < S(L3). Hence, L_3 is the best choice among the alternatives (Tables 3).

Table 3. Best choice alternatives table

Aggregate-D_M	$(L_i\|A_1, A_2, A_3)$
L_1	$[(0.44,0.50,0.51,-0.59,-0.52,-0.45), (0.62,0.34,0.31,-0.34,-0.70,-0.52)]$
L_2	$[(0.45,0.47,0.32,-0.61,-0.59,-0.36), (0.52,0.27,0.24,-0.47,-0.70,-0.49)]$
L_3	$[(0.51,0.54,0.64,-0.54,-0.73,-0.50), (0.72,0.32,0.34,-0.34,-0.80,-0.56)]$
L_4	$[(0.50,0.63,0.42,-0.50,-0.66,-0.41), (0.55,0.41,0.39,-0.27,-0.76,-0.61)]$

7   Comparison of the Proposed $MADM$ Strategies

Now, we compare the results obtained from the MADM strategies presented in Table 4.

Table 4. Comparison of results obtained from the developed strategies

$MADM$ strategies	Ranking of alternatives
$RBNAM$ operator based $MADM$ strategy	$L_2 < L_1 < L_4 < L_3$
$RBNGM$ operator based $MADM$ strategy	$L_1 < L_2 < L_4 < L_3$

From the above comparison table, we can easily observe that L_3 is the most appropriate alternative (E-commerce Company) among the set of alternatives (E-commerce Companies).

8   Conclusions

In this article, we introduce the concept arithmetic mean and geometric mean operator, and establish some results on them under the $RBNS$-environment. Besides, we introduce the score function and accuracy function of RBN-Numbers under the $RBNS$-environment. Further, we develop two $MADM$ strategies based on the $RBNAM$ operator and $RBNGM$ operator respectively under the RBNS environment. Finally, we validate the developed $MADM$ strategies by providing a real-life numerical example, and compare both the $MADM$ strategies. The proposed $MADM$ strategy can also be used to deal with the other decision-making problems such as school choice [33], brick selection [34], logistic center location selection [35], teacher selection [36] etc. Further, we hope that the proposed $MADM$ strategy will open up a new avenue of research in $RBNS$-environment.

Author contributions. All the authors have equal contribution for the preparation of this article.

Conflict of Interest. The authors declare that they have no conflict of interest.

References

1. Broumi, S., Smarandache, F., Dhar, M.: Rough neutrosophic sets. Ital. J. Pure Appl. Math. **32**, 493–502 (2014)
2. Pawlak, Z.: Rough sets. Int. J. Comput. Inf. Sci. **11**, 341–356 (1982)
3. Smarandache, F.: A Unifying Field in Logics, Neutrosophy: Neutrosophic Probability, Set and Logic. American Research Press, Rehoboth (1998)
4. Deli, I., Ali, M., Smarandache, F.: Bipolar neutrosophic sets and their application based on multi-criteria decision making problems. In: Proceedings of the 2015 International Conference on Advanced Mechatronic Systems, Beijing, China, 22–24 August 2015, pp. 249–254 (2015)
5. Lee, K.M.: Bipolar-valued fuzzy sets and their operations. In: Proceedings of International Conference on Intelligent Technologies, Bangkok, Thailand, pp. 307–312 (2000)
6. Salama, A.A., Broumi, S.: Roughness of neutrosophic sets. Elixir Appl. Math. **74**, 26833–26837 (2014)
7. Broumi, S., Smarandache, F.: Interval neutrosophic rough set. Neutrosophic Sets Syst. **7**, 23–31 (2015)
8. Broumi, S., Smarandache, F.: Interval-valued neutrosophic soft rough sets. Int. J. Comput. Math. 232919 (2015). https://doi.org/10.1155/2015/232919
9. Mondal, K., Pramanik, S.: Rough neutrosophic multi-attribute decision-making based on rough accuracy score function. Neutrosophic Sets Syst. **8**, 14–21 (2015)
10. Pramanik, S., Mondal, K.: Some rough neutrosophic similarity measure and their application to multi attribute decision making. Glob. J. Eng. Sci. Res. Manag. **2**(7), 61–74 (2015)
11. Mondal, K., Pramanik, S.: Decision making based on some similarity measures under interval rough neutrosophic environment. Neutrosophic Sets Syst. **10**, 46–57 (2015)
12. Pramanik, S., Mondal, K.: Cotangent similarity measure of rough neutrosophic sets and its application to medical diagnosis. J. New Theory **4**, 90–102 (2015)
13. Mondal, K., Pramanik, S.: Tri-complex rough neutrosophic similarity measure and its application in multi-attribute decision making. Crit. Rev. **11**, 26–40 (2015)
14. Mondal, K., Pramanik, S., Smarandache, F.: Several trigonometric hamming similarity measures of rough neutrosophic sets and their applications in decision making. In: Smarandache, F., Pramanik, S. (eds) New Trends in Neutrosophic Theory and Application, pp. 93–103. Pons Editions, Brussels, Belgium (2016)

15. Mondal, K., Pramanik, S., Smarandache, F.: Multi-attribute decision making based on rough neutrosophic variational coefficient similarity measure. Neutrosophic Sets Syst. **13**, 3–17 (2016)
16. Mondal, K., Pramanik, S., Smarandache, F.: Rough neutrosophic hyper-complex set and its application to multi-attribute decision making. Crit. Rev. **13**, 111–126 (2016)
17. Mondal, K., Pramanik, S., Smarandache, F.: Rough neutrosophic TOPSIS for multi-attribute group decision making. Neutrosophic Sets Syst. **13**, 105–117 (2016)
18. Zhang, C., Zhai, Y., Li, D., Mu, Y.: Steam turbine fault diagnosis based on single-valued neutrosophic multigranulation rough sets over two universes. J. Intell. Fuzzy Syst. **31**(6), 2829–2837 (2016). https://doi.org/10.3233/jifs-169165
19. Pramanik, S., Roy, R., Roy, T.K., Smarandache, F.: Multi criteria decision making using correlation coefficient under rough neutrosophic environment. Neutrosophic Sets Syst. **17**, 29–36 (2017)
20. Yang, H.L., Zhang, C.L., Guo, Z.L., Liu, Y.L., Liao, X.: A hybrid model of single valued neutrosophic sets and rough sets: single valued neutrosophic rough set model. Soft. Comput. **21**, 6253–6267 (2017)
21. Pramanik, S., Roy, R., Roy, T.K.: Multi criteria decision making based on projection and bidirectional projection measures of rough neutrosophic sets. In: Smarandache, F., Pramanik, S. (eds.) New Trends in Neutrosophic Theory and Applications, vol. 2, pp. 175–187. Pons Editions, Brussels (2018)
22. Zhao, X.R., Hu, B.Q.: Three-way decisions with decision theoretic rough sets in multiset-valued information tables. Inf. Sci. **507**, 684–699 (2020). https://doi.org/10.1016/j.ins.2018.08.024
23. Jiao, L., Yang, H.-L., Li, S.-G.: Three-way decision based on decision-theoretic rough sets with single-valued neutrosophic information. Int. J. Mach. Learn. Cybern. **11**(3), 657–665 (2019). https://doi.org/10.1007/s13042-019-01023-3
24. Roy, R., Pramanik, S., Roy, T.K. Interval rough neutrosophic TOPSIS strategy for multi-attribute decision making. In: Abdel-Basset, M., Smarandache, F. (eds.) Neutrosophic Sets in Decision Analysis and Operations Research, pp. 98–118. IGI Global, Hershey (2020). https://doi.org/10.4018/978-1-7998-2555-5.ch005
25. Pramanik, S., Dey, P.P., Giri, B.C., Smarandache, F.: Bipolar neutrosophic projection based models for solving multi-attribute decision making problems. Neutrosophic Sets Syst. **15**, 70–79 (2017)
26. Pramanik, S., Dalapati, S., Alam, S., Roy, T.K.: TODIM method for group decision making under bipolar neutrosophic set environment. In: Smarandache, F., Pramanik, S. (eds.) New Trends in Neutrosophic Theory and Applications, vol. 2, pp. 140–155. Pons Editions, Brussels (2018)
27. Pramanik, S., Dalapati, S., Alam, S., Roy, T.K.: VIKOR based MAGDM strategy under bipolar neutrosophic set environment. Neutrosophic Sets Syst. **19**, 57–69 (2018)
28. Fan, C., Ye, J., Fen, S., Fan, E., Hu, K.: Multi-criteria decision-making method using heronian mean operators under a bipolar neutrosophic environment. Mathematics **7**, 97 (2019)
29. Jamil, M., Abdullah, S., Yaqub Khan, M., Smarandache, F., Ghani, F.: Application of the bipolar neutrosophic hamacher averaging aggregation operators to group decision making: an illustrative example. Symmetry **11**(5), 698 (2019). https://doi.org/10.3390/sym11050698
30. Pramanik, S.: Rough neutrosophic set: an overview. In: Smarandache, F., Broumi, S. (eds.) Neutrosophic Theories in Communication, Management and Information Technology, pp. 275–311. Nova Science Publishers, New York (2020)

31. Pramanik, S., Mondal, K.: Rough bipolar neutrosophic set. Glob. J. Eng. Sci. Res. Manag. **3**(6), 71–81 (2016)
32. Mondal, K., Pramanik, S., Giri, B.C.: Rough neutrosophic aggregation operators for multi-criteria decision-making. In: Kahraman, C., Otay, İ (eds.) Fuzzy Multi-criteria Decision-Making Using Neutrosophic Sets. SFSC, vol. 369, pp. 79–105. Springer, Cham (2019). https://doi.org/10.1007/978-3-030-00045-5_5
33. Mondal, K., Pramanik, S.: Neutrosophic decision making of school choice. Neutrosophic Sets Syst. **7**, 62–68 (2015)
34. Mondal, K., Pramanik, S.: Neutrosophic decision making model for clay-brick selection in construction field based on grey relational analysis. Neutrosophic Sets Syst. **9**, 72–79 (2015)
35. Pramanik, S., Dalapati, S., Roy, T.K.: Logistics center location selection approach based on neutrosophic multicriteria decision making. In: Smarandache, F., Pramanik, S. (eds.) New Trends in Neutrosophic Theory and Application, pp. 161–174. Pons Editions, Brussels (2016)
36. Pramanik, S., Mukhopadhyaya, D.: Grey relational analysis based intuitionistic fuzzy multi criteria group decision-making approach for teacher selection in higher education. Int. J. Comput. Appl. **34**(10), 21–29 (2011)

Single-Valued Neutrosophic Rough Continuous Mapping via Single-Valued Neutrosophic Rough Topological Space

Binod Chandra Tripathy$^{(\boxtimes)}$, Suman Das , and Rakhal Das

Department of Mathematics, Tripura University, Agartala, West Tripura 799022,
India
tripathybc@gmail.com,tripathybc@yahoo.com

Abstract. In this article an attempt is made to introduce and study the notion of single-valued neutrosophic rough continuous mapping, single-valued neutrosophic rough compactness via single-valued neutrosophic rough topological spaces ($SVNRTS$). By defining the concept of single-valued neutrosophic rough continuous function, single-valued neutrosophic rough compactness, we formulate and discuss several interesting results on $SVNRTSs$.

Keywords: SVNRTS · Single-valued neutrosophic rough compactness · Single-valued neutrosophic rough continuous function

1 Introduction

Smarandache [37] grounded the concept of neutrosophic set (NS) theory by extending the notion of fuzzy set (FS) [43] and intuitionistic fuzzy set (IFS) [2] to deal with the uncertainty events having indeterminacy. In the year 2010, Wang et al. [42] studied the notion of single-valued neutrosophic set ($SVNS$). Thereafter, Salama and Alblowi [38] presented the idea of neutrosophic topological space (NTS) via $SVNSs$ by extending the notion of intuitionistic fuzzy topological spaces. Thereafter, many researcher around the globe applied topological concept on neutrosophic set and its extensions [1,4–16,33,34,40]. In the year 1982, Pawlak [25] introduced the concept of rough set for the processing of incomplete information system. Thereafter, Broumi et al. [3] presented the idea of single-valued neutrosophic rough set ($SVNRS$) by extending the notion of fuzzy rough set. Later on, many researchers around the globe applied the concept of $SVNRS$ and its extensions in their practical research [17–24,26–32,36,41]. In the year 2018, Lellis Thivagar et al. [14] applied the topological concept in $SVNRS$ theory and grounded the concept of nano topology on $SVNRS$. Afterwards, Sweety and Arockiarani [39] studied the topological structures of fuzzy neutrosophic rough sets. Later on, Riaz et al. [35] introduced the notion of neutrosophic soft rough topology and presented an application to decision making.

Supported by Organization FRS.

J. F. Peters et al. (Eds.): TRS XXIII, LNCS 13610, pp. 77–95, 2022.
https://doi.org/10.1007/978-3-662-66544-2_6

Recently, Jin et al. [13] grounded the idea of a new *SVNRS* and their related topology.

We illustrate the following example to explain how the memberships are determined.

Example 1. Let a candidate appears in an examination. The paper contains 20 questions. The candidate has to answer 10 out of 20 questions. The candidate knows the answer to 9 questions correctly, 5 he does not know. The rest 6 he can answer but does not know how far the answer to these 6 questions is correct. The memberships are determined with respect to 10 questions. In this case the truth value membership is 0.9, since we can score 90 out of marks. The false membership is 0.5 and similarly the indeterminacy membership is 0.6. In this way, the degree of membership can be determined.

In this article, we introduce and study the concept of single-valued neutrosophic rough continuous function, single-valued neutrosophic rough compactness via *SVNRTSs*. By defining the concept of single-valued neutrosophic rough continuous function, single-valued neutrosophic rough compactness, we formulate and discuss several interesting results on *SVNRTSs*.

The remaining part of this article is designed as follows: In Sect. 2, we recall some definitions and results those are relevant and useful for the preparation of the main results of this article. In Sect. 3, we introduced the notion of single-valued neutrosophic rough continuous function and single-valued neutrosophic rough compactness via *SVNRTSs*. In Sect. 4, we conclude our work done in this article.

2   Some Relevant Results

In this section, we give some definitions and results those are relevant and useful for the preparation of the main results of this article.

Definition 1. *[37] An SVNS V over a fixed set W is defined as follows: $V = \{(r, T_V(r), I_V(r), F_V(r)): r \in W\}$, where, T_V, I_V, $F_V: W \to [0,1]$ are the truth, indeterminacy and falsity membership functions respectively.*

Definition 2. *[37] Suppose that W be a fixed set. Then, the absolute SVNS (1_W) and the null SVNS (0_W) over W are defined as follows:*

1. $1_W = \{(r, 1, 0, 0) : r \in W\}$;
2. $0_W = \{(r, 0, 1, 1) : r \in W\}$.

Definition 3. *[37] Let $M = \{(r, T_M(r), I_M(r), F_M(r)) : r \in W\}$ and $N = \{(r, T_N(r), I_N(r), F_N(r)) : r \in W\}$ be two SVNSs over W. Then, $M \subseteq N$ if and only if $T_M(r) \leq T_N(r), I_M(r) \geq I_N(r), F_M(r) \geq F_N(r), \forall r \in W$.*

Definition 4. *[37] Let* $M = \{(r, T_M(r), I_M(r), F_M(r)): r \in W\}$ *and* $N = \{(r, T_N(r), I_N(r), F_N(r)): r \in W\}$ *be two SVNSs over* W. *Then, the intersection of* M *and* N *is*

$$M \cap N = \{(r, min\{T_M(r), T_N(r)\}, \; max \; \{I_M(r), I_N(r)\}, \; max \; \{F_M(r), F_N(r)\}): r \in W\}.$$

Definition 5. *[37] Let* $M = \{(r, T_M(r), I_M(r), F_M(r)): r \in W\}$ *and* $N = \{(r, T_N(r), I_N(r), F_N(r)): r \in W\}$ *be two SVNSs over* W. *Then, the union of* M *and* N *is*

$$M \cup N = \{(r, max\{T_M(r), T_N(r)\}, \; min \; \{I_M(r), I_N(r)\}, \; min\{F_M(r), F_N(r)\}): r \in W\}.$$

Definition 6. *[37] Let* $M = \{(r, T_M(r), U_M(r), F_M(r)): r \in W\}$ *be an SVNS over a fixed set* W. *Then,* $M^c = \{(r, 1 - T_M(r), \; 1 - I_M(r), \; 1 - F_M(r)): r \in W\}$.

Definition 7. *[3] Suppose that* ρ *be an equivalence relation on a fixed set* W. *Let* $Q = \{(r, T_Q(r), I_Q(r), F_Q(r)) : r \in W\}$ *be an SVNS over* W. *Then, the lower approximation set* $[\underline{N}(Q)]$ *and the upper approximation set* $[\overline{N}(Q)]$ *of* Q *in the approximation space* (W, ρ) *are defined as follows:*

$\underline{N}(Q) = \{(r, T_{\underline{N}(Q)}(r), I_{\underline{N}(Q)}(r), F_{\underline{N}(Q)}(r): p \in [r]_\rho, r \in W\};$

and $\overline{N}(Q) = \{(r, T_{\overline{N}(Q)}(r), I_{\overline{N}(Q)}(r), F_{\overline{N}(Q)}(r): p \in [r]_\rho, \; r \in W\},$

where $T_{\underline{N}(Q)} = \bigwedge_{p \in [r]_\rho} T_Q(r), \; I_{\underline{N}(Q)} = \bigvee_{p \in [r]_\rho} I_Q(r), \; F_{\underline{N}(Q)} = \bigvee_{p \in [r]_\rho} F_Q(r),$ $T_{\overline{N}(Q)} = \bigvee_{p \in [r]_\rho} T_Q(r), \; I_{\overline{N}(Q)} = \bigwedge_{p \in [r]_\rho} I_Q(r), \; F_{\overline{N}(Q)} = \bigwedge_{p \in [r]_\rho} F_Q(r).$ *So,* $0 \leq T_{\underline{N}(Q)}(r) + I_{\underline{N}(Q)}(r) + F_{\underline{N}(Q)}(r) \leq 3$ *and* $0 \leq T_{\overline{N}(Q)}(r) + I_{\overline{N}(Q)}(r) + F_{\overline{N}(Q)}(r) \leq 3.$

Here, the operators "$\bigvee$" *and* "$\bigwedge$" *means* "max" *or* "join" *and* "min" *or* "meet" *operators respectively. Clearly,* $\underline{N}(Q)$ *and* $\overline{N}(Q)$ *are two SVNSs over* W. *The pair* $(\underline{N}(Q), \overline{N}(Q))$ *is called the Single-Valued Neutrosophic Rough Set (SVNRS) in the approximation space* (W, ρ).

Definition 8. *[3] Suppose that* ρ *be an equivalence relation on a fixed set* W. *Let* $Q = \{(r, T_Q(r), I_Q(r), F_Q(r)) : r \in W\}$ *be an SVNS over* W. *Let the lower approximation set* $[\underline{N}(Q)]$ *and the upper approximation set* $[\overline{N}(Q)]$ *of* Q *in the approximation space* (W, ρ). *Then the boundary approximation set is denoted by* $N_B(Q)$ *and is defined by*

$$N_B(Q) = (\overline{N}(Q) \cap \underline{N}^c(Q))$$

Example 2. Let $W = \{r_1, r_2, r_3, r_4, r_5\}$ be a fixed set. Let ρ be an equivalence relation, where its partition of W is given by $W/\rho = \{(r_1, r_3), (r_2, r_5), (r_4)\}$.

Let $Q = \{(r_1, 0.5, 0.2, 0.6), (r_2, 0.8, 0.6, 0.4), (r_3, 0.2, 0.4, 0.6), (r_4, 0.9, 0.7, 0.8), (r_5, 0.6, 0.7, 0.5)\}$ be an SVNS over W. Then, the lower approximation set of the SVNS Q is $\underline{N}(Q) = \{(r_1, 0.2, 0.4, 0.6), (r_2, 0.6, 0.7, 0.5), (r_3, 0.2, 0.4, 0.6), (r_4, 0.9, 0.7, 0.8), \; (r_5, 0.6, 0.7, 0.5)\}$, and the upper approximation set of the SVNS Q is $\overline{N}(Q) = \{(r_1, 0.5, 0.2, 0.6), (r_2, 0.8, 0.6, 0.4), (r_3, 0.5, 0.2, 0.6), (r_4, 0.9, 0.7, 0.8), (r_5, 0.8, 0.6, 0.4)\}$.

Therefore, $(\underline{N}(Q),\ \overline{N}(Q)) = (\{(r_1,0.2,0.2,0.6),\ (r_2,0.6,0.6,0.4),\ (r_3,0.2,$ $0.2,0.6),\ (r_4,0.9,0.7,0.8),\ (r_5,0.6,0.6,0.4)\},\ \{(r_1,0.5,0.4,0.6),\ (r_2,0.8,0.7,0.5),$ $(r_3,0.5,0.4,0.6),\ (r_4,0.9,0.7,0.8),\ (r_5,0.8,0.7,0.5)\})$ is an SVNRS in (W,ρ).

Definition 9. *[3] Let* $N(Q) = (\underline{N}(Q),\overline{N}(Q)) = (\{(r,T_{\underline{N}(Q)}(r),\ I_{\underline{N}(Q)}(r),$ $F_{\underline{N}(Q)}(r)) : p \in [r]_\rho, r \in W\},\ \{(r,T_{\overline{N}(Q)}(r),\ I_{\overline{N}(Q)}(r),\ F_{\overline{N}(Q)}(r)): p \in [r]_(\rho), r \in$ $W\})$ *be an SVNRS in the approximation space* (W,ρ).

Then, $[(T_{\underline{N}(Q)}(r),\ I_{\underline{N}(Q)}(r),\ F_{\underline{N}(Q)}(r)),\ (T_{\overline{N}(Q)}(r),\ I_{\overline{N}(Q)}(r),\ F_{\overline{N}(Q)}(r))]$ *is called a Single-Valued Neutrosophic Rough Number (SVNRN), for all* $r \in W$.

Example 3. Let $N(Q) = (\underline{N}(Q)),\ \overline{N}(Q)))$ be a SVNRS in the approximation space (W,ρ) as it is shown in Example 1. Then, $[(0.2,0.2,0.6),(0.5,0.4,0.6)]$ is a SVNRN in the approximation space (W,ρ).

Definition 10. *[3] Let* $N(Q) = (\underline{N}(Q),\overline{N}(Q))$ *be an SVNRS in the approximation space* (W,ρ). *Then, the complement of* $N(Q) = (\underline{N}(Q),\overline{N}(Q))$ *is defined by* $N(Q)^c = (\underline{N}(Q)^c,\overline{N}(Q)^c)$, *where* $\underline{N}(Q)^c = \{(r,F_{\underline{N}(Q)}(r),\ 1 - I_{\underline{N}(Q)}(r),$ $T_{\underline{N}(Q)}(r)): p \in [r]_\rho,\ r \in W\}$ *and* $\overline{N}(Q)^c = \{(r,F_{\overline{N}(Q)}(r),\ 1 - I_{\overline{N}(Q)}(r),$ $T_{\overline{N}(Q)}(r)) : p \in [r]_\rho, r \in W\}$.

Example 4. Let $N(Q) = (\underline{N}(Q),\overline{N}(Q))$ be a SVNRS in the approximation space (W,ρ) as it is shown in Example 1. Then, the complement of $N(Q)$ is $N(Q)^c = (\underline{N}(Q)^c,\ \overline{N}(Q)^c)$, where $\underline{N}(Q)^c = \{(r_1,0.6,0.8,0.2),\ (r_2,0.4,0.4,0.6),$ $(r_3,0.6,0.8,0.2),\ (r_4,0.8,0.3,0.9),\ (r_5,0.4,0.4,0.6)\}$ and $\overline{N}(Q)^c = \{(r_1,0.6,0.6,0.5),\ (r_2,0.5,0.3,0.8),\ (r_3,0.6,0.6,0.5),\ (r_4,0.8,0.3,0.9),$ $(r_4,0.5,0.3,0.8)\}$.

Definition 11. *[3] Let* $N(Q) = (\underline{N}(Q),\overline{N}(Q))$ *and* $N(V) = (\underline{N}(V),\overline{N}(V))$ *be two SVNRSs in the approximation space* (W,ρ). *Then,* $N(Q) \subseteq N(V)$ *if and only if* $\underline{N}(Q) \subseteq \underline{N}(V)$ *and* $\overline{N}(Q) \subseteq \overline{N}(V)$ *i.e.,* $T_{\underline{N}(Q)}(r) \le T_{\underline{N}(V)}(r),\ I_{\underline{N}(Q)}(r) \ge$ $I_{\underline{N}(V)}(r),\ F_{\underline{N}(Q)}(r) \ge F_{\underline{N}(V)}(r),\ T_{\overline{N}(Q)}(r) \le T_{\overline{N}(V)}(r),\ I_{\overline{N}(Q)}(r) \ge I_{\overline{N}(r)}(r),$ $F_{\overline{N}(Q)}(r) \ge F_{\overline{N}(V)}(r)$, *for all* $r \in W$.

Example 5. Suppose that $N(Q) = (\{(r_1,0.3,0.2,0.6),\ (r_2,0.3,0.6,0.4),\ (r_3,0.2,$ $0.2,0.6),\ (r_4,0.9,0.7,0.8),\ (r_5,0.6,0.6,0.4)\},\ \{(r_1,0.5,0.4,0.6),\ (r_2,0.8,\ 0.7,0.5),\ (r_3,$ $0.5,0.4,0.6),\ (r_4,0.9,0.7,0.8),\ (r_5,0.8,0.7,0.5)\})$ and $N(V) = (\{(r_1,0.3,0.1,0.5),$ $(r_2,0.7,0.4,0.3),\ (r_3,0.5,0.0,0.1),\ (r_4,0.9,0.5,0.3),\ (r_5,0.8,0.2,0.2)\},\ \{(r_1,0.5,$ $0.3,0.6),\ (r_2,0.8,0.5,0.3),\ (r_3,0.5,0.0,0.2),\ (r_4,1.0,0.6,0.4),\ (r_5,0.8,0.4,0.3)\})$ be two SVNRSs in (W,ρ). Clearly, $N(Q) \subseteq N(V)$.

Definition 12. *[3] Let* $N(Q) = (\underline{N}(Q),\overline{N}(Q))$ *and* $N(V) = (\underline{N}(V),\overline{N}(V))$ *be two SVNRSs in the approximation space* (W,ρ). *Then,* $N(Q) = N(V)$ *if and only if* $\underline{N}(Q) = \underline{N}(V)$ *and* $\overline{N}(Q), = \overline{N}(V)$ *i.e.,* $T_{\underline{N}(Q)}(r) = T_{\underline{N}(V)}(r),\ I_{\underline{N}(Q)}(r) =$ $I_{\underline{N}(V)}(r),\ F_{\underline{N}(Q)}(r) = F_{\underline{N}(V)}(r),\ T_{\overline{N}(Q)}(r) = T_{\overline{N}(V)}(r),\ I_{\overline{N}(Q)}(r) = I_{\overline{N}(V)}(r),$ $F_{\overline{N}(Q)}(r) = F_{\overline{N}(V)}(r)$, *for all* $r \in W$.

Example 6. Let $N(Q) = (\{(r_1, 0.2, 0.2, 0.6), (r_2, 0.6, 0.6, 0.4), (r_3, 0.2, 0.2, 0.6),$ $(r_4, 0.9, 0.7, 0.8), (r_5, 0.6, 0.6, 0.4)\}, \{(r_1, 0.5, 0.4, 0.6), (r_2, 0.8, 0.7, 0.5), (r_3, 0.5,$ $0.4, 0.6), (r_4, 0.9, 0.7, 0.8), (r_5, 0.8, 0.7, 0.5)\})$ and $N(V) = (\{(r_1, 0.2, 0.2, 0.6),$ $(r_2, 0.6, 0.6, 0.4), (r_3, 0.2, 0.2, 0.6), (r_4, 0.9, 0.7, 0.8), (r_5, 0.6, 0.6, 0.4)\}, \{(r_1, 0.5,$ $0.4, 0.6), (r_2, 0.8, 0.7, 0.5), (r_3, 0.5, 0.4, 0.6), (r_4, 0.9, 0.7, 0.8), (r_5, 0.8, 0.7, 0.5)\})$ be two SVRPNS in the approximation space (W, ρ). Clearly, $N(Q) = N(V)$.

Definition 13. *[3] Let $N(Q) = (\underline{N}(Q), \overline{N}(Q))$ and $N(V) = (\underline{N}(V), \overline{N}(V))$ be two SVNRSs in the approximation space (W, ρ). Then, the intersection and union of the SVNRSs $N(Q)$ and $N(V)$ are defined as follows:*

$N(Q \cap V) = (\underline{N}(Q \cap V), \overline{N}(Q \cap V))$ *and* $N(Q \cup V) = (\underline{N}(Q \cup V), \overline{N}(Q \cup V))$,

where,

$\underline{N}(Q \cap V) = \{(r, T_{\underline{N}(Q)}(r) \bigwedge T_{\underline{N}(V)}(r), I_{\underline{N}(Q)}(r) \bigvee T_{\underline{N}(V)}(r), F_{\underline{N}(Q)}(r)$ $\bigvee F_{\underline{N}(V)}(r) : p \in [r]_\rho, r \in W\}$;

$\overline{N}(Q \cap V) = \{(r, T_{\overline{N}(Q)}(r) \bigwedge T_{\overline{N}(V)}(r), I_{\overline{N}(Q)}(r) \bigvee T_{\overline{N}(V)}(r), F_{\overline{N}(Q)}(r)$ $\bigvee F_{\overline{N}(V)}(r) : p \in [r]_\rho, r \in W\}$;

$\underline{N}(Q \cup V) = \{(r, T_{\underline{N}(Q)}(r) \bigvee T_{\underline{N}(V)}(r), \ I_{\underline{N}(Q)}(r) \bigwedge I_{\underline{N}(V)}(r), \ F_{\underline{N}(Q)}(r)$ $\bigwedge F_{\underline{N}(V)}(r) : p \in [r]_\rho, r \in W\}$;

and

$\overline{N}(Q \cup V) = \{(r, T_{\overline{N}(Q)}(r) \bigvee T_{\overline{N}(V)}(r), \ I_{\overline{N}(Q)}(r) \bigwedge I_{\overline{N}(V)}(r), \ F_{\overline{N}(Q)}(r)$ $\bigwedge F_{\overline{N}(V)}(r) : p \in [r]_\rho, r \in W\}$

Example 7. Let $N(Q) = (\underline{N}(Q), \ \overline{N}(Q))$ and $N(V) = (\underline{N}(V), \ \overline{N}(V))$ be two SVNRSs in (W, ρ) as they are given in Example 4. Then, $N(Q \cap V) = (\underline{N}(Q \cap V),$ $\overline{N}(Q \cap V)) = (\{(r_1, 0.3, 0.2, 0.6), (r_2, 0.5, 0.6, 0.4), (r_3, 0.2, 0.2, 0.6), (r_4, 0.9, 0.7, 0.8),$ $(r_5, 0.5, 0.6, 0.4)\}, \{(r_1, 0.5, 0.4, 0.6), (r_2, 0.8, 0.7, 0.5), (r_3, 0.5, 0.4, 0.6), (r_4, 0.9, 0.7,$ $0.8), (r_5, 0.8, 0.7, 0.5)\})$, and $N(Q \cup V) = (\underline{N}(Q \cup V), \ \overline{N}(Q \cup V)) = (\{(r_1, 0.3,$ $0.1, 0.5), (r_2, 0.6, 0.4, 0.3), (r_3, 0.5, 0.0, 0.1), (r_4, 0.9, 0.5, 0.3), (r_5, 0.6, 0.2, 0.2)\},$ $\{(r_1, 0.5, 0.3, 0.6), (r_2, 0.8, 0.5, 0.3), (r_3, 0.5, 0.0, 0.2), (r_4, 1.0, 0.6, 0.4), (r_5, 0.8, 0.4,$ $0.3)\})$.

Definition 14. *[14] Suppose that $N(Q) = (\underline{N}(Q), \overline{N}(Q))$ be an SVNRS in the approxima-*
tion space (W, ρ). Then, $\tau_{SVNRS}(\rho) = \{1_W, 0_W, \underline{N}(Q), \overline{N}(Q), N_B(Q)\}$ is said to be an single-valued neutrosophic rough topology (SVNRT) which guarantee the following conditions:

1. 1_W *and* $0_W \in \tau_{RPNT}(\rho)$;
2. *Arbitrary union of members of $\tau_{RPNT}(\rho)$ belongs to $\tau_{RPNT}(\rho)$;*
3. *Finite intersection of members of $\tau_{RPNT}(\rho)$ belongs to $\tau_{RPNT}(\rho)$.*
   The pair $(W, \tau_{RPNT}(\rho))$ is said to be an SVNRTS, if $\tau_{RPNT}(\rho)$ is an SVNRT on (W, ρ).

Example 8. Let $W = \{r_1, r_2, r_3, r_4, r_5\}$ be a fixed set. Let ρ be an equivalence relation, where its partition of W is given by $W/\rho = \{(r_1, r_3), (r_2, r_5), (r_4)\}$. Let $Q = \{(r_1, 0.5, 0.2, 0.6), (r_2, 0.8, 0.6, 0.4), (r_3, 0.2, 0.4, 0.6), (r_4, 0.9, 0.7, 0.8), (r_5, 0.6, 0.7, 0.5)\}$ be an SVNS over W. Then, the lower approximation set of the SVNS Q is

$\underline{N}(Q) = \{(r_1, 0.2, 0.4, 0.6), (r_2, 0.6, 0.7, 0.5), (r_3, 0.2, 0.4, 0.6), (r_4, 0.9, 0.7, 0.8), (r_5, 0.6, 0.7, 0.5)\}$, and the upper approximation set of the SVNS Q is

$\overline{N}(Q) = \{(r_1, 0.5, 0.2, 0.6), (r_2, 0.8, 0.6, 0.4), (r_3, 0.5, 0.2, 0.6), (r_4, 0.9, 0.7, 0.8), (r_5, 0.8, 0.6, 0.4)\}$.

$\underline{N}^c(Q) = \{(r_1, 0.6, 0.6, 0.2), (r_2, 0.5, 0.3, 0.6), (r_3, 0.6, 0.6, 0.2), (r_4, 0.8, 0.3, 0.9), (r_5, 0.5, 0.3, 0.6)\}$, and the boundary approximation set of the SVNS Q is

$N_B(Q) = (\overline{N}(Q) \cap \underline{N}^c(Q))$

$\overline{N}(Q) \cap \underline{N}^c(Q) = \{(r_1, 0.5, 0.6, 0.6), (r_2, 0.5, 0.6, 0.6), (r_3, 0.5, 0.6, 0.6), (r_4, 0.8, 0.7, 0.9), (r_5, 0.5, 0.6, 0.6)\}$.

Clearly $N_B(Q) \subseteq \overline{N}(Q)$ and $N_B(Q) \subseteq \underline{N}^c(Q))$ and $\tau_{SVNRS}(\rho) = \{1_W, 0_W, \underline{N}(Q), \overline{N}(Q), N_B(Q)\}$ satisfies all the property of single-valued neutrosophic rough topological space.

Definition 15. *[14] Let $(W, \tau_{RPNT}(\rho))$ be an SVNRTS. Then, each element of $\tau_{RPNT}(\rho)$ are said to be an single-valued neutrosophic rough open set (SVNROS). The complement of an SVNROS is called an single-valued neutrosophic rough open set (SVNRCS) in $(W, \tau_{RPNT}(\rho))$.*

Proposition 1. *[14] Suppose that $(W, \tau_{RPNT}(\rho))$ be an SVNRTS. Then, the following holds:*

1. *1_W and 0_W are both SVNROS and SVNRCS in $(W, \tau_{RPNT}(\rho))$;*
2. *Arbitrary intersection of SVNRCSs is also an SVNRCS in $(W, \tau_{RPNT}(\rho))$;*
3. *Finite union of SVNRCSs is also an SVNRCS in $(W, \tau_{RPNT}(\rho))$.*

Definition 16. *[14] Suppose that $(W, \tau_{RPNT}(\rho))$ be an SVNRTS such that $\tau_{RPNT}(\rho) = \{1_W, 0_W\}$. Then, $\tau_{RPNT}(\rho)$ is said to be an single-valued neutrosophic rough indiscrete topology (SVNRIT) on W with respect to ρ, and the corresponding space is called an single-valued neutrosophic rough indiscrete topological space (SVNRITS).*

Definition 17. *[14] Suppose that $(W, \tau_{RPNT}(\rho))$ be an SVNRTS w.r.t. ρ, and K be an arbitrary SVNRS over W. Then, the single-valued neutrosophic rough interior of K i.e., $R_{int}(K)$ is the union of all single-valued neutrosophic rough subsets of K. Clearly, $R_{int}(K)$ is the largest SVNROS contained in K.*

Definition 18. *[14] Suppose that $(W, \tau_{RPNT}(\rho))$ be an SVNRTS. Let K be an SVNRS over W. Then, the single-valued neutrosophic rough closure of K i.e., $R_{cl}(K)$ is the intersection of all single-valued neutrosophic rough supersets of K. Clearly, $R_{cl}(K)$ is the smallest SVNRCS which contains K.*

Theorem 1. *[14] Suppose that* $(W, \tau_{RPNT}(\rho))$ *be an SVNRTS w.r.t.* ρ, *and let* M *and* N *be two SVNRSs over* W. *Then, the following holds:*

1. $R_{int}(0_N) = 0_N$, $R_{int}(1_N) = 1_N$, $R_{cl}(0_N) = 0_N$ *and* $R_{cl}(1_N) = 1_N$.
2. $R_{int}(M) \subseteq M$ *and* $M \subseteq R_{cl}(M)$.
3. M *is an SVNROS if and only if* $R_{int}(M) = M$.
4. M *is an SVNRCS if and only if* $R_{cl}(M) = M$.
5. $M \subseteq N \Rightarrow R_{int}(M) \subseteq R_{int}(N)$ *and* $R_{cl}(M) \subseteq R_{cl}(N)$.
6. $R_{int}(M) \cup R_{int}(N) \subseteq R_{int}(M \cup N)$.
7. $R_{int}(M) \cap R_{int}(N) = R_{int}(M \cap N)$.
8. $R_{cl}(M \cup N) = R_{cl}(M) \cup R_{cl}(N)$;
9. $R_{cl}(M \cap N) \subseteq R_{cl}(M) \cap R_{cl}(N)$.

3 Single-Valued Neutrosophic Rough Continuous Function

In this section, we procure the notion of single-valued neutrosophic rough continuous mapping via *SVNRTSs*. Besides, we formulate several interesting results on them via *SVNRTSs*.

Definition 19. *Let* $(W, \tau_{RPNT}(\rho))$ *be an SVNRTS. Then* X, *an SVNRS over* W *is called an*

1. *single-valued neutrosophic rough semi-open set (SVNRSOS) iff*
   $X \subseteq R_{cl}(R_{int}(X))$;
2. *single-valued neutrosophic rough pre-open set (SVNRPOS) iff*
   $X \subseteq R_{int}(R_{cl}(X))$.

Remark 1. The complement of an *SVNRSOS* and an *SVNRPOS* in an *SVNRTS* $(W, \tau_{RPNT}(\rho))$ are called single-valued neutrosophic rough semi-closed set (*SVNRSCS*) and single-valued neutrosophic pre-closed set (*SVNRPCS*) respectively.

Theorem 2. *Let* $(W, \tau_{RPNT}(\rho))$ *be an SVNRTS. Then,*

1. *every SVNROS is an SVNRSOS.*
2. *every SVNROS is an SVNRPOS.*

Proof. 1. Suppose that X be an *SVNROS* in the *SVNRTS* $(W, \tau_{RPNT}(\rho))$. Therefore, $X = R_{int}(X)$. It is known that, $X \subseteq R_{cl}(X)$. This implies, $X \subseteq R_{cl}(R_{int}(X))$. Therefore, X is an *SVNRSOS* in $(W, \tau_{RPNT}(\rho))$.

2. Suppose that X be an *SVNROS* in the *SVNRTS* $(W, \tau_{RPNT}(\rho))$. Therefore, $X = R_{int}(X)$. It is known that, $X \subseteq R_{cl}(X)$. This implies, $R_{int}(X) \subseteq R_{int}(R_{cl}(X))$ i.e., $X = R_{int}(X) \subseteq R_{int}(R_{cl}(X))$. Therefore, $X \subseteq R_{int}(R_{cl}(X))$. Hence, X is an *SVNRPOS* in $(W, \tau_{RPNT}(\rho))$.

Theorem 3. *Suppose that* $(W, \tau_{RPNT}(\rho))$ *be an SVNRTS. Then, the*

1. *union of two SVNRSOSs is also an SVNRSOS;*
2. *union of two SVNRPOSs is also an SVNRPOS.*

Proof. 1. Suppose that X and Y be two *SVNRSOSs* in an *SVNRTS* $(W, \tau_{RPNT}(\rho))$. Therefore, $X \subseteq R_{cl}(R_{int}(X))$ and $Y \subseteq R_{cl}(R_{int}(Y))$. Now, we have $X \cup Y \subseteq R_{cl}(R_{int}(X)) \cup R_{cl}(R_{int}(Y))$
$= R_{cl}(R_{int}(X) \cup R_{int}(Y))$
$\subseteq R_{cl}(R_{int}(X \cup Y))$.
Therefore, $X \cup Y \subseteq R_{cl}(R_{int}(X \cup Y))$. Hence, $X \cup Y$ is an *SVNRSOS* in $(W, \tau_{RPNT}(\rho))$.

2. Let X and Y be two *SVNRPOSs* in an *SVNRTS* $(W, \tau_{RPNT}(\rho))$. Therefore, $X \subseteq R_{int}(R_{cl}(X))$ and $Y \subseteq R_{int}(R_{cl}(Y))$. Now, we have
$X \cup Y$
$\subseteq R_{int}(R_{cl}(X)) \cup R_{int}(R_{cl}(Y))$
$\subseteq R_{int}(R_{cl}(X) \cup R_{cl}(Y))$
$= R_{int}(R_{cl}(X \cup Y))$.
Therefore, $X \cup Y \subseteq R_{int}(R_{cl}(X \cup Y))$. Hence, $X \cup Y$ is an *SVNRPOS* in $(W, \tau_{RPNT}(\rho))$.

Definition 20. *Let $(W, \tau_{RPNT}(\rho))$ be an SVNRTS. Then, an SVNRS X over W is called an single-valued neutrosophic rough α-open set ($SVNR\alpha - OS$) if and only if $X \subseteq R_{int}(R_{cl}(R_{int}(X)))$. The complement of an $SVNR\alpha - OS$ is said to be an single-valued neutrosophic rough α-closed set ($SVNR\alpha - CS$).*

Theorem 4. *In an SVNRTS $(W, \tau_{RPNT}(\rho))$,*

1. *Every SVNROS is also an $SVNR\alpha - OS$;*
2. *Every $SVNR\alpha - OS$ is also an SVNRSOS;*
3. *Every $SVNR\alpha - OS$ is also an SVNRPOS.*

Proof. 1. Suppose that X be an *SVNROS* in the *SVNRTS* $(W, \tau_{RPNT}(\rho))$. Therefore, $X = R_{int}(X)$. It is known that, $X \subseteq R_{cl}(X)$. This implies, $X \subseteq R_{cl}(R_{int}(X))$. Therefore, $R_{int}(X) \subseteq R_{int}(R_{cl}(R_{int}(X)))$, which implies that $X = R_{int}(X) \subseteq R_{int}(R_{cl}(R_{int}(X)))$. Therefore, $X \subseteq R_{int}(R_{cl}(R_{int}(X)))$. Hence, X is an *SVNR\alpha - OS* in $(W, \tau_{RPNT}(\rho))$.

2. Suppose that X be an *SVNR\alpha - OS* in $(W, \tau_{RPNT}(\rho))$. Therefore, $X \subseteq R_{int}(R - cl(R_{int}(X)))$. It is known that, $R_{int}(R - cl(R_{int}(X))) \subseteq R_{cl}(R_{int}(X))$. Thus, we have $X \subseteq R_{cl}(R - int(X))$. Hence, X is an *SVNRSOS*. Therefore, every *SVNR\alpha - OS* is an *SVNRSOS*.

3. Suppose that X be an *SVNR\alpha - OS* in $(W, \tau_{RPNT}(\rho))$. Therefore, $X \subseteq R_{int}(R_{cl}(R_{int}(X)))$. Therefore, $X \subseteq R_{int}(R_{cl}(R_{int}(X)))$. It is known that, $R_{int}(X) \subseteq X$.
This implies, $R_{cl}(R_{int}(X)) \subseteq R_{cl}(X)$, which implies $R_{int}(R_{cl}(R_{int}(X))) \subseteq R_{int}(R_{cl}(X)$. Therefore, $X \subseteq R_{int}(R_{cl}(X)$. Hence, X is an *SVNRPOS*. Therefore, every *SVNR\alpha - OS* is an *SVNRPOS* in $(W, \tau_{RPNT}(\rho))$.

Definition 21. *Let $(W, \tau_{RPNT}(\rho))$ be an SVNRTS. Then, an SVNRS X over W is called an single-valued neutrosophic rough b-open set ($SVNRb - OS$) if* and only if $X \subseteq R_{int}(R_{cl}(X)) \cup R_{cl}(R_{int}(X))$. An SVNRS X of W is called an single-valued neutrosophic rough b-closed set $(SVNRb - CS)$ if and only if X^c is an $SVNRb - OS$ i.e., if and only if $R_{int}(R_{cl}(X)) \cap R_{cl}(R_{int}(X)) \subseteq X$.

Theorem 5. *In an SVNRTS $(W, \tau_{RPNT}(\rho))$, every SVNRPOS $(SVNRSOS)$ is an SVNRb − OS.*

Proof. Let X be an $SVNRPOS$ in an $SVNRTS$ $(W, \tau_{RPNT}(\rho))$. Therefore, $X \subseteq R_{int}(R_{cl}(X))$. This implies, $X \subseteq R_{int}(R_{cl}(X)) \cup R_{cl}(R_{int}(X))$. Hence, X is an $SVNRb - OS$. Therefore, every $SVNRPOS$ is an $SVNRb - OS$. Similarly, it can be shown that every $SVNRSOS$ is an $SVNRb - OS$.

Theorem 6. *The union of two $SVNRb - OSs$ in an $SVNRTS$ $(W, \tau_{RPNT}(\rho))$ is also an $SVNRb − OS$ in $(W, \tau_{RPNT}(\rho))$.*

Proof. Suppose that X and Y be two $SVNRb - OSs$ in $(W, \tau_{RPNT}(\rho))$. Therefore, $X \subseteq R_{int}(R_{cl}(X)) \cup R_{cl}(R_{int}(X))$ and $Y \subseteq R_{int}(R_{cl}(Y)) \cup R_{cl}(R_{int}(Y))$. It is known that, $X \subseteq X \cup Y$ and $Y \subseteq X \cup Y$.
Now, $X \subseteq X \cup Y$ implies $R_{int}(X) \subseteq R_{int}(X \cup Y)$. Therefore, $R_{cl}(R_{int}(X)) \subseteq R_{cl}(R_{int}(X \cup Y))$. Further, $X \subseteq X \cup Y$ implies $R_{int}(R_{cl}(X)) \subseteq R_{int}(R_{cl}(X \cup Y))$. Similarly, it can be shown that, $R_{cl}(R_{int}(Y)) \subseteq R_{cl}(R_{int}(X \cup Y))$ and $R_{int}(R_{cl}(Y)) \subseteq R_{int}(R_{cl}(X \cup Y))$.
From the above, we have
$X \cup Y$
$\subseteq R_{cl}(R_{int}(X)) \cup R_{int}(R_{cl}(X)) \cup R_{cl}(R_{int}(Y)) \cup R_{int}(R_{cl}(Y))$
$\subseteq R_{cl}(R_{int}(X \cup Y)) \cup R_{int}(R_{cl}(X \cup Y)) \cup R_{cl}(R_{int}(X \cup Y)) \cup R_{int}(R_{cl}(X \cup Y))$
$= R_{cl}(R_{int}(X \cup Y)) \cup R_{int}(R_{cl}(X \cup Y))$.
Therefore, $X \cup Y \subseteq R_{cl}(R_{int}(X \cup Y)) \cup R_{int}(R_{cl}(X \cup Y))$. Hence, $X \cup Y$ is an $SVNRb - OS$ in $(W, \tau_{RPNT}(\rho))$. Therefore, the union of two $SVNRb - OSs$ in $(W, \tau_{RPNT}(\rho))$ is also an $SVNRb - OS$ in $(W, \tau_{RPNT}(\rho))$.

Definition 22. *Let $(W, \tau_{RPNT}(\rho))$ be an $SVNRTS$. Then X, an SVNRS over W is called an single-valued neutrosophic rough simply open set (SVNRSO-set) if and only if $R_{int}(R_{cl}(X)) \subseteq R_{cl}(R_{int}(X))$. In that case, the complement of X is called an single-valued neutrosophic rough simply closed set (SVNRSC-set) in $(W, \tau_{RPNT}(\rho))$.*

Remark 2. Let $(W, \tau_{RPNT}(\rho))$ be an $SVNRTS$.
Then, the null $SVNRS$ (0_W) and whole $SVNRS$ (1_W) are both $SVNRSO$-set and $SVNRSC$-set in $(W, \tau_{RPNT}(\rho))$.

Theorem 7. *Let $(W, \tau_{RPNT}(\rho))$ be an $SVNRTS$. If X is an $SVNROS$ in $(W, \tau_{RPNT}(\rho))$, then X is also an SVNRSO-set in $(W, \tau_{RPNT}(\rho))$.*

Proof. Assume that $(W, \tau_{RPNT}(\rho))$ be an $SVNRTS$. Let X be an $SVNROS$ in $(W, \tau_{RPNT}(\rho))$. Therefore, $R_{int}(X) = X$. It is known that, $R_{int}(R_{cl}(X)) \subseteq R_{cl}(X)$ and $R_{cl}(X) = R_{cl}(R_{int}(X))$. This implies, $R_{int}(R_{cl}(X)) \subseteq R_{cl}(R_{int}(X))$. Hence, X is an $SVNRSO$-set in $(W, \tau_{RPNT}(\rho))$.

Theorem 8. *If X is both $SVNRPOS$ and $SVNRSO$-set in an $SVNRTS$ $(W, \tau_{RPNT}(\rho))$, then X is an $SVNRSOS$ in $(W, \tau_{RPNT}(\rho))$.*

Proof. Assume that X be both $SVNRPOS$ and $SVNRSO$-set in an $SVNRTS$ $(W, \tau_{RPNT}(\rho))$. Since X is an $SVNRPOS$, so $X \subseteq R_{int}(R_{cl}(X))$. Further, since X is an $SVNRSO$-set, so $R_{int}(R_{cl}(X)) \subseteq R_{cl}(R_{int}(X))$. This implies, $X \subseteq R_{cl}(R_{int}(X))$. Therefore, X is an $SVNRSOS$ in $(W, \tau_{RPNT}(\rho))$.

Definition 23. *Let $(W, \tau_{RPNT}(\rho))$ and $(M, \tau_{RPNT}(\lambda))$ be two $SVNRTSs$. Then, a one to one and onto mapping $\xi : (W, \tau_{RPNT}(\rho)) \to (M, \tau_{RPNT}(\lambda))$ is called as:*

1. *single-valued neutrosophic rough continuous mapping ($SVNR - C$-mapping) if and only if $\xi^{-1}(L)$ is an $SVNROS$ in W, whenever L is an $SVNROS$ in M;*
2. *single-valued neutrosophic rough semi-continuous mapping ($SVNRS - C$-mapping) if and only if $\xi^{-1}(L)$ is an $SVNRSOS$ in W, whenever L is an $SVNROS$ in M;*
3. *single-valued neutrosophic rough pre-continuous mapping ($SVNRP - C$-mapping) if and only if $\xi^{-1}(L)$ is an $SVNRPOS$ in W, whenever L is an $SVNROS$ in M;*
4. *single-valued neutrosophic rough b-continuous mapping ($SVNRb - C$-mapping) if and only if $\xi^{-1}(L)$ is an $SVNRb - OS$ in W, whenever L is an $SVNROS$ in M;*
5. *single-valued neutrosophic rough simply continuous mapping ($SVNR$-Simply-C-mapping) if and only if $\xi^{-1}(L)$ is an $SVNRSO$-set in W, whenever L is an $SVNROS$ in M.*

Theorem 9. *Suppose that $\xi : (W, \tau_{RPNT}(\rho)) \to (M, \tau_{RPNT}(\lambda))$ and $\zeta : (M, \tau_{RPNT}(\lambda)) \to (Z, \tau_{RPNT}(\kappa))$ be two $SVNR - C$-mappings. Then, the composition mapping $\zeta \circ \xi : (W, \tau_{RPNT}(\rho)) \to (Z, \tau_{RPNT}(\kappa))$ is also an $SVNR - C$-mapping.*

Proof. Let $\xi : (W, \tau_{RPNT}(\rho)) \to (M, \tau_{RPNT}(\lambda))$ and $\zeta : (M, \tau_{RPNT}(\lambda)) \to (Z, \tau_{RPNT}(\kappa))$ be two $SVNR - C$-mappings. Assume that Q be an $SVNROS$ in $(Z, \tau_{RPNT}(\kappa))$. Since, $\zeta : (M, \tau_{RPNT}(\lambda)) \to (Z, \tau_{RPNT}(\kappa))$ is an SVNR-C-mapping, so $\zeta^{-1}(Q)$ is an $SVNROS$ in $(M, \tau_{RPNT}(\lambda))$. Further, since $\xi : (W, \tau_{RPNT}(\rho)) \to (M, \tau_{RPNT}(\lambda))$ is an $SVNR - C$-mapping, so $\xi^{-1}(\zeta^{-1}(Q)) = (\zeta \circ \xi)^{-1}(Q)$ is an $SVNROS$ in $(W, \tau_{RPNT}(\rho))$. Hence, $(\zeta \circ \xi)^{-1}(Q)$ is an $SVNROS$ in $(W, \tau_{RPNT}(\rho))$, whenever Q is an $SVNROS$ in $(Z, \tau_{RPNT}(\kappa))$. Therefore, $\zeta \circ \xi : (W, \tau_{RPNT}(\rho)) \to (Z, \tau_{RPNT}(\kappa))$ is an $SVNR - C$-mapping.

Theorem 10. *Suppose that $\xi : (W, \tau_{RPNT}(\rho)) \to (M, \tau_{RPNT}(\lambda))$ be an $SVNR$-Simply-C-mapping and $\zeta : (M, \tau_{RPNT}(\lambda)) \to (Z, \tau_{RPNT}(\kappa))$ be an $SVNR - C$-mapping. Then, the composition mapping $\zeta \circ \xi : (W, \tau_{RPNT}(\rho)) \to (Z, \tau_{RPNT}(\kappa))$ is an $SVNR$-Simply-C-mapping.*

Proof. Let S be an $SVNROS$ in $(Z, \tau_{RPNT}(\kappa))$.

Since, $\zeta : (M, \tau_{RPNT}(\lambda)) \to (Z, \tau_{RPNT}(\kappa))$ is an $SVNR$-C-mapping, so $\zeta^{-1}(S)$ is an $SVNROS$ in $(M, \tau_{RPNT}(\lambda))$.

Further, since $\xi : (W, \tau_{RPNT}(\rho)) \to (M, \tau_{RPNT}(\lambda))$ is an $SVNR$-Simply-C-mapping, so $\xi^{-1}(\zeta^{-1}(S)) = (\zeta \circ \xi)^{-1}(S)$ is an SVNRSO-set in $(W, \tau_{RPNT}(\rho))$. Hence, $(\zeta \circ \xi)^{-1}(Q)$ is an $SVNRSO$-set in $(W, \tau_{RPNT}(\rho))$, whenever Q is an $SVNROS$ in $(Z, \tau_{RPNT}(\kappa))$. Therefore, $\zeta \circ \xi : (W, \tau_{RPNT}(\rho)) \to (Z, \tau_{RPNT}(\kappa))$ is an $SVNR$-Simply-C-mapping.

Theorem 11. *1. Every $SVNR$-C-mapping is also an $SVNRP$-C-mapping.*

2. Every $SVNR$-C-mapping is also an $SVNRS$-C-mapping.

3. Every $SVNRP$-C-mapping is an $SVNRb$-C-mapping.

4. Every $SVNRS$-C-mapping is also an $SVNRb$-C-mapping.

5. Every $SVNR$-C-mapping is also an $SVNRb$-C-mapping.

6. Every $SVNR$-C-mapping is also an $SVNR$-Simply-C-mapping.

Proof. 1. Suppose that $\xi : (W, \tau_{RPNT}(\rho)) \to (M, \tau_{RPNT}(\lambda))$ be an $SVNR$-C-mapping. Let L be an $SVNROS$ in M. Therefore, $\xi^{-1}(L)$ is an $SVNROS$ in W. Since, every $SVNROS$ is an $SVNRPOS$, so $\xi^{-1}(L)$ is an $SVNRPOS$ in W. Therefore, $\xi^{-1}(L)$ is an $SVNRPOS$ in W, whenever L be an $SVNROS$ in M. Hence, $\xi : (W, \tau_{RPNT}(\rho)) \to (M, \tau_{RPNT}(\lambda))$ is an $SVNRP$-C-mapping.
2. Suppose that $\xi : (W, \tau_{RPNT}(\rho)) \to (M, \tau_{RPNT}(\lambda))$ be an $SVNR$-C-mapping. Let L be an $SVNROS$ in M. Therefore, $\xi^{-1}(L)$ is an $SVNROS$ in W. Since, every $SVNROS$ is an $SVNRSOS$, so $\xi^{-1}(L)$ is an $SVNRSOS$ in W. Therefore, $\xi^{-1}(L)$ is an $SVNRSOS$ in W, whenever L be an $SVNROS$ in M. Hence, $\xi : (W, \tau_{RPNT}(\rho)) \to (M, \tau_{RPNT}(\lambda))$ is an $SVNRS$-C-mapping.
3. Suppose that $\xi : (W, \tau_{RPNT}(\rho)) \to (M, \tau_{RPNT}(\lambda))$ be an $SVNRP$-C-mapping. Let L be an $SVNROS$ in M. Therefore, $\xi^{-1}(L)$ is an $SVNRPOS$ in W. Since, every $SVNRPOS$ is an $SVNRbOS$, so $\xi^{-1}(L)$ is an $SVNRbOS$ in W. Therefore, $\xi^{-1}(L)$ is an $SVNRbOS$ in W, whenever L be an $SVNROS$ in M. Hence, $\xi : (W, \tau_{RPNT}(\rho)) \to (M, \tau_{RPNT}(\lambda))$ is an $SVNRb$-C-mapping.
4. Suppose that $\xi : (W, \tau_{RPNT}(\rho)) \to (M, \tau_{RPNT}(\lambda))$ be an $SVNRS$-C-mapping. Let L be an $SVNROS$ in M. Therefore, $\xi^{-1}(L)$ is an $SVNRSOS$ in W. Since, every $SVNRSOS$ is an $SVNRbOS$, so $\xi^{-1}(L)$ is an $SVNRbOS$ in W. Therefore, $\xi^{-1}(L)$ is an $SVNRbOS$ in W, whenever L be an $SVNROS$ in M. Hence, $\xi : (W, \tau_{RPNT}(\rho)) \to (M, \tau_{RPNT}(\lambda))$ is an $SVNRb$-C-mapping.
5. Suppose that $\xi : (W, \tau_{RPNT}(\rho)) \to (M, \tau_{RPNT}(\lambda))$ be an $SVNR$-C-mapping. Let L be an $SVNROS$ in M. Therefore, $\xi^{-1}(L)$ is an $SVNROS$ in W. Since, every $SVNROS$ is an $SVNRbOS$, so $\xi^{-1}(L)$ is an $SVNRbOS$ in W. Therefore, $\xi^{-1}(L)$ is an $SVNRbOS$ in W, whenever L be an $SVNROS$ in M. Hence, $\xi : (W, \tau_{RPNT}(\rho)) \to (M, \tau_{RPNT}(\lambda))$ is an $SVNRb$-C-mapping.

6. Suppose that ξ : $(W, \tau_{RPNT}(\rho)) \rightarrow (M, \tau_{RPNT}(\lambda))$ be an $SVNR - C$-mapping. Let L be an $SVNROS$ in M. Therefore, $\xi^{-1}(L)$ is an $SVNROS$ in W. Since, every $SVNROS$ is an $SVNRSO$-set, so $\xi^{-1}(L)$ is an $SVNRSO$-set in W. Therefore, $\xi^{-1}(L)$ is an $SVNRSO$-set in W, whenever L be an $SVNROS$ in M. Hence, ξ : $(W, \tau_{RPNT}(\rho)) \rightarrow (M, \tau_{RPNT}(\lambda))$ is an $SVNR - Simply - C$-mapping.

Definition 24. *Let ξ be a mapping from a $SVNRTS$ $(W, \tau_{RPNT}(\rho))$ to another $SVNRTS$ $(M, \tau_{RPNT}(\lambda))$. Then, ξ is called as*

1. *single-valued neutrosophic rough open mapping if and only if $\xi(K)$ is an $SVNROS$ in M, whenever K is an $SVNROS$ in W;*
2. *single-valued neutrosophic rough pre-open mapping if and only if $\xi(K)$ is an $SVNRPOS$ in M, whenever K is an $SVNROS$ in W;*
3. *single-valued neutrosophic rough semi-open mapping if and only if $\xi(K)$ is an $SVNRSOS$ in M, whenever K is an $SVNROS$ in W;*
4. *single-valued neutrosophic rough b-open mapping if and only if $\xi(K)$ is an $SVNRb - OS$ in M, whenever K is an $SVNROS$ in W;*
5. *single-valued neutrosophic rough simply-open mapping if and only if $\xi(K)$ is an $SVNRSO$-set in M, whenever K is an $SVNROS$ in W.*

Theorem 12. *If ξ :$(W, \tau_{RPNT}(\rho)) \rightarrow (M, \tau_{RPNT}(\lambda))$ and ζ :$(M, \tau_{RPNT}(\lambda)) \rightarrow (Z, \tau_{RPNT}(\kappa))$ be two single-valued neutrosophic rough open mappings, then the composition mapping $\zeta \circ \xi$:$(W, \tau_{RPNT}(\rho)) \rightarrow (Z, \tau_{RPNT}(\kappa))$ is also an single-valued neutrosophic rough open mapping.*

Proof. Let Q be an $SVNROS$ in $(W, \tau_{RPNT}(\rho))$. Since, ξ : $(W, \tau_{RPNT}(\rho)) \rightarrow (M, \tau_{RPNT}(\lambda))$ is an single-valued neutrosophic rough open mapping, so $\xi(Q)$ is an $SVNROS$ in $(M, \tau_{RPNT}(\lambda))$. Further, since ζ : $(M, \tau_{RPNT}(\lambda)) \rightarrow (Z, \tau_{RPNT}(\kappa))$ is an single-valued neutrosophic rough open mapping, so $\zeta(\xi(Q)) = (\zeta \circ \xi)(Q)$ is an $SVNROS$ in $(Z, \tau_{RPNT}(\kappa))$. Hence, $(\zeta \circ \xi)(Q)$ is an $SVNROS$ in $(Z, \tau_{RPNT}(\kappa))$, whenever Q is an $SVNROS$ in $(W, \tau_{RPNT}(\rho))$. Therefore, $\zeta \circ \xi$: $(W, \tau_{RPNT}(\rho)) \rightarrow (Z, \tau_{RPNT}(\kappa))$ is an single-valued neutrosophic rough open mapping.

Theorem 13. *Let ξ : $(W, \tau_{RPNT}(\rho)) \rightarrow (M, \tau_{RPNT}(\lambda))$ be an single-valued neutrosophic rough open mapping and ζ : $(M, \tau_{RPNT}(\lambda)) \rightarrow (Z, \tau_{RPNT}(\kappa))$ be an single-valued neutrosophic rough simply open mapping. Then, the composition mapping $\zeta \circ \xi$: $(W, \tau_{RPNT}(\rho)) \rightarrow (Z, \tau_{RPNT}(\kappa))$ is an single-valued neutrosophic rough simply open mapping.*

Proof. Let Q be an $SVNROS$ in $(W, \tau_{RPNT}(\rho))$. Since, ξ : $(W, \tau_{RPNT}(\rho)) \rightarrow (M, \tau_{RPNT}(\lambda))$ is an single-valued neutrosophic rough open mapping, so $\xi(Q)$ is an $SVNROS$ in $(M, \tau_{RPNT}(\lambda))$. Further, since ζ : $(M, \tau_{RPNT}(\lambda)) \rightarrow (Z, \tau_{RPNT}(\kappa))$ is an single-valued neutrosophic rough simply open mapping, so $\zeta(\xi(Q)) = (\zeta \circ \xi)(Q)$ is an $SVNRSO$-set in $(Z, \tau_{RPNT}(\kappa))$. Hence, $(\zeta \circ \xi)(Q)$ is an $SVNRSO$-set in $(Z, \tau_{RPNT}(\kappa))$, whenever Q is an $SVNROS$ in $(W, \tau_{RPNT}(\rho))$. Therefore, the composition mapping $\zeta \circ \xi$: $(W, \tau_{RPNT}(\rho)) \rightarrow (Z, \tau_{RPNT}(\kappa))$ is an single-valued neutrosophic rough simply open mapping.

Theorem 14. *1. Every single-valued neutrosophic rough open mapping is an single-valued neutrosophic rough pre-open mapping.*

2. Every single-valued neutrosophic rough open mapping is an single-valued neutrosophic rough semi-open mapping.

3. Every single-valued neutrosophic rough pre-open mapping is an single-valued neutrosophic rough b-open mapping.

4. Every single-valued neutrosophic rough semi-open mapping is an single-valued neutrosophic rough b-open mapping.

5. Every single-valued neutrosophic rough open mapping is an single-valued neutrosophic rough b-open mapping.

6. Every single-valued neutrosophic rough open mapping is an single-valued neutrosophic rough simply open mapping.

Proof. 1. Suppose that $\xi : (W, \tau_{RPNT}(\rho)) \to (M, \tau_{RPNT}(\lambda))$ be an single-valued neutrosophic rough open mapping. Let L be an $SVNROS$ in W. Therefore, $\xi(L)$ is an $SVNROS$ in M. Since, every $SVNROS$ is an $SVNRPOS$, so $\xi(L)$ is an $SVNRPOS$ in M. Therefore, $\xi(L)$ is an $SVNRPOS$ in M, whenever L be an $SVNROS$ in W. Hence, $\xi : (W, \tau_{RPNT}(\rho)) \to (M, \tau_{RPNT}(\lambda))$ is an single-valued neutrosophic rough pre-open mapping.

2. Suppose that $\xi : (W, \tau_{RPNT}(\rho)) \to (M, \tau_{RPNT}(\lambda))$ be an single-valued neutrosophic rough open mapping. Let L be an $SVNROS$ in W. Therefore, $\xi(L)$ is an $SVNROS$ in M. Since, every $SVNROS$ is an $SVNRSOS$, so $\xi(L)$ is an $SVNRSOS$ in M. Therefore, $\xi(L)$ is an $SVNRSOS$ in M, whenever L be an $SVNROS$ in W. Hence, $\xi : (W, \tau_{RPNT}(\rho)) \to (M, \tau_{RPNT}(\lambda))$ is an single-valued neutrosophic rough semi-open mapping.

3. Suppose that $\xi : (W, \tau_{RPNT}(\rho)) \to (M, \tau_{RPNT}(\lambda))$ be an single-valued neutrosophic rough pre-open mapping. Let L be an $SVNROS$ in W. Therefore, $\xi(L)$ is an $SVNRPOS$ in M. Since, every $SVNRPOS$ is an $SVNRbOS$, so $\xi(L)$ is an $SVNRbOS$ in M. Therefore, $\xi(L)$ is an $SVNRbOS$ in M, whenever L be an $SVNROS$ in W. Hence, $\xi : (W, \tau_{RPNT}(\rho)) \to (M, \tau_{RPNT}(\lambda))$ is an single-valued neutrosophic rough b-open mapping.

4. Suppose that $\xi : (W, \tau_{RPNT}(\rho)) \to (M, \tau_{RPNT}(\lambda))$ be an single-valued neutrosophic rough semi-open mapping. Let L be an $SVNROS$ in W. Therefore, $\xi(L)$ is an $SVNRSOS$ in M. Since, every $SVNRSOS$ is an $SVNRb-OS$, so $\xi(L)$ is an $SVNRb-OS$ in M. Therefore, $\xi(L)$ is an $SVNRb-OS$ in M, whenever L be an $SVNROS$ in W. Hence, $\xi : (W, \tau_{RPNT}(\rho)) \to (M, \tau_{RPNT}(\lambda))$ is an single-valued neutrosophic rough b-open mapping.

5. Suppose that $\xi : (W, \tau_{RPNT}(\rho)) \to (M, \tau_{RPNT}(\lambda))$ be an single-valued neutrosophic rough open mapping. Let L be an $SVNROS$ in W. Therefore, $\xi(L)$ is an $SVNROS$ in M. Since, every $SVNROS$ is an $SVNRb-OS$, so $\xi(L)$ is an $SVNRb-OS$ in M. Therefore, $\xi(L)$ is an $SVNRb-OS$ in M, whenever L be an $SVNROS$ in W. Hence, $\xi : (W, \tau_{RPNT}(\rho)) \to (M, \tau_{RPNT}(\lambda))$ is an single-valued neutrosophic rough b-open mapping.

6. Suppose that $\xi : (W, \tau_{RPNT}(\rho)) \to (M, \tau_{RPNT}(\lambda))$ be an single-valued neutrosophic rough open mapping. Let L be an $SVNROS$ in W. Therefore, $\xi(L)$ is an $SVNROS$ in M. Since, every $SVNROS$ is an $SVNRSO-set$, so $\xi(L)$

is an $SVNRSO-set$ in M. Therefore, $\xi(L)$ is an $SVNRSO-set$ in M, whenever L be an $SVNROS$ in W. Hence, $\xi : (W, \tau_{RPNT}(\rho)) \rightarrow (M, \tau_{RPNT}(\lambda))$ is an single-valued neutrosophic rough simply-open mapping.

Definition 25. *A family* $\{X_\alpha : \alpha \in \Delta\}$, *where* Δ *is an index set and* X_α *is an* $SVNROS$ *in* $(W, \tau_{RPNT}(\rho))$, *for each* $\alpha \in \Delta$, *is called an single-valued neutrosophic rough open cover of an* SVNRS X *if* $X \subseteq \cup_{\alpha \in \Delta} X_\alpha$.

Definition 26. *An* $SVNRTS$ $(W, \tau_{RPNT}(\rho))$ *is called an single-valued neutrosophic rough compact (SVNR-compact) space if each single-valued neutrosophic open cover of* 1_W *has a finite sub-cover.*

Definition 27. *An single-valued neutrosophic rough sub-set* B *of an* $SVNRTS$ $(W, \tau_{RPNT}(\rho))$ *is called an single-valued neutrosophic rough compact relative to* W *if every single-valued neutrosophic rough open cover of* B *has a finite single-valued neutrosophic rough open sub-cover.*

Theorem 15. *Every single-valued neutrosophic rough closed sub-set of an single-valued neutrosophic rough compact space* $(W, \tau_{RPNT}(\rho))$ *is single-valued neutrosophic rough compact relative to* W.

Proof. Assume that $(W, \tau_{RPNT}(\rho))$ be an single-valued neutrosophic rough compact space. Suppose that K be an single-valued neutrosophic rough closed sub-set of $(W, \tau_{RPNT}(\rho))$. Therefore, K^c is an $SVNROS$ in $(W, \tau_{RPNT}(\rho))$. Suppose that $U = \{U_i : i \in \Delta$ and U_i is an $SVNROS$ in $(W, \tau_{RPNT}(\rho))\}$ be an single-valued neutrosophic rough open cover of K. Then, $H = \{K^c\} \cup U$ is an single-valued neutrosophic rough open cover of 1_W. Since, $(W, \tau_{RPNT}(\rho))$ is an single-valued neutrosophic rough compact space, so it has a finite sub-cover say $\{H_1, H_2, H_3, ..., H_n, K^c\}$. This implies, $\{H_1, H_2, H_3, ..., H_n\}$ is a finite single-valued neutrosophic rough open cover of K. Hence, K is an single-valued neutrosophic rough compact set relative to W.

Theorem 16. *If* $\xi : (W, \tau_{RPNT}(\rho)) \rightarrow (M, \tau_{RPNT}(\lambda))$ *be an* $SVNR - C$-*mapping, then for each single-valued neutrosophic rough compact set* Q *relative to* W, $\xi(Q)$ *is an single-valued neutrosophic rough compact set relative to* M.

Proof. Suppose that $\xi : (W, \tau_{RPNT}(\rho)) \rightarrow (M, \tau_{RPNT}(\lambda))$ be an $SVNR - C$-mapping. Let Q be an single-valued neutrosophic rough compact set relative to W. Suppose that $U = \{U_i : i \in \Delta$ and U_i is an $SVNROS$ in $M\}$ be an single-valued neutrosophic rough open cover of $\xi(Q)$. Therefore, $V = \{\xi^{-1}(U_i) : i \in \Delta$ and $\xi^{-1}(U_i)$ is an $SVNROS$ in $W\}$ is an single-valued neutrosophic rough open cover of $\xi^{-1}(\xi(Q)) = Q$. Since, Q be an single-valued neutrosophic rough compact set relative to W.

So there exist a finite sub-cover $\{\xi^{-1}(U_1), \xi^{-1}(U_2), ..., \xi^{-1}(U_n)\}$ such that $Q \subseteq \cup_{i=1}^{n} \xi^{-1}(U_i)$.

This implies, $\xi(Q) \subseteq \xi(\cup_{i=1}^{n} \xi^{-1}(U_i)) = \cup_{i=1}^{n} \xi(\xi^{-1}(U_i)) = \cup_{i=1}^{n} U_i$. Therefore, $\xi(Q) \subseteq \cup_{i=1}^{n} U_i$. Hence, there exist a finite sub-cover $\{U_1, U_2, ..., U_n\}$ such that $\xi(Q) \subseteq \cup_{i=1}^{n} U_i$. Therefore, $\xi(Q)$ is an single-valued neutrosophic rough compact set relative to M, whenever Q be an single-valued neutrosophic rough compact set relative to W.

Theorem 17. *If $\xi : (W, \tau_{RPNT}(\rho)) \to (M, \tau_{RPNT}(\lambda))$ is an single-valued neutrosophic rough open function and $(M, \tau_{RPNT}(\lambda))$ is an single-valued neutrosophic rough compact space, then $(W, \tau_{RPNT}(\rho))$ is also an single-valued neutrosophic rough compact space.*

Proof. Suppose that $\xi : (W, \tau_{RPNT}(\rho)) \to (M, \tau_{RPNT}(\lambda))$ be an single-valued neutrosophic rough open function. Let $(M, \tau_{RPNT}(\lambda))$ be an single-valued neutrosophic rough compact space. Suppose that $U = \{U_i : i \in \Delta$ and U_i is a $SVNROS$ in $W\}$ be an single-valued neutrosophic rough open cover of 1_W. Therefore, $V = \{\xi(U_i) : i \in \Delta$ and $\xi(U_i)$ is an $SVNROS$ in $M\}$ is an single-valued neutrosophic rough open cover of $\xi(1_W)) = 1_M$. Since, $(M, \tau_{RPNT}(\lambda))$ is an single-valued neutrosophic rough compact space, so there exist a finite sub-cover of 1_M say $\{\xi(U_1), \xi(U_2), ..., \xi(U_n)\}$ such that $1_M \subseteq \cup_{i=1}^{n}\xi(U_i)$. This implies, $\xi^{-1}(1_M) \subseteq \xi^{-1}(\cup_{i=1}^{n}\xi(U_i)) = \cup_{i=1}^{n}\xi^{-1}(\xi(U_i)) = \cup_{i=1}^{n}U_i$. Therefore, $1_W = \xi^{-1}(1_M) \subseteq \cup_{i=1}^{n}U_i$. This implies, $1_W \subseteq \cup_{i=1}^{n}U_i$. Hence, there exist a finite sub-cover $\{U_1, U_2, ..., U_n\}$ such that $1_W \subseteq \cup_{i=1}^{n}U_i$. Therefore, $(W, \tau_{RPNT}(\rho))$ is an single-valued neutrosophic rough compact space, whenever $(M, \tau_{RPNT}(\lambda))$ is an single-valued neutrosophic rough compact space.

Definition 28. *A family $\{X_\alpha : \alpha \in \Delta\}$, where Δ is an index set and X_α is an SVNRSO-set in an SVNRTS $(W, \tau_{RPNT}(\rho))$, for each $\alpha \in \Delta$, is called an single-valued neutrosophic rough simply open cover of an SVNRS X if $X \subseteq \cup_{\alpha \in \Delta} X_\alpha$.*

Definition 29. *An SVNRTS $(W, \tau_{RPNT}(\rho))$ is called an single-valued neutrosophic rough simply compact space if each single-valued neutrosophic rough simply open cover of 1_W has a finite sub-cover.*

Definition 30. *An single-valued neutrosophic rough sub-set B of an SVNRTS $(W, \tau_{RPNT}(\rho))$ is called an single-valued neutrosophic rough simply compact relative to W if every single-valued neutrosophic rough simply open cover of B has a finite sub-cover.*

Theorem 18. *Every single-valued neutrosophic rough simply closed subset of an single-valued neutrosophic rough simply compact space $(W, \tau_{RPNT}(\rho))$ is single-valued neutrosophic rough simply compact relative to W.*

Proof. Assume that $(W, \tau_{RPNT}(\rho))$ be an single-valued neutrosophic rough simply compact space. Suppose that K be an single-valued neutrosophic rough simply closed sub-set of $(W, \tau_{RPNT}(\rho))$. Therefore, K^c is an $SVNRSO$-set in $(W, \tau_{RPNT}(\rho))$. Suppose that $U = \{U_i : i \in \Delta$ and U_i is an $SVNRSO$-set in $(W, \tau_{RPNT}(\rho))\}$ be an single-valued neutrosophic rough simply open cover of K. Then, $H = \{K^c\} \cup U$ is an single-valued neutrosophic rough simply open cover of 1_W. Since, $(W, \tau_{RPNT}(\rho))$ is an single-valued neutrosophic rough simply compact space, so it has a finite sub-cover say $\{H_1, H_2, H_3, ..., H_n, K^c\}$. This implies, $\{H_1, H_2, H_3, ..., H_n\}$ is a finite single-valued neutrosophic rough simply open cover of K. Hence, K is an single-valued neutrosophic rough simply compact set relative to W.

Theorem 19. *Every single-valued neutrosophic rough simply compact space is an single-valued neutrosophic compact space.*

Proof. Suppose that $(W, \tau_{RPNT}(\rho))$ be an single-valued neutrosophic rough simply compact space. Therefore, every single-valued neutrosophic rough simply-open cover of 1_W has a finite sub-cover. Suppose that $(W, \tau_{RPNT}(\rho))$ may not be an single-valued neutrosophic compact space. Then, there exists an single-valued neutrosophic rough open cover H (say) of 1_W, which has no finite sub-cover. Since, every $SVNROS$ is an $SVNRSO$-set, so we have an single-valued neutrosophic rough simply-open cover H of 1_W, which has no finite sub-cover. This contradicts the fact that $(W, \tau_{RPNT}(\rho))$ is an single-valued neutrosophic rough simply compact space. Hence, the $SVNRTS$ $(W, \tau_{RPNT}(\rho))$ is an single-valued neutrosophic rough compact space.

Theorem 20. *If $\xi : (W, \tau_{RPNT}(\rho)) \to (M, \tau_{RPNT}(\lambda))$ is an $SVNR$-Simply-C-mapping, then for each single-valued neutrosophic rough simply compact set Q relative to W, $\xi(Q)$ is an single-valued neutrosophic rough compact set relative to M.*

Proof. Suppose that $\xi : (W, \tau_{RPNT}(\rho)) \to (M, \tau_{RPNT}(\lambda))$ be an $SVNR$-Simply-C-mapping. Let Q be an single-valued neutrosophic rough simply compact set relative to W. Suppose that $U = \{U_i : i \in \Delta$ and U_i is an $SVNRSO$-set in $M\}$ be an single-valued neutrosophic rough open cover of $\xi(Q)$. Therefore, $V = \{\xi^{-1}(U_i) : i \in \Delta$ and $\xi^{-1}(U_i)$ is an $SVNRSO$-set in $W\}$ is an single-valued neutrosophic rough simply open cover of $\xi^{-1}(\xi(Q)) = Q$. Since, Q be an single-valued neutrosophic rough simply compact set relative to W, so there exist a finite sub-cover $\{\xi^{-1}(U_1), \xi^{-1}(U_2), ..., \xi^{-1}(U_n)\}$ such that $Q \subseteq \cup_{i=1}^{n}\xi^{-1}(U_i)$. This implies, $\xi(Q) \subseteq \xi(\cup_{i=1}^{n}\xi^{-1}(U_i) = \cup_{i=1}^{n}\xi(\xi^{-1}(U_i)) = \cup_{i=1}^{n}U_i$. Therefore, $\xi(Q) \subseteq \cup_{i=1}^{n}U_i$. Hence, there exist a finite sub-cover $\{U_1, U_2, ..., U_n\}$ such that $\xi(Q) \subseteq \cup_{i=1}^{n}U_i$. Therefore, $\xi(Q)$ is an single-valued neutrosophic rough compact set relative to M, whenever Q be an single-valued neutrosophic rough simply compact set relative to W.

Theorem 21. *If $\xi : (W, \tau_{RPNT}(\rho)) \to (M, \tau_{RPNT}(\lambda))$ is an single-valued neutrosophic rough simply open function and $(M, \tau_{RPNT}(\lambda))$ is an single-valued neutrosophic rough simply compact space, then $(W, \tau_{RPNT}(\rho))$ is an single-valued neutrosophic rough simply compact space.*

Proof. Suppose that $\xi : (W, \tau_{RPNT}(\rho)) \to (M, \tau_{RPNT}(\lambda))$ be an single-valued neutrosophic rough simply open function. Let $(M, \tau_{RPNT}(\lambda))$ be an single-valued neutrosophic rough simply compact space. Suppose that $U = \{U_i : i \in \Delta$ and U_i is an $SVNRSO$-set in $W\}$ be an single-valued neutrosophic rough simply open cover of 1_W. Therefore, $V = \{\xi(U_i) : i \in \Delta$ and $\xi(U_i)$ is an $SVNRSO$-set in $M\}$ is an single-valued neutrosophic rough simply open cover of $\xi(1_W)) = 1_M$. Since, $(M, \tau_{RPNT}(\lambda))$ is an single-valued neutrosophic rough simply compact space, so there exist a finite sub-cover of 1_M say $\{\xi(U_1), \xi(U_2), ..., \xi(U_n)\}$ such that $1_M \subseteq \cup_{i=1}^{n}\xi(U_i)$. This implies, $\xi^{-1}(1_M) \subseteq \xi^{-1}(\cup_{i=1}^{n}\xi(U_i)) = \cup_{i=1}^{n}\xi^{-1}(\xi(U_i)) =$

$\cup_{i=1}^{n} U_i$. Therefore, $1_W = \xi^{-1}(1_M) \subseteq \cup_{i=1}^{n} U_i$. This implies, $1_W \subseteq \cup_{i=1}^{n} U_i$. Hence, there exist a finite sub-cover $\{U_1, U_2, ..., U_n\}$ such that $1_W \subseteq \cup_{i=1}^{n} U_i$. Therefore, $(W, \tau_{RPNT}(\rho))$ is also an single-valued neutrosophic rough simply compact space, whenever $(M, \tau_{RPNT}(\lambda))$ is an single-valued neutrosophic rough simply compact space.

4 Conclusions

In this article, we have introduced the notion of single-valued neutrosophic rough continuous functions and single-valued neutrosophic rough compactness via $SVNRTSs$. Besides, we have formulated several interesting results on them via $SVNRTSs$.

Acknowledgement. The author thanks the reviewer for the careful reading of the paper and the constrictive comments. The comments helped in improving the clarity of presentation of the paper.

Author contributions. All the authors have equal contribution for the preparation of this article.

Conflict of Interest. The authors declare that they have no conflict of interest.

References

1. Arokiarani, I., Dhavaseelan, R., Jafari, S., Parimala, M.: On some new notations and functions in neutrosophic topological spaces. Neutrosophic Sets Syst. **16**, 16–19 (2017)
2. Atanassov, K.: Intuitionistic fuzzy sets. Fuzzy Sets Syst. **20**, 87–96 (1986)
3. Broumi, S., Smarandache, F., Dhar, M.: Rough neutrosophic sets. Neutrosophic Sets Syst. **3**, 60–65 (2014)
4. Das, S., Das, R., Granados, C.: Topology on quadripartitioned neutrosophic sets. Neutrosophic Sets Syst. **45**, 54–61 (2021). https://doi.org/10.5281/zenodo.5485442
5. Das, S., Das, R., Tripathy, B.C.: Topology on rough pentapartitioned neutrosophic set. Iraqi J. Sci. (in Press)
6. Das, S., Pramanik, S.: Generalized neutrosophic b-open sets in neutrosophic topological space. Neutrosophic Sets Syst. **35**, 522–530 (2020)
7. Das, S., Pramanik, S.: Neutrosophic Φ-open sets and neutrosophic Φ-continuous functions. Neutrosophic Sets Syst. **38**, 355–367 (2020)
8. Das, S., Tripathy, B.C.: Neutrosophic simply b-open set in neutrosophic topological spaces. Iraqi J. Sci. (in Press)
9. Ebenanjar, E., Immaculate, J., Wilfred, C.B.: On neutrosophic b-open sets in neutrosophic topological space. J. Phys. Conf. Ser. **1139**(1), 012062 (2018)
10. Iswarya, P., Bageerathi, K.: On neutrosophic semi-open sets in neutrosophic topological spaces. Int. J. Math. Trends Technol. **37**(3), 214–223 (2018)
11. Jayaparthasarathy, G., Arockia Dasan, M., Little Flower, V.F., Ribin Christal, R.: New open sets in N-neutrosophic supra topological spaces. Neutrosophic Sets Syst. **31**, 44–62 (2020)

12. Jayaparthasarathy, G., Little Flower, V.F., Arockia Dasan, M.: Neutrosophic supra topological applications in data mining process. Neutrosophic Sets Syst. **27**, 80–97 (2019)
13. Jin, Q., Hu, K., Bo, C., Li, L.: A new single-valued neutrosophic rough sets and related topology. J. Math. 1–14 (2021). Article ID 5522021. https://doi.org/10.1155/2021/5522021
14. Lellis Thivagar, M., Jafari, S., Sutha Devi, V., Antonysamy, V.: A novel approach to nano topology via neutrosophic sets. Neutrosophic Sets Syst. **20**, 86–94 (2018)
15. Maheswari, C., Sathyabama, M., Chandrasekar, S.: Neutrosophic generalized *b*-closed sets in neutrosophic topological spaces. J. Phys. Conf. Ser. **1139**(1), 012065 (2018)
16. Mohammed Ali Jaffer, I., Ramesh, K.: Neutrosophic generalized pre regular closed sets. Neutrosophic Sets Syst. **30**, 171–181 (2019)
17. Mondal, K., Pramanik, S.: Decision making based on some similarity measures under interval rough neutrosophic environment. Neutrosophic Sets Syst. **10**, 46–57 (2015)
18. Mondal, K., Pramanik, S.: Rough neutrosophic multi-attribute decision-making based on rough accuracy score function. Neutrosophic Sets Syst. **8**, 16–22 (2015)
19. Mondal, K., Pramanik, S.: Rough neutrosophic multi-attribute decision-making based on grey relational analysis. Neutrosophic Sets Syst. **7**, 8–17 (2015)
20. Mondal, K., Pramanik, S., Giri, B.C.: Rough neutrosophic aggregation operators for multi-criteria decision-making. In: Kahraman, C., Otay, İ (eds.) Fuzzy Multi-criteria Decision-Making Using Neutrosophic Sets. SFSC, vol. 369, pp. 79–105. Springer, Cham (2019). https://doi.org/10.1007/978-3-030-00045-5_5
21. Mondal, K., Pramanik, S., Smarandache, F.: Rough neutrosophic TOPSIS for multi-attribute group decision making. Neutrosophic Sets Syst. **13**, 105–117 (2016)
22. Mondal, K., Pramanik, S., Smarandache, F.: Multi-attribute decision making based on rough neutrosophic variational coefficient similarity measure. Neutrosophic Sets Syst. **13**, 3–17 (2016)
23. Mondal, K., Pramanik, S., Smarandache, F.: Rough neutrosophic hyper-complex set and its application to multi attribute decision making. Crit. Rev. **XIII**, 111–123 (2016)
24. Nanda, S., Majumdar, S.: Fuzzy rough sets. Fuzzy Sets Syst. **45**, 157–160 (1992)
25. Pawlak, Z.: Rough sets. Int. J. Inf. Comput. Sci. **11**(5), 341–356 (1982)
26. Pawlak, Z., Sowinski, R.: Rough set approach to multi attribute decision analysis. Eur. J. Oper. Res. **72**(3), 443–459 (1994)
27. Pramanik, S.: Interval quadripartitioned neutrosophic sets. Neutrosophic Sets Syst. (in Press)
28. Pramanik, S., Mondal, K.: Some rough neutrosophic similarity measures and their application to multi attribute decision making. Glob. J. Eng. Sci. Res. Manag. **2**(7), 61–74 (2015)
29. Pramanik, S., Mondal, K.: Rough bipolar neutrosophic set. Glob. J. Eng. Sci. Res. Manag. **3**(6), 71–81 (2016)
30. Pramanik, S., Mondal, K.: Cosine similarity measure of rough neutrosophic sets and its application in medical diagnosis. Glob. J. Adv. Res. **2**(1), 212–220 (2015)
31. Pramanik, S., Roy, R., Roy, T.K., Smarandache, F.: Multi criteria decision making using correlation coefficient under rough neutrosophic environment. Neutrosophic Sets Syst. **17**, 29–36 (2017)
32. Pramanik, S., Roy, R., Roy, T.K.: Multi criteria decision making based on projection and bidirectional projection measures of rough neutrosophic sets. Neutrosophic Sets Syst. **19**, 101–109 (2018)

33. Pushpalatha, A., Nandhini, T.: Generalized closed sets via neutrosophic topological spaces. Malaya J. Matematik **7**(1), 50–54 (2019)
34. Rao, V.V., Srinivasa, R.: Neutrosophic pre-open sets and pre-closed sets in neutrosophic topology. Int. J. ChemTech Res. **10**(10), 449–458 (2017)
35. Riaz, M., Smarandache, F., Karaaslan, F., Hashmi, M.R., Nawaz, I.: Neutrosophic soft rough topology and its applications to multi-criteria decision-making. Neutrosophic Sets Syst. **35**, 198–219 (2020)
36. Roy, R., Pramanik, S., Roy, T.K.: Interval rough neutrosophic TOPSIS strategy for multi-attribute decision making. In: Neutrosophic Sets in Decision Analysis and Operations Research, 21 p (2020). https://doi.org/10.4018/978-1-7998-2555-5.ch005
37. Smarandache, F.: A Unifying Field in Logics, Neutrosophy: Neutrosophic Probability, Set and Logic. Rehoboth, American Research Press (1998)
38. Salama, A.A., Alblowi, S.A.: Neutrosophic set and neutrosophic topological space. ISOR J. Math. **3**(4), 31–35 (2012)
39. Sweety, C.A.C., Arockiarani, I.: Topological structures of fuzzy neutrosophic rough sets. Neutrosophic Sets Syst. **9**, 50–57 (2015)
40. Vijayalakshmi, R., Mookambika, A.P.: Properties of neutrosophic nano semi open sets. Malaya J. Matematik **8**(4), 1851–1858 (2020)
41. Wang, H., Madiraju, P., Zhang, Y.Q., Sunderraman, R.: Interval neutrosophic sets. Int. J. Appl. Math. Stat. **3**(5), 1–18 (2005)
42. Wang, H., Smarandache, F., Zhang, Y.Q., Sunderraman, R.: Single valued neutrosophic sets. Multispace Multistructure **4**, 410–413 (2010)
43. Zadeh, L.A.: Fuzzy sets. Inf. Control **8**, 338–353 (1965)

Regular Paper

The RSDS - Bibliographic Database for Rough Sets and Related Fields

Zbigniew Suraj$^{(\boxtimes)}$ ⓘ and Piotr Grochowalski ⓘ

Institute of Computer Science, University of Rzeszów, Rzeszów, Poland
{zsuraj,pgrochowalski}@ur.edu.pl

Abstract. This paper provides an overview of the Rough Set Database System (the RSDS for short) for creating bibliographies on rough sets and related fields, as well as sharing and analysis. The current version of the RSDS includes a number of modifications, extensions and functional improvements compared to the previous versions of this system. The system was made in the client-server technology. Currently, the RSDS contains over 38 540 entries from nearly 42 860 authors. This system works on any computer connected to the Internet and is available at http://rsds.ur.edu.pl.

Keywords: Rough set · Soft computing · Ontology · Data mining · Knowledge discovery · Pattern recognition · Machine learning · Database system

1 Introduction

The concept of the rough set has its origin in Pawlak's seminal article from 1982 [7]. The rough set theory [8] is a formal theory derived from the fundamental research into the logical properties of information systems [6]. This theory is a simple and effective methodology for database mining or knowledge discovery in relational databases. In its abstract form, it is a new area of soft mathematics [2], related closely to the fuzzy set theory initiated by Lotfi A. Zadeh [29]. Rough and fuzzy sets are complementary generalizations of classical sets. The rapid development of these two approaches formed the basis of the "soft computing" [30] which includes, in addition to rough sets, at least fuzzy logic, neural networks, probabilistic reasoning, belief networks, machine learning, evolutionary computing and chaos theory [1].

For some time we have seen a systematic, global increase of interest in the rough set theory and its applications [5,9,10,27,28]. However, on the other hand, there is a lack of publicly available bibliographic databases that facilitate access to literature and other tools needed by users of rough sets.

The purpose of this paper is to present the latest issue of the RSDS, which to some extent appears to fill this gap.

The system offers a wide range of functional possibilities, including bibliography creation, modifying, downloading, analyzing, visualizing and more capabilities. The bibliography included in the RSDS is formatted according to the

© Springer-Verlag GmbH Germany, part of Springer Nature 2022
J. F. Peters et al. (Eds.): TRS XXIII, LNCS 13610, pp. 99–117, 2022.
https://doi.org/10.1007/978-3-662-66544-2_7

BibTeX specification [31]. It consists of the following publication types: articles, inproceedings, incollections, books, tech reports, proceedings, inbooks, phd theses, master theses, manuals and unpublished. In addition to the bibliography, the system also includes: (1) information about software related to rough sets, (2) bibliographies of famous people working in this field, (3) personal data of the authors of publications available in the system.

The access to the system depends on whether the user is logged in or not. If the user is logged in, he can enter data into the system, edit and classify the data he has entered. However, all users, logged in or not, can download the bibliography from the system and save it to an RTF or BibTeX file. Thanks to this possibility, the user can quickly and easily create a literature list for his or her own needs without having to delve into the structure of available file formats.

An important feature of this system is that it can successfully act as experimental environment for researches related to, inter alia, broadly understood information processing based on methods and techniques in the field of ontology and rough sets as well as advanced data analysis using the methods and techniques of statistics and graph theory. The current version of the RSDS was made in the client-server technology, which is classified as a modern database management technology.

Original and useful functionalities of the RSDS, as far as we know, rather unheard of in other database systems with a similar purpose as our system are:

- Searching according to predefined classifiers [14], as well as using the ontological search method [18,23], thanks to which the search for the desired information becomes more accurate and effective.
- Possibility of searching for information in the system with the use of an interactive world map, illustrating who and where in the world is working on the development of the rough set theory and its applications. To date, we have identified 72 research groups worldwide with 2 416 active members.
- Easy and convenient access to a number of statistics on the data contained in the RSDS and their graphical representation, such as: the number of authors, the number of and types of publications, number and years of publications, information on the percentage of all authors who wrote a certain number of publications at specified intervals, a list of indicators characterizing publishing collaboration between authors [17,19]. The results obtained from such an analysis may be useful in determining the structure of research groups in relation to rough sets, research interests of members of these groups, mutual cooperation between groups and research identification trends. It is also worth adding that the data are analyzed using different statistical methods and graph theory methods [26].
- Determining both two Pawlak numbers and the numbers of individual authors indicating the strength of the publication relationship between the authors represented in the RSDS database [20]–[22].
- Searching for information about software supporting research and experiments based on the rough set methodology and biographies of prominent people dealing with the rough set theory and its applications.

Let's also mention what significant changes have been made to the current version of the system. They include:

- implementation of a new ontological search method based on fuzzy logic,
- extension of the module for calculating Pawlak numbers and individual authors numbers,
- update of the statistical and graphical data analysis module,
- rebuilding the interactive world map,
- addition of the *Help* section,
- update of the system engine,
- rebuilding the user interface,
- modernizing the existing system functionalities,
- increasing the role of system administrator(s),
- increasing amenities for registered users,
- entering data status and modification,
- extension of system-user communication.

Historically speaking, the first version of the RSDS was released almost two decades ago, i.e. in 2003 [11], subsequent editions appeared, among others, in 2005 [13], 2008 [16] and 2014 [24].

In the Table 1 we present the quantitative changes in the system's bibliographic data over the years.

Table 1. The quantitative changes in the RSDS

Year of issue	Number of publications	Number of authors	Source
2003	900	400	[11]
2004	1 400	450	[12]
2005	1 900	815	[13]
2007	3 270	1 670	[15]
2008	3 400	1 900	[16]
2010	3 800	2 250	[19]
2014	4 000	2 380	[24]
2020	38 549	42 859	[33]

This paper is an extended version of our paper [25] presented at the International Joint Conference on Rough Sets held in Bratislava, Slovakia in 2021. The significant changes and extensions in the current version of the paper concern in particular the extension of the introductory part and the addition of two new sections. The first is called "System functionalities" and the second is "The RSDS in scientific research". Additionally, the "Final remarks" and "References" sections were expanded.

The rest of this paper covers the following topics. First, an overview of the main features of the RSDS is presented. Next, the hardware and software requirements are described. Then, we provide three examples of the use of the system in the scientific research. Finally, we discuss the plans for the further development of the RSDS with our concluding remarks.

2   System Functionalities

In this section, we present the main features of the RSDS along with a brief explanation of how to use it.

2.1   Home Page

When the RSDS is activated, its main page appears on the display (see Fig. 1).

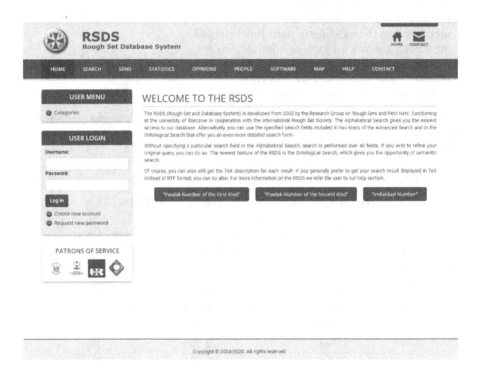

Fig. 1. The home page

There are many options on this page that allow us to navigate the entire system. They are divided into three groups of tasks. The first of them creates the so-called horizontal menu, second vertical and third – informative menu. The horizontal menu includes the following options: *Home, Search, Send, Statistics, Opinions, People, Software, Map, Help, Contact.* The vertical menu allows the user to register in the system. After logging in correctly, the user gains access to a wider range of system capabilities. The central part of the home page displays information about the purpose of the RSDS and the methods of searching information in this system. Additionally, in this part of the page there are three buttons that allow us to designate both Pawlak numbers and the numbers of individual authors [20–22]. The names of options and commands contained in the system are intuitive and their meaning is obvious. Detailed descriptions along with the operating procedures are available in the *Help* option.

2.2   Data Search

To find the required information, use the Search option, which includes following methods: *alphabetical search, advanced search,* predefined *classifier search* and *ontological search.*

For an alphabetical search, the user can select one out of six search keys: *titles, authors, publishers, conferences, journals* or *year of issue* (see Fig. 2).

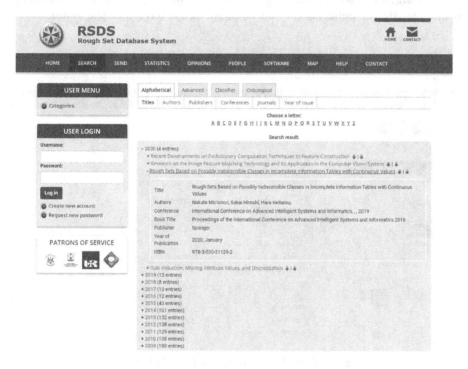

Fig. 2. Search by publication title

When choosing the advanced search method, the user fills into a specially prepared form, the fields which define the details of search criteria. The fields available in the form are equipped with so-called dynamically self-organizing lists, which help in faster and more accurate specification of search criteria. The RSDS also allows you to define complex search criteria in a way that the user can combine individual criteria using the classic logical operators *AND* and *OR*.

In the RSDS, each publication can be assigned to one of the six main thematic sections, which to some extent classify the entire area of research related to the rough set theory and its applications. The classifier available in our system divides all publications into the following groups (see Fig. 3):

- A. Foundations
- B. Applications
- C. Methods
- D. Software systems
- E. History
- F. Didactics

Moreover, each of these groups is divided into an appropriate number of subgroups. The number of subgroups is varied depending on the type of group being divided.

It is worth to mention that each registered user of the RSDS can independently classify the publications that he or she has entered into the system.

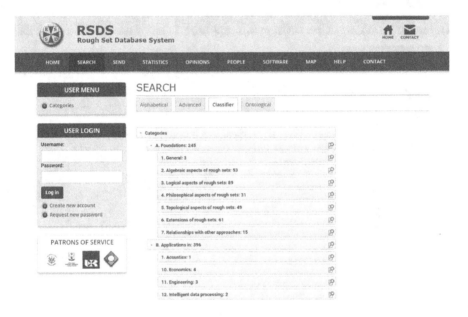

Fig. 3. Search based on the predefined classifier

Search by ontological method is possible by filling in the appropriate form available from the menu level (see Fig. 4). The ontological search form consists of one field in which you can enter a single word or phrase (according to the rules of using logical operators and searching for expressions) or select the appropriate value from the drop-down list. After clicking the *Searching publications* button, we receive grouped search results that meet the given conditions. Information about the number of entries matching each publication category is displayed in brackets.

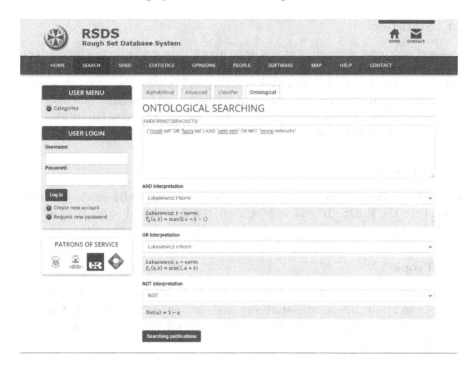

Fig. 4. Ontological search

Publication descriptions are grouped on three levels:

- Level I: types of publications, taking into account the BibTeX specification given in the introduction.
- Level II: grouping the search results depending on how well they match the query as follows: totally matched, extremely matched, matched well, considerable matched, moderately matched, more or less matched, minor matched, minimal matched, mismatched.
- Level III: publication titles in order from the best match to the least under the given conditions.

Details include (if available in the system): title, author(s), journal, number, volume, pages, publisher, year of publication, ISSN, abstracts, keywords, and DOI link.

More information about ontological search can be found in [19,23] and triangular norms in [4].

2.3 Sending Data

It is possible to submit a file with the bibliographic data to the database administrator, who has the software allowing for appending automatically the large data to the database (the *Send* option). User can do it via a special dedicated form. Submissions in the form of *BibTeX* files are preferred.

2.4 Statistics

This section contains a series of statistics about data contained in the system. There are statistics of the number of authors, the number and types of publications, the number and years of publications, information on what percentage of all authors wrote a certain number of publications in specific time intervals, a list of indicators characterizing publishing cooperation between authors and much more. Additional information are provided in Sect. 4.

2.5 Opinions

This section allows to present representative comments and remarks made by users about the RSDS. To add your opinion, the user should select the *Opinion* option.

2.6 People

This section presents two ways to search for the biographies of distinguished people who study the rough set theory and its applications. The first way is to fill in the *Search for people* field, and the second one is to use the *Browse biographies* list.

2.7 Software

By using this option, the user can search for information about research and experiment support software based on the rough set methodology. The first way to search available data here is to fill in the *Search for software* field and the second way is the *Browse software* list. In some cases, this option also provides a link to download the desired software.

2.8 Interactive Map of the World

The possibilities of searching for information in the system have been extended with an interactive world map (the *Map* option), illustrating who and where in the world is working on the development of the rough set theory and its applications. To date, we have identified 72 research groups worldwide with 2 416 active members.

This map is divided into four main parts: (a) the map of the world, (b) maps of the continents, (c) maps of the countries, (d) information about chosen rough set research groups (people).

After selecting the map option, a map of the world with a division into continents appears on the screen, as well as information on how many research groups are dealing with rough sets on a given continent and how many authors come from a given continent.

In this situation, the user can select a specific continent. After doing this, the system will redirect him or her to the continent map with countries marked on it, from which he can choose the country. After selecting a country, the system displays a map with cities in which research groups dealing with rough sets have been identified, showing also information on the number of such groups and the number of authors from the selected country.

Then, after selecting a city, we obtain information about research groups in a given city and who is the group leader. When user selects a specific group, he or she will go to the part with detailed information about this group (see Fig. 5).

Fig. 5. A detailed information for a chosen research group

2.9   Help

The section provides information in what way user can use the RSDS.

2.10   Contact

This section allows you to contact the RSDS administrator.

3   System Requirements

The RSDS can be run on any computer connected to the Internet. The computer must have an operating system with a web browser. These requirements must be met because the system works online and requires constant access to the Internet. Additionally, the web browser in which the system will operate must support JavaScript. The RSDS has been successfully tested with the following browsers: Microsoft Edge, Internet Explorer, Mozilla Firefox and Chrome.

4   The RSDS in Scientific Research

In this section we present three examples of using the RSDS to analyse the data
contained in this system. The results of the analysis show interesting information
about the types of publications, relations between the authors of publications,
and even more. The results can be useful for determining the structure of research
groups in relation to rough sets, the scientific interests by members of these
groups, mutual cooperation between groups and identifying research trends. The
data is analyzed using statistical and graph-theoretical methods [3, 26] as well
as using the so-called Pawlak numbers [20–22].

4.1   Statistical Data Analysis

Currently, 38 549 (3 800) bibliographic entries are available in the system, which
were compiled by 42 859 (2250) authors. (Numbers in brackets apply here and
until the end of this section to 2010 data [19].) Table 2 presents the list of gener-
ical groups of publications and numerical statistics for 2010 and 2020. These
statistics show a significant increase in the number of publications in almost
each group. Some groups have fewer entries. It was a result of the fact that some
entries were redundant and observed duplicates were removed from the system.

Table 2. General breakdown of the publications in the RSDS

Type of publication	Year 2010	Year 2020
article	975	21 552
inproceedings	2 268	15 324
incollection	401	1 232
book	182	176
tech report	148	144
proceedings	73	60
inbook	19	26
phd thesis	16	21
master thesis	12	12
manual	2	2
Total	4 096	38 549

The presented statistics also show that people working on rough sets willingly
present their works in scientific journals or participate in scientific conferences,
which results in a significant increase in the number of publications, both in the
form of articles and conference materials. Moreover, the significant increase in

the number of publications in many categories may also result from the higher interest in the rough set theory. This thesis is also confirmed by a large number of publications in ten-year periods, defined as follows: 1970–1980: 4, 1981–1990: 893 publications, 1991–2000: 7 597 publications, 2001–2010: 19 666 publications, 2011–2020: 10 373 publications (see Fig. 6).

Fig. 6. The data table for a given five-year period

While analysing the data available in the RSDS one can also notice a high interest in the system resources. The global number of all visits to the system in 2020 is 1 156 207. The highest interest this year is observed in Poland–48 083 (2 558), the next places in this category are taken by Germany - 13 959 (62), Russia 7 227 (unknown), China - 5 956 (95), Italy - 2 729 (111), Canada: 299 (86) references etc. There is also a clear increase in interest in the system from the national level.

The data collected in the system made it also possible to detect many interesting relationships between authors, publications and methods of creating publications with the use of statistics. Figure 7 presents the results of the data analysis taking into account the increasing five-year periods, i.e. .: 1981–1985, 1981–1990, ..., in which the changes are rather permanent, not temporary. In this figure we can show an increase in the value of all indicators, except for the number of one and two authors' works. The number of works by one author is clearly decreasing, and the number of two-author works first increased and then decreased. Another observation is the fact that the percentage of multi-author works reaches 31.53%; and this shows an increase in collaboration between authors.

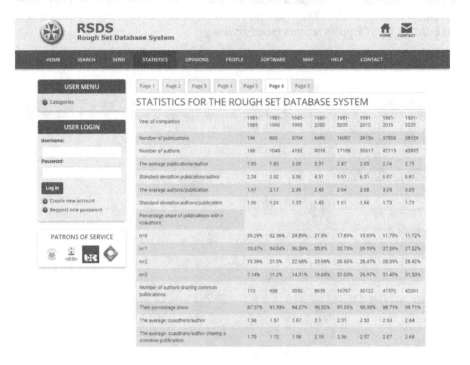

Fig. 7. The cumulative data table to the end of a given five-year period

4.2   Analysis of Collaboration Graph

Data of authors and their publications available in the RSDS allow to define the so-called collaboration graph. Nodes in such a graph represent authors, and edges between nodes represent relationships between authors. Two nodes are connected by an edge if the authors whose nodes they represent in the graph have written a joint work that is available on the system.

The collaboration graph in the system now contains 42 855 (2 414) nodes connected by 117 774 (4 022) edges (see Fig. 8). Among all the nodes of the collaboration graph, there are 565 (140) isolated nodes, i.e. nodes not connected to any other node. This result can be interpreted in such a way that the system includes publications by one author, i.e. their authors do not cooperate in this area.

After removing such nodes, the collaboration graph contains 42 290 (2 274) nodes, and the average degree of each node is 5.57 (3.54). This means that each author in the system collaborated on average with about six authors. Comparing the results of the graph analysis to date with the results of the analysis of the collaboration graph built on the basis of 2020 data, you can notice both an increase in the number of nodes in the graph and an increase in the number of edges in the graph, which means an expansion of cooperation between authors assigned to these nodes.

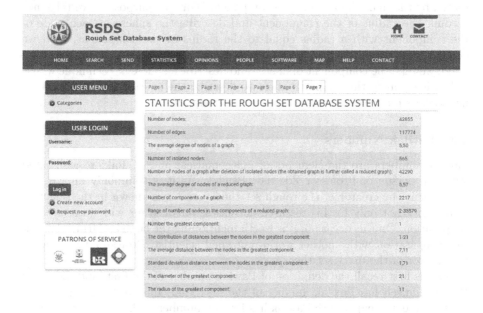

Fig. 8. The collaboration graph

In a collaboration graph, one can distinguish connected components, i.e. subgraphs where there is a link connecting a single node with another node. The defined collaboration graph has one maximum connected component of 33 579 (1 034) nodes. In addition to the maximum connected component, this graph also includes 2 217 (348) other connected components. Each of these components can be interpreted as research groups centered around the group leader. Moreover, on the basis of these components, one can learn about the interests of a given research group. These interests are expressed through the subject of publications written by members of a given group. By analysing the maximum connected component, we can define the average distance between nodes. The distance between two nodes (authors) in a given group means that if we made sphere from a given node (author), with the radius equal to the average distance between two nodes, we would get information about people closely cooperating with that particular author.

In the case of the analysed graph, the average distance is 7.11 (5.15), and the standard deviation is 1.71 (1.68). This distance defines authors who work closely together. The current results of the analysis of the maximum connected component of the collaboration graph show that authors of publications who collaborate in larger and larger groups in order to write joint publications.

The diameter of the maximum connected component, i.e. the maximum distance between two nodes of the component, is 21 (13), and the radius of this component, i.e. the maximum distance from one node to any other node, is 11 (7). These two parameters can be used to characterize the test group represented by the maximum connected component. For this purpose, if we take into account the radius of the component and describe the sphere in each node of the component with a radius equal to the radius of the component, then we can determine the so-called centres representing group leaders. If, on the other hand, we take the component diameter and describe a sphere at each node whose radius is equal to the diameter, then we describe the so-called group satellites, i.e. authors most loosely related to a given group [26].

4.3 Pawlak Numbers

Pawlak numbers are the elements of mathematical and IT folklore associated with the person of Professor Zdzisław Pawlak. As it is commonly known, the professor is the creator of the rough set theory and was known in the scientific community not only for his research activities, but also for wide international cooperation. On this occasion, it is worth mentioning the close and fruitful scientific cooperation with another world-renowned scientist, Professor L.A. Zadeh, the creator of the fuzzy set theory.

Let us first recall the definition of Pawlak number of the first kind. Zdzisław Pawlak himself has Pawlak number of the first kind equal to 0. A person who wrote a research paper with Pawlak has Pawlak number of the first kind equal to 1. In turn, a person who wrote a research paper with someone who has Pawlak number of the first kind equal to 1 has 2 etc. A researcher who writes his works himself receives an infinite Pawlak number of the first kind.

Pawlak number of the first kind among active IT specialists has reached its maximum value of 13, but the weighted arithmetic mean is less than 5.6 and most have Pawlak number the first kind less than 6.

The second kind of Pawlak number is defined in a similar way to Pawlak number of the first kind. An important difference between the first and the second kind is that when defining the second kind of Pawlak number, only those works that were written by two authors, i.e. Pawlak and the person for whom the second kind of Pawlak number are determined, are taken into account.

For the second kind of Pawlak number in the RSDS, similar basic statistics can be observed as for the first kind presented above.

Currently, Pawlak numbers of the above two kinds can be determined using the functions available on the main page of the RSDS (see Fig. 1). In addition, this page includes a function that allows us to specify the numbers of individual authors whose publications are in the RSDS database.

Figure 9 shows the screen of the RSDS during the determination of Pawlak number for Andrzej Skowron.

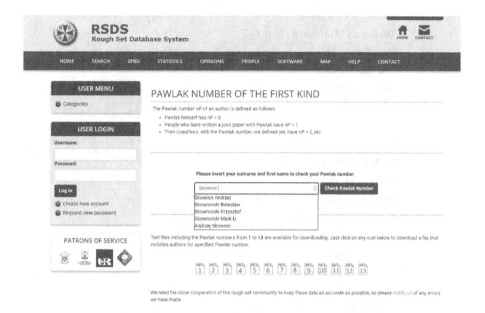

Fig. 9. The screen of the RSDS during the determination of Pawlak number of the first kind

In 2020, scientists with the "one" of the first kind were 22, and 486 scientists who had Pawlak number of the first kind equal to 2, while for Pawlak number of the second kind, there were only 5 with "one" and with "two" 50. For both kinds of numbers, the "two" continues to grow (as new joint works are created), although the number of living scientists with "one" in both cases is still decreasing, because prof. Pawlak died in 2006 and his co-authors are also dying out.

Table 3 presents a sample of both kinds of Pawlak numbers calculated in the RSDS for selected scientists who made a significant contribution to the development of the rough set theory and/or related domains.

The first information about the first kind of Pawlak number appeared in 2013 in [21], while about the second kind of this number in [22]. Both kinds of Pawlak numbers can be interpreted as the strength of publishing cooperation with Pawlak. The so-called Pawlak graphs, built in a similar way to the cooperation graphs described in the previous section, can be used to determine a number of interesting indicators characterizing the rough set community in a multifaceted way, as well as in specific research groups visible, for example, on a world map (see Subsect. 2.8).

Table 3. A sample of both kinds of Pawlak numbers calculated in the RSDS

First name and surname	The first kind	The second kind
Zdzisław Pawlak	0	0
Toshinori Munakata	1	1
Ewa Orłowska	1	1
Cecylia M. Rauszer	1	1
Andrzej Skowron	1	1
Roman Słowiński	1	1
Salvatore Greco	1	2
Jerzy Grzymała-Busse	1	2
Lech Polkowski	1	2
James Peters	1	2
Sheela Ramanna	1	2
Jan H. Komorowski	1	3
Wojciech Ziarko	1	3
Nick Cercone	2	2
Mihir K. Chakraborty	2	2
Stephane Demri	2	2
Ivo Düntsch	2	2
Jouni Järvinen	2	2
Sankar K. Pal	2	2
Helena Rasiowa	2	2
Roman Świniarski	2	2
Ning Zhong	2	2
Masahiro Inuiguchi	2	2
Aboul Hassanien	2	3
Benedetto Matarazzo	2	3
Yiyu Yao	2	3
Tsau Young Lin	2	4
Zbigniew Raś	2	4
Tsumoto Shusaku	2	4
Victor Marek	3	3
Mihinori Nakata	3	5
Piero Pagliani	3	5
Lotfi Zadeh	3	5
Didier Dubois	4	4
Henri Prade	4	5
Gianpiero Cattaneo	4	6

Limitations on the length of the text make it impossible to present detailed experimental results related to these characteristics so they are therefore ignored. They will be presented in detail in our next paper, which will be devoted specifically to a data analysis method based on such a methodology.

5 Further Plans

As mentioned above, the system provides users with a relatively wide range of different functionalities. Of course, they can still be expanded and improved in various ways. In the nearest future, we plan to implement the following tasks:

- develop a method of automatic verification of the correctness of the relationship between the concepts appearing in the general ontology [18],
- increase the efficiency of information retrieval and semantic analysis algorithms in the system,
- expand the scope of automatic data processing in the system,
- add new functionalities regarding automatic user profile detection and data search on the Internet,
- carry out experiments to verify the practical usefulness of our methods of detecting behaviour patterns implemented in the system.

6 Final Remarks

We have presented the RSDS, which collects the different types of computer tools for rough set users. These tools in a natural and effective way support the user in editing, analysing and downloading bibliographies. Additionally, the system provides a wide range of methods of visualizing the results of data analysis contained in the system. A unique feature of the system is the possibility of using it as an experimental platform in the scientific research. An equally important, desirable and practical feature of the RSDS is its expandability: you can easily connect other tools to the system. It is worth to underline that the RSDS was designed and made in accordance with modern design and programming techniques.

Using this system makes it possible to exchange information between academics and practitioners who are interested in the foundations and applications of rough sets. The developers of the RSDS sincerely hope that it will result in a marked increase of interest in rough set theory and related approaches. The research results published in the works represented in this system will stimulate the further development of the foundations, methods and real-life applications of these approaches in intelligent systems [10].

It is not possible to present all aspects of the possibilities and use of the RSDS in one paper with a significant limitation of its volume. Therefore, we plan to prepare separate papers to fully present the benefits of the proprietary methods and algorithms implemented in the RSDS.

The RSDS is being developed by the Rough Set and Petri Net Research Group [34], operating at Rzeszów University in cooperation with the International Rough Set Society [32], among others.

Acknowledgment. We thank Professor Andrzej Skowron for his constructive comments and encouragement to work on the development of the RSDS from its inception to the present. We would also like to thank our students from Rzeszów University for their help in obtaining data and support in expanding the RSDS functionality.

References

1. Chakraverty, S., Sahoo, D.M., Mahato, N.R.: Concepts of Soft Computing: Fuzzy and ANN with Programming, 1st edn. Springer, Singapore (2019). https://doi.org/10.1007/978-981-13-7430-2
2. Devlin, K.: Goodbye Descartes: The End of Logic and the Search for a New Cosmology of the Mind. John Wiley & Sons, New York (1997)
3. Heumann, C., Schomaker, M.: Shalabh: Introduction to Statistics and Data Analysis: With Exercises, Solutions and Applications in R, 1st edn. Springer, Cham (2017). https://doi.org/10.1007/978-3-319-46162-5
4. Klement, E.P., Mesiar, R., Pap, E.: Triangular Norms. Kluwer, Dordrecht (2000)
5. Pal, S.K., Polkowski, L., Skowron, A. (eds.): COGTECH, 1st edn. Springer (2004). https://doi.org/10.1007/978-3-642-18859-6
6. Pawlak, Z.: Information systems - theoretical foundations. Inf. Syst. **6**, 205–218 (1981)
7. Pawlak, Z.: Rough sets. Int. J. Comput. Informat. Sci. **11**, 205–218 (1982)
8. Pawlak, Z.: Rough Sets - Theoretical Aspects of Reasoning About Data. Kluwer, Dordrecht (1991)
9. Pedrycz, W., Skowron, A., Kreinovich, V.: Handbook of Granular Computing. John Wiley & Sons, New York (2008)
10. Skowron, A., Suraj, Z. (eds.): Rough Sets and Intelligent Systems - Professor Zdzisław Pawlak in Memoriam. ISRL, vol. 43. Springer, Heidelberg (2013). https://doi.org/10.1007/978-3-642-30341-8
11. Suraj, Z., Grochowalski, P.: The rough sets database system: an overview. In: Tsumoto, S., Miyamoto, S. (ed.), Bulletin of International Rough Set Society, vol. 7(1/2), pp. 75–81, Shimane, Japan (2003)
12. Suraj, Z., Grochowalski, P.: The rough set database system: an overview. In: Tsumoto, S., Słowiński, R., Komorowski, J., Grzymała-Busse, J.W. (eds.) RSCTC 2004. LNCS (LNAI), vol. 3066, pp. 841–849. Springer, Heidelberg (2004). https://doi.org/10.1007/978-3-540-25929-9_107
13. Suraj, Z., Grochowalski, P.: The rough set database system: an overview. In: Peters, J.F., Skowron, A. (eds.) Transactions on Rough Sets III. LNCS, vol. 3400, pp. 190–201. Springer, Heidelberg (2005). https://doi.org/10.1007/11427834_9
14. Suraj, Z., Grochowalski, P.: Functional extension of the RSDS system. In: Greco, S., Hata, Y., Hirano, S., Inuiguchi, M., Miyamoto, S., Nguyen, H.S., Słowiński, R. (eds.) RSCTC 2006. LNCS (LNAI), vol. 4259, pp. 786–795. Springer, Heidelberg (2006). https://doi.org/10.1007/11908029_81
15. Suraj, Z., Grochowalski, P.: On basic possibilities of RSDS system. In: Wakulicz-Deja, A. (ed.) Systemy Wspomagania Decyzji, pp. 179–185. Wyd. Uniwersytetu Ślaskiego, Katowice (2007)

16. Peters, J.F., Skowron, A.: The Rough Sets Database System. Transactions on Rough Sets VIII. LNCS, vol. 5084. Springer, Heidelberg (2008). https://doi.org/10.1007/978-3-540-85064-9

17. Suraj, Z., Grochowalski, P.: Patterns of collaborations in rough set research. n: Bello, R., Falcon, R., Pedrycz, W., Kacprzyk, J. (eds.) Granular Computing: At the Junction of Rough Sets and Fuzzy Sets. Studies in Fuzziness and Soft Computing, vol. 224, pp. 72–92. Springer, Heidelberg (2008). https://doi.org/10.1007/978-3-540-76973-6_5

18. Suraj, Z., Grochowalski, P.: Toward intelligent searching the rough set database system (RSDS): an ontological approach. Fundam. Inform. **101**, 115–123 (2010)

19. Suraj, Z., Grochowalski, P.: Some comparative analyses of data in the RSDS system. In: Yu, J., Greco, S., Lingras, P., Wang, G., Skowron, A. (eds.) RSKT 2010. LNCS (LNAI), vol. 6401, pp. 8–15. Springer, Heidelberg (2010). https://doi.org/10.1007/978-3-642-16248-0_7

20. Suraj, Z., Grochowalski, P., Lew, Ł: Discovering patterns of collaboration in rough set research: statistical and graph-theoretical approach. In: Yao, J.T., Ramanna, S., Wang, G., Suraj, Z. (eds.) RSKT 2011. LNCS (LNAI), vol. 6954, pp. 238–247. Springer, Heidelberg (2011). https://doi.org/10.1007/978-3-642-24425-4_33

21. Suraj, Z., Grochowalski, P., Lew, Ł: Pawlak collaboration graph and its properties. In: Kuznetsov, S.O., Sezak, D., Hepting, D.H., Mirkin, B.G. (eds.) RSFDGrC 2011. LNCS (LNAI), vol. 6743, pp. 365–368. Springer, Heidelberg (2011). https://doi.org/10.1007/978-3-642-21881-1_56

22. Suraj, Z., Grochowalski, P., Lew, Ł.: Pawlak collaboration graph of the second kind and its properties. In: Proceedings of the Workshop on CS&P 2011, Pułtusk, Poland, September 2011, pp. 512–522. Białystok University of Technology (2011)

23. Suraj, Z., Grochowalski, P., Pancerz, K.: On knowledge representation and automated methods of searching information in bibliographical data bases: a rough set approach. In: kowron, A., Suraj, Z. (eds.) Rough Sets and Intelligent Systems - Professor Zdzislaw Pawlak in Memoriam. Intelligent Systems Reference Library, vol. 43, pp. 515–538. Springer, Heidelberg (2013). https://doi.org/10.1007/978-3-642-30341-8_27

24. Suraj, Z., Grochowalski, P.: About new version of the RSDS system. Fundam. Inform. **135**, 503–519 (2014)

25. Suraj, Z., Grochowalski, P.: The RSDS: a current state and future plans. In: Ramanna, S., Cornelis, C., Ciucci, D. (eds.) IJCRS 2021. LNCS (LNAI), vol. 12872, pp. 97–102. Springer, Cham (2021). https://doi.org/10.1007/978-3-030-87334-9_9

26. Wilson, R.J.: Introduction to Graph Theory. Longman, London (1985)

27. Yao, J.T., Onasanya, A.: Recent development of rough computing: a scientometrics view. Stud. Comput. Intell. **708**, 21–45 (2017)

28. Yao, J.T.: The impact of rough set conferences. In: Mihálydeák, T., Min, F., Wang, G., Banerjee, M., Düntsch, I., Suraj, Z., Ciucci, D. (eds.) IJCRS 2019. LNCS (LNAI), vol. 11499, pp. 383–394. Springer, Cham (2019). https://doi.org/10.1007/978-3-030-22815-6_30

29. Zadeh, L.A.: Fuzzy sets. Inf. Control **8**, 338–353 (1965)

30. Zadeh, L.A.: Fuzzy Logic, Neural Networks and Soft Computing, pp. 77–84. Communication of ACM, March 1994

31. BibTeX. https://www.bibtex.org/

32. International Rough Set Society: https://www.roughsets.org/

33. Rough Set Database System: https://rsds.ur.edu.pl/

34. Rough Set and Petri Net Research Group: https://rspn.ur.edu.pl/

Dissertations

Selected Aspects of Interactive Feature Extraction

Marek Grzegorowski[(⊠)] [iD]

Institute of Informatics, University of Warsaw, Banacha 2, 02-097 Warsaw, Poland
M.Grzegorowski@mimuw.edu.pl

Abstract. In the presented study, the problem of interactive feature extraction, i.e., supported by interaction with users, is discussed, and several innovative approaches to automating feature creation and selection are proposed. The current state of knowledge on feature extraction processes in commercial applications is shown. The problems associated with processing big data sets as well as approaches to process high-dimensional time series are discussed. The introduced feature extraction methods were subjected to experimental verification on real-life problems and data. Besides the experimentation, the practical case studies and applications of developed techniques in selected scientific projects are shown.

Feature extraction addresses the problem of finding the most compact and informative data representation resulting in improved efficiency of data storage and processing, facilitating the subsequent learning and generalization steps. Feature extraction not only simplifies the data representation but also enables the acquisition of features that can be further easily utilized by both analysts and learning algorithms. In its most common flow, the process starts from an initial set of measured data and builds derived features intended to be informative and non-redundant. Logically, there are two phases of this process: the first is the construction of the new attributes based on original data (sometimes referred to as feature engineering), the second is a selection of the most important among the attributes (sometimes referred to as feature selection). There are many approaches to feature creation and selection that are well-described in the literature. Still, it is hard to find methods facilitating interaction with users, which would take into consideration users' knowledge about the domain, their experience, and preferences.

In the study on the interactiveness of the feature extraction, the problems of deriving useful and understandable attributes from raw sensor readings and reducing the amount of those attributes to achieve possibly simplest, yet accurate, models are addressed. The proposed methods go beyond the current standards by enabling a more efficient way to express the domain knowledge associated with the most important subsets of attributes. The proposed algorithms for the construction and selection of features can use various forms of information granulation, problem

The work was supported by the Polish National Science Centre (NCN) grant no. 2018/31/N/ST6/00610.

J. F. Peters et al. (Eds.): TRS XXIII, LNCS 13610, pp. 121–287, 2022.
https://doi.org/10.1007/978-3-662-66544-2_8

decomposition, and parallelization. They can also tackle large spaces of derivable features and ensure a satisfactory (according to a given criterion) level of information about the target variable (decision), even after removing a substantial number of features.

The proposed approaches have been developed based on the experience gained in the course of several research projects in the fields of data analysis and processing multi-sensor data streams. The methods have been validated in terms of the quality of the extracted features, as well as throughput, scalability, and robustness of their operation. The discussed methodology has been verified in open data mining competitions to confirm its usefulness.

Keywords: Feature extraction · Feature selection · Rough set theory

1 Introduction

Every day, the surrounding world is monitored by a still increasing number of sensors. Starting with commonly available sensors from our neighborhood, like mobile phones, automotive sensors, wearables, smart home appliances [17, 203] through medical and telemedical devices [2] for respiratory monitoring [336], auscultation analysis [132], cancer diagnostics [330], or rehabilitation support [204], ending with sensors deployed in factories or coal mines [182, 360] to support diagnostics of manufacturing processes and human staff safety assurance. The variety, variability, and velocity of data have therefore arisen, putting additional pressure on data analysis tools and methods. On the one hand, they should provide the possibility to process various types of data in a multitude of very specialized domains of application. On the other hand, they should seamlessly adapt to drifts, shifts, or the emergence of previously unobserved concepts in data, by interaction with domain experts and data scientists [180].

Interactive data exploration techniques allow analysts to discover interesting dependencies in data due to a fact that it gives the ability to efficiently verify current hypotheses about investigated phenomena and formulate new ones. In practice, this is usually done by conducting various tests on available data and using the results of those tests in consecutive stages of the data exploration process. Very often, the main objective of an analyst is to define such a representation of objects described in the data, that in the future will be the most useful for, e.g., constructing prediction models. There are plenty of methods for automatic feature extraction that are well-described in literature [152, 167, 202, 245, 283]. However, referring to *Judea Pearl*, two more ingredients are needed to move from traditional statistical analysis to causal inference, namely: *"a science-friendly language for articulating causal knowledge, and a mathematical machinery for processing that knowledge, combining it with data and drawing new causal conclusions about a phenomenon"* [293]. The need to embed domain knowledge in data has already been recognized by many scientists as one of the most challenging areas of research [27, 103, 296].

The essential difference between human perceptions and machine-generated measurements, pointed by *Lotfi Zadeh*, is that *"measurements are crisp whereas perceptions are fuzzy"* [424]. Indeed, machine-generated data are often very vague to users. Endless chains of numbers generated by sensor networks are not even close to real-world concepts, problems, and entities [187, 263]. Therefore are particularly hard to interpret. One of the challenges, pointed by *Leslie Valiant*, *is "to characterize the computational building blocks that are necessary for cognition"* [354]. In that context, it is essential to describe the data in a possibly intuitive way. The idea is to make an intelligent use of the information granulation paradigm in the context of aggregating, selecting, and engineering attributes (features/variables/dimensions) that describe the data [141, 422]. For example, we may refer to such straightforward techniques as statistics characterizing granules over sliding time windows. In the case of the underground coal mine sensors, derivation of multivariate series of window-based statistics allows data analysts and experts to deal with noisy and incomplete data sources, better reflect temporal drifts and correlations and reliably describe real situations using higher-level data characteristics [360]. Extracting meaningful features is important in many domains, like medicine or criminal justice [119, 329]. Combining machine-generated data with attributes corresponding to experts' assessments proved to have a positive impact on the quality and robustness of the machine learning models, e.g., in the case of seismic bumps prediction [182].

Feature engineering is recognized as an important but laborious approach [29]. As opposed to the above, representation learning opts to augment artificial intelligence with the capability to autonomously (i.e., without human interaction) identify and disentangle the underlying explanatory factors hidden in the low-level sensory data. The concepts of deep, distributed representations and unsupervised pre-training have recently become a dynamically developing area of research with many successful applications [131, 287, 312, 389]. There are many methods allowing to project or embed the data onto new derived dimensions, forming so-called latent concepts – typically, a combination of (almost) all attributes of the original data [380]. However, regardless of the achievements in the areas, these algorithms generally suffer from the lack of interpretation of the projected dimensions [166, 329] and for that reason are not investigated further in this study. In contrary to feature engineering, which though often labor-intensive, is a way to take advantage of human ingenuity and prior knowledge. Thereby, justifying the effort put into the design of data transformations and preprocessing pipelines when deploying machine learning algorithms.

Let us also emphasize the problems associated with the increasing dimensionality of data, which may exceed human perception. In such cases, users may be still able to interpret attributes' meaning but navigation through their subsets becomes harder. The curse of dimensionality is a well-known problem to machine learning methods, as well. To address this aspect of complexity, one can operate with clusters of attributes inducing similar information (e.g., similar partitions) or employ some techniques of attribute selection, which would replace large attribute sets with their minimal subsets providing comparable information about data [185]. Searching for such subsets is a well-established task within the theory of rough sets (RST) [376]. Given an initial set of attributes, one can search for so called reducts, which induce (almost) the same information as all considered attributes [288]. A number of heuristic methods have been developed to derive the most interesting reducts from large and complex data tables [75, 288].

The task of feature selection may be defined at two levels. Predicting near-future readings of a particular sensor, one could think about it in terms of choosing other relevant sensors and time frames that are enough to start the process of model training. In this case, it is crucial to interact with domain experts and provide appropriate analytic reports/visualizations to let them make the right assessments. This level might be also called an information source selection. The second level considered in the thesis refers to selecting specific features constructed during the sliding time windowing process. The reference to the sliding time window technique is an example of a situation where at a higher level of granularity, essential characteristics need to be defined. Indeed, from the user's point of view, operating on statistics extracted from numerical series covered by a given window interval is definitely more natural than operating on raw numerical data. At this level, one can successfully proceed with relatively simple methods, which yield surprisingly small feature sets in order to establish an efficient framework for deploying, monitoring and tuning the forecasting models embedded into the production system.

In some applications, e.g., related to sensor-based hazard monitoring or medical diagnostics, besides the accuracy, speed, and reliability of a prediction model, no less important is resilience. A single sensor failure (or interruption of signal transmission), which typically causes a missing whole dimension of data, cannot result in the inability to assess the situation. To address this issue, we may refer to the broad studies on missing data imputation techniques [47, 261]. These are often based on univariate series analysis or sampling from original data distribution and may have problems dealing with longer gaps, e.g., resulting in a higher level of uncertainty for subsequent assessment [342, 366]. As an alternative, some researchers study non-imputation methods designed for the classification or regression of incomplete data. Such methods may rely on aggregation techniques and higher-level features (or granules of features) that are less sensitive to missing data [146, 403], ensembles of diverse classification of regression models [182, 402], or enhanced predictive models with additional (redundant) checks, e.g., surrogate splits or verifying cuts in tree-based models [28, 57].

The aforementioned approaches rely on the assumption that features selected for the process of model learning do contain additional (redundant) knowledge, whereas state-of-the-art feature selection techniques attempt to remove redundancy. Our goal is to formulate new constraints, whereby selected feature sets are guaranteed to provide enough information about the considered target variables even if some of those features are dropped. One of the discussed approaches is to rely on a collection of diverse feature subsets with their corresponding prediction models treated as an ensemble. Another approach is to search for feature sets with a guarantee of providing sufficient predictive power even if some of their elements are missing. In the frame of this study, we introduce the idea of resilient feature selection. In particular, we formulate the rough-set-based notion of r-$\mathbb{C}$-reducts – an irreducible subset of features providing a satisfactory level of information about the considered target variable even if up to r features are unavailable.

In the study on the interactiveness of the feature extraction methodologies, we address the problems of deriving useful and understandable parameters (attributes, features) from raw sensor readings and reducing the amount of those parameters in order to achieve possibly simple yet accurate models. In the Sectiontation, a number of innovative approaches to automating feature creation and selection are proposed. The current state of knowledge on feature extraction processes used in commercial applications is shown. The problems associated with processing big data sets and approaches to process high-dimensional time series derived from sensor networks are discussed. Although we rely mainly on RST to specify requirements related to the design and implementation of our approach to interactive attribute selection, the framework presented in this paper can be used together with other well-known methodologies of data analysis.

1.1 Plan of the Paper

In Sect. 2, an overview of the state-of-the-art feature extraction methods is presented. In particular, in Sect. 2.1 a review of feature engineering techniques is provided, with a special emphasis put on sliding window techniques to feature creation. To provide a proper context of other related approaches that are not covered in this study, in Sect. 2.2, we present a comprehensive review of representation learning and dimensionality reduction methods. In Sect. 2.3, we recall relevant approaches to feature selection. In Sect. 2.4, we discuss various approaches to information granulation in feature extraction. In Sect. 3.1, we recall basic concepts from the theory of rough sets (RST).

In Sect. 3, we introduce the idea of resilient feature selection and, accordingly, we introduce r-$\mathbb{C}$-reducts. In Sect. 3.1, we introduce the notion of criterion function $\mathbb{C}$, which enables us to consider various feature selection formulations at a higher level of abstraction. In particular, we show how $\mathbb{C}$ generalizes the RST-based feature selection approaches relying on various definitions of reducts and approximate reducts. In Sect. 3.3, we outline an Apriori-inspired algorithm that generates all r-$\mathbb{C}$-reducts of a given type. In Sect. 3.4, we study the tasks of resilient feature selection from the perspective of their computational complexity. We prove that many NP-hard feature selection/elimination problems remain NP-hard for any arbitrary resilience level r. In Sect. 3.5, we present heuristic DFS algorithms for searching for optimal r-$\mathbb{C}$-reducts, with specific examples of permutation-based and approximation methods.

In Sect. 4, we outline our approach for feature extraction, aimed at processing multivariate time series. In Sect. 4.1, we describe the data in a possibly intuitive way, using statistics characterizing sliding time windows. In Sect. 4.2, we discuss how the concept of granulation can be made useful in selecting and engineering features on large and, possibly, complex data sets. Finally, in Sect. 4.3, the complete framework for linking resilient feature selection and machine learning techniques to build a predictive model resistant to partial data loss is proposed.

In Sect. 5, we provide a comprehensive experimental evaluation of the introduced feature extraction methods over large, multivariate time-series data across significantly different domains. The analysis of potentially dangerous methane concentration and seismic events are presented in Sect. 5.1 and 5.2, respectively. In Sect. 5.3, we evaluate the performance of the introduced framework in the fire and rescue domain that refers to the analysis of data collected from body sensor networks. In this Section, we also evaluate the impact of the developed feature extraction methods on the prediction quality and resilience of various machine learning models. As an important argument for considering interactive feature extraction processes and built-in human-computer interaction into machine learning processes, let us stress out that in the case of seismic data, training models on both sensor readings and domain experts' assessments allowed us to improve the quality of the predictors significantly. In Sect. 5.4, we performed short-term spot prices prediction of many univariate time series collected from the AWS Cloud Spot market. Furthermore, we describe a series of international data mining challenges organized to facilitate this study.

In Sect. 6, we conclude the research and we elaborate on some future research directions.

1.2   Main Contributions

The main contributions of this study are:

In Sect. 2, a broad overview of state-of-the-art feature extraction techniques is provided, with a special emphasis put on sliding window techniques to feature creation and rough set-based granulation and feature selection techniques. Namely, to various extensions of reducts (Definition 1).

In Sect. 3, we introduce a new idea of resilient feature selection – based on so-called r-$\mathbb{C}$-reducts – attribute sets that are well-suited for the investigated problems and provide a level of redundancy that makes these sets more invulnerable with respect to possibly missing or questionable attribute values. The introduced r-$\mathbb{C}$-reducts extend the classical notion of a reduct developed within the rough set theory (which is briefly discussed in Sect. 2). In the provided notion, $\mathbb{C}$ refers to a function encoding the criterion of preserving enough information by the considered sets of attributes. At the same time, r stands for the number of attributes that can be removed from those sets without making them insufficient to build decision models [137] (at the accuracy level corresponding to $\mathbb{C}$). Functions $\mathbb{C}$ (Definition 5) actually enable us to express a number of so-called approximate attribute reduction criteria known from the RST-related literature, based on, e.g., discernibility, entropy, or positive regions. Consequently, by defining the resilience factor r as combinable with an arbitrary $\mathbb{C}$, we generalize all those formulations. The discussed idea of resilience is surely more general, and one may consider it an extension of many other, not necessarily rough-set-based feature selection methods.

The important theoretical contribution in Sect. 3 refers to a broad discussion on the impact of the resilience on the overall complexity of feature selection problems and algorithms. In particular, in Sect. 3.1, we generalize the way of reasoning about attribute subsets by introducing criterion functions, which, for each given decision table $\mathbb{S} = (U, A \cup \{d\})$, return a binary assessment of the candidate attribute subsets. We further use criterion functions to provide a generalization of rough set reducts as *criterion reducts* $\mathbb{C}(R)$ (Definition 6). Lastly, we show some theoretical properties of criterion reducts and we define criterion reducts for a number of well known notions of reducts, i.e., approximate entropy reducts ($\mathbb{C}^{(H,\varepsilon)}$-reducts), γ-reducts ($\mathbb{C}^{\gamma}$-reducts), etc. In Sect. 3.3, we prove that any NP-hard feature selection problem understood as the task of finding – for an input decision table – the minimal $\mathbb{C}$-reduct that may be expressed via so-called monotonic criterion functions $\mathbb{C}$ (e.g., the minimal (H, ε)-approximate reduct problem [356], the minimal (γ, ε)-approximate reduct problem [373], a wide family of discernibility-based approximate/partial reduct optimization problems [273, 279], etc.) retains its NP-hardness for arbitrary resilience level r (Theorem 1).

In order to prove NP-hardness – by providing a polynomial time transformation – in Subsect. 3.4, we introduce a family of artificial attributes, so-called A-attributes, denoted as $\#attr$. In a number of lemmas in Subsect. 3.4, we show some important properties of A-attributes. Among others, we prove that any r A-attributes $\{\#attr_1, \ldots, \#attr_r\}$ form the smallest r-$\mathbb{C}$-reduct (Lemma 3). We show, in Lemma 4, how to construct $\mathbb{C}$-reduct by adding A-attributes to reduct, and we discuss the impact on $\mathbb{S}$ and $\mathbb{C}$ when data representation is extended with A-attributes (Lemma 5). Our study includes also a visual interpretation of the NPH proof (Sect. 3.4), a broad discussion on the meaningfulness of the provided NPH study and the complexity result derivable directly from Theorem 1 (Sect. 3.4). In particular, referring to Theorem 1, we prove NP-hardness of the resilient version of the minimal (H, ε)-approximate reduct problem, and the resilient versions of the minimal α-reduct and (γ, ε)-approximate reduct problems. Let us note that the same mechanism could be easily applied for many other cases as well, in particular, for any formulations of $\mathbb{C}^{(Q,\varepsilon)}$-reducts for which the corresponding measures Q satisfy conditions of Definition 7 [355, 357].

In Sects. 3.3 and 3.5, we discuss opportunities of exhaustive and heuristic search of feature subsets, and we assess computational complexity of proposed algorithms. We elaborate on two generic strategies that follow a popular idea of dynamic exploration of the lattice of feature subsets (i.e., Fig. 2). Namely, breadth first search (BFS) and depth first search (DFS). For BFS, we adapt the well-known Apriori algorithm [335] for the purpose of r-$\mathbb{C}$-reduct search (Sect. 3.3). In Sect. 3.5, we consider two approaches to the depth first search exploration of the lattice, which allow us to identify subsets of attributes that satisfy the resilient version of $test_{\mathbb{C}}$ function: r-$test_{\mathbb{C}}$ (Algorithm 3). Algorithm r-$test_{\mathbb{C}}$ verifies if a given set of attributes $R \subseteq A$ satisfies the resilient criterion

r-$\mathbb{C}$ under the condition that implementation of $test_{\mathbb{C}}$ is given. In Sect. 3.5, we present a novel Algorithm 4 generating r-$\mathbb{C}$-reducts inspired by a permutation-based technique that is common for RST-based approaches [357, 373]. We also discuss the approximation of the permutation-based algorithm for resilient feature selection (Algorithm 5).

In Sect. 4, we outline our approach to feature extraction, aimed at processing data obtained from sensors that provide outputs in the form of time series. In knowledge-based systems, it is common to deploy a potentially large collection of sensors of different types to monitor the environment state and its changes. In such a setting, the gathered data elements can be complex on various levels. Individual readings may take different forms according to the application domain. Values may express continuous phenomena, such as pressure, temperature, or humidity. They can also express a discrete state of the environment, such as an on/off state of a device. Often, data interpretation is possible only in the context of additional knowledge obtained from domain experts. Concerning feature engineering, in Sect. 4.1, we attempt to describe the data in a possibly intuitive way, using statistics characterizing sliding time windows. The proposed approach focuses on extending the sliding window construction process by adding a number of designed statistics and enhancing it with some more static attributes reflecting assessments obtained from domain experts. This brings the opportunity to compare the prediction quality of models trained using derived features with the expert-based assessment and makes it possible to use features derived from experts in ML models training. In the case of the underground coal mine sensors, derivation of multivariate series of window-based statistics allows us to deal with noisy and incomplete data sources, better reflect temporal drifts and correlations, and reliably describe real situations using higher-level data characteristics, which are common problems in time series analysis, reviewed in Sect. 4.1.

The contributions in Sect. 4.2 refer to a general algorithmic framework for performing feature selection on top of a granular representation of attribute space. Our methodology is devised in such a way that it caters to various types of granules and various goals of feature selection. The purpose is to perform a kind of granular attribute selection that exploits to the fullest semantical relationships between variables. Particular contributions in Sect. 4.2 are concentrated around two aspects. First, we put forward a framework for expressing granules in attribute space. Therein, we include original ideas for discovering and managing similarities between attributes for the purpose of constructing granules. Feature granules can be induced by, e.g., hierarchical clustering on attributes or analysis of so-called *heat maps* that convey the knowledge about attribute interchangeability. On the other hand, we show that meaningful granulations can be derived according to such prerequisites as proximity or common functionality of the considered features.

In Sect. 4.2, we discuss how the concept of granulation can be made useful in selecting and engineering features on large and possibly complex data sets. We show how to utilize the intrinsic properties of the data and underlying problem and background/domain knowledge to build a granular representation of attributes. By taking into account a given granulation of attributes, we can configure our algorithms to achieve faster convergence. The proposed methods are designed in a way to deal with large and complex data sets. We present means to make use of efficient, parallelized computational schemes such as MapReduce. Therefore, the provided tools and examples are devised to work with data sets that are very large in terms of the number of objects and the number and complexity of features. Thus, they address some of the challenges posed by the Big Data paradigm.

As a notable aspect and an important contribution in the frame of this research, let us point out the framework for linking resilient feature selection and machine learning techniques to build a predictive model that is resistant to partial data loss (Sect. 4.3). In this section, we focus not only on the extracted features and constructed prediction models but also on data processing stages that are designed to let it work within a big data environment and, particularly, with the high dimensional, multi-stream data. In order to provide high-quality assessments, the Algorithm 9 – for blending the models – is designed in a way, which guarantees that a model can be included if it is accurate enough on validation data and sufficiently different from already selected predictors. The diversity may be achieved by employing a variety of models computed on different subsets of attributes and data samples. For this task, the similarity measures (Sect. 4.2) or resilient attribute subsets (Sect. 3.2) may be applied. As a result of blending diverse models, the final ensemble minimizes the impact of concept drifts and achieves a better prediction quality. The results of conducted experiments (Sect. 5) confirmed that the idea is very promising, and resilient learning may significantly minimize the risk and impact of data loss on predictive analysis.

In Sect. 5, we provide a broad experimental evaluation of learning forecasting models over large multi-sensor data sets, including the steps of sliding window-based feature extraction and rough-set-inspired feature subset ensemble selection. We conducted a series of experiments on data connected to the problems of providing safety of miners working underground, which is the fundamental requirement for the coal mining industry. Analysis and proper assessment of potentially dangerous methane concentration (Sect. 5.1) and seismic events (Sect. 5.2) significantly improve the safety and reduce the costs of underground coal mining.

One of the considered tasks is to construct a model capable of predicting dangerous concentrations of methane at longwalls of a coal mine basing on multivariate time series of sensor readings. The contributions in Sect. 5.1 refer to both the analysis of how the nature of sensor readings influenced the architecture of the developed system and the empirical proof that the designed methods for data processing and analytics turned out to be efficient in practice. We show how the complete mechanism can perform on data collected in an active coal mine and processed with the described framework. We show how the complete mechanism can be built into DISESOR - a decision support system in coal mines. The evaluated feature selection approaches yield excellent results even when combined with the simplest possible prediction techniques like logistic regression. Furthermore, we elaborate on the resilience of the solution in the case of partial data loss, e.g., when particular data sources, sensors are damaged or inactive.

In Sect. 5.2, we investigate how the interactive feature extraction and ensemble blending methods, proposed in Sect. 4, generalize to other problems of multi-stream data analysis. Once again, we address the problem of safety monitoring in underground coal mines. This time, we investigate and compare practical methods for the assessment of seismic hazards using analytical models constructed on both raw multi-stream sensory data and features derived from domain experts. The possibility of representing a problem related to data exploration and analysis with machine-generated features, which are additionally enriched with experts' assessments, is one of the essential aspects from the point of view of interactiveness. Furthermore, in Sect. 5.3, we describe an international data mining challenge organized to facilitate this study. We also demonstrate that the technique used to construct an ensemble of regression models outperformed other approaches used by participants of the described challenge. In Sect. 5.2, we explain how post-competition data was utilized for the purpose of research on the cold start problem in the deployment of decision support systems at new mining sites.

To thoroughly assess the versatility of the developed framework across significantly different domains of application, besides analysis of coal mining-related problems, in Sect. 5.3, we evaluate its performance in the fire and rescue domain that refers to the analysis of data collected from body sensor networks. The aim of this study is to assess how automatic feature extraction and classifier learning (without parameters tuning) can cope with the multi-target learning problem. Furthermore, in Sect. 5.4, we show that, by analyzing spot instance price history and using ARIMA models, it is feasible to perform future spot prices prediction of many univariate time series. The main reason behind the evaluation of ARIMA models on data represented as candlesticks is that both techniques are easy to interpret. Results confirm the quality of the solution, its computational performance, and the versatility of the developed framework resulting in the very short time needed for its adaptation to the significantly different domains.

Some of the partial results of this research were presented at international conferences and workshops. Some were published in conference proceedings and respectable journals. For example, the publications related to the granular and resilient feature selection [137, 141, 145]. Moreover, the research on various applications of feature extraction to improve prediction quality in the field of sensor data analysis in hard coal mining and emergency/firefighting domains [138, 146, 147, 182, 184]. The research on intelligent systems, data models, and processing optimization for interactive feature extraction and data analysis [136, 144, 148, 218, 360–362, 427] are also partial contributions. Some partial results were also published in technical papers and monographs in Polish [142, 181].

2   Feature Extraction

Having in mind the observed variety of possible data representation formats, including text, audio, image, video, relational data, spatio-temporal time series, and many others, it is straightforward that the application of machine learning algorithms and techniques requires a more or less extensive phase of data preparation. Feature extraction (FE) addresses the problem of finding the most compact and informative data representation to improve the efficiency of data storage and processing. The process starts from an initial set of measured data and builds derived features intended to be informative and non-redundant, facilitating the subsequent learning and generalization steps, and leading to better human interpretations. Logically, there are two phases of this process: the first is the construction of new attributes based on original data (sometimes referred to as feature engineering), the second step is a selection of the most important among the attributes (sometimes referred to as feature selection). In this chapter, we provide a broad overview of the state-of-the-art feature extraction methods, including feature construction, selection, granulation, and selected methods from rough set theory. We also briefly present some other related topics, like dimensionality reduction and representation learning.

2.1   Feature Engineering

Feature engineering (FE) is the process of using domain knowledge of the data to create features that make machine learning algorithms work [202]. The importance of feature engineering was aptly identified by *Pedro Domingos: "At the end of the day, some machine learning projects succeed and some fail. What makes the difference? Easily the most important factor is the features used"* [94]. Indeed, this process is fundamental to the application of machine learning resulting in simpler and more effective predictive models, improved models' robustness and resilience, reduced computation time and resources needed, and foremost, better interpretability of the results. In this section, we present a comprehensive review of the state-of-the-art in the area with a strong emphasis on the structural, relational, and time series data extraction methods.

Some machine learning algorithms, like a decision tree model, can handle various data representations. However, most have fairly restrictive limitations and usually require a specific data format. For example, rough set reudcts construction methods can work with categorical data only. In the case of numerical attributes, those algorithms require a discretization to transform continuous attributes before the data reduction may be performed. On the other hand, some algorithms, like neural networks, require numeric inputs [379]. The typical data preprocessing steps that allow adapting attributes' format to the requirements of selected learning methods include: normalization, standardization, discretization, categorical encoding, imputation of missing values, outliers detection, and user-defined custom transformations, e.g., min/max values, percentiles, or generating polynomial features [112, 147].

There are several data preprocessing techniques to be used for encoding categorical variables [305]. One hot encoding is the most widely used scheme to transform a single variable with 'd' distinct values to 'd' binary variables indicating the presence (1) or absence (0) of the particular category. In ordinal coding, an integer value is assigned to each category (assuming that the number of categories is known). Polynomial coding is a form of trend analysis that looks for trends in the categorical variable. Leave-one-out is an example of the target-based encoder that calculates mean target of a given category for each observation, supposing that this observation is removed from the data set. We may also mention sum coding, Helmert and James-Stein encoders, etc. [305]. All leading to converting categorical features to binary, integer, or continuous ones, as expected by ML algorithms' inputs.

In the case of some machine learning algorithms, objective functions may not work at all, or may perform less effectively, without proper feature scaling [10]. For example, we may recall stochastic gradient descent and its variants, which are recognized as an effective way of training deep networks [172, 375]. The need for normalization and standardization arises naturally when dealing with clustering [175], in the case of experiments involving multiple arrays [39], or whenever data are collected from various sources [360]. In some applications of high-density oligonucleotide arrays, the goal is to learn how RNA populations differ in expression in response to genetic and environmental differences. For example, large expression of a particular gene or genes may cause an illness resulting in variation between diseased and normal tissue. The obscuring sources of variation can have many different effects on data. Unless arrays are appropriately normalized, comparing data from different arrays can lead to misleading results [173]. Let us now briefly recall some common approaches to feature scaling.

Given a lower bound $min(a)$ and an upper bound $max(a)$ for an attribute "a", the min-max normalization is one of the elementary methods to scale the range in $[0, 1]$.

$$\hat{a} = \frac{a - min(a)}{max(a) - min(a)}$$

The general formula to rescale a range between values $[\hat{a}^{inf}, \hat{a}^{sup}]$ is given as:

$$\hat{a} = \hat{a}^{inf} + \frac{(a - min(a)) * (\hat{a}^{sup} - \hat{a}^{inf})}{max(a) - min(a)}$$

Mean normalization refers to the average $avg(a)$ values of a feature "a":

$$\hat{a} = \frac{a - avg(a)}{max(a) - min(a)}$$

Standardization (or Z-score normalization) is a technique used to scale the data such that the mean of the data becomes zero and the standard deviation becomes one. Here the values are not restricted to a particular range. We can use standardization in the case of large differences between input data attributes' ranges. Standardization is widely used in many machine learning algorithms, e.g., support vector machines, logistic regression, or deep learning [253]. The general method of calculation is to determine the distribution mean μ and standard deviation σ for each feature a and to replace it with the following formula:

$$\Phi^{\mathrm{M}}_{candidate} = \frac{a - \mu}{\sigma}$$

Furthermore, we may scale features according to a given norm $\|.\|$, i.e., Euclidean length, L_1 (city-block length), or any other user-defined norm. We may also mention rank and quantile normalizations, and their applications in the image processing and genetics [10, 39]. There are also decoupling and Gaussian normalization that are successfully applied in collaborative filtering [193, 194]. In [228] was proposed an interesting framework to handle some special cases when standard normalization techniques are not capable of eliminating technical bias due to skewed distribution of variables. We may also recall a variety of methods adjusted to the given feature distribution [10], which obviously do not close the range.

Many algorithms, like Apriori or Naive Bayes, can handle only nominal or discrete attributes [416]. Even in the case of algorithms, which are able to deal with continuous attributes, learning is far less efficient and effective. Thus an embedded or an external discretization of data is often required [408]. The main goal of discretization is to transform a set of continuous attributes into discrete ones, e.g., by associating categorical values to intervals and thus transforming quantitative data into qualitative data [417]. In this manner, symbolic data mining algorithms can be applied over continuous data, and the representation of information is more concise and specific.

Assuming that the data is represented by a set of objects (instances, observations) U, set of attributes (features, variables) A, and (in the case of supervised problems) a set of classes D, a discretization algorithm would split the continuous attribute $a \in A$ in this data set into k discrete, non-overlapping intervals:

$$A_{discr} = \{[a_0, a_1], (a_1, a_2], .., (a_{k-1}, a_k]\}$$

where $a_0 = min(a)$ is the minimal observed value of the attribute a (or $-\infty$, if attribute values are not bounded), $a_k = max(a)$ is the maximal value (constant or ∞), and $\forall_{0 \leq i < j \leq k} a_i \leq a_j$. Such a discrete result A_{discr} is called a discretization scheme on attribute $a \in A$, and $A_{cuts} = \{a_1, .., a_{k-1}\}$ is the set of cut points of attribute $a \in A$. Let us briefly present selected, common discretization methods.

Equal-frequency (EFB), equal-width (EWB), and fixed-frequency (FFB) binnings are the simplest and most straightforward discretization methods. All those methods involve sorting of the observed values V_a of a continuous feature a. For the given k number of intervals, EFB divides the sorted values into k intervals, so that each contains approximately the same number of training instances [97]. Let U refer to the set of observations (objects), then each interval contains $\frac{|U|}{k}$ training instances with adjacent values. Note that training instances with identical values are placed in the same interval, thus it is not always possible to generate k equal frequency intervals.

If a continuous variable is observed to have values bounded by $min(a)$ and $max(a)$, EWB aims to divide the range of observed values into k equally sized intervals (bins), where k is a given constant parameter. The width δ_a of each interval is computed as:

$$\delta_a = \frac{max(a) - min(a)}{k}$$

the cuts (boundaries of the intervals) are defined as: $min(a) + i * \delta_a$, where $i = 1, .., k - 1$.

For a predefined interval frequency k, FFB discretizes the sorted values into intervals so that each interval has approximately the same number k of training instances with adjacent values [417]. All above mentioned methods are applied to each continuous feature independently, hence all are classified as univariate. They also make no use of class information (unsupervised).

The scientific literature provides numerous proposals of discretization techniques, and there are many different axes by which they can be classified, e.g., univariate vs. multivariate, supervised vs. unsupervised, global vs. local, static vs. dynamic, etc. [120]. The most common evaluation measures used by the discretizer to assess the best discretization scheme are derived from information theory (Gini index, entropy), statistics (χ^2, ChiMerge), or rough set theory (RST) [108, 281, 385]. Furthermore, some methods utilize wrapper approach, like ID3 [311], Bayesian approach [43], fuzzy functions [326], and many other techniques [34, 272]. It is also important to stress out that obtaining the optimal discretization is a NP-complete problem [68].

When analyzing real data sets, one may face a broad spectrum of problems related to data, varying around: missing values, anomalies, exceptions, discordant observations, or contaminants [58, 60]. Missing values imputation has been studied for several decades being the basic solution for incomplete data problems, specifically those where some data samples contain one or more missing attribute values [242, 435]. Outlier detection techniques strive to solve the problem of discovering patterns that do not fit to expected behaviors [396]. This is a particularly challenging and important problem in the case of big sensor networks and multidimensional time series data analysis [214, 360]. The problems related to missing attributes, noisy data, or outliers refer more to data quality aspects and data cleaning rather than to feature engineering. Therefore, some selected approaches to imputation of missing values and outlier detection methods are discussed concisely further in Sect. 4.1.

The widespread growth of Big Data and the evolution of Internet of Things enable various entities to continuously generate and collect streams of data [277].

Stream data analysis is essential for many fields of application where processes are typically monitored by a number of sensor devices [203], such as: logistics, mining industry, health-care, medicine, and even agriculture [76, 143, 386, 430], Proper understanding of data collected from many sensors and application of machine learning methods are very challenging and time-consuming tasks that usually require particular feature engineering [6, 115]. Feature extraction approaches, which output interpretable and dimensionally consistent features, are still in big demand and are considered as an important research topic [76]. Among them, techniques based on sliding window segmentation are considered as one of the simplest yet very effective for constructing easily interpretable features from time series derived from data streams [360, 430].

Deriving statistics from sliding time windows can be regarded as a crucial FE stage in all knowledge discovery process investigating sensor readings and (multivariate) time series [146, 238]. A sliding window is defined by a *length* and an *offset*. The length determines the size of a window, whether it is a fixed number of readings contained in a window or a fixed time interval. The offset is the extent to which the consecutive windows overlap to each other. In Fig. 1, we provide four examples of possible sliding window set-ups. The example marked in red shows the situation when the length of a sliding window is equal to the offset. The green and blue examples show the consecutive positions of a sliding window when the offset is equal to $\frac{1}{2}$ and $\frac{1}{3}$ of the length. The example marked in cyan illustrates the situation when the offset is twice as large as the length (or in general just greater) of a sliding window [147].

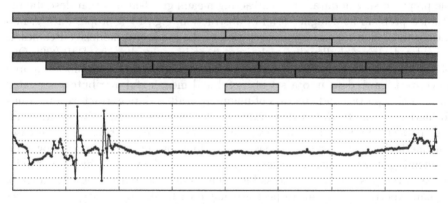

Fig. 1. A set of possible sliding windows set-ups.

Sliding time windows are represented by various statistics computed over their values [138, 143]. In practice, such methods require an extensive feature engineering step, which often needs to be domain-specific [45, 426]. For a comprehensive study on efficient maintenance of basic statistics derived from sliding windows, we may refer to [85]. For an example how to integrate multi-sensor analysis with external sources of spatio-temporal information, let us refer to

[332]. An example of utilization of such statistics as higher-level features can be found in [401]. Time series can be also filtered or smoothed (using, e.g., running averages) in order to reduce its complexity while maintaining its important characteristics [238]. Series of data points can be approximated using methods, such as: piecewise constant approximation [211], or piecewise linear representation [371]. Selection of an appropriate time series representation is the fundamental aspect of an efficient analysis of sequential data. For a more detailed overview of approaches to time series representation, one may refer to [115].

In a broader sense, algorithms and systems for the on-line prediction based on sensor readings can be placed within the scope of research on time series data mining [299], or pattern recognition from multivariate time series [113,304]. In this field, a lot of research was conducted on topics such as searching for similar subsequences [84,215], or time series segmentation and dimensionality reduction [210]. In tasks such as subsequence matching, a sliding time window approach is used in a combination with series compression techniques, e.g., the symbolic aggregate approximation [241]. In many domains, series transformations such as the discrete Fourier transforms (DFT) are also often applied in this context [425]. Obviously, sliding time windows approaches may vary depending on the area of application [254]. Nevertheless, the overall mechanism of computing time-window-based representations can be treated as a universal approach.

There are numerous well-known automatic feature engineering methods. In some approaches, they are tightly integrated with the modeling process, e.g., hidden layers of a deep neural network model internal representations in a way analogous to constructed features. In other approaches, they are limited to simple preprocessing of data. Still, extracting meaningful features that describe the studied problem at a higher level of abstraction, e.g., by a proper data granulation, thus allowing easier interpretation of the predictive models' outcomes, is considered a very challenging task, important in many domains [119,329]. One of the possible approaches to feature engineering, which is sometimes required to convert "raw" data into a set of useful and meaningful attributes, is related to human expertise and creation of manually crafted data extractors and transformations. Despite the evident value of the features obtained this way, leading to easily interpretable and well suited data representation, in some cases this method may be far too expensive and time-consuming. Therefore, by complementing it with automatic methods, one can achieve a viable compromise between the possibility to process big volumes of data and taking advantage of human expertise. With this respect, we may refer to already mentioned statistics characterizing granules over sliding time windows, which may be easily defined or interpreted by users. Furthermore, the process of sliding window-based feature creation may be automated, and the derived data representation may be complemented by experts' assessments.

2.2 Representation Learning and Dimensionality Reduction

Representation learning (or feature learning) allows to automatically discover the representations from raw data. This approach is an established alternative to classical feature engineering. There are many feature learning methods that can be either supervised (e.g., neural networks) or unsupervised (e.g., matrix factorization, auto-encoders), linear (e.g., linear discriminant analysis) or non-linear (e.g., kernel methods). However as the number of features increases, the model training takes far more time, and consumes more compute resources and storage. Trained predictors may become more complex, and may relay on misleading, redundant, or noisy information. This may lead to decreased models' accuracy and over-fitting. There are many methods allowing to project or embed the data into a lower dimensional space while retaining as much information as possible. Classical examples are singular value decomposition, principal component analysis, kernel principal component analysis, independent component analysis, multidimensional scaling, word embeddings, auto-encoders, deep learning, etc. [51, 391]. In this section, we provide an overview of the state-of-the-art in the area.

Singular value decomposition (SVD) and principal component analysis (PCA) are two commonly used dimensionality reduction methods that attempt to find linear combinations of features in the original high dimensional data matrix to construct a meaningful, yet compressed representation of a data set. They are preferred by different fields of application. PCA is often used for biomedical data, or in genetics [102, 126]. Meanwhile, SVD is more popular when the investigated problem is related to sparse representations, e.g., in (mechanical) faults diagnosis, or in the case of complex chemical processes analysis [160, 414].

SVD is a factorization of a real (or complex[1], i.e., $\mathbb{Z}$) matrix that generalizes the eigen-decomposition of a square matrix (i.e., $n \times n$) to a rectangular one (i.e., $m \times n$). More formally, with SVD any real matrix $\mathbf{A} \in \mathbb{R}^{m \times n}$:

$$\mathbf{A} = \begin{pmatrix} a_{1,1} & a_{1,2} & \cdots & a_{1,n} \\ a_{2,1} & a_{2,2} & \cdots & a_{2,n} \\ \vdots & \vdots & \ddots & \vdots \\ a_{m,1} & a_{m,2} & \cdots & a_{m,n} \end{pmatrix}$$

is decomposed into the product of two unitary[2] matrices $\mathbf{U} \in \mathbb{R}^{m \times m}$ and $\mathbf{V}^T \in \mathbb{R}^{n \times n}$, and a diagonal rectangular matrix of singular values $\mathbf{\Sigma} \in \mathbb{R}^{n \times m}$. The general formula $\mathbf{A} = \mathbf{U} \times \mathbf{\Sigma} \times \mathbf{V}^T$ is shown in-detail below:

$$\mathbf{A} = \begin{pmatrix} \overrightarrow{u_1} & \overrightarrow{u_2} & \cdots & \overrightarrow{u_m} \end{pmatrix} \times \begin{pmatrix} \sigma_1 & & & \\ & \ddots & & 0 \\ & & \sigma_r & \\ 0 & & & 0 \end{pmatrix} \times \begin{pmatrix} \overrightarrow{v_1} & \overrightarrow{v_2} & \cdots & \overrightarrow{v_n} \end{pmatrix}^T$$

---

[1] Later in this section, we discuss real matrices, as they are more relevant for real-life data sets.

[2] We call a matrix $X \in \mathbb{Z}^{n \times n}$ unitary iff $XX^H = X^H X = \mathbb{I}$. For a real matrix $X \in \mathbb{R}^{n \times n}$, we have $X^H = X^T$, and we say that a matrix is orthogonal, i.e., $XX^T = X^T X = \mathbb{I}$.

where vectors $\vec{u_i} \in \mathbb{R}^m$, $\vec{v_i} \in \mathbb{R}^n$, and the singular values σ_i on the diagonal of the matrix Σ are non-negative and ordered according to their importance, i.e., $\sigma_1 \geq \sigma_2 \geq \cdots \geq \sigma_r \geq 0$, where $r \leq min(m, n)$ is the rank of the matrix $\mathbf{A}$. Naturally, we may compress all the matrices in the above formula with the rank r of the original matrix $\mathbf{A}$. In such a case: $\mathbf{U} \in \mathbb{R}^{m \times r}$, $\Sigma \in \mathbb{R}^{r \times r}$, and $\mathbf{V}^T \in \mathbb{R}^{r \times n}$.

The central idea of principal component analysis (PCA) is to reduce the dimensionality of a data set $\mathbf{A} \in \mathbb{R}^{m \times n}$ (consisting of a potentially large number of interrelated variables) retaining as much as possible of the variation present in the data. This is achieved by transforming original data representation to a new set of variables, so-called principal components, which are uncorrelated, and which are ordered so that the first few retain most of the variation present in all of the original variables. I the first step, we center the data in matrix $\mathbf{A} \in \mathbb{R}^{m \times n}$ by subtracting it with matrix $\mathbf{A}_{mean}$ consisting of mean vectors for each column in matrix $\mathbf{A}$:

$$\mathbf{A}_{mean} = \begin{pmatrix} mean(a_{1,1}, .., a_{m,1}) & \cdots & mean(a_{1,n}, .., a_{m,n}) \\ mean(a_{1,1}, .., a_{m,1}) & \cdots & mean(a_{1,n}, .., a_{m,n}) \\ \vdots & \ddots & \vdots \\ mean(a_{1,1}, .., a_{m,1}) & \cdots & mean(a_{1,n}, .., a_{m,n}) \end{pmatrix}$$

This way, every column in matrix $\mathbf{B} = \mathbf{A} - \mathbf{A}_{mean}$ has a zero (0) mean. The next step is to calculate the co-variance[3] matrix $\mathbf{C} \in \mathbb{R}^{n \times n}$ for the columns (features) in table $\mathbf{B} = \left(\vec{b_1} \cdots \vec{b_n} \right)$. Since every column in $\mathbf{B}$ has a zero mean (i.e.,$\forall_{1 \leq i \leq n} E[\vec{b_i}] = 0$) then co-variance between features:

$$Cov[\vec{b_x}, \vec{b_y}] = \frac{1}{m} \sum_{1 \leq i \leq m} (x_i - E[\vec{b_x}])(y_i - E[\vec{b_y}]) = \frac{1}{m} \sum_{1 \leq i \leq m} x_i * y_i$$

where x_i, y_i correspond to i-th observations (rows) in $\vec{b_x}$ and $\vec{b_y}$, respectively. Hence, we may express a co-variance matrix as follows:

$$\mathbf{C} = \frac{1}{m} \mathbf{B}^T \mathbf{B}$$

Here, we can calculate eigenvectors, and the corresponding eigenvalues, for matrix $\mathbf{C}$, such as:

$$\mathbf{CV} = \mathbf{V\Sigma}$$

where matrix $\mathbf{V}$ contains eigenvectors, a diagonal matrix Σ contains eigenvalues[4]. For the purpose of dimensionality reduction, we can project the data points

---

[3] $Cov[X, Y] = E[XY] - E[X]E[Y]$, or $Cov[X, Y] = \frac{1}{m} \sum_{1 \leq i \leq m} (x_i - E[X])(y_i - E[Y])$.

[4] In literature, matrix $\mathbf{V}$ is often denoted as $\mathbf{W}$, whereas Σ as Λ. We, however, continue with the notation as introduces with SVD example above.

onto the first k principal components, i.e., truncating matrix $\mathbf{V}$ to only k most significant features ($\mathbf{V}_k$) and projecting the original data $\mathbf{A_k} = \mathbf{A V}_k$ retaining enough variance. The first principal component is the direction in feature space along which projections of observations have the largest variance. The second principal component is the direction which maximizes variance among all directions orthogonal to the first one. The k-th component is the variance-maximizing direction orthogonal to the previous k-1 components.

It is also worth mentioning a few other dimensionality reduction methods [18]. Fisher's linear discriminant allows to find a linear combination of features that separates two or more classes of objects. Linear discriminant analysis is a generalized version of Fisher's linear discriminant, typically used for compressing supervised data [369]. This technique projects data in a way to maximize the target class separability. In independent component analysis, the original inputs are linearly transformed into features which are mutually statistically independent. Robust principal component analysis is proposed since the standard PCA is very sensitive to noise or outliers, and the estimated values obtained by PCA can be arbitrarily far from the true value [250]. Kernel principal component analysis (KPCA) is an extension of conventional PCA that is capable of constructing nonlinear mappings that maximize the variance in the data. Multilinear principal component analysis (MPCA) is a multilinear subspace learning algorithm. Compared with other commonly used dimensionality reduction algorithms, MPCA has proven performance for the tensor data [150]. Autoencoders are a specific type of neural networks that uses an adaptive encoder to transform high-dimensional data into a low-dimensional code to then reconstruct the output from this representation [163, 247]. In [18, 369, 388], a comprehensive review of more related methods can be found.

Reduction techniques, like PCA, are useful for 2D or 3D visualizations of high-dimensional datasets [439]. Given a matrix $D^{m \times m}$ with distances between each pair of m objects form the original set, and a number of dimensions (typically, 2 or 3, for 2D or 3D output), multidimensional scaling (MDS) places each object into low-dimensional space in a way that preserves (as well as possible) pairwise distances between object. In genetics and microbiology, typical data analysis pipelines include a dimensionality reduction step for visualising the data in two dimensions, frequently performed with t-distributed stochastic neighbour embedding (t-SNE) [216]. A self-organizing map (SOM) is a type of artificial neural network used to produce a low-dimensional, nonlinear approximation of data. This makes it an appealing instrument for visualizing and exploring high-dimensional data, with a wide range of applications [308]. In addition to already mentioned, there are many more methods that can be used for a similar purpose, including: locally linear embeddings (LLE), isomap, or Laplacian eigenmaps [236].

In the case of texts, the raw data, i.e., a sequence of symbols with variable length, cannot be used directly to the already mentioned algorithms as most of them expect numerical feature vectors with a fixed size. Extraction of text features is an important matter for information retrieval (IR) or natural language processing (NLP) [144]. The standard methods derived form IR refer to tokenization, lemmatization, removal of stop words, Tf-Ifd term weighting, or building various n-gram representations for document corpus, etc. [259]. Below, we present 5 exemplary documents $\{D_1, D_2, D_3, D_4, D_5\}$ to better depict some of the reviewed concepts.

D_1: *"Role of granulation in feature selection"*
D_2: *"On resilient feature selection with r-$\mathbb{C}$-reducts"*
D_3: *"Interactive attribute selection with reducts"*
D_4: *"Predicting seismic events"*
D_5: *"Forecasting seismic events"*

Word embedding is one of the core feature learning techniques in NLP, where documents are mapped to vectors of real numbers [13]. In its simple form, the embedding may be represented as a term-document incidence matrix $M^{m \times n}$, where rows refer to m documents in corpus, columns refer to the n unique terms constituting the vocabulary of the document corpus, and cells $m_{i,j} \in M$ may determine whether i-th document contains j-th term, or a number of times each term occurs per document. There are many variants of this technique, e.g., by combining it with Tf-Ifd or word co-occurrence[5]. In Table 1, a simple term-document incidence matrix for exemplary documents $D_1, \ldots, D_5$ is presented.

Basing on co-occurrence, we may discover hidden similarities between words. Latent semantic indexing (LSI) relies on SVD to identify relationships between terms and hidden topics[6] contained in text. LSI assumes that words which are close in meaning often occur in a similar context. For example, cosine similarity for vectors representing terms "attribute" and "feature", or "predict" and "forecast" in Table 1 would indicate that those terms are dissimilar, whereas LSI would discover their similarity since both appear in a similar contexts. There are many other methods for topic modeling, besides LSI, it is important to mention latent Dirichlet allocation (LDA), which is one of the most popular in this field of study [190]. On the other hand, explicit semantic analysis (ESA) augments text representations with concept-based features, which are automatically extracted from massive human knowledge repositories such as Wikipedia. This way, it is

Table 1. A term-document incidence matrix for the exemplary documents.

Doc\Term	Attribute	Event	Feature	Forecast	Granule	Interact	Predict	Reduct	Resilient	Role	Seismic	Select
D_1	0	0	1	0	1	0	0	0	0	1	0	1
D_2	0	0	1	0	0	0	0	1	1	0	0	1
D_3	1	0	0	0	0	1	0	1	0	0	0	1
D_4	0	1	0	0	0	0	1	0	0	0	1	0
D_5	0	1	0	1	0	0	0	0	0	0	1	0

---

[5] For a given corpus, the co-occurrence of two words is the number of times they appear together (and are close enough, e.g., no more than 30 words separates them in text) in documents.

[6] The main topic for documents $\{D_1, D_2, D_3\}$ could be related to "feature selection".

possible to assign a human-readable name for hidden topics, or even to automatically generate a short substitute summary for documents [269]. Global vectors for word representation (GloVe), is a global log-bilinear regression model for the unsupervised learning of word vectors that is also basing on word co-occurrences [298]. GloVe combines the advantages of the two major model families, i.e., global matrix factorization and local context window methods.

Computing distributed word representations using neural networks is yet another very interesting technique because the learned vectors encode many linguistic regularities and patterns [131]. The skip-gram model is an efficient method for learning high-quality distributed vector representations that capture a large number of syntactic and semantic word relationships. The continuous bag of words (CBOW) model attempts to predict the current target word (the center word) basing on its context (surrounding words) [264]. For the exemplary document D_4: *"Predicting seismic events"*, for the context window of size 3, the task would be to predict the central word *"seismic"* having the context words: *"predict"* and *"event"*[7]. In [69], authors observe that – in the case of statistical machine translation – adding features computed by neural networks consistently improves the performance.

Recurrent neural network (RNN), and particularly long short term memory networks (LSTM), form a broad group of architectures that handle sequential data such as natural language, and hence are particularly useful for NLP. Transformers use attention mechanism to gather information about the relevant context of a given word, and to encode that context in the vector representation [389]. Likewise many other techniques, attention mechanism, which was initially invented for machine translation, has found applications in many other tasks, and currently, can help understanding objects' inter-relations in an image just as well as it supports machine translation tasks [287]. Bidirectional encoder representations from transformers (BERT) is designed to pre-train deep bidirectional representations from unlabeled text by jointly conditioning on both left and right context in all layers [89]. Natural language processing comprises a much wider range of diverse tasks, such as: part-of-speech tagging, chunking, named entity recognition, textual entailment, question answering, or semantic role labeling, and is supported by a vast amount of divers representation learning techniques [73, 313, 382].

The recent advances in objects recognition and image classification were achieved mainly due to convolutional neural networks (CNNs) [64, 221]. The term convolution refers to the mathematical combination of two functions. In the case of CNN, convolution is a specialized type of linear operation used for feature extraction, typically represented as $N \times N$ matrix, which is sometimes referred to as kernel, mask, convolution matrix, or filter [379]. It is used to enhance an image representation via blurring, sharpening, embossing, edge detection, etc. A convolution layer is a fundamental component of the CNN architecture that performs

---

[7] It is worth mentioning that the neural network input is a numeric vector embedding for each word (typically, word vectorization is performed after the initial preprocessing).

feature extraction, which consists of a combination of a convolution operation and an activation function. Typically, CNNs include also pooling layers to reduce dimensions of data and to provide effective controls for over-fitting. Still, automatic learning of high quality features is considered as a challenge also in this field of study [233]. To improve the process of feature engineering from sequential data performed by traditional CNNs, the convolutional recurrent neural network model extracts features from hidden states or outputs of the recurrent layer [212]. Along to their unquestioned role in image classification, CNNs were successfully applied in many other domains, including natural language processing [73], time series analysis and algorithmic financial trading [340], human activity recognition using multiple accelerometer and gyroscope sensors [154], or in radiology where a deep convolutional neural network was designed to detect COVID-19 cases from chest X-ray images [397].

In the case of images, CNNs practically outperformed all other approaches to feature engineering. However for videos, the well crafted features play a major role. There exists a large number of approaches for extracting local spatio-temporal features, including histograms of oriented gradients (HOG), histograms of optical flow (HOF), and combination of those two [229]. Another popular descriptors are: SIFT [337], and motion boundary histograms (MBH), which rely on differential optical flow [80]. Spatio-temporal interest points encode video information at a given location in space and time. In contrast, dense trajectories track a given spatial point over time to capture motion information [395].

Recently, a variety of model designs and methods have blossomed. There is, however, a hidden catch: the reliance of these models on hand-labeled training data. It is easy to collect and store a large amount of data, however it is difficult and time-consuming to label data, since interaction with human experts is usually essential for this process [180]. Deep hierarchical representations carry some interesting advantages with that respect. On the one hand, they promote the re-use of features, e.g., by unsupervised learning of intermediate representations, which can be used on a variety of supervised learning tasks. On the other hand, deep architectures can potentially lead to more abstract features at higher layers of representations [29]. Shared representations are useful to handle multiple modalities or domains, or to transfer learned knowledge to tasks for which few or no examples are given but a task representation exists. Learning reusable feature representations from large unlabeled data sets has been an area of active research. For example, one way to build good image representations is by training generative adversarial networks (GANs), and later reusing parts of the generator and discriminator networks as feature extractors for supervised tasks [312]. Much research has been dedicated to learning, understanding, and evaluating the representations of both supervised and unsupervised pre-training methods. With that respect, unsupervised multi-task learning is a promising area of research [328]. For example, generative pre-trained transformer trained on general language data sets can be fine-tuned to specific language tasks [313]. There are more areas of data science, which consider similar problems, e.g., transfer learning [314], weak supervised learning [318], or active learning [180].

As discussed in this section, modern approaches to representation learning and deep neural networks (DNN) enable performing feature extraction with various network architectures [29,33]. The feature extraction and selection is often performed as an implicit phase in training of network's hidden layers. We can think of DNNs trained by supervised learning as performing a kind of representation learning. The last layer of the network is typically a linear classifier, such as a softmax. Whereas, the hidden layers of the network learn to provide a representation to this classifier. In many applications, features extracted form hidden layers are processed directly [212], whereas in statistical machine translation this is a natural model behavior [69].

Deep learning methods employ multiple processing layers to learn hierarchical representations of data, and have preeminent results in many domains [89,379, 397]. Naturally, there are many more topics related to dimensionality reduction and representation learning [59], including: restricted Boltzmann machines, deep belief networks, or graph neural networks, which have shown high capability in handling relational dependencies behind multivariate time series forecasting where variables depend on one another [409]. This section provides only a high level overview of those techniques, which, in many cases, can be considered as a foremost alternative to the feature extraction methods. However, regardless of the unquestioned achievements in this area, these algorithms generally suffer from the lack of interpretation of the projected dimensions [166,432], and for that reason, they are not studied further in this paper.

2.3   Feature Selection

Feature selection (FS), sometimes referred to as variable elimination, or attribute subset selection, is the process of determining those attributes that potentially contribute to the predictive models. Along to dimensionality reduction, discussed in Sect. 2.2, FS is one of the most popular approaches defying the curse of dimensionality, by removing irrelevant and redundant attributes from data [61,372]. There are many benefits of eliminating surplus variables. On the one hand, the excessive amount of features increases the time and compute resources required to train models. On the other hand, training models on a large number of features may lead to over-fitting, resulting in their lower performance. Furthermore, FS is facilitating data visualization, providing a better understanding of the underlying process that generated the data [151]. Feature selection not only simplifies the obtained data representation, but also allows to acquire features that can be easily utilized by both analysts and learning algorithms [197]. FS can be designed at different levels spanning from a standard tabular data scenario, whereby features take a form of the existing columns/attributes, toward determination of data sources that can be used to extract features in further steps [280]. Feature selection mechanisms can be also combined with other approaches to machine learning and knowledge discovery, e.g., by means of analyzing components of neural network structures (interpreted as features) in order to achieve compact hybrid data representations [327]. FS has become increasingly important for data analysis with numerous successful applications in real life machine learning problems in various domains [54,176,192].

Due to the large search space, FS is a difficult combinatorial problem, i.e., for a data with n features the number of possible solutions is 2^n [353,398]. Searching for a (near)optimal subset of features is a challenging optimization problem, for which many meta-heuristics, including: bee or ant colony optimization [123], simulated annealing and whale optimization [255], Harris hawks [161] or grey wolf optimizers [1], have been successfully applied. We may also distinguish several search strategies to select a subset of variables from the input data, including: exhaustive or heuristic search algorithms, genetic algorithms, evolutionary computation techniques, forward propagation and backward elimination strategies, or various hybrid strategies combining the above [95,413]. The forward propagation (sometimes referred to as sequential forward selection or addition) strategy starts with the empty set and consecutively adds one attribute at a time until certain criteria are met [61]. On the other hand, the backward elimination strategy starts with the full set (or relatively large set of attributes that satisfies required criteria). In each iteration, one attribute is removed - as long as the reduced set satisfies given criteria. Those algorithms which aim to obtain the possibly minimal set of attributes usually combine the heuristic search or forward selection with the subsequent phase of backward elimination [186,419]. We may also refer to a number of studies on parallelization of feature selection algorithms, e.g., by exploiting the computational capabilities of modern heterogeneous systems that contain several CPUs and GPUs [129], or by using Map Reduce paradigm and Spark framework [286]. Big Data aspects of attribute granulation and selection are discussed further in Sect. 11.

Depending on whether the training set is labeled or not, feature selection algorithms can be categorized into supervised, unsupervised, and semi-supervised. Given the input data as a table with m samples and n features $A = \{a_1, .., a_n\}$, and the target variable d, the supervised feature selection problem is to find a sub-set of features $R \subseteq A$ that "optimally" characterizes d [297]. Unsupervised feature selection is a less constrained search problem (without class labels), often depending on clustering quality measures [438], statistical and information measures [437], or on various hierarchical and granular structures – as briefly discussed further in Sect. 2.4. A comprehensive review of unsupervised methods can be found in [12]. It is, however, quite common to have a data set with huge dimensionality but a small labeled sample size. Under the assumption that, both, labeled and unlabeled data are sampled from the same population generated by the target concept, the semi-supervised feature selection methods make use of both labeled and unlabeled data to estimate the relevance of evaluated features [345]. One way to do this is to transform the partially labeled data into completely labeled. Whereas, the other approach is to construct a measurement to cover both labeled and unlabeled data. For this purpose, one may use ensemble selectors, for example based on rough set based local neighborhood decision error rate [248], or may incorporate additional knowledge, like graph-based structures, into semi-supervised FS methods [346].

Feature selection methods can be further categorized into three main groups: wrapper, embedded, and filter methods [40,151]. Wrapper methods make a selection of attributes based on the results of a preliminary data analysis. Wrappers use the learning algorithm as a part of the feature subsets evaluation, i.e., classification (or regression) model is used as a black box for assessing the feature subsets usefulness in terms of the error (or fitness) rate obtained by a wrapped model on a testing set. Wrapper methods include simple approaches, like greedy sequential searches, but also more elaborate algorithms like recursive feature elimination, or evolutionary and swarm intelligence algorithms [171,257]. Although these techniques may lead to feature subsets well corresponding to the analyzed problem, they require training a model for a combinatorial number of times, hence the computational cost becomes prohibitive for high dimensional data sets. Embedded methods are nested in machine learning algorithms, and incorporate knowledge about the specific structure of the class of functions used by a certain learner, e.g., bounds on the leave-one-out error of SVMs [31]. Other examples are: Lasso regression, classification and regression trees, or gradient boosting [36]. Embedded methods are usually less computationally expensive, still are much slower than filters. Same as in the case of wrappers, the selected features are dependent on the learning machine.

In contrast to the above-discussed methods, filters carry out the attributes selection regardless of the chosen model, since for the assessment of feature subsets, they use evaluation metrics independent of the induction learning algorithm [232]. This strategy is particularly useful because of its efficiency. Typically, attributes are ranked according to various types of scores, and those with the highest scores are used to train the model, with an implicit assumption that choosing appropriate attributes improves the accuracy and efficiency of classification or regression. By applying statistical measures, one can find columns that do not contribute to the accuracy of a model (or might in fact decrease its accuracy) and remove them before the final training phase. Filter methods can be roughly classified further by the filtering measures they employ to heuristically determine the subset of attributes with the highest predictive power, i.e., information, distance, dependence, consistency, similarity, or statistical measures. Examples of which include univariate criteria like: correlation between evaluated features and a target variable [155,176], entropy, chi-square, analysis of variance (ANOVA), or other statistical tests [40], as well as multivariate tests like various attribute dependency measures from the rough set theory (RST) [78,186,373], which are discussed in detail in Sect. 3.1.

Using the correlation coefficient is a simple yet effective approach to FS [155,176]. For the attribute $a \in A$ and the decision d, Pearson's correlation coefficient $r_{a,d}$ (or $r_{a,b}$ in the case of correlation between two attributes $a, b \in A$), for data with m samples, is defined as:

$$r_{a,d} = \frac{\sum_{i=1}^{m}(a_i - \bar{a})(d_i - \bar{d})}{\sqrt{\sum_{i=1}^{m}(a_i - \bar{a})^2}\sqrt{\sum_{i=1}^{m}(d_i - \bar{d})^2}}$$

where $\bar{a}$ and $\bar{d}$ are the mean values for the investigated attribute and the decision, respectively. Whereas, a_i and d_i are the values of the attribute a and the decision d for the i-th sample.

Another simple yet very popular test to maximize the relevance of selected features is mutual information. Given two random variables X and Y, and their probabilistic density functions p(x), p(y), and p(x,y), mutual information is defined as:

$$I(X;Y) = \int_Y \int_X p(x,y) \log \frac{p(x,y)}{p(x),p(y)} dxdy$$

Given m samples, we may approximate density functions $\hat{p}$ for the attribute $a \in A$ as:

$$\hat{p}(a) = \frac{1}{|A|} \sum_{i=1}^m \delta(a - a_i, h)$$

where a_i is the value of the attribute a for the i-th sample, $\delta(.)$ is the density estimator (e.g., Parzen window function for which h is the window length) [101]. Naturally, we may consider mutual-information-based feature selection for both discrete and continuous data. For discrete (categorical) variables probability tables can be estimated from data samples with the following formula:

$$I(X;Y) = \sum_{y \in Y} \sum_{x \in X} p(x,y) \log \frac{p(x,y)}{p(x),p(y)}$$

Mutual information may be equivalently expressed with entropy[8] as:

$$I(X;Y) = H(X) - H(X|Y)$$

where $H(X|Y)$ is the conditional entropy. Other well-known feature ranking strategies are Fisher Score that optimize between-class variance and the within-class variance, or Relief-based algorithms that order features based on the nearest neighbor distance [387].

Among filter-based feature selection methods, the most interesting from our viewpoint are multivariate algorithms [383]. Such approaches rely on inter-feature dependencies when selecting a feature subset. Most of multivariate filtering algorithms attempt to avoid including unnecessary features by measuring redundancy within the selected subset. Such methods iteratively select features that provide the most relevant information regarding dependent variable values (e.g., are highly correlated or have a high value of mutual information index) and, on the other hand, are less dependent on the already-selected features [246].

---

[8] Entropy is one of the basic measures of information contained in data. For a discrete random variable X with possible values $\{x_1, .., x_m\}$ is defined as: $H(X) = -\sum_{i=1}^m p(x_i) log(p(x_i))$.

As a result, they produce quite compact feature sets – what is a big advantage in practice [360]. For example, in [176], authors search for features which have strong correlations with a target class, yet uncorrelated mutually. This way implementing the correlation-based multivariate FS method to identify the most prognostic genes to classify biological samples of binary and multi-class cancers.

In terms of mutual information, feature selection algorithms aim to find a feature set $R \subseteq A$, containing features that jointly have the largest dependency on the target variable d. One of the most prominent examples of the multivariate FS based on mutual information is minimum redundancy maximum relevance algorithm (mRMR) [91, 297]. The main objective of mRMR is to find the subset of features $R \subseteq A$ that maximizes the following criterion:

$$\max_{R \subseteq A} \Phi(R, d) = \frac{1}{|R|} \sum_{a_i \in R} I(a_i; d) - \frac{1}{|R|^2} \sum_{a_i, a_j \in R} I(a_i; a_j)$$

Algorithm 1: Hybrid FS combining mRMR filter with a wrapper model

---

Data: U – samples, A – set of features, d – target variable;
N – max number of features to be evaluated during forward propagation
$\mathbb{M}^{\Phi}$ – selected ML algorithm for wrapper model
Φ^{M} – selected criterion to assess model quality
k – cross validation parameter
Result: $R \subseteq A$ – selected attribute subset

1 /* Initialization   */
2 $R \leftarrow argmax_{a \in A}(I(a; d))$
3 $i \leftarrow 1$
4 $R_{best} \leftarrow R$

5 $\Phi^{M}_{best} \leftarrow$ evaluation of $\mathbb{M}^{\Phi}(R, U)$ with Φ^{M} under k-fold cross validation

6 $\Phi^{M}_{candidate} \leftarrow 0$

7 /* Forward propagation   */
8 **while** $(i < N)$ **do**
9 $\quad$ $i++$
10 $\quad$ /* incremental mRMR criterion   */
11 $\quad$ $a = argmax_{a_i \in A \setminus R}(I(a_i; d) - \frac{1}{i} \sum_{a_j \in R} I(a_i; a_j))$
12 $\quad$ $R \leftarrow R \cup \{a\}$

13 $\quad$ /* evaluation with wrapper model   */
14 $\quad$ $\Phi^{M}_{candidate} \leftarrow$ eval. of $\mathbb{M}^{\Phi}(R, U)$ with Φ^{M} under k-fold cross validation

15 $\quad$ **if** $(\Phi^{M}_{candidate} > \Phi^{M}_{best})$ **then**
16 $\quad\quad$ $R_{best} \leftarrow R$
17 $\quad\quad$ $\Phi^{M}_{candidate} \leftarrow \Phi^{M}_{best}$
18 $\quad$ **end**

19 **end**
20 *return* R_{best};

The objective of mRMR algorithm is to maximizes relevance between selected features and the decision (the left factor of the above subtraction), and to minimize the redundancy among selected features (the right factor of the above subtraction) [91]. In practice, we may use an incremental search strategy to find the near-optimal solution as shown in line 11 in Algorithm 1. As proposed in [297], mRMR criterion may be combined with wrapper FS method. In each iteration of the forward propagation the wrapper model $\mathbb{M}^{\Phi}$ is evaluated with k-fold cross validation on the given data sample and so far selected features, i.e., $R^{(1)} \subset R^{(2)} \subset R^{(3)} \subset \cdots \subseteq A$ to assess the predictive quality of candidate feature set – as presented in Algorithm 1.

To provide more examples of multivariate methods, we may further refer to N-MRMCR-MI method based on the normalization of maximum relevance and minimum common redundancy for the nonlinear optimization problems [63]. There are more approaches that rely on mutual information, e.g., maximum relevance minimum multicollinearity (MRmMC) [338], double input symmetrical relevance filter (DISR), or normalized joint mutual information maximization (NJMIM) [32]. We may also recall a linear feature selection method called dynamic change of selected feature with the class (DCSF) that employs both mutual information and conditional mutual information [117], which eliminates irrelevant and redundant features by introducing the dynamic information change of already-selected features with the class. In [158], authors propose two FS algorithms and evaluation criterion inspired by mutual information, ReliefF, and Fisher score. Naturally, there are many more multivariate filters [30,54], or combinations of filters and wrappers [123,157]. In [40,390,398,413], a comprehensive review of more related methods can be found.

Most of feature selection approaches are focused on achieving possibly compact data representation to perform efficiently on large data volumes [381], or to scale with respect to high dimensionality [246]. However, as in real life applications data may be processed continually over time, and some features may become temporarily unavailable or unreliable, it is also worth to study various extensions of standard feature selection algorithms, including such aspects as: incomplete data handling [79], dynamic and incremental data processing [195], or feature cost analysis [266]. To some extent, the ideas presented in this study could be compared to the notion of stability (or robustness) of selected feature subsets [201,285]. Stability of feature selection techniques can be expressed as a variation in feature selection results due to changes in the data, e.g., when training samples are added or removed [333]. If the FS algorithm produces a significantly different subsets for any perturbations in the training data, then that algorithm becomes unreliable. Measuring stability of selected features is particularly important in biological and medical research, indicating whether the selected features are likely to be a real clinical signals worth further investigation, or not [127]. There are two popular approaches to assess the stability of particular FS algorithm: a similarity-based approach and the frequency-based approach. In both cases, we may measure the stability of a given feature selection algorithm as the variability of its output with respect to data sampling [284].

Let $\mathbb{R} = \{R_1, .., R_M\}$ be the set containing M feature subsets $R_i \subseteq A$ being results of M consecutive runs of the evaluated FS algorithm, e.g., on different data sub-samples. In the frequency-based approach, we interpret the feature selection results $\mathbb{R} = \{R_1, .., R_M\}$ as a binary embedding, where 1 means that feature a_i has been selected, whereas 0 means the opposite. For a given data containing $|A| = n$ features, we may represent $\mathbb{R}$ in a tabular form as:

$$\mathbb{R} = \begin{pmatrix} selected(a_1, 1) & \cdots & selected(a_n, 1) \\ selected(a_1, 2) & \cdots & selected(a_n, 2) \\ \vdots & \ddots & \vdots \\ selected(a_1, M) & \cdots & selected(a_n, M) \end{pmatrix}$$

where i-th row corresponds to subset $R_i \in \mathbb{R}$ selected in i-th algorithm run, and the $selected(.,.)$ function for an attribute $a \in A$ and i-th algorithm iteration is defined as:

$$selected(a, i) = \begin{cases} 1 & \text{if } a \text{ was selected in i-th FS algorithm run} \\ 0 & \text{opposite} \end{cases}$$

The observed frequency of selection of a feature $a \in A$ after M algorithm runs may be defined as:

$$\hat{p}(a) = \frac{1}{M} \sum_{i=1}^{M} selected(a, i)$$

Here, one can define the stability measure by the selection frequencies of each feature after M algorithm runs, e.g., as the frequency of selection averaged over all features [127]:

$$\hat{\Phi}(\mathbb{R}) = \frac{1}{|A|} \sum_{a \in A} \hat{p}(a)$$

Naturally, there are more frequency-based approaches to assess FS stability, for example: relative weighted consistency, or entropy of feature sets [284].

In the similarity-based approach, we define stability of algorithms as the average pairwise similarity between the possible pairs of feature sets in $\mathbb{R}$:

$$\hat{\Phi}(\mathbb{R}) = \frac{1}{|\mathbb{R}|(|\mathbb{R}| - 1)} \sum_{R_i \in \mathbb{R}} \sum_{\substack{R_j \in \mathbb{R} \\ R_j \neq R_i}} Sim^{\phi}(R_i, R_j)$$

where the stability measure $\hat{\Phi}$ depends on the similarity measure Sim^{ϕ} of a choice, which may be a Hamming ($Sim^{Hamming}$) or Dice-Sørensen (Sim^{Dice}) index, fuzzy similarity measures, e.g., generalized weighted Jaccard similarity (Sim^{GWJS}) [347], or RST based similarity measures, e.g., based on discernibility Sim^{Disc} (24) as presented later in Sect. 4.2, and many others [178]. For example, given the Jaccard index as the similarity measure $(Sim^{Jaccard})$, the stability measure $\hat{\Phi}$, defined as the average pairwise similarity between the possible pairs of features, is as follows:

$$\hat{\Phi}(\mathbb{R}) = \frac{1}{|\mathbb{R}|(|\mathbb{R}| - 1)} \sum_{R_i \in \mathbb{R}} \sum_{\substack{R_j \in \mathbb{R} \\ R_j \neq R_i}} \frac{|R_i \cap R_j|}{|R_i \cup R_j|}$$

Stability was investigated from various perspectives, e.g., by means of avoiding over-fitting [285], or minimizing an impact of data noise [14]. In the case of our study, the meaning of stability may refer to minimizing a risk of information insufficiency subject to a loss of access to some of pre-selected features.

Another thread of research that corresponds to robust/stable FS [227] is related to ensemble-based feature selection [46,333]. An ensemble (sometimes referred to as a committee) is collection of single classification (or regression) models whose predictions are aggregated, e.g., by majority voting [220]. To address this aspect while building classifier ensembles, meta-learning algorithms, such as boosting or bagging, can be used. Ensembles for feature selection might be further classified following many diverse criteria [38], but the most simple division is into the homogeneous ones, in the case of which the base selectors are all of the same kind, and the heterogeneous ensembles that combine outputs of diverse FS methods. Diversity of ensamble-based FS may be investigated from many perspective, as thoroughly discussed further in Sect. 4. Yielding in an improved prediction performance, as confirmed in a detailed evaluation on real data sets presented further in Sects. 5.1 and 5.2, is one of the main reasons to use an ensemble method with divers components. It would not make any sense to build an ensemble in which all the components offered the same result.

Ensembles have been shown to be an efficient way of improving predictive accuracy or/and decomposing a complex learning problem into easier sub-problems. The ensemble feature selection may be interpreted two-fold. On the one hand, several feature selection processes may be carried out (either using different training sets, different FS methods, or both), with the final goal to produce a single feature set as a combination of particular ensemble components [3]. In this approach, the aggregated predictions are expected to obtain more accurate and stable results, hence reducing the risk of choosing an unstable subsets, what is especially important for non-stationary environments [220], such as imbalanced data streams [50], or in the presence of concept drifts [24]. Indeed, merging multiple feature subsets obtained using ensemble techniques can yield results that are robust (or stable) from the above viewpoint. This kind of merging can actually lead toward establishing feature sets that induce high-quality prediction models [138]. However, robustness with regard to small data changes is not the same as robustness with regard to losing some of data dimensions. This latter aspect is specially relevant in knowledge discovery, and even more in those cases in which data dimensionality is very high, but the number of samples is not such, as they are more sensible to generalization problems [223].

On the other hand, we may apply FS for several times in order to produce the diversity of subsets for the purpose of subsequent ensemble learning methods. Here, ensemble-based methods in feature selection can be considered by means of creating multiple prediction models [90], whereby each model is built over a different subset of features, e.g., by constructing a rule-based classifier for each selected feature subset and aggregating results of such predictors [365]. In [302], it was noted that ensemble construction based on random subspace selection can partially solve the missing feature problem, which is exactly what we want to address in our resilient feature selection framework presented in Sect. 3. Analogous approaches can be found also in the rough set literature, with respect to both, standard reducts [168] and approximate reducts [404]. Actually, using a variety of approximate reducts to construct an ensemble of diversified models can be efficient in many areas [360]. Still, when comparing to the approach to resilient feature selection, introduced further in Sect. 3, those aforementioned ideas – based either on merging multiple reducts [363], or on treating them as an ensemble [365] – are heuristic methods that miss explicit mathematical formulation of the properties of resilient feature sets and explicit optimization goals for algorithms aimed at searching for such sets in the data.

2.4 Information Granulation in Feature Extraction

Granular Computing (GrC) arose as a synthesis of insights into human-centred information processing that mimics human, intelligent synthesis of knowledge from information [23,423]. Currently, information granulation plays an important role in modern machine learning and knowledge discovery algorithms, with a number of successful applications in various domains [140,359,400,433]. In this section, we focus on feature space granulation approaches introduced by now. Our objective is to provide a general overview of GrC, and to identify the main items on its agenda associating their usage in the setting of feature extraction. This way, we lay foundation for our approach by explaining how the granules can be formed, interpreted, and utilized by feature extraction algorithms.

Decision support in solving problems related to complex systems requires relevant computation models as well as methods for reasoning [352]. In recent years, one can observe a growing interest in the area of GrC as a methodology for modeling and conducting complex computations, in various domains of information technology, machine learning, and feature extraction, in particular [141]. On the other hand, human-centricity comes as an inherent feature of intelligent systems [296]. It is anticipated that a two-way human-machine communication is imperative, and interactive communication of intelligent systems with users becomes substantial [177].

The possibility to take advantage of additional domain knowledge provided by human experts relies on the observation that human thinking and perception in general, and their reasoning while performing data exploration tasks in particular, can comprise different levels of abstraction, display a natural ability to switch focus from one level to another, or operate on several levels simultaneously [282]. Human, however, perceives the world, reasons, and communicates at some level of abstraction that, unlike information systems and algorithms, comes hand in hand with non-numeric constructs. Those embrace collections of entities characterized by some notions of closeness, proximity, functionality, resemblance, or similarity, referred to as *information granules* (*granules* or *infogranules*, for short) [179, 295, 420].

The construction of a granular system for a given data set is frequently portrayed as a procedure of zooming in and out on the data or, in other words, changing the data "resolution". Depending on the chosen level of granularity, some data items (objects, cases, instances) become indistinguishable. Hence, the "length" of the data is altered, which corresponds to possible reduction of the storage and processing resources. By employing compact descriptions of granules – defined as collections of original data elements gathered together – one can accelerate computations and, moreover, make the results of those computations more meaningful for domain experts. It is also worth mentioning that the idea of zooming in and out – i.e., switching between different levels of information granularity – is popular in the area of analytical processing in databases [359]. However, one should remember that *data granularity* can have different meanings. In traditional databases, by *granular* data one usually means the most detailed, low level, exact data representation [11].

The granular approach to dealing with information systems does not have to be limited to just the length/volume dimension of the data set. It can also be used to modify, reduce and transform the "width" and "depth" of information. In GrC this is sometimes called *variable granulation* and *concept granulation*. Just like in a case of the "classical" granulation, where data objects are combined into more complex entities, attributes in data can be granulated by using similarity, distance or correlation between them. In particular, by constructing granules over the space of attributes in the data set it is possible to reduce dimensionality. In the simplest form it can be used to replace multiple features/dimensions by just the representative one of the corresponding granule. A more complex, yet still similar approach is represented by a reduction based on an information function and discernibility, typical for the theory of rough sets, where the original set of attributes is replaced by a reduct, i.e., a subset that carries the same amount of important information.

In the context of attribute granulation, two attributes are usually regarded as similar if they convey similar information about objects described in the data. For instance, one may consider similar two attributes whose values in the data are highly correlated. In fact, Pearson and Spearman correlation coefficients are commonly used as measures of attribute similarity for the purpose of attribute clustering [370]. There are, however, some other possibilities as well. For instance, we examine an idea of building similarity of attributes by means of their ability to replace each other in the constructed decision models. Namely, if an attribute can be replaced by another without losing important information about investigated objects, it means that they complement in the same way the remaining attributes. The more generic approach is to search for whole feature sets with a guarantee of providing sufficient predictive power even if some of their elements are dropped [137, 145].

The proximity of attributes may have a few meanings as well. Typically, this term is used as a synonym of similarity. However, when it comes to granules of attributes, it may also be understood as a "physical proximity". For example, in coal mines, there are many sensors monitoring the safety of miners, which constantly gather data about the conditions underground [182]. When analyzing this type of data, it is important to consider locations of sensors, since readings from closely co-located devices are inherently correlated [15]. Moreover, events observed by one group of sensors are detected by other groups after some time and the delay, as well as the order, in which different sensors denote the event, often corresponds to the ventilation scheme of the mine. For this reason, as noted in [280], it is often worth to consider the whole chunks of attributes corresponding to such proximate sensors. In this way, it is not only possible to improve readability of the resulting decision models, but also increase the performance of the whole data processing chain due to a more efficient utilization of local buffers for reading data streams [136, 147]. Another practical consideration is the aspect of model robustness and fault tolerance. In this context, proximity of attributes may be regarded as a degree of dependency on a specific hardware equipment [182]. For instance, if one sensor is faulty, all attributes whose values are dependent on its readings will be unreliable [145].

It may also be desirable to consider granules of attributes that share some higher-level properties or that are tied by constraints imposed by a given application area [130]. Typically, domain experts associate such attributes with similar functionalities of investigated objects. Let us consider an example of the brain MRI data set investigated in [400], whereby features derived using some parameterized image processing procedures may be associated with groups of attributes that take different values for particular objects (these values depend on particular parameter settings) but describe the same aspect of the data. Another example of this type of situation is apparent in the analysis of a stock market. Many financial experts use technical indices to describe the behavior of stock prices in time. Such indices (e.g., moving averages, moving variance, RSI, TDI, stochastic oscillators, etc.) have many parameters, such as the considered time window size. Over long periods, the accuracy of time series model forecasting is invariably affected by interval length, and formulating effective interval partitioning methods can be very difficult. In [65], an interdisciplinary review of the idea of granularity in economic decision making from different perspectives, including: psychology, cognitive science, complex science, as well as behavioral and experimental economics is discussed.

The above considerations lead toward several observations. To begin, the spaces of features/attributes that require to be granulated can be more complex than a set of columns in a tabular data. The above considerations lead toward the observation that the spaces of features/attributes A[9] that require to be granulated can be more complex than a set of columns in a tabular data. In some real-life scenarios, the set A may require granulation because of its high cardinality. An example of such situation can be found, e.g., in [134], where an interactive

---

[9] Typically the set of all features/attributes is denoted with A [291].

GUI-based approach for grouping genes-attributes was introduced. However, in other scenarios the set A may not exist in a materialized form. We can rather think about a set A^* gathering all derivable attributes/features, e.g., wavelet coefficients in the case of EEG signal analysis [378] or JSON-driven aggregates defined for a semi-structured data set [189]. Thus, one could think about A^* as a space of all outcomes of the feature engineering/extraction techniques applied in a given application area. We shall treat A^* (sometimes taking a simple form of A) as our granulation domain.

The second observation is about the meaning of granules built over A^* (or A) from the perspective of data understanding and decision model construction, including feature selection. With respect to data understanding, it is implicitly assumed that features dropping into the same granules should be assessed by domain experts as having some kind of common background, by means of physical, functional or information-specific comparability. In particular, the information level of comparability may correspond to the way, in which particular features contribute to decision models aimed at classifying or distinguishing between different states of target variables. This aspect, as previously mentioned, seems to be close to the ideas of adapting various data clustering methods for the purpose of grouping together similarly acting or replaceable/interchangeable attributes [3,180]. However, we also need to remember that all of the above flavors of similarity need to be coupled with some tangible criteria for assessing the quality of pre-defined or produced granules, especially in the context of feature extraction [188].

Identification of subgroups of similar variables is especially important for high-dimensional data exploration [41,149]. In this context it is frequently useful to apply the modern algorithms aimed at big data clustering. Several instance clustering algorithms, like the expectation maximization or k-means, have already been implemented in the scalable environments [20,81]. There are also some prior results reported on the feature clustering algorithms that are of particular interest in this paper. The hierarchies of granules/groups of features can be constructed using some interactive clustering methods as well [134]. It is also important to realize that the feature similarity measures employed in the above clustering approaches should somehow correspond to the ultimate goal of finding the groups of attributes that can play mutually comparable roles in the constructed decision models [3].

The demand for efficiency and effectiveness in Big Data scenarios resulted in a number of approaches to massively parallel feature reduction [310,436], as well as highly scalable instance selection and deduplication [377,384]. Popular code libraries like Spark or Mahout provide parallel implementations of well-known feature selection methods [104]. There are also approximate implementations of standard algorithms, which derive heuristic feature evaluation scores from granulated data summaries [62]. The speed of the feature and instance selection processes becomes especially important in interactive approaches [362], whereby, additionally, granular hierarchies of attributes may help the users to navigate through rich feature spaces. Introducing approximate computations into the feature selection processes is – in combination with making them highly parallel – an example of a more general trend in machine learning and knowledge discovery [7].

It is noteworthy that, just as for other popular feature selection methods, there were some interesting attempts to perform RST reduct derivation within the MapReduce framework [239]. The ideas of scalable performance of feature extraction, in particular reduct calculation [138], are most commonly related to decomposing computations with respect to rows/instances [165]. However, by introducing the elements of granulation into the feature spaces we can additionally scale up the algorithms in an "attribute-oriented" fashion. Surely, such granulation-related ideas could be considered – besides the algorithms originating from the theory of rough sets – within the scope of other popular feature selection/engineering solutions as well [151,297].

Besides the so-far-mentioned rough sets [288,291], there are numerous formal frameworks of information granules [296]. Let us recall some selected alternatives. Among the most popular ones we may point out the set theory, interval analysis [271] and fuzzy sets which deliver an important conceptual and algorithmic generalization of sets by admitting partial membership of an element to a given information granule [422]. Shadowed sets distinguish among elements, which (i) fully belong to the concept, (ii) are excluded from it, and (iii) their belongingness is completely unknown [294]. The list of formal frameworks is quite extensive [290], interesting examples are also rough-fuzzy and fuzzy-rough sets [100], probabilistic sets [164], probabilistic rough sets [251], axiomatic fuzzy sets [249], or three-way decisions [418] under dynamic granulation [309], and many more [71,434]. However, we mainly focus on the interactive feature extraction methods related to the theory of rough sets [141,145]. In the next sections, we discuss the advantages of pre-grouping of attributes from the perspective of feature selection, with the reduct-based decision models originating from the theory of rough sets [373].

2.5   Rough Set Methods for Feature Selection

One of the data exploration methodologies where a large emphasis is put on the granulation of attribute space and multivariate feature selection is rough set theory (RST) [252,351]. RST as a whole provides a formalism for reasoning about imperfect data, handling such problems as data veracity, uncertainty, or incompleteness [156,159,222]. Its fundamental concept related to feature selection – and particularly dimensionality reduction – is a decision reduct, which is an irreducible subset of attributes (features, columns) that determines a target variable (so-called decision attribute) at the same level as the whole set of considered attributes.

In RST, we assume that the whole available information about an object $u \in U$ is represented in a structure called an information system [291] – a tuple (U, A), where U is a finite, non-empty set of objects, and A is a finite, non-empty set of attributes. Let us distinguish a decision attribute (class feature, target variable), which defines a partitioning of U into disjoint sets representing decision classes (or categories) that we want to describe using other attributes.

An information system with specified decision attribute is called a decision table (or decision system) and is denoted by $\mathbb{S} = (U, A \cup \{d\})$, $A \cap \{d\} = \emptyset$.

For a given decision system $\mathbb{S} = (U, A \cup \{d\})$, one considers functions $a : U \rightarrow V_a$, $a \in A$, where V_a is the set of values of a. Such functions allow us to represent $\mathbb{S}$ as a table with rows labeled by objects, columns labeled by attributes, and cells corresponding to pairs (u, a) assigned with values $a(u) \in V_a$ (see Table 2). Obviously, this kind of tabular representation is one of many equivalent formats of representing the data [405].

Table 2. A decision table $\mathbb{S}$ that is used in further illustrations in the frame of this study.

$U \backslash^A$	a_1	a_2	a_3	a_4	a_5	a_6	d
u_1	false	'a'	▽	○	□	'x'	good
u_2	false	'b'	▽	⊙	●	'x'	good
u_3	false	'c'	△	⊘	◇	'x'	good
u_4	false	'd'	▽	⊗	◁	'x'	good
u_5	false	'e'	▽	⊖	★	'y'	good
u_6	false	'f'	▽	⊘	▷	'y'	good
u_7	true	'g'	△	⊗	□	'y'	bad
u_8	true	'h'	△	⊖	●	'z'	bad
u_9	true	'i'	▽	⊕	◇	'z'	bad

The indiscernibility relation (IND) expresses the fact that due to a lack of information (or knowledge) we are unable to discern some objects employing available information. In general, we are unable to deal with each particular object but we have to consider granules (clusters) of indiscernible objects as a fundamental basis for RST. Let us define indiscernibility relation $IND(R)$: $U \times U$, for any $R \subseteq A$, as follows [291, 307]:

$$IND(R) = \{(u_i, u_j) : \forall a \in R, a(u_i) = a(u_j)\} \tag{1}$$

after considering the decision:

$$IND(R) = \{(u_i, u_j) : \forall a \in R, a(u_i) = a(u_j) \wedge d(u_i) \neq d(u_j)\} \tag{2}$$

By analogy, we can define a discernibility (or more precise R-discernibility) relation $DIS(R)$, as:

$$DIS(R) = \{(u_i, u_j) : \exists a \in R, a(u_i) \neq a(u_j)\} \tag{3}$$

after considering the decision:

$$DIS(R) = \{(u_i, u_j) : \exists a \in R, a(u_i) \neq a(u_j) \wedge d(u_i) \neq d(u_j)\} \tag{4}$$

(In)discernibility relations enable us to express dependencies among attributes at a more universal level. We may notice that indiscernibility (1) is an equivalence relation. We denote an equivalence class of each object $u \in U$ as $[u]_A$.

An excessive amount of attributes in A provides a great potential for data-driven reasoning. However, many of those attributes may be dispensable, or could be irrelevant from the point of view of a given problem corresponding to d. In such situations, A-based information about objects in U needs to be simplified. Selecting informative sets of attributes is conducted by referring to the notion of a reduct [291].

Definition 1 (Reduct). *Let* $\mathbb{S} = (U, A \cup \{d\})$ *be a decision table. Subset* $R \subseteq A$ *is called a superreduct, if and only if it determines d within U, denoted as* $R \Rightarrow d$. *Superreduct R is called a reduct, if and only if there is no proper subset* $R' \subsetneq R$, *which holds the superreduct condition.*

From a formal point of view, we should write $\Rightarrow_\mathbb{S}$ instead of $\Rightarrow$, as the requirement of determining d by R is data-specific. However, we use a simplified notation whenever it does not lead to misunderstandings. Analogously, one may think about the usage of various heuristic measures while evaluating (subsets of) attributes in filter-based feature selection algorithms. There are plenty of interpretations of the reduct Definition 1 that correspond to several other concepts and theorems like: (in)discernibility relations, the (in)discernibility matrix, or the positive region. Below we provide a short review of several significant and representative reduct interpretations.

The first example refers to the indiscernibility relation (Eq. 2), which enables us to express dependencies among attributes.

Definition 2 (Reduct - by IND relation). *Let* $\mathbb{S} = (U, A \cup \{d\})$ *be a decision table. Subset* $R \subseteq A$ *is called a superreduct, if and only if* $IND(R) \subseteq IND(A)$. *Superreduct R is called a reduct, if and only if there is no proper subset* $R' \subsetneq R$, *which holds the superreduct condition.*

Equivalently, we may say that $R \subseteq A$ is a decision reduct, if and only if it is an irreducible subset of attributes such that each pair of objects $u_i, u_j \in U$ satisfying the inequality $d(u_i) \neq d(u_j)$ is discerned by R.

Another reduct definition is related to the indiscernibility relation (1) and its quotient set (i.e., is constructed by the equivalence classes of IND). A subset of features $R \subseteq A$ is called a decision superreduct iff for any object $u \in U$ the indiscernibility class of u relative to A is a subset of some decision class, its indiscernibility class relative to R should also be a subset of that decision class.

Definition 3 (Reduct - by equivalence classes of IND). *Let* $\mathbb{S} = (U, A \cup \{d\})$ *be a decision table. Subset* $R \subseteq A$ *is called a superreduct, if and only if* $[u]_A \subseteq [u]_d \Rightarrow [u]_R \subseteq [u]_d$. *Superreduct R is called a reduct, if and only if there is no proper subset* $R' \subsetneq R$, *which holds the superreduct condition.*

The next example refers to the discernibility relation. The numeric $Disc$ measure is based on the arity of discernibility relation:

$$Disc(R) = |\{(u, u') : \exists a \in R, a(u) \neq a(u') \wedge d(u) \neq d(u')\}| \qquad (5)$$

The definition of the decision reduct by $Disc$ measure would be very similar to the above listed. The only difference would refer to the superreduct condition, which, for $Disc$ measure, would be defined with the following equation:

$$Disc(R) = Disc(A) \qquad (6)$$

Another popular reducts formulation refers to the notion of function $\gamma :$ $\mathcal{P}(A) \rightarrow [0, 1]$, which is commonly used to express a degree of dependence between a subset of attributes and the decision:

$$\gamma(R) = \frac{|POS(R)|}{|U|} \qquad (7)$$

where POS denotes the positive region induced by R [291]:

$$POS(R) = \{u \in U : \forall_{u' \in U} d(u) \neq d(u') \Rightarrow \exists_{a \in R} a(u) \neq a(u')\} \qquad (8)$$

For a decision system $\mathbb{S} = (U, A \cup \{d\})$, where cardinality of A and U is: $|A| = m$, $|U| = n$ we can define a discernibility matrix $\mathbb{M}(R)$. Discernibility matrices are useful for deriving possibly small subsets of attributes, still keeping the knowledge encoded within a decision system [353]. Each cell $c_{i,j}$ of $\mathbb{M}(R)$ for $i, j = 1..n$, $1 \leq i < j \leq n$ contains a list of attributes in $R \subseteq A$, which are discerning objects u_i, u_j with different decisions, or more formally:

$$c_{i,j} = \{a \in R \subseteq A : u_i, u_j \in U, u_i \neq u_j, a(u_i) \neq a(u_j) \wedge d(u_i) \neq d(u_j)\} \qquad (9)$$

Among the presented variety of extensions of decision reducts, let us also discuss their approximate interpretations. Criteria for calculating approximate decision reducts are usually based on functions evaluating degrees of decision information induced by attribute subsets and thresholds for values of those functions' specifying which of those subsets are good enough. Such an approach may lead us to obtain subsets of attributes that are less accurate than exact reducts but could be preferred in some real-life applications to deal with large or noisy data, ultimately leading to smaller data representations.

For example, we may refer to α-approximations of reducts, where $\alpha \in (0, 1]$ is a real parameter [279]. The set of attributes $R \subseteq A$ is called α-reduct iff it is minimal in sense of set-inclusion, intersecting at least $\alpha \cdot 100\%$ of pairs of objects that are necessary to be discerned with respect to decision. More formally, we may define α-reduct with the following equation:

$$\frac{|\{c_{i,j} : R \cap c_{i,j} \neq \emptyset\}|}{|\{(u_i, u_j) : d(u_i) \neq d(u_j)\}|} \geq \alpha \qquad (10)$$

We may also easily introduce the approximation threshold ε for many reduct criteria. For example, let us introduce it into criteria based on $Disc$ measure (6):

$$Disc(R) \geq (1 - \varepsilon) * Disc(A) \tag{11}$$

As a yet another significant example, we may point out approximate entropy reducts [356], in the case of which, the superreduct criterion relay on the conditional entropy $H(d|R) = H(R \cup \{d\}) - H(R)$. In the following specification, H plays the role of a penalty measure, which, with the given approximation threshold ε, corresponds to (H, ε)-approximate reducts introduced in [356].

Definition 4 (Reduct - by conditional entropy). *Let* $\mathbb{S} = (U, A \cup \{d\})$ *be a decision table. Subset* $R \subseteq A \subseteq A$ *is called a* (H, ε)-*approximate superreduct, if and only if* $H(d|R) \leq H(d|A) - \log_2(1 - \varepsilon)$. *Superreduct* R *is called a* (H, ε)-*approximate reduct, if and only if there is no proper subset* $R' \subsetneq R$, *which holds the superreduct condition.*

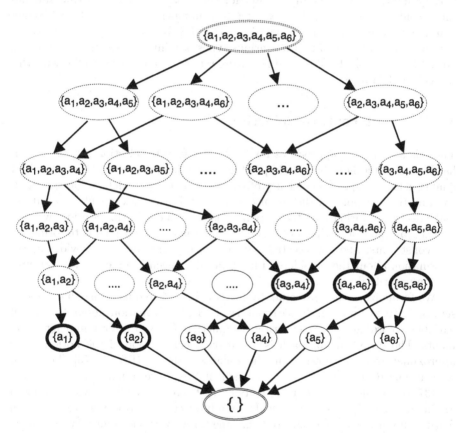

Fig. 2. The lattice for the data in Table 2. Each oval corresponds to a single attribute subset, starting from the empty set at the bottom, ending with the whole set of attributes at the top. Bold ovals correspond to reducts, dotted ovals correspond to superreducts, arrows correspond to the set inclusion relation $\subset$.

There are a lot more extensions for approximate and exact reducts, often hybridized with other methods [307,373]. For instance, in [186] a combination of iterative filter-based feature selection with statistical significance tests based on random probes and a typical RST-based redundant feature elimination was applied to calculate dynamically adjusted approximate reducts (DAAR). To provide a wider range of reasoning strategies, we could easily refer to the *rough membership function* $\mu_{d/R} : V_d \times V_R^U \to [0,1]$ [289], *majority decision function* $m_{d/B} : V_R^U \to 2^{V_d}$ [355], and many others [54,256,344]. A similar mixture of iterative feature selection and reduction was considered in [83]. Although in that latter case authors did not refer to the rough set literature, their feature reduction phase actually follows the same criteria as RST-based reduct calculation methods referring to the discernibility of (almost all) pairs of objects having different target variable values.

Figure 2 presents the attribute lattice for the data in Table 2. In this context, reduct computation means the search through the lattice. A minimal reduct may be interpreted as the first subset for every path from $\emptyset$ to A that satisfies the considered criterion. We may refer to a number of well-established reduct search algorithms with this respect [118,323,419]. It is also important to stress out that the problems of finding various types of minimal reducts are known to be NP-hard and the RST-related literature reports a number of studies with this respect [93,353,364].

3   Resilient Feature Selection

There is a variety of approaches for automatic feature selection [61]. Still, it is hard to find a method that would put together different aspects of feature subset quality, such as expected efficiency of the corresponding model, its interpretability from the viewpoint of the end-users, a risk of loss or lack of sufficient data to make decisions during long-term operations, and so on. In this Section, we concentrate on the last of the above-listed quality aspects. Our goal is to formulate new constraints, whereby selected feature sets are guaranteed to provide enough information about the considered target variables even if some of those features are temporarily dropped.

We formalize such constraints by introducing r-$\mathbb{C}$-reducts – irreducible subsets of features providing a satisfactory level of information about the target variable according to a given criterion function $\mathbb{C}$, even after removing r elements. The proposed approach is based on generalization of the notion of an approximate reduct known from the rough set theory (RST) [373]. This way, we continue RST-based research on resilient feature selection that was started in [137] by extending standard reducts [291]. However, the framework proposed in this paper embraces a much wider family of criteria specifying that a given feature subset is good enough to determine target variable values. We are actually able to refer to the whole realm of filter-based feature selection strategies [82], now defining a satisfactory feature set as the one whose evaluation function exceeds a certain threshold even after removing its r elements, $r \geq 0$.

In the feature selection process based on r-$\mathbb{C}$-reducts, an analyst should be able to control the level of resilience occurring in generated subsets of features while maintaining their relevance to the analyzed problem. In other words, the idea is to let an analyst achieve a relatively compact representation of the data tuned for the investigated problem, whereby the selected feature set should preserve its relevance even in case of partial data loss. However, to make this approach feasible, we need to investigate computational complexity of the corresponding search tasks. Then, we should also design efficient algorithms deriving meaningful r-$\mathbb{C}$-reducts from the data. The rough set literature is a good source of inspiration for both above aspects.

In RST, there is a lot of attention paid to NP-hardness of finding various versions of reducts in the data [273]. For example, in [355] it was proposed to evaluate feature subsets using a measure modeling accuracy of rule-based classifiers induced by those subsets over the training data. Then, the problem of finding minimal – in terms of cardinality – subset of features providing ε-almost the same value of such measure as the whole set of features was shown to be NP-hard for an arbitrary fixed threshold $\varepsilon \in [0, 1)$. Problems related to r-$\mathbb{C}$-reducts are analogous. As the important theoretical contribution in this Section, we show that any NP-hard problem of finding minimal attribute subset that yields satisfactory level of information according to a given $\mathbb{C}$ remains NP-hard for an arbitrary resilience level r. As a special case, the task of finding minimal subset of features providing ε-almost the same level of the aforementioned accuracy measure as the whole set even after removing arbitrary r elements is NP-hard.

The second contribution is a broad study on algorithmic aspects of searching for minimal r-$\mathbb{C}$-reducts. By following a popular idea of dynamic exploration of the lattice of feature subsets, whereby some of its elements turn out to be labeled as satisfying the criteria for providing enough information while others do not, we elaborate on two generic strategies, namely, breadth first search (BFS) and depth first search (DFS). For BFS, we adapt the well-known Apriori algorithm [335] for the purpose of r-$\mathbb{C}$-reduct search. For DFS, we extend standard reduct construction methods [357] to incorporate resilience of generated feature sets. Our study includes also some illustrative examples of data sets, as well as the analysis of computational cost of particular algorithms.

The rest of the Section is organized as follows. In Sect. 3.1, we introduce the notion of criterion function $\mathbb{C}$, which enables us to consider various feature selection formulations at a higher level of abstraction with $\mathbb{C}$-reducts. In Sect. 3.2, we discuss the idea of resilient feature selection and, accordingly, we introduce r-$\mathbb{C}$-reducts. In Sect. 3.3, we outline an Apriori-inspired algorithm that generates all r-$\mathbb{C}$-reducts of a given type. In Sect. 3.4, we study the tasks of resilient feature selection from the perspective of their computational complexity. We prove that many NP-hard feature selection/elimination problems remain NP-hard for any arbitrary resilience level r. In Sect. 3.5, we present heuristic DFS algorithms for searching for optimal r-$\mathbb{C}$-reducts, with specific examples of permutation-based and approximation methods.

3.1   ℂ-reducts

In this section, we take a step towards a generalization of feature selection methods as a process of achieving a feature subset that satisfies expected criteria. In many cases, especially in data analysis, it is much more interesting whether the given feature subset complies with respect to the defined function that is verifying some specified criteria rather than the exact value of a quality (or error) measure. Below, we generalize this way of reasoning about attribute subsets by introducing criterion functions, which, for each given decision table $\mathbb{S} = (U, A \cup \{d\})$, return a binary assessment of the candidate attribute subsets.

Definition 5 (Criterion Function). *A criterion function $\mathbb{C}$ is a function, which assigns, for any $\mathbb{S} = (U, A \cup \{d\})$, values 0 and 1 to the subsets of A (i.e., $\mathbb{C} : \mathcal{P}(A) \rightarrow \{0, 1\}$, where $\mathcal{P}(A)$ denotes the set of all subsets of A) in such a way that, for any $X, Y \subseteq A$, if $X \Rightarrow Y$ then $\mathbb{C}(X) \geq \mathbb{C}(Y)$.*

We write $\mathbb{C} : \mathcal{P}(A) \rightarrow \{0, 1\}$ instead of $\mathbb{C}_{\mathbb{S}} : \mathcal{P}(A) \rightarrow \{0, 1\}$. However, we go back to explicit data-specific notation in Sect. 3.4. Having this in mind, let us note that for any $X, Y \subseteq A$, if $X \supseteq Y$ then $X \Rightarrow Y$, hence $\mathbb{C}(X) \geq \mathbb{C}(Y)$. Such monotonicity of $\mathbb{C}$ is illustrated in the attribute subset lattice in Fig. 2. The above definition allows us to consider a very broad range of criteria, not all of which could be anyhow reasonable for feature selection. Still, once we conclude the particular approach does have a sense and is compliant to the presented generic definition (as we see in the following sections, there is a number of so far developed approaches that do comply), we may easily formulate its resilient versions and appraise their complexity.

Besides the constraint expressed for the selected R by $\mathbb{C}$, the proposed approach follows the very common for RST (but not only for RST – see, e.g., [83]) objective to achieve the smallest feature subsets – reducts. Below we extend the notion of a reduct with respect to $\mathbb{C}$.

Definition 6 (Criterion Reduct). *Let $\mathbb{S} = (U, A \cup \{d\})$ and $\mathbb{C}$ be given. Subset $R \subseteq A$ is called a $\mathbb{C}$-superreduct, if and only if $\mathbb{C}(R) = 1$. We call R a $\mathbb{C}$-reduct, if and only if, additionally, there is no proper subset $R' \subsetneq R$ such that $\mathbb{C}(R') = 1$.*

In relation to the notions introduced in Definition 1, we may notice that they can be easily rephrased using specific criterion function, namely:

$$\mathbb{C}^{\Rightarrow}(R) = \begin{cases} 1 & \text{if } R \Rightarrow d \\ 0 & \text{otherwise} \end{cases} \tag{12}$$

Indeed, reducts and $\mathbb{C}^{\Rightarrow}$-reducts are equivalent to each other. It is also worth adding that there are decision tables $\mathbb{S} = (U, A \cup \{d\})$ for which $\mathbb{C}^{\Rightarrow}(A) = 0$. In RST, they are called inconsistent. In such cases, there are no (super)reducts in terms of Definition 1. Definition 6 is surely far more general than Definition 1, subject to a choice of $\mathbb{C}$. In the literature, one can find many variants of reduct definitions. Below, we recall some of the popular extensions reviewed in Sect. 2.5, and re-formulate them using their corresponding criterion functions.

The criterion function that encapsulates the reduct definition based on $Disc$ measure (Eq. 6) may be constructed as follows:

$$\mathbb{C}^{Disc}(R) = \begin{cases} 1 & \text{if } Disc(R) = Disc(A) \\ 0 & \text{otherwise} \end{cases} \tag{13}$$

The criterion function that encapsulates the reduct definition based on discernibility matrix $\mathbb{M}$ may be constructed as follows:

$$\mathbb{C}^{\mathbb{M}}(R) = \begin{cases} 1 & \forall_{1 \leq i < j \leq n}, c_{i,j} \in \mathbb{M}(A), c'_{i,j} \in \mathbb{M}(R), \text{ if } |c_{i,j}| > 0 \Rightarrow |c'_{i,j}| > 0 \\ 0 & \text{otherwise} \end{cases} \tag{14}$$

The criterion function defining so-called γ-reducts is as follows:

$$\mathbb{C}^{\gamma}(R) = \begin{cases} 1 & \text{if } \gamma(R) = \gamma(A) \\ 0 & \text{otherwise} \end{cases} \tag{15}$$

The above examples can be generalized using the notion of a quality measure $Q : \mathcal{P}(A) \to \Theta$, where $(\Theta, \succeq)$ refers to a partially ordered set in which every two elements have a unique lower and upper bound [145, 278, 357]:

Definition 7 (Quality Measure). *Function Q is called a quality measure if, for any $\mathbb{S} = (U, A \cup \{d\})$, it assigns the subsets of A with the elements of Θ (i.e., $Q : \mathcal{P}(A) \to \Theta$) in such a way that for any $X, Y \subseteq A$, if $X \Rightarrow Y$ then $Q(X) \succeq Q(Y)$.*

The above property of Q will be further referred to as the monotonicity with respect to functional dependencies, which yields in particular – like in the case of Definition 5 – the monotonicity with respect to set inclusion (cf. the lattice in Fig. 4). In practice, the most commonly used specification of Θ are $\mathbb{R}$, $\mathbb{N}$, or $(0, 1]$ with $\geq$ relation. Similarly, as in the case of the criterion function Definition 5, Definition 7 is intended to cover essential properties of feature subset measures to generalize the further discussion, not to implement a feature selection by itself.

The following criterion functions correspond to a number of Q-based definitions of so-called approximate reducts. A general mechanism is to use measures $Q : \mathcal{P}(A) \to [0, +\infty)$ together with an approximation threshold $\varepsilon \in [0, 1)$, which is responsible for the allowed degree of losing information while removing attributes from A:

$$\mathbb{C}^{(Q,\varepsilon)}(R) = \begin{cases} 1 & Q(R) \geq (1 - \varepsilon) * Q(A) \\ 0 & \text{otherwise} \end{cases} \tag{16}$$

Proposition 1. *For every Q satisfying conditions of Definition 7, for every $\varepsilon \in [0, 1)$, criterion function $\mathbb{C}^{(Q,\varepsilon)}$ satisfies conditions of Definition 5.*

Proof. Straightforward.

For example, let us note that RST-based functions $Disc$ and γ can be considered as special cases of the above framework [278,373]. The corresponding criterion function for $Disc$ measure would be defined as follows:

$$\mathbb{C}^{(Disc,\varepsilon)}(R) = \begin{cases} 1 & Disc(R) \geq (1-\varepsilon) * Disc(A) \\ 0 & \text{otherwise} \end{cases} \tag{17}$$

The criterion function defining γ-reducts with the approximation threshold is as follows:

$$\mathbb{C}^{(\gamma,\varepsilon)}(R) = \begin{cases} 1 & \gamma(R) \geq (1-\varepsilon) * \gamma(A) \\ 0 & \text{otherwise} \end{cases} \tag{18}$$

Proposition 2. *Functions $Disc$ and γ satisfy conditions of Definition 7.*

Proof. It is known that both considered functions are monotonic with respect to set inclusion. The property related to functional dependencies can be shown analogously.

Proposition 3. *Function $\mathbb{C}^{\mathbb{M}}(R)$ (Eq. 14) satisfies conditions of Definition 7.*

Proof. Straightforward. Conditions of discernibility matrix are constructed basing on discernibility relation DIS and has similar behaviour as $Disc$. Compare Proposition 2.

The above simple facts will be important while considering examples of $\mathbb{C}^{(Disc,\varepsilon)}$-reducts and $\mathbb{C}^{(\gamma,\varepsilon)}$-reducts. Let us refer to [191,291,355] for more examples of quality measures that can be utilized to specify $\mathbb{C}^{(Q,\varepsilon)}$-reducts.

Yet another option to utilize Definition 6 to express various variations of reducts is to consider functions modeling a lack of information about the decision attribute, such as, e.g., conditional entropy $H(d|R) = H(R \cup \{d\}) - H(R)$. In the following specification, H plays the role of a penalty measure, which, with the given approximation threshold ε, corresponds to (H,ε)-approximate reducts introduced in [356].

$$\mathbb{C}^{(H,\varepsilon)}(R) = \begin{cases} 1 & H(d|R) \leq H(d|A) - \log_2(1-\varepsilon) \\ 0 & \text{otherwise} \end{cases} \tag{19}$$

Surely, as in the case of $\Rightarrow$ and $\mathbb{C}$, functions H, $Disc$, γ, etc., could be marked with additional subscript corresponding to specific $\mathbb{S}$, though we omit it for simplicity.

Proposition 4. *Criterion function $\mathbb{C}^{(H,\varepsilon)}$ satisfies conditions of Definition 5.*

Proof. Straightforward, like in the case of Proposition 2.

Let us finish this section with several remarks on computational aspects of deriving reducts from the data. In some cases, when $\mathbb{C}(\emptyset) = \mathbb{C}(A)$, reduct computation is trivial. This can happen if there is either no subset of attributes satisfying the given criterion ($\mathbb{C}(A) = 0$), or every subset does it ($\mathbb{C}(\emptyset) = 1$). Nevertheless, the problems of finding various types of minimal reducts are known to be NP-hard. For instance, let us consider the problem of finding minimal already-mentioned α-reducts [279], which are actually equivalent to $\mathbb{C}^{(Disc,\varepsilon)}$-reducts for $\alpha = 1 - \varepsilon$ (see Eq. (16)). As another example, let us mention NP-hardness of the problem of finding minimal (H,ε)-approximate reducts (or $\mathbb{C}^{(H,\varepsilon)}$-reducts using the terminology of Eq. (19)) proved in [356]. For further formulations of NP-hard problems related to the search of ε/α-related approximate reducts let us refer to [273,373]. We may also refer to a number of well-established search algorithms with this respect [186,310,323,419]. Accordingly, those well-known methods of finding reducts can be reconsidered for the purposes of $\mathbb{C}$-reducts as well. This interpretation can be also compared to other search strategies applied in the area of feature selection [83].

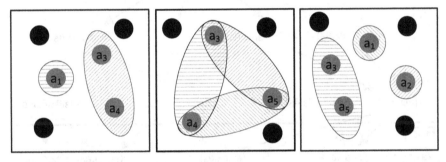

Fig. 3. Three examples of r-reducts. Points represent attributes from Table 2, ovals are grouping attributes into standard reducts, the set of all labelled attributes included in reducts on each figure forms an r-reduct. The leftmost and the middle present examples of 1-reducts, since after removal of any attribute, the remaining attributes still form a superreduct. Analgously, the rightmost figure presents 2-reducts.

3.2 r-$\mathbb{C}$-reducts

In applications such as threat detection or recommendation systems, classification models often have to work on incomplete data. Most of the studies on feature extraction focus on deriving useful and understandable parameters (variables, attributes, features) in order to achieve possibly simplest, yet accurate models [197]. However, almost none of the available methods takes into account that the data may be lost or temporarily unavailable for the analysis. In this regard, let us recall r-reducts [137] – one of the approaches to resilient feature selection that extends the concept of reduct to enable the governance of the redundancy level and, hence, to improve the resilience of the analysis. In Fig. 3, a graphical interpretation of r-reducts is shown.

Definition 8 (r-reduct). *Let* $\mathbb{S} = (U, A \cup \{d\})$ *be a decision table. Subset* $\breve{R} \subseteq A$ *is called an r-superreduct, if and only if, after removing any $1 \leq n \leq r$ attributes $a_1, \ldots, a_n$ from $\breve{R}$, the remaining $R = \breve{R} \setminus \{a_1, \ldots, a_n\}$ is a superreduct. We say that $\breve{R}$ is an r-reduct, if and only if it is an r-superreduct and there is no proper subset $\breve{R}' \subsetneq \breve{R}$, which is an r-superreduct.*

To emphasize the meaning of resilience let us consider the following scenario. Let us assume, for simplicity, that for each attribute in $R \subseteq A$, the probability that it is missing in the data during application of a prediction model is independent and equal to $p \in (0, \frac{1}{q * |A|})$, where $q > 1$ (q may refer to, e.g., the quality or price of utilized sensors). Then, for a standard version of (approximate) reducts the risk that the expected quality measure will not be satisfied is equal to p, while for r-reducts it is p^{r+1}.

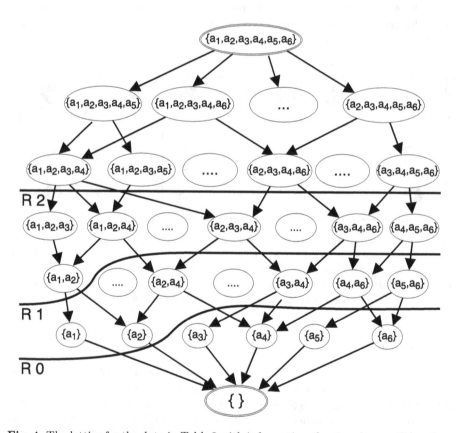

Fig. 4. The lattice for the data in Table 2 with information about various resilience levels guaranteed by r-reducts. The line marked as R0 visualizes the border above which attribute subsets discern all objects in Table 2 with respect to d. The R1 line corresponds to the border above which each subset is a 1-$\mathbb{C}$-superreduct. That is, after the removal of any attribute the remaining attributes hold enough information to discern all the objects in Table 2. The R2 line visualizes subsets that are 2-$\mathbb{C}$-superreducts.

To give a better understanding of r-reducts let us present all the subsets of attributes from the decision table in Table 2 as a lattice – starting with the empty set $\emptyset$ and ending with the full attribute set $A = \{a_1, a_2, a_3, a_4, a_5, a_6\}$ – see Fig. 4. We may imagine that some subsets of A have special properties that are retained by supersets – an example of such a property is discernibility of objects in a decision table $\mathbb{S} = (U, A \cup \{d\})$. Subsets: $\{a_1\}, \{a_2\}, \{a_3, a_4\},$ $\{a_4, a_6\}, \{a_5, a_6\}$ in Fig. 4 correspond to reducts (compare Table 2) and the line marked as R0 corresponds to the border above which attribute subsets discern all the objects in the decision table.

Subsets $\{a_1, a_2\}, \{a_2, a_3, a_4\}, \{a_3, a_4, a_6\}, \{a_4, a_5, a_6\}$ presented in Fig. 4 are 1-reducts. We may notice, that removal of any attribute from those sets guaranty discernibility of all objects, however if we remove two attributes it is not guaranteed – e.g., removal of $\{a_4, a_5\}$ from $\{a_4, a_5, a_6\}$. Thus, the line marked as R1 corresponds to the border above which set guarantee that after the removal any attribute the remaining ones discern all the objects. Similarly, the line marked as R2 corresponds to 2-reducts.

Using the criterion functions we may express the notion of, both, r-reducts and approximate r-reducts. Having defined the criterion function $\mathbb{C}$ for approximate reducts, e.g., $\mathbb{C}^{(Q, \varepsilon)}$ in Eq. (16), we may define a resilient criterion function r-$\mathbb{C}^{(Q, \varepsilon)}$ as shown in Eq. (20). Approximate r-reducts may be defined exactly the same way as presented in Definition 9. Therefore, in order to provide background for further discussions, let us reformulate r-reducts with the notion of criterion function. First, let us define the resilient version of criterion function r-$\mathbb{C} : \mathcal{P}(A) \rightarrow \{0, 1\}$ as:

$$r\text{-}\mathbb{C}(R) = \begin{cases} 1 & \text{if } \min_{R' \subseteq R : |R'| \geq max(|R|-r, 0)} \mathbb{C}(R') = 1 \\ 0 & \text{otherwise} \end{cases} \quad (20)$$

Given the above, the resilient criterion reduct (r-$\mathbb{C}$-reduct) formulation is straightforward:

Definition 9 (r-$\mathbb{C}$-reduct). *Let $\mathbb{S} = (U, A \cup \{d\})$, $\mathbb{C}$ and the expected resilience level r be given. A subset of attributes $\check{R}$ is called an r-$\mathbb{C}$-superreduct, if and only if r-$\mathbb{C}(\check{R}) = 1$. We say that $\check{R}$ is an r-$\mathbb{C}$-reduct, if and only if it is an r-$\mathbb{C}$-superreduct and there is no proper subset $\check{R}' \subsetneq \check{R}$, which is an r-$\mathbb{C}$-superreduct.*

Below we elaborate on some interesting properties of r-$\mathbb{C}$-reducts. Before doing this, let us just mention that in some special cases there is no risk of losing information, e.g., when $R = \emptyset$ is a reduct. Then we assume that $\check{R} = \emptyset$ is an r-reduct for any r.

Proposition 5. *For every non-empty r-$\mathbb{C}$-reduct $\check{R}$ there exist at least $r + 1$ reducts R for which $\check{R} \cap R = R \wedge \check{R} \cup R = \check{R}$ (which means that r-$\mathbb{C}$-reduct may be expressed as a union of at least $r + 1$ $\mathbb{C}$-reducts).*

Proof. Let $\breve{R}$ be an r-$\mathbb{C}$-reduct in $\mathbb{S} = (U, A \cup \{d\})$. We put $\mathcal{R} = \{R \subseteq \breve{R} \mid R$ is a $\mathbb{C}$-reduct$\}$. Let $|\mathcal{R}| = k$ and $k \leq r$. Consider a set $X = \{a_1, \ldots, a_k\}$, such that for each $\mathbb{C}$-reduct $R_i \in \mathcal{R}$ there is $X \cap R_i \neq \emptyset$. Let us remove attributes $a_1, \ldots, a_k$ from $\breve{R}$. For the remaining set $R' = \breve{R} \setminus X$, for each $\mathbb{C}$-reduct $R_i \in \mathcal{R}$, there is $R' \not\supseteq R_i$. So, for every $R \subseteq R'$, R is not a $\mathbb{C}$-reduct. Hence, $R' = \breve{R} \setminus \{a_1, \ldots, a_k\}$ is not a $\mathbb{C}$-superreduct, whereas should be because $k \leq r$ and $\breve{R}$ is an r-$\mathbb{C}$-reduct. Contradiction.

Proposition 6. *If in a given decision table $\mathbb{S}$ there exists a non-empty r-$\mathbb{C}$-reduct $\breve{R}$, for $r > 0$, then for each $a \in \breve{R}$ there exists $\mathbb{C}$-reduct R such that $a \in R$ and $R \subsetneq \breve{R}$.*

Proof. Let $\breve{R}$ be an r-$\mathbb{C}$-reduct and $\breve{R} = \breve{R}' \cup \{b\}$, where b is such that for each $R \subseteq \breve{R}$, if R is a $\mathbb{C}$-reduct, then $b \notin R$. For any r attributes $a_1, \ldots, a_r$ that satisfy $\{a_1, \ldots, a_r\} \cap \{b\} = \emptyset$, if we remove $\{a_1, \ldots, a_r\}$ from $\breve{R}$, then $R' = \breve{R}' \setminus \{a_1, \ldots, a_r\} \cup \{b\}$ meets the $\mathbb{C}$-superreduct condition. However, we know that b does not contribute to any $\mathbb{C}$-reduct. Hence, b is superfluous in R' because $R'' = R' \setminus \{b\}$ also meets the $\mathbb{C}$-superreduct condition. So, $\breve{R}' = \breve{R} \setminus \{b\}$ is also an r-$\mathbb{C}$-reduct and $|\breve{R}'| < |\breve{R}|$. Contradiction.

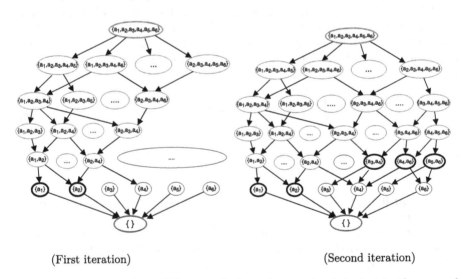

(First iteration)   (Second iteration)

Fig. 5. The interpretation of 'downward closure' property of Apriori in the case of reduct computation. The bold ovals correspond to reducts, meanwhile the dashed ovals correspond to sets that will not be explored, since in every iteration we remove all so-far-found superreducts from the F_k.

Remark 1 *(Zero redundancy).* A 0-$\mathbb{C}$-reduct R is a $\mathbb{C}$-reduct.

Remark 2 *(Redundant attributes removal).* After the removal of any attribute a from a non-empty r-$\mathbb{C}$-reduct $\breve{R}$, the remaining set $\breve{R}' = \breve{R} \setminus \{a\}$ satisfies the following:

1. $\check{R}'$ is a $(r-1)$-$\mathbb{C}$-superreduct
2. $\exists \check{R}^{(r\text{-}1)} \subseteq \check{R}'$, where $\check{R}^{(r\text{-}1)}$ is a $(r-1)$-$\mathbb{C}$-reduct.

3.3 Breadth First Search Algorithms

There are a lot of feature selection methods described in the literature (see Sect. 2.3), however it is hard to find those which would take into account not only the quality and relevance of selected attribute subsets but also their resilience to partial loss or lack of the data. Although there are some preliminary studies of algorithmic approaches that allow to construct resilient subsets of attributes basing on controlled redundancy in generated subsets [137], or relying on ensemble techniques [360], in general, it is not straightforward how to provide an expected resilience level to the feature selection.

In this section, we introduce a novel approach to perform resilient feature selection. It is inspired by the well-known Apriori algorithm that was adapted in many ways in both, RST-based [405] and non-RST-based [205] feature selection frameworks. The presented mechanism generates r-$\mathbb{C}$-reducts for a given implementation of a test function $test_C(R)$ for a criterion $\mathbb{C}$. The $test_C(R)$ allows us to verify whether a given subset of attributes R satisfies the examined criterion.

Apriori-Based Algorithm. Originally, the Apriori algorithm was supposed to discover association rules between items in a database of sales transactions. Given a set of transactions, the problem is to generate all association rules that have support and confidence no less than the user-specified thresholds (called *minsup* and *minconf*, respectively). Apriori is characterized as a level-wise complete search (breadth first search) algorithm using anti-monotonicity of itemsets: "If an itemset is not frequent, any of its superset is never frequent", which is also called the downward closure.

For resilient attribute subsets, the downward closure property refers to the monotonicity of $\mathbb{C}$ (cf. Definition 5) – that is, if the subset R of attributes satisfies function $\mathbb{C}$, then every superset $\bar{R} \supseteq R$ does it too. Hence, we do not need to explore supersets of R. Moreover, for the optimization problems considered in this paper, it is enough to find minimal sets satisfying $\mathbb{C}$, so that the algorithm could stop (see Fig. 5). Going further, in the case of resilient attribute subsets, we know that each r-$\mathbb{C}$-reduct may be reached only by adding an attribute to an $(r-1)$-$\mathbb{C}$-superreduct (see Remark 2).

The resilient version – r-$apriori\_gen(F_{k-1})$ – of the original *apriori_gen* procedure, takes as an argument F_{k-1} – the set of all frequent $(k-1)$-items (in our case, attribute subsets of size $k-1$), and returns a superset 'C_k' containing all frequent k-itemsets. First, in the '*from*' part of the SQL implementation below, F_{k-1} is joined with all attributes from A (A may be represented as a single-column table with attributes in rows). In the *'group by'* phase, the set is compacted and some additional meta-data is created, e.g., *is-1-Superreduct, ...,* *is-r-Superreduct* properties are generated – which means that candidate R has particular resilience level r. Actually, one can compare this kind of SQL-based approach with some other Apriori-style SQL implementations [335], as well as SQL-based RST-related calculations [405].

The candidate set $R \in C_k$ is *is-r-Superreduct*, if and only if all subsets $R' \subset R$ that $|R'| = |R| - 1$ are $r - 1$ resilient (see Remark 2). Moreover, we rely on monotonicity of $test_{\mathbb{C}}$ – if any subset of candidate $R' \subset R$ satisfies $test_{\mathbb{C}}(R') = 1$ then R as well – this way, in some cases, we may omit necessity to perform $test_{\mathbb{C}}$ calculations. In the *'having'* part of SQL, we discard candidates corresponding to subsets that were left out earlier (*'count'* is less than k, if and only if some of subsets of R are missing in F_{k-1}). The *sort()* function is responsible for sorting attributes according to, e.g., a lexicographical order. Below is the SQL implementation of the r-apriori_gen(F_{k-1}):

```
INSERT into C_k
select sort(p.item_1, p.item_2, ..., p.item_{k-1}) as candidate,
       max(testC) as testC, min(testC) as is-1-Superreduct,
       min(is-1-Superreduct) as is-2-Superreduct, ...,
       min(is-(r - 1)-Superreduct) as is-r-Superreduct, count(*)
from F_{k-1} p, A a
group by candidate
having count(*) = k
```

To better illustrate the proposed SQL implementation of r-apriori_gen, in Fig. 6, we present two iterations of the procedure on the limited set A containing three attributes $A = \{a_1, a_2, a_3\}$. In the preliminary iteration (left snippet in Fig. 6) it is necessary to apply $test_{\mathbb{C}}(a)$ for each attribute $a \in \{a_1, a_2, a_3\}$. The cost of such operation is $\mathcal{O}(|A|) \times \mathcal{O}(test_{\mathbb{C}})$. The result confirmed that attributes $\{a_1\}$ and $\{a_2\}$ satisfy $\mathbb{C}$ (let us call them $\mathbb{C}$-reducts), that is $test_{\mathbb{C}}(\{a_1\}) = test_{\mathbb{C}}(\{a_2\}) = 1$, however $\{a_3\}$ does not – $test_{\mathbb{C}}(\{a_3\}) = 0$. In the first iteration (right snippet in Fig. 6) there is no need to execute $test_{\mathbb{C}}$ at all, since all the sets: $\{a_1, a_2\}$, $\{a_1, a_3\}$, $\{a_2, a_3\}$ has direct connection in the lattice to at least one $\mathbb{C}$-reduct, hence all satisfy $test_{\mathbb{C}} = 1$ because of the monotonicity of $\mathbb{C}$ (in the presented SQL implementation of r-apriori_gen it is interpreted as *'max(testC) as testC'*). Moreover, we know that the set $\{a_1, a_2\}$ is 1-$\mathbb{C}$-reduct, since every edge down in the lattice ends in a $\mathbb{C}$-reduct (interpreted as *'min(testC) as is-1-Superreduct'* in SQL). This short discussion shows that bottom-up approach based on *apriori_gen* allowed to conclude information about given superset basing only on properties of its subset, without necessity to perform additional calculations.

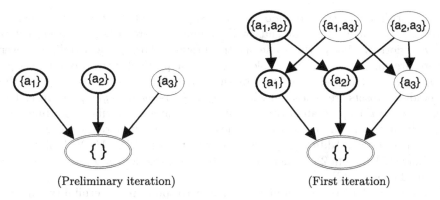

(Preliminary iteration)   (First iteration)

Fig. 6. The preliminary (left) and the first (right) iteration of the *r-apriori_gen* procedure.

Algorithm 2: *r*-apriori for resilient feature selection

Data: F_{k-1}; $\mathbb{S} = (U, A \cup \{d\})$; $Flag \in \{ALL_{\breve{R}}, MIN_{\breve{R}}\}$
Result: $\mathbb{R}$ – a set of all *r*-$\mathbb{C}$-reducts or all smallest *r*-$\mathbb{C}$-reducts

1   $k \leftarrow 1$; $r \leftarrow 0$; $F_k \leftarrow A$; $\mathbb{R} \leftarrow \emptyset$
2   **for** $k = 2; F_{k-1} \neq \emptyset; k++$ **do**
3   /* Generate candidates with *r-apriori_gen*   */
4      $C_k \leftarrow r\text{-}apriori\_gen(F_{k-1})$
5   **foreach** *candidate* $R \in C_k$ **do**
6   **if** $!R.test_C$ **then**
7          $R.test_C \leftarrow test_C(R)$
8   **end**
9   **end**
10     $\mathcal{R}_r \leftarrow \{R \in C_k | R \text{ is } r\text{-}\mathbb{C}\text{-superreduct}\}$
11     $\mathbb{R} \leftarrow \mathbb{R} \cup \mathcal{R}_r$
12   **if** $Flag = MIN_{\breve{R}} \ AND \ |\mathbb{R}| > 0$ **then**
13   break;
14   **end**
15     $F_k \leftarrow C_k \setminus \mathcal{R}_r$
16 **end**
17 *return* $\mathbb{R}$

Table 3. Summary of experiments for *r*-Apriori.

k	$\mathcal{R}_r$	$\mathbb{R}$	Comments
1	$\emptyset$	$\emptyset$	Preliminary step; $k = 1$
2	$\{a_1, a_2\}, \{a_1, a_6\}, \{a_2, a_6\}$	$\{a_1, a_2\}, \{a_1, a_6\}, \{a_2, a_6\}$	minimal $1\text{-}\mathbb{C}^{(H,\ 0.2)}$-reducts
3	$\{a_1, a_4, a_5\}, \{a_2, a_4, a_5\}, \{a_3, a_4, a_5\}, \{a_3, a_4, a_6\}$	$\{a_1, a_2\}, \{a_1, a_6\}, \{a_2, a_6\}, \{a_1, a_4, a_5\}, \{a_2, a_4, a_5\}, \{a_3, a_4, a_5\}, \{a_3, a_4, a_6\}$	all $1\text{-}\mathbb{C}^{(H,\ 0.2)}$-reducts

Algorithm 2 presents pseudo-code for r-apriori that, for the given r, generates all r-$\mathbb{C}$-superreducts or ends with minimal r-$\mathbb{C}$-reducts. The overall flow of r-apriori is almost the same as the original Apriori algorithm, however there are differences in the implementation of particular functions like, e.g., *r-apriori_gen*. In every iteration of the outer *'for'* loop in Algorithm 2, the candidate subsets are generated with the resilient *r-apriori_gen* procedure. The inner *'foreach'* loop iterates over generated candidates and verifies $test_{\mathbb{C}}$ function. F_k is built on candidate set without those sets, which are already recognized as r-$\mathbb{C}$-superreducts. Additionally, there are two flags $\{ALL_{\breve{R}}, MIN_{\breve{R}}\}$ that allow to control Algorithm 2 in order to generate all r-$\mathbb{C}$-reducts in $\mathbb{S}$ ($ALL_{\breve{R}}$), or just minimal ones ($MIN_{\breve{R}}$).

For a standard reduct problem, $\mathbb{C}$ may correspond to discernibility whereas $test_C(R)$ may correspond to function *isReduct* [405]. As other examples, $test_C$ may be implemented as correlation with a decision attribute, a constraint for conditional entropy (19), the RST-related function γ (15), and many others [141].

Table 4. The course of experiments for *r-apriori_gen*.

k	C_k	
1	!testC	$\{a_3\}, \{a_4\}, \{a_5\}$
	testC	$\{a_1\}, \{a_2\}, \{a_6\}$
	is-1-Superreduct	$\emptyset$
	is-2-Superreduct	$\emptyset$
2	!testC	$\emptyset$
	testC	$\{a_1, a_3\}, \{a_1, a_4\}, \{a_1, a_5\}, \{a_2, a_3\}, \{a_2, a_4\}, \{a_2, a_5\},$ $\{a_3, a_6\}, \{a_4, a_6\}, \{a_5, a_6\}, \{a_3, a_4\}, \{a_3, a_5\}, \{a_4, a_5\}$
	is-1-Superreduct	$\{a_1, a_2\}, \{a_1, a_6\}, \{a_2, a_6\}$
	is-2-Superreduct	$\emptyset$
3	!testC	$\emptyset$
	testC	$\emptyset$
	is-1-Superreduct	$\{a_1, a_4, a_5\}, \{a_2, a_4, a_5\}, \{a_3, a_4, a_5\}, \{a_3, a_4, a_6\}$
	is-2-Superreduct	$\{a_1, a_2, a_6\}$

Algorithm Working Example. To confirm that Algorithm 2 generates all (smallest) r-$\mathbb{C}$-reducts (depending on the control flag) let us perform illustrative experiments for the data in Table 2. Let us refer to the conditional entropy H as the penalty measure and consider the criterion function $\mathbb{C}^{(H,\ 0.2)}$. The resilient version of $\mathbb{C}^{(H,\ 0.2)}$ may be constructed for a given r as described in Eq. (20). Let us now present a concise experiment for the data in Table 2, aimed at finding all/the smallest 1-$\mathbb{C}^{(H,\ 0.2)}$-reducts.

In Table 3, we summarize the states after every iteration of *'for loop'* in Algorithm 2. In Table 4, each row corresponds to the call of the *r-apriori_gen* procedure, whereby the generated attribute subsets are assigned to one of four groups: *!testC*, *testC*, *is-1-Superreduct* and *is-2-Superreduct*. Note that for $k = 2$ (the second iteration of *r-apriori_gen*) minimal 1-$\mathbb{C}^{(H,\ 0.2)}$-reducts were found, with candidate sets $\{a_1, a_2\}$, $\{a_1, a_6\}$, $\{a_2, a_6\}$ assigned to *is-1-Superreduct* group. Thus, in the case of '*Flag*' set to $MIN_{\check{R}}$, the algorithm would stop its execution (compare the if condition in line 12).

3.4   Computational Complexity Study

A natural question arises whether the most meaningful, particularly minimal r-$\mathbb{C}$-reducts can be derived from the data in a more efficient way than by using the aforementioned breath first search techniques. Intuitively, having in mind the already-published NP-hardness results corresponding to minimal r-reducts [137] – somewhat similar to (multi)set multicover problems studied, e.g., in [169] – we should not expect the existence of fast deterministic algorithms with this respect. Still, one might think that this kind of complexity could depend on the choice of criterion function, i.e., although the problem of finding minimal r-reducts (Definition 8) is known to be NP-hard, the analogous problems of finding minimal r-$\mathbb{C}$-reducts (Definition 9) could be computationally more feasible at least for some functions $\mathbb{C}$.

In this section, we prove that every NP-hard attribute reduction problem P that may be expressed via an appropriately defined criterion function $\mathbb{C}^P$ remains NP-hard even for its resilient version $\check{P}$, where r refers to the resilience level and means that any r attributes of the examined set R may be unavailable without any impact on the criterion $\mathbb{C}^P$. The presented NP-hardness proof mechanism works for any functions $\mathbb{C}$ that meet the requirements of Definition 5. On the one hand, one may say that it overlaps with other complexity studies. For instance, let us refer to NP-hardness of partial multi-cover problems [317], which might be used as a prerequisite for proving NP-hardness of a resilient version of the aforementioned minimal α-reduct problem [279]. On the other hand, our theoretical result is broader as it allows us to deal with a far wider family of formulations of attribute reduction problems [355,373].

A-Attributes. For further needs, let us consider a family of artificial attributes, so-called *A*-attributes, denoted as *#attr*. The values of *#attr* are constructed, for each $u \in U$, as concatenations of all values $a(u)$, $a \in A^{10}$. Polynomial reduction presented later in Subsect. 3.4 is based on the following properties of *A*-attributes.

Lemma 1 (*A*-attribute *#attr*). *For a given decision table* $\mathbb{S} = (U, A \cup \{d\})$ *and* $\mathbb{C} : \mathcal{P}(A) \to \{0,1\}$ *we may generate an arbitrary number of A-attributes #attr such that:*

1. *For any* $n \in \mathbb{N}$, *for all* i, j *such that* $1 \leq i, j \leq n$, *there is* $\mathbb{C}(\{\#attr_i\}) = \mathbb{C}(\{\#attr_i, \#attr_{i+1}, \ldots, \#attr_j\}) = \mathbb{C}(A) = \mathbb{C}(A \cup \{\#attr_i, \#attr_{i+1}, \ldots, \#attr_j\})$.

---

[10] If attribute domains are overlapping, i.e., there exist $a_i, a_j \in A$ for which $V_{a_i} \cap V_{a_j} \neq \emptyset$, then concatenation may include a delimiter $|_A$ such that for each $a \in A$ we have $|_A \notin V_a$.

2. *Singleton sets $\{\#attr_i\}$ are the smallest non-empty subsets of attributes satisfying $\mathbb{C}$ in the extended decision table $\mathbb{S}' = (U, A \cup \{\#attr_{1 \leq i \leq n}\} \cup \{d\})$.*

Proof. Item 1: Since every $\#attr$ is generated as concatenation of all attribute values, there are the following functional dependencies ($\Rightarrow$) between A and each $\#attr$: For any $n \in \mathbb{N}$, for every i, j, such that $1 \leq i \leq n$, there is $A \Rightarrow \#attr_i \wedge \#attr_i \Rightarrow A$. So, $\mathbb{C}(\{\#attr\}) \leq \mathbb{C}(A) \wedge \mathbb{C}(\{\#attr\}) \geq \mathbb{C}(A)$. Thus, given $1 \leq i, j \leq n$, $\mathbb{C}(\{\#attr\}) = \mathbb{C}(\{\#attr_i, \#attr_{i+1}, \ldots, \#attr_j\}) = \mathbb{C}(A) = \mathbb{C}(A \cup \{\#attr_i, \#attr_{i+1}, \ldots, \#attr_j\})$.

Item 2: A singleton set $\{\#attr\}$ satisfies $\mathbb{C}$ (that is: $\mathbb{C}(\{\#attr\}) = 1$) and $\{\#attr\}$ is a single attribute, hence it is the smallest one – what ends the proof.

Let us strengthen the meaning of Lemma 1 with the following remarks:

Lemma 2 (Only one $\mathbb{C}$-reduct contains $\#attr$). *If $\mathbb{C}(\emptyset) \neq \mathbb{C}(A)$, then, for each i, $R = \{\#attr_i\}$ is the smallest $\mathbb{C}$-superreduct in decision table $\mathbb{S}' = (U, A \cup \{\#attr_{1 \leq i \leq n}\} \cup \{d\})$ and, in particular, there are no other $\mathbb{C}$-reducts containing $\#attr_i$.*

Proof. Straightforward. In particular, we know that $\{\#attr_i\}$ is a $\mathbb{C}$-reduct. Therefore, for any set $R = \{\#attr_i, a\}$, a is dispensable.

Lemma 3 ($\#ATTRs$ forms the smallest r-$\mathbb{C}$-reduct). *If $\mathbb{C}(\emptyset) \neq \mathbb{C}(A)$, then the set $\#ATTRs = \{\#attr_1, \ldots, \#attr_r, \#attr_{r+1}\}$ is the smallest set of attributes that satisfies r-$\mathbb{C}$ in $\mathbb{S}' = (U, A \cup \{\#attr_{1 \leq i \leq n}\} \cup \{d\})$.*

Proof. We need to show the two following things. First, $\#ATTRs$ is an r-$\mathbb{C}$-reduct. Second, it is the smallest one.

1. After removal of any r elements from $\#ATTRs$ there is still one $\#attr$ attribute left. From Lemma 1, we know that such attribute satisfies $\mathbb{C}$. Hence, $\#ATTRs$ is the r-$\mathbb{C}$-superreduct.
2. Assume that there is $\check{R}'$ that satisfies r-$\mathbb{C}$ and $|\check{R}'| \leq r < r+1 = |\#ATTRs|$. If so, after removal of r attributes from $\check{R}'$ there is no attribute. Since $\mathbb{C}(\emptyset) \neq \mathbb{C}(A)$, i.e., $\mathbb{C}(\emptyset) = 0$ and $\mathbb{C}(A) = 1$, $\check{R}'$ is not an r-$\mathbb{C}$-reduct. Contradiction.

Lemma 4 (Reducts and $\#ATTRs$). *Let R be a non-empty $\mathbb{C}$-reduct in $\mathbb{S}' = (U, A \cup \{\#attr_{1 \leq i \leq n}\} \cup \{d\})$, such that $R \cap \{\#attr_1, \ldots, \#attr_r\} = \emptyset$. Then, the set $\check{R} = R \cup \{\#attr_1, \ldots, \#attr_r\}$ is an r-$\mathbb{C}$-reduct.*

Proof. From Lemma 3 we know that $\{\#attr_1, \ldots, \#attr_r\}$ is the smallest $(r-1)$-$\mathbb{C}$-reduct. From Lemma 2 we know that for each $\mathbb{C}$-reduct $R \subset A$ and $\#attr$, if $R \neq \{\#attr\}$, then $R \cap \{\#attr\} = \emptyset$. To show that $\check{R} = R \cup \{\#attr_1, \ldots, \#attr_r\}$ is an r-$\mathbb{C}$-reduct, we need to prove that for any $R' \subset \check{R}$ such that $|R'| \leq r$ the remaining set $\check{R} \setminus R'$ is a $\mathbb{C}$-superreduct, thus it satisfies the condition $\mathbb{C}(\check{R} \setminus R') = 1$.

There are two cases to be considered. First, $R' = \{\#attr_1, \ldots, \#attr_r\}$. In that case the remaining set after $\check{R} \setminus R'$ is R. Hence, it is a $\mathbb{C}$-reduct. Second, $R' \neq \{\#attr_1, \ldots, \#attr_r\}$. In that case the remaining set $\check{R} \setminus R'$ contains at least one $\#attr$ attribute. So, from Lemma 1 it satisfies $\mathbb{C}(\check{R} \setminus R') = 1$. Thus, it is a $\mathbb{C}$-superreduct.

Let us continue with the study of the impact of $\#attr$ attributes on the properties of $\mathbb{S}$ and $\mathbb{C}$, with an emphasis on the extended data representation $\mathbb{S}'$. In order to make a proper distinction between $\mathbb{S}$ and $\mathbb{S}'$, we go back to the aforementioned explicit data-specific notation $\mathbb{C}_\mathbb{S}$ and $\mathbb{C}_{\mathbb{S}'}$, respectively.

Lemma 5 (Impact of $\#attr$ on $\mathbb{S}$ and $\mathbb{C}$). *Let $\mathbb{S} = (U, A \cup \{d\})$ be a decision table, $\mathbb{C} : \mathcal{P}(A) \rightarrow \{0,1\}$ be a given criterion. Let $\mathbb{S}' = (U, A \cup \{\#attr_1, \ldots, \#attr_r\} \cup \{d\})$ be an extended data representation. Then, the following properties hold:*

1. *$\forall_{R \subseteq A}$ if $\mathbb{C}_\mathbb{S}(R) = 0$ then $\mathbb{C}_{\mathbb{S}'}(R) = 0$;*
2. *$\forall_{R \subseteq A}$ if $\mathbb{C}_\mathbb{S}(R) = 1$ then $\mathbb{C}_{\mathbb{S}'}(R) = 1$;*
3. *$\forall_{R \subseteq A} \; \mathbb{C}_\mathbb{S}(R) = \mathbb{C}_{\mathbb{S}'}(R)$*
4. *If $\forall_{R' \subseteq A \cup \{\#attr_1, \ldots, \#attr_r\}} \; \mathbb{C}_{\mathbb{S}'}(R') = 0$ then $\forall_{R \subseteq A} \mathbb{C}_\mathbb{S}(R) = 0$;*
5. *If $\exists_{R' \subseteq A \cup \{\#attr_1, \ldots, \#attr_r\}} \; \mathbb{C}_{\mathbb{S}'}(R') = 1$ then $\exists_{R \subseteq A} \mathbb{C}_{\mathbb{S}'}(R) = \mathbb{C}_\mathbb{S}(R) = 1$.*

Proof. Item 1: If $\mathbb{C}_\mathbb{S}(A) = 0$ then from Lemma 1 $\mathbb{C}_\mathbb{S}(A) = \mathbb{C}_{\mathbb{S}'}(A \cap \{\#attr_1, \ldots, \#attr_r\}) = 0$. So, from monotonicity $\mathbb{C}_{\mathbb{S}'}(R) = 0$. If $\mathbb{C}_\mathbb{S}(A) = 1$ then $\mathbb{C}_\mathbb{S}(R) < \mathbb{C}_\mathbb{S}(A)$. From Lemma 1 we have $\mathbb{C}_\mathbb{S}(A) = \mathbb{C}_{\mathbb{S}'}(A \cap \{\#attr_1, \ldots, \#attr_r\})$. So, $\mathbb{C}_{\mathbb{S}'}(R) < \mathbb{C}_{\mathbb{S}'}(A \cap \{\#attr_1, \ldots, \#attr_r\})$.

Item 2: Directly from Lemma 1.

Item 3: Directly from items (1) and (2).

Item 4: By contradiction. We have that for each $R' \subseteq A \cup \{\#attr_1, \ldots, \#attr_r\}$ there is $\mathbb{C}_{\mathbb{S}'}(R') = 0$. Let $R \subseteq A$ and $\mathbb{C}_\mathbb{S}(R) = 1$. Then, directly from (2), $\mathbb{C}_{\mathbb{S}'}(R) = 1$.

Item 5: If $R \cap \{\#attr_1, \ldots, \#attr_r\} = \emptyset$ then $R' \subseteq A$. Otherwise, if $R \cap \{\#attr_1, \ldots, \#attr_r\} \neq \emptyset$ then from Lemma 1, we have $\mathbb{C}_\mathbb{S}(A) = \mathbb{C}_{\mathbb{S}'}(A) = \mathbb{C}_{\mathbb{S}'}(\{\#attr_1, \ldots, \#attr_r\}) = \mathbb{C}_{\mathbb{S}'}(A \cup \{\#attr_1, \ldots \#attr_r\}) = 1$. Thus, A is the solution, i.e., $\mathbb{C}_\mathbb{S}(A) = \mathbb{C}_{\mathbb{S}'}(A) = 1$.

Resilient NP-Hardness. In this section, we concentrate on showing that every NP-hard attribute reduction problem P expressed by means of criterion function $\mathbb{C}^P$ remains NP-hard also for its resilient variant $\check{P}$ expressed by r-$\mathbb{C}^P$.

Theorem 1 *(NP-hardness of minimal r-$\mathbb{C}^P$). Let P be a problem of finding the minimal set R satisfying condition expressed via a criterion $\mathbb{C}^P : \mathcal{P}(A) \rightarrow \{0,1\}$. If P is NP-hard, then finding minimal set $\check{R}$ satisfying r-$\mathbb{C}^P$ (see Eq. (20)) is also NP-hard.*

To prove Theorem 1, we will show that P can be polynomially reduced to $\check{P}$. The reduction is as follows. Given a problem input, i.e., a decision table

$\mathbb{S} = (U, A \cup \{d\})$, the reduction comes to creation of a new data representation $\mathbb{S}'$ that contains additional r conditional A-attributes, where r corresponds to the expected resilience level. Obviously, the whole reduction is polynomial:

1. Given the original data representation $\mathbb{S} = (U, A \cup \{d\})$ and integer r
2. Add r #attr attributes and, this way, create the extended data representation $\mathbb{S}' = (U, A \cup \{\#attr_1, \ldots, \#attr_r\} \cup \{d\})$
3. Solve the problem $\check{P}$ defined via $r\text{-}\mathbb{C}^P$ (20)
4. Extract the solution of P as described below.

To show that the reduction is applicable, we need to show two things. First, if there is a solution of P, then there must be a solution of $\check{P}$ ($P \rightarrow \check{P}$). Second, if there is a solution of $\check{P}$, then there must be a solution of P ($\check{P} \rightarrow P$).

In order to distinguish applications of functions $\mathbb{C}^P$ and $r\text{-}\mathbb{C}^P$ on the original decision table $\mathbb{S}$ and the extended data representation $\mathbb{S}'$, we use the following notation: $\mathbb{C}_{\mathbb{S}}$, $r\text{-}\mathbb{C}_{\mathbb{S}}$, $\mathbb{C}_{\mathbb{S}'}$ and $r\text{-}\mathbb{C}_{\mathbb{S}'}$, respectively.

Proof ($P \rightarrow \check{P}$). First, let us discuss the boundary condition $\mathbb{C}_{\mathbb{S}}(\emptyset) = \mathbb{C}_{\mathbb{S}}(A)$: a) If $\mathbb{C}_{\mathbb{S}}(A) = 0$ there is no solution of P in $\mathbb{S}$. So, from Lemma 1 we have $\mathbb{C}_{\mathbb{S}'}(A \cup \#ATTRs) = 0$. So, there is no solution of P in $\mathbb{S}'$. Thus, the Eq. (20) is never met. Hence, there is no solution of $\check{P}$ in $\mathbb{S}'$. b) If $\mathbb{C}_{\mathbb{S}}(\emptyset) = 1$ everything is a solution of P and $\mathbb{C}_{\mathbb{S}'}(\emptyset) = 1$ – directly from Lemma 5. The Eq. (20) is always met so $r\text{-}\mathbb{C}_{\mathbb{S}'}(\emptyset) = 1$. Hence, everything is a solution of $\check{P}$ in $\mathbb{S}'$. Below, we consider the more complex case when $\mathbb{C}_{\mathbb{S}}(\emptyset) \neq \mathbb{C}_{\mathbb{S}}(A)$:

Let $R, |R| = k$, be a solution of P, in particular, $\mathbb{C}_{\mathbb{S}}(R) = 1$. Let $\#ATTRs = \{\#attr_1, \ldots, \#attr_r\}$ be a set of r #attr attributes $|\#ATTRs| = r$. Let $\check{R} = R \cup \#ATTRs$. Directly from Lemma 4, we have $r\text{-}\mathbb{C}_{\mathbb{S}'}(R \cup \#ATTRs) = 1$.

Now, we need to show that $\check{R} = R \cup \#ATTRs$ of size $|\check{R}| = k + r$ is minimal. Assume that there exists $\check{R}' \subseteq A \cup \{\#attr_1, \ldots, \#attr_r\}$ such that $r\text{-}\mathbb{C}_{\mathbb{S}'}(\check{R}') = 1$ and $|\check{R}'| < |\check{R}|$. From Eq. (20), we know that after removing any r attributes from $\check{R}'$ we have R' of size $l < k$ ($|R'| < |R|$) that satisfies $\mathbb{C}_{\mathbb{S}'}$ (in $\mathbb{S}'$). From Lemma 5, we know that R' satisfies also $\mathbb{C}_{\mathbb{S}}$ (in $\mathbb{S}$). Whereas, R of size k is a solution of P for $\mathbb{S}$. Contradiction.

In order to prove $\check{P} \rightarrow P$, let us introduce an auxiliary lemma for $\#ATTRs$:

Lemma 6 (#ATTRs and $\check{P}$). *If there is a non-empty solution $\check{R}$ to $\check{P}$ in $\mathbb{S}' = (U, A \cup \{\#attr_1, \ldots, \#attr_r\} \cup \{d\})$, then there exists a solution $\check{R}'$ to $\check{P}$ in $\mathbb{S}'$ that satisfies:*

1. $|\check{R}'| = |\check{R}|$
2. $\{\#attr_1, \ldots, \#attr_r\} \subseteq \check{R}'$

As a proof, we present a constructive algorithm that transforms a given solution $\check{R}$ to the problem $\check{P}$ into $\check{R}'$, whereby $|\check{R}'| = |\check{R}|$ and $\{\#attr_1, \ldots, \#attr_r\} \subseteq \check{R}'$. Let $\check{R}$ be a solution to $\check{P}$ in $\mathbb{S}' = (U, A \cup \{\#attr_1, \ldots, \#attr_r\} \cup \{d\})$.

In the first step, we remove all $\#attr$ attributes from $\breve{R}$, where $|\breve{R} \cap \{\#attr_1, \ldots, \#attr_r\}| = m$ and $0 \leq m \leq r$. Next, we remove any $r - m$ other attributes. The remaining set R of size $|\breve{R}| - r$ is satisfying $\mathbb{C}_{\mathbb{S}'}(R) = 1$ (see Eq. (20)).

In the second step, we create a solution $\breve{R}' = R \cup \{\#attr_1, \ldots, \#attr_r\}$. We know that the solution $\breve{R}'$ constructed this way satisfies $r\text{-}\mathbb{C}_{\mathbb{S}'}(\breve{R}') = 1$ and is minimal because $|\breve{R}'| = |\breve{R}|$. The complexity of the above algorithm is obviously polynomial.

Proof ($\breve{P} \rightarrow P$). First, let us discuss the case of $\mathbb{C}_{\mathbb{S}'}(\emptyset) = \mathbb{C}_{\mathbb{S}'}(A)$: a) If $r\text{-}\mathbb{C}_{\mathbb{S}'}(A \cup \{\#attr_1, \ldots, \#attr_r\}) = 0$ there is no solution of $\breve{P}$ in $\mathbb{S}'$. Hence, from Eq. (20), for any $R \subseteq A \cup \{\#attr_1, \ldots, \#attr_r\}$, if $|R| \geq r$, then $\mathbb{C}_{\mathbb{S}'}(A \cup \{\#attr_1, \ldots, \#attr_r\} \setminus R) = 0$. Thus, for $R = \{\#attr_1, \ldots, \#attr_r\}$ we have $\mathbb{C}_{\mathbb{S}'}(A) = 0$. So, from monotonicity of $\mathbb{C}$, we have that for any $R \subseteq A$ there is $\mathbb{C}_{\mathbb{S}}(A) = 0$. b) If $r\text{-}\mathbb{C}_{\mathbb{S}'}(\emptyset) = 1$ then from Eq. (20) $\mathbb{C}_{\mathbb{S}'}(\emptyset) = 1$. Hence, from Lemma 5 $\mathbb{C}_{\mathbb{S}}(\emptyset) = 1$.

Now, we consider the case of $\mathbb{C}_{\mathbb{S}}(\emptyset) \neq \mathbb{C}_{\mathbb{S}}(A)$. Let $\breve{R}$ be a solution to $\breve{P}$ in $\mathbb{S}' = (U, A \cup \{\#attr_1, \ldots, \#attr_r\} \cup \{d\})$. Without loosing generality, we may assume that $\{\#attr_1, \ldots, \#attr_r\} \subseteq \breve{R}$ (see Lemma 6). Now, we must show that $R = \breve{R} \setminus \{\#attr_1, \ldots, \#attr_r\}$ is a solution of P in $\mathbb{S} = (U, A \cup \{d\})$.

Table 5. An exemplary decision table extended with two A-attributes.

$U \backslash {}^A$	a_1	a_2	a_3	a_4	a_5	a_6	$\#attr_1$	$\#attr_2$	d
u_1	f	'a'	$\triangledown$	$\circ$	$\square$	'x'	f'a'$\triangledown\circ\square$'x'	f'a'$\triangledown\circ\square$'x'	*good*
u_2	f	'b'	$\triangledown$	$\odot$	$\bullet$	'x'	f'b'$\triangledown\odot\bullet$'x'	f'b'$\triangledown\odot\bullet$'x'	*good*
u_3	f	'c'	$\triangle$	$\oslash$	$\diamond$	'x'	f'c'$\triangle\oslash\diamond$'x'	f'c'$\triangle\oslash\diamond$'x'	*good*
u_4	f	'd'	$\triangledown$	$\otimes$	$\triangleleft$	'x'	f'd'$\triangledown\otimes\triangleleft$'x'	f'd'$\triangledown\otimes\triangleleft$'x'	*good*
u_5	f	'e'	$\triangledown$	$\ominus$	$\star$	'y'	f'e'$\triangledown\ominus\star$'y'	f'e'$\triangledown\ominus\star$'y'	*good*
u_6	f	'f'	$\triangledown$	$\oslash$	$\triangleright$	'y'	f'f'$\triangledown\oslash\triangleright$'y'	f'f'$\triangledown\oslash\triangleright$'y'	*good*
u_7	t	'g'	$\triangle$	$\otimes$	$\square$	'y'	t'g'$\triangle\otimes\square$'y'	t'g'$\triangle\otimes\square$'y'	*bad*
u_8	t	'h'	$\triangle$	$\ominus$	$\bullet$	'z'	t'h'$\triangle\ominus\bullet$'z'	t'h'$\triangle\ominus\bullet$'z'	*bad*
u_9	t	'i'	$\triangledown$	$\oplus$	$\diamond$	'z'	t'i'$\triangledown\oplus\diamond$'z'	t'i'$\triangledown\oplus\diamond$'z'	*bad*

From Eq. (20), we know that after removing any r attributes from $\breve{R}$ the remaining set R satisfies $\mathbb{C}_{\mathbb{S}'}(R) = 1$. Hence, it is a solution to P in $\mathbb{S}'$. From Lemma 5, we know that $\mathbb{C}_{\mathbb{S}}(R) = \mathbb{C}_{\mathbb{S}}(A) = 1$. So R is a solution to P in $\mathbb{S}$.

The last thing is to show that $R = \breve{R} \setminus \{\#attr_1, \ldots, \#attr_r\}$ constructed this way is minimal in $\mathbb{S}$. Suppose that there is a solution $R' \subseteq A$ that $\mathbb{C}_{\mathbb{S}}(R') = 1$ in $\mathbb{S}$ and $|R'| < |R|$. Thus, we may construct set $\breve{R}' = R' \cup \{\#attr_1, \ldots, \#attr_r\}$ that satisfies $r\text{-}\mathbb{C}_{\mathbb{S}'}(\breve{R}') = 1$ and $|\breve{R}'| < |\breve{R}|$. Contradiction, because $\breve{R}$ is minimal in $\mathbb{S}'$, what ends the proof.

Visual Interpretation. In order to provide a better understanding of the proof of Theorem 1, let us present a visual interpretation of the aforementioned reduction. In Table 5, we may find the exemplary decision table extended with two additional A-attributes created as described in Subsect. 3.4. That is, for each object $u \in U$, we put $\#attr_1(u) = \#attr_2(u) = concat(a_1(u), a_2(u), a_3(u), a_4(u), a_5(u), a_6(u))$.

In the lattice in Fig. 7, we can see the additional A-attributes. For simplicity, let us consider the most standard case of $(r\text{-})$reducts introduced in Definitions 1 and 8. We remember that each $\#$attribute forms a row identifier that maintains the same functional dependencies with the decision attribute d as the full attribute set $A = \{a_1, a_2, a_3, a_4, a_5, a_6\}$. Hence, singleton sets $\{\#attr_1\}$ and $\{\#attr_2\}$ are decision reducts. (As we know, for the considered data set the empty set of attributes is not a reduct.) Furthermore, once we consider attribute sets presented above $R1$ line in Fig. 7, we may notice that every two-element combination of attributes a_1, a_2, $\#attr_1$, $\#attr_2$ is a 1-reduct because after removal of any attribute from $\{a_1, a_2\}$, $\{a_1, \#attr_1\}$, $\{a_1, \#attr_2\}$, $\{\#attr_1, \#attr_2\}$ the remaining singletons constitute reducts. Similarly, every combination of three attributes out of a_1, a_2, $\#attr_1$, $\#attr_2$ is a 2-reduct, etc.

Let us discuss briefly the polynomial reduction basing on an exemplary visualization of decision Table 2. In the lattice in Fig. 2, we may easily notice that there are two minimal reducts: $\{a_1\}$ and $\{a_2\}$ in the original decision Table 2. A presented reduction is as follows: in the first step, we add two A-attributes, this way we create extended decision Table 5 - obviously the computational complex-

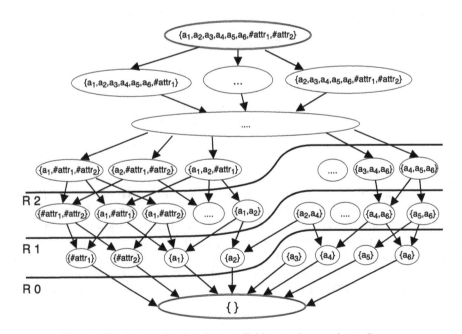

Fig. 7. The lattice for the data in Table 5 with two A-attributes.

ity of this step is polynomial with respect to the original decision table size. In Fig. 7, we present the impact of the aforementioned extension on the lattice.

Let us assume that there is a polynomial algorithm $\breve{Alg}^2()$ that solves the minimal 2-reduct problem. Hence, the result of Algorithm $\breve{Alg}^2()$ executed on the decision Table 5 is one of the following sets: $\{a_1, a_2, \#attr_1\}$, $\{a_1, a_2, \#attr_2\}$, $\{a_1, \#attr_1, \#attr_2\}$, $\{a_2, \#attr_1, \#attr_2\}$ - without losing generality let it be $\{a_1, a_2, \#attr_2\}$. According to the reduction presented in Subsect. 3.4, we should remove all A-attributes from the resulting set, this way we have: $\{a_1, a_2\}$. Now, we may remove any attribute. Without loosing generality let it be a_1. We may easily notice that the remaining set $\{a_2\}$ is a decision reduct in the original decision Table 2.

Impact of Complexity Study. Let us briefly discuss the impact of Theorem 1. For example, consider the resilient version of the minimal (H, ε)-approximate reduct problem or equivalently minimal $\mathbb{C}^{(H,\varepsilon)}$-reduct problem – using the nomenclature of Eq. (19). We may define r-$\mathbb{C}^{(H,\varepsilon)}$ as shown in Eq. (20). We obtain the following:

Theorem 2 *(Minimal r-$\mathbb{C}^{(H,\varepsilon)}$-reduct problem is NP-hard). For each $\varepsilon \in [0, 1)$ and $r \in \mathbb{N}$, the minimal r-$\mathbb{C}^{(H,\varepsilon)}$-reduct problem – i.e., the problem of finding, for an input decision table $\mathbb{S}$ the minimal (in the sense of the number of elements) subset $\breve{R} \subseteq A$ such that r-$\mathbb{C}^{(H,\varepsilon)}(\breve{R}) = 1$ – is NP-hard.*

Proof. From Proposition 4, we know that $\mathbb{C}^{(H,\varepsilon)}$ satisfies conditions of Definition 5. As already mentioned, the minimal (H, ε)-approximate reduct problem is NP-hard [356]. Thus, directly from Theorem 1, the minimal r-$\mathbb{C}^{(H,\varepsilon)}$-reduct problem is NP-hard too.

Similarly, we may refer to the minimal α-reduct problem [279], which – as already discussed – corresponds to the $Disc$ measure (5).

Theorem 3 *(Minimal r-$\mathbb{C}^{(Disc,\varepsilon)}$-reduct problem is NP-hard). For each $\varepsilon \in [0, 1)$ and $r \in \mathbb{N}$, the minimal r-$\mathbb{C}^{(Disc,\varepsilon)}$-reduct problem – i.e., the problem of finding, for an input decision table $\mathbb{S}$ the minimal (in the sense of the number of elements) subset $\breve{R} \subseteq A$ such that r-$\mathbb{C}^{(Disc,\varepsilon)}(\breve{R}) = 1$ – is NP-hard.*

Proof. From Proposition 2, we know that $\mathbb{C}^{(Disc,\varepsilon)}$ satisfies conditions of Definition 5. Since we know that the minimal α-reduct problem is NP-hard [279] and the criterion r-$\mathbb{C}^{(Disc,\varepsilon)}(\breve{R}) = 1$ is equivalent for the one formulated for α-reducts for $\alpha = 1 - \varepsilon$, thus from Theorem 1 we know that the minimal r-$\mathbb{C}^{(Disc,\varepsilon)}$-reduct problem is also NP-hard.

The last example of the complexity result derivable directly from Theorem 1 is the following. However, let us note that the same mechanism could be applied for many other cases as well, in particular, for any formulations of $\mathbb{C}^{(Q,\varepsilon)}$-reducts for which the corresponding measures Q satisfy conditions of Definition 7 [355, 357].

Theorem 4 *(Minimal r-$\mathbb{C}^{(\gamma,\varepsilon)}$-reduct problem is NP-hard). For each $\varepsilon \in [0,1)$ and $r \in \mathbb{N}$, the minimal r-$\mathbb{C}^{(\gamma,\varepsilon)}$-reduct problem – i.e., the problem of finding, for an input decision table $\mathbb{S}$ the minimal (in the sense of the number of elements) subset $\check{R} \subseteq A$ such that r-$\mathbb{C}^{(\gamma,\varepsilon)}(\check{R}) = 1$ – is NP-hard.*

Proof. From Proposition 2, we know that $\mathbb{C}^{(\gamma,\varepsilon)}$ satisfies conditions of Definition 5. Since the minimal (γ,ε)-approximate reduct problem is NP-hard [373], thus from Theorem 1 we know that the minimal r-$\mathbb{C}^{(\gamma,\varepsilon)}$-reduct problem is also NP-hard.

3.5   Depth First Search Algorithms

The task of heuristic search of reducts in a given data set is broadly investigated in the literature. For instance, in [360] a combination of iterative filter-based feature selection algorithm with a statistical significance stop criterion and an RST-based redundant feature elimination was applied. A similar mixture of iterative feature selection and reduction was suggested in [83]. In [419], three groups of algorithms based on the deletion, addition-deletion and addition strategies were discussed. There are also many other approaches to algorithmic reduct construction that refer, e.g., to randomized search [98], ensembles [404], or various methods of feature granulation [141].

Algorithm 3: *r-test$_\mathbb{C}$*

Data: $\check{R}$ – the examined subset of attributes; r – resilience level;
$\mathbb{S} = (U, A \cup \{d\})$– decision table; $test_\mathbb{C}()$ – verifies criterion $\mathbb{C}$
Result: *true – if a given set of attributes $\check{R}$ satisfies selected criteria test$_\mathbb{C}$ with expected resilience level r, false – otherwise*
1 *candidate$\_$subsets $\leftarrow \{X \subset \check{R}$ such that $|X| = r\}$*
2 **foreach** *$X \in candidate\_subsets$* **do**
3   **if** *!test$_\mathbb{C}(\check{R} \setminus X)$* **then**
4 | *return* false;
5   **end**
6 **end**
7 *return* true;

In the following subsections, we consider two approaches to the depth first search exploration of the lattice, which allow us to identify subsets of attributes that satisfy the resilient version of *test$_\mathbb{C}$* function: *r-test$_\mathbb{C}$* (Algorithm 3). Algorithm *r-test$_\mathbb{C}$* verifies if a given set of attributes $R \subseteq A$ satisfies the resilient criterion r-$\mathbb{C}$ under the condition that implementation of *test$_\mathbb{C}$* is given.

In Subsect. 11, we present a novel algorithm generating r-$\mathbb{C}$-reducts inspired with a permutation-based technique that is common for RST-based approaches [357,373]. In Subsect. 11, we follow with discussing the approximation of the permutation-based algorithm for resilient feature selection.

Permutation-Based Algorithm. The function $test_{\mathbb{C}}(R)$ verifies if a given set of attributes $R \subseteq A$ satisfies the specified criterion $\mathbb{C}$ in a given decision table $\mathbb{S} = (U, A \cup \{d\})$. Let us assume that we have an implementation of $test_{\mathbb{C}}$ for $\mathbb{C}$. Algorithm 3 shows how to introduce function $r\text{-}test_{\mathbb{C}}(\breve{R})$ that verifies whether the given set $\breve{R}$ satisfies $test_{\mathbb{C}}$ after removal of any r attributes.

Algorithm 4 ($genRed_{\breve{R}}$) generates an r-$\mathbb{C}$-reduct $\breve{R}$ for a given criterion r-$\mathbb{C}$ with the expected resilience level r. The pessimistic computational complexity of $genRed_{\breve{R}}$ with respect to $r\text{-}test_{\mathbb{C}}(\breve{R})$ is $\mathcal{O}(|A|)$, since both loops – 'while' and 'foreach' – are iterated at most $|A|$ times. Thus, the computational complexity of the $r\text{-}test_{\mathbb{C}}(\breve{R})$ implementation has a crucial impact on the complexity of $genRed_{\breve{R}}$.

We may notice that for relatively big values of r, e.g., $r = \frac{|A|}{2}$ or $r = \frac{|A|}{3}$, the $r\text{-}test_{\mathbb{C}}(\breve{R})$ may iterate $test_{\mathbb{C}}(\breve{R})$ exponentially many times. However, for any constant r, the algorithm is polynomial. Comparing to the current market standards/defaults with regard to security, resilience and high availability of services and the data, we may notice that data replication levels offered out-of-the-box by database management systems vary near relatively low values (2 to 6) as a reasonable compromise between resilience and storage costs. For example, the default data replication level in most of MapReduce implementations like, e.g., Hadoop[11] is set to 3. Cloud services that offer very high level of durability and availability of stored data usually use between 3 and 6 data replicas, depending on the service pricing level[12]. Having that in mind, we may calculate that

---

Algorithm 4: $genRed_{\breve{R}}$

  Data: r – expected resilience level; $\mathbb{S} = (U, A \cup \{d\})$– decision table; $test_{\mathbb{C}}$– the function that verifies if a given set of attributes $R \subseteq A$ satisfies the specified criteria $\mathbb{C}$

  Result: $\breve{R}$ – a subset of attributes that satisfies resilient criterion r-$\mathbb{C}$

1   $\breve{R} \leftarrow \emptyset$; Permute set A;

2   /* Forward propagation   */

3   **while** $r < |\breve{R}| \wedge !r\text{-}test_C(\breve{R})$ **do**

4       $a \leftarrow removeNextAttribute(A)$

5       $\breve{R} \leftarrow \breve{R} \cup a$

6   **end**

7   /* Backward elimination   */

8   **foreach** $a \in \breve{R}$ **do**

9   **if** $r\text{-}test_C(\breve{R} \setminus a)$ **then**

10          $\breve{R} \leftarrow \breve{R} \setminus a$

11   **end**

12   **end**

13   *return* $\breve{R}$;

---

[11] http://hadoop.apache.org.

[12] https://docs.microsoft.com/en-us/azure/storage/storage-redundancy.

in the case of $r = 3$ (or $r = 6$ in a very restrictive case) the complexity of $r\text{-}test_\mathbb{C}(\check{R})$ relies mostly on the complexity of $test_\mathbb{C}(\check{R})$ and the size of $\check{R}$. With limited constant r, the function remains polynomial and may be estimated as $\mathcal{O}(|A|^r) \times \mathcal{O}(test_\mathbb{C})$ where $\mathcal{O}(test_\mathbb{C})$ refers to the complexity of $test_\mathbb{C}$. In that case, the complexity of $genRed_{\check{R}}$ depends on the specifics of $\mathbb{C}$. In the case of, e.g., the classical discernibility criterion $\mathbb{C}^{(Disc,0)}$, the pessimistic complexity of $isReduct$ is $\mathcal{O}(|U| \times |A|^2)$.

Approximation Algorithm. Once we have defined the straightforward permutation algorithm, we may elaborate on possible approximations in order to improve the overall feature selection performance. There are plenty of approximation methods that may be adopted to this case like, for example, the DAAR heuristics introduced in [186].

Algorithm 5 ($approximateRed_{\check{R}}$) follows the idea that $r+1$ disjoint attribute subsets, which individually satisfy $\mathbb{C}$, constitute a set that satisfies $r\text{-}\mathbb{C}$ after being merged together. In the presented pseudo-code we rely on the permutation-based algorithm to construct disjoint $r + 1$ $\mathbb{C}$-reducts. We merge them together to form an $r\text{-}\mathbb{C}$-superreduct $\check{R}$. The set $\check{R}$ constructed this way, is for sure a $r\text{-}\mathbb{C}$-superreduct, since we may remove any r attributes and at least one $\mathbb{C}$-reduct will be untouched. The size of the output is no more than r times bigger than an $r\text{-}\mathbb{C}$-reduct could be, which may be still acceptable for highly multidimensional real life data sets.

Algorithm 5: $approximateRed_{\check{R}}$

Data: r – expected redundancy level; $\mathbb{S} = (U, A \cup \{d\})$– decision table;
$\qquad genRed_C()$ – the function that generates the set of attributes
$\qquad R \subseteq A$, which satisfies the specified criteria $\mathbb{C}$
Result: $\check{R}$ – *a subset of attributes that is a r-$\mathbb{C}$-superreduct*

1 $\check{R} \leftarrow \emptyset$
2 Permute set A
3 **for** $i = 0; i < r + 1; i + +$ **do**
4 $\quad$ $R \leftarrow genRed_C(A)$;
5 $\quad$ $A \leftarrow A \setminus R$;
6 $\quad$ $\check{R} \leftarrow \check{R} \cup R$;
7 $\quad$ **if** $A = \emptyset$ **then**
8 $\quad\quad$ | break; /* If we tested all attributes $\qquad\qquad$ */
9 $\quad$ **end**
10 **end**
11 *return* $\check{R}$;

As a conclusion, in the case of resilient feature selection, the analyst should elaborate on the required level of resilience from the perspective of importance and sensitivity of the problem. Nevertheless, one should have in mind the impact of resilience level on the performance of feature selection and should adjust it with respect to the aforementioned factors. On the other hand, minimal subsets of attributes are not always desired. In some situations, it is worth combining the groups of approximate reducts in order to improve performance of prediction models [138]. This shows that properly managed redundancy in selected attribute sets may not only increase the resilience of the solution but also may have a positive impact on the quality of trained models.

Algorithm Working Example. In order to provide a better understanding of the presented algorithms and to verify the quality of the proposed approach, let us experiment with the size of r-$\mathbb{C}$-reducts acquired by Algorithm 4 for the data in Table 2 and conditional entropy H, namely the criterion function $\mathbb{C}^{(H,\ 0.2)}$. The resilient version of criterion r-$\mathbb{C}^{(H,\ 0.2)}$ may be constructed for a given resilience level $r = 1$ as described in Eq. (20).

Below, we present the step-by-step description of a single execution of Algorithm 4. Afterwards, we present a summary of 10 independent executions and the statistical analysis of the expected size of 1-$\mathbb{C}^{(H,\ 0.2)}$-reducts for the data in Table 2.

Table 6. Summary of the experiments for Algorithm 4.

i	σ_i	1-$\mathbb{C}^{(H,\ 0.2)}$-reduct	size
1	$\sigma_1 : a_6, a_3, a_1, a_4, a_5, a_2$	$\{a_1, a_6\}$	2
2	$\sigma_2 : a_2, a_3, a_4, a_1, a_5, a_6$	$\{a_2, a_3, a_4\}$	3
3	$\sigma_3 : a_5, a_3, a_2, a_1, a_4, a_6$	$\{a_2, a_3, a_5\}$	3
4	$\sigma_4 : a_1, a_3, a_2, a_4, a_5, a_6$	$\{a_1, a_2\}$	2
5	$\sigma_5 : a_1, a_5, a_3, a_2, a_4, a_6$	$\{a_1, a_3, a_5\}$	3
6	$\sigma_6 : a_6, a_4, a_2, a_5, a_3, a_1$	$\{a_2, a_6\}$	2
7	$\sigma_7 : a_2, a_5, a_1, a_6, a_3, a_4$	$\{a_1, a_2\}$	2
8	$\sigma_8 : a_4, a_3, a_2, a_1, a_5, a_6$	$\{a_2, a_3, a_4\}$	3
9	$\sigma_9 : a_6, a_2, a_1, a_5, a_4, a_3$	$\{a_1, a_6\}$	2
10	$\sigma_{10} : a_5, a_3, a_6, a_2, a_4, a_1$	$\{a_3, a_5, a_6\}$	3

During the first execution of the experiment, we sorted the set A (line 2 in Algorithm 4) with the permutation σ. Within the *'while'* loop (lines 3-6), the algorithm iterated over A according to $\sigma_1 : a_6, a_3, a_1, a_4, a_5, a_2$. After consecutive draws of attributes, we evaluated whether the condition 1-$\mathbb{C}^{(H,\ 0.2)}$ was met. Below we enumerate each iteration of the *'while'* loop during the first execution of the experiment:

– 1st iteration
  1. $removeNextAttribute(A)$ returns a_6, hence $\check{R} = \{a_6\}$
  2. 1-$test_{\mathbb{C}}(\{a_6\}) = 0$, because $\check{R} \setminus \{a_6\} = \emptyset$
– 2nd iteration

1. *removeNextAttribute*(A) returns a_3, hence $\check{R} = \{a_6, a_3\}$
2. 1-*test*$_{\mathbb{C}}$($\{a_6, a_3\}$) $= 0$ because for the subset $\{a_3\}$, $H(d|\{a_3\}) = 0.74$, hence $H(d|\{a_3\}) > -\log_2(1 - 0.2)$
- 3rd iteration
    1. *removeNextAttribute*(A) returns a_1, hence $\check{R} = \{a_6, a_3, a_1\}$
    2. 1-*test*$_{\mathbb{C}}$($\{a_6, a_3, a_1\}$) $= 1$ because for all subsets $\check{R}' \in \{\{a_1, a_3\}, \{a_1, a_6\}, \{a_3, a_6\}\}$, we have $H(d|\check{R}') \leq -\log_2(1 - 0.2)$

In the first execution, in the 'foreach' loop (lines 8–12), we iterated over $\check{R} = \{a_6, a_3, a_1\}$ trying to eliminate superfluous attributes. The attribute a_6 could not be removed because 1-*test*$_{\mathbb{C}}$($\{a_1, a_3\}$) $= 0$. The attribute a_3 was removed with no impact on 1-$\mathbb{C}^{(H, \ 0.2)}$ because $H(d|\{a_1\}) \leq -\log_2(1 - 0.2)$ and $H(d|\{a_6\}) \leq -\log_2(1 - 0.2)$, thus 1-*test*$_{\mathbb{C}}$($\{a_1, a_6\}$) $= 1$. The last attribute a_1 could not be removed. This way, we reached 1-$\mathbb{C}^{(H, \ 0.2)}$-reduct $\check{R} = \{a_1, a_6\}$ of size $|\check{R}| = 2$ – which is the minimal possible.

To provide higher reliability, the experiment was repeated 10 times for 10 randomly chosen permutations. Table 6 summarizes each iteration 'i', including permutations $\sigma_1, \ldots, \sigma_{10}$ and the derived 1-$\mathbb{C}^{(H, \ 0.2)}$-reducts $\check{R}_1, \ldots, \check{R}_{10}$ with their size.

Lastly, let us elaborate on the expected size of 1-$\mathbb{C}^{(H, \ 0.2)}$-reducts that may be generated with Algorithm 4. The minimal size of 1-$\mathbb{C}^{(H, \ 0.2)}$-reducts is 2, whereas the maximal size of 1-$\mathbb{C}^{(H, \ 0.2)}$-reducts in the analyzed data is 3, because for each two-element set $R \subset A, H(d|R) \leq -\log_2(1 - 0.2)$. There are 6! possible permutations σ of A. Let us estimate the number of permutations that would result in 1-$\mathbb{C}^{(H, \ 0.2)}$-reducts of size 2. Such permutations should have two of attributes $\{a_1, a_2, a_6\}$ within the first 3 attributes. There are $\binom{3}{2} * \binom{3}{1} = 9$ three-element combinations that contain two attributes of $\{a_1, a_2, a_6\}$ and one that contains all. Each of them may be permuted in 3! ways and the rest of σ may be arranged in 3! ways. So, $\frac{10*3!*3!}{6!} = \frac{1}{2}$. The rest half of permutations would result in 1-$\mathbb{C}^{(H, \ 0.2)}$-reducts of size 3. Thus, the expected size of 1-$\mathbb{C}^{(H, \ 0.2)}$-reduct generated by Algorithm 4 for the data in Table 2 is equal to $\frac{1}{2} * 2 + \frac{1}{2} * 3 = 2.5$.

4   Technical Aspects of Interactive Feature Engineering

Several useful techniques have already been applied for decision making in the domains of data warehousing (DW), business intelligence (BI) and machine learning (ML) [394]. However, before one may apply machine learning algorithms on the collected data, several steps need to be performed in the first place [87,431]. Among them, feature extraction and selection are recognized as the most challenging, time-consuming and computationally cost-full [77]. Optimistically, the similarity of the nature of sensory and machine generated data provides an opportunity to construct generic, reusable mechanisms for interactive data processing, exploration and analysis. In this chapter, we introduce a new approach for learning forecasting models over large multi-sensor data sets, including the steps of sliding window-based feature extraction and ensemble-based feature selection.

4.1   Sliding Window-Based Feature Engineering

In this section, we outline our approach to feature extraction, aimed at processing data obtained from sensors that monitor certain changes in the environment and provide outputs in a form of time series. Individual readings may take different forms according to the application domain [270,303]. Values may express continuous phenomena, such as pressure, humidity, or the level of methane concentration in a longwall of a coal mine [200]. They can also express a discrete state of the environment, such as an on/off state of a device or vehicle movement direction. To acquire knowledge about environment state and its changes, it is common to set up a collection of sensors, potentially of different types. Therefore, the gathered data elements can be complex on various levels and sometimes their interpretation is possible only in a context of additional knowledge obtained from domain experts. This, in turn, requires appropriate mechanisms for human-system interaction. No less important is the ability to properly parallelize data processing in order to deal with various challenges related to Big Data [16,92].

Prerequisites and Data Preprocessing. Information systems - especially in the field related to reporting, business intelligence, machine learning, and decision support - integrate data from many other systems and sensors [360]. Having data from various sources in a single place provides a valuable opportunity for pattern discovery. However, those systems and sensors are usually provided by many different corporations, and are developed using various technologies with different data formats. Information provided by each can be textual, categorical, numeric, etc. All may also refer to different data dictionaries and taxonomies. Therefore, prior to feature extraction, let us outline some typical challenges related to data integrity and quality as well as basic steps of data preprocessing which becomes particularly important when real streams of sensor readings are involved.

In the first place, the analysis of a big variety of data representations requires some kind of unification protocol. With that respect, let us refer to so called *sensor card* - an information template created on the basis of investigation of a large variety of sensors that can be applied whenever heterogeneous data sources need to be systematized, so that could be integrated. An example of a sensor card - a common interface that allows to describe in a consistent way a variety of types of sensors and devices that are used in the domain of underground mining - is presented in [360]. This way, the sensor readings from different data sources may not only be integrated but also further enhanced with some specific features highly depended on the analyzed problem, like: the organization of the shift work, shift schedule and plan, information about bank holidays, recent local events, etc.

The industrial monitoring systems usually produce multidimensional streams of sensor readings for which performing standard preprocessing steps such as data integration, data cleaning, feature extraction and selection, etc. is quite challenging. Measurements recorded by sensor devices tend to be noisy. Because of faults and errors that may have place in real environments, it is also difficult to maintain decision models that should be used in an on-line fashion. Thus, the goal of data preparation and cleaning is to translate data to a form acceptable by the forecasting model construction methods. This phase is focused on the preparation of the training sets for further analysis and, once the models are ready, becomes responsible for feeding them with new inputs. Let us outline some typical issues connected to the data acquisition in real life environments:

1. *Unsynchronized readings*: Reading frequencies differ for different sensors.
2. *Missed readings*: Sensors may stop delivering in a given time interval.
3. *Outlier readings*: Sensor readings are frequently imprecise or unreliable.
4. *Rare readings*: Usually, the most critical events occur in data very sparsely.

The first task is to adjust sensor readings that are collected at different frequencies. Also, some systems collect a new reading only in the case of a sufficiently significant change of the measured value. Although some ML algorithms can fit models on incomplete data [72], the main objectives of subsequent tasks are imputation of missing values, and outliers detection [342, 366].

These are particularly complex tasks for multivariate time series and spatio-temporal data environments [15, 47]. With this respect, one can, e.g., create a logical expression that defines value replacements (for instance, to replace values < 1 with "low state"), use a default value, follow the last valid reading, take an average of the neighboring readings, or apply linear regression based on the preceding values. Imputation techniques are often based on univariate series analysis or sampling from original data distribution [342, 366]. There are also approaches based on auto-regressive models constructed, e.g., by combining an expectation-minimization algorithm with the prediction error minimization method, or based on Bayesian models [25]. The maximum allowed number of consecutive missing values that can be imputed should be set up too. On the other hand, the missing value imputation (or outlier replacement) is not always a requirement. It may depend on a context of a given sensor and knowledge about its operation conditions, or simply on further preprocessing steps which can deal with incomplete data on the fly.

In the proposed approach, a sliding time window method allows us to overcome, or at least to minimize the impact of, some of the above-mentioned issues related to time series data. A viable approach to deal with missing or unreliable attributes - the resilient feature selection with r-$\mathbb{C}$-reducts – was proposed in Sect. 3. Later in Sect. 4.2, we discuss another approach to overcome problem with missing data basing on feature selection over granular attribute space due to attribute interchangeability. Whereas, in Sect. 4.3, we propose ensemble-based feature selection method which is also a practical approach to introduce a certain redundancy of information.

A special case of data preprocessing corresponds to the creation of a dependent variable. This is a crucial aspect for any supervised learning approach. For this purpose, we may use a dedicated operator which allows us to define a dependent variable as the maximal value measured for a given sensor within a specified time interval (e.g., three to six minutes into the future). This can be considered as an example of a broader window-based methodology described below. Such a style of specifying a dependent variable may also decrease sparsity of its critical values. This is because a single high measurement influences a score of the whole time interval. Given the above steps of data preprocessing, we are now ready to go to the topic of feature extraction.

Sliding Windows. A sliding window travels through the time series from the beginning to the end and replaces a sequence of raw sensor readings with some of its derived statistics, accordingly to the predefined aggregation functions (cf. Table 7). The range of aggregation can be chosen by the users by means of, e.g., a time unit that defines windows containing sensor readings to be grouped together. For each outcome of aggregation, we can calculate a weight corresponding to the quality of the original data that is, e.g., inversely proportional to a number of missing values or outliers involved. This approach allows us to reduce the number of missing values in data, and to introduce weights that can be utilized further by analytical methods. It is also worth mentioning that such aggregation operations can work on multiple sensor readings with unsynchronized frequencies [360].

Table 7. Examples of features that represent each time window.

Feature type	Description
Basic meta information	e.g., a data source identifier, ID or a name of the sensor, a total number of readings "n", a number of valid readings "nValid", etc.
Quality assurance and reliability of a given windows	e.g., a ratio of correct readings in the window $= \frac{nValid}{n}$, or just a number of identified outliers or missing values
Time-range of a window and time related data	e.g., year, month, day of month, day of week, hour, time-range, etc.
Basic summary of all readings in a given time window	Statistics: mean, min, max, stdDev, median and pecentiles: $5th$, $10th$, $25th$, $75th$, $90th$, and $95th$, etc.
Transformations and measures of values in time window	e.g., selected Fourier transform coefficients, skewness, Kurtosis measure, etc.
Summaries of consecutive sub-windows	The same statistics as above, computed for sub-windows of a given window
Trends related to recent readings in a time window	Differences between the last reading and min/max values, differences between last/first readings in a window
Statistics for differentiated values in a time window	Mean, median, min, max, stdDev and percentiles of differences between two consecutive readings
Measures derived from summaries of a time window	Differences between min and max, mean and median, max and percentiles, etc.
Measures derived from summaries of sub-windows	Differences between quantities of mean, median, min, and max values representing consecutive sub-windows
Indicators of extreme readings	Position in a time window of a reading with min/max value, position of a min/max value in the latest sub-window
Indicators of extreme transformed readings	Position of a maximum difference between consecutive readings, etc.
A set of values that express the trend between statistics in consecutive sliding windows	Inter window data, e.g., a difference between min, max, $mean$ or between Xth percentiles, where: $X \in \{5, 10, 25, 50, 75, 90, 95, etc.\}$

Processing of a single time series requires two parameters: *length* and *offset*. The first of them defines the size of a sliding time window, e.g., the number of readings to be involved, or a corresponding time interval. The alignment of processed time windows is controlled by the second parameter. It defines a degree to which two consecutive windows overlap each other. Let us here recall Fig. 1 – presented in a preliminary Sect. 2.1 – which highlights four examples of sliding window set-ups. The first example, marked in red, shows the situation when the length of a sliding window is equal to the offset. The green and blue examples show the consecutive positions of a sliding window when the offset is equal to $\frac{1}{2}$ and $\frac{1}{3}$ of the length, respectively. The system is also capable to express the situation when the offset is greater than the length – the example marked in cyan.

A collection of sensors that perform readings produces a corresponding collection of time series data. A single time series is an ordered sequence of readings associated with the timestamp at which it was collected. A collection of aggregated values created from a time series may be organized arbitrarily at a higher level. For instance, if a time window covers one minute, then we may be interested in five consecutive windows that cover five minutes of data in order to analyze various trends over derived statistics within that period. It is quite different than aggregations over a single five-minutes time window [138,146]. Going further, sensors may correspond to many different time series processed independently. To obtain a complete description of the environment at a given time point, time series collected from different sensors should be combined together to form a larger set of aggregated values [147].

During the process of moving a time window through a time series, each of its fixed positions defines so-called *basic window* (cf. Fig. 8). For such a basic slice of data, a predefined aggregation functions are applied. Each aggregation function can be seen as a new window feature. This step may be adjusted to a specific data domain by supplying different aggregation implementations [147,182,360]. The proposed set of features which are calculated to represent the data in a basic window are presented in Table 7. If we consider more than one consecutive basic window to represent the environment state at a given time point, then we can extract so-called *inter window features* expressing trends and changes between pairs of basic windows. Some examples of inter window features that are calculated to represent the changes between two consecutive basic windows are presented in Table 7. A schematic view of extracted basic window features, inter window differential statistics as well as their relation to the processed time series is shown in Fig. 5.

Yet another very specific approach is the extension of the sliding window construction process by adding some more static attributes reflecting assessments obtained from domain experts [182]. This brings the opportunity to compare the prediction quality of models trained using derived features with the expert-based assessment as well as makes it possible to use features derived from experts in ML models training. Furthermore, the expert assessment helps to address the so-called *cold start* problem, when a decision support system is installed, e.g., in

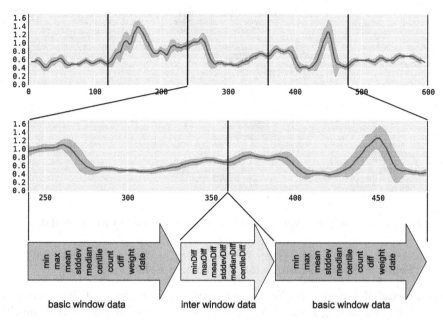

Fig. 8. Overview of the proposed sliding window approach. The topmost time series is split into five non-overlapping basic windows of equal length. Two of them are zoomed in the middle of the above diagram. The statistics are computed for each window separately. Inter window features express the dynamics of changes of basic window statistics.

a new location and it does not yet have a sufficient amount of data to fit into the new environment [368].

The proposed framework is capable of operating on multidimensional time series derived from a number of sensors. The default method of processing multiple series is hierarchical, i.e., each time series is processed independently and then the results are combined according to specified settings. Afterwards, depending on configuration, features corresponding to basic windows and inter window features derived for selected sensors create so-called composite windows which represent the overall state for a given time point.

This way we provide a comprehensive data preprocessing and feature extraction framework that can be used for constructing informative and robust representation of multidimensional time series data, as visible in Fig. 9. The overall mechanism of computing time-window-based representations can be treated as a universal approach. It is worth noticing that due to a diversity of extracted features and a high number of considered sensors this representation of data may be highly dimensional. Hence, it may require feature selection before forecasting model construction techniques could be applied.

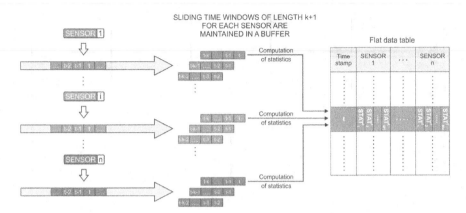

Fig. 9. Window-based features calculated for a portion of time series data.

4.2   Feature Space Granulation in Feature Selection

The process of feature selection aims at exploring the given attribute space A (or A^*) and extracting a relatively small subset $R \subseteq A$ of attributes that, on the one hand, are the most relevant and, on the other hand, are sufficient to solve the investigated problem. Such selection/extraction process is often conducted by applying statistical tests in order to determine, which attributes contribute to the constructed decision model [152]. Although the standard feature selection algorithms are not configured for attributes that are structured or bound by relationships, the knowledge about attribute granulation can have an important impact on the final subset composition.

Feature Space Granulation. The attribute granules can take various forms. It is possible to group or cluster features on the basis of their relationship, and it can be done in a parameterized manner. For example, we can produce various versions of granulations depending on the choice of cutoff value after the original attributes are hierarchically clustered [141]. In this context, it is important to have the means of assessment of the resulting granules, similar to those developed for standard data clustering. By making the feature selection process aware of the underlying granular structure of attribute space one can make better use of the knowledge contained therein. This in turn may lead to selecting the sets of features that are not only optimal from the perspective of some mathematical criteria but are also more useful for interpreting knowledge hidden in the data.

Let us now present two specific examples of the granulation based on the attribute interchangeability. The first approach is centered around the notion of *explicit interchangeability* of features in attribute subsets that are small in size but sufficient to model the target decision classes/labels. In the theory of rough sets, such attribute subsets are usually referred as decision reducts (Definition 1). Intuitively, if two attributes rarely belong to the same subset but they both often appear together with similar groups of other attributes, they may be considered interchangeable. In the opposite situation, when two attributes often belong to the same subset or appear in a company of completely different features, it seems reasonable to assume that they convey different information and thus are not similar. More formally, this type of attribute interchangeability can be measured using a co-occurrence frequency matrix F, whose entry in i-th row and j-th column equals $f_{i,j}$:

$$f_{i,j} = \frac{|\{k : a_i \in R_k \wedge a_j \in R_k\}|}{|\{k : a_i \in R_k\}|} \tag{21}$$

where a_i, a_j are attributes, $i \neq j$ and R_k is the k-th pre-computed attribute subset (reduct). All values at diagonal of F are set to 0. The final values of attribute interchangeability can be computed as a difference between the similarity of corresponding feature sets and the frequency, with which the given features co-occur, e.g.:

$$I(a_i, a_j) = cosine(f_{i,\cdot}, f_{j,\cdot}) - f_{i,j} \tag{22}$$

In this formula, $f_{i,\cdot}$ and $f_{j,\cdot}$ are vectors of values from i-th and j-th rows of F, respectively. Such an approach was successfully applied in [183]. Figure 11 depicts a heat map of a distance matrix that was used for identifying key risk factors for firefighters during fire&rescue actions. Analogous heat maps could be computed using, e.g., ensembles of possibly diverse *approximate* decision reducts that preserve information about the decision classes only to some extent and, thus, they can utilize different groups of attributes to concentrate on different aspects of approximate data dependencies [357].

A slightly different approach is centered around the attribute similarity function sim^{Disc}, which refers to the discernibility relation (4) and its numeric representation - measure $Disc$ (5). For a given decision system $\mathbb{S} = (U, A \cup \{d\})$, we may define $sim^{Disc} : A \times A \to \mathbb{R}$ as follows:

$$sim^{Disc}(a, a') = \frac{|\{(u, u') : d(u) \neq d(u') \wedge a(u) \neq a(u') \wedge a'(u) \neq a'(u')\}|}{|\{(u, u') : d(u) \neq d(u') \wedge (a(u) \neq a(u') \vee a'(u) \neq a'(u'))\}|} \tag{23}$$

where $(u, u') \in U \times U$ and $a, a' \in A$. So defined attribute similarity measure expresses a ratio between a number of pairs of objects from different decision classes that are discerned by *exactly one* attribute from the considered pair, to a number of such objects discerned by *at least one* of the compared attributes.

We may extend the definition of the above attribute similarity measure sim^{Disc} (23), so that it operates on subsets of attributes instead of individual attributes. For a decision system $\mathbb{S} = (U, A \cup \{d\})$, the attribute subsets similarity function $Sim^{Disc} : \mathcal{P}(A) \times \mathcal{P}(A) \to \mathbb{R}$ is defined as $Sim^{Disc}(R, R')$:

$$\frac{|\{(u, u') : \exists_{a \in R, a' \in R'} \; d(u) \neq d(u') \wedge a(u) \neq a(u') \wedge a'(u) \neq a'(u')\}|}{|\{(u, u') : \exists_{a'' \in R \cup R'} \; d(u) \neq d(u') \wedge a''(u) \neq a''(u')\}|} \tag{24}$$

Feature Selection Algorithms with Attribute Granules. In this section, we examine to what extent feature granulation can guide the process of choosing the most appropriate collections of attributes. We argue that it should influence the order, in which we investigate attributes. We discuss the meaning of similarity, proximity and functionality while considering the granules of physically existing, or potentially derivable attributes in the feature extraction process. We also propose several approaches to utilize granulation structures defined over the feature spaces in feature selection algorithms. In particular, we consider the algorithms developed within the theory of rough sets, aimed at finding irreducible subsets of attributes that are sufficient to distinguish between the cases belonging to different target decision classes.

For the purpose of further discussion, let us concentrate on the approach to conducting feature selection proposed in Sect. 3.1. Certainly, we do not claim that all possible methods follow the scheme below. Nevertheless, it is sufficiently general to explain the benefits of working with attribute granules. For a given decision system $\mathbb{S} = (U, A \cup \{d\})$ and input set of attributes A, let us consider a criterion function $\mathbb{C} : \mathcal{P}(A) \rightarrow \{0, 1\}$ (cf. Definition 5) that indicate which subsets of attributes are already rich enough to serve as the outcomes of the selection process. In practice, $\mathbb{C}$ may correspond to a collection of criteria reflecting different requirements. Additionally, let us consider an arbitrary heuristic quality function $Q : \mathcal{P}(A) \rightarrow \Theta$ (cf. Definition 7) that can be utilized iteratively to add the most "promising" elements to the constructed feature subset. Let us note that Q can combine various aspects of relationships between the selected attributes and a target variable [83,117,176,297]. Let us also mention that the last item of the following procedure has strong roots in the theory of rough sets, where there is a particular focus on the simplification of decision models [78,376,419].

1. While the criterion $\mathbb{C}(R)$ is not met by the selected feature subset R continue the following:
   (a) Select candidate subsets of features $B_1, .., B_k$ to be added to R
   (b) Evaluate $B_1, .., B_k$ with the desired attribute subset quality measure Q
   (c) If the best B_x contributes to R, then $R \leftarrow R \cup B_x$
   (d) Verify if the criterion $\mathbb{C}(R)$ is met
2. Eliminate superfluous attributes from R

Algorithm 6 reflects our generic idea of embedding the additional knowledge about attribute granulation into the above-described feature selection process. In each iteration of the main loop, in order to limit the attribute space A, the subset of granules $\{G_1, .., G_m\} \subseteq \mathbb{G}$ is selected with respect to the granulation preferences expressed by, e.g., a permutation $\sigma_\mathbb{G} : \{G_{\sigma(1)}, G_{\sigma(2)}, ..\}$ (which means that the granule $G_{\sigma(1)}$ is most preferred to draw attributes from). By limiting the search space using the additional knowledge about attribute granulation, we may quickly generate a set of candidates $\{B_1, .., B_k\}$. After the evaluation of candidates with the correlation, Gini index, or other implementation of the function Q, the feature subset R may be extended if only the selected B contributes

---

Algorithm 6: General framework for granular feature selection

---

Data: $\mathbb{G}$ – set of granules, A – attribute space,
$\mathbb{C}$– criterion function, $\sigma_{\mathbb{G}}$ – granule preferences
Result: R – selected attribute subset

1 /* Initialization   */
2 $R \leftarrow \emptyset$
3 **while** R *does not satisfy* $\mathbb{C}(R)$ **do**
4      $B \leftarrow \emptyset$
5   Select granules $\{G_1, .., G_m\} \subseteq \mathbb{G}$ with respect to $\sigma_{\mathbb{G}}$
6   Limit attribute space $A_G \leftarrow A \cap \bigcup_{1 \leq i \leq m} G_i$
7   Generate candidates $B_1, .., B_k \subseteq A_G$
8   Evaluate candidates $\{B_1, .., B_k\}$
9      $B \leftarrow selectBestCandidate(\{B_1, .., B_k\}, ...)$
10   **if** B *contributes to* R **then**
11 | $R \leftarrow R \cup B$
12   **end**
13 **end**
14 $R \leftarrow eliminateSuperfluousAttributes(R)$
15 *return* R;

---

to R. The loop continues until a "good enough" R is collected or all combinations/candidates are explored. Finally, we conduct a backward elimination of superfluous attributes.

The presented framework does not enforce any particular interpretation of the information granules and, thus, different implementations may vary in a way of their utilization. In some cases, it may be preferred to select features that belong to only one, specific granule. For example, the analysis of coal mine sensor readings may be oriented on the one, particular mine shaft [182]. In that case, the analyst could generate granules on the basis of a sensor location and introduce a constraint that the finally selected attributes should/must belong to the particular ones. In other applications, it may be convenient to generate an attribute subset that contains attributes from multiple granules in order to provide higher robustness [3]. Regardless of the way that we use the attribute granulation, the general framework is still the same.

Attribute granulation may also influence a feature selection process with respect to the expected robustness and resilience of decision models. In real-life applications, we may observe various anomalies in explored data sets, which cause a model over-fitting. Some researchers emphasize the role of appropriate granulation of attributes during feature engineering in achieving higher stability of the created models. With that respect, we may refer to several techniques using, e.g., clustering or histograms [429]. During the decision model construction, there are also some non-functional factors that could impact the continuity of analysis like, e.g., temporal or permanent unavailability of some sources during on-line data collection [137]. From this perspective, it is advisable to use diverse feature subsets and ensemble methods, whereby each of separate decision mod-

---

Algorithm 7: Full-granule-oriented version of Algorithm 6

Data: $\mathbb{G}$ – set of granules, A – attribute space,
Q– quality function, $\mathbb{C}$– criterion function
Result: R – selected attribute subset

```
1  /* Initialization                                                    */
2  R ← ∅
3  while R does not satisfy ℂ(R) = 1 do
4  │   Select granules {G₁,..,Gₘ} ⊆ 𝔾
5  │   Evaluate granules {G₁,..,Gₘ} and pick the best Gₓ
6  │   if Gₓ contributes to R then
7  │   │   R ← R ∪ Gₓ
8  │   end
9  end
10 R ← eliminateSuperfluousGranules(R)
11 return R;
```

---

els is based on a few attributes but, overall, many attributes are involved [185]. Thus, it is important to combine the feature selection approaches relying on the attribute granulation with some feature subset diversification methods.

In this context, the objective is to achieve more robust and resilient results due to, e.g., exploitation of attributes extracted from diverse sources. In particular, the method outlined by Algorithm 6 could be used to compose an attribute subset R as a collection of features from diverse granules. In this case, the attribute reduction algorithm should aim at achieving feature subsets of minimal cardinality $|R|$ and also ensure the diversity of granules by, e.g., maximization of $|\{G \in \mathbb{G} : R \cap G \neq \emptyset\}|$. Accordingly, a specialized configuration of the main loop in the presented framework can take into account, both, the so-far-selected features and the granules that are used less often, i.e., granules G_i that minimize the quantity of $|G_i \cap R|$. The presented approach may be considered as a practical solution to the problem of resilient feature selection introduced in Sect. 3.

The feature selection methods should be also able to operate on the whole granules or their subsets instead of individual attributes. To some extent, it corresponds to the idea of so-called *decision systems with constraints* – the enriched data representation proposed in [280]. The goal of this approach is not only to record the presence of granules (called constraints) but also to make it possible to apply various computational methods that make use of them. Let us consider Algorithm 7, where the overall scheme is aligned with Algorithm 6, though one can notice some simplifications like selecting particular granules $G_1,..,G_m$ as the candidate subsets B_i. Similarly, the backward elimination concerns removal of the whole granules instead of individual attributes. In such approaches, as it was observed also by other researchers, the properties of selected attribute subsets can depend a lot on coarsening or refining granules [196]. Therefore, there is a need for a framework allowing the domain experts and algorithm design-

Attributes ordering approach - toward maximal number of granules in the selected feature subset

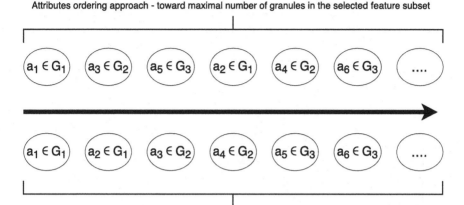

Attributes ordering approach - toward minimal number of granules in the selected feature subset

Fig. 10. A diagram with two significantly different attribute ordering strategies that take into account granulation of attributes.

ers to assess the results of feature selection/granulation processes from different perspectives.

As we could see above, Algorithm 6 can be treated as a general umbrella for various approaches aiming at utilization of the attribute space granulation for the purpose of enhancing the feature selection process. Surely, there are still several details to be discussed. First, it is useful to look at different strategies of validating whether a given attribute sufficiently *contributes* to the result R [186]. Second, it is interesting to compare the proposed framework with methods based on attribute orderings. The main idea behind this class of methods is to iterate along diversified permutations σ_A over A. Such permutations can be induced partially with respect to some heuristic function Q, or they can be generated fully randomly [373]. In the latter case, the procedure is repeated a number of times and the best of the obtained attribute subsets (or a bigger ensemble of subsets) is eventually selected.

Figure 10 shows how we can use the knowledge about granules to influence permutations of attributes, e.g., by arranging the elements of the same granule within consecutive subsequences, or mixing them together as much as possible (by following a "preference" permutation $\sigma_{\mathbb{G}} : \{G_{\sigma(1)}, G_{\sigma(2)}, ..\}$). It is important to note that such two semi-randomized strategies are in a correspondence to the ideas of operating with regular granules (Algorithm 7) and maximally diverse attribute subsets, respectively. This shows that the attribute granulation is easily applicable to the ordering-based feature selection algorithms, without a necessity to modify their code. On the one hand, the described scenarios of "granular ordering" are conceptually aligned with Algorithm 6. On the other hand, the phase of selecting granules/candidates can be performed implicitly at a level of generation of attribute permutations.

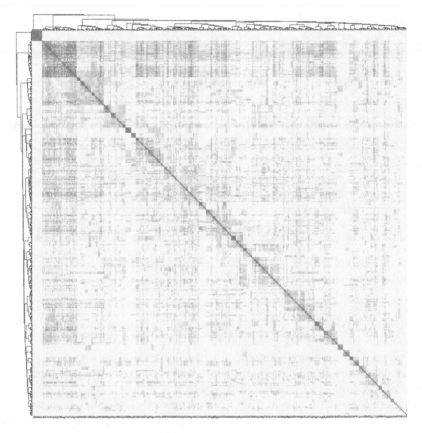

Fig. 11. A heat map expressing interchangeability of risk factors (represented as attributes) taken from the AAIA'14 Data Mining Competition [183]. Granules of attributes are arranged along the diagonal of the matrix [141].

While the "case-oriented" granulation is a way to cope with ever-growing amounts of the data, the "attribute-oriented" granulation may turn out to be useful for high-dimensional data problems, whereby the amounts of possible features become difficult to handle. This is visible at the stage of feature selection that is aimed at deriving compact sets of attributes that can be an appropriate input to construct the final decision models. Computational complexity of typical feature selection algorithms depends heavily on the number of potentially useful and derivable features, therefore, any ideas how to reasonably introduce granulation into the feature space are essential for the Big Data.

BigData Aspects of Attribute Granulation. In this section, we discuss how the concept of granulation can be made useful in selecting and engineering features on big and possibly complex data sets. We show how to utilize the intrinsic properties of the data and underlying problem as well as background/domain knowledge for the purpose of building granular representation of attributes. All the provided tools and examples are devised to work with data sets that are very large in terms of the number of objects, as well as the number and complexity of features. Thus, we address at least some of the challenges posed by the Big Data paradigm.

Big Data is often characterized by presence of "Five Vs" – *Volume, Variety, Velocity, Variability,* and *Veracity* – reflecting in the enormous complexity that directly impacts the aforementioned data processing [16, 92]. They make it a challenge to represent the task at hand in a way that is at the same time computationally useful and comprehensible for the user. Domain experts expect intuitive data representation, whereas machine generated data collected from, e.g., large sensor networks may have all of the required technical properties but may be hardly understandable and detached from the real-life phenomenon that it is meant to record. The variety of incompatible data formats and non-aligned data structures spanning across photographs, sensor data, tweets, text documents, encrypted packets, etc., can make it hard to perform data analytics. The possible reduction and transformation of the data set provided by "classical" object-wise granulation mostly addresses the *Volume,* with some additional, lesser impact on *Velocity* and *Variability.* With granular feature selection and construction it is possible to take care of the other "Vs", in particular *Variability* and *Veracity.*

The high velocity and volume of still-incoming records are often a curse of storage systems and machine learning algorithms. Furthermore, raw records are often insufficient for the purpose of predictive analysis and the process of feature engineering is commonly employed to construct more relevant attributes [9]. The massively parallel feature engineering methods may be efficiently performed via the MapReduce programming model what, in turn, may multiply the initial number of explored attributes [138]. Still, the question remains how to choose which attributes should be evaluated. As suggested in [400], the actual feature selection process can be performed at a level of general labels of some attribute granules, whereby specific elements of those granules are not materialized prior to the algorithm's start. This style of hierarchical feature space exploration fits perfectly Algorithm 6 and its specific configurations.

From the perspective of Big Data, an introduction of some hierarchies of granularity into the spaces of investigated attributes can make the feature selection and extraction processes more efficient. Tackling the complexity of large data sets is an issue noticed by many researchers [107, 121]. The typical challenges associated with Big Data, as symbolized by the presence of "Five Vs", make things even more complicated. Besides the complexity and scale of calculations that affect the required amounts of resources, the superfluous features may negatively influence the understanding of the data by the analysts, therefore, affecting their ability to monitor and tune the knowledge discovery processes [182, 321].

Models and frameworks for parallel computing focus on various aspects of data processing [141]. Some of them respond to high velocity of the data, which makes them closer to incremental stream processing [147]. Others concentrate on

---

Algorithm 8: Granular feature selection with iterative MapReduce. In each of ℓ-phases the following program is executed.

Map(Key: $a \in A$, **Value:** G_a, V_a)   :

1 Given $\mathbb{R} = \{R_1, .., R_n\}$
2 **foreach** $R_i \in \mathbb{R}$ **do**
3 | **if** a *is relevant to* R_i **then**
4 | | $R_i \leftarrow R_i \cup \{a\}$
5 | | **emit**(sortAttributes(R_i), $\sigma_{\mathbb{G}}^i$, score)
6 | **end**
7 **end**

Reduce(Key: $R_i, \sigma_{\mathbb{G}}^i$, **Value:** $\{score, score, ..\}$):
1 **emit(** $R_i, \sigma_{\mathbb{G}}^i$, score)

---

batch processing models and adapt well-known mechanisms, such as the apriori-based breadth first exploration of a feature space [410]. Herein, the MapReduce paradigm seems to be a good choice to consider [234,239]. We may distinguish two popular approaches in this field. One of them implements the solution as a single job, whereas the other – iterative MapReduce – encompasses ℓ consecutive job runs that may be controlled automatically or manually [70,105]. One can think about parallelization of the discussed granular feature selection methods using both of these approaches.

Let us outline one of possible implementations of a massively parallel granular feature selection process as an iterative MapReduce program. Consider ℓ consecutive iterations, where each of them is based on Algorithm 8. We propose to work on the transmuted data, i.e., the mappers are executed on attributes a assigned to a granule G_a and having a vector V_a of values for objects/records in the analyzed data set. The outcome of a single iteration is a sorted set of candidate attribute subsets, whereas only n best intermediate outputs $\mathbb{R} = \{R_1, .., R_n\}$ are passed to the subsequent phase. The map functions are provided with the collection $\mathbb{R}$ and the vector V_d containing values of the decision attribute d. To each subset R_i there has been assigned granulation preferences $\sigma_{\mathbb{G}}^i$, whereby the diversification of granule-level permutations may play a similar role as for the previously discussed attribute-level permutations. During the evaluation of a, we verify its relevance to every considered R_i, with respect to a quality function Q, preferences $\sigma_{\mathbb{G}}^i$, or any other factor of interest. If the performed assessment reveals that a is relevant for R_i (where relevance may be expressed as a mixture of preference, contribution, etc.), then the set $R_i \cup \{a\}$ is emitted. The role of reducers is then to aggregate subsets R_i and sort them according to their score. The whole process ends when the expected number of feature subsets satisfies $\mathbb{C}$.

The main objective of the above illustrative example of a MapReduce program is to evaluate a possibly large number of attribute subsets, in order to reach a higher quality, compactness and/or diversification of the produced out-

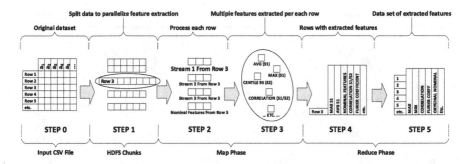

Fig. 12. The overview of feature extraction process split into individual steps. The labels above the curly braces (at the top of the diagram) indicate objectives in each processing step. The labels below the curly braces (at the bottom of the diagram) indicate how the individual processing steps were implemented.

comes. Obviously, parallel programming models allow to implement the granular feature selection framework in many other ways [239,310]. Let us also mention DiReliefF, a distributed version of the well-known ReliefF [286], or fast-mRMR algorithm for high-dimensional data [316].

The idea of operating on attribute granules – regardless of their origin – is worth combining with the principles of parallelization of feature selection methods with respect to complex spaces of derivable features and their subsets, yet fitting the iterative nature of most ML algorithms [325]. In [350], granular computing was utilized to discretize M-factors time series data to obtain granular intervals. Information granules naturally emerge when dealing with data, including those coming in the form of data streams [296]. However, regardless the particular application the ultimate objective is to describe the underlying phenomenon in an easily understood way and at a certain level of abstraction to enable human-system interaction.

4.3 Framework for Multi-stream Data Analysis

In this section, we focus not only on the extracted features and constructed prediction models but also on data processing stages that are designed to let it work within a big data processing environment [141,315], and particularly with high dimensional, multi-stream data [184,187]. In order to provide high quality assessments, the presented solution requires constructing an ensemble of diverse models [94,182]. The diversity may be obtained by employing a variety of models computed on different subsets of attributes and data samples. For this task, the granular similarity measures (Sect. 4.2) or resilient attribute subsets (Sect. 3.2) may be applied. As a result of blending diverse models, the final ensemble minimizes the impact of a concept drift [45], and achieves a better prediction quality [182]. The proposed architecture can be used both in the

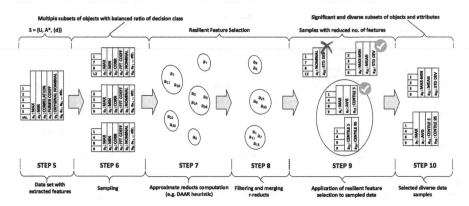

Fig. 13. The overview of the feature selection process.

incremental, stream processing model [37,136], and in highly scalable, batch processing model, i.e., MapReduce [22,88].

In Fig. 12, a high-level overview of the feature extraction process divided into individual steps is presented. The'original data set' in STEP 0 corresponds to a collection of historical data provided as a training set for a machine learning task, where features: $a_1, a_2, a_3, ..$ correspond to attributes in the data. STEP 1 is designed to partition the original set into individual rows (objects) in order to parallelize calculations - this step may be implemented within, e.g., MapReduce framework [138]. The purpose of STEP 2 is to split each row into static data, e.g., features reflecting assessments obtained interactively from domain experts (as discussed in Sects. 2.4, 4.2), and time series data (Sect. 4.1). In STEP 3, the feature extraction framework is applied to each time series in the data, e.g., to a numerical time series containing consecutive values expressing the average energy of the most active geophones at a longwall in a coal mine [182], and all features derived from time series are constructed (as described in Sect. 4.1). In STEPs 4 and 5, all attributes are combined together.

The process of feature engineering is performed basing on the sliding time-window approach that is designed to process data sets containing multiple time series (Sect. 4.1). During the process of moving a time window through the series, aggregating functions are applied. Table 7 presents the overview of features that may be extracted from individual time series. As emphasized in Sect. 4.1, standard statistics extracted form a sliding window may be supplemented by more sophisticated ones, e.g., correlations between pairs of time series. Furthermore, since more than one window is generated per time series, we may extract inter window statistics - as depicted in Fig. 8. That is, a set of values which express changes between the same statistics obtained in consecutive sliding windows. The inter window stats are presented in Table 7.

During the feature extraction process, a large number of potentially relevant, however very often redundant, data characteristics are generated [146,426].

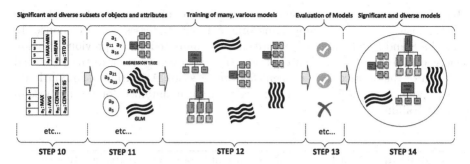

Fig. 14. Ensemble blending.

Therefore, after the construction of features, an attribute subset selection algorithm is applied to reduce the attribute space [61]. In the case of designing decision support systems, the scope of feature selection is twofold, related to both interaction with domain experts and analysts while running the system online, as well as off-line exploration of gathered data in order to find out the best feature selection algorithms and prepare the best possible feature sets for further processing [56, 362]. It is herein worth noting that the step of feature selection - conducted independently or in an iterative fashion - is often taken into account in combination with various machine learning methods such as neural networks or support vector machines while building decision systems, e.g., aimed at equipment and environment monitoring in coal mines [237].

In Fig. 13, the overview of the feature selection process, split into individual steps, is presented. In STEP 6, the random samples of objects with a more balanced distribution of classes are drawn. The generated samples are randomly divided into two disjoint groups. First one serves for the purpose of feature selection, whereas the second group of samples is used for training predictive models. It should be noted that, in order to use the feature selection algorithms derived from RST (selected RST based FS methods are surveyed in Sect. 2.5, the novel ones are introduced in Sect. 3), numerical attributes in data should be subjected to discretization (discretization methods are surveyed in Sect. 2.1).

In STEP 7, the reduced attribute subsets are calculated. With regard to the proposed architecture (Fig. 13), we decided to focus mainly on filter-based methods, which (comparing to wrapper and embedded techniques - Sect. 2.3) assure relatively high computational efficiency, as well as independence of the resulting feature sets from a particular model. This last property allows the obtained feature sets to be used in combination with various types of forecasting approaches. Among filter-based feature selection methods, we pay special attention to multivariate algorithms. One of the most prominent examples of this approach are methods based on the mRMR framework. Another popular approach that is implemented in the presented framework refers to computation of approximate decision reducts developed within the theory of rough sets, e.g., dynamically adjusted approximate reducts (DAAR), where a statistical test based on random probes is used to avoid selection of features that are likely to distinguish data records supporting different target classes only by chance. In STEPs 8 and 9, a number of feature subsets computed in the previous step are merged into several larger subsets. For this purpose, we refer to the presented version of the approximate resilient feature selection Algorithm 5. In STEP 10, only significantly different attribute subsets are maintained for the purpose of model training.

Algorithm 9: Construction of the ensemble of diverse regression models.

Data:
- *attSubsets* - pre-calculated subsets of attributes (e.g., approximate reducts)
- *objectSamples* - pre-calculated samples of objects
- *testSet* - a test set
- *regressionAlgorithms*, e.g., { rPart, SVM, glm }
- *allowedAttempts* and *minQuality* - parameters governing quality of models

Result: *ensemble of regression models*

1 /* Initialization of variables   */
2 $ensemble \leftarrow \emptyset$; $weakAttempts \leftarrow 0$
3 $alg \leftarrow regressionAlgorithms.removeFirst$
4 **while** *TRUE* **do**
5     $a_1, a_2 \leftarrow attSubsets.drawAndRemoveTwo$
6     $b_1, b_2 \leftarrow objectSamples.drawAndRemoveTwo$
7   /* Models are trained and validated on different samples   */
8     $model \leftarrow alg.trainAndEvaluate(a_1, b_1, a_2, b_2)$
9     $score \leftarrow model.score(testSet)$
10   /* The ensemble is expanded only if the newly trained model meets
        the specified quality criteria and there is no other similar
        model in the ensemble.   */
11   **if** $score > minQuality \wedge \neg ensemble.containsSimilar(model, score)$ **then**
12 | $ensemble \leftarrow ensemble \cup \{model \oplus score\}$
13   **else**
14         $weakAttempts \leftarrow weakAttempts + 1$
15   **if** $weakAttempts < allowedAttempts$ **then**
16 | $continue$;
17   **end**
18   **if** $regressionAlgorithms \neq \emptyset$ **then**
19 | $alg \leftarrow regressionAlgorithms.removeFirst$
20 | $weakAttempts \leftarrow 0$
21   **else** /* end of ensemble blending */
22 | $break$;
23   **end**
24   **end**
25 **end**
26 $return \sum_{s \in ensemble.scores} s$;

In order to provide a good clarity of the presentation, in the subsequent steps of the framework (in Fig. 14), let us focus on the well known task of regression analysis[13]. The final solution is an ensemble of diverse regression models. The diversity is achieved by training models on different subsets of attributes and objects (STEP 11). The Algorithm 9 for blending models refers to STEPs: 11 to 14 in Fig. 14, and is designed in a way which guarantees that a model can be included only if it is accurate enough on validation data, and sufficiently different from already selected predictors (e.g., correlation of its prediction with predictions of other models is small enough). This could be seen as a method for increasing robustness of predictions in the case of noisy and heterogeneous data. An additional advantage of using different features for different models is that it may reduce the influence on the final ensemble of a concept drift between the training and test cases. This approach is also expected to protect the model against over-fitting and, hence, a decrease in the prediction quality on the final test set.

5   Evaluation, Practical Applications

The goal of this Section is to present examples of a data mining investigation, which illustrate the performance of the framework presented in Sect. 4 and the resilient feature selection methods described in Sect. 3 when handling real-life problems related to coal mining and fire & rescue domains. The following study complements the so-far presented research with the experimental evaluation of the proposed feature extraction and selection methods.

5.1   Methane Outbreaks

In this section, we provide a broad experimental evaluation of learning forecasting models over large multi-sensor data sets, including the steps of sliding window feature extraction and rough-set-inspired feature subset ensemble selection. The considered task is to construct a model capable of predicting dangerous concentrations of methane at longwalls of a coal mine basing on multivariate time series of sensor readings. We show how the described framework performed on data collected from a sensor network in an active coal mine and, how the complete mechanism can be built into DISESOR - a particular decision support system.

The contributions in this section refer to both the analysis of how the nature of sensor readings influenced the architecture of the developed solution and the empirical proof that the designed methods turned out to be efficient in practice. Furthermore, we elaborate on the resilience of the solution in the case of partial data loss, e.g., when particular data sources (e.g., sensors) are damaged or inactive.

---

[13] The description for the classification tasks would differ mainly in the training algorithms, and model evaluation criteria used.

Natural Hazards Monitoring in Coal Mines. Coal mining requires working in hazardous conditions. Miners in an underground coal mine can face several threats, such as methane explosions, rock-bursts or seismic tremors, etc. [49, 237,260]. To provide protection for people working underground, systems for active monitoring of production processes are typically used [219]. One of their fundamental applications is screening dangerous gas concentrations (methane in particular) in order to prevent spontaneous explosions [184]. For that purpose, the ability to predict dangerous concentrations of gases in the nearest future can be even more important than monitoring the current sensor readings [360].

Coal mines are well equipped with monitoring, supervising, and dispatching systems connected with machines, devices, and transport facilities. There are also systems for monitoring natural hazards. Such systems are provided by many different companies, hence some problems with data quality, integration, and interpretation may be observed (Sect. 4.1). Once someone is able to overcome these issues, the collected data can be used for ongoing visualization of conditions in particular places of a mine [274]. Moreover, by utilizing the domain knowledge and patterns derived from integrated historical data [113], one can construct forecasting models to enrich the upcoming sensory data with additional predictions. This way, it is possible to considerably improve both the safety of miners and work efficiency [268]. For example, thanks to short-term prognoses related to methane concentrations combined with information regarding the location and work intensity of a cutter loader, it is possible to prevent emergency energy shutdowns and maintain continuity of mining [184]. This, in turn, allows for increasing the production volume and reducing the wear of electrical elements whose exploitation time largely depends on a number of switch-ons and switch-offs [349]. Furthermore, a decision support system should be easily comprehended by the experts and end-users who, not only, need access to its outcomes, but also to arguments or causes that were taken into account (Sect. 4.2).

DISESOR is a decision support system designed for monitoring potential threats in coal mines [360], which processes data from sensors of various types, like: $CO2$, methanometers, machine monitoring devices and many others (Table 20, and Appendices A.1, A.2). In Fig. 15, a draft architecture of DISESOR system is presented [306]. Basing on a vast amount of sensor readings collected via integration of data from various monitoring systems [142], e.g., THOR or Zefir, DISESOR provides predictive analytics of mine conditions and threats. As the most important use cases of the system we can indicate: Assessment of seismic hazard probabilities in the vicinity of the mine; Forecasting dangerous increase in the methane concentration in the mine shafts; Forecasting of the possible ranges of the sensor readings in advance; Detection of endogenous fires and conveyor belts fires; Detecting anomalies in the consumption of media such as electricity [181].

From data processing point of view, a decision support system that aid in controlling the coal mining process requires efficient methods that can handle large volumes of data from many sensors enriched with features provided by

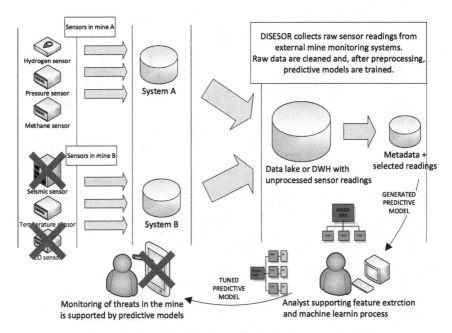

Fig. 15. DISESOR system architecture.

domain experts (Sects. 11, 4.3). The continuous collection and analysis of multiple streams of readings from a large network of sensors located underground raises certain problems with providing expected resilience (Sect. 3) in business continuity plans. In Figs. 15 and 16, a potential impact of missing data sources is shown. Figure 16 outlines – in a schematic manner – the window based processing of two data streams from methane and CO_2 sensors. We may notice on the drafts above that the impacted stream (in red) propagates the problem forward and blocks the subsequent processing tasks. That is especially harmful when we combine a number of sources – i.e., joining data streams in Fig. 15. The majority of predictive models are very sensitive to the (in)completeness of input data.

Predictive models providing a proper assessment of potentially dangerous methane concentrations (Sect. 5.1) and seismic events (Sect. 5.2), which are resilient to missing data sources and are able to interact with domain experts to use their assessments, could significantly improve the safety and reduce the costs of underground coal mining.

IJCRS'15 Data Challenge. Based on the sensor readings collected in an active Polish coal mine, a data mining competition was organized at the international conference IJCRS'15 [184]. By publishing this data set and defining the corresponding problem in the form of a competition task, we obtained and analyzed 1,676 solutions submitted by 90 registered research teams from 18 different countries. Additionally, 40 teams provided reports describing their approach.

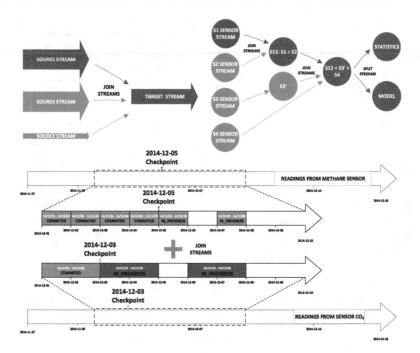

Fig. 16. Sliding window data flow and missing data impact.

Altogether, these solutions can be regarded as state-of-the-art in the predictive analysis of multivariate time series data and as the reference in our research.

Data prepared for the competition correspond to a mining period between March 2, 2014, and June 16, 2014. Among the thousands of sensors located over tens of kilometers of underground corridors, 28 sensors monitoring the work in the immediate vicinity of the shearer workplace were selected. Prepared data records were composed of raw sensor readings arranged in 10-minutes time series, with measurements taken every second. Hence, each record consisted of 16,800 numerical features, i.e., 600 values per sensor. The detailed information about all the sensors can be found in Table 20. In Fig. 17, a detailed location of all sensors, as well as a workplace of a longwall shearer, on a fragment of the coal mine plan is shown.

Each row of the training data was tagged with three labels, each from the set $\{0, 1\}$, where 0 and 1 corresponded to "normal" and "warning" labels, respectively. Labels indicated whether a warning threshold had been reached in a period between three and six minutes after a given measurement, for three methane meters denoted as $MM263$, $MM264$ and $MM256$ (Fig. 17). If a given data row corresponded to a time period between t_{-599} and t_0, then its dependent variable value for a meter MM was "warning", if and only if $\max\{MM(t_{181}), \ldots, MM(t_{360})\} \geq \rho$, where ρ is a safety threshold. The value of this threshold may vary for different longwalls, however, it is usually set between

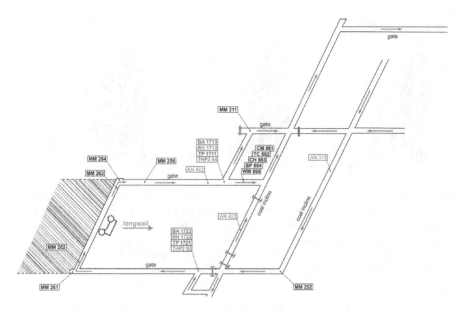

Fig. 17. A scheme of the mining process corresponding to the data set considered in [184]. A shearer moves along the wall of coal extraction between the sensors MM261 and MM264. The progress of the coal extraction is unveiled by an arrow described with "longwall". Thin arrows depict flow direction of the air in the mine sidewalks which is enforced by a ventilation system.

1% and 1.5% on the basis of interviews with experts (see, e.g., a sensor card in [360]) and the national regulations on hazard estimation [135].

The training set contained sensor readings registered within 51,700 time periods. Periods in the training set were overlapping and given in a chronological order. However, periods included in the testing set did not overlap and they were given in a random order. Figure 18 presents frequency distributions of values for the three sensors $MM263$, $MM264$ and $MM256$. The vast majority of the observations stored in the data set are below the warning threshold. Table 8 presents the amount of "warnings" observed for each investigated sensor in the training data and may be used as a premise to realize the decision class imbalance in the context of methane concentration monitoring. Selected, more in-depth insights into the methane-related data are provided in Appendix A.1. Data sets are available online on the KnowledgePit platform.

The task of the data challenge was to predict the likelihood of the label 1 ("warning") for the threshold $\rho = 1\%$. The solutions were evaluated with the Area Under the ROC Curve (AUC) measure, which was computed separately for each of the target sensors. The final score corresponded to the average AUC for a submitted solution "s":

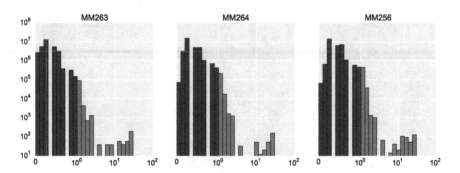

Fig. 18. Frequency distributions for sensors MM263, MM264 and MM256 in the training data set. The majority of readings are in the $[0, 1.2]$ range and a relatively small number is spread in the $(1.2, 30]$ range. The dark blue bars drawn on a linear scale with a step of 0.1 correspond to the readings below the warning threshold the threshold $\rho = 1\%$. The light blue bars drawn on a logarithmic scale with a step of 1 represent hazardous situations as well as outliers. (Color figure online)

$$score(s) = \frac{AUC_{MM263}(s) + AUC_{MM264}(s) + AUC_{MM256}(s)}{3} \qquad (25)$$

Table 9 shows the scores of top-ranked teams together with the score of our methods. Baseline model was created by averaging 10 simple rule-based models computed out-of-the-box using the *RoughSets* package [323]. The Zagorecki approach [425] assumed generation of a large number of variables characterizing sensor measurements and operating with the time series derived from those measurements. Separate random-forest-based models were then created to predict the "warning" states for each of three considered methane meters. On the other hand Boullè [44] focused on a problem of distribution drift between the training and testing data sets. Informativeness of each considered feature with respect to both classification and drift detection was evaluated. As a result, the training data set was reduced to a single sensor per target class. The prediction model was then generated by the Naïve Bayes classifier. Grzegorowski and Stawicki [146] provided a logistic regression model based on the linear combination of selected three features – extracted with the sliding window framework (presented in Sect. 4.1). The Ruta and Cen [331] method was also based on a logistic regression model computed over a small subset of sensor observations. The authors utilized their self-organizing framework to choose this particular model out of a number of other solutions including decision trees, support vector machines, etc. Among other successful approaches used by participants of the competition, there were also deep learning models using the LSTM networks [292].

Based on the analysis of the most successful solutions submitted to the competition (Table 9), we reached a conclusion that a robust prediction of methane concentration levels can be achieved even when a small subset of features is used for constructing the model. Although the Zagorecki [425] solution used nearly

Table 8. Occurrences of the labels in the training data set.

	MM263	MM264	MM256	count
Label values	Normal	Normal	Normal	48695
			Warning	1208
		Warning	Normal	1258
			Warning	74
	Warning	Normal	Normal	435
			Warning	24
		Warning	Normal	2
			Warning	4

Table 9. A comparison of the logistic regression performance for the competition data set between the implemented feature extraction methods and the top ranked solutions of IJCRS'15 Data Challenge.

Method	Macro averaged AUC (25)
Zagorecki [425]	0.9592
FE+DAAR	0.9545
Grzegorowski and Stawicki [146]	0.9473
Boullè [44]	0.9439
Ruta and Cen [331]	0.9436
FE+mRMR	0.9413
Pawłowski and Kurach [292]	0.94
Baseline	0.9004

5,000 features in the learning process, several of the other top-ranked teams achieved similar results with models considering far fewer features. Another interesting outcome was that a vast majority of solutions followed the ideas of producing sliding window aggregations and that such aggregations, treated as low-level features, were useful while learning various prediction models [226].

Evaluation of Multi-stream Framework. In the second part of our experiments, we considered exactly the same sensor readings as those used in Sect. 5.1, now, taking into account our multi-stream feature extraction framework with ensemble blending (Sect. 4.3). We applied two different feature selection methods into our framework and we examined the AUC scores of prediction outcomes obtained using ensembles of simple models. Both cases were following the general framework presented in Sect. 4.3, in particular, in Algorithm 9. In both cases the training algorithm was used independently for each of three dependent variables – implementing a particular transformation for the original multi-target problem [42]. Subsequently, we extended the analysis with two new multi-stream

---

Algorithm 10: Implemented version of mRMR feature selection method.

Input: set of features A and dependent variable d;
$\phi : A \times A \cup \{d\} \to \mathbb{R}^+$ function for measuring dependency;
$N \in \mathbb{N}; \varepsilon \in [0, 1)$;
Output: subset of features $R \subseteq A$

1 **begin**
2      $stopFlag \leftarrow FALSE$;
3      $R \leftarrow \arg\max_{a \in A} \phi(a, d)$;
4      $A \leftarrow A \setminus R$;
5   **while** $stopFlag == FALSE$ **do**
6          $\bar{a} \leftarrow \arg\max_{a \in A} \left(\phi(a, d) - \max_{b \in A'} \phi(a, b) \right)$;
7   **foreach** $i \in 1, ..., N$ **do**
8              $\bar{p}_i \leftarrow$ random permutation of $\bar{a}$;
9   **end**
10   **if** $\frac{|\{i : |\phi(\bar{p}_i, d)| > |\phi(\bar{a}, d)|\}| + 1}{N + 2} > \varepsilon$ **then**
11             $stopFlag \leftarrow TRUE$;
12   **else**
13             $R \leftarrow R \cup \bar{a}$
14   **end**
15   **end**
16 **end**

---

data sets, both related to hard coal mining, to verify how effectively the discussed approach could transfer to new data in the same domain with implementation changes limited to adaptation of the sliding window feature extraction layer only [334].

In the first setup of the framework, we applied our version of the minimum redundancy maximum relevance (mRMR) method [316]. Comparing to the standard mRMR, provided modifications are related to criteria for selecting the best feature in each iteration and to a stopping condition – outlined in Algorithm 10. First, we select a feature that maximizes the difference between its relevance (the dependency score $\phi(a, d)$) and its maximal dependency on features selected before. Second, we stop the algorithm if the feature selected in a given iteration does not pass the random probe test, i.e., the estimation of the probability that a randomly generated feature obtains a higher score than the selected feature exceeds an allowed threshold [374]. Thus, we guarantee compactness and the relatively high independence of the resulting feature set.

We obtained three small subsets of features containing three, six and seven elements, respectively. With these feature sets, we trained three independent logistic regression models and utilized them to make predictions for the testing cases. Although we used a very simple prediction method and a completely automated feature selection, the average AUC of this solution for the testing set was 0.9413 – see FE+mRMR in Table 9.

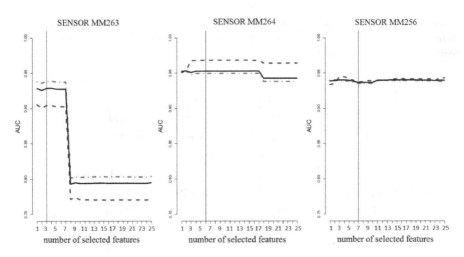

Fig. 19. AUC values obtained for simple linear regression models trained on features selected in subsequent iterations of mRMR procedure for each of dependent variable values in the data: dashed blue lines show the scores on the preliminary testing set; dashed-dotted red lines show the scores on the final testing set; thick black lines show the scores on the joined testing set; thin vertical lines mark the iteration on which the stopping condition of our implementation of mRMR algorithm was triggered. (Color figure online)

In order to verify the stopping criteria we repeated the experiment with the probe condition switched off. Figure 19 shows the results of the forecasting model trained using features selected in 25 consecutive iterations of the mRMR procedure. It can be seen that, for each of dependent variable values, the results obtained for our stopping method were close to optimal. Moreover, by virtue of the random probe test, the resulting feature subsets were compact. Overall, they contained the lowest number of features among the solutions submitted by the top-ranked teams in the competition.

In the second setup, we implemented the granular algorithmic schema outlined in Algorithm 7 using – as a granulation technique – a DAAR heuristic [186]. For each of three dependent variables, we computed 10 different feature sets. Then, for each set, we computed a logistic regression model using the same training algorithm as for features selected with mRMR. Finally, we created a simple ensemble by averaging their prediction. The final score achieved by the ensemble of DAAR-based logistic regression models was 0.9545 – see FE+DAAR in Table 9.

In order to thoroughly investigate the performance of the proposed framework with particular interes on feature extraction and selection part, we evaluated it on two additional data sets that were obtained from different coal mines. The first of those sets contained sensor readings from over a month (November 2007 – December 2007). Similarly as the first set there were three target methane meters (MM532, MM533 and MM534) and the sensor reading frequency was one

Table 10. A comparison of logistic regression performance (AUC measure) for individual target methane meters from the three data sets considered in our study (macro averaged AUC – in the last row).

Target variable	FE+mRMR	FE+DAAR	Zagorecki [425]
MM256	0.9432	**0.9579**	0.9439
MM263	0.9374	0.9564	**0.9760**
MM264	0.9433	0.9492	**0.9579**
MM532	**0.9176**	0.9170	0.8968
MM533	0.8501	**0.8681**	0.8283
MM534	0.9276	**0.9321**	0.9299
MWR116	0.9389	0.9431	**0.9575**
Macro averaged AUC	0.9226	**0.9320**	0.9272

per second. The data set contained readings from eight sensors plus additional information regarding the coal shearer status. After the initial preprocessing, the data consisted of 51,329 records corresponding to 10-min periods of sensor readings. These records were divided into two disjoint sets - one for training the compared models and the second one for validation. The testing set corresponded to the last two weeks of sensor readings.

In the second of the considered new data sets, there was only one target methane meter (denoted by MWR116) and the data spanned across five months (September 2013 – January 2014). In this set, there were readings from 14 sensors, sampled once per minute. After the initial preprocessing, the set consisted of 204,465 records. They were divided into separate training and testing sets as well. The test period corresponded to the last two months of sensor readings. The results obtained for each of the target sensors from all three data sets are presented in Table 10. In addition to our own models, we include there the results obtained using implementation of the model reported by the IJCRS'15 competition winner [425].

It is also worth to notice that the multi-stream framework setup with DAAR feature selection algorithm achieved the highest macro average AUC on all seven target sensors (Table 10). The paired Wilcoxon test did not reveal statistically significant differences in the results between the AutoML multi-stream framework results (Sect. 4.3) and the best of the fine-tuned solutions constructed on nearly 5000 features. It is, therefore, sufficient to conclude that the proposed framework can successfully replace more complicated and hardly interpretable machine learning approaches. Moreover, computation time required to train our model was an order of magnitude lower. For instance, for the third data set, our model was constructed in 19 min, whereas the construction of the random forest model took nearly five hours.

The aforementioned analysis clearly shows the accuracy obtained using our approach, taking into account its subsequent layers of feature creation and selection (Figs. 12 and 13) and the forecasting models training and ensemble blending (Fig. 14). Both evaluated feature selection approaches – mRMR and DAAR methods – yield very good results even when combined with the simplest possible prediction techniques – the logistic regression. From the prediction accuracy perspective, they perform comparably to the model developed by the competition winner which was manually tuned for over two months. Moreover, they are easy to maintain, efficient to compute and, what is maybe the most important aspect from the point of view of interactiveness, they are understandable for the system users and domain experts by means of operating with small subsets of intuitively defined features. In particular, this is why the DAAR-based method was deployed in a production system responsible for processing sensor readings collected from multiple monitoring and dispatching systems deployed in different coal mines. [360].

Impact of Feature Extraction on Resilience. Let us evaluate the impact of the sliding window-based feature extraction on the quality and resilience of methane prediction. For that purpose, we refer to the same data set of sensor readings collected from an active coal mine in Poland between March 2, 2014, and June 16, 2014 - as described in Sect. 5.1. In the following study, we performed a series of experiments on both raw, unprocessed data and on the data after performing feature extraction (as described in Sect. 4). The problem was to predict maximal methane concentrations in a six minutes time horizon for three selected methane meters, denoted as: $MM263$, $MM264$ and $MM256$ (Fig. 17) – similarity as in JCRS'15 Data Challenge – see Sect. 5.1. This time, however, evaluations were performed with three error measures designed to assess regression problems: mean absolute error (MAE), root mean squared error (RMSE), and root relative squared error (RRSE).

In the frame of the experiment, we focused on those methods that can predict the methane concentration even in the absence of selected conditional attributes, including linear regression (lm), two implementations of regression trees (rpart, ctree), regression rules (cubist), and gradient boosting (gbm). As in most short-term regression problems, the last know value (last val.) is usually a reasonably good naive approach, commonly used as a baseline forecast. Apart from that, we also used as a predictor a very simple statistic, that is an average calculated on the training set (train μ). Since the purpose of the experiment was to assess the relative impact of the performed feature extraction vs. using raw data, we did not enforce any parameter tuning of the models. The results are summarized in Table 11.

The first column in Table 11 indicates the target variable. The second one provides information about the prediction method used. Three consecutive columns provide information about the prediction error under the assumption that all the sensors from the training set were available during the assessment - the best results per each target variables in bold. We may notice that the results of simple approaches like the last value were quite often the best. Regression threes (rpart) performed very well in the case of $MM256$, however by far the

Table 11. Impact of missing features on prediction quality. No feature extraction applied.

Target Attr. MM256

Method	− MAE	RMSE	RRSE	MM256 MAE	RMSE	RRSE	MM263, MM264 MAE	RMSE	RRSE	MM256, MM263, MM264 MAE	RMSE	RRSE
train μ	0.1756	0.3532	1	0.1756	0.3532	1	0.1756	0.3532	1	0.1756	0.3532	1
last val	**0.0733**	0.2971	0.841	–	–	–	**0.0733**	**0.2971**	**0.841**	–	–	–
lm	0.1335	0.3198	0.9052	0.2166	0.3834	1.0855	0.0969	0.3033	0.8586	0.19	0.3652	1.0339
rpart	0.0906	**0.2941**	**0.8327**	0.1688	0.3796	1.0747	0.1179	0.7232	2.0474	0.1846	0.5273	1.4928
ctree	0.0918	0.3297	0.9334	**0.144**	**0.3395**	**0.9611**	0.0883	0.305	0.8635	**0.1529**	**0.3491**	**0.9883**
gbm	0.1623	0.3459	0.9793	0.1719	0.3508	0.993	0.1623	0.3459	0.9792	0.1738	0.3516	0.9955
cubist	0.1328	0.3892	1.1017	0.2198	0.4047	1.1456	0.1238	0.3384	0.958	0.3555	0.632	1.7891

Target Attr. MM263

Method	− MAE	RMSE	RRSE	MM263 MAE	RMSE	RRSE	MM256, MM264 MAE	RMSE	RRSE	MM256, MM263, MM264 MAE	RMSE	RRSE
train μ	0.1331	0.3247	1	**0.1331**	**0.3247**	1	0.1331	0.3247	1	**0.1331**	**0.3247**	1
last val	**0.0498**	**0.2905**	**0.8945**	–	–	–	**0.0498**	**0.2909**	**0.8957**	–	–	–
lm	0.0612	0.2909	0.8957	0.176	0.3484	1.073	0.0601	0.291	0.8961	0.1772	0.3483	1.0725
rpart	1.1592	3.0303	9.3317	0.1348	0.3285	1.0115	1.1592	3.0303	9.3317	0.1337	0.3263	1.005
ctree	0.2502	1.2467	3.8391	0.1792	0.3577	1.1015	0.2508	1.246	3.8369	0.1826	0.3591	1.1058
gbm	0.1245	0.3198	0.9849	0.1341	0.3255	1.0024	0.1245	0.3198	0.9849	0.1341	0.3255	1.0023
cubist	0.1796	0.6836	2.105	0.2517	0.4375	1.3474	0.1005	0.331	1.0192	0.1869	0.3683	1.134

Target Attr. MM263

Method	− MAE	RMSE	RRSE	MM264 MAE	RMSE	RRSE	MM256. MM263 MAE	RMSE	RRSE	MM256. MM263. MM264 MAE	RMSE	RRSE
train μ	0.165	0.3267	1	0.165	0.3267	1	0.165	0.3267	1	0.165	0.3267	1
last val	**0.0632**	**0.2704**	**0.8277**	–	–	–	**0.0632**	0.2704	0.8277	–	–	–
lm	0.0823	0.2714	0.8308	0.165	0.3267	1	0.0715	**0.2667**	**0.8165**	0.1711	0.3316	1.0151
rpart	0.0738	0.3411	1.0441	0.1799	0.341	1.044	0.0819	0.3981	1.2187	0.2256	0.3857	1.1808
ctree	0.0892	0.3231	0.9892	0.2328	0.413	1.2643	0.0722	0.269	0.8236	**0.1534**	**0.3217**	**0.985**
gbm	0.1551	0.3192	0.9773	**0.1623**	**0.3247**	**0.9939**	0.1552	0.3192	0.9773	0.1636	0.3249	0.9946
cubist	0.0857	0.2894	0.886	0.4085	0.6752	2.0671	0.1056	0.303	0.9276	0.2559	0.4242	1.2986

Table 12. Impact of sliding window feature extraction on resilience of methane prediction.

Target Attr.: MM256

Method	Missing: -			Missing: MM256			Missing: MM263, MM264			Missing: MM256, MM263, MM264		
	MAE	RMSE	RRSE	MAE	RMSE	RRSE	MAE	RMSE	RRSE	MAE	RMSE	RRSE
rpart without FE	0.0906	0.2941	0.8327	0.1688	0.3796	1.0747	0.1179	0.7232	2.0474	0.1846	0.5273	1.4928
rpart+FE(l:1/o:1)	0.0713	0.2661	0.691	0.1146	0.3126	0.7248	0.0567	0.274	0.6888	0.1139	0.3067	0.7711
rpart+FE(l:3/o:1)	0.0737	0.2543	0.6603	0.1163	0.2905	0.7542	0.0591	0.2897	0.7284	0.1204	0.316	0.8206
rpart+FE(l:6/o:1)	0.0733	0.2582	0.6491	0.1141	0.224	0.5817	0.0701	0.2945	0.7403	0.1153	0.2628	0.6823
best rpart+FE	**0.0713**	**0.2543**	**0.6491**	**0.1141**	**0.224**	**0.5817**	**0.0567**	**0.274**	**0.6888**	**0.1139**	**0.2628**	**0.6823**
best without FE	0.0733	0.2941	0.8327	0.144	0.3395	0.9611	0.0733	0.2971	0.841	0.1529	0.3491	0.9883

Target Attr.: MM263

Method	Missing: -			Missing: MM263			Missing: MM256, MM264			Missing: MM256, MM263, MM264		
	MAE	RMSE	RRSE	MAE	RMSE	RRSE	MAE	RMSE	RRSE	MAE	RMSE	RRSE
rpart without FE	1.1592	3.0303	9.3317	0.1348	0.3285	1.0115	1.1592	3.0303	9.3317	0.1337	0.3263	1.005
rpart+FE(l:1/o:1)	0.0521	0.2708	0.7395	0.137	0.3335	0.911	0.0457	0.2598	0.686	0.1234	0.3161	0.8348
rpart+FE(l:3/o:1)	0.0753	0.2368	0.6467	0.1356	0.2862	0.7816	0.0544	0.2569	0.6785	0.1204	0.316	0.8206
rpart+FE(l:6/o:1)	0.0521	0.2596	0.6459	0.12	0.2438	0.6658	0.0511	0.2526	0.667	0.1338	0.3004	0.8204
best rpart+FE	**0.0521**	**0.2368**	**0.6459**	**0.12**	**0.2438**	**0.6658**	**0.0457**	**0.2526**	**0.667**	**0.1204**	**0.3004**	**0.8204**
best without FE	**0.0498**	0.2905	0.8945	0.1331	0.3247	1	0.0498	0.2909	0.8957	0.1331	0.3247	1

Target Attr.: MM264

Method	Missing: -			Missing: MM264			Missing: MM256, MM263			Missing: MM256, MM263, MM264		
	MAE	RMSE	RRSE	MAE	RMSE	RRSE	MAE	RMSE	RRSE	MAE	RMSE	RRSE
rpart without FE	0.0738	0.3411	1.0441	0.1799	0.341	1.044	0.0819	0.3981	1.2187	0.1534	0.3217	0.985
rpart+FE(l:1/o:1)	0.0698	0.2555	0.665	0.1227	0.2886	0.7337	0.0635	0.3262	0.8492	0.1257	0.328	0.8537
rpart+FE(l:3/o:1)	0.07	0.2351	0.5978	0.1295	0.2651	0.6901	0.0497	0.2894	0.7533	0.1245	0.2925	0.7614
rpart+FE(l:6/o:1)	0.0719	0.2433	0.6333	0.1241	0.3113	0.8102	0.0619	0.2952	0.7684	0.1294	0.2613	0.68
best rpart+FE	0.0698	**0.2351**	**0.5978**	0.1227	**0.2651**	**0.6901**	**0.0497**	**0.2894**	**0.7533**	**0.1245**	**0.2613**	**0.68**
best without FE	**0.0632**	0.2704	0.8277	0.1623	0.3247	0.9939	0.0632	**0.2667**	0.8165	0.1534	0.3217	0.985

worst in the case of $MM263$ – we could observe similar behavior whenever the $MM263$ was available in the conditional attribute set - evidently, the cause of the observed over-fitting. This raised the question of whether the rpart model on different data representation (i.e., with sliding window-based feature engineering applied) could result in a more robust behavior of this method.

Due to hazardous events or harsh conditions prevailing in mines, sensors or wires transmitting data may be damaged. As presented in Fig. 15, this may cause gaps in the collected readings, resulting in missing values of particular attributes. Thus, all the predictive models utilizing affected features from failed sensors may become impacted (Fig. 15). To verify the resilience of the methods to missing attributes, we performed the following three experiments: we had been removing from the test data the most important attributes: $MM256$, $MM263$, and $MM264$. The *"Missing:"* keywords in Table 11 indicate rows with sensor symbols that were excluded from the conditional attributes in the test set. Columns 6–8 indicate the error measured when the historical values of the target variable were not present - such a situation obviously disables forecasts with the *last value* predictor. The last three columns (12–14) provide an assessment of the methane concentration forecasts when all $MM256$, $MM263$ and $MM264$ sensors would be, for some reason, disabled. We may notice that tree-based methods handled relatively well that crisis-scenario.

In Table 12, we provide results of the same experiments, this time, however, performed on a slightly different representation of data - after the sliding window feature extraction was applied. For that study, we picked one of the tree-based models - *rpart* - and applied three different sliding window setting with respect to window length and offset (Recall, e.g., Fig. 1) designated with "+FE" in Table 12. For each $rpart+FE(l : \ldots /o : \ldots)$ method, information in brackets indicates a particular sliding window setting, where "l:" indicates *a length of a time window* (in minutes) and "o:" refers to *an offset of a time window* (in minutes). To emphasize the quality gain achieved with sliding window feature extraction, we additionally added "rpart without FE" to the comparison. The last two rows per each target variable contain the best results achieved with feature extraction - that is $rpart+FE = MIN(rpart+FE(l : 1/o : 1), rpart+FE(l : 3/o : 1), rpart+FE(l : 6/o : 1))$ vs. the best results without FE - that is with minimal error in Table 11. We may notice that rpart with feature extraction outperformed the best of the evaluated methods 33 out of 36 times - the best results in bold.

The results confirm that the developed feature extraction methods have a positive impact on prediction of methane concentration. An important contribution of this research is the evaluation of the impact of the developed feature extraction methods not only on the quality of prediction but also on the resilience of various machine learning models in the case of partial data loss, i.e., missing attributes. The performed experimental study confirms that it is feasible to assure a proper resilience level of methane concentration prediction. Hence, would allow us to immunize decision support systems in case of data loss.

5.2   Seismic Events

In this section, we investigate how the interactive feature extraction and ensemble blending methods, proposed in Sect. 4, generalize to other problems of multistream data analysis. Once again, we address the problem of safety monitoring in underground coal mines. This time, we investigate and compare practical methods for the assessment of seismic hazards using analytical models constructed on both multi-stream sensory data and features derived from domain experts. The possibility to represent a problem related to data exploration and analysis with machine-generated features enriched with expert assessments, we consider as one of the essential aspects from the point of view of interactiveness.

For our case study, we use a rich data set collected during a period of over five years from several active Polish coal mines. We focus on comparing the prediction quality between expert methods, which serve as a standard in the coal mining industry, and state-of-the-art machine learning methods for mining high-dimensional time series data. We describe an international data mining challenge organized to facilitate our study. We also demonstrate that the technique, which we employed to construct an ensemble of regression models (presented in Sect. 4.3) together with the sliding window feature extraction framework (Sect. 4.1) were able to outperform other approaches used by participants of the challenge. Finally, we explain how we utilized the data obtained during the competition for the purpose of research on the cold start problem in deploying decision support systems at new mining sites.

Seismic Hazards in Coal Mines. Coal mining is one of the most important branches of heavy industry in the world, with the employment level exceeding 3.5M people worldwide [182]. As briefly outlined in Sect. 5.1, there are many threats that may be encountered by miners working in underground coal mines. An important aspect of safe and efficient coal mining is the prediction of seismic hazards, particularly those related to high-energy destructive tremors, which may result in rock-bursts [49]. Safety refers to saving workers from accidents and injuries, while efficiency refers to unplanned shut-downs of longwall systems. From this perspective, proper prognosis of potentially dangerous methane concentrations [349] and seismic events [109] constitutes one of the most important challenges that should lead toward improving the safety and reducing the costs of underground coal mining.

More and more advanced seismic and seismoacoustic monitoring systems allow for a better understanding of rock mass processes [109] and defining seismic hazard prediction methods [124]. Seismic hazard assessment and prediction methods, among others, include: probabilistic analysis [231, 275] that predicts the energy of future tremors or a linear prediction method [217], which can be used to predict aggregated seismic and seismoacoustic energy emitted from a mining longwall. Both methods perform analysis in a given time horizon. An application of data clustering techniques to seismic hazard assessment was presented in [235]. There are also approaches to the prediction of seismic tremors by means of artificial neural networks [198] and rule-based systems [200]. The accuracy of the methods created so far is, however, far from perfect. These methods often require a special, non-standard measuring apparatus and that is the main reason why some of them are not applied in mining practice.

Microseismic monitoring and multi-parameter indices may be also considered as a tool for the early warning of rock-bursts [96]. In the context of dealing with uncertainties in geomechanical underground works, particularly interesting are techniques that apply the Bayesian modeling approach [267]. Rule induction and decision tree construction techniques were also applied for this purpose [348]. There are also applications of machine learning methods to diagnostics of mining equipment and machinery [393]. The issue of mining devices diagnostics was raised among others in [268]. Still, expert systems are currently the most popular method of natural hazard prediction in the area of underground coal mining.

Two basic methods are routinely used by experts for the assessment of seismic hazards in Polish coal mines. These methods are often called *seismic* and *seismoacoustic*, respectively [200]. In Appendix B.1, we briefly describe both methods of seismic hazard assessment. The seismic and seismoacoustic methods are the result of the work of many domain experts and serve as a current standard in the Polish mining industry. Therefore, estimating the accuracy of those expert methods for natural hazard assessment and comparing their reliability with automatic prediction models constructed using statistical and machine learning techniques is of the utmost importance. This was one of the objectives of the presented research.

Processes related to the seismic activity are often considered the hardest types of natural hazards to predict. In this respect, they are comparable to earthquakes. Seismic activity in underground coal mines occurs in the case of a specific structure of geological deposits and due to the excavation of coal. Factors which influence the nature of seismic hazards are diverse. Relationships between those factors are very complex and still insufficiently recognized. To provide protection for people working underground, systems for active monitoring of coal extraction processes are typically used. One of their fundamental applications is to screen seismic activity in order to minimize the risk of severe mining incidents. Such a situation occurs in the Upper Silesian Coal Basin, where the additional conditions are related to the multi-seam structure of the coal deposit [170].

In almost all mines in this region, there are systems that detect and assess seismic activity degrees. The current industry standard in this regard (and the regulations imposed by Polish law) involves manual assessment of hazards by mining experts. However, the question remains whether the existing systems and expert methods take full advantage of the available data in order to provide their users with the maximum possible prediction accuracy. Moreover, it is important to design seismic hazard prediction methods that can adapt to new conditions. There is also a question of whether the way in which currently deployed systems work is sufficiently clear and comprehensible, so the users can properly interpret their results and react in case of possible false emergencies.

Table 13. Basic characteristics for data obtained from different working sites. The first column shows working sites ids, whereas the subsequent ones present information regarding initial expert assessments of the working site's safety, the number of data samples in the training and test sets, and the percentage of cases with the *'warning'* decision label.

Main working site ID	Initial assessment	Number of training cases	Number of test cases	Training warnings (percent)	Testwarnings (percent)
146	a	5591	98	0.0014	0.0000
149	b	4248	98	0.0718	0.0018
155	b	3839	98	0.1681	0.0094
171	a	0	49	0.0000	0.0000
264	b	20533	0	0.0039	0.0000
373	b	31236	0	0.0113	0.0000
437	b	11682	0	0.0041	0.0000
470	c	0	258	0.0000	0.0078
479	a	2488	35	0.0000	0.0000
490	a	0	160	0.0000	0.0500
508	a	0	58	0.0000	0.0172
541	b	6429	5	0.0087	0.0000
575	b	4891	253	0.0045	0.0012
583	b	3552	215	0.0021	0.0029
599	a	1196	363	0.0148	0.0289
607	b	2328	209	0.0000	0.0000
641	a	0	97	0.0000	0.0103
689	b	0	83	0.0000	0.1205
703	a	0	145	0.0000	0.0069
725	b	14777	330	0.0920	0.0021
765	a	4578	329	0.0000	0.0022
777	b	13437	330	0.0000	0.0009
793	b	2346	330	0.0000	0.0045
799	a	0	317	0.0000	0.0000
Total	-	133151	3860	0.0226	0.0508

AAIA'16 Data Challenge. The complexity of seismic processes and the imbalanced distribution of the positive (e.g., *'warning'*) and negative (*'normal'*) examples is a serious difficulty in seismic hazard prediction. Commonly used statistical methods are still insufficient to achieve good sensitivity and specificity of the predictions. Therefore, it is essential to search for new and more efficient techniques of natural hazard assessment, including methods derived from the field of machine learning. By organizing an international data challenge related to seismic hazards assessment as an open, on-line competition we aimed to conveniently review and evaluate the performance of the available state-of-the-art methods. Furthermore, this allowed us to verify not only the viability of the predictive models but also whole analytic processes, including preprocessing, feature extraction, model construction, and post-processing of predictions (e.g., ensemble approaches).

AAIA'16 Data Mining Challenge: Predicting Dangerous Seismic Events in Active Coal Mines took place between October 5, 2015, and February 27, 2016. It was organized at the KnowledgePit platform, under auspices of 11th International Symposium on Advances in Artificial Intelligence and Applications (AAIA'16) which is a part of the FedCSIS conference series. The task in this competition was related to the assessment of safety conditions in underground coal mines with regard to seismic activity and early detection of seismic hazards.

The data set provided to participants was composed of readings from sensors that monitor the seismic activity perceived at longwalls of different coal mines and measure the energy released by so-called seismic bumps. Each case in the data was described by a series of hourly aggregated sensor readings from a 24 h period. The provided data also contained information regarding the intensity of recent mining activities at the corresponding working site, coupled with the latest assessments of the safety conditions made by mining experts (for instance, ratings obtained using the seismic and seismoacoustic methods – described in Appendix B.1). To further enrich the available data, for each working site that occurs in the data set, some additional meta-data were made available, such as identifiers of the mine and region where the working site is located or a working site's height. The detailed list of all data attributes is available in Table 21 in Appendix A.2.

Participants of the competition were asked to design a prediction model which would be capable of accurately detecting periods of increased seismic activity. In particular, the target attribute in the provided data (the decision) indicated cases for which the total energy of seismic bumps observed in a following 8 hour period exceeded the warning level of $5 \cdot 10^4$ Joules (i.e., the energy released in the period starting after the last hour of aggregated readings describing the case and ending 8 hours later). In total, the provided data was described by 541 main attributes and 6 additional features related to particular working sites. Most of the attributes were numeric, but there were also a few symbolic (qualitative) ones, e.g., assessments made by experts. The competition's data correspond to over 5 years of readings which, to the best of our knowledge, makes this research the most comprehensive study related to this domain, conducted anywhere in the world.

The data set was divided into a training part, which was made available to participants along with the corresponding decision labels, and a test part. The labels for the test set were hidden from participants. The division of cases between the training and test sets was made based on time stamps. In particular, the training data set corresponding to a period between May 5, 2010 and March 6, 2014. It consisted of 133151 data rows, each corresponding to a different 24-hour period, overlapping for consecutive cases. The test data covered the period between March 7, 2014 and June 24, 2015. Unlike the training set, to facilitate the objective evaluation of solutions and to prevent a common problem with so-called data leakage [207], the test cases were not overlapping and provided in random order. For this

Table 14. Final results and number of submissions from the selected, top ranked teams in the AAIA'16 Data Challenge. The last row shows results obtained solely from assessments made by mining experts that were available in the data (see Appendix B.1).

Method	No. of submissions	AUC
Grzegorowski M. [138]	2	0.9396
Milczek et al. (Deepsense Inc.) [265]	111+31	0.9393
Tabandeh Y. (Golgohar Inc.[a])	54	0.9342
Podlodowski [301]	71	0.9336
Kurach & Pawłowski [224]	32	0.9312
Başak et al. [153]	30	0.9297
...	...	...
Experts assessment (18th place)	–	0.9196

[a] Golgohar Mining & Industrial Company, Iran.

reason, the test set used in the challenge was much smaller than the training data but still covered a period of nearly 16 months.

Table 13 shows some basic data characteristics from each working site that was used in the competition. It is worth noting that not all working sites present in the training data also appeared in the test set and there were a few working sites that were present in the test data but not in the training set. Such a situation reflects a real-life problem when the exploration of coal shifts to a new site for which there is no data available. A similar issue can also be identified within other domains, e.g., recommender systems, and is commonly referred to as the *cold start* problem [368]. A fact worth noticing is also that the distribution of cases with a *'warning'* decision label is quite uneven for different working sites.

Solutions submitted by participants had a form of scores assigned to the test cases (i.e., real numbers, which could be interpreted as likelihoods of 'warning' signals). In practical applications related to the monitoring of safety conditions, such a form of predictions is more valuable than the exact decision labels because it allows for tuning the sensitivity of the utilized model. Due to imbalanced distributions of decision labels, the quality of each submission was measured using Area Under the ROC (AUC). The AUC measure explicitly relates the true alarm rate to the false alarm rate and, thus, is appropriate for measuring the performance of prediction models in a situation when underestimating the risk of a minority binary class (i.e., a seismic event) is significantly worse (in our case in terms of safety) than overestimating the risk.

Among the registered teams, 106 were active, i.e., submitted at least one solution. Table 14 shows scores achieved by the selected, top-ranked teams. In total, we received 3236 solutions, of which 3135 were correctly formatted and successfully passed the evaluation procedure. In Sect. 5.2, we explain how we used those submissions in our post-competition analysis of the cold start problem in the deployment of predictive models for new working sites. Additionally, 50 of the participating teams provided reports describing their approach, e.g., [138, 265, 428]. These reports turned out to be a valuable source of knowledge regarding the state-of-the-art solutions in the predictive analysis of time series data related to early detection of seismic hazards.

Construction of a Seismic Hazard Assessment Model. In this section, we focus not only on the constructed prediction models but also on data processing stages that were designed to let it work within a big data processing environment, and particularly with multi-sensor data streams. The highest result in the final evaluation of AAIA'16 Data Challenge was obtained out-of-the-box by the solution trained with the framework described in Sect. 4.3. The Grzegorowski M. [138] method (Table 14) is based on the framework presented in Sect. 4 – and had been successfully applied to a similar problem, namely, the prediction of dangerous methane concentration levels in corridors of coal mines (Sect. 5.1). The fact that we were able to reuse this approach confirms its attractive generality. The overall work we spend on the framework configuration, data preprocessing, feature extraction and selection, models training, and ensemble blending for the purpose of seismic data prediction did not exceed 2 h.

Let us now take a closer look at the best performing solution – Grzegorowski M. [138] (see Table 14) – of the AAIA'16 Data Challenge. In order to provide high quality assessments, this solution constructed an ensemble of diverse logistic regression models (Sect. 4.3). The diversity is obtained by employing a variety of models computed on various subsets of attributes and examples (Sect. 4.1). By aggregating predictions of those models using the Algorithm 9, we were able to obtain robust performance even for new mining sites. As a result, the final ensemble minimized the impact of a concept drift [24, 322] and achieved a better quality and robustness of prediction than models used by all other teams participating in the competition.

A scheme of the whole feature extraction process is the same as depicted in Fig. 12. In the *'Map'* phase (compare steps 2 and 3 in Fig. 12), each data row was divided into sub-series of numerical values from various sensors, a set of static and aggregated features and, in the case of the training set, also a label. The labels, as well as the static attributes from experts, were transferred to the *'Reduce'* phase unchanged while the time series were subjected to the feature extraction process described in Sect. 4.1. In the *'Reduce'* phase (steps: 4 and 5 in Fig. 12), all the attributes obtained for each row were combined again.

The design of our model and particularly the method for construction of an ensemble was largely affected by the imbalanced distribution of *'warning'* cases in the data (only about 2.3% of all objects). Firstly, we drew a number of random samples that contained between 10000 and 20000 objects from the training set. The samples differed in the number of objects from the *'warning'* class – it was assured that each contained a minimum of 1000 and a maximum of 2000 such cases. Objects within a particular sample were unique, but they could be repeated between different samples. Such prepared samples supported the ensemble feature selection technique to yield more robust results [333]. All steps of the feature selection, models training, and the construction of the final ensemble are presented in Figs. 13 and 14.

It is worth noticing that the selection of attribute subsets was carried out using a technique that originated within the theory of rough sets (cf. Sect. 2.5). Similarly, as in the FE+DAAR (Table 9) approach presented in Sect. 5.1, the DAAR heuristic for computation of approximate decision reducts was applied to find relatively small subsets of relevant features. Furthermore, we decided to combine a few approximate reducts into a single attribute subset to extend the feature space for each logistic regression model while increasing their diversity (compare steps 8–10 in Fig. 13) – the combined reducts were randomly matched. Only significantly different subsets were maintained for the purpose of model training, the rest were filtered out.

In the next steps (compare Fig. 14), the obtained subsets of attributes and objects with a more balanced distribution of classes (see, e.g., SMOTE oversampling technique for sensor readings and ensemble learning [240]) were used to train logistic regression models using pre-selected algorithms. The most important models were used to form an ensemble. Criteria for selecting models for the ensemble considered a quality of individual regressors as well as the degree of diversity of a resulting collection of models. The course of the experiment is presented in Algorithm 9 and in Fig. 14. All the processing steps were implemented in R environment for statistical computing using additional libraries, e.g., *rpart, e1071* and *RoughSets* [323].

The final solution was an ensemble of diverse logistic regression models which interpret the *'warning'* label as 1 and the *'normal'* label as 0. The diversity was achieved by training the models on different subsets of attributes and objects. Algorithm 9 guarantees that a model can be included only if it is accurate enough on validation data and sufficiently different from already selected predictors (in that case, correlation of its prediction with predictions of other models is small enough). This could be seen as a method for increasing the robustness of predictions in the case of noisy and heterogeneous data.

The final ensemble consisted of 8 different regression models which were calculated using three various algorithms, namely: regression trees (calculated using the implementation from the *rpart* library), SVM regression, and a generalized linear model. These particular models were selected:

- five regression trees calculated with rpart (default settings),
- two SVM models with different kernel functions:
    - SVM_1 - linear kernel, cost: 1, eps: 0.1
      (the number of support vectors: 2968),
    - SVM_2 - radial kernel, cost: 1, gamma: 0.07143, eps: 0.1
      (the number of support vectors: 7171),
- one logistic regression model computed with glm.

A comparison of the selected solutions to predictions that were based solely on assessments made by experts revealed that more complex models were able to quickly attain significantly higher scores for working sites with available training data. In the case of the remaining working sites, the advantage of complex prediction models was not that clear. The average results for selected models in phase 6 were only slightly higher, however, for a part of the investigated solutions, the difference was much more favorable than for others.

Table 14 shows scores achieved by the selected, top-ranked teams. The second highest result, achieved by Milczek et al. solution [265], was based on a mixture of multiple gradient-boosted trees [66,114], extremely randomized trees [35,122] multi-task logistic regression and linear discriminant analysis models [276]. Interestingly, when we computed the Spearman's correlation between predictions of our model and the Milczek et al. (Deepsense Inc.) [265] model, it turned out to be relatively small (≈ 0.77). When we combined these predictions by averaging ranks of predicted values for the test cases, we obtained a higher AUC than those of the individual models (0.9421 vs. 0.9396 and 0.9393). This result highlights the benefit of using diverse prediction models for constructing ensembles.

In general, an overview of the most successful approaches in the competition suggests that the key steps to achieve a good result in this task included:

1. Extracting relevant features (computing a new data representation) that aggregate time series data and are robust with regard to a concept drift.
2. Designing an appropriate evaluation procedure for testing performance of used prediction models and tuning their parameters.
3. Using ensemble learning techniques for blending predictions of simpler models.

Such a general approach, which is strongly dependent on feature engineering, was employed by eight out of the ten top-ranked teams. A slightly different methods utilized deep neural networks (DNN) to automatically learn a representation of data [421]. Some details regarding DNN approaches may be also found in [224].

It is here very important to stress out that a solution based solely on domain experts' assessments achieved the 18th position. It confirmed that data mining techniques may outperform experts in forecasting seismic hazards. It also showed that this problem is not easy to solve even using state-of-the-art ML methods. Furthermore, a closer analysis of the reports submitted by the most successful teams revealed that the attributes corresponding to experts' assessments were commonly used by their models. Interestingly, all submitted solutions performed significantly better when the set of input data included the most recent evaluations provided by experts. It clearly shows how important the domain knowledge is for the efficient assessment of seismic hazards in coal mines. Furthermore, as we show in the subsequent section, expert knowledge allowed us to successfully transfer the knowledge to the new working sites.

The Cold Start Problem. The cold start problem is an important practical issue that is related to real-world applications of many decision support systems. In the case of coal mining, it typically appears when a system for monitoring natural hazards is being deployed for new, previously unexplored longwalls. One of the research objectives, motivating the organization of AAIA'16 Data Mining

Challenge (described in Sect. 5.2), was to investigate the severity of this problem in the context of systems for early detection of periods of increased seismic activity.

To gather comprehensive data about the impact of the size of available data on the quality of predictions for a given working site, the training set in the competition was divided into five separate parts and the challenge was split into six phases. Table 15 shows some basic statistics related to the consecutive phases, including the maximum size of the data set available to participants in each phase of the challenge. After the start of the competition, only the first part of the training data was revealed. The four consecutive parts were made available in monthly intervals. In the sixth phase, which lasted for the last two weeks of the competition, all training data parts were revealed to all participating teams. Since in each phase a new subset of training data was made available to participants, we were able to verify the impact of this additional information by examining the quality of solutions submitted in consecutive phases.

Table 15. Basic statistics for each phase of the challenge, including: training set size, number of submissions (cases) as well as best, mean and standard deviation of AUC scores.

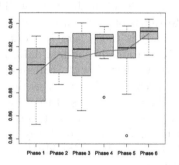

Phase	Training set size	No. of uploads	Best score	Mean score	Std. dev.
1	79893	99	0.9290	0.8251	0.0672
2	93211	278	0.9320	0.8851	0.0587
3	106527	1377	0.9405	0.8307	0.1058
4	119839	363	0.9375	0.8772	0.0831
5	133151	513	0.9379	0.8857	0.0625
6	133151	505	0.9439	0.8942	0.0696

Fig. 20. Distribution of the best AUC scores per team and phase.

A detailed analysis of the distribution of scores in time reveals some interesting observations. For the analysis, we only used valid solutions with a reasonable quality (we disregarded 'random' submissions and those which obtained a preliminary score lower than 0.65). Table 15 shows the mean and standard deviation of evaluation scores for each of the competition phases.

As an interesting observation related to the analysis of the results shown in Table 15, we may point that, starting at some point in time, the use of additional training data has a diminishing impact on the performance of prediction models. For instance, if we compare the average results from the second phase with the results from the fourth or fifth phase, we see that the difference is minimal. Even though in these phases we received a comparable number of submissions, and the available training set in the phase 5 was larger than in the phase 2 by nearly 43%. This was even less expected due to the fact that the data available in the phase 2 contained information about only 9 out of 21 main working sites present in the test data (these sites corresponded to ≈ 45% of observations in the test set), whereas in the phase 5 this number was much higher (13 out of 21 sites; ≈ 70% of observations). This suggests that models relatively quickly became saturated with the training examples.

To further confirm this observation, we analyzed the best solutions in each phase taking into account only the submissions from well-performing teams – that obtained AUC scores higher than 0.85 – results of such teams better reflect the performance of the state-of-the-art models. Figure 20 visualizes basic statistics (min, max, quantiles, and the mean values) of the best AUC scores of those submissions. The average scores slightly increase from phase to phase. However, when we checked the statistical significance of the changes, it turned out that a significant difference (p-value lower than 0.01) occurred only between results from the fifth and sixth phases. For other consecutive phases, the p-values of the Wilcoxon test were always higher than 0.175.

Let us now thoroughly investigate the performance of top-ranked solutions submitted in each phase, with regard to individual working sites. For this purpose, we disregarded working sites for which there were no examples with the 'warning' label in the test set. The reason for that was the inability to compute values of AUC on such data subsets. This way, for the remaining part of our analysis there were 15 working sites left, which corresponded to ≈ 81.5% of observations in the test data. From solutions submitted in each competition phase, we chose 6 with scores in the top 10% for a given phase. Table 16 shows their average AUC values with respect to individual working sites. Additionally, the last two rows of the table give average values of AUC for working sites that were present in the training set and for those which were unavailable in the training data, respectively. Finally, the last column of Table 16 shows AUC values obtained for individual working sites using only the assessments made by experts.

For most of the working sites there is a statistically significant improvement (tested using t-test with a confidence level of 0.95) of results from the later competition phases in comparison to the first phase. However, in nearly all cases the improvement between the second and later phases becomes marginal (one exception is the working site with ID:599). Interestingly, there are working sites (e.g., ID:689, ID:777) for which there is a noticeable drop in the average quality of solutions between the second phase and phases 3, 4 and 5. Our partial explanation of this phenomenon is that in the case of site 689, there is no training data for this particular site, while in the test set, it is characterized by the highest percentage of 'warning' cases (over 12%). In the case of site ID:777, the situation is the opposite. For this site, there were many examples in the training data and their number was increasing in consecutive phases. However, all examples in the training set belonged to 'normal' decision class, whereas the test set contained a few observations from the 'warning' class (see Table 13). Such a distribution of labels could trick the models into thinking that all cases from this mining site should be 'normal', and as a result, decreased their performance. Interesting is also the fact that the top solutions obtained consistently higher scores for working sites that were not present in the training data.

Table 16. Average scores of top solutions for individual working sites, in different phases of the competition. Evaluations of expert assessments are given for comparison in the last column. Additionally, the last two rows display aggregated values (averages) for working sites with some data in the training set and the working sites without any available training data.

Working site ID	Phase 1	Phase 2	Phase 3	Phase 4	Phase 5	Phase 6	Expert's assessment
149	0.8984	0.9056	0.8523	0.9062	0.8766	0.9005	0.9306
155	0.6578	0.7328	0.7492	0.7393	0.7242	0.7487	0.6845
470	0.9749	0.9922	0.9876	0.9935	0.9922	0.9964	0.9707
490	0.8013	0.8122	0.8340	0.8021	0.7892	0.8289	0.8109
508	0.9825	0.9971	1.0000	0.9942	0.9854	1.0000	1.0000
575	0.9348	0.9845	0.9859	0.9826	0.9820	0.9825	0.9723
583	0.9000	0.9419	0.9363	0.9388	0.9370	0.9401	0.9280
599	0.8391	0.8585	0.8678	0.8445	0.8670	0.8710	0.8020
641	0.9809	0.9983	1.0000	0.9965	1.0000	1.0000	1.0000
689	0.7723	0.8812	0.8523	0.8685	0.8582	0.8938	0.8884
703	0.9346	0.9792	0.9826	0.9699	0.9873	0.9722	0.9722
725	0.8968	0.9188	0.9251	0.9151	0.9176	0.9099	0.8955
765	0.7989	0.7911	0.7367	0.7608	0.7423	0.7808	0.7587
777	0.9118	0.9354	0.9242	0.9252	0.9175	0.9408	0.9444
793	0.9499	0.9545	0.9585	0.9538	0.9361	0.9468	0.8868
Avail. in training	0.8653	0.8915	0.8818	0.8852	0.8778	0.8912	0.8670
Unavail. in training	0.9077	0.9433	0.9428	0.9374	0.9354	0.9486	0.9404

The above observations show that having a sufficiently large data set, it is possible to construct efficient prediction models for the assessment of seismic hazards. The created models can outperform the currently used expert methods even for completely new working sites, as long as these sites have comparable geophysical properties and the same methodology is used for collecting new data. At this point, it is worth emphasizing that all the evaluated models were trained on both sensory data and domain experts' assessments. Such an approach allowed to significantly improve the quality of machine learning models since it encapsulated part of the knowledge about the particular conditions of each working site not covered by sensors. This is an important argument for considering interactive feature extraction processes and built-in human-computer interaction into machine learning processes.

5.3   Tagging Firefighter Posture and Activities

A fire ground is considered to be one of the most challenging decision-making environments. In dynamically changing situations, such as those occurring at a fire scene, all decisions need to be taken in a very short time. Since wrong decisions might have severe consequences, a commander of the response team is forced to act under huge pressure [183]. Based on several thousands of carefully analyzed reports, experts identified the "lack of situational awareness" as the main factor associated with major accidents among firefighters [133]. According to studies on causes of mortal accidents during actions of firefighters conducted by the Department of Homeland Security of the United States [5] over 43% of deaths at a fire scene was caused by stress or overexertion. Therefore, another critical way of increasing firefighter safety is by monitoring their kinematics and psycho-physical condition during the course of fire & rescue actions.

The computer systems for human activity recognition may help to reduce unsafe events, improving communication, and increasing the efficacy of incident management. Human activity recognition using Body Sensor Networks (BSN) is a non-invasive system that is able to deliver information about person motion patterns, posture and specific actions performed [154,230]. A network of sensors located on a firefighter body are used to gather kinematic (motion) data from different parts of the body that may be additionally complemented with physiological data sensors for vital function monitoring. Afterward, sensor readings are transferred, pre-processed and various machine learning classification techniques may be applied in order to estimate the current activity. In this section, we present a practical application of the presented framework for interactive feature extraction in the fire and rescue domain that refers to BSN data analysis.

Additional Constraints and Requirements. The aim of our research was to assess how the automatic feature extraction and classifiers learning (without parameters tuning) can cope with the multi-target learning problem [19,433]. The evaluated mechanism was previously applied for the purpose of processing multiple streams of readings generated by sensor networks in coal mines (Sects. 5.1 and 5.2). Hence, one of our objectives was to assess the versatility of the developed framework across significantly different domains of application. Working on the solution, we imposed a few additional constraints and requirements that are essential for the emergency and threats detection domains:

1. The overall time spent on solving the problem must not exceed a total of 2MD (two man days - that is 16 h).
2. The overall computation time required to prepare data and train the classifiers must not exceed the total of 10 minutes.
3. The total time required to pre-process a single row of data to a format accepted by a classifier and assignment of both labels must not exceed one second.

The first of the imposed restrictions was intended to evaluate the possibility to adapt quickly to a new domain with a satisfactory quality of the model. We put a requirement that the time for configuring the framework, processing data, training, and applying the model should not exceed 2 MD which has been recognized as sufficient for researchers to become familiar with the task and to adjust original data representation to formats accepted by the evaluated feature extraction mechanisms.

The second point posed a constraint on the time necessary to re-train the model on new data. After a certain time, the quality may fall below the predetermined threshold due to, e.g., concept shift or drift [128,243]. We assumed that the time required for re-training the model should not exceed 10 min.

The last point, we consider as the most important because it imposed limits on the permissible delay in the operation of the pre-processor and classifiers when acting in a production environment. According to the assumptions, maximum delay between data collection, complete processing, and labeling of a single row of data should not exceed one second. This was one of the main reasons for excluding from consideration all object-based methods as well as heavy classifier ensembles.

AAIA'15 Data Challenge. The competition: Tagging Firefighter Activities at a Fire Scene [263] – organized within the frame of the International Symposium on Advances in Artificial Intelligence and Applications[14] – concerned the problem of an automatic assignment of labels (activities) to short series of readings from sensors that monitor activities and movements of firefighters during an action. The aim of the competition was to maximize a balanced accuracy measure which is defined as an average accuracy within all decision classes. It was computed separately for the labels describing the posture and main activities of firefighters. The final score is a weighted average of balanced accuracies computed for those two sets of labels and is defined as follows:

$$score(s) = \frac{BAC_p(s) + 2 \cdot BAC_a(s)}{3} \tag{26}$$

where BAC_p is the balanced accuracy for labels describing the posture and BAC_a for the main activity. Recall the definition of the balanced accuracy (BAC):

$$BAC(preds, labels) = \frac{\sum_{1 < i < l} ACC_i(preds, labels)}{l}$$

$$ACC_i(preds, labels) = \frac{|\, j : preds_j = labels_j = i \,|}{|\, j : labels_j = i \,|}$$

The data provided in the competition were obtained during training exercises conducted by a group of eight firefighters from the Main School of Fire Service. The sensors placed on a chest were registering vital functions, while the sensors placed on the torso, hands, arms, and legs were registering movements of a firefighter. Along with recording the data from sensors, all training sessions were also filmed. The video recordings, firstly synchronized with the sensor readings, were presented to experts who manually labeled them with actions performed

---

[14] AAIA'15, https://fedcsis.org/aaia.

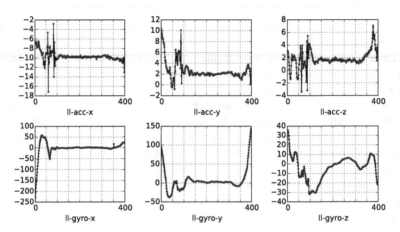

Fig. 21. Time series form an accelerometer (measured in m/s^2) and a gyroscope (measured in deg/s).

during the exercises. The training and test data sets contain 20000 rows and 17242 columns each. The data are available online on the KnowledgePit platform as CSV files. The considered task was even more challenging since the training and test data sets consist of recordings from disjoint groups of firefighters.

Each single row in data sets corresponds to several short time series with length equal to approximately 1.8 s. The first 42 columns contain basic statistics (like mean, standard deviation, maximum, minimum, etc.) of data from sensors monitoring a firefighter's vital functions over the given, fixed time period. The raw readings for the vital functions were recorded using Equivital Single Subject Kit (EQ-02-KIT-SU-4) fitted with two medical-quality ECG units, heart rate and breath rate units, and thermometers for measuring skin temperature. The remaining columns contain readings from a set of kinetic sensors that were attached to seven places on a body, i.e., left leg, right leg, left hand, right hand, left arm, right arm, and torso. The enumerated body areas correspond to the following name prefixes in data: $ll, rl, lh, rh, la, ra, torso$, respectively. An infix – acc or gyro – refers to an accelerometer (dynamic bandwidth: ± 16G) or gyroscope (scale up to 2000 deg/s), respectively. Each sensor of both types produced three readings corresponding to the three dimensions. A suffix x, y, or z indicates the axis readings came from. An average time difference between consecutive sensory readings in the data is 4.5 ms. Eventually, time series are divided into 400 chunks that represent consecutive points in time. Figure 21 contains exemplary, six time series (each consists of 400 values that correspond to approximately 1.8 s) from a set of sensors placed on a left hand of a firefighter performing an exercise.

The above description shows the details of the values arrangement in the data. In the frame of the experiment, we considered each row as a separate data set containing readings from many sensors. Values from the vital sensors were aggregated externally but the kinetic ones were provided in the raw form of

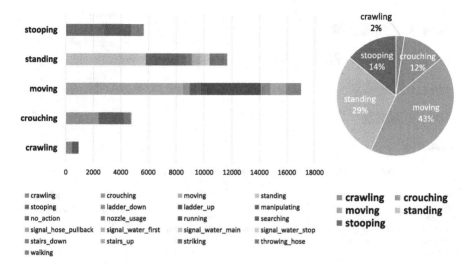

Fig. 22. Inter-dependencies between posture and activity labels in training data.

time series - this setup was out-of-the-box covered by the framework - compare Fig. 12 in Sect. 4.3. However, the preliminary data analysis revealed two interesting characteristics of the investigated data. The posture and activity labels were not independent [206] – see Figs. 22 and 23). Furthermore, the analysis revealed an imbalance in label distribution [399, 411].

Feature Extraction. For the purpose of posture and activity recognition, we processed the data with three configurations of a sliding window mechanism (Sect. 4.1). As presented in Fig. 24, every time series were split into 1, 2, and 5 consecutive, non-overlapping sliding windows, respectively. If there were more than one window generated for the time series we extracted so-called inter window statistics (in addition to those included in a basic window) – that is a set of features expressing changes of attribute's values between a pair of consecutive windows (Sect. 4.1).

According to the task description, the kinetic sensors (accelerometers and gyroscopes) used during the exercises have symmetric scales with 0 as their neutral reading. The specificity of the firefighter activities like walking, running, moving up the stairs or ladder, could cause the readings to be more significant when considered as a group – e.g., a whole tuple (x, y, z) from a given accelerometer rather than separate readings x, y, and z. For that reason, we introduced a concept of so-called *virtual sensors*. Besides applying the aggregate functions to the original time series available in the delivered files, we implemented an idea of creating artificial time series derived from the original ones. The virtual sensors were created on the basis of one or more time series from other sensors (whether original or virtual) by applying a particular function. In our solution, we decided to create virtual sensors for readings from all accelerometers and

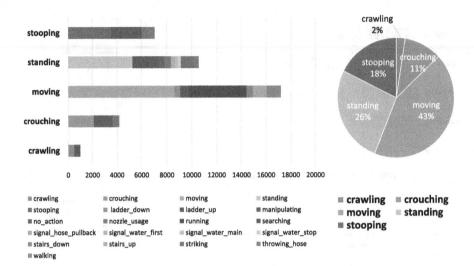

Fig. 23. Inter-dependencies between posture and activity labels in test data set.

gyroscopes' axes separately, applying an *abs* (absolute value) function. We created also virtual sensors for readings grouped in tuples (x, y, z) for each kinetic sensor – computing the Manhattan and Euclidean norms for the (x, y, z) vectors. An example that illustrates the concept of virtual sensors used in our solution is presented in Fig. 25.

Along with so far mentioned attributes, we additionally extracted more domain-specific features, e.g., a sum of the selected features for the left and right hand or a sum for the left and right leg – to exclude the symmetry of right- and left-handed people. This was important because the training samples were created based on the behavior of different people than the test samples. Moreover, training and test set data were acquired during observation of a small group of firefighters, hence the training sample could not contain all possible patterns. All extracted statistics were joined together, in a sense of appending all their values in a data table, and served as an input for the further steps of data analysis and experiments (see the subsequent steps – 5 and above – in Fig. 13 and in Fig. 14).

The above-mentioned adjustments and re-configurations of the presented FE framework were performed in stages in an interactive way. In Fig. 26, the ultimate schema of feature extraction process is presented. The extracted data sets had a total of 27177 attributes, due to 3 different sliding windows configurations: 2199 – one sliding window per time series; 6315 – two sliding windows per time series, and 18663 – five sliding windows. All identifiers and all constant attributes were removed from data. In the process of feature selection, we employed a wrapper approach. In the forward propagation phase, we had been progressively enlarging the number of candidate features and making periodic evaluations with SVM

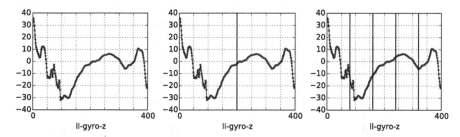

Fig. 24. Illustration of the sliding window configurations. Time series were processed with varied granularity, ranging from the statistics computed for the whole time series, to calculate them for 2 or 5 shorter, non-overlapping sliding windows which divided the time series to the parts of equal length.

model after each step. Ultimately, 163 most significant attributes were selected for the purpose of model training.

Model Training. The main multi-target learning problem of labeling multi-variate time series with many labels that are interdependent (Figs. 22 and 22) may be modeled and solved in a number of ways [19,433]. One of the options could be transformation to a typical classification problem by training classifiers to solve posture and activity recognition independently. Another option could be creation of new labels corresponding to couples of posture and activity, e.g., for posture: "moving" and action "stairs_up" the combined label would be "mov-ing_stairs_up". Such an approach would incorporate the additional knowledge about dependencies between labels, on the other hand, the number of cases for niche activities would be relatively small. The way in which the assessment of the solutions is defined (Eq. 5.3), that is uneven importance of labels for posture and activity, encourages to consider various concepts like a multi-label classification with label ranking [116] or a graded multilabel classification [67]. The experimentation with label power-set methods [320,415], however, did not provide satisfactory results.

Ultimately, we decided to follow ensembles of classifier chains (ECC) [319]. This approach involves linking together classifiers in a chain structure [412], such that posture label predictions become features for activity classifiers. Class imbalance is an intrinsic characteristic of analyzed multi-label data (Compare Figs. 22 and 22). Some of the labels in data were associated with a small number of training examples. In general, class imbalance poses a key challenge that plagues most multi-label learning methods [244]. Classifier Chains [319] – one of the most prominent multi-label learning methods – is no exception to this rule, as each of the models it builds is trained on all positive and negative examples of each label. To make a ECC resilient to class imbalance, we coupled it with over- and under-sampling (recall STEP 6 in Fig. 13) [367].

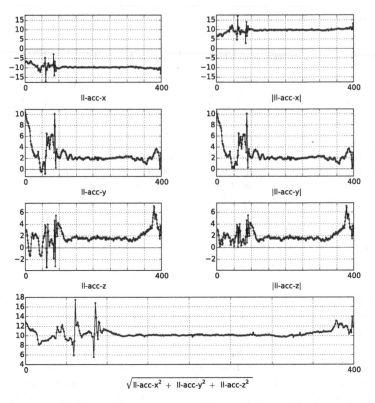

Fig. 25. An example of virtual sensors extracted by applying an absolute value function and the Euclidean norm to the original time series.

Experiments were implemented and carried out in the R software environment with additional packages. We experimented with decision trees, random forest, and SVM models [206]. To align with agreed constraints, that is: not to exceed 10 min of model training and at most 1 s for single input processing and labeling, the final solution was based on two SVM models with radial kernel set up in a classification chain. In Fig. 27, the schema of classifier chain training – carried out in order to solve the problem of labeling sensor time series with posture and main activity of a firefighter – is presented. A model responsible for recognizing a firefighter's posture was trained on the 163 attributes. The (second in the chain) SVM model which classified the data with a main activity had one additional attribute – the prediction for posture label. Although constrained effort in solving the problem (limited with a maximum of 16 h), the final evaluation reached $BAC = 0.72$ and significantly exceeded the organizer's baseline $BAC = 0.6$.

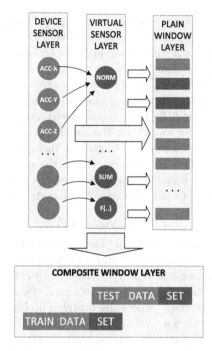

Fig. 26. Pre-processing and feature extraction.

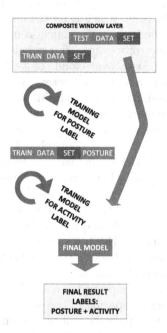

Fig. 27. Classifier chain.

In the final embodiment, the total time of training classifiers did not exceed 7 min. Extraction of all the features, including those for both: raw and virtual-sensors readings, took approximately 450 ms per a single csv file row. The post-processing, including assignment of the labels, was performed in R software environment for statistical computing and consisted of: importing data 1.5 ms per record (overall 30 seconds per 20000 rows of test data set), feature selection 0.5 ms per record (overall 10 s per 20000 rows of test set) and labeling 3.5 ms per record (classification with SVM of both labels for 20000 rows took in total 70 s).

5.4 Spot Instances Price Prediction

The ability to analyze the available data is a valuable asset for any successful business, especially when the analysis yields meaningful knowledge. Analytical data processing has become the cornerstone of today's businesses success, and it is facilitated by Big Data platforms that offer virtually limitless scalability. The storage technologies with high level of compression that support stream data collection and analytics [21,358] as well as the data processing and integration tools [52,324,427], which can scale up to thousands of compute resources [88, 136,174] allowed companies to store and analyze data collected from ubiquitous sensors. Furthermore, cloud computing has emerged as an important paradigm offering a variety of low-cost hardware and software in pay-as-you-go pricing model [406], which is particularly convenient for Big Data analytics [343].

Cloud computing offers a number of Big Data solutions related to scalable storage, processing, and sophisticated business analytics. Due to the growth of Big Data over cloud, cost-effective allocation of appropriate resources has become a significant research problem [208,343]. Minimizing the total cost of ownership (TCO) for the infrastructure supporting Big Data is considered a very challenging task. The number of available pricing models on the cloud markets is overwhelming, but it is worth paying special attention to two of them, in particular: the on-demand and spot markets. The first one represents the pay-as-you-go cloud model, and today is the most common way the resources are provisioned. The second one allows customers to save up to 90% of costs by using the cloud data centers' idle servers.

In this section, we show that, by analyzing spot instance price history and using ARIMA models, it is feasible to leverage the discounted prices of the cloud spot market [148]. In particular, we evaluate savings opportunities when using Amazon EC2 spot instances comparing to on-demand resources. The performed experiments confirmed the feasibility of short-term future spot prices prediction, which can improve the cost-effectiveness of any cloud processing bringing up big savings comparing to the on-demand prices. This way, we provide a significantly different application of the presented framework for multi-stream feature extraction and analysis. Instead of referring to multi-dimensional data representation, we performed univariate analysis of many time series independently where the feature extraction part was limited to extracting candlesticks from sliding windows over the spot price bidding data collected from AWS Cloud [8,148]. The main reason behind the evaluation of ARIMA models on data represented as candlestick is that both techniques are very popular and easily interpretable by experts.

Introduction. Proper allocation of cloud resources is a challenging task, particularly for computationally cumbersome tasks like data processing. There are quite a few examples of cluster size optimizations for Big Data analytics that focus on resource management for sustainable and reliable cloud computing [125]. One of the approaches could rely on initial estimations of data stream characteristics expressed in a vector termed Characteristics of Data (CoD). Clusters of cloud resources could then be created dynamically with the help of, e.g., Self-Organizing Maps [208,258]. Another approach – presented in [106] – focuses on the optimization of short-running jobs. Authors in [162] propose a query-like environment where developers can query for the required cluster size. The proposed approach requires, however, implementation-specific details. The evaluation of historic executions and metrics is considered as one of the prominent methods that leads to proper optimization, resulting in the timely processing of data [427].

Cloud providers aim to optimize server utilization to avoid idle capacity and significant peaks [209]. This led to the emergence of cloud spot markets on which service providers and customers can trade computation power in near real-time. One of the evident concerns regarding the spot model is that prices fluctuate along with changes in supply and demand. Furthermore, cloud providers may terminate provisioned instances with a minute notice due to outbidding. The ability to forecast future spot prices in a time horizon necessary to complete the data processing tasks would be a game-changer allowing to decrease total costs of operation of data processing pipelines, and to minimize the risk of resource terminations. In practice, typical data processing tasks, and in particular feature engineering tasks, have a degree of temporal flexibility - they need to be fulfilled before a specified deadline. However, it is often possible to defer the computations if it could lead to overall cost reduction due to price fluctuation. With a reliable forecasting model that provides accurate spot price prediction for a given time horizon, and reliable estimation of resources required to perform the task, one could recommend an efficient and cost-effective cluster configuration.

Some of the frameworks for cluster size optimization, to minimize the deployment cost, consider allocating server time to spot cloud resources. For that purpose, a fine-tuned heuristic to automate application deployment, and a Markov model that describes the stochastic evolution of the spot price and its influence on virtual machine reliability are proposed [99]. In [407], the authors describe an integral framework for sharing time on servers between on-demand and spot services. This is one way to guarantee that on-demand users can be served quickly while spot users can stably use servers for an appropriately long period. This is a critical feature in making both on-demand and spot services accessible. However, guaranteeing timely cloud job execution on a spot instance is a very challenging task, and existing strategies may not fulfill requests in case of outbidding.

Changes in supply and demand are the primary factor that impacts the price of a given service. This behavior is well-known in the stock exchange or commodity markets [55,110]. Among many available methods for time series regression [341] – which are the most suitable for modeling the problem of price prediction – one of the most popular and broadly used are autoregressive integrated moving average (ARIMA) models [4]. Results obtained in this study confirmed that ARIMA has a strong potential for short-term spot prediction.

The ability to accurately forecast future spot prices is essential to minimize the risk of resource terminations. A number of models have already been applied for that task [209]. For example, in [26], the authors evaluated a model to predict EC2 spot prices based on long/short-term memory recurrent neural networks. The problem of forecasting EC2 spot prices one day and one week ahead was also evaluated with random forest regressors [213]. Because of the similarity between cloud spots and financial markets [110], we decided to assess ARIMA models [74], which are known to be robust and efficient in short-term time series forecasting on stock exchanges or commodity markets [55,86].

Cloud Spot Market. Spot instances can be regarded as spare compute capacity in cloud data centers. They are offered as one of the three ways cloud providers sell their computing capacity - the other two are on-demand and reserved instances. In terms of the servers, there is no difference between the three. The difference is in the business model. On-demand instances represent the pay-as-you-go model, while reserved instances facilitate long-term renting of computing resources with a discount. However, spot instances allow customers to save up to 90% of costs by using the cloud's unused servers (cf. Fig. 28). The two most popular cloud providers, Amazon AWS[15] and Windows Azure[16], have such spot instance offerings. Even though both Windows Azure and Amazon AWS offer spot instances, there are years of spot instance price history available for AWS. Therefore, most of the discussion later in this section revolves around AWS types of spot instances.

With spot instances, customers never pay more than the maximum price specified in the bid. However, the evident concern with the spot model is that the cloud provider may terminate these instances with literally last-minute notice. AWS offers various options to configure interruption behavior of Spot instances and Spot fleets (a set of spot instances), including hibernation and automatic restarting. When the AWS Spot service determines to hibernate a Spot Instance, an interruption notice is issued as a CloudWatch event, but the customer does not have time before the Spot Instance is interrupted, and hibernation begins immediately. To prevent interruptions, the best practices suggest using the on-demand price for bidding, storing necessary data regularly at persistent storage (e.g., Amazon S3, Amazon EBS, or DynamoDB), and dividing the work into small tasks while using checkpoints. The more advanced techniques consider future spot price prediction this, in turn, allow to either avoid resource termination by bidding higher, or to store partial results before interruption is triggered by cloud provider.

Univariate Prediction Methods

Naive Predictions. As in many short-term forecast problems, the last known value is a reasonably good indicator of the next value. Thus, such predictions are commonly used as a baseline. In this section, we refer to it as *Naive* prediction model – that refers to the last known spot price of a particular instance type in the given availability zone. With this approach, at the time when we need to predict future spot price of a particular instance, we simply use the current price and predict that all future prices are going to be equal to it. Obviously, this is a very naive assumption, completely ignoring the dynamic demand for spot instances. As the prediction is about values further away in the future, the expectation is that the quality of such a forecast would significantly decrease. Still, the motivation to include this approach in the evaluation is derived from the exploratory analysis of the dataset showing that the spot prices of some instances were infrequently changing.

---

[15] aws.amazon.com/ec2/spot/.

[16] azure.microsoft.com/en-us/pricing/spot/.

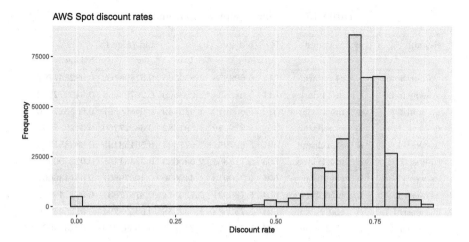

Fig. 28. Histogram of discount rates for the Linux/UNIX spot machines compared to on-demand pricing. The values are computed based on the data described in Sect. 5.4

Autoregressive Integrated Moving Average. ARIMA form a class of time series models that are widely applicable in the field of time series forecasting. ARIMA models are known to be robust and efficient in short-term time series forecasting, with some prominent results in financial and commodity markets, or for anomaly detection in IoT environments [74, 86]. In the ARIMA model, the future value of a variable is a linear combination of past values and errors after removing the trend – by differencing. Given a time series data Y_t where t is an integer index, an ARMA(p,q) model is given by:

$$Y_t = c + \varepsilon_t + \sum_{i=1}^{p} \varphi_i Y_{t-i} + \sum_{i=1}^{q} \theta_i \varepsilon_{t-i}$$

where Y_t and ε_t are the actual value and error at time period t, respectively. Whereas, c is a constant, θ_i and φ_i are model parameters to be estimated in the process of model training. ARIMA(p,d,q) model is an extension of ARMA that aims to model non-stationary processes. When the observed time series has a trend, the difference between consecutive observations is computed d times until the observed process becomes stationary.

To provide high-quality spot price forecasts, we trained ARIMA models separately for each AWS instance type in each availability zone. To adjust price prediction to the still changing environment on the AWS spot market, hence minimizing the effect of so-called concept drifts [243], models were iteratively re-trained after each day. The more detailed analysis of spot price prediction is provided further in Sect. 5.4.

Table 17. The most popular spot machines.

Region	az	Machine	N	bid freq.(h)		bid price ($)	
				Avg	StDev	Avg	StDev
ap-northeast-2	a	r4.8xlarge	513	5.609858	1.850305	0.61898109	0.032847797
ap-south-1	c	m5.4xlarge	511	5.646654	2.098600	0.26793268	0.049972972
ap-south-1	a	m4.10xlarge	510	5.664042	1.919549	0.64866667	0.065327196
us-west-1	a	r4.8xlarge	509	5.679289	1.691382	0.68227269	0.042938466
us-west-1	a	r5.4xlarge	507	5.697051	1.577393	0.56153195	0.082851776
sa-east-1	c	m4.4xlarge	505	5.718381	2.680653	0.33331168	0.022590959
us-west-2	b	c5n.4xlarge	504	5.720918	1.442650	0.40579861	0.031436667
us-west-1	b	m5d.4xlarge	504	5.730931	1.736511	0.37877837	0.071149562
eu-central-1	c	c5d.9xlarge	504	5.732888	1.632358	0.64598294	0.032545920
ap-south-1	b	c5.2xlarge	504	5.740978	1.712241	0.19892956	0.022324986

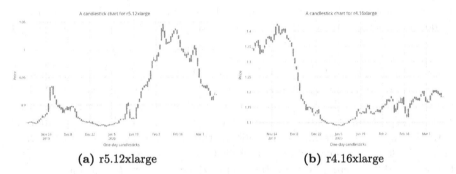

(a) r5.12xlarge   (b) r4.16xlarge

Fig. 29. Candlestick charts for two popular machine types in N. Virginia region, both in AZ: 'b'

Dataset. Contributing to the popularity in industry and research community of the Amazon Web Services (AWS), and the hardware heterogeneity offered in various instance types [262], AWS was used for the experimental evaluation. AWS cloud consists of geographically dispersed regions around the world, each with multiple availability zones (AZ's)[17]. In each of the regions, AWS offers a broad number of cloud services, among which the Elastic Cloud Compute (EC2) is the essential one.

The analyzed spot price data were collected from 11 AWS regions[18]: Tokyo (*ap-northeast-1*), Seoul (*ap-northeast-2*), Mumbai (*ap-south-1*), Singapore (*ap-southeast-1*), Sydney (*ap-southeast-2*), Canada (*ca-central-1*), Frankfurt (*eu-central-1*), Ireland (*eu-west-1*), São Paulo (*sa-east-1*), N. Virginia (*us-east-1*), N. California (*us-west-1*), and Oregon (*us-west-2*) over the period between

---

[17] See AWS global cloud infrastructure at aws.amazon.com/about-aws/global-infrastructure.
[18] AWS code-names for regions in brackets.

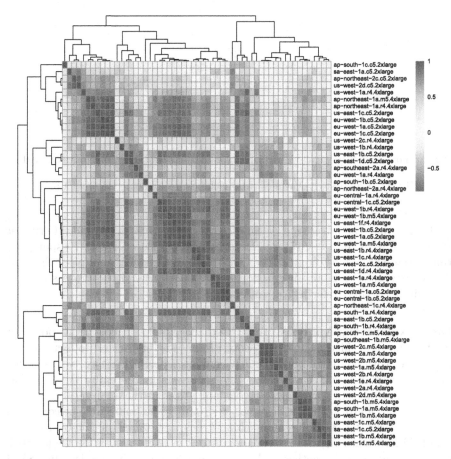

Fig. 30. Correlation heatmap for the three popular spot machine types: r4.4xlarge, m5.4xlarge, c5.2xlarge.

November 11, 2019 and March 11, 2020. After the preliminary data filtering, we left only those records, which referred to the EMR compatible EC2 machine types – i.e., dedicated for Big Data processing[19] – working with *Linux/UNIX* operating system. For spot instances there is also a constraint that the root volume must be an Elastic Block Store (EBS) volume, not an instance store volume, which eliminated some of the instances from this study. In our study, we were interested in ETL related servers, that is: the memory oriented machine types – m and r series; the computation oriented – c series, and g and p series which are popular for the data analysis and machine learning. The remaining instance families were ignored.

Concerning the savings that could be made with spot instances compared to the corresponding on-demand prices, the prelimynary data exploration results

---

[19] See docs.aws.amazon.com/emr/latest/ManagementGuide/emr-supported-instance-types.

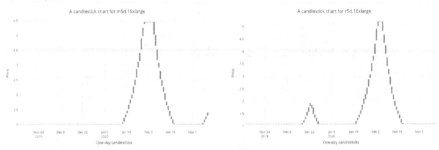

(a) ap-northeast-2(b), machine:m5d.16xl   **(b)** us-west-1(b), machine:r5d.16xl

Fig. 31. Candlestick charts of two machines troublesome for prediction.

confirmed that the advertised claim of savings up to 90% was indeed true, as shown in Fig. 28. In Table 17, we also present a brief overview of 10 spot price time series for the most popular (i.e., with the most frequent spot price changes) machine types. For more information about the data and the process of data acquisition, we may refer to Appendix A.4.

Data Exploratory Analysis. To verify the feasibility of short term spot price prediction, we decided to limit the scope of analysis further and to focus on the time series with non-trivial price change characteristics. Therefore, we discarded machines with infrequent bids (less than 100 bids in the entire data set), as well as the time series with almost constant prices in the analyzed time range (with the standard deviation of prices $\sigma < 0.01$). The final data set contained 854 time series for 85 different machine types aggregated in non-overlapping candlesticks – a standard tool in financial stock market analysis [8]. Candlestick charts are often used together with various machine learning models, like SVM or DNN [225]. In the performed experiments, each candlestick contained volume of operations as well as open, high, low, and close price during one day. The exemplary candlestick charts for two popular machine types in North Virginia (*us-east-1*) region are depicted in Fig. 29. In Fig. 32, the schematic flow of the entire data collection and machine learning process is presented.

Spot Price Predictions. Having the data aggregated in one-day candlesticks, for a given day (t_x), we aim to predict the highest price during the next day (t_{x+1}). The models are evaluated with two commonly used error metrics, namely, root mean square error (RMSE) and mean absolute percentage error (MAPE). However, to make the prediction and obtained error rates comparable between various machines, the prices were scaled – they were divided by the on-demand price of the same machine type in the corresponding region. This allows providing an estimation of a budget needed for the data processing task. The preliminary analysis showed that even the *Naive* model, which used as a prediction the last day price, achieved a relatively good quality. The highest MAPE of 5.49% was recorded for the *m5d.16xlarge* machine type in *ap-northeast-2* region (See

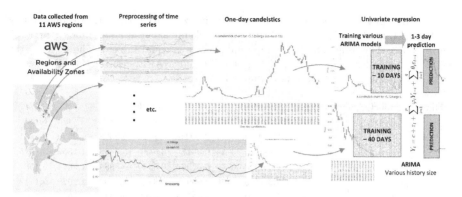

Fig. 32. A schematic flow of the entire data collection and machine learning process.

Fig. 31(a)). The median and macro average of MAPE over all 854 evaluated time series was 0.66% and 0.87%, respectively. These results mean that in the worst case, we can expect a budget overrun of ca. 5.49% in the event of a rapid price change (as shown in Fig. 31). However, on average, the error will be much smaller. The results of Naive model performance aggregated over all 854 time series are presented in Table 18 in the row signed *Naive*. The median RMSE, in Table 18, refers to median value of 854 experiments. Similarly, in the table, we also report macro average, 3rd quantile, and max value for RMSE and MAPE.

In the performed study, we trained ARIMA models for each of 854 time series in data. Similar to the Naive model's case, the evaluation was performed on the last two months in data (60 days). Before each assessment (at time t_x) of next day price, the ARIMA model was re-trained on all available historical data ($t_0...t_x$). The aggregated performance of the ARIMA model trained on all available history is presented in Table 18 in the row marked as "ARIMA(All)". To further verify the optimal history size for the estimation of model's parameters, that allows more dynamically respond to the shifts in characteristics of AWS spot prices, we repeated the experiments for various length of training data history to fit the ARIMA model (from 10 days to 50 days). In the case of 40 day long history the maximal MAPE error of 3.84% was recorded in *us-west-1* region for *r5d.16xlarge* machine (see Fig. 31(b)). This provides us with the worst-case estimation of cost under- or over-run of spot resource allocation. Still, in the (macro) averaged or mean case, we would be far more accurate. In Table 18, the aggregated analysis for the Naive model and all ARIMA settings is presented.

For the more in-depth analysis, we decided to select Naive, ARIMA(All), ARIMA(40) models. The first one presents a baseline, the second achieved lowest average errors, whereas the ARIMA(40) minimized the maximal MAPE error, which assures the lowest worst-case budget misestimation. The three main parameters to be estimated in the ARIMA(p, d, q) model are the number of time lags of the auto-regressive model p, degree of differencing d, and the order of the moving average model (q). In our experiments, these parameters were estimated

Table 18. Spot price prediction in one day horizon. ARIMA with various sliding window settings (training history length in brackets) and the naive model with last known value. The minimal error for all classifiers was equal to zero. The 1st quantile was always smaller than 0.001 and 0.15 for R2 and MAPE, respectively. Prediction evaluated on the last two months in data (60 days).

Model	Hist. size	RMSE				MAPE			
		Median	MacAvg	3^{rd}Qu	Max	Median	MacAvg	3^{rd}Qu	Max
ARIMA	10	0.00354	0.00624	0.00651	**0.04195**	0.7069	0.9088	1.2520	5.1401
	15	0.00315	**0.0055**	0.0071	0.0617	0.6486	0.8008	1.1552	3.8915
	20	0.00325	0.00643	0.00759	0.05964	0.6620	0.8884	1.2644	5.1581
	25	0.00313	0.00630	0.00750	0.05926	0.6322	0.8597	1.1928	6.4112
	30	0.00305	0.00598	0.00714	0.05909	0.6199	0.8240	1.2005	5.5383
	35	0.00299	0.00606	0.00715	0.08033	0.6101	0.8197	1.1622	4.7605
	40	0.00293	0.00577	0.00688	0.05909	0.5987	0.8020	1.1433	**3.842**
	45	0.00286	0.00577	0.00704	0.07563	0.5954	0.8032	1.1157	4.0203
	50	0.00286	0.00557	0.00692	0.10255	0.5893	**0.789**	1.0645	4.9812
	All	**0.0028**	0.00551	0.00739	0.07884	**0.5866**	0.8051	**1.0335**	6.5449
Naive	1	0.00314	0.00607	**0.0059**	0.04648	0.6640	0.8782	1.2556	5.4873

using the Box-Jenkins approach. The analysis of the selected models revealed that for various time series and length of available training data, different p and q parameter values were chosen. In the performed experiments, the series was most often differenced once - trend components for the trained ARIMA models were usually $d = 1$. The AR parameters were typically equal to 1 or 2, whereas MA parameters varied from 0 to 4. The seasonality test was negative in all examined cases. Hence, the trained ARIMA models were of a form ARIMA$(1 - 2, 1, 0 - 4)$.

To validate the statistical significance of observed differences between the performance of the selected models, we decided to employ the Wilcoxon signed rank test – due to a very low p-value observed during Shapiro-Wilk normality test on both RMSE and MAPE distributions achieved during the tests. In all the cases, p-value of Shapiro-Wilk normality test was: $p\text{-}value < 1.0e\text{-}15$. In the case of RMSE, the Wilcoxon signed rank test, with the null hypothesis that the errors of the Naive model are not greater than those of ARIMA(40) did not allow to reject this hypothesis ($p\text{-}value = 0.1954$). However, when the ARIMA(All) model was compared to ARIMA(40) and Naive, the p-values of both tests were very low, i.e., $3.786e\text{-}08$ for Naive and $4.649e-06$ for ARIMA(40), respectively. It allowed us to reject the null hypothesis, hence showing the statistical significance of differences between the models. Slightly different observations were made for the MAPE measure. In this case, the Wilcoxon test revealed that ARIMA(40) model was significantly better than Naive ($p\text{-}value = 3.053e\text{-}07$). However, the ARIMA(All) again performed significantly better than both ARIMA(40) ($p\text{-}value = 0.005515$) and Naive ($p\text{-}value = 3.396e\text{-}11$) models.

In the last part of our study, we attempted to validate the feasibility of spot price prediction in a bit longer horizon of two and three days ahead. We examined the performance of the three selected models from our previous test: Naive,

Table 19. A summary of the prediction errors of the selected models for various forecasting horizons (1–3 days).

Model	Pred. day	RMSE				MAPE			
		Median	MacAvg	3^{rd}Qu	Max	Median	MacAvg	3^{rd}Qu	Max
ARIMA (All)	1	0.0028	0.0055	0.0074	0.0788	0.5866	0.8051	1.034	6.545
	2	0.0053	0.01	0.0118	0.095	1.15	1.537	2.029	10.73
	3	0.0076	0.0146	0.0152	0.142	1.647	2.28	2.948	16.32
ARIMA (40)	1	0.0029	0.0058	0.0069	0.0591	0.599	0.802	1.143	3.842
	2	0.0056	0.0105	0.0123	0.079	1.187	1.57	2.18	7.96
	3	0.0079	0.015	0.017	0.136	1.72	2.36	3.24	13.98
Naive	1	0.00314	0.0061	0.006	0.0465	0.664	0.878	1.256	5.487
	2	0.00547	0.0112	0.011	0.0887	1.1841	1.6459	2.335	11.01
	3	0.0073	0.0157	0.014	0.126	1.609	2.345	3.317	16.48

ARIMA(All) and ARIMA(40). The results - presented in Table 19 - showed that the observed drop of each model performance is significant, and the maximal MAPE error exceeds 16% for both ARIMA(All) and Naive models. However, we may conclude that prediction is still feasible two and three days ahead, with a median of MAPE errors only slightly exceeding 1.6% for ARIMA(All).

An interesting approach to further investigation would be to use multivariate methods, mainly due to the observed correlations between various time series in multiple regions and availability zones, as shown in Fig. 30. This figure is a heatmap with a dendrogram added to the left side and to the top where the colour of each cell represents the correlation between the price of a pair of instance types in different availability zones. A dendrogram is a tree-structured graph that visualizes the result of a hierarchical clustering calculation. For the dendrogram on the left side of the heatmap, the individual rows in the clustered data are represented by the right-most nodes (i.e., the leaf nodes). Each node in the dendrogram represents a cluster of all rows from the connected leaves. The left-most node in the dendrogram is therefore a cluster that contains all rows.

6   Concluding Remarks and Future Works

This section concludes the presented study and indicates some possible research directions for the future development of the interactive feature extraction methods and points out some interesting application areas.

6.1   Summary

In this work, we discuss interactive feature extraction, and we propose several innovative approaches to automating feature creation and selection processes. In the study on the interactiveness of the feature extraction methodologies, we address the problems of deriving relevant and understandable attributes from raw sensor readings and reducing the amount of those attributes to achieve possibly simplest yet accurate models. The proposed algorithms for the construction and selection of features can use various forms of granulation, problem decomposition, and parallelization. Consequently, they respond to the requirements of expressing complex concepts intuitively and efficiently, which are essential for the feasibility of feature selection.

Feature selection is crucial for constructing prediction and classification models, resulting in their higher quality and interpretability. However, the selected features may become temporarily unavailable in a long-term time frame, which can disable a pre-trained model and cause a severe impact on business continuity. The presented methods go beyond the current standards. Accordingly, we formalize the notion of resilient feature selection by introducing r-$\mathbb{C}$-reducts – irreducible subsets of attributes providing a satisfactory level of information about the target variable according to a given criterion function $\mathbb{C}$, even after removing arbitrary r elements. The proposed approach is based on a generalization of (approximate) reducts known from the rough set theory (RST) [373]. The framework proposed in this paper embraces a much wider family of criteria specifying that a given feature subset is good enough to determine target variable values. We are actually able to refer to the whole realm of filter-based feature selection strategies [82], now defining a satisfactory feature set as the one whose evaluation function exceeds a certain threshold even after removing its arbitrary r elements, $r \geq 0$.

We proved that any NP-hard problem of finding a minimal attribute subset that yields a satisfactory level of information according to a given criterion function $\mathbb{C}$ remains NP-hard for an arbitrary resilience level r. As a special case, the task of finding a minimal subset of features providing ε-almost the same level of the aforementioned accuracy measure as the whole set even after removing arbitrary r elements is NP-hard. We also discuss opportunities of the exhaustive and heuristic search of r-$\mathbb{C}$-reducts. By following a popular idea of dynamic exploration of the lattice of feature subsets, whereby some of its elements turn out to be labeled as satisfying the criteria for providing enough information while others do not, we elaborate on two generic algorithmic strategies, namely: breadth first search (BFS), and depth first search (DFS). For BFS, we adapted the well-known Apriori algorithm [335] for the purpose of r-$\mathbb{C}$-reduct search (Sect. 3.3). For DFS, we extended standard reduct construction methods [357] to incorporate resilience of generated feature sets (Sect. 11). The presented results confirm that the idea is very promising, and resilient feature selection may significantly minimize the risk and impact of data loss on predictive analysis.

With regard to feature engineering, we present a particular take on the challenge of devising a more effective and efficient feature extraction methodology. The main idea behind our approach is to make intelligent use of the information granulation paradigm in the context of aggregating, selecting, and engineering attributes (features/variables/dimensions) that describe the data. The gist is to operate on attribute granules that are formed through the use of various knowledge discovery algorithms, such as, e.g., clustering or interchangeability analysis through heat maps. In many instances, as exemplified by the use cases discussed, granules built over the attribute space may represent semantic relationships that are important for domain experts. The proposed framework facilitates discovering meaningful knowledge from the underlying data, which may be further leveraged in order to obtain a more comprehensible and user-friendly representation that is described in a possibly intuitive way, i.e., using statistics characterizing sliding time windows (Sect. 4.1). In the case of the underground coal mine sensors, derivation of multivariate series of window-based statistics allowed us to deal with noisy and incomplete data sources, better reflected temporal drifts and correlations, and reliably described real situations using higher-level data characteristics.

As a notable aspect and an important research field addressed in the frame of this study, let us point out the framework for linking sliding window-based feature creation, resilient feature selection, and machine learning techniques to build predictive models that are understandable for experts and resistant to partial data loss (Sect. 4). The solution conveys the granular knowledge in the data to the final decision model. At the same time, it is designed to deal with enormous amounts of information that needs to be processed when facing the kinds of tasks typical for Big Data. The proposed methods for feature extraction are easy to maintain and efficient to compute. They are understandable for the system users and domain experts by means of operating with small subsets of intuitively defined features, which is an important aspect from the point of view of interactiveness.

The proposed approaches to interactive feature extraction have been developed based on the experience gained in the course of several research projects in the fields of processing mutli-sensory streams in various domains, but also textual data analysis [144]. The experimental study in Sect. 5 confirms the quality of the proposed framework, taking into account its subsequent layers of feature creation, selection (Figs. 12, and 13), forecasting models training and ensemble blending (Fig. 14). The methods have been validated in terms of the quality of the obtained features, throughput, scalability, and resilience of their operation. The discussed methodology has been successfully applied in several real-life problems related to the time series data [136, 144, 182, 219, 360, 427]. Furthermore, we describe a series of international data mining challenges organized to facilitate this study.

In this work, we addresses a number of challenges related, among others, to the comprehensible and concise representation of the analyzed data or the possibility of embedding domain knowledge into the data. The investigated problems have been thoroughly considered both from the theoretical and practical sides. The developed solutions have been meticulously evaluated in terms of various qualitative aspects like the diversity of the solution or its resilience to data deficiencies. We provide a comprehensive rationale for this research direction, building a solid theoretical foundation for further considerations related to the interactivity of the feature extraction and machine learning process.

6.2   Future Works

The next steps towards practical use of the outlined methodology would be to devise methods and tools that automate this process and, at the same time, maintain an acceptable level of transparency and human readability in a possibly visual way while taking into account various constraints [53, 280]. Further research on the interactive incorporation of domain knowledge into feature extraction is also a desired direction. For example, in one of the possible scenarios, an analyst collaborates with a feature selection algorithm through a specially designed user interface. In an iterative way, the analyst passes feedback on the relevance of attributes proposed by the algorithm. Therefore, allows to limit the scope of the analysis and improve the quality of the obtained features. A complete system capable of flexible, comprehensible, and extensible interaction with a data scientist who analyzes massive data sets and provides their input by interruption to the extraction process would be an invaluable tool [362].

No less important is integration with the existing technologies. In the presented study, we have shown how the MapReduce principles can be employed. There are, however, many more other techniques that were developed over the years with Big Data in mind. For example, it could be helpful to integrate the proposed methods with the existing tools for the management of massive relational data sets (such as Apache Hive or some approximate database engines) [359]. This way, we could embed the "zoom in/out" operations on attributes into a convenient RDBMS environment.

It would be valuable to continue the study on problems related to monitoring natural hazards in underground coal mines. In particular, to continue work on extending and better utilizing information registered while adding new data sources. There are many places where such information could be useful to configure the steps of consistent data ingestion, preprocessing, and learning forecasting models. It is commonly known that the stage of feature extraction should take into account the semantics of both the overall forecasting task and particular inputs that may help to build its solution. One of the potentially valuable capabilities would be to use it to customize window-based and inter window-based aggregations applied for particular sensors and groups of sensors during the process of feature extraction. In the case of multi-sensor analytics in the domain of underground coal mining, it may mean applying different feature extraction strategies for different types of sensors, as well as constructing higher-level features basing on spatial information represented by the corresponding mining schemes, such as the one shown in Fig. 17.

As for future research in this area, it is important to perform a deeper analysis of errors made by different models to identify factors influencing wrong predictions for different mining sites. It is important to continue research on the problem of reliability of prediction models in cases when some of the sensory devices which gather the input data are malfunctioning. This is a very common situation in active mines due to a very harsh working environment. In order to construct prediction models which are robust and insensitive to gaps in incoming data, we would like to develop methods for automatic detection of exchangeable features (i.e., attributes whose values come from different sources but express very similar information). We plan to further investigate feature subset selection methods that keep a controllable degree of information redundancy.

In this particular area, the notion of r-$\mathbb{C}$-reduct can be regarded as a feature selection method derived from the theory of rough sets that allows constructing small subsets of features while maintaining the discernibility of objects in a data set, even in a case when we suspect that a part of attributes may not be available in future. It is, however, worth remembering that there are also some cases of criterion functions $\mathbb{C}$, which model data-based information encoded by attribute subsets in a not (strictly) monotonic way, which means that smaller attribute sets can potentially yield a higher level of information. As a special case, we could consider classes of functions with relaxed monotonicity conditions. In particular, it would be valuable to study functions providing weak, quasi, and directional monotonicity [300, 339]. We could also consider various aggregation functions (called also pre-aggregation functions), directionally increasing conjunctors and implications [48], or mixture functions - a type of weighted averages for which the corresponding weights are calculated by means of appropriate continuous functions of their inputs, which need not be monotone increasing [392].

It is particularly important to investigate all the above aspects in both theoretical and empirical study, including the assessments in real-life environments, taking into account simulations based on pessimistic and random attribute removal scenarios that model temporary, partial unavailability of data sources. For the same reason, it is needed to empirically compare the proposed framework with other (both RST-based and not RST-based) approaches to stable and robust feature selection [3, 201].

We do believe that further research results may have a significant impact on the development of feature extraction. The presented future research directions have a solid practical motivation. Deriving meaningful features, which are interpretable for human experts, is important in many domains, as medicine, criminal justice, or industrial processes monitoring [119, 329, 360]. Such methods may increase the safety of people working in, e.g., underground mines or participating in firefighting rescue operations. Furthermore, through analysis of variables' importance [111] and co-predictive mechanisms between interpretable features [118], they may foster the understanding of the root causes of modeled phenomena.

Acknowledgements. I would like to thank my advisors Dominik Ślęzak and Andrzej Janusz for their guidance and invaluable support in conducting this research.

I would also like to sincerely thank all from the faculty of Mathematics, Informatics, and Mechanics, University of Warsaw, who I consider not only my colleagues but also my friends. I have learnt a lot from their knowledge and experience. Moreover, I would like to thank all my co-authors: Cas Apanowicz, Ace Dimitrievski, Łukasz Grad, Andrzej Janusz, Mateusz Kalisch, Michał Kijowski, Marcin Kowalski, Michał Kozielski, Zdzisław Krzystanek, Petre Lameski, Marcin Michalak, Sinh Hoa Nguyen, Przemysław Wiktor Pardel, Marek Sikora, Sebastian Stawicki, Krzysztof Stencel, Marcin Szczuka, Dominik Ślęzak, Piotr Wojtas, Łukasz Wróbel, Eftim Zdravevski, whose excellent ideas have taught me to stay open-minded, and have been a motivation and inspiration for my research and writing.

My deepest gratitude goes to my Lovely Wife and Kids for their support in all my efforts and understanding during many years of research activities. None of the things that I achieved would be possible without you. Thank You!

Partial contributions of the study were influenced by the author's work in several research projects. The content of this article comes from the author's Ph.D. dissertation [139].

A   Data Insights

A.1   Methane Data

The appendix presents selected, more in-depth insights into the methane-related data set, reffed to in Sect. 5.1. The data correspond to a mining period between March 2, 2014, and June 16, 2014, and is collected from selected 28 (out of thousands) sensors located in an active Polish coal mine, which were located in the vicinity of the coal extraction area. The data has been made available on KnowlegePit platform for the purpose of organizing IJCRS'15 Data Challenge: Mining Data from Coal Mines [184].

Table 20. Sensors related to methane data.

Sensor	Type	Unit	Type	Additional Info
AN311	Anemometer	m/s	Alarming	Threshold A: none, Threshold B: $<= 0.3\,\text{m/s}$
AN422	Anemometer	m/s	Switching off	Threshold A: $<= 1.1\,\text{m/s}$, Threshold B: $<= 1.3\,\text{m/s}$
AN423	Anemometer	m/s	Switching off	Threshold A: $<= 1.0\,\text{m/s}$, Threshold B: $<= 1.2\,\text{m/s}$
TP1721	Thermometer	$°C$	Registering	Tri-constituent sensor THP2/93
RH1722	Humidity	%RH	Registering	Tri-constituent sensor THP2/93
BA1723	Barometer	hPa	Registering	Tri-constituent sensor THP2/93
TP1711	Thermometer	$°C$	Registering	Tri-constituent sensor THP2/94
RH1712	Humidity	%RH	Registering	Tri-constituent sensor THP2/94
BA1713	Barometer	hPa	Registering	Tri-constituent sensor THP2/94
MM252	Methanometer	$\%CH_4$	Switching off	Threshold A: 2.0%, Threshold B: 1.5%
MM261	Methanometer	$\%CH_4$	Switching off	Threshold A: 1.5%, Threshold B: 1.0%
MM262	Methanometer	$\%CH_4$	Switching off	Threshold A: 1.0%, Threshold B: 0.6%
MM263	Methanometer	$\%CH_4$	Switching off	Threshold A: 1.5%, Threshold B: 1.0%
MM264	Methanometer	$\%CH_4$	Switching off	Threshold A: 1.5%, Threshold B: 1.0%
MM256	Methanometer	$\%CH_4$	Switching off	Threshold A: 1.5%, Threshold B: 1.0%
MM211	Methanometer	$\%CH_4$	Switching off	Threshold A: 2.0%, Threshold B: 1.5%
CM861	Methanometer	$\%CH_4$	Registering	Measures high concentrations of methane
CR863	Pressure difference	Pa	Registering	Sensor is placed on the demethanisation orifice
P_864	Barometer	kPa	Registering	Pressure inside the pipeline for methane drainage
TC862	Temperature	$°C$	Registering	Temperature inside the pipeline for methane drainage
WM868	Methane expense	m^3/min	Registering	Methane expense calculated by CM, CR, P, TC
AMP1	Ammeter	A	Registering	Current in the motor in the left arm of the shearer
AMP2	Ammeter	A	Registering	Current in the motor in the right arm of the shearer
DMP3	Ammeter	A	Registering	Current in the motor in the left tractor of the shearer
DMP4	Ammeter	A	Registering	Current in the motor in the right tractor of the shearer
AMP5	Ammeter	A	Registering	Current in the hydraulic pump motor of the shearer
F_SIDE	Drive direction	Left, right	Registering	The driving direction of the shearer
V	Shearer speed	Hz	Registering	Work frequency, 100 Hz means ca 20 m/min

In Table 20, three groups of sensors are presented (groups are separated with horizontal lines). The first group is responsible for the monitoring of the mine atmosphere, the second group monitors the methane drainage flange, the third group monitors the operating status of a longwall shearer. The fifth column provides additional information about security thresholds assigned to the selected sensors. After crossing the threshold A the "switching off" sensors cut off the electricity supply. After crossing the threshold B both the "alarming" and "switching off" sensors display a predefined warning message. All of sensor recordings are collected and stored for the purpose of the further analysis. A detailed location of all sensors, as well as a workplace of a longwall shearer, on a fragment of the coal mine plan, is shown in Fig. 17, in Sect. 5.1.

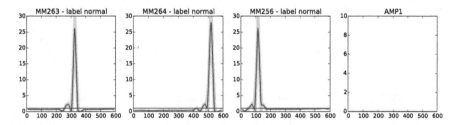

Fig. 33. Examples of outliers in methane concentration time series.

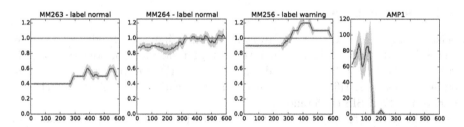

Fig. 34. Methane indications oscillating near the warning threshold. On the rightmost plot the current cut off after exceeding the methane concentration threshold.

The plots show that, apart from outliers (Fig. 33), the variation of methanometer indications is relatively small (Fig. 35). Hence, indications oscillating near the warning threshold, like in Fig. 34, are the most difficult cases. Furthermore, "warnings" were rarely indicated by more than one sensor (see example in Fig. 36) what could potentially affect on, e.g., multi-target approaches related to classifier chains technique [319].

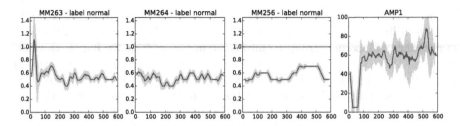

Fig. 35. Relatively small dynamics of changes in methane concentration for three methane detectors. On the rightmost plot we can observe current consumption of the cutter loader that corresponds to ongoing coal mining.

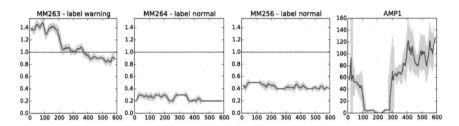

Fig. 36. Three methane detectors and ammeter located on the cutter loader. The sensor MM263 has exceeded the warning threshold.

A.2   Seismic Data

Seismic data is related to AAIA'16 Data Mining Challenge: Predicting Dangerous Seismic Events in Active Coal Mines that took place between October 5, 2015, and February 27, 2016, under auspices of 11th International Symposium on Advances in Artificial Intelligence and Applications (AAIA'16) which is a part of the FedCSIS conference.

All the attributes of the data set are described in Table 21. Test and train data sets are available online at the competition's web page – AAIA'16 Data Mining Challenge at the KnowledgePit platform. To access the data it is necessary to register.

Table 21. Attributes of the seismic data. The experts assessments (a/b/c/d) corresponds to: a - no hazard; b - moderate hazard; c - high hazard; d - dangerous.

Attribute no	Description
1	ID of the main working site where the measurements were taken
2	Total energy of seismic bumps registered in the last 24 h
3	Total energy of major seismic bumps registered in the last 24 h
4	Total energy of destressing blasts in the last 24 h
5	Total seismic energy of all types of bumps
6	Latest progress in the mining from, both, left and right side
7	Latest seismic hazard assessment made by experts (a/b/c/d)
8	Latest seismoacoustic hazard assessment by experts (a/b/c/d)
9	Latest (alternative) seismoacoustic hazard assessment (a/b/c/d)
10	Latest comprehensive hazard assessment made by experts (a/b/c/d)
11	Maximum yield from the last meter of the small-diameter drilling
12	Depth at which the maximum yield was registered
13–37	Time series containing number of seismic bumps with energy in range $(0, 10^2]$ per hour (1..24)
38–61	Time series containing number of seismic bumps with energy in range $(10^2, 10^3]$ per hour (1..24)
62–85	Time series containing number of seismic bumps with energy in range $(10^3, 10^4]$ per hour (1..24)
86–109	Time series containing number of seismic bumps with energy in range $(10^4, 10^5]$ per hour (1..24)
110–133	Time series containing number of seismic bumps with energy in range $(10^5, Inf)$ per hour (1..24)
134–157	Time series containing sum of energy of seismic bumps with energy in rang $(0, 10^2]$ per hour (1..24)
158–181	Time series containing sum of energy of seismic bumps with energy in rang $(10^2, 10^3]$ per h (1..24)
182–205	Time series containing sum of energy of seismic bumps with energy in rang $(10^3, 10^4]$ per h (1..24)
206–229	Time series containing sum of energy of seismic bumps with energy in rang $(10^4, 10^5]$ per h (1..24)
230–253	Time series containing sum of energy of seismic bumps with energy in rang $(10^5, Inf)$ per h (1..24)
254–277	Time series containing number of seismic bumps per hour (1..24)
278–301	Time series containing number of rock bursts per hour (1..24)
302–325	Time series containing number of destressing blasts per hour (1..24)
326–349	Time series containing energy of the strongest seismic bump per hour (1..24)
350–373	Time series containing max activity of the most active geophone per hour (1..24)
374–397	Time series containing max energy of the most active geophone per hour (1..24)
398–421	Time series containing avg activity of the most active geophone per hour (1..24)
422–445	Time series containing avg energy of the most active geophone per hour (1..24)
446–469	Time series containing maximum difference in activity of the most active geophone per hour (1..24)
470–493	Time series containing maximum difference in energy of the most active geophone per hour (1..24)
494–517	Time series containing average difference in activity of the most active geophone per hour (1..24)
518–541	Time series containing average difference in energy of the most active geophone per hour (1..24)

A.3   Firefighter Data

The data used in the AAIA'15 data mining competition: Tagging Firefighter Activities at a Fire Scene were collected during training exercises conducted by a group of firefighters from the Main School of Fire Service in Warsaw. All cadets participating in the experiment were equipped with several sensors located on their body, including seven inertial measurement units (IMU) Polulu AltIMU-9 rev-4 with 3–axis (horizontal, vertical, and altitudinal) accelerometers with $\pm16g$ dynamic range, and 3–axis gyroscopes with $\pm2000 \circ /s$ maximum angular rate, and a physiological data sensor - Equivital Single Subject Kit (EQ-02-KIT-SU-4). Sensors were integrated with a data acquisition unit (DAU) on Odroid-U3+ with an external battery, additional Bluetooth, and a Wi-Fi module. The data acquisition process was further supported by XBee-PRO 868 communication nodes and Arduino micro prototype platform connected via USB to DAU.

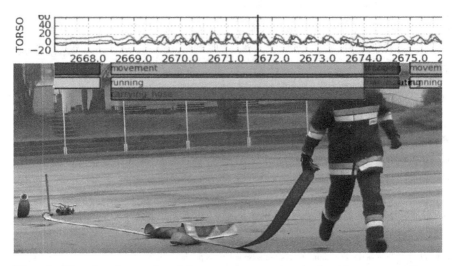

Fig. 37. Filmed and tagged training exercises synchronized with sensor readings.

During the exercise, cadets were simulating typical actions related to a fire incident. The video recordings of the experiment were synchronized with sensor readings and provided to domain experts, who tagged videos with the observed posture and activity. In Fig. 37, we provide a demonstrative frame from the video recording of the exercises with sample labels and selected sensor readings[20]. The data set achieved this way contains 20,000 rows and 17,242 columns, and is available on-line at KnowlegePit platform. Each data row corresponds to a short time series (approximately 1.8 s long) of sensor readings; hence an average time difference between consecutive sensory readings is 4.5 ms. The descriptions of data attributes and two labels are provided in Table 22. For further reading about the measuring apparatus or the recording procedure, we may refer to the summary of AAIA'15 data mining competition [263].

---

[20] The video recording is available on-line on KnowlegePit platform.

Table 22. Attributes of the firefighter data.

Attribute	Description
1-42	42 attributes with aggregated statistics, including: average, standard deviation, median, skew, and avgDiff collected from physiological sensor measuring: ECG, breath rate, heart beat rate, heart rate interval (RR), and body temperature. Each attribute name ends with: ECG1, ECG2, HRBR1, HRBR2, RR, and TEMP, indicating the measured vital parameter, and each one is prefixed with the type of statistics, e.g., min_TEMP, max_ECG1, or avg_HRBR2.
time1..400	relative time of each measure (1..400)
ll_acc_x_1..400	400 readings from x-axis of accelerometers attached to each firefighter's left leg
ll_acc_y_1..400	400 readings from y-axis of accelerometers attached to each firefighter's left leg
ll_acc_z_1..400	400 readings from z-axis of accelerometers attached to each firefighter's left leg
ll_gyro_x_1..400	400 readings from x-axis of gyroscopes attached to each firefighter's left leg
ll_gyro_y_1..400	400 readings from y-axis of gyroscopes attached to each firefighter's left leg
ll_gyro_z_1..400	400 readings from z-axis of gyroscopes attached to each firefighter's left leg
rl_acc_x_1..400	400 readings from x-axis of accelerometers attached to each firefighter's right leg
rl_acc_y_1..400	400 readings from y-axis of accelerometers attached to each firefighter's right leg
rl_acc_z_1..400	400 readings from z-axis of accelerometers attached to each firefighter's right leg
rl_gyro_x_1..400	400 readings from x-axis of gyroscopes attached to each firefighter's right leg
rl_gyro_y_1..400	400 readings from y-axis of gyroscopes attached to each firefighter's right leg
rl_gyro_z_1..400	400 readings from z-axis of gyroscopes attached to each firefighter's right leg
lh_acc_x_1..400	400 readings from x-axis of accelerometers attached to each firefighter's left hand
lh_acc_y_1..400	400 readings from y-axis of accelerometers attached to each firefighter's left hand
lh_acc_z_1..400	400 readings from z-axis of accelerometers attached to each firefighter's left hand
lh_gyro_x_1..400	400 readings from x-axis of gyroscopes attached to each firefighter's left hand
lh_gyro_y_1..400	400 readings from y-axis of gyroscopes attached to each firefighter's left hand
lh_gyro_z_1..400	400 readings from z-axis of gyroscopes attached to each firefighter's left hand
rh_acc_x_1..400	400 readings from x-axis of accelerometers attached to each firefighter's right hand
rh_acc_y_1..400	400 readings from y-axis of accelerometers attached to each firefighter's right hand
rh_acc_z_1..400	400 readings from z-axis of accelerometers attached to each firefighter's right hand
rh_gyro_x_1..400	400 readings from x-axis of gyroscopes attached to each firefighter's right hand
rh_gyro_y_1..400	400 readings from y-axis of gyroscopes attached to each firefighter's right hand
rh_gyro_z_1..400	400 readings from z-axis of gyroscopes attached to each firefighter's right hand
la_acc_x_1..400	400 readings from x-axis of accelerometers attached to each firefighter's left arm
la_acc_y_1..400	400 readings from y-axis of accelerometers attached to each firefighter's left arm
la_acc_z_1..400	400 readings from z-axis of accelerometers attached to each firefighter's left arm
la_gyro_x_1..400	400 readings from x-axis of gyroscopes attached to each firefighter's left arm
la_gyro_y_1..400	400 readings from y-axis of gyroscopes attached to each firefighter's left arm
la_gyro_z_1..400	400 readings from z-axis of gyroscopes attached to each firefighter's left arm
ra_acc_x_1..400	400 readings from x-axis of accelerometers attached to each firefighter's right arm
ra_acc_y_1..400	400 readings from y-axis of accelerometers attached to each firefighter's right arm
ra_acc_z_1..400	400 readings from z-axis of accelerometers attached to each firefighter's right arm
ra_gyro_x_1..400	400 readings from x-axis of gyroscopes attached to each firefighter's right arm
ra_gyro_y_1..400	400 readings from y-axis of gyroscopes attached to each firefighter's right arm
ra_gyro_z_1..400	400 readings from z-axis of gyroscopes attached to each firefighter's right arm
torso_acc_x_1..400	400 readings from x-axis of accelerometers attached to each firefighter's torso
torso_acc_y_1..400	400 readings from y-axis of accelerometers attached to each firefighter's torso
torso_acc_z_1..400	400 readings from z-axis of accelerometers attached to each firefighter's torso
torso_gyro_x_1..400	400 readings from x-axis of gyroscopes attached to each firefighter's torso
torso_gyro_y_1..400	400 readings from y-axis of gyroscopes attached to each firefighter's torso
torso_gyro_z_1..400	400 readings from z-axis of gyroscopes attached to each firefighter's torso
posture	a label describing a posture of a firefighter
activity	a label describing a main activity of a firefighter

A.4   AWS Spot Data

The data were iteratively collected over the period between November 11, 2019 and March 11, 2020, from 11 AWS regions[21]: *ap-northeast-1* (427309), *ap-northeast-2* (303113), *ap-south-1* (322593), *ap-southeast-1* (459592), *ap-southeast-2* (382703), *ca-central-1* (206218), *eu-central-1* (453457), *eu-west-1* (546474), *sa-east-1* (280076), *us-east-1* (1061429), *us-west-1* (285975), and *us-west-2* (661369) using AWS command line interface (CLI v2) with the following command:

```
aws2 ec2 describe-spot-price-history
--region <e.g., us-east-1>
--start-time  <e.g., 2019-11-11T12:00:00>
--end-time <e.g., 2019-11-18T12:00:00>
--output text
```

The raw data consisted of a total of $5.4M$ unique records, each corresponding to a bid for one of the AWS spot instances, in a form as presented in Table 23. The preliminary data exploration revealed that the spot price time series for a given instance type differ between regions and availability zones.

Table 23. Exemplary spot price bids collected from AWS.

SPOTPRICEHISTORY	Region & AZ	Instance type	System	Bid price	Bid date & time
SPOTPRICEHISTORY	sa-east-1c	m4.xlarge	Linux/UNIX	0.076300	2020-02-11T14:50:42+00:00
SPOTPRICEHISTORY	sa-east-1b	m5.large	Linux/UNIX	0.052800	2020-02-11T14:25:45+00:00
SPOTPRICEHISTORY	sa-east-1c	m5.24xlarge	Linux/UNIX	2.055900	2020-02-11T14:11:04+00:00
SPOTPRICEHISTORY	sa-east-1c	m5.24xlarge	Windows	6.471900	2020-02-11T14:10:52+00:00
SPOTPRICEHISTORY	sa-east-1a	r3.2xlarge	SUSE Linux	0.262800	2020-02-11T14:08:50+00:00

After the initial filtering and pre-processing, the data were aligned into 854 time series. One per each *region, AZ, instance type* triple, and aggregated daily as presented in Table 24. *Volume* column presents the number of price changing bids recorded in a given time window. *Open* refers to the first bid in a given time window. Whereas *High, Low,* and *Close* columns refer to highest, lowest, and the last bid in each time window, respectively. If no bids were recorded, i.e., *Volume* equals to zero (cf. last row in Table 24), *Open, High, Low,* and *Close* rates were assigned the same value as the *Close* price in the previous window. Such a data format allowed us to represent time series as candlestick charts (cf. Fig. 29).

---

[21] Numbers in brackets indicate the amount of unique bids in data for each region.

Table 24. Data aggregated in 24 h long time windows starting at *Window Begin*.

Region	AZ	Instance type	Window Begin	Open ($)	High ($)	Low ($)	Close ($)	Vol.
us-east-1	d	r5.12xlarge	2020-03-07 12:00:00	0.8790	0.8790	0.8763	0.8763	3
us-east-1	d	r5.12xlarge	2020-03-08 12:00:00	0.8761	0.8761	0.8757	0.8757	4
us-east-1	d	r5.12xlarge	2020-03-09 12:00:00	0.8767	0.8810	0.8767	0.8810	3
us-east-1	d	r5.12xlarge	2020-03-10 12:00:00	0.8799	0.8882	0.8799	0.8863	4
us-east-1	d	r5.12xlarge	2020-03-11 12:00:00	0.8863	0.8863	0.8863	0.8863	0

B

B.1   Expert Methods for Classifications of Seismic Hazards in Coal Mines

Two basic methods are routinely used by experts for the assessment of seismic hazards in Polish coal mines. These methods are often called *seismic* and *seismoacoustic*, respectively [200].

The essence of the seismic method is the analysis of tremor occurrences in mines. Table 25 presents the basis for quantitative hazard assessment using this method. This type of assessment is performed routinely every shift. As shown in Table 25, very simple and intuitive rules are used in order to model the relationship between the energy of tremors and rock bursts. These rules were designed by experts based on their experience and common sense.

The seismoacoustic method is based on an analysis of seismoacoustic emissions recorded at a given longwall. The seismoacoustic emission is described by its intensity, understood as the number of registered events and their total energy. The dependency between the seismoacoustic emission seismic hazards was often observed in practice by mining experts. In this type of assessment, the following factors are considered as crucial:

- registered seismoacoustic emissions,
- the number of pulses recorded by geophones, which is converted into so-called conventional seismic energy using an appropriate formula.

Available studies on the effectiveness of the seismic and seismoacoustic methods are limited to those conducted by the Polish Central Mining Institute [199]. Its analysis of selected cases of rock-bursts showed that the seismic method correctly predicted these dangerous events in only about 17% of cases. When the seismic method was coupled with the seismoacoustic approach (i.e., a hazardous state is predicted when any of the methods indicate the state 'd'), the prediction accuracy increased to about 20%. However, the data set used for the purpose of that experiment was relatively small and did not cover coal mines located in different geographical locations. The used data set is not publicly available, hence the results of this evaluation were difficult to reproduce.

Table 25. Quantitative assessments of seismic hazards based on the observed seismic activity, as outlined in [200].

Rockburst hazard	Caved faces	Roadways
a no hazard	1. No tremors or single tremors with energies E of the order of 10^2 J – 10^3 J 2. $E_{max} \leq 10^4$ J 3. $\Sigma E < 10^5$ J per 5m of longwall advance	1. No tremors or single tremors with energies E of the order of 10^2 J 2. $E_{max} \leq 10^3$ J 3. $\Sigma E < 10^3$ J per 5m of longwall advance
b low hazard	1. Occurrence of tremors with energies E of the order of 10^2 J – 10^5 J 2. 10^4 J $< E_{max} \leq 10^5$ J 3. $10^5 \leq \Sigma E < 10^6$ J per 5m of longwall advance	1. Occurrence of single tremors or single tremors with energies E of the order of 10^2 – 10^3 J 2. $E_{max} \leq 5 \cdot 10^3$ J 3. 10^3 J $\leq \Sigma E < 10^4$ J per 5m of longwall advance
c moderate hazard	1. Occurrence of tremors with energies E of the order of 10^2 J – 10^6 J 2. $5 \cdot 10^5$ J $< E_{max} \leq 5 \cdot 10^6$ J 3. $10^6 \leq \Sigma E < 10^7$ J per 5m of longwall advance	1. Occurrence of single tremors or single tremors with energies E of the order of 10^2 – 10^4 J 2. $5 \cdot 10^3$ J $\leq E_{max} \leq 5 \cdot 10^5$ J 3. 10^4 J $\leq \Sigma E < 10^5$ J per 5m of longwall advance
d high hazard	1. Occurrence of tremors with energies E of the order of 10^2 J – 10^6 J 2. $E_{max} > 5 \cdot 10^6$ J 3. $\Sigma E > 10^7$ J per 5m of longwall advance	1. Occurrence of single tremors or single tremors with energies E of the order of 10^2 – 10^5 J 2. $E_{max} > 10^5$ J 3. $\Sigma E > 10^5$ J per 5m of longwall advance

References

1. Abdel-Basset, M., El-Shahat, D., El-Henawy, I.M., de Albuquerque, V.H.C., Mirjalili, S.: A new fusion of grey wolf optimizer algorithm with a two-phase mutation for feature selection. Expert Syst. Appl. **139**, 112824 (2020). https://doi.org/10.1016/j.eswa.2019.112824

2. Abedjan, Z., et al.: Data science in healthcare: benefits, challenges and opportunities. In: Consoli, S., Recupero, D.R., Petkovic, M. (eds.) Data Science for Healthcare, pp. 3–38. Springer, Cham (2019). https://doi.org/10.1007/978-3-030-05249-2_1

3. Abeel, T., Helleputte, T., de Peer, Y.V., Dupont, P., Saeys, Y.: Robust biomarker identification for cancer diagnosis with ensemble feature selection methods. Bioinformatics **26**(3), 392–398 (2010)

4. Adebiyi, A.A., Adewumi, A.O., Ayo, C.K.: Comparison of ARIMA and artificial neural networks models for stock price prediction. J. Appl. Math. **2014**, 614342:1–614342:7 (2014). https://doi.org/10.1155/2014/614342

5. U. S. Fire Administration: Annual report on firefighter fatalities in the United States. http://apps.usfa.fema.gov/firefighter-fatalities/

6. Aggarwal, C. (ed.): Managing and Mining Sensor Data. Springer, New York (2013). https://doi.org/10.1007/978-1-4614-6309-2

7. Agrawal, A., et al.: Approximate computing: challenges and opportunities. In: IEEE International Conference on Rebooting Computing, ICRC 2016, San Diego, CA, USA, 17–19, October 2016, pp. 1–8. IEEE Computer Society (2016). https://doi.org/10.1109/ICRC.2016.7738674

8. Ahmadi, E., Jasemi, M., Monplaisir, L., Nabavi, M.A., Mahmoodi, A., Jam, P.A.: New efficient hybrid candlestick technical analysis model for stock market timing on the basis of the support vector machine and heuristic algorithms of imperialist competition and genetic. Expert Syst. Appl. **94**, 21–31 (2018). https://doi.org/10.1016/j.eswa.2017.10.023

9. Ahmed, F., Samorani, M., Bellinger, C., Zaïane, O.R.: Advantage of integration in big data: feature generation in multi-relational databases for imbalanced learning. In: Proceedings of IEEE Big Data, pp. 532–539 (2016)

10. Aksoy, S., Haralick, R.M.: Feature normalization and likelihood-based similarity measures for image retrieval. Pattern Recogn. Lett. **22**(5), 563–582 (2001). https://doi.org/10.1016/S0167-8655(00)00112-4

11. Al-Ali, H., Cuzzocrea, A., Damiani, E., Mizouni, R., Tello, G.: A composite machine-learning-based framework for supporting low-level event logs to high-level business process model activities mappings enhanced by flexible BPMN model translation. Soft. Comput. **24**(10), 7557–7578 (2019). https://doi.org/10.1007/s00500-019-04385-6

12. Alelyani, S., Tang, J., Liu, H.: Feature selection for clustering: a review. In: Aggarwal, C.C., Reddy, C.K. (eds.) Data Clustering: Algorithms and Applications, pp. 29–60. CRC Press, Boca Raton (2013)

13. Almeida, F., Xexéo, G.: Word embeddings: a survey. CoRR abs/1901.09069 (2019). http://arxiv.org/abs/1901.09069

14. Altidor, W., Khoshgoftaar, T.M., Napolitano, A.: Measuring stability of feature ranking techniques: a noise-based approach. Int. J. Bus. Intell. Data Min. **7**(1–2), 80–115 (2012)

15. Appice, A., Guccione, P., Malerba, D., Ciampi, A.: Dealing with temporal and spatial correlations to classify outliers in geophysical data streams. Inf. Sci. **285**, 162–180 (2014)

16. Assunção, M.D., Calheiros, R.N., Bianchi, S., Netto, M.A., Buyya, R.: Big data computing and clouds: trends and future directions. J. Parallel Distrib. Comput. **79**, 3–15 (2015)

17. Augustyniak, P., Smoleń, M., Mikrut, Z., Kańtoch, E.: Seamless tracing of human behavior using complementary wearable and house-embedded sensors. Sensors **14**(5), 7831–7856 (2014). https://doi.org/10.3390/s140507831

18. Ayesha, S., Hanif, M.K., Talib, R.: Overview and comparative study of dimensionality reduction techniques for high dimensional data. Inf. Fusion **59**, 44–58 (2020). https://doi.org/10.1016/j.inffus.2020.01.005

19. Azad, M., Moshkov, M.: Minimization of decision tree average depth for decision tables with many-valued decisions. Procedia Comput. Sci. **35**, 368–377 (2014). https://doi.org/10.1016/j.procs.2014.08.117. Knowledge-Based and Intelligent Information & Engineering Systems 18th Annual Conference, KES-2014 Gdynia, Poland, September 2014 Proceedings

20. Bahmani, B., Moseley, B., Vattani, A., Kumar, R., Vassilvitskii, S.: Scalable K-Means++. Proc. VLDB Endow. **5**(7), 622–633 (2012)

21. Bałazińska, M., Zdonik, S.: Databases meet the stream processing era, pp. 225–234. Association for Computing Machinery and Morgan and Claypool (2018)

22. Bansal, A., Jain, R., Modi, K.: Big data streaming with spark. In: Mittal, M., Balas, V.E., Goyal, L.M., Kumar, R. (eds.) Big Data Processing Using Spark in Cloud. SBD, vol. 43, pp. 23–50. Springer, Singapore (2019). https://doi.org/10.1007/978-981-13-0550-4_2

23. Bargiela, A., Pedrycz, W.: The roots of granular computing. In: 2006 IEEE International Conference on Granular Computing, pp. 806–809. IEEE (2006)

24. de Barros, R.S.M., de Carvalho Santos, S.G.T.: An overview and comprehensive comparison of ensembles for concept drift. Inf. Fusion **52**, 213–244 (2019). https://doi.org/10.1016/j.inffus.2019.03.006

25. Bashir, F., Wei, H.L.: Handling missing data in multivariate time series using a vector autoregressive model-imputation (VAR-IM) algorithm. Neurocomputing **276**, 23–30 (2018). https://doi.org/10.1016/j.neucom.2017.03.097. Machine Learning and Data Mining Techniques for Medical Complex Data Analysis

26. Baughman, M., Haas, C., Wolski, R., Foster, I., Chard, K.: Predicting amazon spot prices with LSTM networks. In: Proceedings of the 9th Workshop on Scientific Cloud Computing, ScienceCloud 2018, p. 7. Association for Computing Machinery, New York (2018). https://doi.org/10.1145/3217880.3217881

27. Bazan, J.G.: Hierarchical classifiers for complex Spatio-temporal concepts. In: Peters, J.F., Skowron, A., Rybiński, H. (eds.) Transactions on Rough Sets IX. LNCS, vol. 5390, pp. 474–750. Springer, Heidelberg (2008). https://doi.org/10.1007/978-3-540-89876-4_26

28. Bazan, J.G., Bazan-Socha, S., Buregwa-Czuma, S., Dydo, Ł, Rząsa, W., Skowron, A.: A classifier based on a decision tree with verifying cuts. Fundam. Informaticae **143**(1–2), 1–18 (2016). https://doi.org/10.3233/FI-2016-1300

29. Bengio, Y., Courville, A.C., Vincent, P.: Representation learning: a review and new perspectives. IEEE Trans. Pattern Anal. Mach. Intell. **35**(8), 1798–1828 (2013). https://doi.org/10.1109/TPAMI.2013.50

30. Benítez-Caballero, M.J., Medina, J., Ramírez-Poussa, E., Ślęzak, D.: A computational procedure for variable selection preserving different initial conditions. Int. J. Comput. Math. **97**(1–2), 387–404 (2020). https://doi.org/10.1080/00207160.2019.1613530

31. Benítez-Peña, S., Blanquero, R., Carrizosa, E., Ramírez-Cobo, P.: Cost-sensitive feature selection for support vector machines. Comput. Oper. Res. **106**, 169–178 (2019). https://doi.org/10.1016/j.cor.2018.03.005
32. Bennasar, M., Hicks, Y., Setchi, R.: Feature selection using joint mutual information maximisation. Expert Syst. Appl. **42**(22), 8520–8532 (2015). https://doi.org/10.1016/j.eswa.2015.07.007
33. Benoit, F., van Heeswijk, M., Miche, Y., Verleysen, M., Lendasse, A.: Feature selection for nonlinear models with extreme learning machines. Neurocomputing **102**, 111–124 (2013). https://doi.org/10.1016/j.neucom.2011.12.055
34. Berrado, A., Runger, G.C.: Supervised multivariate discretization in mixed data with random forests. In: 2009 IEEE/ACS International Conference on Computer Systems and Applications, pp. 211–217, May 2009. https://doi.org/10.1109/AICCSA.2009.5069327
35. Berrouachedi, A., Jaziri, R., Bernard, G.: Deep extremely randomized trees. In: Gedeon, T., Wong, K.W., Lee, M. (eds.) ICONIP 2019. LNCS, vol. 11953, pp. 717–729. Springer, Cham (2019). https://doi.org/10.1007/978-3-030-36708-4_59
36. Biau, G., Cadre, B., Rouvière, L.: Accelerated gradient boosting. Mach. Learn. **108**(6), 971–992 (2019). https://doi.org/10.1007/s10994-019-05787-1
37. Bifet, A., Holmes, G., Pfahringer, B., Kirkby, R., Gavaldà, R.: New ensemble methods for evolving data streams. In: IV, J.F.E., Fogelman-Soulié, F., Flach, P.A., Zaki, M.J. (eds.) Proceedings of the 15th ACM SIGKDD International Conference on Knowledge Discovery and Data Mining, Paris, France, 28 June–1 July 2009, pp. 139–148. ACM (2009). https://doi.org/10.1145/1557019.1557041
38. Bolón-Canedo, V., Alonso-Betanzos, A.: Ensembles for feature selection: a review and future trends. Inf. Fusion **52**, 1–12 (2019). https://doi.org/10.1016/j.inffus.2018.11.008
39. Bolstad, B.M., Irizarry, R.A., Åstrand, M., Speed, T.P.: A comparison of normalization methods for high density oligonucleotide array data based on variance and bias. Bioinformatics **19**(2), 185–193 (2003). https://doi.org/10.1093/bioinformatics/19.2.185
40. Bommert, A., Sun, X., Bischl, B., Rahnenführer, J., Lang, M.: Benchmark for filter methods for feature selection in high-dimensional classification data. Comput. Stat. Data Anal. **143**, 106839 (2020). https://doi.org/10.1016/j.csda.2019.106839
41. Bondell, H.D., Reich, B.J.: Simultaneous regression shrinkage, variable selection, and supervised clustering of predictors with OSCAR. Biometrics **64**(1), 115–123 (2008)
42. Borchani, H., Varando, G., Bielza, C., Larrañaga, P.: A survey on multi-output regression. Wiley Int. Rev. Data Min. and Knowl. Disc. **5**, 216–233 (2015). https://doi.org/10.1002/widm.1157
43. Boullé, M.: MODL: a Bayes optimal discretization method for continuous attributes. Mach. Learn. **65**(1), 131–165 (2006). https://doi.org/10.1007/s10994-006-8364-x
44. Boullé, M.: Prediction of methane outbreak in coal mines from historical sensor data under distribution drift. In: Yao, Y., Hu, Q., Yu, H., Grzymala-Busse, J.W. (eds.) RSFDGrC 2015. LNCS (LNAI), vol. 9437, pp. 439–451. Springer, Cham (2015). https://doi.org/10.1007/978-3-319-25783-9_39
45. Boullé, M.: Predicting dangerous seismic events in coal mines under distribution drift. In: Ganzha, M., Maciaszek, L.A., Paprzycki, M. (eds.) Proceedings of FedCSIS 2016, pp. 227–230. IEEE (2016)
46. Brahim, A.B., Limam, M.: Robust ensemble feature selection for high dimensional data sets. In: Proceedings of HPCS 2013, pp. 151–157 (2013)

47. Bruni, R., Daraio, C., Aureli, D.: Imputation techniques for the reconstruction of missing interconnected data from higher educational institutions. Knowl.-Based Syst. **212**, 106512 (2021). https://doi.org/10.1016/j.knosys.2020.106512
48. Bustince, H., et al.: On some classes of directionally monotone functions. Fuzzy Sets Syst. **386**, 161–178 (2020). https://doi.org/10.1016/j.fss.2019.01.024
49. Cai, W., et al.: A new seismic-based strain energy methodology for coal burst forecasting in underground coal mines. Int. J. Rock Mech. Min. Sci. **123**, 104086 (2019). https://doi.org/10.1016/j.ijrmms.2019.104086
50. Cano, A., Krawczyk, B.: Kappa updated ensemble for drifting data stream mining. Mach. Learn. **109**(1), 175–218 (2019). https://doi.org/10.1007/s10994-019-05840-z
51. Cao, L., Chua, K.S., Chong, W.K., Lee, H.P., Gu, Q.M.: A comparison of PCA, KPCA and ICA for dimensionality reduction in support vector machine. Neurocomputing **55**(1), 321–336 (2003). https://doi.org/10.1016/S0925-2312(03)00433-8
52. Carbone, P., Ewen, S., Fóra, G., Haridi, S., Richter, S., Tzoumas, K.: State management in apache Flink®: consistent stateful distributed stream processing. Proc. VLDB Endow. **10**(12), 1718–1729 (2017). https://doi.org/10.14778/3137765.3137777
53. Carrizosa, E., Guerrero, V., Morales, D.R.: On mathematical optimization for the visualization of frequencies and adjacencies as rectangular maps. Eur. J. Oper. Res. **265**(1), 290–302 (2018). https://doi.org/10.1016/j.ejor.2017.07.023
54. Çekik, R., Uysal, A.K.: A novel filter feature selection method using rough set for short text data. Expert Syst. Appl. **160**, 113691 (2020). https://doi.org/10.1016/j.eswa.2020.113691
55. Cen, Z., Wang, J.: Crude oil price prediction model with long short term memory deep learning based on prior knowledge data transfer. Energy **169**, 160–171 (2019). https://doi.org/10.1016/j.energy.2018.12.016
56. Cerrada, M., Sánchez, R., Cabrera, D., Zurita, G., Li, C.: Multi-stage feature selection by using genetic algorithms for fault diagnosis in gearboxes based on vibration signal. Sensors **15**(9), 23903–23926 (2015). https://doi.org/10.3390/s150923903
57. Cevallos Valdiviezo, H., Van Aelst, S.: Tree-based prediction on incomplete data using imputation or surrogate decisions. Inf. Sci. **311**, 163–181 (2015). https://doi.org/10.1016/j.ins.2015.03.018
58. Chakraborty, D., Narayanan, V., Ghosh, A.: Integration of deep feature extraction and ensemble learning for outlier detection. Pattern Recognit. **89**, 161–171 (2019). https://doi.org/10.1016/j.patcog.2019.01.002
59. Chalapathy, R., Chawla, S.: Deep learning for anomaly detection: a survey. CoRR abs/1901.03407 (2019)
60. Chalapathy, R., Khoa, N.L.D., Chawla, S.: Robust deep learning methods for anomaly detection. In: Proceedings of the 26th ACM SIGKDD International Conference on Knowledge Discovery & Data Mining, KDD 2020, pp. 3507–3508. Association for Computing Machinery, New York (2020). https://doi.org/10.1145/3394486.3406704
61. Chandrashekar, G., Sahin, F.: A survey on feature selection methods. Comput. Electr. Eng. **40**(1), 16–28 (2014)
62. Chądzyńska-Krasowska, A., Betliński, P., Ślęzak, D.: Scalable machine learning with granulated data summaries: a case of feature selection. In: Kryszkiewicz, M., Appice, A., Ślęzak, D., Rybinski, H., Skowron, A., Raś, Z.W. (eds.) ISMIS 2017. LNCS (LNAI), vol. 10352, pp. 519–529. Springer, Cham (2017). https://doi.org/10.1007/978-3-319-60438-1_51

63. Che, J., Yang, Y., Li, L., Bai, X., Zhang, S., Deng, C.: Maximum relevance minimum common redundancy feature selection for nonlinear data. Inf. Sci. **409–410**, 68–86 (2017). https://doi.org/10.1016/j.ins.2017.05.013

64. Chen, L., Papandreou, G., Kokkinos, I., Murphy, K., Yuille, A.L.: DeepLab: semantic image segmentation with deep convolutional nets, atrous convolution, and fully connected CRFs. IEEE Trans. Pattern Anal. Mach. Intell. **40**(4), 834–848 (2018). https://doi.org/10.1109/TPAMI.2017.2699184

65. Chen, S.-H., Du, Y.-R.: Granularity in economic decision making: an interdisciplinary review. In: Pedrycz, W., Chen, S.-M. (eds.) Granular Computing and Decision-Making. SBD, vol. 10, pp. 47–71. Springer, Cham (2015). https://doi.org/10.1007/978-3-319-16829-6_3

66. Chen, T., Guestrin, C.: XGBoost: a scalable tree boosting system. In: Proceedings of the 22nd ACM SIGKDD International Conference on Knowledge Discovery and Data Mining, pp. 785–794. Association for Computing Machinery, New York (2016). https://doi.org/10.1145/2939672.2939785

67. Cheng, W., Dembczyński, K., Hüllermeier, E.: Graded multilabel classification: the ordinal case. In: Fürnkranz, J., Joachims, T. (eds.) Proceedings of the 27th International Conference on Machine Learning (ICML-10), 21–24 June 2010, Haifa, Israel, pp. 223–230. Omnipress (2010)

68. Chlebus, B.S., Nguyen, S.H.: On finding optimal discretizations for two attributes. In: Polkowski, L., Skowron, A. (eds.) RSCTC 1998. LNCS (LNAI), vol. 1424, pp. 537–544. Springer, Heidelberg (1998). https://doi.org/10.1007/3-540-69115-4_74

69. Cho, K., et al.: Learning phrase representations using RNN encoder-decoder for statistical machine translation. In: Moschitti, A., Pang, B., Daelemans, W. (eds.) Proceedings of the 2014 Conference on Empirical Methods in Natural Language Processing, EMNLP 2014, 25–29 October 2014, Doha, Qatar, A meeting of SIGDAT, A Special Interest Group of the ACL, pp. 1724–1734. ACL (2014)

70. Chu, C.T., et al.: Map-reduce for machine learning on multicore. In: Proceedings of NIPS, pp. 281–288 (2006)

71. Ciucci, D., Yao, Y.: Synergy of granular computing, shadowed sets, and three-way decisions. Inf. Sci. **508**, 422–425 (2020). https://doi.org/10.1016/j.ins.2019.09.003

72. Clark, P.G., Grzymała-Busse, J.W., Hippe, Z.S., Mroczek, T., Niemiec, R.: Complexity of rule sets mined from incomplete data using probabilistic approximations based on generalized maximal consistent blocks. Procedia Comput. Sci. **176**, 1803–1812 (2020). https://doi.org/10.1016/j.procs.2020.09.219. Knowledge-Based and Intelligent Information & Engineering Systems: Proceedings of the 24th International Conference KES2020

73. Collobert, R., Weston, J., Bottou, L., Karlen, M., Kavukcuoglu, K., Kuksa, P.P.: Natural language processing (almost) from scratch. J. Mach. Learn. Res. **12**, 2493–2537 (2011). http://dl.acm.org/citation.cfm?id=2078186

74. Cook, A.A., Misirli, G., Fan, Z.: Anomaly detection for IoT time-series data: a survey. IEEE Internet Things J. **7**(7), 6481–6494 (2020). https://doi.org/10.1109/JIOT.2019.2958185

75. Cornelis, C., Jensen, R., Martín, G.H., Ślęzak, D.: Attribute selection with fuzzy decision reducts. Inf. Sci. **180**(2), 209–224 (2010)

76. Crochepierre, L., Boudjeloud-Assala, L., Barbesant, V.: Interpretable dimensionally-consistent feature extraction from electrical network sensors (2020). https://bitbucket.org/ghentdatascience/ecmlpkdd20-papers/raw/master/ADS/sub_795.pdf

77. Lakshmipadmaja, D., Vishnuvardhan, B.: Classification performance improvement using random subset feature selection algorithm for data mining. Big Data Res. **12**, 1–12 (2018). https://doi.org/10.1016/j.bdr.2018.02.007

78. Chelly Dagdia, Z., Zarges, C., Beck, G., Lebbah, M.: A scalable and effective rough set theory-based approach for big data pre-processing. Knowl. Inf. Syst. **62**(8), 3321–3386 (2020). https://doi.org/10.1007/s10115-020-01467-y

79. Dai, J., Xu, Q.: Approximations and uncertainty measures in incomplete information systems. Inf. Sci. **198**, 62–80 (2012)

80. Dalal, N., Triggs, B., Schmid, C.: Human detection using oriented histograms of flow and appearance. In: Leonardis, A., Bischof, H., Pinz, A. (eds.) ECCV 2006. LNCS, vol. 3952, pp. 428–441. Springer, Heidelberg (2006). https://doi.org/10.1007/11744047_33

81. Das, A.S., Datar, M., Garg, A., Rajaram, S.: Google news personalization: scalable online collaborative filtering. In: Proceedings of WWW, pp. 271–280 (2007)

82. Das, S.: Filters, wrappers and a boosting-based hybrid for feature selection. In: Proceedings of ICML 2001, pp. 74–81 (2001)

83. Dash, M., Liu, H.: Consistency-based search in feature selection. Artif. Intell. **151**(1–2), 155–176 (2003)

84. Dash, P.K., Nayak, M., Senapati, M.R., Lee, I.W.C.: Mining for similarities in time series data using wavelet-based feature vectors and neural networks. Eng. Appl. Artif. Intell. **20**(2), 185–201 (2007). https://doi.org/10.1016/j.engappai.2006.06.018

85. Datar, M., Gionis, A., Indyk, P., Motwani, R.: Maintaining stream statistics over sliding windows. SIAM J. Comput. **31**(6), 1794–1813 (2002)

86. David, S.A., Machado, J.A.T., Trevisan, L.R., Inácio, C.M.C., Lopes, A.M.: Dynamics of commodities prices: integer and fractional models. Fundam. Inform. **151**(1–4), 389–408 (2017). https://doi.org/10.3233/FI-2017-1499

87. Dayal, U., Castellanos, M., Simitsis, A., Wilkinson, K.: Data integration flows for business intelligence. In: Proceedings of the 12th International Conference on Extending Database Technology: Advances in Database Technology, EDBT 2009, pp. 1–11. ACM, New York (2009). https://doi.org/10.1145/1516360.1516362

88. Dean, J., Ghemawat, S.: MapReduce: simplified data processing on large clusters. Commun. ACM **51**(1), 107–113 (2008). https://doi.org/10.1145/1327452.1327492

89. Devlin, J., Chang, M., Lee, K., Toutanova, K.: BERT: pre-training of deep bidirectional transformers for language understanding. CoRR abs/1810.04805 (2018)

90. Dietterich, T.G.: Ensemble methods in machine learning. In: Kittler, J., Roli, F. (eds.) MCS 2000. LNCS, vol. 1857, pp. 1–15. Springer, Heidelberg (2000). https://doi.org/10.1007/3-540-45014-9_1

91. Ding, C.H.Q., Peng, H.: Minimum redundancy feature selection from microarray gene expression data. J. Bioinform. Comput. Biol. **3**(2), 185–206 (2005). https://doi.org/10.1142/S0219720005001004

92. Dobre, C., Xhafa, F.: Parallel programming paradigms and frameworks in big data era. Int. J. Parallel Prog. **42**(5), 710–738 (2013). https://doi.org/10.1007/s10766-013-0272-7

93. Doherty, P., Szalas, A.: Rough set reasoning using answer set programs. Int. J. Approx. Reason. **130**, 126–149 (2021). https://doi.org/10.1016/j.ijar.2020.12.010

94. Domingos, P.: A few useful things to know about machine learning. Commun. ACM **55**(10), 78–87 (2012)

95. Dong, H., Li, T., Ding, R., Sun, J.: A novel hybrid genetic algorithm with granular information for feature selection and optimization. Appl. Soft Comput. **65**, 33–46 (2018). https://doi.org/10.1016/j.asoc.2017.12.048

96. Dou, L., Cai, W., Cao, A., Guo, W.: Comprehensive early warning of rock burst utilizing microseismic multi-parameter indices. Int. J. Min. Sci. Technol. **28**(5), 767–774 (2018). https://doi.org/10.1016/j.ijmst.2018.08.007

97. Dougherty, J., Kohavi, R., Sahami, M.: Supervised and unsupervised discretization of continuous features. In: Proceedings of the Twelfth International Conference on International Conference on Machine Learning, ICML 1995, pp. 194–202. Morgan Kaufmann Publishers Inc., San Francisco (1995)

98. Dramiński, M., Rada-Iglesias, A., Enroth, S., Wadelius, C., Koronacki, J., Komorowski, H.J.: Monte Carlo feature selection for supervised classification. Bioinformatics **24**(1), 110–117 (2008)

99. Dubois, D.J., Casale, G.: OptiSpot: minimizing application deployment cost using spot cloud resources. Clust. Comput. **19**(2), 893–909 (2016). https://doi.org/10.1007/s10586-016-0568-7

100. Dubois, D., Prade, H.: Rough fuzzy sets and fuzzy rough sets. Int. J. Gen. Syst. **17**, 191–209 (1990). https://doi.org/10.1080/03081079008935107

101. Duda, P., Rutkowski, L., Jaworski, M., Rutkowska, D.: On the Parzen kernel-based probability density function learning procedures over time-varying streaming data with applications to pattern classification. IEEE Trans. Cybern. **50**(4), 1683–1696 (2020). https://doi.org/10.1109/TCYB.2018.2877611

102. Duforet-Frebourg, N., Luu, K., Laval, G., Bazin, E., Blum, M.G.: Detecting genomic signatures of natural selection with principal component analysis: application to the 1000 genomes data. Mol. Biol. Evol. **33**(4), 1082–1093 (2015). https://doi.org/10.1093/molbev/msv334

103. Dutta, S., Jankowski, A., Rozenberg, G., Skowron, A.: Linking reaction systems with rough sets. Fundam. Informaticae **165**(3–4), 283–302 (2019). https://doi.org/10.3233/FI-2019-1786

104. Eiras-Franco, C., Bolón-Canedo, V., Ramos, S., González-Domínguez, J., Alonso-Betanzos, A., Touriño, J.: Multithreaded and Spark Parallelization of Feature Selection Filters. J. Comput. Sci. **17**, 609–619 (2016)

105. Ekanayake, J., et al.: Twister: a runtime for iterative mapreduce. In: Proceedings of HPDC, pp. 810–818 (2010)

106. Elmeleegy, K.: Piranha: optimizing short jobs in Hadoop. Proc. VLDB Endow. **6**(11), 985–996 (2013)

107. Fan, J., Lv, J.: A selective overview of variable selection in high dimensional feature space. Stat. Sin. **20**(1), 101–148 (2010)

108. Fayyad, U.M., Irani, K.B.: Multi-interval discretization of continuous-valued attributes for classification learning. In: IJCAI, pp. 1022–1029 (1993)

109. Feng, J., Wang, E., Ding, H., Huang, Q., Chen, X.: Deterministic seismic hazard assessment of coal fractures in underground coal mine: a case study. Soil Dyn. Earthq. Eng. **129**, 105921 (2020). https://doi.org/10.1016/j.soildyn.2019.105921

110. Fischer, T., Krauss, C.: Deep learning with long short-term memory networks for financial market predictions. Eur. J. Oper. Res. **270**(2), 654–669 (2018). https://doi.org/10.1016/j.ejor.2017.11.054

111. Fisher, A., Rudin, C., Dominici, F.: All models are wrong, but many are useful: learning a variable's importance by studying an entire class of prediction models simultaneously. J. Mach. Learn. Res. **20**(177), 1–81 (2019). http://jmlr.org/papers/v20/18-760.html

112. Florescu, D., England, M.: Algorithmically generating new algebraic features of polynomial systems for machine learning. CoRR abs/1906.01455 (2019). http://arxiv.org/abs/1906.01455

113. Fontes, C.H., Pereira, O.: Pattern recognition in multivariate time series - a case study applied to fault detection in a gas turbine. Eng. Appl. Artif. Intell. **49**, 10–18 (2016). https://doi.org/10.1016/j.engappai.2015.11.005

114. Friedman, J.H.: Greedy function approximation: a gradient boosting machine. Ann. Stat. **29**, 1189–1232 (2000)

115. Fu, T.C.: A review on time series data mining. Eng. Appl. Artif. Intell. **24**(1), 164–181 (2011)

116. Fürnkranz, J., Hüllermeier, E., Loza Mencía, E., Brinker, K.: Multilabel classification via calibrated label ranking. Mach. Learn. **73**(2), 133–153 (2008). https://doi.org/10.1007/s10994-008-5064-8

117. Gao, W., Hu, L., Zhang, P.: Class-specific mutual information variation for feature selection. Pattern Recogn. **79**, 328–339 (2018). https://doi.org/10.1016/j.patcog.2018.02.020

118. Garbulowski, M., et al.: R.ROSETTA: an interpretable machine learning framework. BMC Bioinform. **22**(1), 110 (2021). https://doi.org/10.1186/s12859-021-04049-z

119. Garbulowski, M., et al.: Interpretable machine learning reveals dissimilarities between subtypes of autism spectrum disorder. Front. Genet. **12**, 73 (2021). https://doi.org/10.3389/fgene.2021.618277

120. García, S., Luengo, J., Sáez, J.A., López, V., Herrera, F.: A survey of discretization techniques: taxonomy and empirical analysis in supervised learning. IEEE Trans. Knowl. Data Eng. **25**(4), 734–750 (2013). https://doi.org/10.1109/TKDE.2012.35

121. García-Torres, M., Gómez-Vela, F., Melián-Batista, B., Moreno-Vega, J.M.: High-dimensional feature selection via feature grouping. Inf. Sci. **326**, 102–118 (2016)

122. Geurts, P., Ernst, D., Wehenkel, L.: Extremely randomized trees. Mach. Learn. **63**(1), 3–42 (2006). https://doi.org/10.1007/s10994-006-6226-1

123. Ghosh, M., Guha, R., Sarkar, R., Abraham, A.: A wrapper-filter feature selection technique based on ant colony optimization. Neural Comput. Appl. **32**(12), 7839–7857 (2019). https://doi.org/10.1007/s00521-019-04171-3

124. Gibowicz, S.J., Lasocki, S.: Seismicity induced by mining: 10 years later. In: Advances in Geophysics, pp. 81–164 (2001)

125. Gill, S.S., et al.: Holistic resource management for sustainable and reliable cloud computing: an innovative solution to global challenge. J. Syst. Softw. **155**, 104–129 (2019). https://doi.org/10.1016/j.jss.2019.05.025

126. Giuliani, A.: The application of principal component analysis to drug discovery and biomedical data. Drug Discov. Today **22**(7), 1069–1076 (2017). https://doi.org/10.1016/j.drudis.2017.01.005

127. Goh, W.W.B., Wong, L.: Evaluating feature-selection stability in next-generation proteomics. J. Bioinform. Comput. Biol. **14**(5), 1–23 (2016). https://doi.org/10.1142/S0219720016500293

128. Goldenberg, I., Webb, G.I.: Survey of distance measures for quantifying concept drift and shift in numeric data. Knowl. Inf. Syst. **60**(2), 591–615 (2018). https://doi.org/10.1007/s10115-018-1257-z

129. González-Domínguez, J., Expósito, R.R., Bolón-Canedo, V.: CUDA-JMI: acceleration of feature selection on heterogeneous systems. Future Gener. Comput. Syst. **102**, 426–436 (2020). https://doi.org/10.1016/j.future.2019.08.031

130. Govindan, P., Chen, R., Scheinberg, K., Srinivasan, S.: A scalable solution for group feature selection. In: Proceedings of IEEE Big Data 2015, pp. 2846–2848 (2015)

131. Grave, E., Bojanowski, P., Gupta, P., Joulin, A., Mikolov, T.: Learning word vectors for 157 languages. CoRR abs/1802.06893 (2018). http://arxiv.org/abs/1802.06893

132. Grochala, D., Kajor, M., Kucharski, D., Iwaniec, M., Kańtoch, E.: A novel approach in auscultation technology - new sensors and algorithms. In: Bujnowski, A., Kaczmarek, M., Ruminski, J. (eds.) 11th International Conference on Human System Interaction, HSI 2018, Gdansk, Poland, 4–6 July 2018, pp. 240–244. IEEE (2018). https://doi.org/10.1109/HSI.2018.8431339

133. Grorud, L.J., Smith, D.: The national fire fighter near-miss reporting. Annual Report 2008. An Exclusive Supplement to Fire & Rescue Magazine, pp. 1–24 (2008)

134. Gruźdź, A., Ihnatowicz, A., Ślęzak, D.: Interactive gene clustering - a case study of breast cancer microarray data. Inf. Syst. Front. 8(1), 21–27 (2006). https://doi.org/10.1007/s10796-005-6100-x

135. Grychowski, T.: Hazard assessment based on fuzzy logic. Arch. Min. Sci. 53(4), 595–602 (2008)

136. Grzegorowski, M.: Scaling of complex calculations over big data-sets. In: Ślezak, D., Schaefer, G., Vuong, S.T., Kim, Y.-S. (eds.) AMT 2014. LNCS, vol. 8610, pp. 73–84. Springer, Cham (2014). https://doi.org/10.1007/978-3-319-09912-5_7

137. Grzegorowski, M.: Governance of the redundancy in the feature selection based on rough sets' reducts. In: Flores, V., et al. (eds.) IJCRS 2016. LNCS (LNAI), vol. 9920, pp. 548–557. Springer, Cham (2016). https://doi.org/10.1007/978-3-319-47160-0_50

138. Grzegorowski, M.: Massively parallel feature extraction framework application in predicting dangerous seismic events. In: Proceedings of FedCSIS 2016, pp. 225–229 (2016)

139. Grzegorowski, M.: Selected aspects of interactive feature extraction. Ph.D. thesis, University of Warsaw (2021)

140. Grzegorowski, M., Janusz, A., Lazewski, S., Swiechowski, M., Jankowska, M.: Prescriptive analytics for optimization of FMCG delivery plans. In: Ciucci, D., et al. (eds.) IPMU 2022. Computer and Information Science, vol. 1602, pp. 44–53. Springer, Cham (2022). https://doi.org/10.1007/978-3-031-08974-9_4

141. Grzegorowski, M., Janusz, A., Ślęzak, D., Szczuka, M.S.: On the role of feature space granulation in feature selection processes. In: Nie, J., et al. (eds.) 2017 IEEE International Conference on Big Data, BigData 2017, Boston, MA, USA, 11–14 December 2017, pp. 1806–1815. IEEE Computer Society (2017). https://doi.org/10.1109/BigData.2017.8258124

142. Grzegorowski, M., Kalisch, M., Kozielski, M., Wróbel, Ł.: Hurtowania danych i procesy ETL. In: Przystałka, P., Sikora, M. (eds.) Zintegrowany, szkieletowy system wspmagania decyzji dla systemów monitorowania procesów, urządzeń i zagrożeń, chap. 3, pp. 31–40. Monograficzna Seria Wydawnicza Instytututu Technik Innowacyjnych EMAG (2017). http://disesor.ibemag.pl/www/disesor.ibemag.pl/data/Image/hurtownia.pdf

143. Grzegorowski, M., Litwin, J., Wnuk, M., Pabis, M., Marcinowski, L.: Survival-based feature extraction - application in supply management for dispersed vending machines. IEEE Trans. Industr. Inform. (2022). https://doi.org/10.1109/TII.2022.3178547

144. Grzegorowski, M., Pardel, P.W., Stawicki, S., Stencel, K.: SONCA: scalable semantic processing of rapidly growing document stores. In: Pechenizkiy, M., Wojciechowski, M. (eds.) New Trends in Databases and Information Systems. Advances in Intelligent Systems and Computing, vol. 185, pp. 89–98. Springer, Heidelberg (2012). https://doi.org/10.1007/978-3-642-32518-2_9

145. Grzegorowski, M., Ślęzak, D.: On resilient feature selection: computational foundations of r-C-reducts. Inf. Sci. **499**, 25–44 (2019). https://doi.org/10.1016/j.ins.2019.05.041

146. Grzegorowski, M., Stawicki, S.: Window-based feature engineering for prediction of methane threats in coal mines. In: Yao, Y., Hu, Q., Yu, H., Grzymala-Busse, J.W. (eds.) RSFDGrC 2015. LNCS (LNAI), vol. 9437, pp. 452–463. Springer, Cham (2015). https://doi.org/10.1007/978-3-319-25783-9_40

147. Grzegorowski, M., Stawicki, S.: Window-based feature extraction framework for multi-sensor data: a posture recognition case study. In: Ganzha, M., Maciaszek, L.A., Paprzycki, M. (eds.) 2015 Federated Conference on Computer Science and Information Systems, FedCSIS 2015, Lódz, Poland, 13–16 September 2015, pp. 397–405. IEEE (2015). https://doi.org/10.15439/2015F425

148. Grzegorowski, M., Zdravevski, E., Janusz, A., Lameski, P., Apanowicz, C., Ślęzak, D.: Cost optimization for big data workloads based on dynamic scheduling and cluster-size tuning. Big Data Res. **25**, 100203 (2021). https://doi.org/10.1016/j.bdr.2021.100203

149. Gu, B., Liu, G., Huang, H.: Groups-keeping solution path algorithm for sparse regression with automatic feature grouping. In: Proceedings of the KDD, pp. 185–193 (2017)

150. Guo, Y., Zhou, Y., Zhang, Z.: Fault diagnosis of multi-channel data by the CNN with the multilinear principal component analysis. Measurement **171**, 108513 (2020). https://doi.org/10.1016/j.measurement.2020.108513

151. Guyon, I., Elisseeff, A.: An introduction to variable and feature selection. J. Mach. Learn. Res. **3**, 1157–1182 (2003)

152. Guyon, I., Nikravesh, M., Gunn, S., Zadeh, L.A. (eds.): Feature Extraction. Studies in Fuzziness and Soft Computing, vol. 207. Springer, Heidelberg (2006). https://doi.org/10.1007/978-3-540-35488-8

153. Güzel, B.E.K., Karaçalı, B.: Fisher's linear discriminant analysis based prediction using transient features of seismic events in coal mines. In: Ganzha, M., Maciaszek, L., Paprzycki, M. (eds.) Proceedings of the 2016 Federated Conference on Computer Science and Information Systems. Annals of Computer Science and Information Systems, vol. 8, pp. 231–234. IEEE (2016). https://doi.org/10.15439/2016F116

154. Ha, S., Choi, S.: Convolutional neural networks for human activity recognition using multiple accelerometer and gyroscope sensors. In: 2016 International Joint Conference on Neural Networks (IJCNN), pp. 381–388 (2016). https://doi.org/10.1109/IJCNN.2016.7727224

155. Hall, M.: Correlation-based feature selection for machine learning. Ph.D. thesis, University of Waikato (1999)

156. Hamed, A., Sobhy, A., Nassar, H.: Distributed approach for computing rough set approximations of big incomplete information systems. Inf. Sci. **547**, 427–449 (2021). https://doi.org/10.1016/j.ins.2020.08.049

157. Hancer, E.: Differential evolution for feature selection: a fuzzy wrapper–filter approach. Soft. Comput. **23**(13), 5233–5248 (2018). https://doi.org/10.1007/s00500-018-3545-7

158. Hancer, E., Xue, B., Zhang, M.: Differential evolution for filter feature selection based on information theory and feature ranking. Knowl. Based Syst. **140**, 103–119 (2018). https://doi.org/10.1016/j.knosys.2017.10.028

159. Hariri, R.H., Fredericks, E.M., Bowers, K.M.: Uncertainty in big data analytics: survey, opportunities, and challenges. J. Big Data **6**(1), 1–16 (2019). https://doi.org/10.1186/s40537-019-0206-3

160. He, Y.L., Tian, Y., Xu, Y., Zhu, Q.X.: Novel soft sensor development using echo state network integrated with singular value decomposition: application to complex chemical processes. Chemometr. Intell. Lab. Syst. **200**, 103981 (2020)

161. Heidari, A.A., Mirjalili, S., Faris, H., Aljarah, I., Mafarja, M.M., Chen, H.: Harris hawks optimization: algorithm and applications. Future Gener. Comput. Syst. **97**, 849–872 (2019). https://doi.org/10.1016/j.future.2019.02.028

162. Herodotou, H., Dong, F., Babu, S.: No one (cluster) size fits all: automatic cluster sizing for data-intensive analytics. In: Proceedings of the 2nd ACM Symposium on Cloud Computing, p. 18. ACM (2011)

163. Hinton, G.E., Salakhutdinov, R.R.: Reducing the dimensionality of data with neural networks. Science **313**(5786), 504–507 (2006). https://doi.org/10.1126/science.1127647. http://science.sciencemag.org/content/313/5786/504

164. Hirota, K.: Concepts of probabilistic sets. Fuzzy Sets Syst. **5**(1), 31–46 (1981). https://doi.org/10.1016/0165-0114(81)90032-4

165. Hońko, P.: Attribute reduction: a horizontal data decomposition approach. Soft. Comput. **20**(3), 951–966 (2016). https://doi.org/10.1007/s00500-014-1554-8

166. Hosseini, B., Hammer, B.: Interpretable discriminative dimensionality reduction and feature selection on the manifold. In: Brefeld, U., Fromont, E., Hotho, A., Knobbe, A., Maathuis, M., Robardet, C. (eds.) ECML PKDD 2019. LNCS (LNAI), vol. 11906, pp. 310–326. Springer, Cham (2020). https://doi.org/10.1007/978-3-030-46150-8_19

167. Hu, L., Zhang, Z. (eds.): EEG Signal Processing and Feature Extraction. Springer, Singapore (2019). https://doi.org/10.1007/978-981-13-9113-2

168. Hu, X.: Ensembles of classifiers based on rough sets theory and set-oriented database operations. In: Proceedings of IEEE GrC 2006, pp. 67–73 (2006)

169. Hua, Q.-S., Yu, D., Lau, F.C.M., Wang, Y.: Exact algorithms for set multicover and multiset multicover problems. In: Dong, Y., Du, D.-Z., Ibarra, O. (eds.) ISAAC 2009. LNCS, vol. 5878, pp. 34–44. Springer, Heidelberg (2009). https://doi.org/10.1007/978-3-642-10631-6_6

170. Huang, Q., Cao, J.: Research on coal pillar malposition distance based on coupling control of three-field in shallow buried closely spaced multi-seam mining, China. Energies **12**(3), 462 (2019). https://doi.org/10.3390/en12030462

171. Huang, X., Zhang, L., Wang, B., Li, F., Zhang, Z.: Feature clustering based support vector machine recursive feature elimination for gene selection. Appl. Intell. **48**(3), 594–607 (2017). https://doi.org/10.1007/s10489-017-0992-2

172. Ioffe, S., Szegedy, C.: Batch normalization: accelerating deep network training by reducing internal covariate shift. In: Proceedings of the 32nd International Conference on International Conference on Machine Learning - Volume 37, ICML 2015, pp. 448–456. JMLR.org (2015). http://dl.acm.org/citation.cfm?id=3045118.3045167

173. Irizarry, R.A., et al.: Exploration, normalization, and summaries of high density oligonucleotide array probe level data. Biostatistics **4**(2), 249–264 (2003). https://doi.org/10.1093/biostatistics/4.2.249

174. Islam, M.T., Srirama, S.N., Karunasekera, S., Buyya, R.: Cost-efficient dynamic scheduling of big data applications in apache spark on cloud. J. Syst. Softw. **162**, 110515 (2020). https://doi.org/10.1016/j.jss.2019.110515
175. Jain, A.K., Dubes, R.C.: Algorithms for Clustering Data. Prentice-Hall Inc, New Jersey (1988)
176. Jain, I., Jain, V.K., Jain, R.: Correlation feature selection based improved-binary particle swarm optimization for gene selection and cancer classification. Appl. Soft Comput. **62**, 203–215 (2018). https://doi.org/10.1016/j.asoc.2017.09.038
177. Jankowski, A., Skowron, A., Swiniarski, R.W.: Interactive complex granules. Fundam. Inform. **133**(2–3), 181–196 (2014). https://doi.org/10.3233/FI-2014-1070
178. Janusz, A.: Algorithms for similarity relation learning from high dimensional data. Ph.D. thesis, University of Warsaw (2014)
179. Janusz, A.: Algorithms for similarity relation learning from high dimensional data. In: Peters, J.F., Skowron, A. (eds.) Transactions on Rough Sets XVII. LNCS, vol. 8375, pp. 174–292. Springer, Heidelberg (2014). https://doi.org/10.1007/978-3-642-54756-0_7
180. Janusz, A., Grad, Ł., Grzegorowski, M.: Clash Royale challenge: how to select training decks for win-rate prediction. In: Ganzha, M., Maciaszek, L.A., Paprzycki, M. (eds.) Proceedings of the 2019 Federated Conference on Computer Science and Information Systems, FedCSIS 2019, Leipzig, Germany, 1–4 September 2019. Annals of Computer Science and Information Systems, vol. 18, pp. 3–6 (2019). https://doi.org/10.15439/2019F365
181. Janusz, A., et al.: Przykłady zastosowania systemu DISESOR w analizie i predykcji zagrożeń. In: Przystałka, P., Sikora, M. (eds.) Zintegrowany, szkieletowy system wspmagania decyzji dla systemów monitorowania procesów, urządzeń i zagrożeń, chap. 11, pp. 31–40. Monograficzna Seria Wydawnicza Instyututu Technik Innowacyjnych EMAG (2017). http://disesor.ibemag.pl/www/disesor.ibemag.pl/data/Image/przyklad-predykcja.pdf
182. Janusz, A., Grzegorowski, M., Michalak, M., Wróbel, Ł, Sikora, M., Ślęzak, D.: Predicting seismic events in coal mines based on underground sensor measurements. Eng. Appl. Artif. Intell. **64**, 83–94 (2017)
183. Janusz, A., Krasuski, A., Stawicki, S., Rosiak, M., Ślęzak, D., Nguyen, H.S.: Key risk factors for polish state fire service: a data mining competition at knowledge pit. In: Ganzha, M., Maciaszek, L.A., Paprzycki, M. (eds.) Proceedings of the 2014 Federated Conference on Computer Science and Information Systems, Warsaw, Poland, 7–10 September 2014. Annals of Computer Science and Information Systems, vol. 2, pp. 345–354 (2014). https://doi.org/10.15439/2014F507
184. Janusz, A., et al.: Mining data from coal mines: IJCRS'15 data challenge. In: Yao, Y., Hu, Q., Yu, H., Grzymala-Busse, J.W. (eds.) RSFDGrC 2015. LNCS (LNAI), vol. 9437, pp. 429–438. Springer, Cham (2015). https://doi.org/10.1007/978-3-319-25783-9_38
185. Janusz, A., Ślęzak, D.: Rough set methods for attribute clustering and selection. Appl. Artif. Intell. **28**(3), 220–242 (2014). https://doi.org/10.1080/08839514.2014.883902
186. Janusz, A., Ślęzak, D.: Computation of approximate Reducts with dynamically adjusted approximation threshold. In: Esposito, F., Pivert, O., Hacid, M.-S., Raś, Z.W., Ferilli, S. (eds.) ISMIS 2015. LNCS (LNAI), vol. 9384, pp. 19–28. Springer, Cham (2015). https://doi.org/10.1007/978-3-319-25252-0_3

187. Janusz, A., Ślęzak, D., Sikora, M., Wróbel, Ł.: Predicting dangerous seismic events: AAIA'16 data mining challenge. In: Ganzha, M., Maciaszek, L.A., Paprzycki, M. (eds.) Proceedings of the 2016 Federated Conference on Computer Science and Information Systems, FedCSIS 2016, Gdańsk, Poland, 11–14 September 2016. Annals of Computer Science and Information Systems, vol. 8, pp. 205–211. IEEE (2016). https://doi.org/10.15439/2016F560
188. Janusz, A., Szczuka, M.S.: Assessment of data granulations in context of feature extraction problem. In: Proceedings of IEEE GrC, pp. 116–120 (2014)
189. Janusz, A., Tajmajer, T., Świechowski, M.: Helping AI to play hearthstone: AAIA'17 data mining challenge. In: Proceedings of FedCSIS, pp. 121–125 (2017)
190. Jelodar, H., et al.: Latent Dirichlet allocation (LDA) and topic modeling: models, applications, a survey. Multimed. Tools Appl. **78**(11), 15169–15211 (2018). https://doi.org/10.1007/s11042-018-6894-4
191. Jia, X., Shang, L., Zhou, B., Yao, Y.: Generalized attribute Reduct in rough set theory. Knowl. Based Syst. **91**, 204–218 (2016). https://doi.org/10.1016/j.knosys.2015.05.017
192. Jiménez, F., Palma, J.T., Sánchez, G., Marín, D., Ortega, F.P., López, M.D.L.: Feature selection based multivariate time series forecasting: an application to antibiotic resistance outbreaks prediction. Artif. Intell. Med. **104**, 101818 (2020)
193. Jin, R., Si, L.: A study of methods for normalizing user ratings in collaborative filtering. In: Proceedings of the 27th Annual International ACM SIGIR Conference on Research and Development in Information Retrieval, SIGIR 2004, pp. 568–569. ACM, New York (2004). https://doi.org/10.1145/1008992.1009124. http://doi.acm.org/10.1145/1008992.1009124
194. Jin, R., Si, L., Zhai, C., Callan, J.: Collaborative filtering with decoupled models for preferences and ratings. In: Proceedings of the Twelfth International Conference on Information and Knowledge Management, CIKM 2003, pp. 309–316. ACM, New York (2003). https://doi.org/10.1145/956863.956922. http://doi.acm.org/10.1145/956863.956922
195. Jing, Y., Li, T., Fujita, H., Wang, B., Cheng, N.: An incremental attribute reduction method for dynamic data mining. Inf. Sci. **465**, 202–218 (2018). https://doi.org/10.1016/j.ins.2018.07.001
196. Jing, Y., Li, T., Luo, C., Horng, S.J., Wang, G., Yu, Z.: An incremental approach for attribute reduction based on knowledge granularity. Knowl. Based Syst. **104**, 24–38 (2016)
197. Jovic, A., Brkic, K., Bogunovic, N.: A review of feature selection methods with applications. In: Proceedings of MIPRO 2015, pp. 1200–1205 (2015)
198. Kabiesz, J.: Effect of the form of data on the quality of mine tremors hazard forecasting using neural networks. Geotech. Geol. Eng. **24**(5), 1131–1147 (2006). https://doi.org/10.1007/s10706-005-1136-8
199. Kabiesz, J.: The justification and objective to modify methods of forecasting the potential and assess the actual state of rockburst hazard. In: Methods for Assessment of Rockburst Hazard in Coal Mines' Excavations, vol. 44, pp. 44–48 (2010). (in Polish)
200. Kabiesz, J., Sikora, B., Sikora, M., Wróbel, Ł: Application of rule-based models for seismic hazard prediction in coal mines. Acta Montanistica Slovaca **18**(3), 262–277 (2013)
201. Kalousis, A., Prados, J., Hilario, M.: Stability of feature selection algorithms: a study on high-dimensional spaces. Knowl. Inf. Syst. **12**(1), 95–116 (2007)
202. Kang, M., Tian, J.: Machine Learning: Data Pre-processing, pp. 111–130 (2019). https://doi.org/10.1002/9781119515326.ch5

203. Kańtoch, E., Augustyniak, P., Markiewicz, M., Prusak, D.: Monitoring activities of daily living based on wearable wireless body sensor network. In: 36th Annual International Conference of the IEEE Engineering in Medicine and Biology Society, EMBC 2014, Chicago, IL, USA, 26–30 August 2014, pp. 586–589. IEEE (2014). https://doi.org/10.1109/EMBC.2014.6943659
204. Kántoch, E., Grochala, D., Kajor, M., Kucharski, D.: The prototype of wearable sensors system for supervision of patient rehabilitation using artificial intelligence methods. In: IBE 2017. AISC, vol. 623, pp. 205–214. Springer, Cham (2018). https://doi.org/10.1007/978-3-319-70063-2_22
205. Karabatak, M., Ince, M.C.: A new feature selection method based on association rules for diagnosis of erythemato-squamous diseases. Expert Syst. Appl. **36**(10), 12500–12505 (2009)
206. Kasinikota, A., Balamurugan, P., Shevade, S.: Modeling label interactions in multi-label classification: a multi-structure SVM perspective. In: Phung, D., Tseng, V.S., Webb, G.I., Ho, B., Ganji, M., Rashidi, L. (eds.) PAKDD 2018. LNCS (LNAI), vol. 10937, pp. 43–55. Springer, Cham (2018). https://doi.org/10.1007/978-3-319-93034-3_4
207. Kaufman, S., Rosset, S., Perlich, C., Stitelman, O.: Leakage in data mining: formulation, detection, and avoidance. TKDD **6**(4), 15 (2012). https://doi.org/10.1145/2382577.2382579. http://doi.acm.org/10.1145/2382577.2382579
208. Kaur, N., Sood, S.K.: Efficient resource management system based on 4Vs of big data streams. Big Data Res. **9**, 98–106 (2017). https://doi.org/10.1016/j.bdr.2017.02.002
209. Keller, R., Häfner, L., Sachs, T., Fridgen, G.: Scheduling flexible demand in cloud computing spot markets. Bus. Inf. Syst. Eng. **62**(1), 25–39 (2019). https://doi.org/10.1007/s12599-019-00592-5
210. Keogh, E., Lin, J., Fu, A.: Hot sax: efficiently finding the most unusual time series subsequence. In: Proceedings of the Fifth IEEE International Conference on Data Mining, ICDM 2005, pp. 226–233. IEEE Computer Society, Washington, DC (2005). https://doi.org/10.1109/ICDM.2005.79
211. Keogh, E.J., Pazzani, M.J.: Scaling up dynamic time warping for datamining applications. In: Proceedings of the Sixth ACM SIGKDD International Conference on Knowledge Discovery and Data Mining, KDD 2000, pp. 285–289. ACM, New York (2000). https://doi.org/10.1145/347090.347153
212. Keren, G., Schuller, B.W.: Convolutional RNN: an enhanced model for extracting features from sequential data. In: 2016 International Joint Conference on Neural Networks, IJCNN 2016, Vancouver, BC, Canada, 24–29 July 2016, pp. 3412–3419. IEEE (2016). https://doi.org/10.1109/IJCNN.2016.7727636
213. Khandelwal, V., Chaturvedi, A.K., Gupta, C.P.: Amazon EC2 spot price prediction using regression random forests. IEEE Trans. Cloud Comput. **8**(1), 59–72 (2020)
214. Kieu, T., Yang, B., Guo, C., Jensen, C.S.: Outlier detection for time series with recurrent autoencoder ensembles. In: Kraus, S. (ed.) Proceedings of the Twenty-Eighth International Joint Conference on Artificial Intelligence, IJCAI 2019, Macao, China, 10–16 August 2019, pp. 2725–2732. ijcai.org (2019). https://doi.org/10.24963/ijcai.2019/378
215. Kin-Pong Chan, F., Wai-chee Fu, A., Yu, C.: Haar wavelets for efficient similarity search of time-series: with and without time warping. IEEE Trans. Knowl. Data Eng. **15**(3), 686–705 (2003). https://doi.org/10.1109/TKDE.2003.1198399
216. Kobak, D., Berens, P.: The art of using t-SNE for single-cell transcriptomics. Nature Commun. **10**, 1–14 (2019). https://doi.org/10.1038/s41467-019-13056-x

217. Kornowski, J.: Linear prediction of aggregated seismic and seismoacoustic energy emitted from a mining longwall. Acta Montana Ser. A **22**(129), 5–14 (2003)
218. Kowalski, M., Ślęzak, D., Stencel, K., Pardel, P.W., Grzegorowski, M., Kijowski, M.: RDBMS model for scientific articles analytics. In: Bembenik, R., Skonieczny, L., Rybiński, H., Niezgodka, M. (eds.) Intelligent Tools for Building a Scientific Information Platform. Studies in Computational Intelligence, vol. 390, pp. 49–60. Springer, Heidelberg (2012). https://doi.org/10.1007/978-3-642-24809-2_4
219. Kozielski, M., Sikora, M., Wróbel, Ł.: DISESOR - decision support system for mining industry. In: Ganzha, M., Maciaszek, L.A., Paprzycki, M. (eds.) 2015 Federated Conference on Computer Science and Information Systems, FedCSIS 2015, Lódz, Poland, 13–16 September 2015. Annals of Computer Science and Information Systems, vol. 5, pp. 67–74. IEEE (2015). https://doi.org/10.15439/2015F168
220. Krawczyk, B., Minku, L.L., Gama, J., Stefanowski, J., Woźniak, M.: Ensemble learning for data stream analysis: a survey. Inf. Fusion **37**, 132–156 (2017). https://doi.org/10.1016/j.inffus.2017.02.004
221. Krizhevsky, A., Sutskever, I., Hinton, G.E.: Imagenet classification with deep convolutional neural networks. In: Pereira, F., Burges, C.J.C., Bottou, L., Weinberger, K.Q. (eds.) Advances in Neural Information Processing Systems 25, pp. 1097–1105. Curran Associates, Inc. (2012)
222. Kryszkiewicz, M.: Rough set approach to incomplete information systems. Inf. Sci. **112**(1), 39–49 (1998). https://doi.org/10.1016/S0020-0255(98)10019-1
223. Kuncheva, L.I., Diez, J.J.R.: On feature selection protocols for very low-sample-size data. Pattern Recognit. **81**, 660–673 (2018). https://doi.org/10.1016/j.patcog.2018.03.012
224. Kurach, K., Pawłowski, K.: Predicting dangerous seismic activity with recurrent neural networks. In: Ganzha, M., Maciaszek, L., Paprzycki, M. (eds.) Proceedings of the 2016 Federated Conference on Computer Science and Information Systems. Annals of Computer Science and Information Systems, vol. 8, pp. 239–243. IEEE (2016). https://doi.org/10.15439/2016F134
225. Kusuma, R.M.I., Ho, T.T., Kao, W.C., Ou, Y.Y., Hua, K.L.: Using deep learning neural networks and candlestick chart representation to predict stock market (2019)
226. Lameski, P., Zdravevski, E., Mingov, R., Kulakov, A.: SVM parameter tuning with grid search and its impact on reduction of model over-fitting. In: Yao, Y., Hu, Q., Yu, H., Grzymala-Busse, J.W. (eds.) RSFDGrC 2015. LNCS (LNAI), vol. 9437, pp. 464–474. Springer, Cham (2015). https://doi.org/10.1007/978-3-319-25783-9_41
227. Lan, G., Hou, C., Nie, F., Luo, T., Yi, D.: Robust feature selection via simultaneous sapped norm and sparse regularizer minimization. Neurocomputing **283**, 228–240 (2018)
228. Landfors, M., Philip, P., Rydén, P., Stenberg, P.: Normalization of high dimensional genomics data where the distribution of the altered variables is skewed. PLOS ONE **6**(11), 1–11 (11 2011). https://doi.org/10.1371/journal.pone.0027942. https://doi.org/10.1371/journal.pone.0027942
229. Laptev, I., Marszalek, M., Schmid, C., Rozenfeld, B.: Learning realistic human actions from movies. In: 2008 IEEE Computer Society Conference on Computer Vision and Pattern Recognition (CVPR 2008), 24–26 June 2008, Anchorage, Alaska, USA. IEEE Computer Society (2008). https://doi.org/10.1109/CVPR.2008.4587756

230. Lara, O.D., Labrador, M.A.: A survey on human activity recognition using wearable sensors. IEEE Commun. Surv. Tutorials **15**(3), 1192–1209 (2013). https://doi.org/10.1109/SURV.2012.110112.00192

231. Lasocki, S.: Probabilistic analysis of seismic hazard posed by mining induced events. In: Proceedings of Sixth International Symposium on Rockburst and Seismicity in Mines, pp. 151–156 (2005)

232. Lazar, C., et al.: A survey on filter techniques for feature selection in gene expression microarray analysis. IEEE ACM Trans. Comput. Biol. Bioinform. **9**(4), 1106–1119 (2012). https://doi.org/10.1109/TCBB.2012.33

233. LeCun, Y., Kavukcuoglu, K., Farabet, C.: Convolutional networks and applications in vision. In: ISCAS, pp. 253–256. IEEE (2010)

234. Lee, K.H., Lee, Y.J., Choi, H., Chung, Y.D., Moon, B.: Parallel data processing with mapreduce: a survey. SIGMOD Rec. **40**(4), 11–20 (2012)

235. Leśniak, A., Isakow, Z.: Space-time clustering of seismic events and hazard assessment in the Zabrze-Bielszowice coal mine, Poland. Int. J. Rock Mech. Min. Sci. **46**(5), 918–928 (2009). https://doi.org/10.1016/j.ijrmms.2008.12.003

236. Levada, A.L.: Parametric PCA for unsupervised metric learning. Pattern Recogn. Lett. **135**, 425–430 (2020). https://doi.org/10.1016/j.patrec.2020.05.011

237. Li, C., Ai, D.: Automatic crack detection method for loaded coal in vibration failure process. PLOS ONE **12**(10), 1–21 (2017). https://doi.org/10.1371/journal.pone.0185750

238. Li, M., Hinnov, L., Kump, L.: Acycle: time-series analysis software for paleoclimate research and education. Comput. Geosci. **127**, 12–22 (2019). https://doi.org/10.1016/j.cageo.2019.02.011

239. Li, P., Wu, J., Shang, L.: Fast approximate attribute reduction with MapReduce. In: Proceedings of RSKT 2013, pp. 271–278 (2013)

240. Lin, C.C., Deng, D.J., Kuo, C.H., Chen, L.: Concept drift detection and adaption in big imbalance industrial IoT data using an ensemble learning method of offline classifiers. IEEE Access **7**, 56198–56207 (2019). https://doi.org/10.1109/ACCESS.2019.2912631

241. Lin, J., Vlachos, M., Keogh, E., Gunopulos, D.: Iterative incremental clustering of time series. In: Bertino, E., et al. (eds.) EDBT 2004. LNCS, vol. 2992, pp. 106–122. Springer, Heidelberg (2004). https://doi.org/10.1007/978-3-540-24741-8_8

242. Lin, W.-C., Tsai, C.-F.: Missing value imputation: a review and analysis of the literature (2006–2017). Artif. Intell. Rev. **53**(2), 1487–1509 (2019). https://doi.org/10.1007/s10462-019-09709-4

243. Liu, A., Lu, J., Liu, F., Zhang, G.: Accumulating regional density dissimilarity for concept drift detection in data streams. Pattern Recogn. **76**, 256–272 (2018). https://doi.org/10.1016/j.patcog.2017.11.009

244. Liu, B., Tsoumakas, G.: Dealing with class imbalance in classifier chains via random undersampling. Knowl.-Based Syst. **192**, 105292 (2020). https://doi.org/10.1016/j.knosys.2019.105292

245. Liu, H., Motoda, H. (eds.): Feature Extraction, Construction and Selection. Springer, Heidelberg (1998). https://doi.org/10.1007/978-1-4615-5725-8

246. Liu, H., Wu, X., Zhang, S.: A new supervised feature selection method for pattern classification. Comput. Intell. **30**(2), 342–361 (2014)

247. Liu, J., Wang, S., Yang, W.: Sparse autoencoder for social image understanding. Neurocomputing **369**, 122–133 (2019). https://doi.org/10.1016/j.neucom.2019.08.083

248. Liu, K., Yang, X., Yu, H., Mi, J., Wang, P., Chen, X.: Rough set based semi-supervised feature selection via ensemble selector. Knowl. Based Syst. **165**, 282–296 (2019). https://doi.org/10.1016/j.knosys.2018.11.034
249. Liu, X., Pedrycz, W.: The development of fuzzy decision trees in the framework of axiomatic fuzzy set logic. Appl. Soft Comput. **7**(1), 325–342 (2007). https://doi.org/10.1016/j.asoc.2005.07.003
250. Liu, Y., Gao, X., Gao, Q., Shao, L., Han, J.: Adaptive robust principal component analysis. Neural Netw. **119**, 85–92 (2019). https://doi.org/10.1016/j.neunet.2019.07.015
251. Luo, C., Li, T., Yao, Y.: Dynamic probabilistic rough sets with incomplete data. Inf. Sci. **417**, 39–54 (2017). https://doi.org/10.1016/j.ins.2017.06.040
252. Luo, J., Fujita, H., Yao, Y., Qin, K.: On modeling similarity and three-way decision under incomplete information in rough set theory. Knowl.-Based Syst. **191**, 105251 (2020). https://doi.org/10.1016/j.knosys.2019.105251
253. Luts, J., Ojeda, F., de Plas, R.V., Moor, B.D., Huffel, S.V., Suykens, J.A.: A tutorial on support vector machine-based methods for classification problems in chemometrics. Anal. Chim. Acta **665**(2), 129–145 (2010). https://doi.org/10.1016/j.aca.2010.03.030
254. Ma, C., Li, W., Cao, J., Du, J., Li, Q., Gravina, R.: Adaptive sliding window based activity recognition for assisted livings. Inf. Fusion **53**, 55–65 (2020). https://doi.org/10.1016/j.inffus.2019.06.013
255. Mafarja, M.M., Mirjalili, S.: Hybrid whale optimization algorithm with simulated annealing for feature selection. Neurocomputing **260**, 302–312 (2017). https://doi.org/10.1016/j.neucom.2017.04.053
256. Mafarja, M.M., Mirjalili, S.: Hybrid binary ant lion optimizer with rough set and approximate entropy reducts for feature selection. Soft. Comput. **23**(15), 6249–6265 (2018). https://doi.org/10.1007/s00500-018-3282-y
257. Maldonado, S., López, J.: Dealing with high-dimensional class-imbalanced datasets: Embedded feature selection for SVM classification. Appl. Soft Comput. **67**, 94–105 (2018). https://doi.org/10.1016/j.asoc.2018.02.051
258. Malondkar, A., Corizzo, R., Kiringa, I., Ceci, M., Japkowicz, N.: Spark-GHSOM: growing hierarchical self-organizing map for large scale mixed attribute datasets. Inf. Sci. **496**, 572–591 (2019). https://doi.org/10.1016/j.ins.2018.12.007
259. Manning, C.D., Raghavan, P., Schütze, H.: Introduction to Information Retrieval. Cambridge University Press, Cambridge (2008). https://doi.org/10.1017/CBO9780511809071. https://nlp.stanford.edu/IR-book/pdf/irbookprint.pdf
260. Mark, C.: Coal bursts in the deep longwall mines of the United States. Int. J. Coal Sci. Technol. **3**(1), 1–9 (2016)
261. Mason, A.J.: Bayesian methods for modelling non-random missing data mechanisms in longitudinal studies. Ph.D. thesis, Imperial College London (2009)
262. Mathew, S.: Overview of Amazon Web Services, April 2017. Accessed 04 June 2019
263. Meina, M., Janusz, A., Rykaczewski, K., Ślęzak, D., Celmer, B., Krasuski, A.: Tagging firefighter activities at the emergency scene: summary of AAIA'15 data mining competition at knowledge pit. In: Ganzha, M., Maciaszek, L.A., Paprzycki, M. (eds.) 2015 Federated Conference on Computer Science and Information Systems, FedCSIS 2015, Lódz, Poland, 13–16 September 2015. Annals of Computer Science and Information Systems, vol. 5, pp. 367–373. IEEE (2015). https://doi.org/10.15439/2015F426

264. Mikolov, T., Chen, K., Corrado, G., Dean, J.: Efficient estimation of word representations in vector space. In: Bengio, Y., LeCun, Y. (eds.) 1st International Conference on Learning Representations, ICLR 2013, Scottsdale, Arizona, USA, 2–4 May 2013, Workshop Track Proceedings (2013). http://arxiv.org/abs/1301.3781

265. Milczek, J.K., Bogucki, R., Lasek, J., Tadeusiak, M.: Early warning system for seismic events in coal mines using machine learning. In: Ganzha, M., Maciaszek, L., Paprzycki, M. (eds.) Proceedings of the 2016 Federated Conference on Computer Science and Information Systems. Annals of Computer Science and Information Systems, vol. 8, pp. 213–220. IEEE (2016). https://doi.org/10.15439/2016F420

266. Min, F., Hu, Q., Zhu, W.: Feature selection with test cost constraint. Int. J. Approx. Reason. **55**(1), 167–179 (2014)

267. Miranda, T., Correia, A.G., e Sousa, L.R.: Bayesian methodology for updating geomechanical parameters and uncertainty quantification. Int. J. Rock Mech. Mining Sci. **46**(7), 1144–1153 (2009). https://doi.org/10.1016/j.ijrmms.2009.03.008

268. Moczulski, W., Przystałka, P., Sikora, M., Zimroz, R.: Modern ICT and mechatronic systems in contemporary mining industry. In: Rough Sets - International Joint Conference, IJCRS 2016, Santiago de Chile, Chile, 7–11 October 2016, Proceedings, pp. 33–42 (2016). https://doi.org/10.1007/978-3-319-47160-0_3

269. Mohamed, M., Oussalah, M.: SRL-ESA-TextSum: a text summarization approach based on semantic role labeling and explicit semantic analysis. Inf. Process. Manag. **56**(4), 1356–1372 (2019). https://doi.org/10.1016/j.ipm.2019.04.003

270. Mönks, U., Dörksen, H., Lohweg, V., Hübner, M.: Information fusion of conflicting input data. Sensors **16**(11), E1798 (2016)

271. Moore, R.E., Kearfott, R.B., Cloud, M.J.: Introduction to Interval Analysis. Society for Industrial and Applied Mathematics (2009)

272. Mörchen, F., Ultsch, A.: Optimizing time series discretization for knowledge discovery. In: Proceedings of the Eleventh ACM SIGKDD International Conference on Knowledge Discovery in Data Mining, KDD 2005, pp. 660–665. ACM, New York (2005). https://doi.org/10.1145/1081870.1081953

273. Moshkov, M.J., Piliszczuk, M., Zielosko, B.: On construction of partial reducts and irreducible partial decision rules. Fund. Inform. **75**(1–4), 357–374 (2007)

274. Mu, L., Ji, Y.: Integrated coal mine safety monitoring system. In: Zhang, W. (ed.) SEKEIE 2012, pp. 365–371. Springer, Heidelberg (2012). https://doi.org/10.1007/978-3-642-29455-6_51

275. Mulargia, F., Stark, P.B., Geller, R.J.: Why is probabilistic seismic hazard analysis (PSHA) still used? Phys. Earth Planet. Inter. **264**, 63–75 (2017). https://doi.org/10.1016/j.pepi.2016.12.002

276. Murphy, K.P.: Machine Learning: A Probabilistic Perspective. The MIT Press, Cambridge (2012)

277. Nasiri, H., Nasehi, S., Goudarzi, M.: Evaluation of distributed stream processing frameworks for IoT applications in Smart Cities. J. Big Data **6**(1), 1–24 (2019). https://doi.org/10.1186/s40537-019-0215-2

278. Nguyen, H.S.: Approximate boolean reasoning: foundations and applications in data mining. Trans. Rough Sets **5**, 334–506 (2006). https://doi.org/10.1007/11847465_16

279. Nguyen, H.S., Ślęzak, D.: Approximate reducts and association rules. In: Zhong, N., Skowron, A., Ohsuga, S. (eds.) RSFDGrC 1999. LNCS (LNAI), vol. 1711, pp. 137–145. Springer, Heidelberg (1999). https://doi.org/10.1007/978-3-540-48061-7_18

280. Nguyen, S.H., Szczuka, M.: Feature selection in decision systems with constraints. In: Flores, V., Gomide, F., Janusz, A., Meneses, C., Miao, D., Peters, G., Ślęzak, D., Wang, G., Weber, R., Yao, Y. (eds.) IJCRS 2016. LNCS (LNAI), vol. 9920, pp. 537–547. Springer, Cham (2016). https://doi.org/10.1007/978-3-319-47160-0_49

281. Nguyen, S.H., Skowron, A.: Quantization of real value attributes - rough set and boolean reasoning approach. In: Proceedings of the Second Joint Annual Conference on Information Sciences, Wrightsville Beach, North Carolina, 28 September–1 October 1995, pp. 34–37 (1995)

282. Nguyen, T.T., Skowron, A.: Rough-Granular Computing in Human-Centric Information Processing. In: Bargiela, A., Pedrycz, W. (eds.) Human-Centric Information Processing Through Granular Modelling. Studies in Computational Intelligence, vol. 182, pp. 1–30. Springer, Heidelberg (2009). https://doi.org/10.1007/978-3-540-92916-1_1

283. Nixon, M.S., Aguado, A.S.: Feature Extraction and Image Processing for Computer Vision, 4th edn. Academic Press (2020)

284. Nogueira, S.: Quantifying the stability of feature selection. Ph.D. thesis, University of Manchester (2018)

285. Nogueira, S., Sechidis, K., Brown, G.: On the stability of feature selection algorithms. J. Mach. Learn. Res. **18**, 174:1–174:54 (2017)

286. Palma-Mendoza, R.-J., Rodriguez, D., de-Marcos, L.: Distributed ReliefF-based feature selection in Spark. Knowl. Inf. Syst. **57**(1), 1–20 (2018). https://doi.org/10.1007/s10115-017-1145-y

287. Parmar, N., Ramachandran, P., Vaswani, A., Bello, I., Levskaya, A., Shlens, J.: Stand-alone self-attention in vision models. In: Wallach, H.M., Larochelle, H., Beygelzimer, A., d'Alché-Buc, F., Fox, E.B., Garnett, R. (eds.) Advances in Neural Information Processing Systems 32: Annual Conference on Neural Information Processing Systems 2019, NeurIPS 2019, 8–14 December 2019, Canada, Vancouver, BC, pp. 68–80 (2019)

288. Pawlak, Z.: Rough Sets: Theoretical Aspects of Reasoning about Data, System Theory, Knowledge Engineering and Problem Solving, vol. 9. Kluwer (1991)

289. Pawlak, Z., Skowron, A.: Rough membership functions. In: Advances in the Dempster-Shafer Theory of Evidence, pp. 251–271. Wiley, New York (1994)

290. Pawlak, Z., Skowron, A.: Rough sets: some extensions. Inf. Sci. **177**(1), 28–40 (2007)

291. Pawlak, Z., Skowron, A.: Rudiments of rough sets. Inf. Sci. **177**(1), 3–27 (2007)

292. Pawłowski, K., Kurach, K.: Detecting methane outbreaks from time series data with deep neural networks. In: Yao, Y., Hu, Q., Yu, H., Grzymala-Busse, J.W. (eds.) RSFDGrC 2015. LNCS (LNAI), vol. 9437, pp. 475–484. Springer, Cham (2015). https://doi.org/10.1007/978-3-319-25783-9_42

293. Pearl, J.: Causal inference in statistics: an overview. Stat. Surv. **3**, 96–146 (2009). https://doi.org/10.1214/09-SS057

294. Pedrycz, W.: Interpretation of clusters in the framework of shadowed sets. Pattern Recogn. Lett. **26**(15), 2439–2449 (2005). https://doi.org/10.1016/j.patrec.2005.05.001

295. Pedrycz, W.: Granular Computing: Analysis and Design of Intelligent Systems. CRC Press, Boca Raton (2013)

296. Pedrycz, W.: Granular computing for data analytics: a manifesto of human-centric computing. IEEE CAA J. Autom. Sinica **5**(6), 1025–1034 (2018). https://doi.org/10.1109/JAS.2018.7511213

297. Peng, H., Long, F., Ding, C.H.Q.: Feature selection based on mutual information: criteria of max-dependency, max-relevance, and min-redundancy. IEEE Trans. Pattern Anal. Mach. Intell. **27**(8), 1226–1238 (2005). https://doi.org/10.1109/TPAMI.2005.159

298. Pennington, J., Socher, R., Manning, C.D.: GloVe: global vectors for word representation. In: Moschitti, A., Pang, B., Daelemans, W. (eds.) Proceedings of the 2014 Conference on Empirical Methods in Natural Language Processing, EMNLP 2014, 25–29 October 2014, Doha, Qatar, A meeting of SIGDAT, a Special Interest Group of the ACL, pp. 1532–1543. ACL (2014). https://doi.org/10.3115/v1/d14-1162

299. Perez-Benitez, J.A., Padovese, L.R.: A system for classification of time-series data from industrial non-destructive device. Eng. Appl. Artif. Intell. **26**(3), 974–983 (2013). https://doi.org/10.1016/j.engappai.2012.09.006

300. Persson, L.E., Samko, N., Wall, P.: Quasi-monotone weight functions and their characteristics and applications. Math. Inequalities Appl. **15**, 685–705 (2012). https://doi.org/10.7153/mia-15-61

301. Podlodowski, Ł.: Utilizing an ensemble of SVMs with GMM voting-based mechanism in predicting dangerous seismic events in active coal mines. In: Ganzha, M., Maciaszek, L., Paprzycki, M. (eds.) Proceedings of the 2016 Federated Conference on Computer Science and Information Systems. Annals of Computer Science and Information Systems, vol. 8, pp. 235–238. IEEE (2016). https://doi.org/10.15439/2016F122

302. Polikar, R., DePasquale, J., Mohammed, H.S., Brown, G., Kuncheva, L.I.: Learn++.MF: a random subspace approach for the missing feature problem. Pattern Recognit. **43**(11), 3817–3832 (2010)

303. Ponciano, V., et al.: Mobile computing technologies for health and mobility assessment: research design and results of the timed up and go test in older adults. Sensors **20**(12), 3481 (2020). https://doi.org/10.3390/s20123481

304. Popieul, J.C., Loslever, P., Todoskoff, A., Simon, P., Rotting, M.: Multivariate analysis of human behavior data using fuzzy windowing: example with driver-car-environment system. Eng. Appl. Artif. Intell. **25**(5), 989–996 (2012). https://doi.org/10.1016/j.engappai.2011.11.011

305. Potdar, K., Pardawala, T., Pai, C.: A comparative study of categorical variable encoding techniques for neural network classifiers. Int. J. Comput. Appl. **175**, 7–9 (2017). https://doi.org/10.5120/ijca2017915495

306. Przystałka, P., Sikora, M. (eds.): Zintegrowany, szkieletowy system wspomagania decyzji dla systemów monitorowania procesów, urządzeń i zagrożeń. Monograficzna Seria Wydawnicza Instytututu Technik Innowacyjnych EMAG (2017)

307. Qian, J., Miao, D., Zhang, Z., Li, W.: Hybrid approaches to attribute reduction based on indiscernibility and discernibility relation. Int. J. Approx. Reason. **52**(2), 212–230 (2011). https://doi.org/10.1016/j.ijar.2010.07.011

308. Qian, J., et al.: Introducing self-organized maps (SOM) as a visualization tool for materials research and education. Results Mater. **4**, 100020 (2019). https://doi.org/10.1016/j.rinma.2019.100020

309. Qian, J., Dang, C., Yue, X., Zhang, N.: Attribute reduction for sequential three-way decisions under dynamic granulation. Int. J. Approx. Reason. **85**, 196–216 (2017). https://doi.org/10.1016/j.ijar.2017.03.009

310. Qian, J., Lv, P., Yue, X., Liu, C., Jing, Z.: Hierarchical attribute reduction algorithms for big data using MapReduce. Knowl.-Based Syst. **73**, 18–31 (2015)

311. Quinlan, J.R.: C4.5: Programs for Machine Learning. Morgan Kaufmann Publishers Inc., San Francisco (1993)

312. Radford, A., Metz, L., Chintala, S.: Unsupervised representation learning with deep convolutional generative adversarial networks. In: Bengio, Y., LeCun, Y. (eds.) 4th International Conference on Learning Representations, ICLR 2016, San Juan, Puerto Rico, 2–4 May 2016, Conference Track Proceedings (2016)

313. Radford, A., Wu, J., Child, R., Luan, D., Amodei, D., Sutskever, I.: Language models are unsupervised multitask learners. OpenAI Blog (2019)

314. Raffel, C., et al.: Exploring the limits of transfer learning with a unified text-to-text transformer. J. Mach. Learn. Res. **21**, 140:1–140:67 (2020)

315. Rakthanmanon, T., et al.: Addressing big data time series: mining trillions of time series subsequences under dynamic time warping. ACM Trans. Knowl. Discov. Data **7**(3), 10:1–10:31 (2013). https://doi.org/10.1145/2500489

316. Ramírez-Gallego, S., et al.: Fast-mRMR: fast minimum redundancy maximum relevance algorithm for high-dimensional big data. Int. J. Intell. Syst. **32**, 134–152 (2017)

317. Ran, Y., Shi, Y., Zhang, Z.: Local ratio method on partial set multi-cover. J. Comb. Optim. **34**(1), 302–313 (2017)

318. Ratner, A., Hancock, B., Dunnmon, J., Sala, F., Pandey, S., Ré, C.: Training complex models with multi-task weak supervision. In: AAAI 2019, Honolulu, Hawaii, USA, 27 January–1 February 2019, pp. 4763–4771. AAAI Press (2019). https://doi.org/10.1609/aaai.v33i01.33014763

319. Read, J., Pfahringer, B., Holmes, G., Frank, E.: Classifier chains: a review and perspectives. CoRR abs/1912.13405 (2019). http://arxiv.org/abs/1912.13405

320. Read, J., Puurula, A., Bifet, A.: Multi-label classification with meta-labels. In: Kumar, R., Toivonen, H., Pei, J., Huang, J.Z., Wu, X. (eds.) 2014 IEEE International Conference on Data Mining, ICDM 2014, Shenzhen, China, 14–17 December 2014, pp. 941–946. IEEE Computer Society (2014). https://doi.org/10.1109/ICDM.2014.38

321. Rehman, M.H., Chang, V., Batool, A., Wah, T.Y.: Big data reduction framework for value creation in sustainable enterprises. Int. J. Inf. Manag. **36**(6), 917–928 (2016)

322. dos Reis, D.M., Flach, P.A., Matwin, S., Batista, G.E.A.P.A.: Fast unsupervised online drift detection using incremental kolmogorov-smirnov test. In: Krishnapuram, B., Shah, M., Smola, A.J., Aggarwal, C.C., Shen, D., Rastogi, R. (eds.) Proceedings of the 22nd ACM SIGKDD International Conference on Knowledge Discovery and Data Mining, San Francisco, CA, USA, 13–17 August 2016, pp. 1545–1554. ACM (2016). https://doi.org/10.1145/2939672.2939836

323. Riza, L.S., et al.: Implementing algorithms of rough set theory and fuzzy rough set theory in the R package 'RoughSets'. Inf. Sci. **287**, 68–89 (2014)

324. Röger, H., Mayer, R.: A comprehensive survey on parallelization and elasticity in stream processing. ACM Comput. Surv. **52**(2), 1–37 (2019). https://doi.org/10.1145/3303849

325. Rosen, J., et al.: Iterative MapReduce for Large Scale Machine Learning. CoRR abs/1303.3517 (2013)

326. Roy, A., Pal, S.K.: Fuzzy discretization of feature space for a rough set classifier. Pattern Recogn. Lett. **24**(6), 895–902 (2003). https://doi.org/10.1016/S0167-8655(02)00201-5

327. Roy, D., Murty, K.S.R., Mohan, C.K.: Feature selection using deep neural networks. In: Proceedings of IJCNN 2015, pp. 1–6 (2015)

328. Ruder, S.: An overview of multi-task learning in deep neural networks. CoRR abs/1706.05098 (2017). http://arxiv.org/abs/1706.05098

329. Rudin, C.: Please stop explaining black box models for high stakes decisions. CoRR abs/1811.10154 (2018). http://arxiv.org/abs/1811.10154

330. Menasalvas Ruiz, E., et al.: Profiling lung cancer patients using electronic health records. J. Med. Syst. **42**(7), 1–10 (2018). https://doi.org/10.1007/s10916-018-0975-9

331. Ruta, D., Cen, L.: Self-organized predictor of methane concentration warnings in coal mines. In: Yao, Y., Hu, Q., Yu, H., Grzymala-Busse, J.W. (eds.) RSFDGrC 2015. LNCS (LNAI), vol. 9437, pp. 485–493. Springer, Cham (2015). https://doi.org/10.1007/978-3-319-25783-9_43

332. Rzeszótko, J., Nguyen, S.H.: Machine learning for traffic prediction. Fund. Inform. **119**(3–4), 407–420 (2012)

333. Saeys, Y., Abeel, T., Van de Peer, Y.: Robust feature selection using ensemble feature selection techniques. In: Daelemans, W., Goethals, B., Morik, K. (eds.) ECML PKDD 2008. LNCS (LNAI), vol. 5212, pp. 313–325. Springer, Heidelberg (2008). https://doi.org/10.1007/978-3-540-87481-2_21

334. Salaken, S.M., Khosravi, A., Nguyen, T., Nahavandi, S.: Seeded transfer learning for regression problems with deep learning. Expert Syst. Appl. **115**, 565–577 (2019). https://doi.org/10.1016/j.eswa.2018.08.041

335. Sarawagi, S., Thomas, S., Agrawal, R.: Integrating association rule mining with relational database systems: alternatives and implications. Data Min. Knowl. Disc. **4**(2–3), 89–125 (2000)

336. Schaefer, M., Eikermann, M.: Contact-free respiratory monitoring using bed-wheel sensors: a valid respiratory monitoring technique with significant potential impact on public health. J. Appl. Physiol. **126**, 1430–1431 (2019). https://doi.org/10.1152/japplphysiol.00198.2019

337. Scovanner, P., Ali, S., Shah, M.: A 3-dimensional sift descriptor and its application to action recognition. In: Proceedings of the 15th ACM International Conference on Multimedia, MM 2007, pp. 357–360. ACM, New York (2007). https://doi.org/10.1145/1291233.1291311

338. Senawi, A., Wei, H., Billings, S.A.: A new maximum relevance-minimum multi-collinearity (MRmMC) method for feature selection and ranking. Pattern Recogn. **67**, 47–61 (2017)

339. Sesma-Sara, M., Mesiar, R., Bustince, H.: Weak and directional monotonicity of functions on Riesz spaces to fuse uncertain data. Fuzzy Sets Syst. **386**, 145–160 (2020). https://doi.org/10.1016/j.fss.2019.01.019

340. Sezer, O.B., Ozbayoglu, A.M.: Algorithmic financial trading with deep convolutional neural networks: time series to image conversion approach. Appl. Soft Comput. **70**, 525–538 (2018). https://doi.org/10.1016/j.asoc.2018.04.024

341. Shah, D., Isah, H., Zulkernine, F.: Stock market analysis: a review and taxonomy of prediction techniques. Int. J. Financ. Stud. **7**(2), 26 (2019). https://doi.org/10.3390/ijfs7020026

342. Shah, J.S.: Novel statistical approaches for missing values in truncated high-dimensional metabolomics data with a detection threshold. Ph.D. thesis, University of Louisville (2017)

343. Shawi, R.E., Sakr, S., Talia, D., Trunfio, P.: Big data systems meet machine learning challenges: towards big data science as a service. Big Data Res. **14**, 1–11 (2018). https://doi.org/10.1016/j.bdr.2018.04.004

344. She, Y.H., Qian, Z.H., He, X.L., Wang, J.T., Qian, T., Zheng, W.L.: On generalization reducts in multi-scale decision tables. Inf. Sci. **555**, 104–124 (2021). https://doi.org/10.1016/j.ins.2020.12.045

345. Sheikhpour, R., Sarram, M.A., Gharaghani, S., Chahooki, M.A.Z.: A survey on semi-supervised feature selection methods. Pattern Recognit. **64**, 141–158 (2017). https://doi.org/10.1016/j.patcog.2016.11.003

346. Sheikhpour, R., Sarram, M.A., Gharaghani, S., Chahooki, M.A.Z.: A robust graph-based semi-supervised sparse feature selection method. Inf. Sci. **531**, 13–30 (2020). https://doi.org/10.1016/j.ins.2020.03.094

347. Shishavan, S.A.S., Gündogdu, F.K., Farrokhizadeh, E., Donyatalab, Y., Kahraman, C.: Novel similarity measures in spherical fuzzy environment and their applications. Eng. Appl. Artif. Intell. **94**, 103837 (2020). https://doi.org/10.1016/j.engappai.2020.103837

348. Sikder, I.U., Munakata, T.: Application of rough set and decision tree for characterization of premonitory factors of low seismic activity. Expert Syst. Appl. **36**(1), 102–110 (2009). https://doi.org/10.1016/j.eswa.2007.09.032

349. Sikora, M., Sikora, B.: Improving prediction models applied in systems monitoring natural hazards and machinery. Int. J. Appl. Math. Comput. Sci. **22**(2), 477–491 (2012). https://doi.org/10.2478/v10006-012-0036-3

350. Singh, P., Dhiman, G.: A hybrid fuzzy time series forecasting model based on granular computing and bio-inspired optimization approaches. J. Comput. Sci. **27**, 370–385 (2018). https://doi.org/10.1016/j.jocs.2018.05.008

351. Skowron, A., Dutta, S.: Rough sets: past, present, and future. Nat. Comput. **17**(4), 855–876 (2018). https://doi.org/10.1007/s11047-018-9700-3

352. Skowron, A., Jankowski, A., Dutta, S.: Interactive granular computing. Granular Comput. **1**(2), 95–113 (2015). https://doi.org/10.1007/s41066-015-0002-1

353. Skowron, A., Rauszer, C.: The discernibility matrices and functions in information systems. In: Słowiński, R. (ed.) Intelligent Decision Support. Theory and Decision Library, vol. 11, pp. 331–362. Springer, Dordrecht (1992). https://doi.org/10.1007/978-94-015-7975-9_21

354. Skowron, A., Wasilewski, P.: Interactive information systems: toward perception based computing. Theor. Comput. Sci. **454**, 240–260 (2012). https://doi.org/10.1016/j.tcs.2012.04.019

355. Ślęzak, D.: Normalized decision functions and measures for inconsistent decision tables analysis. Fund. Inform. **44**(3), 291–319 (2000)

356. Ślęzak, D.: Approximate entropy reducts. Fund. Inform. **53**(3–4), 365–390 (2002)

357. Ślęzak, D.: Rough sets and functional dependencies in data: foundations of association reducts. Trans. Comput. Sci. **5**, 182–205 (2009)

358. Ślęzak, D.: Compound analytics of compound data within RDBMS framework – Infobright's perspective. In: Kim, T., Lee, Y., Kang, B.-H., Ślęzak, D. (eds.) FGIT 2010. LNCS, vol. 6485, pp. 39–40. Springer, Heidelberg (2010). https://doi.org/10.1007/978-3-642-17569-5_5

359. Ślęzak, D., Glick, R., Betliński, P., Synak, P.: A new approximate query engine based on intelligent capture and fast transformations of granulated data summaries. J. Intell. Inf. Syst. **50**(2), 385–414 (2017). https://doi.org/10.1007/s10844-017-0471-6

360. Ślęzak, D., et al.: A framework for learning and embedding multi-sensor forecasting models into a decision support system: a case study of methane concentration in coal mines. Inf. Sci. **451–452**, 112–133 (2018)

361. Ślęzak, D., Grzegorowski, M., Janusz, A., Stawicki, S.: Interactive Data Exploration with Infolattices. Abstract Materials of BAFI 2015 (2015)

362. Ślęzak, D., Grzegorowski, M., Janusz, A., Stawicki, S.: Toward interactive attribute selection with infolattices – a position paper. In: Polkowski, L., et al. (eds.) IJCRS 2017. LNCS (LNAI), vol. 10314, pp. 526–539. Springer, Cham (2017). https://doi.org/10.1007/978-3-319-60840-2_38

363. Ślęzak, D., Janusz, A.: Ensembles of bireducts: towards robust classification and simple representation. In: Kim, T., et al. (eds.) FGIT 2011. LNCS, vol. 7105, pp. 64–77. Springer, Heidelberg (2011). https://doi.org/10.1007/978-3-642-27142-7_9

364. Ślęzak, D., Stawicki, S.: The problem of finding the simplest classifier ensemble is NP-hard – a rough-set-inspired formulation based on decision bireducts. In: Bello, R., Miao, D., Falcon, R., Nakata, M., Rosete, A., Ciucci, D. (eds.) IJCRS 2020. LNCS (LNAI), vol. 12179, pp. 204–212. Springer, Cham (2020). https://doi.org/10.1007/978-3-030-52705-1_15

365. Ślęzak, D., Widz, S.: Evolutionary inspired optimization of feature subset ensembles. In: Takagi, H., Abraham, A., Köppen, M., Yoshida, K., de Carvalho, A.C.P.L.F. (eds.) Second World Congress on Nature & Biologically Inspired Computing, NaBIC 2010, 15–17 December 2010, Kitakyushu, Japan, pp. 437–442. IEEE (2010). https://doi.org/10.1109/NABIC.2010.5716365

366. Smuk, M.: Missing data methodology: sensitivity analysis after multiple imputation. Ph.D. thesis, University of London (2015)

367. Sobhani, P., Viktor, H., Matwin, S.: Learning from imbalanced data using ensemble methods and cluster-based undersampling. In: Appice, A., Ceci, M., Loglisci, C., Manco, G., Masciari, E., Ras, Z.W. (eds.) NFMCP 2014. LNCS (LNAI), vol. 8983, pp. 69–83. Springer, Cham (2015). https://doi.org/10.1007/978-3-319-17876-9_5

368. Son, L.H.: Dealing with the new user cold-start problem in recommender systems: a comparative review. Inf. Syst. **58**, 87–104 (2016). https://doi.org/10.1016/j.is.2014.10.001

369. Sorzano, C.O.S., Vargas, J., Montano, A.P.: A survey of dimensionality reduction techniques (2014)

370. de Souto, M.C.P., Costa, I.G., de Araujo, D.S.A., Ludermir, T.B., Schliep, A.: Clustering cancer gene expression data: a comparative study. BMC Bioinform. **9**, 1–14 (2008)

371. Sripada, S.G., Reiter, E., Hunter, J., Yu, J., Davy, I.P.: Modelling the task of summarising time series data using ka techniques. In: Macintosh, A., Moulton, M., Preece, A. (eds.) Applications and Innovations in Intelligent Systems IX, pp. 183–196. Springer, London (2002). https://doi.org/10.1007/978-1-4471-0149-9_14

372. Stańczyk, U., Zielosko, B., Jain, L.C.: Advances in feature selection for data and pattern recognition: an introduction. In: Stańczyk, U., Zielosko, B., Jain, L.C. (eds.) Advances in Feature Selection for Data and Pattern Recognition. ISRL, vol. 138, pp. 1–9. Springer, Cham (2018). https://doi.org/10.1007/978-3-319-67588-6_1

373. Stawicki, S., Ślęzak, D., Janusz, A., Widz, S.: Decision bireducts and decision reducts - a comparison. Int. J. Approx. Reason. **84**, 75–109 (2017)

374. Stoppiglia, H., Dreyfus, G., Dubois, R., Oussar, Y.: Ranking a random feature for variable and feature selection. J. Mach. Learn. Res. **3**, 1399–1414 (2003). http://jmlr.org/papers/v3/stoppiglia03a.html

375. Sutskever, I., Martens, J., Dahl, G., Hinton, G.: On the importance of initialization and momentum in deep learning. In: Proceedings of the 30th International Conference on International Conference on Machine Learning, ICML 2013, vol. 28. pp. 1139–1147. JMLR.org (2013). http://dl.acm.org/citation.cfm?id=3042817.3043064

376. Świniarski, R.W., Skowron, A.: Rough set methods in feature selection and recognition. Pattern Recogn. Lett. **24**(6), 833–849 (2003)

377. Szczuka, M.S., Ślęzak, D.: How deep data becomes big data. In: Proceedings of IFSA/NAFIPS 2013, pp. 579–584 (2013)

378. Szczuka, M.S., Wojdyłło, P.: Neuro-wavelet classifiers for EEG signals based on rough set methods. Neurocomputing **36**(1–4), 103–122 (2001)

379. Sze, V., Chen, Y., Yang, T., Emer, J.S.: Efficient processing of deep neural networks: a tutorial and survey. Proc. IEEE **105**(12), 2295–2329 (2017). https://doi.org/10.1109/JPROC.2017.2761740

380. Taguchi, Y.H.: Unsupervised Feature Extraction Applied to Bioinformatics. Springer, Cham (2020). https://doi.org/10.1007/978-3-030-22456-1

381. Teixeira de Souza, J., Matwin, S., Japkowicz, N.: Parallelizing feature selection. Algorithmica **45**(3), 433–456 (2006)

382. Tenney, I., Das, D., Pavlick, E.: BERT rediscovers the classical NLP pipeline. In: Korhonen, A., Traum, D.R., Màrquez, L. (eds.) Proceedings of the 57th Conference of the Association for Computational Linguistics, ACL 2019, Florence, Italy, 28July–2 August 2019, Volume 1: Long Papers, pp. 4593–4601. Association for Computational Linguistics (2019). https://doi.org/10.18653/v1/p19-1452

383. Tran, T.N., Afanador, N.L., Buydens, L.M., Blanchet, L.: Interpretation of variable importance in partial least squares with significance multivariate correlation (SMC). Chemom. Intell. Lab. Syst. **138**, 153–160 (2014)

384. Triguero, I., Peralta, D., Bacardit, J., García, S., Herrera, F.: MRPR: a MapReduce solution for prototype reduction in big data classification. Neurocomputing **150**, 331–345 (2015)

385. Tsai, C.F., Chen, Y.C.: The optimal combination of feature selection and data discretization: an empirical study. Inf. Sci. **505**, 282–293 (2019). https://doi.org/10.1016/j.ins.2019.07.091

386. Tsakiridis, N.L., et al.: Versatile internet of things for agriculture: an explainable AI approach. In: Maglogiannis, I., Iliadis, L., Pimenidis, E. (eds.) AIAI 2020. IAICT, vol. 584, pp. 180–191. Springer, Cham (2020). https://doi.org/10.1007/978-3-030-49186-4_16

387. Urbanowicz, R.J., Meeker, M., Cava, W.G.L., Olson, R.S., Moore, J.H.: Relief-based feature selection: introduction and review. J. Biomed. Inform. **85**, 189–203 (2018). https://doi.org/10.1016/j.jbi.2018.07.014

388. Van Der Maaten, L., Postma, E., Van den Herik, J.: Dimensionality reduction: a comparative review. Tilburg University Technical Report, TiCC-TR 2009 (2009)

389. Vaswani, A., et al.: Attention is all you need. In: Guyon, I., et al. (eds.) Advances in Neural Information Processing Systems 30: Annual Conference on Neural Information Processing Systems 2017, 4–9 December 2017, Long Beach, CA, USA, pp. 5998–6008 (2017)

390. Vergara, J.R., Estévez, P.A.: A review of feature selection methods based on mutual information. Neural Comput. Appl. **24**(1), 175–186 (2013). https://doi.org/10.1007/s00521-013-1368-0

391. Vincent, P., Larochelle, H., Bengio, Y., Manzagol, P.A.: Extracting and composing robust features with denoising autoencoders. In: Proceedings of the 25th International Conference on Machine Learning, ICML 2008, pp. 1096–1103. ACM, New York (2008). https://doi.org/10.1145/1390156.1390294. http://doi.acm.org/10.1145/1390156.1390294

392. Špirková, J., Beliakov, G., Bustince, H., Fernandez, J.: Mixture functions and their monotonicity. Inf. Sci. **481**, 520–549 (2019). https://doi.org/10.1016/j.ins.2018.12.090. http://www.sciencedirect.com/science/article/pii/S002002551831048X

393. Wachla, D., Moczulski, W.A.: Identification of dynamic diagnostic models with the use of methodology of knowledge discovery in databases. Eng. Appl. Artif. Intell. **20**(5), 699–707 (2007). https://doi.org/10.1016/j.engappai.2006.11.002

394. Wang, H., Xu, Z., Fujita, H., Liu, S.: Towards felicitous decision making: an overview on challenges and trends of big data. Inf. Sci. **367–368**, 747–765 (2016). https://doi.org/10.1016/j.ins.2016.07.007

395. Wang, H., Kläser, A., Schmid, C., Liu, C.L.: Dense trajectories and motion boundary descriptors for action recognition. Int. J. Comput. Vision **103**(1), 60–79 (2013). https://doi.org/10.1007/s11263-012-0594-8

396. Wang, H., Bah, M.J., Hammad, M.: Progress in outlier detection techniques: a survey. IEEE Access **7**, 107964–108000 (2019). https://doi.org/10.1109/ACCESS.2019.2932769

397. Wang, L., Lin, Z.Q., Wong, A.: COVID-Net: a tailored deep convolutional neural network design for detection of COVID-19 cases from chest X-ray images. Sci. Rep. **10**, 1–12 (2020). https://doi.org/10.1038/s41598-020-76550-z

398. Wang, L., Wang, Y., Chang, Q.: Feature selection methods for big data bioinformatics: a survey from the search perspective. Methods **111**, 21–31 (2016). https://doi.org/10.1016/j.ymeth.2016.08.014

399. Wang, X., Liu, X., Japkowicz, N., Matwin, S.: Resampling and cost-sensitive methods for imbalanced multi-instance learning. In: 2013 IEEE 13th International Conference on Data Mining Workshops, pp. 808–816 (2013)

400. Widz, S., Ślęzak, D.: Granular attribute selection: a case study of rough set approach to MRI segmentation. In: Proceedings of PReMI 2013, pp. 47–52 (2013)

401. Wieczorkowska, A., Wróblewski, J., Synak, P., Ślęzak, D.: Application of temporal descriptors to musical instrument sound recognition. J. Intell. Inf. Syst. **21**(1), 71–93 (2003)

402. Wójtowicz, A.: Ensemble classification of incomplete data - a non-imputation approach with an application in ovarian tumour diagnosis support. Ph.D. thesis, University in Poznań (2017)

403. Wójtowicz, A., Żywica, P., Stachowiak, A., Dyczkowski, K.: Solving the problem of incomplete data in medical diagnosis via interval modeling. Appl. Soft Comput. **47**, 424–437 (2016). https://doi.org/10.1016/j.asoc.2016.05.029

404. Wróblewski, J.: Ensembles of classifiers based on approximate reducts. Fund. Inform. **47**(3–4), 351–360 (2001)

405. Wróblewski, J., Stawicki, S.: SQL-based KDD with infobright's RDBMS: attributes, reducts, trees. In: Kryszkiewicz, M., Cornelis, C., Ciucci, D., Medina-Moreno, J., Motoda, H., Raś, Z.W. (eds.) RSEISP 2014. LNCS (LNAI), vol. 8537, pp. 28–41. Springer, Cham (2014). https://doi.org/10.1007/978-3-319-08729-0_3

406. Wu, C., Buyya, R., Ramamohanarao, K.: Cloud pricing models: taxonomy, survey, and interdisciplinary challenges. ACM Comput. Surv. **52**(6), 108:1–108:36 (2020). https://doi.org/10.1145/3342103

407. Wu, X., Pellegrini, F.D., Gao, G., Casale, G.: A framework for allocating server time to spot and on-demand services in cloud computing. TOMPECS **4**(4), 20:1–20:31 (2019). https://doi.org/10.1145/3366682
408. Wu, X., et al.: Top 10 algorithms in data mining. Knowl. Inf. Syst. **14**(1), 1–37 (2007). https://doi.org/10.1007/s10115-007-0114-2
409. Wu, Z., Pan, S., Long, G., Jiang, J., Chang, X., Zhang, C.: Connecting the dots: multivariate time series forecasting with graph neural networks. In: Proceedings of the 26th ACM SIGKDD International Conference on Knowledge Discovery & Data Mining, KDD 2020, pp. 753–763. Association for Computing Machinery, New York (2020). https://doi.org/10.1145/3394486.3403118
410. Xie, J., Wu, J., Qian, Q.: Feature selection algorithm based on association rules mining method. In: Proceedings of ICIS 2009, pp. 357–362 (2009)
411. Xioufis, E.S., Spiliopoulou, M., Tsoumakas, G., Vlahavas, I.: Dealing with concept drift and class imbalance in multi-label stream classification. In: Proceedings of the Twenty-Second International Joint Conference on Artificial Intelligence - Volume Two, IJCAI 2011, pp. 1583–1588. AAAI Press (2011). https://doi.org/10.5591/978-1-57735-516-8/IJCAI11-266
412. Spyromitros-Xioufis, E., Tsoumakas, G., Groves, W., Vlahavas, I.: Multi-target regression via input space expansion: treating targets as inputs. Mach. Learn. **104**(1), 55–98 (2016). https://doi.org/10.1007/s10994-016-5546-z
413. Xue, B., Zhang, M., Browne, W.N., Yao, X.: A survey on evolutionary computation approaches to feature selection. IEEE Trans. Evol. Comput. **20**(4), 606–626 (2016). https://doi.org/10.1109/TEVC.2015.2504420
414. Yang, H., Lin, H., Ding, K.: Sliding window denoising k-singular value decomposition and its application on rolling bearing impact fault diagnosis. J. Sound Vib. **421**, 205–219 (2018). https://doi.org/10.1016/j.jsv.2018.01.051
415. Yang, Y., Gopal, S.: Multilabel classification with meta-level features in a learning-to-rank framework. Mach. Learn. **88**(1–2), 47–68 (2012). https://doi.org/10.1007/s10994-011-5270-7
416. Yang, Y., Webb, G.I.: Discretization for Naive-Bayes learning: managing discretization bias and variance. Mach. Learn. **74**(1), 39–74 (2009). https://doi.org/10.1007/s10994-008-5083-5
417. Yang, Y., Webb, G.I., Wu, X.: Discretization methods. In: Maimon, O., Rokach, L. (eds.) The Data Mining and Knowledge Discovery Handbook, pp. 113–130. Springer, Boston (2005). https://doi.org/10.1007/0-387-25465-X_6
418. Yao, Y.: Three-way decision and granular computing. Int. J. Approx. Reason. **103**, 107–123 (2018). https://doi.org/10.1016/j.ijar.2018.09.005
419. Yao, Y., Zhao, Y., Wang, J.: On reduct construction algorithms. Trans. Comput. Sci. **2**, 100–117 (2008)
420. Yao, Y., Zhong, N.: Granular computing. In: Wah, B.W. (ed.) Wiley Encyclopedia of Computer Science and Engineering. Wiley, Hoboken (2008)
421. Yin, J., Zhao, W.: Fault diagnosis network design for vehicle on-board equipments of high-speed railway: a deep learning approach. Eng. Appl. Artif. Intell. **56**, 250–259 (2016). https://doi.org/10.1016/j.engappai.2016.10.002
422. Zadeh, L.A.: Fuzzy sets. Inf. Control **8**(3), 338–353 (1965). https://doi.org/10.1016/S0019-9958(65)90241-X
423. Zadeh, L.A.: Toward a theory of fuzzy information granulation and its centrality in human reasoning and fuzzy logic. Fuzzy Sets Syst. **90**(2), 111–127 (1997)

424. Zadeh, L.A.: From computing with numbers to computing with words—from manipulation of measurements to manipulation of perceptions. In: Azvine, B., Nauck, D.D., Azarmi, N. (eds.) Intelligent Systems and Soft Computing. LNCS (LNAI), vol. 1804, pp. 3–40. Springer, Heidelberg (2000). https://doi.org/10.1007/10720181_1

425. Zagorecki, A.: Prediction of methane outbreaks in coal mines from multivariate time series using random forest. In: Yao, Y., Hu, Q., Yu, H., Grzymala-Busse, J.W. (eds.) RSFDGrC 2015. LNCS (LNAI), vol. 9437, pp. 494–500. Springer, Cham (2015). https://doi.org/10.1007/978-3-319-25783-9_44

426. Zagorecki, A.: A versatile approach to classification of multivariate time series data. In: Ganzha, M., Maciaszek, L.A., Paprzycki, M. (eds.) 2015 Federated Conference on Computer Science and Information Systems, FedCSIS 2015, Lódz, Poland, 13–16 September 2015, pp. 407–410. IEEE (2015)

427. Zdravevski, E., Lameski, P., Dimitrievski, A., Grzegorowski, M., Apanowicz, C.: Cluster-size optimization within a cloud-based ETL framework for Big Data. In: 2019 IEEE International Conference on Big Data (Big Data), Los Angeles, CA, USA, 9–12 December 2019, pp. 3754–3763. IEEE (2019). https://doi.org/10.1109/BigData47090.2019.9006547

428. Zdravevski, E., Lameski, P., Kulakov, A.: Automatic feature engineering for prediction of dangerous seismic activities in coal mines. In: Ganzha, M., Maciaszek, L., Paprzycki, M. (eds.) Proceedings of the 2016 Federated Conference on Computer Science and Information Systems. Annals of Computer Science and Information Systems, vol. 8, pp. 245–248. IEEE (2016). https://doi.org/10.15439/2016F152

429. Zdravevski, E., Lameski, P., Mingov, R., Kulakov, A., Gjorgjevikj, D.: Robust histogram-based feature engineering of time series data. In: Proceedings of Fed-CSIS 2015, pp. 381–388 (2015)

430. Zdravevski, E., et al.: Improving activity recognition accuracy in ambient-assisted living systems by automated feature engineering. IEEE Access 5, 5262–5280 (2017). https://doi.org/10.1109/ACCESS.2017.2684913

431. Zhang, S., Zhang, C., Yang, Q.: Data preparation for data mining. Appl. Artif. Intell. 17(5–6), 375–381 (2003). https://doi.org/10.1080/713827180

432. Zhang, X., Qian, B., Cao, S., Li, Y., Chen, H., Zheng, Y., Davidson, I.: Inprem: an interpretable and trustworthy predictive model for healthcare. In: Proceedings of the 26th ACM SIGKDD International Conference on Knowledge Discovery & Data Mining, KDD 2020, pp. 450–460. Association for Computing Machinery, New York (2020). https://doi.org/10.1145/3394486.3403087

433. Zhang, Y., Miao, D., Pedrycz, W., Zhao, T., Xu, J., Yu, Y.: Granular structure-based incremental updating for multi-label classification. Knowl. Based Syst. 189, 105066 (2020). https://doi.org/10.1016/j.knosys.2019.105066

434. Zhao, X.R., Yao, Y.: Three-way fuzzy partitions defined by shadowed sets. Inf. Sci. 497, 23–37 (2019). https://doi.org/10.1016/j.ins.2019.05.022

435. Zhao, Y., Udell, M.: Missing value imputation for mixed data via gaussian copula. In: Proceedings of the 26th ACM SIGKDD International Conference on Knowledge Discovery & Data Mining, KDD 2020, pp. 636–646. Association for Computing Machinery, New York (2020). https://doi.org/10.1145/3394486.3403106

436. Zhao, Z., Zhang, R., Cox, J., Duling, D., Sarle, W.: Massively parallel feature selection: an approach based on variance preservation. Mach. Learn. 92(1), 195–220 (2013)

437. Zheng, W., Zhu, X., Wen, G., Zhu, Y., Yu, H., Gan, J.: Unsupervised feature selection by self-paced learning regularization. Pattern Recogn. Lett. **132**, 4–11 (2020). https://doi.org/10.1016/j.patrec.2018.06.029
438. Zhu, P., Zhu, W., Hu, Q., Zhang, C., Zuo, W.: Subspace clustering guided unsupervised feature selection. Pattern Recogn. **66**, 364–374 (2017). https://doi.org/10.1016/j.patcog.2017.01.016
439. Zong, W., Chow, Y., Susilo, W.: Interactive three-dimensional visualization of network intrusion detection data for machine learning. Future Gener. Comput. Syst. **102**, 292–306 (2020). https://doi.org/10.1016/j.future.2019.07.045

A Study of Algebraic Structures and Logics Based on Categories of Rough Sets

Anuj Kumar More[(✉)] [iD]

Department of Mathematics and Statistics, Indian Institute of Technology Kanpur,
Kanpur 208016, UP, India
more39@gmail.com

Abstract. The theory of rough sets has been studied extensively, both from foundation and application points of view, since its introduction by Pawlak in 1982. On the foundations side, a substantial part of work on rough set theory involves the study of its algebraic aspects and logics. The present work is in this direction, initiated through the study of categories of rough sets.

Starting from two categories RSC and $ROUGH$ of rough sets, it is shown that they are equivalent. Moreover, RSC, and thus $ROUGH$, are found to be a quasitopos, a structure slightly weaker than topos. The construction is then lifted to a more general set-up to give the category RSC($\mathscr{C}$) with an arbitrary non-degenerate topos $\mathscr{C}$ serving as a 'base', just as sets constitute a base for defining rough sets.

The category-theoretic study gives rise to two directions of work. In one direction of work, a particular example of RSC($\mathscr{C}$) when $\mathscr{C}$ is the topos of monoid actions on sets is considered. It yields the monoid actions on rough sets and that of transformation semigroups (ts) for rough sets, leading to decomposition results. A semiautomaton for rough sets is also defined.

In the other direction, we incorporate Iwinski's notion of 'relative rough complementation' in the internal algebra of the quasitopos RSC($\mathscr{C}$). This results in the introduction of two new classes of algebraic structures with two negations, namely *contrapositionally complemented pseudo-Boolean algebra* (*ccpBa*) and *contrapositionally $\vee$ complemented pseudo-Boolean algebra* (*c$\vee$cpBa*). Examples of *ccpBa*s and *c$\vee$cpBa*s are developed, comparison with existing algebras is done and representation theorems are established.

The logics ILM and ILM-$\vee$ corresponding to *ccpBa*s and *c$\vee$cpBa*s respectively are defined, and different relational semantics are obtained. It is shown that ILM is a proper extension of a variant JP$'$ of Peirce's logic, defined by Segerberg in 1968. The inter-relationship between relational semantics and the algebraic semantics of ILM and ILM-$\vee$ are investigated. Lastly, in the line of Dunn's study of logics, the two negations are expressed without the help of the connective of implication, and the resulting logical and algebraic structures are also studied.

© Springer-Verlag GmbH Germany, part of Springer Nature 2022
J. F. Peters et al. (Eds.): TRS XXIII, LNCS 13610, pp. 288–507, 2022.
https://doi.org/10.1007/978-3-662-66544-2_9

Keywords: Rough sets · Quasitopos · Intuitionistic logic · Minimal logic · Relational semantics · Compatibility frames

1 Introduction

Rough sets were first introduced by Pawlak in 1982 to deal with the concept of vagueness and incomplete information, and a resulting indiscernibility among objects in the universe of discourse. Incompleteness or vagueness could arise due to some missing values of attributes of the objects in the universe. The theory of rough sets has been studied from both foundation and application points of view. It has been used in various fields of computer science like data mining, medical data analysis and automata theory. Algebraic and logical properties of rough sets have interested researchers since the inception of the theory. Pawlak himself studied basic operations of rough sets such as *intersection* and *union*. Iwiński's work [48] may be considered as the beginning of algebraic studies on rough sets; he used the term *rough algebras* for the first time. Since then, based on different definitions and properties of rough sets, various known algebras have been related to rough sets and new algebras have been proposed, such as MV-algebras, quasi-Boolean, topological quasi-Boolean, Kleene, Stone, double Stone and Nelson algebras (cf. e.g. [7,15,56,73], to name a few). The new structures such as topological quasi-Boolean algebras and pre-rough algebras arose specifically in the context of rough sets [5].

In 1993, Banerjee and Chakraborty started the study of categories of rough sets [4]. The aim was to investigate the nature of inter-translation between the indiscernibility relations and the rough sets. They showed that a category (*ROUGH*) of rough sets does not behave like a (elementary) *topos* - the categorial generalization of the category *SET* of sets. Since then, studying categorial properties of rough sets has been an active research area [6,8,25,26], in directions such as categories of rough sets with multiple relations [60], or connections of rough sets with partially ordered monads [35]. In 2008, Li and Yuan [57] defined a category (*RSC*) of rough sets, based on Iwiński's I-*rough sets* [48], and showed that it forms a *weak topos*. Most recently, a category based on rough sets has been studied in [13].

An elementary topos has an algebraic structure on the class of 'subobjects' of any object, called its *internal* algebra. For example, in *SET*, this class and the associated algebra is just the power set and the Boolean algebra respectively. A structure weaker than elementary topos, which exhibits this property on a subclass of subobjects is (elementary) *quasitopos*. Li and Yuan [57] briefly discuss the algebra obtained through the weak topos property of *RSC*. In this work, we shall explore in detail, the internal algebras of *RSC*, *ROUGH*, related categories and their generalizations.

An interesting example of a topos is the category **M-Set** of monoid actions on sets. Monoid or semigroup actions on sets have been classically studied as transformation semigroups, with applications in automata theory [21,22,40]. The algebra of these transformation semigroups is developed by defining basic constructions such as resets, coverings, (direct and wreath) products and admissible partitions for the structures. Just like the well-known Jordan Holder decomposition for groups, there are two major decomposition results, namely Krohn-Rhodes decomposition and Holonomy decomposition result for transformation semigroups [34,46] in literature. While generalizing the categories of rough sets, we shall encounter *monoid actions on rough sets* leading to development of the theory of transformation semigroups for rough sets.

As mentioned, transformation semigroups have natural connections with automata theory. Having defined transformation semigroups for rough sets, we move our attention to 'rough' automata. Rough sets have been previously connected to automata theory and transformation semigroups [9,79,81,82], with notions of *rough semi-automaton* and *rough transformation semigroups*. The basis of both structures lies in non-deterministic automata. We propose semi-automata for rough sets, and do a comparison with the existing structures.

The study of algebraic structures (with negation) in lattice theory has a vast literature [75]. Some of these that get directly related to our work are Boolean algebras, pseudo-Boolean algebras, contrapositionally complemented lattices and Nelson algebras (cf. [39,75,76,84]). Representation theorems connecting algebras and topological spaces were first given by Stone (cf. [75]). Since then, various representation and duality results have been obtained (cf. [12,16,20,24]). We shall use Priestley and Esakia dualities to give representations for new algebraic structures obtained in this dissertation.

Studies on classes of algebras have led to the study of logics whose semantics *correspond* to the concerned classes of algebras. The validity of a formula in an algebraic class in such cases coincides with the provability of the formula in the logic. Semantics of any logic are just not restricted to algebras. Kripke [54,55] studied relational semantics for modal logics, as well as intuitionistic logic. He defines validity of a formula at a world w in a relational frame (W, R), depending on the worlds related to w. For propositional logics with negation, different relational semantics have been introduced by Došen, Segerberg and others [27, 28,33,44,78,85]. One given by Woodruff [87], called *sub-normal* frames, is used in our work. Došen also studied the inter-translation between frames of two different semantics giving the validity correspondence between two semantics. We shall look at these in the context of the logics that we define. As done for algebras, we shall compare the proposed logics with known ones, namely with Johansson's minimal, Peirce's, Glivenko's, and intuitionistic logics [49,70,71,78].

Negations in rough sets have a special place in the algebraic and logical studies of the theory. We next turn to logics with negations. Studying negation independently, particularly in the absence of other connectives, constitutes an important area of work on logics. Dunn [29–31] and Vakarelov [85] have independently studied logics with negations, in the absence of *implication* [32,33,77,80].

A basic feature of these logics introduced by Dunn is that the logical consequence, in the absence of the implication, is defined as a pair of formulas (ϕ, ψ). This is called a *sequent*, written as $\phi \vdash \psi$. Then negation is taken as a fundamental connective, and properties are introduced in the logics as axioms or rules, thus defining *pre-minimal, minimal, intuitionistic* or other negations. Relational semantics of such logics are, given by 'compatibility' frames, and negation is treated as a modal 'impossibility' operator – which is also done by Došen. We shall study two new logics and their semantics based on Dunn's logical frameworks.

Major directions of research, not mutually exclusive, that we shall be focusing on in this dissertation are as follows:

(1) categorial aspects of rough sets (Sects. 3 and 5),
(2) transformation semigroups in the context of rough sets (Sect. 4),
(3) algebras from rough sets (Sect. 5), and
(4) study of logics corresponding to particular classes of algebras (Sects. 6 and 7).

Section 2 covers the preliminaries required to proceed with our work. Before proceeding any further, let us first formally define rough sets. Consider the pair (U, R), where U is a set (universe of discourse) and R is an equivalence relation on U (parameter for vagueness or indiscernibility). (U, R) is called an (Pawlakian) *approximation space*. For any subset $X \subseteq U$, $\overline{\mathcal{X}}$ denotes the collection of equivalence classes in U properly intersecting X; and $\underline{\mathcal{X}}$ denotes the collection of equivalence classes in U contained in X. Then the pair $(\underline{X}, \overline{X})$, where $\overline{X} = \bigcup \overline{\mathcal{X}}$ (*upper approximation*) and $\underline{X} = \bigcup \underline{\mathcal{X}}$ (*lower approximation*), is called a *rough set*. Observe that $\underline{X} \subseteq \overline{X}$. The set $\overline{X}$, $\underline{X}$ and $\overline{X} \backslash \underline{X}$ are called the *possible, definite* and *boundary* regions of the rough set.

Section 3 starts by studying the relationship between the categories RSC, $ROUGH$, and $\xi\text{-}ROUGH$ of rough sets. In the very first result, we observe that $ROUGH$ and RSC are in fact equivalent, in category-theoretic terminology. Thus studying any one of them is sufficient, as equivalent categories share all category-theoretic properties. Moreover, RSC is a full subcategory of the arrow category $SET^{\rightarrow}$, with objects as inclusions. Since these categories of rough sets have SET as their 'base' category, we see that the categories of rough sets can indeed be lifted to such a general setting using elementary topos. We define a category $RSC(\mathscr{C})$, where $\mathscr{C}$ is an arbitrary non-degenerate elementary topos. Objects of $RSC(\mathscr{C})$ are represented as (A, B, m), where $m : A \to B$ is a monic arrow in $\mathscr{C}$. An arrow in $RSC(\mathscr{C})$ with domain (X_1, X_2, m) and codomain (Y_1, Y_2, m') is a pair (f', f), where $f' : X_1 \to Y_1$ and $f : X_2 \to Y_2$ are arrows in $\mathscr{C}$ such that $f \circ m = m' \circ f'$. RSC is then just a full subcategory equivalent to $RSC(SET)$, the latter a special case of $RSC(\mathscr{C})$, when $\mathscr{C}$ is SET. In this sense, we conclude that $RSC(\mathscr{C})$ is indeed a generalization of RSC.

We observe that $RSC(\mathscr{C})$ forms a *quasitopos*, a structure slightly weaker than topos. This observation assumes significance from the perspective of foundations of rough sets. Just as an elementary topos is an abstraction of the category of sets, the quasitopos $RSC(\mathscr{C})$ becomes an abstraction of the category (RSC or $ROUGH$) of rough sets. That $RSC(\mathscr{C})$ is a quasitopos, is also a consequence of a general result by Johnstone [52]. In this thesis, we explicitly derive all the category-theoretic constructs that make $RSC(\mathscr{C})$ a quasitopos. We also define ξ-RSC and ξ-$RSC(\mathscr{C})$, the categories of rough sets, in which the arrows preserve the *information* in the boundary region. It is shown that ξ-$RSC(\mathscr{C})$ forms a topos if and only if $\mathscr{C}$ is a Boolean topos.

Further, these are not the only categories of rough sets possible. Based on the various definitions of rough sets and also the *information* region required to be preserved, one can define objects and arrows suitably, and give rise to different categories of rough sets.

Next, we look at an example of $RSC(\mathscr{C})$, when $\mathscr{C}$ is not SET. In particular, we focus on the case when $\mathscr{C}$ is the topos **M-Set**, the category of actions of a monoid **M** on sets. The construction of $RSC(\textbf{M-Set})$-objects yields the definition of *monoid actions on rough sets*, which motivates *transformation semigroups in rough sets*. In Sect. 4 we propose and study a theory of transformation semigroups in the context of rough sets, where a rough set is taken as an object of category $ROUGH$, i.e. of the form (U, R, X).

A *transformation semigroup for a rough set* $\mathcal{U} := (U, R, X)$ is a triple $\mathcal{A} := (\mathcal{U}, S)$, where $(\overline{\mathcal{X}}, S)$ is a ts such that $\underline{\mathcal{X}}s \subseteq \underline{\mathcal{X}}$, for all $s \in S$. Alternatively, this can be viewed as a faithful action of S on $\overline{\mathcal{X}}$ such that its restriction is an action of S on $\underline{\mathcal{X}}$. $(\overline{\mathcal{X}}, S)$ is called the *upper ts* and $(\underline{\mathcal{X}}, S/\sim)$ the *lower ts* for $\mathcal{A}$, where $s_1 \sim s_2$ if and only if $qs_1 = qs_2$, for all $q \in \underline{\mathcal{X}}$. Basic constructions like resets, products, coverings, admissible partitions in ts for sets are extended to ts for rough sets. A comparison for each construction has been made with those for classical ts for sets. Further, we present a decomposition theorem for reset ts for rough sets. Some results in decomposition for general ts for rough sets have been obtained for special cases. Continuing the study in this section, we also define an appropriate notion of *semiautomaton for rough sets* corresponding to ts for rough sets, using the standard notion of substructures of an automaton.

In Sect. 5, we study the algebraic properties of (strong) subobjects of an object in RSC and $RSC(\mathscr{C})$. As $RSC(\mathscr{C})$ is a quasitopos, the set $\mathcal{M}((U_1, U_2, m))$ of strong monics of an $RSC(\mathscr{C})$-object (U_1, U_2, m) forms a pseudo-Boolean algebra (Heyting algebra). We derive the operators on the algebra $\mathcal{M}((U_1, U_2, m))$ explicitly. In case of RSC, the algebra of strong subobjects of an RSC-object forms a Boolean algebra. This runs counter to the known notions of algebras from rough sets, because the latter are non-classical in nature. We decide to focus on the negation operation in the Boolean algebra. The relative complement of the subobject (A_1, A_2) of an RSC-object (U_1, U_2) in the algebra is the pair $(U_1 \backslash A_1, U_2 \backslash A_2)$. We note that the possible region $U_2 \backslash A_2$ of the relative complement does not include the boundary region $A_2 \backslash A_1$ which is part of the 'non-definite' region of the subobject. We address this point by considering

Iwiński's *relative rough complementation* of rough sets, by which the relative complement of (A_1, A_2) would be the pair $(U_1 \backslash A_1, U_2 \backslash A_1)$. We adopt this new negation. In $RSC(\mathscr{C})$, the new 'rough' negation $\sim$ on $\mathcal{M}((U_1, U_2, m))$ is defined as

$$\sim : \ \sim (f', f) := (\neg f', \neg(m \circ f')),$$

where (f', f) is a strong monic with codomain (U_1, U_2, m). We study the properties of connectives of the internal algebra in RSC and $RSC(\mathscr{C})$, involving both the old and new negations. As a consequence of this study we obtain two new algebraic structures with *two negations*, as briefly described below. The structures are an amalgamation of contrapositionally complemented lattices and pseudo-Boolean algebras.

An abstract algebra $\mathcal{A} := (A, 1, 0, \cap, \cup, \rightarrow, \neg, \sim)$ is said to be a *contrapositionally complemented pseudo-Boolean algebra* (*ccpBa*) if $(A, 1, 0, \cap, \cup, \rightarrow, \neg)$ forms a pseudo-Boolean algebra and for all $a \in A$, $\sim a = a \rightarrow (\neg\neg \sim 1)$. If, in addition, for all $a \in A$, $a \vee \sim a = 1$, we call $\mathcal{A}$ a *c$\vee$c-pseudo-Boolean algebra* (*c$\vee$cpBa*). Bringing the new negation into the generalized scenario of $RSC(\mathscr{C})$, we define a new 'rough' negation $\sim$ on $\mathcal{M}((U_1, U_2, m))$ as

$$\sim : \ \sim (f', f) := (\neg f', \neg(m \circ f')),$$

where (f', f) is a strong monic with codomain (U_1, U_2, m). For each $RSC(\mathscr{C})$-object (U_1, U_2, m), $\mathcal{A} := (\mathcal{M}((U_1, U_2, m)), (1_{U_1}, 1_{U_2}), (i_{U_1}, i_{U_2}), \cap, \cup, \rightarrow, \neg, \sim)$ is a *ccpBa*. In the special case of RSC, it forms a *c$\vee$cpBa*. We make a detailed study of *ccpBas* and *c$\vee$cpBas* in this section. On abstracting the constructions in the algebra of $RSC(\mathscr{C})$-subobjects, we obtain a class of examples for *ccpBa* and *c$\vee$cpBa*. A comparison is made with known and close structures such as pseudo-Boolean algebras and Nelson algebras. Every *ccpBa* $\mathcal{A} := (A, 1, 0, \cap, \cup, \rightarrow, \neg, \sim)$ is proved to be embeddable into a certain *ccpBa* of all open subsets of a topological space X. Further representation theorems for both the algebraic classes are obtained with respect to topological spaces that are certain restrictions of Esakia spaces, the topological spaces corresponding to pseudo-Boolean algebras.

In Sect. 6, we study in detail, ILM (Intuitionistic Logic with Minimal Negation) and ILM$-\vee$, propositional logics corresponding to the algebraic classes of *ccpBas* and *c$\vee$cpBas* respectively. The language of ILM is given by the scheme

$$p_i \mid \top \mid \bot \mid \alpha \wedge \beta \mid \alpha \vee \beta \mid \alpha \rightarrow \beta \mid \neg\alpha \mid \sim\alpha$$

The axioms and rules of ILM are those of intuitionistic logic (IL) along with the axiom:

$$\sim\alpha \leftrightarrow (\alpha \rightarrow (\neg\neg\sim\top)).$$

We compare ILM with IL and minimal logic (ML), the logic corresponding to contrapositionally complemented lattices. We utilize notions of 'interpretation' as given in [14, 37, 38, 75]. We compare ILM with IL, ML, and Peirce's logic. The axioms and rules of Peirce's logic (JP) are those of ML along with the axiom $((\alpha \to \beta) \to \alpha) \to \alpha$. A special case is Glivenko's logic JP$'$, for which the axioms and rules are those of ML along with the axiom $((\sim\top \to \beta) \to \sim\top) \to \sim\top$. A strict extension of JP$'$ with the propositional constant $\perp$ and axiom $\perp \to \alpha$, is seen to be equivalent with ILM.

The next focus is on finding the Kripke-style 'relational' semantics for ILM and ILM$-\vee$. We obtain two different relational semantics for ILM. We then investigate the inter-relationship between these relational semantics and the algebraic semantics of ILM and ILM-V. It is shown that any $ccpBa$ $(c\vee cpBa)$ $\mathcal{A}$ can be embedded into the $ccpBa$ $(c\vee cpBa)$ formed by the 'canonical' frame $\mathcal{F}$ of $\mathcal{A}$ (the 'complex algebra' of $\mathcal{F}$). Similarly, a sub-normal frame $\mathcal{F}$, satisfying certain conditions, can be embedded into the canonical frame of the complex algebra of $\mathcal{F}$.

In both the logics ILM and ILM$-\vee$, the properties of the two negations depend on other connectives. Specifically recall that in the algebraic counterpart of $ccpBas$, the negation $\sim$ satisfies the property $\sim a = a \to (\neg\neg\sim 1)$. In Sect. 7, we study the negation $\sim$, in the absence of the connective of implication in the language. In our case, we consider bounded *distributive lattice logic* along with an intuitionistic negation $\neg$ as base, and introduce the negation $\sim$ by defining these two axioms.

1. $\sim\alpha \vdash \neg(\alpha \wedge \neg\sim\top)$
2. $\neg(\alpha \wedge \neg\sim\top) \vdash \sim\alpha$

These axioms give rise to two logics K_{im} and $K_{im-\vee}$. Note that the language of these logics has no implication and is given by the scheme

$$p_i \mid \top \mid \perp \mid \alpha \wedge \beta \mid \alpha \vee \beta \mid \neg\alpha \mid \sim\alpha$$

The algebras corresponding to K_{im} and $K_{im-\vee}$ are defined, and it is shown that the class of K_{im}-algebras is not the same as the class of $ccpBas$. More precisely, the reduct of any $ccpBa$ obtained by removing implication $\to$, is a K_{im}-algebra. However, not every K_{im}-algebra can be extended to a $ccpBa$. We also give relational semantics for K_{im} and $K_{im-\vee}$ in the same lines as those obtained for ILM and ILM$-\vee$. We end this section by studying the inter-relationship between the relational semantics and algebraic semantics of K_{im} and $K_{im-\vee}$. Any K_{im}-algebra ($K_{im-\vee}$-algebra) $\mathcal{A}$ can be embedded into the K_{im}-algebra ($K_{im-\vee}$-algebra) formed by the 'canonical' frame $\mathcal{F}$ of $\mathcal{A}$ (the 'complex algebra' of $\mathcal{F}$). Similarly, a compatibility frame $\mathcal{F}$, satisfying certain conditions, can be embedded into the canonical frame of the complex algebra of $\mathcal{F}$.

We conclude the thesis in Sect. 8 by giving a summary of the work and indicating some future directions.

2   Preliminaries

Let us first introduce the notion of rough sets. Consider the pair (U, R), where U is a set (universe of discourse) and R is an equivalence relation on U, called the *indiscernibility relation*. Elements of U inside equivalence classes due to R are regarded as indiscernible, due to lack of complete information about the universe. The pair (U, R) is called an (Pawlakian) *approximation space*. Given an approximation space (U, R), an equivalence class of U under R, denoted by $[x]_R$, is called an *elementary set* of (U, R). The collection of all elementary sets, i.e. the quotient set due to R, is denoted by U/R.

Definition 1 (Definable set).
For an approximation space (U, R), a subset $X \subseteq U$ is called definable *if it is a union of elementary sets of (U, R).*

Note that not all subsets of U are definable. However, we can always 'approximate' a subset X of U from outside and within, with the help of equivalence classes properly intersecting X, and those contained in X. More precisely, we have the following.

Definition 2 (R-lower approximation and R-upper approximation).
For any subset X of U in the approximation space (U, R), the R-lower approximation of X is $\underline{X}_R := \{x \in U \mid [x]_R \subseteq X\}$ and the R-upper approximation of X is $\overline{X}_R := \{x \in U \mid [x]_R \cap X \neq \emptyset\}$. The pair $(\underline{X}_R, \overline{X}_R)$ is called a rough set *in the approximation space (U, R).*
$\overline{X}_R \backslash \underline{X}_R$ is called the boundary *of X and $U \backslash \overline{X}_R$ is called the* negative *region with respect to X (Fig. 1).*

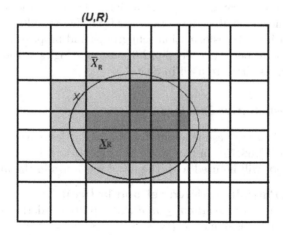

Fig. 1. A rough set

Any rough set $(\underline{X}_R, \overline{X}_R)$ thus divides the universe U into four definable regions, namely 'definite' $\underline{X}_R$ (elements here are definitely inside X), 'possible' $\overline{X}_R$ (elements are possibly within X), 'boundary' $\overline{X}_R \backslash \underline{X}_R$ (elements could be within or outside X), and 'negative' $U \backslash \overline{X}_R$ (elements are definitely outside X) regions. Note that $\underline{X} \subseteq \overline{X}$. The following notation shall be useful in this work [4].

Notation 1. *Let $\overline{\mathcal{X}}_R$ and $\underline{\mathcal{X}}_R$ denote the collections of equivalence classes in U contained in the R-upper approximation and R-lower approximation of set X ($\subseteq U$) respectively. So*

$$\overline{\mathcal{X}}_R := \{[x]_R \in U/R \mid [x]_R \cap X \neq \emptyset\}, \ and$$
$$\underline{\mathcal{X}}_R := \{[x]_R \in U/R \mid [x]_R \subseteq X\}.$$

Note that the R-upper approximation of X, $\overline{X}_R = \bigcup \overline{\mathcal{X}}_R$ and the R-lower approximation of X, $\underline{X}_R = \bigcup \underline{\mathcal{X}}_R$, where union is taken over elements of the equivalence classes. Here, $\underline{X}_R \subseteq \overline{X}_R$.

We can drop the lower suffix R from the notation, if the approximation space is obvious from the context.

There are equivalent definitions of rough sets in literature. Pawlak (1991) called the set X a rough set in the approximation space (U, R), if its boundary is non-empty. More generally, any triple (U, R, X) may be called a rough set [4]. However, the pair $(\underline{X}_R, \overline{X}_R)$ is most commonly taken as a rough set in (U, R). It may be noted that any triple (U, R, X) gives rise, uniquely, to the pair $(\underline{X}, \overline{X})$, and conversely, for any pair $(\underline{X}, \overline{X})$, there is an underlying triple (U, R, X).

This section consists of basic notions and results that have been used in this thesis. There are five sections in the section. The first section introduces basic algebraic notions that shall be required in Sect. 5. Part of this section is also used in Sect. 2.2, which contains the preliminaries for the category-theoretic study of rough sets presented in Sect. 3. Section 2.3 gives a few categories of rough sets in literature. Our work starts from these results. Sections 2.4 and 2.5 cover the basics of transformation semigroups and propositional logics, used in Sects. 4 and 6 respectively. Section 2.6 describes logics without implication and their semantics that are used in Sect. 7.

2.1   Some Distributive Lattices

The basic definitions and results about distributive lattices covered in this section can be found in [12,68,75]. In this work, the symbols $\forall$, $\exists$, $\Rightarrow$, $\Leftrightarrow$, & (*and*), *or*, *if ... then* and *not* will be used with the usual meanings in the metalanguage.

Definition 3 (Partially ordered set and lattice).
Let A be a set. A binary relation $\leq$ is called a partial ordering *on A if $\leq$ is a reflexive ($x \leq x$), antisymmetric (($x \leq y$ & $y \leq x$) $\Rightarrow x = y$) and transitive relation (($x \leq y$ & $y \leq z$) $\Rightarrow x \leq z$) in A.*
The pair $(A, \leq)$ is called a partially ordered set *(abbreviation - poset).*
A lattice *is taken as a poset such that every pair a, b of elements in A has a*

greatest lower bound $(a \wedge b)$ *and a lowest upper bound* $(a \vee b)$ *with respect to the underlying partial order. In the denotation of a lattice, the partial order will not be included, i.e. it will be a tuple of the form* $(A, \vee, \wedge)$.

Definition 4 (Embedding of posets).
Given two posets $(A, \leq)$ *and* $(A', \leq')$. *The mapping* $\phi : A \to A'$ *is an embedding if it satisfies the following: for all* $a, b \in A$, $a \leq b \Rightarrow \phi(a) \leq' \phi(b)$ *(Order-preserving), and* $\phi(a) \leq' \phi(b) \Rightarrow a \leq b$ *(Order-reflecting).*
If an embedding ϕ *is onto, then* ϕ *is called an* order-isomorphism.

Consider a lattice $(A, \vee, \wedge)$ and a binary operator 'implication' $\to$ on A.

Definition 5 (Relatively pseudo-complemented lattice).
The lattice $(A, \vee, \wedge, \to)$ *is said to be a* relatively pseudo-complemented *(abbreviation - rpc) lattice if* $a \wedge x \leq b \Leftrightarrow x \leq a \to b$ *holds for any* $a, b, x \in A$.
The element $a \to b$ *is called the* pseudo-complement of a relative to b.

Some properties of *rpc* lattices which shall be used in our work are as follows [43, 75].

Proposition 1. [75]

(A) *Any rpc lattice is distributive.*
(B) *In any rpc lattice* $(A, \vee, \wedge, \to)$, *for all* $a, b, c \in A$,

(1) $b \leq a \to a$.
(2) $a \to (b \to c) = b \to (a \to c) = (a \wedge b) \to c$.
(3) $a \wedge c \leq b \Leftrightarrow c \leq a \to b$.
(4) $1 \to a = a$.
(5) $(a \to b) \wedge (a \to c) = (a \to (b \wedge c))$.
(6) $(a \to c) \wedge (b \to c) = (a \vee b) \to c$.
(7) $a \to b = 1 \Leftrightarrow a \leq b$.
(8) $b \leq c \Rightarrow a \to b \leq a \to c$.
(9) $a \wedge (a \to b) = a \wedge b \leq b$.
(10) $b \wedge (a \to b) = b$.
(11) $(a \to b) \leq (a \wedge c) \to (b \wedge c)$.
(12) $a \to (b \to c) \leq (a \to b) \to (a \to c)$.

Property (1) in the above proposition tells that any *rpc* lattice $(A, \vee, \wedge, \to)$ has a top element $a \to a$ (where $a \in A$), denoted by 1. Thus, hereafter any *rpc* lattice will be represented as $(A, 1, \vee, \wedge, \to)$. Let us now see a few sub-classes of *rpc* lattices involving negation that will be fundamental in our work.

Definition 6 (Contrapositionally Complemented Lattices).

1. [75] Consider an *rpc* lattice $(A, 1, \vee, \wedge, \to)$ with a unary 'negation' operator $\neg$ satisfying the contraposition law: $x \to \neg y = y \to \neg x$ for all $x, y \in A$. Then $\mathcal{A} := (A, 1, \vee, \wedge, \to, \neg)$ is called a *contrapositionally complemented* (abbreviation - cc) lattice.
   Equivalently, $(A, 1, \vee, \wedge, \to, \neg)$ is a cc lattice if and only if $(A, 1, \vee, \wedge, \to)$ is an *rpc* lattice, and for any $a \in A$, $\neg a = a \to \neg 1$.

2. [68] A *contrapositionally* ∨ *complemented* (abbreviation - *c∨c*) lattice is a *cc* lattice $(A, 1, \vee, \wedge, \rightarrow, \neg)$ such that $x \vee \neg x = 1$ holds for all $x \in A$.

Note that none of the above lattices necessarily has a bottom (i.e. least) element 0 in the lattice structure. If a *cc* (or *c∨c*) lattice $(A, 1, 0, \vee, \wedge, \rightarrow, \neg)$ has a bottom element 0, then we shall call it *cc* (or *c∨c*) *lattice with* 0. An important sub-class of *cc* lattices with 0 is one in which the lattices satisfy the condition $\neg 1 = 0$.

Definition 7 (Pseudo-Boolean and Boolean algebras [75]).

1. A *pseudo-Boolean algebra* (abbreviation - *pBa*) $(A, 1, 0, \vee, \wedge, \rightarrow, \neg)$ *is a lattice such that the reduct* $(A, 1, \vee, \wedge, \rightarrow)$ *is an rpc lattice with the bottom element* 0, *and the negation* ¬ *satisfies* $\neg a := a \rightarrow 0$ *for all* $a \in A$.
   Equivalently, $(A, 1, 0, \vee, \wedge, \rightarrow, \neg)$ *is a pBa if and only if* $(A, 1, 0, \vee, \wedge, \rightarrow, \neg)$ *is a cc lattice with* 0, *and for any* $a \in A$, $\neg a = a \rightarrow 0$.
2. A *Boolean algebra is a pBa* $(A, 1, 0, \vee, \wedge, \rightarrow, \neg)$ *such that the negation* ¬ *satisfies the* law of excluded middle*:* $a \vee \neg a = 1$ *for all* $a \in A$.

A *pBa* is also called a *Heyting* algebra, and a Boolean algebra is also called a *classical algebra*. Let us now see how to compare two algebraic structures. Given any algebra $\mathcal{A}$, there is always a set A *associated* with $\mathcal{A}$, on which the n-ary operators are defined. Note that constants are nothing but operators of zero order. We say two algebras $\mathcal{A}$ and $\mathcal{B}$ are of the *same type* if for each n-ary operator $\circ_{\mathcal{A}}$ in $\mathcal{A}$, there exists an n-ary operator $\circ_{\mathcal{B}}$ in $\mathcal{B}$ and vice-versa $(n \in \mathbb{N})$. The operator $\circ_{\mathcal{B}}$ is said to *correspond* to the operator $\circ_{\mathcal{A}}$.

Definition 8 (Homomorphism and isomorphism).
Consider two algebras $\mathcal{A}$ and $\mathcal{B}$ of the same type. We say that $\mathcal{A}$ is homomorphic *to $\mathcal{B}$ if there exists a map* $\Phi : A \rightarrow B$, *where A and B are the sets associated with the algebras $\mathcal{A}$ and $\mathcal{B}$ respectively, such that for every n-ary operator* $\circ_{\mathcal{A}}$ *in $\mathcal{A}$,* $a_i \in A$ *and* $b_i \in B$, $1 \le i \le n$ $(n \in \mathbb{N})$, $\Phi(\circ_{\mathcal{A}}(a_1, a_1, \ldots, a_n)) = \circ_{\mathcal{B}}(\Phi(a_1), \Phi(a_2), \ldots, \Phi(a_n))$.
If Φ is injective (surjective), then $\mathcal{A}$ is monomorphic (epimorphic) *to $\mathcal{B}$.*
If Φ is bijective, then $\mathcal{A}$ is isomorphic *to $\mathcal{B}$.*

Let us now see some properties and examples of *cc* lattices and *pBa* that we shall use in the work.

Proposition 2. [75] *In any cc lattice* $\mathcal{A} := (A, 1, \vee, \wedge, \rightarrow, \neg)$, *the following hold for all* $a, b \in A$.

(1) $a \le b \Rightarrow \neg b \le \neg a$.
(2) $a \le \neg \neg a$.
(3) $\neg a = \neg \neg \neg a$.
(4) $\neg(a \vee b) = \neg a \wedge \neg b$.
(5) $a \wedge b \le c \Rightarrow a \wedge \neg c \le \neg b$.

In any pBa $\mathcal{A} := (A, 1, 0, \vee, \wedge, \rightarrow, \neg)$, *in addition to the above conditions, the following also hold for all* $a, b \in A$.

(6) $a \wedge \neg a = 0$.
(7) $\neg 0 = 1$.
(8) $\neg 1 = 0$.
(9) $\neg a \leq \neg a \vee b \leq a \rightarrow b$.
(10) $(a \rightarrow b) \wedge (a \rightarrow \neg b) \leq \neg a$.
(11) $\neg(a \vee \neg a) = 0$.

If $\mathcal{A}$ is a Boolean algebra, then the following are true in addition to the above.

(12) $\neg a \vee b = a \rightarrow b$,
(13) $\neg(a \wedge b) = \neg a \vee \neg b$.
(14) $\neg a \vee a = 1$.
(15) $\neg b \rightarrow \neg a \leq a \rightarrow b$.

Example 1 (Example of rpc lattice, cc lattice and pBa). Consider the following 6-element bounded lattice $\mathcal{H}_6 := (H_6, 1, 0, \vee, \wedge, \rightarrow)$, for which the Hasse diagram is given by Fig. 2. Since $\mathcal{H}_6$ does not have any occurrence of the lattices - M_3 (Fig. 3a) or N_5 (Fig. 3b) as its sublattice, $(H_6, \vee, \wedge)$ is distributive [24]. Moreover, any finite distributive lattice is an *rpc* lattice and so $\mathcal{H}_6$ is an *rpc* lattice. Table 1 gives the values for the implication '$\rightarrow$' in $\mathcal{H}_6$.

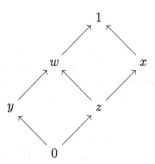

Fig. 2. $\mathcal{H}_6$ - a 6-element *rpc* lattice.

If we define negation $\neg$ on H_6 as $\neg a := a \rightarrow \neg 1$ for all $a \in H_6$, where $\neg 1 = c_0$ for a fixed $c_0 \in H_6$, then the resulting algebra $(H_6, 1, 0, \wedge, \vee, \rightarrow, \neg)$ is a *cc* lattice with 0. In particular, when $c_0 = 0$ and $\neg a = a \rightarrow 0$ for all $a \in H_6$, then $(H_6, 1, 0, \wedge, \vee, \rightarrow, \neg)$ is a *pBa*.

Example 2 (Example of Boolean algebras).

(1) Consider the 4-element lattice $\mathcal{B}_4 := (B_4, 1, 0, \wedge, \vee, \rightarrow, \neg)$, for which the Hasse diagram is given by Fig. 4. Clearly, $(B_4, \vee, \wedge)$ is a finite distributive lattice. The negation and implication operators in H_4 are defined as follows: $\neg 0 := 1$, $\neg 1 := 0$, $\neg a := b$, $\neg b := a$, and for all $a, b \in B_4$, $a \rightarrow b := \neg a \vee b$. The algebra $\mathcal{B}_4$ is a Boolean algebra.

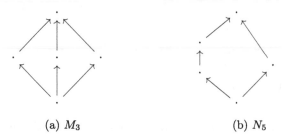

(a) M_3 (b) N_5

Fig. 3. Non-distributive lattices

Table 1. Implication in $\mathcal{H}_6$

$\rightarrow$	0	y	z	w	x	1
0	1	1	1	1	1	1
y	x	1	x	1	x	1
z	y	y	1	1	1	1
w	0	y	x	1	x	1
x	y	y	w	w	1	1
1	0	y	z	w	x	1

(2) For any set A, $(\mathcal{P}(A), A, \emptyset, \cap, \cup, \rightarrow, \neg)$ forms a Boolean algebra, called *Power set Boolean algebra*. Here, $\cup$ and $\cap$ are the usual set intersections and unions; $\neg$ is the relative complement with respect to A, that is, for $B \subseteq A$, $\neg B = A \backslash B$; and $\rightarrow$ is defined for $B, C \subseteq A$ as $B \rightarrow C := \neg B \cup C = (A \backslash B) \cup C$.

An important direction in the study of a class of algebras is establishing representation results for the given class. One of the first representations was given for the class of Boolean algebras by M. H. Stone in 1936. He gave the algebra-topology duality by showing that any Boolean algebra is isomorphic to an algebra of sets, more specifically to the Boolean algebra formed by the clopen (closed and open) subsets of a Stone space (cf. [75]). Let us first see the representation theorems for the class of *rpc* lattices, *cc* lattices, and *pBa* [75]. For the basic notions of topology required here, we refer to [66,75].

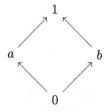

Fig. 4. An example of a Boolean algebra B_4

Definition 9 (Set lattices).
Consider a topological space X with an interior operator I. Let $G(X)$ be the class of open subsets of X. Then $\mathcal{G}(X) := (G(X), X, \cup, \cap, \rightarrow)$ forms an rpc lattice, where $\rightarrow$ is defined as follows:

$$Y \rightarrow Z := I((X \backslash Y) \cup Z) \text{ for all } Y, Z \in G(X).$$

(1) $\mathcal{G}(X)$ *is called a* relatively pseudo complemented set lattice *(abbreviation - rpc set lattice) of all open subsets of the topological space X.*
(2) *If we define a unary operator on $G(X)$ as follows: $\neg Z := Z \rightarrow \neg X$, where $\neg X := Y_0$ for some $Y_0 \in G(X)$, then $(G(X), X, \cup, \cap, \rightarrow, \neg)$ forms a cc lattice, and is called* contrapositionally complemented set lattice *(abbreviation - cc set lattice).*
(3) *If $Y_0 = \emptyset$ in (2), then $(G(X), X, \emptyset, \cup, \cap, \rightarrow, \neg)$ forms a pBa, and is called a* pseudo-field *of all open subsets of the topological space X.*

Proposition 3. [75] *For every pBa (cc lattice) $\mathcal{A}$, there exists a monomorphism h from $\mathcal{A}$ into the pseudo-field (cc set lattice) of all open subsets of some topological space.*

Another representation theorem was given by H. Priestley (1970) and L. Esakia (1974) for the class of distributive lattices and pBa. This was obtained using a particular kind of partially ordered topological space - Priestley space and Esakia space (cf. [16,24]), called dual topological space with respect to distributive lattices and pBa. We define these topological spaces; let us first give the concepts of upsets in any poset, filters in a distributive lattice and some basic results about them [75].

Definition 10 (Upsets and downsets).
Consider a poset $(X, \leq)$.

(1) *A subset $Y \subseteq X$ is called an* upset *if for all $x \in Y$ and $y \in X$, $x \leq y$ implies $y \in Y$.*
(2) *A subset $Y \subseteq X$ is called a* downset *if for all $x \in Y$ and $y \in X$, $y \leq x$ implies $y \in Y$.*
(3) *For a subset $Z \subseteq X$,*

$$\uparrow Z := \{x \in X \mid \text{ there exists } y \in Z \text{ satisfying } y \leq x\}, \text{ and}$$
$$\downarrow Z := \{x \in X \mid \text{ there exists } y \in Z \text{ satisfying } x \leq y\}$$

are the upset and downset *generated by Z respectively.*

Definition 11 (Filters).
Consider a distributive lattice $\mathcal{A} := (A, \vee, \wedge)$.

(1) *A subset $F \subseteq A$ is called a* filter *in $\mathcal{A}$ if for all $a, b \in A$, $a \wedge b \in F$ if and only if $a \in F$ and $b \in F$.*

(2) *A filter F in A is* proper *if $F \neq A$.*

(3) *A proper filter F in A is* prime *if for all $a, b \in A$, $a \vee b \in F$ if and only if $a \in F$ or $b \in F$.*

(4) *A proper filter F in A is* maximal *if there does not exist any proper filter F' such that $F \subsetneq F'$ i.e. F is a proper subset of F'.*

(5) *For a non-empty subset $B \subseteq A$, the filter generated by B (notation - $\langle B \rangle$) is the set of all elements $a \in A$ such that there exist $a_1, \ldots, a_n \in B$ satisfying $a_1 \wedge \ldots \wedge a_n \leq a$.*

Lemma 1. *The following hold for any filter F in a distributive lattice $\mathcal{A} := (A, \vee, \wedge)$.*

(1) *F is an upset in A.*

(2) *If F is maximal then F is prime.*

(3) *For all $a \in A$, $b \in \langle F \cup \{a\} \rangle$ if and only if there exists $b' \in F$ such that $b' \wedge a \leq b$.*

(4) *The property in (3) can be extended: for all filters F_1, F_2 in $\mathcal{A}$, $x \in \langle F_1 \cup F_2 \rangle$ if and only if there are $y \in F_1$ and $z \in F_2$ such that $y \wedge z \leq x$.*

(5) *For all $a, b \in A$ such that $a \nleq b$, there exists a prime filter Q in $\mathcal{A}$ such that $a \in Q$ and $b \notin Q$.*

(6) *The property in (5) can be extended: let F be proper. Suppose $\Gamma \subseteq A$ is such that $\Gamma \cap F = \emptyset$, and Γ is $\vee$-closed, i.e. for all $a, b \in \Gamma$, $a \vee b \in \Gamma$. Then there exists a prime filter P in $\mathcal{A}$ such that $F \subseteq P$ and $\Gamma \cap P = \emptyset$.*

We shall also be making use of

Lemma 2 (Zorn's Lemma). *Suppose a partially ordered set $(X, \leq)$ (X non-empty) has the property that 'every chain in X has an upper bound in X', i.e. for every $\{x_i\}_{i \in \mathbb{N}} \subseteq X$ such that $x_1 \leq x_2 \leq \ldots$, there exists $x \in X$ satisfying $x_i \leq x$ for all $i \in \mathbb{N}$.*
Then the set X contains a maximal element $m \in X$, i.e. if there exists $m' \in X$ such that $m \leq m'$ then $m = m'$.

Definition 12 (Priestley and Esakia spaces (cf. [16,24])).

A Priestley space *is a tuple $(X, \tau, \leq)$ where $(X, \leq)$ is a poset and τ ($\neq \emptyset$) is a compact topological space on X satisfying the following property: for every $x, y \in X$, if $x \nleq y$, then there exists a clopen upset Y of X such that $x \in Y$ and $y \notin Y$.*

Additionally, if a Priestley space $(X, \tau, \leq)$ satisfies the property that for any $U \subseteq X$, U is clopen implies $\downarrow U$ is clopen, then it is called an Esakia space.

Let $CpUp(X)$ be the set of clopen upsets of τ in a Priestley space $(X, \tau, \leq)$. Then $\mathcal{D}(X) := (CpUp(X), X, \emptyset, \cup, \cap)$ forms a bounded distributive lattice. Define the operator $\rightarrow$ on $CpUp(X)$, for any $U, V \in CpUp(X)$:

$$U \rightarrow V := X \backslash \downarrow (U \backslash V). \tag{2.1}$$

The operator $\rightarrow$ is closed in $CpUp(X)$ if and only if $(X, \tau, \leq)$ is an Esakia space (cf. [16]). In this case, $\mathcal{D}(X) := (CpUp(X), X, \emptyset, \cup, \cap, \rightarrow, \neg)$ forms a *pBa*, where

$\neg U := U \to \emptyset$.

Now consider a pBa $\mathcal{A} := (A, 1, 0, \vee, \wedge, \to, \neg)$, and let X_A denote the set of prime filters in $\mathcal{A}$. Define the topology τ_A on X_A generated by the subbasis

$$\{\sigma(a) \mid \sigma(a) \subseteq X_A \ \& \ a \in A\} \cup \{X_A \backslash \sigma(a) \mid \sigma(a) \subseteq X_A \ \& \ a \in A\},$$

where $\sigma(a)$ is the set of prime filters containing $a \in A$. Then $(X_A, \tau_A, \subseteq)$ forms an Esakia space. $\sigma(a)$ and $X_A \backslash \sigma(a)$ are the only clopen upsets in τ_A. These definitions lead to the representation theorem for pBa's.

Theorem 1 (Duality result for pBa). [12]

(1) *The pBa $\mathcal{A}$ is isomorphic to $\mathcal{D}(X_A)$, through the map $\Phi : A \to CpUp(X_A)$ defined as $\Phi(a) := \sigma(a)$, for any $a \in A$.*

(2) *Given any Esakia space $(X, \tau, \leq)$, τ is homeomorphic to $\tau_{CpUp(X)}$ and the poset $(X, \leq)$ is order-isomorphic to the poset $(X_{CpUp(X)}, \subseteq)$.*

2.2 Basic Category Theory

This section comprises the basic definitions in category theory [3,43,88] that we shall require, en route to the notions of elementary topos and quasitopos.

Definition 13 (Category and subcategory).

(1). *A category $\mathscr{C}$ consists of two classes - the class $Obj(\mathscr{C})$ of objects and the class $Arr(\mathscr{C})$ of arrows such that for each arrow $f \in Arr(\mathscr{C})$, there are two objects in $Obj(\mathscr{C})$ - domain $A := dom(f)$ of f and codomain $B := cod(f)$ of f (notation - $f : A \to B$) satisfying the following properties.*

(a) *Given two arrows $f : A \to B$ and $g : B \to C$ in $Arr(\mathscr{C})$, there exists an arrow $g \circ f : A \to C$ in $Arr(\mathscr{C})$.*

(b) *For each $A \in Obj(\mathscr{C})$, there exists an identity arrow $1_A : A \to A$ in $Arr(\mathscr{C})$.*

(c) *For arrows $f : A \to B$, $g : B \to C$ and $h : C \to D$ in $Arr(\mathscr{C})$, $h \circ (g \circ f) = (h \circ g) \circ f$.*

(d) *For arrow $f : A \to B$ in $Arr(\mathscr{C})$, $f \circ 1_A = f = 1_B \circ f$.*

The collection of all arrows between any two objects A and B in $Obj(\mathscr{C})$ is denoted as $Hom_{\mathscr{C}}(A, B)$.

(2). *By a subcategory $\mathscr{C}'$ of the category $\mathscr{C}$, we shall mean that $\mathscr{C}'$ is itself a category such that an object in $Obj(\mathscr{C}')$ is an object in $Obj(\mathscr{C})$ and an arrow in $Arr(\mathscr{C}')$ is an arrow in $Obj(\mathscr{C})$.*

(3). *A subcategory $\mathscr{C}'$ of $\mathscr{C}$ is called full if for any two objects A and B in $Obj(\mathscr{C}')$, $Hom_{\mathscr{C}'}(A, B) = Hom_{\mathscr{C}}(A, B)$.*

Arrows are also referred to as *morphisms*. Any object $A \in Obj(\mathscr{C})$ is called a $\mathscr{C}$-object (or any object in $\mathscr{C}$) and any arrow $f \in Arr(\mathscr{C})$ is called a $\mathscr{C}$-arrow (or an arrow in $\mathscr{C}$). If the category is clear from the context, we shall drop the prefix $\mathscr{C}$ and just say A is an object and f is an arrow. For our purpose, we shall restrict our study to those categories for which $Hom_{\mathscr{C}}(A, B)$ is a set for all $\mathscr{C}$-objects A and B. Such categories are called '*locally small*'. Let us see some examples of categories.

Example 3.

(1) Category *SET* of sets - with objects as sets, and arrows as set functions.
(2) Category of a fixed poset (poset) $(P, \leq)$ - with objects as elements of P, and for any $p, q \in P$ such that $p \leq q$, there is a unique arrow $p \to q$; and these are the only arrows in $(P, \leq)$.
(3) The category **M-Set**, where **M** is a monoid. Let $*$ be the binary operator, and e be the identity element in the monoid **M**. An **M-Set**-object is a pair (X, λ) such that $\lambda : \mathbf{M} \times X \to X$ is a function satisfying
    (a) $\lambda(e, x) = x$ for all $x \in X$, and
    (b) $\lambda(m, \lambda(n, x)) = \lambda(m * n, x)$ for all $m, n \in \mathbf{M}$ and $x \in X$.
    An **M-Set**-arrow $f : (X, \lambda) \to (Y, \mu)$ is an action-preserving function $f : X \to Y$, i.e. $\mu(m, f(x)) = f(\lambda(m, x))$ for all $m \in \mathbf{M}$ and $x \in X$. Composition of arrows is functional composition.
(4) The category $\mathscr{C}^{\to}$, where $\mathscr{C}$ is itself a category. $\mathscr{C}^{\to}$-Objects are arrows in $\mathscr{C}$ and, $\mathscr{C}^{\to}$-arrows with domain f and codomain g are the pairs (u, v) such that $g \circ u = v \circ f$, whenever the compositions $(g \circ u)$ and $(v \circ f)$ are defined for $u, v, f, g \in Arr(\mathscr{C})$.
(5) The category $\mathscr{C} \times \mathscr{D}$, where $\mathscr{C}$ and $\mathscr{D}$ are itself categories. $(\mathscr{C} \times \mathscr{D})$-objects are taken to be pairs (X_1, X_2), where X_1 and X_2 are $\mathscr{C}$-object and $\mathscr{D}$-object respectively; and $(\mathscr{C} \times \mathscr{D})$-arrows are pairs $(h, k) : (X_1, X_2) \to (Y_1, Y_2)$, where $h : X_1 \to Y_1$ and $k : X_2 \to Y_2$ are $\mathscr{C}$-arrow and $\mathscr{D}$-arrow respectively. If $\mathscr{C}$ and $\mathscr{D}$ are the same category, then $\mathscr{C} \times \mathscr{D}$ is denoted by $\mathscr{C}^2$.

Definition 14 (Monics, epics, isos and strong monics).
Consider a category $\mathscr{C}$.

(1) *A $\mathscr{C}$-arrow $f : A \to B$ is called a* monic *(or monomorphism) if given any $\mathscr{C}$-arrows $g, h : C \to A$, $f \circ g = f \circ h$ implies $g = h$.*
(2) *A $\mathscr{C}$-arrow $f : A \to B$ is called an* epic *(or epimorphism) if given any $\mathscr{C}$-arrows $g, h : B \to D$, $g \circ f = h \circ f$ implies $g = h$.*
(3) *A $\mathscr{C}$-arrow $f : A \to B$ is called an* iso*(or isomorphism) if there exists a $\mathscr{C}$-arrow $g : B \to A$ such that $g \circ f = 1_A$ and $f \circ g = 1_B$. If there exists an iso f between two objects A and B, then we say that A is isomorphic to B, denoted as $A \cong B$.*
    Any iso arrow is always monic and epic.
(4) *A monic arrow m is called* strong monic *if every commutative square $m \circ u = v \circ e$ with e epic has a 'diagonal' in $\mathscr{C}$, that is, there is a $\mathscr{C}$-arrow t such that $u = t \circ e$ and $v = m \circ t$ in $\mathscr{C}$ (Fig. 5).*

Note that in the category *SET*, the monics, epics and isos correspond to the injective, surjective and bijective functions respectively. The strong monics in *SET* are the same as the monics. In a poset category $(P, \leq)$, every arrow is monic as well as epic. However, an arrow is strong monic in $(P, \leq)$ if and only if it is an identity arrow.

The following definition gives a 'mapping' between two categories.

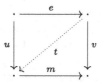

Fig. 5. Strong monics

Definition 15 (Functor).

A functor $F : \mathscr{C} \to \mathscr{D}$ between two categories $\mathscr{C}$ and $\mathscr{D}$ consists of two mappings - one from $Obj(\mathscr{C})$ to $Obj(\mathscr{D})$ and another from $Arr(\mathscr{C})$ to $Arr(\mathscr{D})$ such that

(1) *for any $\mathscr{C}$-arrow $f : A \to B$, $F(dom(f)) = dom(F(f))$ and $F(cod(f)) = cod(F(f))$,*
(2) *for any $\mathscr{C}$-arrows $f : A \to B$ and $g : B \to C$, $F(g \circ f) = F(g) \circ F(f)$, and*
(3) *for any identity $\mathscr{C}$-arrow 1_A, $F(1_A) = 1_{F(A)}$.*

In a functor $F : \mathscr{C} \to \mathscr{D}$, both the mappings - the object map and the arrow map from $\mathscr{C}$ to $\mathscr{D}$ - are denoted by F itself.

Definition 16 (Full, faithful and essentially surjective functors).

Consider a functor $F : \mathscr{C} \to \mathscr{D}$. For any two $\mathscr{C}$-objects A and B, define the map:

$$\Phi_{A,B} \; : \; Hom_{\mathscr{C}}(A, B) \to Hom_{\mathscr{D}}(FA, FB)$$

given by $\Phi_{A,B}(f) = F(f)$.

(1) *F is faithful if $\Phi_{A,B}$ is injective for all $\mathscr{C}$-objects A and B.*
(2) *F is full if $\Phi_{A,B}$ is surjective for all $\mathscr{C}$-objects A and B.*
(3) *F is essentially surjective if for every $\mathscr{D}$-object D, there exists a $\mathscr{C}$-object C such that $F(C) \cong D$.*

When do we say that two categories are 'same', i.e. have same category-theoretic properties? If in the functor $F : \mathscr{C} \to \mathscr{D}$, both the mappings - object map and the arrow map are one-one and onto, then we say that $\mathscr{C}$ is *isomorphic* to $\mathscr{D}$. However, to ask for isomorphism between two categories is too strong for most purposes in category theory. Instead the concept of 'equivalence' between two categories is sufficient to show that the two categories have same category-theoretic properties.

Definition 17 (Equivalence between two categories).

A category $\mathscr{C}$ is equivalent to $\mathscr{D}$, denoted as $\mathscr{C} \simeq \mathscr{D}$, if there exists a functor $F : \mathscr{C} \to \mathscr{D}$ such that F is full, faithful and essentially surjective.

Definition 18 (Diagram).

A diagram D in a category $\mathscr{C}$ consists of a collection of $\mathscr{C}$-objects $A_1, A_2, \ldots$ along with some $\mathscr{C}$-arrows $g : A_i \to A_j$ between these objects.

Definition 19 (Cone and cocone).
A cone *for diagram D in a category $\mathscr{C}$ consists of a $\mathscr{C}$-object C together with $\mathscr{C}$-arrows $f_i : C \to A_i$ for each object $A_i \in D$ such that for every $g : A_i \to A_j$ in D, we have $f_j = g \circ f_i$. A cone is denoted as $\{f_i : C \to A_i\}$ for the diagram D. A* cocone *for diagram D in a category $\mathscr{C}$ consists of a $\mathscr{C}$-object C together with $\mathscr{C}$-arrows $f_i : A_i \to C$ for each object $A_i \in D$ such that for every $g : A_i \to A_j$ in D, $f_i = f_j \circ g$. A cocone is denoted as $\{f_i : A_i \to C\}$ for the diagram D.*

Definition 20 (Limits and colimits).
A limit *of a diagram D in a category $\mathscr{C}$ is a cone $\{f_i : C \to A_i\}$ for D such that for any other cone $\{f_i' : C' \to A_i\}$ for D, there exists a unique arrow $f : C' \to C$ such that $f_i' = f_i \circ f$ for every object $A_i \in D$. A* colimit *of a diagram D in a category $\mathscr{C}$ is a cocone $\{f_i : A_i \to C\}$ for D such that for any other cone $\{f_i' : A_i \to C\}$ for D, there exists a unique arrow $f : C \to C'$ such that $f_i' = f \circ f_i$ for every object $A_i \in D$.*

Remark 1. In the above definitions of limits and colimits, the property 'existence of a unique arrow' is called *universal* property of limits and colimits.

Let us see some fundamental examples of limits and colimits of a diagram D. Consider a category $\mathscr{C}$, and a diagram D in $\mathscr{C}$.

(1) **Terminal and initial objects -** *If D is empty, that is, it has no objects and no arrows.* Any object A is a cone for D. The limit of the empty diagram is an object C such that there exists exactly one arrow $f : A \to C$ for every object A. In this case, D is called the *terminal object*, denoted by 1 and the arrow f is denoted by $!_A$ (Fig. 6a). The colimit of an empty diagram is called *initial* object, denoted by 0. The arrow f is denoted by i_A (Fig. 6b). The arrow $0 \to 1$ is denoted just by ! (or i).
In *SET*, the 1-element set $\{0\}$ is a terminal object, and empty set $\emptyset$ is an initial object.

$$A \xrightarrow{\quad !_A \quad} 1 \qquad\qquad 0 \xrightarrow{\quad i_A \quad} A$$

(a) Terminal object 1 in $\mathscr{C}$ (b) Initial object 0 in $\mathscr{C}$

Fig. 6. Initial and terminal objects in $\mathscr{C}$

(2) **Products and coproducts -** *If D consists of two $\mathscr{C}$-objects A and B, and no arrows between them.* The limit, in this case, is called *products*, denoted by $(A \times B, p_1, p_2)$, where $A \times B$ is a $\mathscr{C}$-object and $p_1 : A \times B \to A$ and $p_2 : A \times B \to B$ are $\mathscr{C}$-arrows (Fig. 7a). The object $A \times B$ is called the *product* of A and B. The colimit is called *coproduct*, denoted by $(A + B, i_1, i_2)$, and the $\mathscr{C}$-object $A + B$ is called *coproduct* of A and B (Fig. 7b).
In *SET*, the product of two sets A and B is the standard cross product of

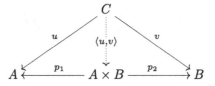

(a) Products in $\mathscr{C}$

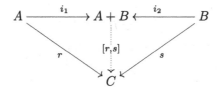

(b) Coproducts in $\mathscr{C}$

Fig. 7. Products and coproducts in $\mathscr{C}$

two sets $A \times B = \{(x, y) \mid x \in A,\ y \in B\}$. The co-product of two sets A and B is their disjoint union $A + B = \{(a, 0) \mid a \in A\} \cup \{(b, 1) \mid b \in B\}$.

(3) **Equalizers and coequalizers** - *If D consists of two $\mathscr{C}$-objects A and B, and two $\mathscr{C}$-arrows $f, g : A \to B$, the limit of D is called equalizer, denoted by $e : E \to A$ (Fig. 8a). The colimit of D is called coequalizer, denoted by $c : B \to C$ (Fig. 8b).*

(4) **Pullbacks and pushouts** - *If D consists of $\mathscr{C}$-objects A, B and C, and $\mathscr{C}$-arrows $f : A \to C$ and $g : B \to C$. The limit of D is called pullback, denoted by (T, p_A, p_B), where T is a $\mathscr{C}$-object, and $p_A : T \to A$ and $p_B : T \to B$ are $\mathscr{C}$-arrows (Fig. 9a). We say that p_A and p_B are pullbacks of f with g. The colimit of D is called pushout (Fig. 9b).*

We shall call the inner squares in the Figs. 9a and 9b as pullbacks and pushouts squares respectively.

Remark 2.

(A) Given two $\mathscr{C}$-arrows $f : A \to C$ and $g : B \to D$, using the universal property of the product $C \times D$ of C and D, we obtain a unique arrow $\langle f \circ p_1, g \circ p_2 \rangle : A \times B \to C \times D$, where p_1 and p_2 are arrows associated with the product $A \times B$ of A and B. This arrow $\langle f \circ p_1, g \circ p_2 \rangle$ is denoted by $f \times g$.

(B) Given two $\mathscr{C}$-arrows $f : A \to C$ and $g : B \to D$, using the universal property of the coproduct $A + B$ of A and B, we obtain a unique arrow $[i_3 \circ f, i_4 \circ g] : A + B \to C + D$, where i_3 and i_4 are arrows associated with the coproduct $C + D$. This arrow $[i_3 \circ f, i_4 \circ g]$ is denoted by $f + g$.

(C) Every equalizer is a monic, and every co-equalizer is an epic.

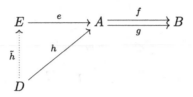

(a) Equalizers in $\mathscr{C}$

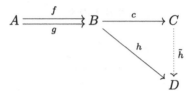

(b) Coequalizers in $\mathscr{C}$

Fig. 8. Equalizers and coequalizers in $\mathscr{C}$

(D) Let $g : E \to A$, $f : A \to B$, $k : E' \to C$ and $h : C \to D$ be $\mathscr{C}$-arrows. Then
$(f \times h) \circ (g \times k) = (f \circ g) \times (h \circ k) : E \times E' \to B \times D$.

Proposition 4 (The pullback lemma [43]).
Consider the commutative diagram in the Fig. 10.

(1) If the two inner squares (the left hand side square and the right hand side square) are pullbacks, then the outer square (it is a rectangle) is also a pullback.

(2) If the outer square and the left hand side square are pullbacks, then the right hand side square is also a pullback.

Proposition 5. [43] *In the diagram below if the square is a pullback such that f is monic, then r is also a monic.*

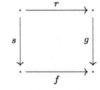

Definition 21 (Finite complete and finite cocomplete category).
If the limits (colimits) of all finite diagrams exist in $\mathscr{C}$, then $\mathscr{C}$ is called a finite complete (finite cocomplete) *category.*

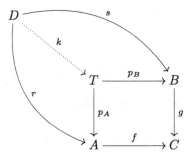

(a) Pullbacks in $\mathscr{C}$

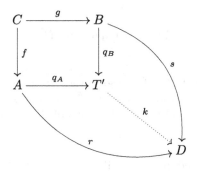

(b) Pushouts in $\mathscr{C}$

Fig. 9. Pullbacks and pushouts in $\mathscr{C}$

All the categories that we shall encounter in this work are finitely complete and cocomplete. The following proposition tells that it is not necessary to always check the existence of all the limits for finite completeness and all the colimits for finite cocompleteness of a category.

Proposition 6. *For a category $\mathscr{C}$, the following are equivalent.*

(1) *$\mathscr{C}$ is finitely complete (finitely cocomplete).*
(2) *$\mathscr{C}$ has terminal object (initial object) and pullbacks (pushouts).*
(3) *$\mathscr{C}$ has products (coproducts) and equalizers (coequalizers).*

Definition 22 (Cartesian closed).
Consider two $\mathscr{C}$-objects A and B in the category $\mathscr{C}$. A $\mathscr{C}$-object B^A is called an exponent if there exists a $\mathscr{C}$-arrow $ev : B^A \times A \to B$ (called evaluation map) such that for any $\mathscr{C}$-object C and $\mathscr{C}$-arrow $g : C \times A \to B$, there is a unique $\mathscr{C}$-arrow $\hat{g} : C \to B^A$ such that $ev \circ (\hat{g} \times 1_A) = g$ (Fig. 11).
If for any two $\mathscr{C}$-objects A and B, the exponent B^A and evaluation map $ev : B^A \times A \to B$ exist in $\mathscr{C}$, then $\mathscr{C}$ is called a cartesian closed category.

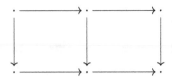

Fig. 10. The pullback Lemma

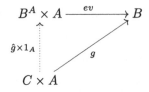

Fig. 11. Exponents in $\mathscr{C}$

In SET, the exponent B^A of two sets A and B is the set of all the functions from A to B, and the evaluation map $ev : B^A \times A \to B$ is given as $ev(g, a) = g(a)$, where $g : A \to B$, $a \in A$ and $g(a)$ is the value of the set function g at a.

Remark 3. [43]

(1) A category $\mathscr{C}$ is called *degenerate* if for any two $\mathscr{C}$-objects A and B, $A \cong B$. Moreover, for a cartesian closed category with initial object 0 and terminal object 1, if there exists an arrow $f : 1 \to 0$ in $\mathscr{C}$, then $\mathscr{C}$ is degenerate.

(2) Let $\mathscr{C}$ be a cartesian closed category with initial object 0 and terminal object 1. Then (i) any arrow $i_A : 0 \to A$ is monic, and (ii) if there exists an arrow $f : A \to 0$, then $A \cong 0$.

The collection of all monics with a fixed codomain exhibit interesting properties in some categories. For example, in SET, the domain A of a monic arrow $m : A \to B$ is isomorphic to the subset $m(A) \subseteq B$. Moreover, given any subset $C \subseteq B$, there exists an inclusion function $m : C \to B$ such that m is a monic arrow. This motivates the definition of 'subobjects' in a category. Consider a category $\mathscr{C}$ and a $\mathscr{C}$-object D. Let T denote the set of all monics in $\mathscr{C}$ with codomain D. Define the following ordering '$\subseteq$' on T: for any $f : A \to D$ and $g : B \to D$ in T, $f \subseteq g$ if and only if there exists an arrow $h : A \to B$ such that $f = g \circ h$. This ordering is reflexive and transitive, but is not a partial order (Definition 3). To obtain a partial ordering, a relation $\simeq$ on T is defined as follows: $f \simeq g$ if and only if $f \subseteq g$ and $g \subseteq f$. $\simeq$ is an equivalence relation on T. Now, define an ordering $\subseteq$ on the quotient set $T/\simeq$: $[f] \subseteq [g]$ if and only if $f \subseteq g$. This ordering is well-defined and a partial order. Notice that for any $f_1, f_2 \in [f]$, $dom(f_1) \cong dom(f_2)$.

Definition 23 (Subobjects and strong subobjects).
A subobject (strong subobject) of D *is defined as an element of the set*

$$Sub(D) := \{[f] \mid f \text{ is a monic (strong monic) with codomain } D\}.$$

Notation 2. *The following notations for subobjects and strong subobjects are used in this work, adopting standard practice [88].*
(1). The set of subobjects and the set of strong subobjects for a $\mathscr{C}$-object D - both are denoted by $Sub(D)$.
(2). An arrow f, or sometimes the domain $dom(f)$ of f, is called the subobject (strong subobject) of D, where f is a representative element from an equivalence class in $T/\simeq$.

Definition 24 (Subobject classifier for monics and strong monics).
Consider a category $\mathscr{C}$ with the terminal object 1. A subobject classifier for a class of monics (strong monics) is a $\mathscr{C}$-object Ω along with the $\mathscr{C}$-arrow 'true' $\top : 1 \to \Omega$ such that for each monic (strong monic) $m : A \to D$, there exists a unique $\mathscr{C}$-arrow $\chi_m : D \to \Omega$ such that the diagram in Fig. 12 is a pullback square.
For the monic (strong monic) $m : A \to D$, the arrow $\chi_m : D \to \Omega$ is called the characteristic arrow of m.

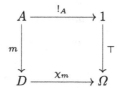

Fig. 12. Subobject classifier for class of monics (strong monics)

Observation 1. *In SET, the set of subobjects of a set D is in bijection with the power set $\mathcal{P}(D)$ of D. The subobject classifier is $\Omega = \{0, 1\}$ with the set function $\top : \{0\} \to \{0, 1\}$ mapping 0 to 1.*

A special class of categories called *elementary toposes* was defined by W. Lawvere in 1964 to generalize the category *SET* of sets (cf. [3, 43]). These will be of particular relevance in this work.

Definition 25 (Elementary topos [43]).
A category $\mathscr{C}$ is an elementary topos if $\mathscr{C}$ is cartesian closed, finitely complete, finitely cocomplete and has a subobject classifier for the class of monics.

In the above definition, some conditions can be inferred from the others. Therefore, the following result is very helpful to show that a category $\mathscr{C}$ is an elementary topos.

Proposition 7. *For a category $\mathscr{C}$, the following are equivalent.*

(1) $\mathscr{C}$ *is an elementary topos.*
(2) $\mathscr{C}$ *is cartesian closed and has products, equalizers, and subobject classifier for class of monics.*
(3) $\mathscr{C}$ *is cartesian closed and has terminal objects, pullbacks, and a subobject classifier for class of monics.*

Proposition 8. *Consider a topos $\mathscr{C}$, where the subobject classifier is given by Ω along with the arrow $\top : 1 \to \Omega$. Then we have the following.*

(1) $\top : 1 \to \Omega$ *is always a monic.*
(2) *Given any arrow $\chi : D \to \Omega$, there exists a monic $m : A \to D$ such that $\chi_m = \chi$.*
(3) *There is a bijection between the set of subobjects with codomain D and the set of arrows $f : D \to \Omega$.*

Proposition 9. [88] *In any topos $\mathscr{C}$, we have the following.*

(1) *Any arrow which is epic and monic is iso.*
(2) *The class of equalizers is equal to the class of monics, which is equal to the class of strong monics.*
(3) *For two objects A and B, consider the coproduct $(A + B, i_1, i_2)$. Then the following diagram is a pullback and pushout.*

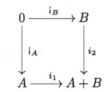

That is, (a) i_1 and i_2 are pushouts of i_A and i_B, and (b) i_A and i_B are pullbacks of i_1 and i_2. Moreover, the arrows i_1 and i_2 are monics.

Apart from SET, we have the following examples of elementary toposes.

Example 4 (Example of toposes).

(1) The category **M-Set**, where **M** is a monoid, is an elementary topos.
(2) $\mathscr{C}^2$ and $\mathscr{C}^{\to}$, for any elementary topos $\mathscr{C}$.

There have been various efforts to weaken the definition of topos (cf. [51]). One particular definition which stands out is that of an 'elementary quasitopos'. The definition is based on the following fact: $\mathscr{C}$ may not have the subobject classifier for the class of monics, but has one for a subclass of the class of monics, namely the class of strong monics (Definition 14). Some of the categories we study in this work will be quasitoposes. Let us first see a few more terminologies, before giving the definition of an elementary quasitopos.

Definition 26 (Relation and partial morphisms).
Consider $\mathscr{C}$-arrows $u : C \to A$ and $v : C \to B$. The pair (u, v) is called a relation with respect to the class of monics (strong monics) in $\mathscr{C}$ if the $\mathscr{C}$-arrow $\langle u, v \rangle : C \to A \times B$ in Fig. 7a is a monic (strong monic).
The pair (u, v) is called a partial morphism with respect to the class of monics (strong monics) in $\mathscr{C}$ if u is a monic (strong monic).
The relation or partial morphism is denoted as $(u, v) : A \to B$, where A is called the domain and B is called the codomain of the partial morphism.

Remark 4. Any partial morphism is a relation, i.e. if u is a monic (or strong monic), then $\langle u, v \rangle : C \to A \times B$ is also a monic (or strong monic).

In the category SET, a partial morphism $(u, v) : A \to B$ is nothing but a partial function from A to B.

Definition 27 (Representation of a partial morphism).
A monic (or strong monic) $\nu_B : B \to \tilde{B}$ in $\mathscr{C}$ represents partial morphisms in $\mathscr{C}$ with codomain B if for every partial morphism $(m, f) : A \to B$ in $\mathscr{C}$ with codomain B there exists a unique $\mathscr{C}$-arrow $\bar{f} : A \to \tilde{B}$ such that the diagram in Fig. 13 is a pullback square.

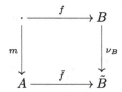

Fig. 13. Representation of partial morphisms with codomain B

Definition 28 (Elementary quasitopos (cf. [88])**).**
A category $\mathscr{C}$ is an elementary quasitopos if $\mathscr{C}$ is cartesian closed, finitely cocomplete and all the partial morphisms with respect to the class of strong monics are represented.

Note: We shall drop the word 'elementary' hereafter, and just refer to elementary topos and elementary quasitopos as topos and quasitopos respectively.

Proposition 10. (cf. [88]) *Any quasitopos $\mathscr{C}$ has the following properties.*

(1) *$\mathscr{C}$ is finitely complete.*
(2) *$\mathscr{C}$ has a subobject classifier for the class of strong monics.*
(3) *Any arrow which is epic and strong monic is iso.*

Moreover, any topos is a quasitopos.

Example 5.

(1) The toposes *SET* and **M-Set** are quasitoposes.
(2) The poset category $(P, \leq)$ $(|P| > 1)$ is a quasitopos, and it is not a topos.

Proposition 11. *Consider two categories $\mathscr{C}_1$ and $\mathscr{C}_2$.*

(1) *If $\mathscr{C}_1$ and $\mathscr{C}_2$ are toposes (quasitoposes) then the product $\mathscr{C}_1 \times \mathscr{C}_2$ forms a topos (quasitopos).*
(2) *If $\mathscr{C}_1 \simeq \mathscr{C}_2$ and $\mathscr{C}_1$ is a topos (quasitopos) then $\mathscr{C}_2$ also forms a topos (quasitopos).*

Another concept that we shall be using in our work is 'image factorization'. Consider a set function $f : A \to B$. We know that f can be factored into two functions $e : A \to Im(f)$ and $m : Im(f) \to B$, such that $f = m \circ e$, where e is a surjective map and m is an injective map. This construction can be generalized in any topos. Let $\mathscr{C}$ be a topos, and $f : A \to B$ be a $\mathscr{C}$-arrow. Let $p : B \to T'$ and $q : B \to T'$ be the pushout of f with f. Further, let $Im(f) : E \to B$ be the equalizer of p and q. Since $q \circ f = p \circ f$, there exists a unique arrow $f^* : A \to f(A)$ such that the diagram in Fig. 14 commutes. Here, f^* is epic [43]. Therefore, we have obtained the factorization of f into two arrows - one epic and one equalizer. Further, in a topos any equalizer is a monic (and vice versa), therefore we can factor f into two arrows - one epic and one monic. Such a 'factorization' of f can also be obtained by taking the coequalizer of the pullback of f with f [43].

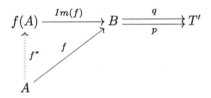

Fig. 14. Image factorization of f

Definition 29 (Image factorization).
Let $\mathscr{C}$ be a topos, and $f : A \to B$ be a $\mathscr{C}$-arrow. There exist an epic e and a monic m such that $f = m \circ e$. This is called the image factorization *of f, and m is called the* image *of f.*

2.2.1   Internal Algebra of a Topos or Quasitopos

An important property of any topos (quasitopos) $\mathscr{C}$ is that on the set of subobjects (strong subobjects) of any $\mathscr{C}$-object A, we can define a unary operator $\neg$ and binary operators $\cup$, $\cap$ and $\to$ such that $(Sub(A), 1_A, i_A, \cap, \cup, \to, \neg)$ forms a *pBa* (Definition 7). For example, in SET, we know that given any set A, the power set $\mathcal{P}(A)$ forms a Boolean algebra along with the operators intersection ($\cap$), union ($\cup$), implication ($\rightarrow$) and complement ($\neg$). $\emptyset$ (empty set) and A are the bottom and top elements respectively. We have already seen that the equivalent version of the concept of 'subsets of a set' in the topos (quasitopos) is the 'subobjects (strong subobjects) of an object'. Let us now see how we can define the standard operators $\cup$, $\cap$, $\rightarrow$ and $\neg$ on the subobjects of a $\mathscr{C}$-object.

Note: The definition below is given for monics in a topos $\mathscr{C}$. However, all the constructions in the definition also hold when $\mathscr{C}$ is a quasitopos, with the only difference that 'monics' and 'subobjects' are replaced by 'strong monics' and 'strong subobjects' respectively.

Definition 30 (Algebra on $Sub(C)$).

Consider a topos $\mathscr{C}$, where $\emptyset$ and 1 are initial and terminal objects in $\mathscr{C}$ respectively, and the subobject classifier for the class of monics is denoted by the $\mathscr{C}$-object Ω along with the $\mathscr{C}$-arrow 'true' $\top : 1 \rightarrow \Omega$. Consider a $\mathscr{C}$-object C and the set $Sub(C)$ of subobjects of C.

(1) **Intersection:** Given two monic arrows $f : A \rightarrow C$ and $g : B \rightarrow C$ in $\mathscr{C}$ (i.e. $f, g \in Sub(C)$), the intersection of f and g is defined as the monic arrow $f \cap g : A \cap B \rightarrow C$ whose characteristic morphism is $\cap \circ \langle \chi_f, \chi_g \rangle$, where χ_f and χ_g are the characteristic morphisms of f and g respectively; and $\cap : \Omega \times \Omega \rightarrow \Omega$ is the characteristic morphism of $\langle \top, \top \rangle : 1 \rightarrow \Omega \times \Omega$ (Fig. 15).

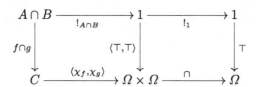

Fig. 15. Intersection of f and g in a topos or quasitopos $\mathscr{C}$

(2) **Union:** Given two monic arrows $f : A \rightarrow C$ and $g : B \rightarrow C$ in $\mathscr{C}$ (i.e. $f, g \in Sub(C)$), the union of f and g is defined as the monic arrow $f \cup g : A \cup B \rightarrow C$ whose characteristic morphism is $\cup \circ \langle \chi_f, \chi_g \rangle$, where χ_f and χ_g are the characteristic morphisms of f and g respectively; and $\cup : \Omega \times \Omega \rightarrow \Omega$ is the characteristic morphism of $[\langle \top_\Omega, 1_\Omega \rangle, \langle 1_\Omega, \top_\Omega \rangle] : \Omega + \Omega \rightarrow \Omega \times \Omega$ (Fig. 16). Here, $\top_\Omega := \top \circ !_\Omega$.

(3) **Implication:** Given two monic arrows $f : A \rightarrow C$ and $g : B \rightarrow C$ in $\mathscr{C}$ (i.e. $f, g \in Sub(C)$), the implication $f \rightarrow g$ is defined as the monic arrow $f \rightarrow g : A \rightarrow B \rightarrow C$ whose characteristic morphism is $\rightarrow \circ \langle \chi_f, \chi_g \rangle$, where χ_f and χ_g are the characteristic morphisms of f and g respectively; and $\rightarrow : \Omega \times \Omega \rightarrow \Omega$ is the characteristic morphism of $e : E \rightarrow \Omega \times \Omega$ (Fig. 17).

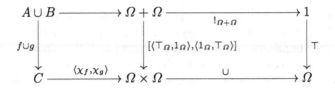

Fig. 16. Union of f and g in a topos or quasitopos $\mathscr{C}$

Here, e is the equalizer of $\Omega \times \Omega \overset{\cap}{\underset{p_1}{\rightrightarrows}} \Omega$, where $\cap$ is the intersection arrow as defined in (1), and p_1 is one of the arrows associated with the product $\Omega \times \Omega$.

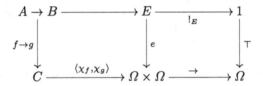

Fig. 17. $f \to g$ in a topos or quasitopos $\mathscr{C}$

(4) **Negation:** Given a monic arrow $f : A \to C$ in $\mathscr{C}$ (i.e. $f \in Sub(C)$), the negation of f is defined as the monic arrow $\neg f : \neg A \to C$ whose characteristic morphism is $\neg \circ \chi_f$, where χ_f is the characteristic morphism of f; $\neg : \Omega \to \Omega$ is the characteristic morphism of 'false' $\bot$; and $\bot : 1 \to \Omega$ is the characteristic morphism of $! : 0 \to 1$ (Fig. 18).

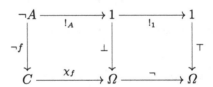

Fig. 18. Negation of f in a topos or quasitopos $\mathscr{C}$

Observation 2. Intersection of $f, g \in Sub(C)$ can also be obtained by taking the pullback of $f : A \to C$ with $g : B \to C$ (diagram below). Moreover, let $r : D \to A$ and $s : D \to B$ be the pullback of two monics $f : A \to C$ and $g : B \to C$ with the same codomain C, then $f \cap g \simeq f \circ r = g \circ s$ and $A \cap B \simeq D$.

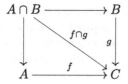

Proposition 12. *Consider a topos $\mathscr{C}$ and a $\mathscr{C}$-object A. Let r and s be the pullback of monic $f : X \to A$ with monic $g : Y \to A$. Then $dom(r) = dom(s) = X \cap Y$, $g \circ s = f \circ r = f \cap g$, and the inner square in the following diagram is a pushout. Moreover, $g = (f \cup g) \circ v$ and $f = (f \cup g) \circ u$.*

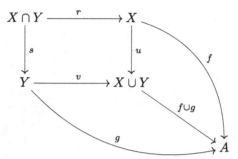

The following theorem gives the algebraic structure that is formed by the set of subobjects (strong subobjects) of an object in a topos (quasitopos).

Proposition 13.
In a topos (quasitopos) $\mathscr{C}$, $(Sub(A), 1_A, i_A, \cup, \cap, \to, \neg)$ forms a pBa for all $A \in Obj(\mathscr{C})$.

Observation 3. Consider the category SET and a set A. Using Observation 1, we have $Sub(A) \cong \mathcal{P}(A)$, where 1_A is identified with A and i_A is identified with $\emptyset$. Therefore, $(\mathcal{P}(A), A, \emptyset, \cup, \cap, \to, \neg)$ forms a *pBa*. However, we know that $(\mathcal{P}(A), A, \emptyset, \cup, \cap, \to, \neg)$ forms a Boolean algebra (Example 2). Since a topos is defined as an abstraction of the category SET, this difference between a general topos $\mathscr{C}$ and the topos SET is notable. While the topos has 'good' properties like having subobjects and exponents, the Boolean-ness property '$f \cup \neg f = 1_A$ for all $f \in Sub(A)$' is lost.

Definition 31 (Boolean topos).
A topos (quasitopos) $\mathscr{C}$ is called Boolean if $(Sub(A), 1_A, i_A, \cup, \cap, \to, \neg)$ is a Boolean algebra for every $\mathscr{C}$-object A.

2.3   Some Categories of Rough Sets

In [4] the category $ROUGH$ was proposed, with objects as rough sets in Pawlak approximation spaces. More precisely, objects of the category are triples of the form $\langle U, R, X \rangle$, where R is an equivalence relation on the set U, and X a subset of U. The R-upper and R-lower approximations of X are used to define arrows in the category, with the basic requirement that arrows should map upper approximations in one space to those in another, *preserving information present in the lower approximations*. Let us formally define this.

Definition 32 (*ROUGH* **category** [4]).
Objects of ROUGH are of the form (U, R, X) *(as above). An arrow in ROUGH with domain* (U, R, X) *and codomain* (V, S, Y) *is a map* $f : \overline{\mathcal{X}}_R \to \overline{\mathcal{Y}}_S$ *such that* $f(\underline{\mathcal{X}}_R) \subseteq \underline{\mathcal{Y}}_S$ *(Fig. 19).*

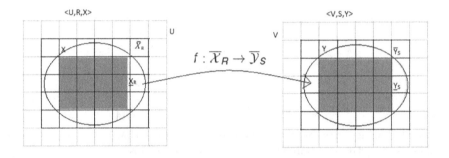

Fig. 19. An arrow in *ROUGH*

In [4], another category on rough sets is defined capturing the idea that during 'communications' (being represented by arrows) between rough sets, the boundary region may be another invariant. On imposing a further condition in the definition of arrows, namely that *information in boundaries also be preserved*, one obtained a subcategory ξ-*ROUGH*.

Definition 33 (ξ-*ROUGH* **category** [4]).
Objects are (U, R, X), *that is, same as those in ROUGH. An arrow in* ξ-*ROUGH with domain* (U, R, X) *and codomain* (V, S, Y) *is a map* $f : \overline{\mathcal{X}}_R \to \overline{\mathcal{Y}}_S$ *such that* $f(\underline{\mathcal{X}}_R) \subseteq \underline{\mathcal{Y}}_S$ *and* $f(\overline{\mathcal{X}}_R \backslash \underline{\mathcal{X}}_R) \subseteq \overline{\mathcal{Y}}_S \backslash \underline{\mathcal{Y}}_S$ *(Fig. 20).*

The arrows of ξ-*ROUGH* thus preserve the lower approximation *and* the boundary. It is clear that ξ-*ROUGH* is a subcategory of *ROUGH* containing the same collection of objects.
In [4], it was shown that *ROUGH* is finitely complete and is not a topos, using the fact that in any topos, a monic and epi arrow must be an iso arrow.

Theorem 2. *ROUGH is not a topos.*

Another category of rough sets was introduced by Li and Yuan in 2008 [57]. In this, the objects are not collection of equivalence classes, but sets obtained from some 'rough universes'. The approach follows that of Iwiński,

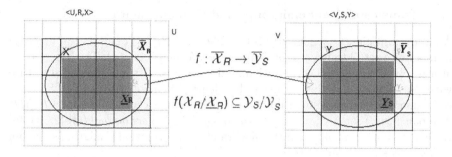

Fig. 20. An arrow in ξ-*ROUGH*

who gave an interpretation of rough sets based on a Boolean algebra [48]. A pair $(U, \mathcal{B})$, called a *rough universe*, was considered, where U is the universe and $\mathcal{B} := (B, U, \emptyset, \cap, \cup, \rightarrow, \neg)$ is a subalgebra of the power set Boolean algebra $(\mathcal{P}(U), U, \emptyset, \cap, \cup, \rightarrow, \neg)$. Any pair (A_1, A_2), where $A_1, A_2 \in B$ and $A_1 \subseteq A_2$, is called an *I-rough set* of $(U, \mathcal{B})$. Li and Yuan defined the category taking these *I-rough sets* as objects.

Definition 34 (*RSC* category [57,88]).
Objects of RSC are all I-rough sets, that is objects are of the form (X_1, X_2). An arrow in RSC with domain (X_1, X_2) and codomain (Y_1, Y_2) is a map $f : X_2 \rightarrow Y_2$ such that $f(X_1) \subseteq Y_1$.
Composition of two arrows is composition of two mappings. Identity arrow on (X_1, X_2) is the identity map on X_2 (Fig. 21).

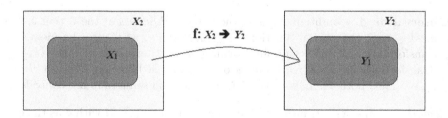

Fig. 21. An arrow in *RSC*

In [57], it was shown that *RSC* has limits and exponents. Further, there is no subobject classifier for the class of monics. Hence, the following is concluded.

Theorem 3. *RSC is not a topos.*

Moreover, *RSC* is a weak topos, a category-theoretic structure weaker than the topos.

2.4   Transformation Semigroups and Semiautomata

An important class of semigroups is the collection $PF(Q)$ of all partial functions from a finite set Q to itself, representing *transformations of Q*. The binary operator involved is function composition, and in fact, results in a monoid structure, with the identity function on Q as the identity element. Any subset S of this collection that is closed under function composition is a subsemigroup of $PF(Q)$. The pair (Q, S) for such S, is called a *transformation semigroup (ts)* [22,34,46]. In this section, we shall recall the algebra of transformation semigroups, leading to decomposition results [34,46].

One of the reasons to study transformation semigroups has been a natural connection with automata theory [46]. We shall also mention some basic structures in algebraic automata [46].

2.4.1   Transformation Semigroups

Semigroup *actions* give an alternative and equivalent way of viewing transformation semigroups [22]. Recall that an action of a semigroup S on the set Q is a function $\delta : Q \times S \to Q$ satisfying $\delta(\delta(q, s_1), s_2) = \delta(q, s_1 s_2)$, for all $q \in Q$ and $s_1, s_2 \in S.$, where $s_1 s_2$ denotes the application of the binary operator of S on s_1 and s_2. If the function δ is partial, δ is called a *partial semigroup action* of S on the set Q. Then we have the following definition.

Definition 35 (Transformation semigroups [34]).
A transformation semigroup *is a pair* $\mathcal{A} := (Q, S)$ *consisting of a finite set Q, a finite semigroup S, along with a partial semigroup action δ of S on Q that satisfies:*

$$\text{for any } s_1, s_2 \in S, \text{ if } \delta(q, s_1) = \delta(q, s_2) \text{ for all } q \in Q, \text{ then } s_1 = s_2. \qquad (2.2)$$

Observation 4. Condition 2.2 is termed the *faithfulness* of the action δ. For a fixed $s \in S$, the partial function $\delta_s := \delta(-, s) : Q \to Q$ can be viewed as a transformation of the set Q, and δ can also be interpreted as a set $\{\delta_s\}_{s \in S}$ of transformations of Q. Faithfulness of δ ensures a bijection between S and $\{\delta_s\}_{s \in S}$. Thus both the definitions of transformation semigroup are equivalent.

Remark 5. Consider a transformation semigroup $\mathcal{A} := (Q, S)$ with δ as the corresponding semigroup action. If S is a monoid and $\delta(q, 1) = q$ for all $q \in Q$, then $\mathcal{A}$ is called a *transformation monoid*. This shall be generally represented as (Q, M) where M denotes the monoid.

One can extend the semigroup S in the transformation semigroup (Q, S) to obtain a monoid $S^{\bullet}$ by adding the identity function 1_Q to S. Then the pair $(Q, S^{\bullet})$ forms a transformation monoid, denoted by $\mathcal{A}^{\bullet}$.

If Q is non-empty and S forms a group such that S contains the identity transformation then $\mathcal{A}$ is called *transformation group*, represented as (Q, G) where G is a group. Observe that in a transformation group (Q, G), every transformation in G has to be invertible.

Notation 3. *Hereafter, $\delta(q,s)$ shall be denoted by 'qs' and for any $P \subseteq Q$,*

$$Ps := \{qs \mid q \in P \text{ and } qs \text{ is defined}\}.$$

A transformation semigroup of a set shall be abbreviated as 'ts'. Similarly, 'tm' and 'tg' will be used for a transformation monoid and a transformation group respectively. Plurals of each of them shall also be denoted by the same abbreviation.

Constant functions and permutations motivate the definition of two special kinds of *ts*.

Definition 36 (Reset ts).
A ts $\mathcal{A} := (Q, S)$ is called reset *if $|Qs| \leq 1$ for any $s \in S$.*

Definition 37 (Permutation ts).
$\mathcal{A} := (Q, S)$ is called a permutation *ts if $Qs = Q$ for all $s \in S$.*

Definition 38 (Closure of a ts).
Given a ts $\mathcal{A} := (Q, S)$, the closure *$\overline{\mathcal{A}}$ of $\mathcal{A}$ is the subsemigroup of $PF(Q)$ that is generated by the set $S \cup \{\bar{q} \mid q \in Q\}$, where $\bar{q}$ represents a constant function on Q mapping any element of Q to q. In notation, $\overline{\mathcal{A}} := (Q, \langle S \cup \{\bar{q} \mid q \in Q\}\rangle)$.*

Example 6. A trivial example of a reset *ts* is the pair $(Q, \emptyset)$. If $|Q| = n$, the *ts* $(Q, \emptyset)$ is denoted as $\boldsymbol{n}$. Then $\overline{\boldsymbol{n}} := (Q, \{\bar{q} \mid q \in Q\})$, which is again a reset *ts*. For $n = 2$, the closure $\overline{\boldsymbol{2}} = (\{0,1\}, \{\bar{0}, \bar{1}\})$ can be diagrammatically represented as follows:

where $\bar{1}$ and $\bar{0}$ are the constant functions to 1 and 0 respectively. If we include only $\bar{0}$ and not $\bar{1}$ in the semigroup, then we have the following *ts* $(\{0,1\}, \{\bar{0}\})$, denoted by C.

Observation 5. Let Q be a finite set, S a finite semigroup and δ a partial semigroup action of Q on S. Does (Q, S) form a *ts*? Not necessarily, as δ may not be faithful. One then defines a relation $\sim$ on S by $s \sim s' \Leftrightarrow qs = qs'$ for all $q \in Q$. $\sim$ is a congruence relation on S and $S/\sim$ is a quotient semigroup of S. The pair $(Q, S/\sim)$ forms a *ts* with the action defined by $q[s] := qs$, for all $q \in Q$, $[s] \in S/\sim$.
If $Q = \emptyset$ then $S/\sim$ is the singleton $\{S\}$.

Example 7. Consider a non-empty semigroup S.

(A) For each $s \in S$ define a function $\mu_s : S \to S$ mapping x to xs for all $x \in S$. The collection S' of all μ_s's forms a semigroup under composition, and gives a semigroup action on S. Note that S' may not be isomorphic to S. By defining a congruence relation as in Observation 5, we can obtain a ts $(S, S'/\!\!\sim)$.

(B) Adjoin an identity element 1 in S to get $S^{\cdot} := S \cup \{1\}$. $\delta : S^{\cdot} \times S \to S^{\cdot}$ mapping (x, s) to xs, where $x \in S^{\cdot} = S \cup \{1\}$ and $s \in S$. δ is a semigroup action. The pair $(S^{\cdot}, S)$ forms a ts. If S is a monoid, then $S^{\cdot} = S$ and (S, S) forms a tm. Similarly for a group G, (G, G) forms a tg. By abuse of notation, we shall denote $(S^{\cdot}, S)$ and (G, G) by S and G respectively.

Observation 6. Two different definitions of *restriction* of a given ts are found in literature [34]. Consider a ts $\mathcal{A} := (Q, S)$, $P \subseteq Q$ and the inclusion function $i : P \to Q$.

(a) Define a subsemigroup $T := \{s \mid s \in S \text{ and } Ps \subseteq P\}$. Using part (5) of this observation, $\mathcal{A}_P := (P, T/\!\!\sim)$ forms a ts.

(b) Define a partial function i^{-1} from Q to P given by $i^{-1}(q) = q$ for all $q \in P$ and not defined for $q \notin P$. Let S' be the semigroup generated by the partial functions $s' = isi^{-1} : P \to P$ for all $s \in S$. Then $\mathcal{A}|P := (P, S')$ is also a ts.

In some cases, these definitions coincide [34]:

Proposition 14. *For a ts $\mathcal{A} := (Q, S)$ and $P \subseteq Q$, if $Ps \subseteq P$ for all $s \in S$, then $\mathcal{A}|P = \mathcal{A}_P$.*
Note that '=' here is used to express isomorphism between the associated semigroups.

Definition 39 (Transformation Semigroup Homomorphism).
Consider two ts $\mathcal{A} := (Q, S)$ and $\mathcal{B} := (P, T)$. Let $\alpha : Q \to P$ be a set function and $\beta : S \to T$ a semigroup homomorphism such that $\alpha(qs) = \alpha(q)\beta(s)$, whenever qs is defined for any $q \in Q$ and $s \in S$. The pair (α, β) is called a ts homomorphism from $\mathcal{A}$ to $\mathcal{B}$.
If α and β are both one-one, onto or bijections, (α, β) is called a ts monomorphism, epimorphism or isomorphism of ts respectively.
If $(\alpha, \beta) : \mathcal{A} \to \mathcal{B}$ is an isomorphism then $\mathcal{A}$ is said to be equivalent to $\mathcal{B}$ and this is denoted by $\mathcal{A} \cong \mathcal{B}$.

Observation 7. *If there is a bijection $f : Q \to P$, and (Q, S) is a ts, then (P, S) becomes a ts with the same semigroup action as on (Q, S). Further, $(Q, S) \cong (P, S)$.*

One can easily see that ts constitute a category.

Definition 40 (The category TS of transformation semigroups).
*Objects of **TS** are ts and arrows of **TS** are ts homomorphisms.*

Note: The object class of **TS** in the above Definition is different from the class **TS** defined in [34].

Definition 41 (Direct and wreath products [46]).
Consider ts $\mathcal{A} := (Q, S)$ and $\mathcal{B} := (P, T)$.

(1) $\mathcal{A} \times \mathcal{B} := (Q \times P, S \times T)$ is a ts, called the direct product of $\mathcal{A}$ and $\mathcal{B}$. The semigroup operation/action involved is defined component-wise.
(2) The wreath product of $\mathcal{A}$ and $\mathcal{B}$, denoted by $\mathcal{A} \circ \mathcal{B}$, is given by $(Q \times P, S^P \times T)$, where S^P is the set of all functions from P to S. The semigroup operation on $S^P \times T$ is defined as $(f, t) \cdot (g, t') := (f * g, tt')$, where $f, g \in S^P$, $t, t' \in T$ and $f * g$ is a map from P to S given by $(f * g)(p) := f(p)g(pt')$. $S^P \times T$ then acts on $(Q \times P)$ in the following way:

$$(q, p)(f, t) = (qf(p), pt)$$

for $(q, p) \in Q \times P$ and $(f, t) \in S^P \times T$.

Definition 42 (Admissible partitions and quotients).
Consider a ts $\mathcal{A} := (Q, S)$ and $\pi := \{H_i\}_{i \in I}$ a set of non-empty subsets of Q. π is called a partition of Q if $\bigcup_{i \in I} H_i = Q$ and $H_i \cap H_j = \emptyset$ for any $i, j \in I$. π is called admissible if for every $H_i \in \pi$ and $s \in S$, if $H_i s$ is non-empty then there exists $H_j \in \pi$ such that $H_i s \subseteq H_j$. Note that such a choice for H_j would be unique for the H_i. Then a partial semigroup action $*$ of S on π can be defined as follows: For any $H_i \in \pi$, $s \in S$,
(1) $H_i * s := H_j$ if $H_i s \subseteq H_j$; (2) $H_i * s$ is not defined if $H_i s = \emptyset$.
This action may not be faithful. However as discussed in Observation 5, one can obtain the quotient ts $\mathcal{A}/\langle \pi \rangle := (\pi, S/\sim)$, using the congruence relation $\sim$. When $|Q| > 2$, π is said to be non-trivial if $1 < |H_i| < |Q|$ for some $i \in I$.

Note: When the context is clear, we shall use the same notation '$\sim$' to denote equivalence relations on different domains.

Definition 43 (Orthogonal partitions).
For a non-trivial admissible partition $\pi := \{H_i\}_{i \in I}$ on ts (Q, S), if there exists another non-trivial admissible partition $\tau := \{K_j\}_{j \in J}$ such that $|H_i \cap K_j| \leq 1$ for all $i \in I$ and $j \in J$, then π is called an orthogonal partition on Q [46]. The condition '$|H_i \cap K_j| \leq 1$ for all $i \in I$ and $j \in J$' is denoted as '$\pi \cap \tau = 1_Q$'.

Definition 44 (Coverings [46]).
A ts $\mathcal{B} := (P, T)$ covers the ts $\mathcal{A} := (Q, S)$, written as $\mathcal{A} \preccurlyeq \mathcal{B}$, if there exists a surjective partial function $\eta : P \to Q$ such that for each $s \in S$, there is $t_s \in T$ satisfying $\eta(p)s = \eta(pt_s)$ whenever $\eta(p)s$ is defined for any $p \in P$. η is called a covering of $\mathcal{A}$ by $\mathcal{B}$, or $\mathcal{B}$ is said to cover $\mathcal{A}$ by η. t_s is said to cover s (Fig. 22).

Using the definition and the fact that any element in the semigroup $\langle S \rangle$ generated by S can be written as a finite product of elements of S, one gets

Proposition 15. *For ts $(Q, \langle S \rangle)$ and (P, T), the following are equivalent.*

(a) $(Q, \langle S \rangle) \preccurlyeq (P, T)$.

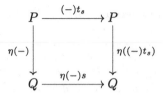

Fig. 22. A covering of ts $\mathcal{A} := (Q, S)$ by $\mathcal{B} := (P, T)$

(b) *There exists a surjective partial function $\eta : P \to Q$ satisfying the following property: for each $s \in S$ there exists a $t_s \in T$ such that for any $p \in P$, if $\eta(p)s$ is defined then $\eta(p)s = \eta(pt_s)$.*

In algebra, expressing a general structure in terms of simpler parts is always studied. For example, any finite group is isomorphic to the 'products' of simple subgroups using Jordan-Holder's theorem. Similarly, in ts theory, a ('useful') decomposition of a ts $\mathcal{A}$ is a covering of $\mathcal{A}$ by products of some $\overline{\mathcal{A}_i}$'s where each $\mathcal{A}_i$ is 'smaller' than $\mathcal{A}$ – in terms of cardinality of components in the pairs constituting the ts. Products involved in the decomposition may not always be direct products; they may be wreath products or others (such as cascade products). Here we say (Q_1, S_1) 'smaller' than (Q_2, S_2) when $(|S_1|, |Q_1|) < (|S_2|, |Q_2|)$ where the order '<' is lexicographical, i.e. $(m, n) < (r, s)$ if and only if either (a). $m < r$ or (b). $m = r$ and $n < s$.

The decomposition result obtained for reset ts is the following. $\prod^k \mathcal{A}_i$ denotes the direct product of $\mathcal{A}_i$, $i = 1, \ldots, k$.

Proposition 16. [46] *Any reset ts can be covered by $\prod^k \mathbf{2}$.*

For the general ts, we shall use the following results [34]. Recall the notation used in Example 7(B).

Proposition 17. [34] *Consider a ts $\mathcal{A} := (Q, S)$. We have the following decomposition results.*

1. *If S is a monoid and G denotes the maximal subgroup of S, then $\mathcal{A} \preccurlyeq (Q, S\backslash G)^{\cdot} \circ G$.*
2. *Let $p \in Q$ be such that $Qs \subseteq Q\backslash\{p\} \neq \emptyset$ for all $s \in S$. Then we have $\mathcal{A} \preccurlyeq \overline{\mathcal{B}} \circ C$, where $\mathcal{B} = \mathcal{A}|(Q\backslash\{p\})$ as in Observation 6 and $C := (\{0, 1\}, \{\overline{0}\})$ as in Example 6.*
3. *If L is a left ideal in S, and T a subsemigroup of S such that $L \cup T = S$, then $\mathcal{A} \preccurlyeq (Q, L)^{\cdot} \circ \overline{\mathcal{B}}$, where $\mathcal{B} := (T \cup \{1_Q\}, T)$.*

The above proposition is used to prove the following theorem, which in turn proves Krohn-Rhodes decomposition theorem (stated below).

Theorem 4. [34] *For a ts $\mathcal{A} := (Q, S)$ with $S \neq \emptyset$, we have the following decomposition*

$$\mathcal{A} \preccurlyeq \overline{\mathcal{A}_1} \circ \ldots \circ \overline{\mathcal{A}_n}$$

where $\mathcal{A}_i := (Q_i, G_i)$ is a tg with $G_i \preccurlyeq S$ for all $1 \leq i \leq n$.

Theorem 5 (Krohn Rhodes decomposition for transformation semi-groups [34]).
Every ts $\mathcal{A} := (Q, S)$ with $S \neq \emptyset$ can be decomposed in the following way:

$$\mathcal{A} \preccurlyeq \mathcal{A}_1 \circ \ldots \circ \mathcal{A}_n$$

where $\mathcal{A}_i$ is either $\overline{\mathbf{2}}^{\boldsymbol{\cdot}}$ or is a simple group satisfying $\mathcal{A}_i \preccurlyeq S$ for all $1 \leq i \leq n$.

2.4.2 Algebraic Automata Theory

Automata theory (cf. [47] was developed to model the concept of a machine. 'Machine' may represent switching circuits, or natural language [18], or a cellular system [58]. We now focus on connections of transformation semigroups with *semiautomata.*

The basic algebraic concepts in this context may be found in [34] and [1]. For all practical purposes here, we work with finite automata. Semiautomata are automata without outputs, defined in the following way [46]. Note that in literature, a semiautomaton is sometimes referred to as an 'automaton' or as a 'state machine'. Here, we shall use the term 'semiautomaton' only.

Definition 45 (Semiautomaton).
A semiautomaton is a triple $\mathcal{M} := (Q, \Sigma, \delta)$, where Q and Σ are finite sets, and $\delta : Q \times \Sigma \to Q$ is a partial function.

A semiautomaton $\mathcal{M}$ can be associated with the free semigroup Σ^*, and the partial function δ can be extended to define a semigroup action of Σ^* on the set of states Q. So Σ^* can be seen as a collection of transformations of Q. The relation between semiautomata and transformation semigroups of finite sets is given as follows.

Remark 6. Given any semiautomaton $\mathcal{M} := (Q, \Sigma, \delta)$, one can obtain a *ts* by forcing the action of the free semigroup Σ^* on Q to be faithful, as done in Observation 5 by defining a congruence relation $\sim$ on Σ^*. The pair $TS(\mathcal{M}) := (Q, \Sigma^*/\!\!\sim)$ forms a *ts*. Conversely, given a *ts* $\mathcal{A} := (Q, S)$, the triple $SM(\mathcal{A}) := (Q, S, \delta)$ is a semiautomaton, where δ is the semigroup action associated with the *ts* $\mathcal{A}$.

Definition 46 (Semiautomaton homomorphism [46]).
For semiautomata (Q, Σ, δ) and (P, Λ, γ), consider the functions $\alpha : Q \to P$ and $\beta : \Sigma \to \Lambda$ such that if $\alpha(\delta(q, s))$ is defined then

$$\alpha(\delta(q, s)) = \gamma(\alpha(q), \beta(s)) \text{ for any } q \in Q \text{ and } s \in \Sigma.$$

The pair (α, β) is called a semiautomaton homomorphism.

It can be shown that for a *ts* $\mathcal{A}$ and a semiautomaton $\mathcal{M}$, $TS(SM(\mathcal{A}))$ is isomorphic to $\mathcal{A}$, while there is a homomorphism from $\mathcal{M}$ to $SM(TS(\mathcal{M}))$.

In our work, we shall use the concept of a 'subautomaton'. Substructures of an automaton were first defined by Ginsburg [40], and studied extensively by others – the literature contains various definitions of subautomata depending on the applications. A discussion can be found in [61]. We consider the following definition in our work.

Definition 47 (Subautomaton [10]).
The tuple $\mathcal{M}' := (Q', \Sigma, \nu)$ *is a* subautomaton *of the semiautomaton* $\mathcal{M} :=$ (Q, Σ, δ) *if* $Q' \subseteq Q$ *and* $\nu = \delta$ *on* $Q' \times \Sigma$.

2.5   The Logic

The basic definitions covered here can be found in [75].

Definition 48 (Formalized language).
An alphabet of a formalized language *is the tuple* $\mathcal{L} := (V, L_0, L_1, L_2, U)$, *where*

(a) *V, L_0, L_1, L_2, and U are disjoint sets,*
(b) *V is an infinite set,*
(c) *L_0, L_1 and L_2 are finite sets,*
(d) *L_0 contains a propositional constant* $\top$,
(e) *L_2 contains a binary connective* $\rightarrow$, *called the* implication, *and*
(f) *U contains parentheses '(' and ')'.*

Elements of V are called propositional variables *(PV), denoted by $p, q, r, \dots$.*
Elements in L_0, L_1 and L_2 are called propositional constants, unary proposi-tional connectives, *and* binary propositional connectives *respectively.*
The set F of well-formed formulas *over the alphabet is defined in the standard way. The pair comprising the alphabet and set of formulas is said to be a* for-malized language $\mathcal{L}$.

Hereafter, a formalized language $\mathcal{L}$ shall be referred to as just a *language*. A *logic* L on a language is given by a set of axioms and rules of inference, and a 'consequence relation' $\vdash_L \subseteq \mathcal{P}(F) \times F$. $\vdash_L$ is defined as follows.

Definition 49 (Consequence relation $\Gamma \vdash_L \alpha$).
We say that α is provable from Γ in L, *denoted as $\Gamma \vdash_L \alpha$, where $\Gamma \cup \{\alpha\} \subseteq F$, if there exists a finite sequence of formulas $\{\alpha_i\}_{i=1}^n \subseteq F$, such that $\alpha := \alpha_n$, and each α_i, $1 \le i \le n$, satisfies either of these:*

(a) *$\alpha_i \in \Gamma$, or*
(b) *α_i is one of the axioms of* L, *or*
(c) *α_i is obtained from previous members of the sequence using the rules of inference in* L.

We write $\vdash_L \alpha$ if Γ is empty. The sequence of formulas $(\alpha_1, \alpha_2, \dots, \alpha_{n-1}, \alpha_n = \alpha)$ is called a proof of α.

Note: A logic L is called *consistent* if there exists an L-formula α such that $\nvdash_L \alpha$. In this thesis, we shall be dealing with consistent logics only.

An algebra $\mathcal{A}$, consisting of a set A and some operators, is said to be *associ-ated with* the language $\mathcal{L}$ if for each connective $\circ$ in $\mathcal{L}$, there exists an operator $\circ_A$ in $\mathcal{A}$ of the same order. When $\circ$ is a propositional constant, the operator $\circ_A$ is a constant element in A. The operator $\circ_A$ is said to be *associated* with $\circ$; and A is the set associated with the algebra $\mathcal{A}$.

We have already seen in Definition 8, the mappings between two algebras having same set of constants and n-ary operators, i.e. the two algebras have the same language $\mathcal{L}$. Let us now see the connection between the logic L and the algebra $\mathcal{A}$ associated with the language $\mathcal{L}$ of the logic L.

Definition 50 (Valuation).

A valuation v_0 of $\mathcal{L}$ on an algebra $\mathcal{A}$ associated with $\mathcal{L}$, is a map from the set PV of propositional variables in $\mathcal{L}$ to the set A, where A is the underlying set in the algebra $\mathcal{A}$.

Every valuation v_0 of $\mathcal{L}$ on the algebra $\mathcal{A}$ can be uniquely extended to give a map v from the set F of formulas to the set A as follows:

(1) For every propositional variable $p \in F$, $v(p) := v_0(p)$;
(2) For every propositional constant $\circ$, $v(\circ) := \circ_{\mathcal{A}}$, where $\circ_{\mathcal{A}}$ is the constant element associated with the unary connective $\circ$;
(3) For every unary connective $\circ$, $v(\circ\alpha) := \circ_{\mathcal{A}}v(\alpha)$, where $\circ_{\mathcal{A}}$ is a the unary operator associated with the unary connective $\circ$, $\alpha \in F$, and $v(\alpha)$ $(\in A)$ is already known; and
(4) For every binary connective $\circ$, $v(\alpha_1 \circ \alpha_2) := (v(\alpha_1)\circ_{\mathcal{A}}v(\alpha_2))$, where $\circ_{\mathcal{A}}$ is the binary operator associated with the binary connective $\circ$, for $i = 1, 2$, $\alpha_i \in F$ and $v(\alpha_i)$ $(\in A)$ are already known.

Note: Hereafter, we shall drop the suffix $\mathcal{A}$, and denote a connective in the logic and the operator corresponding to the connective by the same notation. For example, $\rightarrow$ shall denote the binary operator in $\mathcal{A}$ associated with the binary connective $\rightarrow$ in $\mathcal{L}$. $1 \in \mathcal{A}$ shall denote the constant associated with the constant $\top$ in $\mathcal{L}$. We shall also use the same notation for a valuation on propositional variables and its extension to the set of all formulas.

Definition 51 (Truth and validity). *A formula $\alpha \in F$ is said to be* true *for the valuation v on $\mathcal{A}$ (notation - $\mathcal{A}, v \vDash \alpha$), if $v(\alpha) = 1$.*
If for every valuation v on $\mathcal{A}$, $\mathcal{A}, v \vDash \alpha$, then we say that α is valid *in $\mathcal{A}$ (notation - $\mathcal{A} \vDash \alpha$).*

Definition 52 (L-algebras [75]).
Consider an algebra $\mathcal{A}$ associated with the language $\mathcal{L}$ of the logic L. $\mathcal{A}$ is an L-algebra *provided that*

(1) *if α is an axiom in L, then for every valuation v of $\mathcal{L}$ on $\mathcal{A}$, we have $v(\alpha) = 1$;*
(2) *if a rule (r) in L assigns to premises $\alpha_1, \alpha_2, \ldots, \alpha_n$ the conclusion α, then for every valuation v on $\mathcal{L}$ on $\mathcal{A}$, $v(\alpha_i) = 1$ (for all $1 \leq i \leq n$) implies $v(\alpha) = 1$;*
(3) *for all $a, b, c \in \mathcal{A}$, $a \rightarrow b = 1$ and $b \rightarrow c = 1$ imply $a \rightarrow c = 1$; and*
(4) *for all $a, b \in \mathcal{A}$, $a \rightarrow b = 1$ and $b \rightarrow a = 1$ imply $a = b$.*

Definition 53 (L-valid).
A formula $\alpha \in F$ is called L-valid in a class $\mathcal{C}$ of algebras, if $\mathcal{A} \vDash \alpha$ for every algebra $\mathcal{A}$ in $\mathcal{C}$.

For providing a logic L with algebraic semantics with respect to a class $\mathcal{C}$ of algebras, one proves, for any $\alpha \in F$,

(A). (Soundness) $\vdash_L \alpha$ implies α is L-valid in $\mathcal{C}$, and

(B). (Completeness) α is L-valid in $\mathcal{C}$ implies $\vdash_L \alpha$.

To establish (A), one shows that any algebra $\mathcal{A} \in \mathcal{C}$ is an L-algebra. For (B), one constructs the 'Lindenbaum-Tarski algebra' $\mathcal{U}(L)$ on the set F of L-formulas as follows.

Definition 54 (Lindenbaum-Tarski algebra $\mathcal{U}(L)$).

Define a relation on the set F of formulas in the language $\mathcal{L}$ as follows: $\alpha \simeq \beta$ if and only if $\vdash_L (\alpha \to \beta)$ and $\vdash_L (\beta \to \alpha)$. We assume that the logic L has axioms and rules of inference such that $\simeq$ becomes an equivalence relation on F. The following operators are then defined on the quotient set $F/\simeq$. Let $[\alpha], [\beta] \in F/\simeq$.

(1) *For every unary connective $\circ$ in $\mathcal{L}$, $\circ[\alpha] := [\circ\alpha]$.*

(2) *For every binary connective $*$ in $\mathcal{L}$, $[\alpha] * [\beta] := [(\alpha * \beta)]$.*

Then $(F/\sim, [\top], [e_i], \dots, [e_r], \circ_1, \dots, \circ_s, \to, *_1, \dots, *_t)$ is called the Lindenbaum-Tarski algebra of L, *where e_i's, $\circ_i$'s and $*_i$'s are propositional constants, unary connectives and binary connectives in $\mathcal{L}$ respectively.*

For completeness of L with respect to the class $\mathcal{C}$, one has to ensure that (i) the operations in $\mathcal{U}(L)$ are well-defined and $[\top]$ is the top element consisting of exactly L-theorems, and (ii) $\mathcal{U}(L)$ belongs to $\mathcal{C}$. Note that $\mathcal{U}(L)$ is an L-algebra, and as shown in [75], L is sound and complete with respect to the class of L-algebras. For the kind of algebraic structures that we use in our work, if we have soundness and completeness of L with respect to the class of L-algebras as well as a class $\mathcal{C}$ of algebras, the two classes become identical.

Let us now look at the ways to connect two different logics. Consider any two logics L and L' over languages $\mathcal{L}$ and $\mathcal{L}'$ respectively, such that $\mathcal{L}$ and $\mathcal{L}'$ have the same set PV of propositional variables and $\{\top, \to\} \in \mathcal{L} \cap \mathcal{L}'$. The sets of formulas are denoted by F and F' respectively. *Equivalence* between the logics is defined as follows.

Definition 55 (Equivalence between logics).

The logic L is equivalent to the logic L', denoted as $L \cong L'$, if there exist two maps $\theta : F \to F'$ and $\rho : F' \to F$ such that for all $\alpha, \alpha_1, \dots, \alpha_n \in F$ and $\beta, \beta_1, \dots, \beta_n \in F'$, the following hold.

(1) $\vdash_L \alpha \Leftrightarrow \vdash_{L'} \theta(\alpha)$.

(2) $\vdash_{L'} \beta \Leftrightarrow \vdash_L \rho(\beta)$.

(3) $\vdash_L \alpha \leftrightarrow \rho(\theta(\alpha))$ *and* $\vdash_{L'} \beta \leftrightarrow \theta(\rho(\beta))$.

(4) θ *and* ρ *preserve all propositional variables, that is, $\theta(p) = p = \rho(p)$ for all $p \in PV$.*

(5) θ *and* ρ *preserve all common constants and n-ary connectives, that is, for all constants $c \in \mathcal{L} \cap \mathcal{L}'$ and n-ary connectives $* \in \mathcal{L} \cap \mathcal{L}'$,*

i. $\theta(c) = c = \rho(c)$,

ii. $\theta(*(\alpha_1, \ldots, \alpha_n)) = *(\theta(\alpha_1), \ldots, \theta(\alpha_n))$ *and*
$\rho(*(\beta_1, \ldots, \beta_n)) = *(\rho(\beta_1), \ldots, \rho(\beta_n))$.

(6) *For constants* $c \in \mathcal{L}$ *and* $c' \in \mathcal{L}'$ *not in the intersection of the languages,* $\theta(c)$ *and* $\rho(c')$ *consist entirely of n-ary connectives and constants, that is,* $\theta(c)$ *and* $\rho(c')$ *do not contain any propositional variable.*

Note that if $\mathcal{L} = \mathcal{L}'$, that is, $F = F'$, both θ and ρ are just the identity maps. Conditions (1)–(5) give *definition equivalence* between two logics as in [69]. For equivalence of logics that may have constants outside the intersection of their languages, Condition (6) is added to ensure that the translation of any such constant also behaves like a constant.

Observation 8. Using the definition of the consequence relation $\vdash$, if we show that θ images of the axioms in the logic L are theorems in the logic L$'$ and the rules in L are derivable in L$'$, and ρ images of the axioms in the logic L$'$ are theorems in the logic L and the rules in L$'$ are derivable in L, then we have L $\cong$ L$'$.

We may not always get the equivalence between two logics. However, there are other ways to relate two different logics. Various definitions of mappings from one logical system to another can be found in literature. The very first studies of logic connections were done by Kolmogorov in 1925 and Glivenko in 1929 (cf. [74]). Some definitions of mappings from one logic into another, called translations, can be found in [14,37]. A detailed study of connections between various propositional logics can be found in [74], which has defined the term 'interpretable'. We shall use a more general definition of interpretation.

Definition 56 (Interpretation [64]).
Consider two logics L *and* L$'$. *A mapping* $r : F \to F'$, *from the set* F *of formulas in* L *to the set* F' *of formulas in* L$'$, *is called an* interpretation *of* L *in* L$'$, *if for any formula* $\alpha \in F$, *we have the following condition:*

$$\vdash_{\mathrm{L}} \alpha \text{ if and only if } \Delta_\alpha \vdash_{\mathrm{L}'} r(\alpha),$$

where $\Delta_\alpha \subseteq F'$ *is a finite set corresponding to* α.

The mapping r *is called an* interpretation *of* L *in* L$'$ *with respect to derivability, if for any set* $\Gamma \cup \{\alpha\} \subseteq F$, *we have*

$$\Gamma \vdash_{\mathrm{L}} \alpha \text{ if and only if } r(\Gamma) \cup \Delta_\alpha \vdash_{\mathrm{L}'} r(\alpha),$$

where $\Delta_\alpha \subseteq F'$ *is a finite set corresponding to* α.

Definition 57 (Embedding of logics).
Let L *and* L$'$ *be two logics, and* F *and* F' *be the sets of formulas in the languages of* L *and* L$'$ *respectively. We say that* L *is* embedded *in* L$'$ *(notation -* L $\prec$ L$'$*) provided there exists a map* $r : F \to F'$ *such that for any* $\alpha \in F$, $\vdash_{\mathrm{L}} \alpha$ *if and only if* $\vdash_{\mathrm{L}'} r(\alpha)$.
In case r *is the inclusion map,* L$'$ *is called an* extension *of* L.

2.5.1   Positive Logic and Its Extensions

[71,75] The alphabet of positive logic (PL) consists of propositional variables $p, q, r, \ldots$, and binary connectives $\wedge, \vee, \rightarrow$. The class F of well-formed formulas is defined recursively and given by the scheme:

$$p \mid \alpha \wedge \beta \mid \alpha \vee \beta \mid \alpha \rightarrow \beta$$

Abbreviation: $\alpha \leftrightarrow \beta := (\alpha \rightarrow \beta) \wedge (\beta \rightarrow \alpha)$, where $\alpha, \beta \in F$.

Definition 58 (Positive logic).
The axiomatic scheme for PL *is as follows:*

> (A1). $\alpha \rightarrow (\beta \rightarrow \alpha)$
> (A2). $(\alpha \rightarrow (\beta \rightarrow \gamma)) \rightarrow ((\alpha \rightarrow \beta) \rightarrow (\alpha \rightarrow \gamma))$
> (A3). $\alpha \rightarrow (\alpha \vee \beta)$
> (A4). $\beta \rightarrow (\alpha \vee \beta)$
> (A5). $(\alpha \rightarrow \gamma) \rightarrow ((\beta \rightarrow \gamma) \rightarrow ((\alpha \vee \beta) \rightarrow \gamma))$
> (A6). $(\alpha \wedge \beta) \rightarrow \alpha$
> (A7). $(\alpha \wedge \beta) \rightarrow \beta$
> (A8). $(\alpha \rightarrow \beta) \rightarrow ((\alpha \rightarrow \gamma) \rightarrow (\alpha \rightarrow (\beta \wedge \gamma)))$

*Rule of Inference-Modus Ponens (*MP*):*

$$\frac{\alpha \rightarrow \beta, \quad \alpha}{\beta}$$

With the above axioms and the rule of inference, we can define the 'consequence (binary) relation' $\vdash_{PL} \subseteq \mathcal{P}(F) \times F$ using Definition 49. By virtue of this definition, the following holds in PL.

Proposition 18. *For any set* $\Delta \cup \Delta' \cup \{\alpha, \beta\}$ *of formulas in the language of positive logic, we have the following.*

(1) $\alpha \in \Delta \Rightarrow \Delta \vdash_{PL} \alpha$.
(2) $\Delta \vdash_{PL} \alpha \Rightarrow \Delta \cup \Delta' \vdash_{PL} \alpha$.
(3) $\Delta \vdash_{PL} \alpha$ *&* $\Delta' \cup \{\alpha\} \vdash_{PL} \beta \Rightarrow \Delta \cup \Delta' \vdash_{PL} \beta$.

Note that a constant $\top$ (top) can be defined in this language, given as $\top := \alpha \rightarrow \alpha$, where α is any (fixed) formula in the language. Then, we can obtain the following formula as a theorem in PL:

(A9). $\beta \rightarrow \top$.   ($\top$- top)

Therefore, if we take β as a theorem in PL, i.e. $\vdash_{PL} \beta$, then using MP again on $\vdash_{PL} \beta \rightarrow \top$, we have $\vdash_{PL} \top$ as a theorem in PL.
In fact, we can include $\top$ in the language of PL itself as a constant and take (A9) as an axiom in the axiomatic system for PL (Definition 58). In this case, the following is obtained as a theorem in PL.

$$\vdash_{\mathrm{PL}} \top \leftrightarrow (\alpha \to \alpha)$$

Based on the above discussion, we shall assume the language of PL as given by the scheme:

$$\top \mid p \mid \alpha \wedge \beta \mid \alpha \vee \beta \mid \alpha \to \beta$$

The following two results (Propositions 19 and 20) are very useful in proving theorems in PL. Let us first recall the concept of the set $sub(\alpha)$ of subformulas of a formula α. The definition is obtained recursively.

Definition 59 (Subformula of a formula).
For a formula α, the set of subformulas of α, denoted as $\mathrm{sub}(\alpha)$, is defined recursively as follows:

(1) *If α is a propositional variable or a constant, $\mathrm{sub}(\alpha) := \{\alpha\}$;*
(2) *For any unary connective !, $\mathrm{sub}(!\alpha) := \mathrm{sub}(\alpha) \cup \{!\alpha\}$; and*
(3) *For a binary connective $\circ$, $\mathrm{sub}(\alpha \circ \beta) := \mathrm{sub}(\alpha) \cup \mathrm{sub}(\beta) \cup \{\alpha \circ \beta\}$.*

Any formula in the set $\mathrm{sub}(\alpha)$ is called a subformula of α.

Proposition 19 (Deduction Theorem in PL [75]).
For any set of formulas $\Delta \cup \{\alpha, \beta\}$ in the language of positive logic, we have the following.

$$\Delta \cup \{\alpha\} \vdash_{\mathrm{PL}} \beta \Rightarrow \Delta \vdash_{\mathrm{PL}} (\alpha \to \beta).$$

The converse also holds in positive logic, i.e. $\Delta \vdash_{\mathrm{PL}} (\alpha \to \beta) \Rightarrow \Delta \cup \{\alpha\} \vdash_{\mathrm{PL}} \beta$.

Proposition 20 (Equivalence Theorem in PL [75]).
Let α, β, γ be formulas in PL such that $\beta \in \mathrm{sub}(\alpha)$. Suppose α' is a formula obtained by replacing some (or all) occurrences of β in α by γ. Then we have

$$\vdash_{\mathrm{PL}} \alpha \ \& \ \vdash_{\mathrm{PL}} \beta \leftrightarrow \gamma \ \Rightarrow \ \vdash_{\mathrm{PL}} \alpha'.$$

Note: All the logics considered in this section satisfy the corresponding versions of Propositions 18, 19 and 20, where PL is replaced by the concerned logic in the discussion.

Let us now move to the logics with negation - by adding a unary connective $\neg$ to the language of PL.

Definition 60 (Minimal logic (ML) [75]).
Define a logic with the axiomatic system same as that of PL (Axioms (A1)-(A9) and MP) along with the following additional axiom:

(A10). $(\alpha \to \beta) \to ((\alpha \to \neg\beta) \to \neg\alpha)$ *($\neg$ reductio ad absurdum)*

The resulting logic obtained is called minimal logic (ML).
The negation operator $\neg$ is called the minimal negation.

Proposition 21. [75] *The following hold in ML, for any ML-formulas* α, β.

1. $\vdash_{\text{ML}} (\alpha \to \beta) \to (\neg\beta \to \neg\alpha)$
2. $\vdash_{\text{ML}} (\alpha \to \neg\beta) \to (\beta \to \neg\alpha)$
3. $\vdash_{\text{ML}} (\alpha \to \neg\neg\alpha)$
4. $\vdash_{\text{ML}} (\top \to \alpha) \leftrightarrow \alpha$
5. $\vdash_{\text{ML}} \neg\alpha \leftrightarrow (\alpha \to \neg\top)$
6. $\vdash_{\text{ML}} (\neg\alpha \leftrightarrow \neg\neg\neg\alpha)$

Observation 9. In ML, a new constant $\bot$ (bottom) can be defined as $\bot :=$ $\neg\top$. Further, motivated by the above Proposition 21(5), we have an equivalent definition of ML. The alphabet is the same as that of PL with the addition of a propositional constant $\bot$; the axiomatic system has (A1)-(A9) as axioms and MP. Note that there is no negation $\neg$ in this case. However, we can define a unary connective '$\neg$' as $\neg\alpha := \alpha \to \bot$. Then (A10) can be obtained as a theorem in this system (cf. [71]). Recall Definition 55 and Observation 8. Define a map θ over the appropriate set of formulas such that for any ML-formula α, $\theta(\alpha)$ is obtained from α by replacing every subformula of the form $\neg\beta$ by the formula $\beta \to \bot$. Define another map ρ over the appropriate set of formulas by mapping every instance of $\bot$ in any subformula to $\neg\top$. Then by Observation 8, we have the equivalence of both the logical systems for ML. A detailed study of ML can be found in [71, 75].

Note that unlike $\top$ for which $\vdash_{\text{ML}} \beta \to \top$, the following is *not* a theorem in ML.

(A11). $\bot \to \beta$. ($\bot$- bottom)

Addition of the above formulas in the axiomatic system gives us *Intuitionistic logic* (IL).

Definition 61 (Intuitionistic logic (IL) [75]**).**
The alphabet of IL *is same as that of* ML, *and formulas are given by the scheme:*

$$\top \mid p \mid \alpha \wedge \beta \mid \alpha \vee \beta \mid \alpha \to \beta \mid \neg\alpha$$

The axiomatic system for IL *has Axioms* (A1) − (A10), *the rule of inference* MP, *and the following axiom:*

(A12). $\neg\alpha \to (\alpha \to \beta)$

If negation $\neg$ *satisfies* (A10) *and* (A12), *then it is called* intuitionistic negation *(or* pseudo-complemented negation*).*

Observation 10.

(1) In this case, if we define $\bot := \neg\top$ then (A11) is a theorem in IL.
(2) As observed for ML, IL can also be defined by an alphabet without the unary connective $\neg$. Consider the alphabet of PL with the addition of the constant $\bot$. Define the axiomatic system by taking the Axioms (A1)–(A9) and (A11); and rule of inference MP. If we define $\neg$ as $\neg\alpha := \alpha \to \bot$, then (A10) and (A12) can be obtained as a theorem in IL. Thus, using the mappings defined in Observation 9, both the logical systems for IL are equivalent.

Note: For our convenience, in this work, we shall include both the connectives - a propositional constant $\perp$ and a unary connective $\neg$ - in the alphabet of IL, and the axiomatic system for IL will be given by the Axioms (A1)–(A12) and the rule MP.

A distinguishing feature between ML and IL is the following formula, which is a theorem in IL but not in ML.

$$(\alpha \wedge \neg\alpha) \rightarrow \beta \qquad\qquad (\neg \text{ Explosion})$$

This can be proved in IL using (A12). Any logic with negation where $\neg$-Explosion is not a theorem is called a *Paraconsistent logic*. ML is an example of such a logic. A detailed discussion on such logics can be found in [71]. Another formula which is important in the study of propositional logic is the following.

(A13). $\alpha \vee \neg\alpha$ $\qquad\qquad\qquad\qquad\qquad\qquad$ ($\neg$ LEM)

Definition 62 (Classical logic (CL)).
The alphabet of CL is same as that of IL, and formulas given by the scheme:

$$\top \mid \perp \mid p \mid \alpha \wedge \beta \mid \alpha \vee \beta \mid \alpha \rightarrow \beta \mid \neg\alpha$$

The axiomatic system for CL has Axioms (A1)-(A13) and the rule of inference MP. If negation $\neg$ satisfies (A11), (A12) and (A13), then we call it classical negation.

The following correspondence between the above logical systems and the classes of algebras are important in our work.

Theorem 6. *The logics - PL, ML, IL and CL are sound and complete with respect to the class of rpc lattices, cc lattices, pBa and Boolean algebras respectively.*

2.5.2 Relational Semantics for the Logics
Kripke defined a semantics for IL [55], where frames are partially ordered sets $(W, \leq)$. The relation $\leq$ is the 'intuitionistic' accessibility relation. Using the posets $(W, \leq)$, Segerberg [78] constructed a natural semantics for ML and its various extensions. To capture ML, he added to the frames $(W, \leq)$ a hereditary set Y_0 of 'queer' worlds at each of which $\perp$ holds [78]. ML is the weakest logic that can be captured by models of the type $(W, \leq, Y_0)$ [27]. More recent work on extensions of minimal logic and its semantics can be found in [71].

Meanwhile, Došen [27] and Vakarelov [85] gave different Kripke-style relational semantics for ML and IL, treating negation as an impossibility (modal) operator. In fact, their approach was applied to logics with negation weaker than minimal negation [27].

Let us first see the Kripke semantics for ML and its extensions as given by Segerberg.

2.5.2.1 The j-frames

Recall the definition of an *upset* on a poset $(W, \leq)$ (Definition 10).

Definition 63 (*j*-frame and normal frame).
The triple $(W, \leq, Y_0)$, *where* $(W, \leq)$ *is a poset with* $W \neq \emptyset$ *and* Y_0 *is an upset of* W, *is called a* j-frame.
The j-frame $(W, \leq, \emptyset)$ *is called a* normal *frame, denoted by the pair* $(W, \leq)$.
The sets W, Y_0 *and* $W \backslash Y_0$ *are called the sets of* worlds, non-normal worlds *and* normal worlds *respectively.*

The simplest examples of a j-frame are of the form $\mathcal{F} := (X, \leq, Y_0)$, where Y_0 is either empty or equal to X.

We have seen the definition of a valuation on an algebra (Definition 50); and the truth and validity (Definition 51) of a formula in the algebra. Let us now define the valuation of a logic on a j-frame; and truth and validity of a formula in j-frames. Consider the language $\mathcal{L}$ and the logic ML (Definition 60). Let F denote the set of formulas in $\mathcal{L}$.

Definition 64 (Extension of ML).
Consider a logic S *such that the language of* S *is* $\mathcal{L}$; S *is deductively closed under* MP; *and all the theorems of* ML *are theorems of* S. *Then* S *is called an* extension *of* ML.

For any extension S of ML, we can define $\Gamma \vdash_S \alpha$ using Definition 49, where $\Gamma \cup \{\alpha\} \subseteq F$. Let us see the definition of valuations, truth and validity of a formula α on j-frames.

Definition 65 (Valuation on a *j*-frame).
A valuation $v_0 : PV \to \mathcal{P}(W)$ *of* $\mathcal{L}$ *on a* j-frame $\mathcal{F} := (W, \leq, Y_0)$ *is defined as a mapping from the set of propositional variables (PV) in* $\mathcal{L}$ *to the power set* $\mathcal{P}(W)$ *of* W *such that* $v_0(p)$ *is an upset of* W *for any* $p \in PV$.

Definition 66 (Model on a *j*-frame).
Consider a valuation v_0 *of* $\mathcal{L}$ *on a* j-frame $\mathcal{F} := (W, \leq, Y_0)$. *The pair* $\mathcal{M} := (\mathcal{F}, v_0)$ *is called a* model *on the* j-frame $\mathcal{F}$.

Definition 67 (Truth of a formula).
The truth *of a formula* α *at a world* $w \in W$ *in the model* $\mathcal{M} := (\mathcal{F}, v_0)$ *(notation* - $\mathcal{M}, w \vDash \alpha$*) is defined by extending the map* $v_0 : PV \to \mathcal{P}(W)$ *to the set* F *of formulas as follows:*

1. $\mathcal{M}, w \vDash p \Leftrightarrow w \in v_0(p)$, *for all propositional variables* p.
2. $\mathcal{M}, w \vDash \phi \wedge \psi \Leftrightarrow \mathcal{M}, w \vDash \phi$ *and* $\mathcal{M}, w \vDash \psi$.
3. $\mathcal{M}, w \vDash \phi \vee \psi \Leftrightarrow \mathcal{M}, w \vDash \phi$ *or* $\mathcal{M}, w \vDash \psi$.
4. $\mathcal{M}, w \vDash \phi \to \psi \Leftrightarrow$ *for all* $w' \in W$, *if* $w \leq w'$ *and* $\mathcal{M}, w' \vDash \phi$ *then* $\mathcal{M}, w' \vDash \psi$.
5. $\mathcal{M}, w \vDash \neg\phi \Leftrightarrow$ *for all* $w' \in W$, *if* $w \leq w'$ *and* $\mathcal{M}, w' \vDash \phi$ *then* $w' \in Y_0$.
6. $\mathcal{M}, w \vDash \top$.

$\mathcal{M}, w \vDash \phi$ *is read as '*ϕ *is true at the world* w *in the model* $\mathcal{M}$'.
A formula α *is said to be* true in a model $\mathcal{M}$ *(notation* - $\mathcal{M} \vDash \alpha$*) if* $\mathcal{M}, w \vDash \alpha$ *for all* $w \in W$.

The notation '$\mathcal{M}, w \not\models \phi$' denotes that ϕ is not true at the world $w \in W$ in the model $\mathcal{M}$.

Definition 68 (Validity of a formula).
A formula $\alpha \in F$ is valid in the j-frame $\mathcal{F}$ (notation - $\mathcal{F} \models \alpha$) if $\mathcal{M} \models \alpha$ for all models $\mathcal{M}$ on the j-frame $\mathcal{F}$.
A formula $\alpha \in F$ is valid in a class $\mathcal{C}$ of j-frames if for any j-frame $\mathcal{F} \in \mathcal{C}$ we have $\mathcal{F} \models \alpha$.

Definition 69.
A logic L is determined by the class $\mathcal{C}$ of frames or $\mathcal{C}$ characterizes the logic L, if it is sound and complete with respect to $\mathcal{C}$, i.e. for any formula $\alpha \in F$, $\vdash_L \alpha$ if and only if α is valid in $\mathcal{C}$.

Segerberg gave the characterization of ML and its various extensions [78].

Proposition 22. [78]

(A) ML *is determined by the class of all j-frames.*
(B) IL *is determined by the class of all normal frames.*

Observation 11. Consider the logic IL and the truth of a formula ϕ at a world $w \in W$ in the normal model $((W, \leq), v)$. Since Y_0 is empty, Definition 67(5) can be re-expressed as follows:

7. $\mathcal{M}, w \models \neg\phi \Leftrightarrow$ for all $w' \in W$, if $w \leq w'$ then $\mathcal{M}, w' \not\models \phi$.

In IL, we have $\bot := \neg\top$. Thus, the validity definition for $\bot$ is:

8. $\mathcal{M}, w \not\models \bot$.

Following is an extension of ML and its characterizing class of j-frames which shall be used in our work.

Proposition 23. [78] *Consider the logic L in the language $\mathcal{L}$ such that L satisfies all the axioms and rules of ML, along with the axiom (A13) : $(\alpha \vee \sim\alpha)$. Then the logic L is determined by the class of all j-frames $(W, \leq, Y_0)$ satisfying the following condition.*

$$\forall x, y \notin Y_0 \Rightarrow (x \leq y \Rightarrow y \leq x).$$

The above condition can be interpreted as follows: in any j-frame $(W, \leq, Y_0)$, the relation $\leq$ is the identity relation on the set of normal worlds $W \backslash Y_0$. We know that if we add (A13) as an axiom to IL, we obtain CL (Definition 62). In the characterizing class of j-frames $(W, \leq, Y_0)$ of IL, Y_0 is empty, i.e. $W \backslash Y_0 = W$. Thus, we have the characterizing class of j-frames for CL.

Proposition 24. [78] *CL is determined by the class of all normal frames $(W, \leq)$ such that $\leq$ is the identity relation on W.*

2.5.2.2 The N-frames

In the above semantics with respect to j-frames, negation has been expressed in terms of implication. A major drawback of j-frames is that the class of j-frames cannot be used to characterize logics with negation weaker than minimal negation. This problem was addressed by Došen [27] and Vakarelov [85] independently. In their work, taking motivation from Kripke frames for modal logic and its modal accessibility relation, the negation is considered as an impossibility (modal) operator. Thus to define the semantics one refers to the relational semantics for modal logic given by Kripke [55]. For a unary modal operator $\Diamond$ in basic modal logic, the semantics in the model (W, R) is defined for any formula α and possible world $w \ (\in W)$ as:

$$w \vDash \Diamond\alpha \Leftrightarrow \exists v \in W(wRv \ \& \ v \vDash \alpha).$$

Treating negation as an 'impossibility' modal operator, represented as $\not\Diamond$, means: Whenever w satisfies $\not\Diamond\alpha$, the right side of the above equation should never be true, that is, α should not be true at any world 'accessible' from w.

$$w \vDash \not\Diamond\alpha \Leftrightarrow \forall v \in W(wRv \Rightarrow v \nvDash \alpha).$$

A detailed discussion on negation as modal operator can be found in [28]. Let us see what are the frames in Došen's semantics.

Definition 70 (N-frames, condensed and strictly condensed N-frames [27]).

(1) An N-frame is a triple $\mathcal{F} := (X, R_I, R_N)$ satisfying the following properties:
   (a) X is a non-empty set,
   (b) $R_I \subseteq X \times X$ such that R_I is reflexive and transitive, and
   (c) $R_N \subseteq X \times X$ such that $R_I R_N \subseteq R_N R_I^{-1}$.
(2) An N-frame $\mathcal{F} := (X, R_I, R_N)$ such that $R_I R_N \subseteq R_N$ is called a condensed N-frame.
(3) A condensed N-frame $\mathcal{F} := (X, R_I, R_N)$ such that $R_N R_I^{-1} \subseteq R_N$ is called a strictly condensed N-frame.

Here $R_I R_N$ or $R_N R_I^{-1}$ are composition of relations and R_I^{-1} represents the inverse of R_I.
The set X is called the set of worlds.

Consider a language $\mathcal{L}_\sim$ given by the following scheme for formulas:

$$\mathcal{L}_\sim := p \mid \alpha \wedge \beta \mid \alpha \vee \beta \mid \alpha \to \beta \mid \sim\alpha$$

Let F denote the set of formulas in $\mathcal{L}_\sim$. Consider a logic S on the language $\mathcal{L}_\sim$. Let us see the definition of valuations, truth and validity of a formula α on N-frames.

Definition 71 (Valuation on an N-frame).

A valuation v_0 of $\mathcal{L}_\sim$ on an N-frame $\mathcal{F} := (X, R_I, R_N)$ is a mapping from the set of propositional variables in $\mathcal{L}_\sim$ to the power set $\mathcal{P}(X)$ of X such that for any propositional variable p, $v_0(p)$ satisfies the following: for every $x, y \in X$,

$$(xR_Iy \ \& \ x \in v_0(p)) \Rightarrow y \in v_0(p).$$

Definition 72 (Model on an N-frame).
Consider a valuation v_0 of $\mathcal{L}$ on an N-frame $\mathcal{F} := (X, R_I, R_N)$. The pair $\mathcal{M} := (\mathcal{F}, v_0)$ is called a model on the N-frame $\mathcal{F}$.

Definition 73 (Truth of a formula).
The truth of a formula α at a world $x \in X$ in the model $\mathcal{M} := (\mathcal{F}, v_0)$ (notation - $\mathcal{M}, x \vDash \alpha$) is defined by extending the valuation map $v_0 : PV \to \mathcal{P}(X)$ to the set F of formulas as follows:

(1) $\mathcal{M}, x \vDash p \Leftrightarrow x \in v_0(p)$ *for all propositional variables p.*
(2) $\mathcal{M}, x \vDash \phi \wedge \psi \Leftrightarrow \mathcal{M}, x \vDash \phi$ *and* $\mathcal{M}, x \vDash \psi$.
(3) $\mathcal{M}, x \vDash \phi \vee \psi \Leftrightarrow \mathcal{M}, x \vDash \phi$ *or* $\mathcal{M}, x \vDash \psi$.
(4) $\mathcal{M}, x \vDash \phi \to \psi \Leftrightarrow$ *for all $y \in X$, if xR_Iy and $\mathcal{M}, y \vDash \phi$ then $\mathcal{M}, y \vDash \psi$.*
(5) $\mathcal{M}, x \vDash {\sim}\phi \Leftrightarrow$ *for all $y \in X$, if xR_Ny then $\mathcal{M}, y \nvDash \phi$.*

$\mathcal{M}, x \vDash \phi$ *is read as 'ϕ is true at the world x in the model $\mathcal{M}$'.*
A formula α is said to be true *in the model $\mathcal{M}$ (notation - $\mathcal{M} \vDash \alpha$) if $\mathcal{M}, x \vDash \alpha$ for all $x \in X$.*

Note:
(1). Validity is defined as in Definition 68, where j- frame is replaced by N-frame.
(2). The definition of soundness, completeness and characterizing classes of N-frames are as in Definition 69.

Došen gave characterizing conditions on N-frames for various extensions of logics with negation by putting conditions on R_N [27].

Definition 74 (J-frames and H-frames).
An N-frame $\mathcal{F} := (X, R_I, R_N)$ is called a J-frame, if it satisfies the following conditions.

(1) $R_N R_I^{-1}$ *is symmetric, and*
(2) $\forall x, y \in X(xR_Ny \Rightarrow \exists z \in X(xR_Iz \,\&\, yR_Iz \,\&\, xR_Nz))$;

A J-frame $\mathcal{F} := (X, R_I, R_N)$ is called an H-frame if the relation $R_N R_I^{-1}$ is reflexive.

The following characterizations for ML and IL are obtained.

Proposition 25. [27] *The class of strictly condensed J-frames and the class of strictly condensed H-frames, where R_I is a partial order, characterize ML and IL respectively.*

In our work, we shall utilize strictly condensed J-frames and H-frames, where R_I is a partial order, denoted by $\leq$. The intuitive explanation of R_I and R_N is given by Došen [28] as follows: for any N-frame $\mathcal{F} := (X, R_I, R_N)$,
(1). xR_Iy is read as 'y extends x',
(2). xR_Ny is read as 'x and y are *compatible*'.
In the reading of xR_Ny, R_N can be interpreted as a symmetric relation. This may not be true in general. However, in all the frames considered in this work,

R_N is considered to be symmetric. We can interpret properties of R_I and R_N in terms of the above understanding. For example, in Definition 74, the property

$$\forall x, y \in X(x R_N y \Rightarrow \exists z \in X(x R_I z \ \& \ y R_I z \ \& \ x R_N z))$$

can be read as 'if x is compatible with y then there is an extension of x and y, which is compatible with x.
In strictly condensed J-frame $\mathcal{F} := (X, R_I, R_N)$, R_N can be defined using R_I as follows: $x R_N y \Leftrightarrow \exists z \in X(x R_I z \ \& \ y R_I z)$. This can be read as follows: x is compatible with y if and only if there exists a common extension of x and y. Detailed discussion on this can be found in [28].

2.5.2.3 Connection Between Different Semantics

Došen had shown that models on the strictly condensed J-frames are 'inter-translatable' with j-frames for ML, preserving the truth of any formula $\alpha \in F$ at any world w of the respective frame. Let us see the translation.

Theorem 7. [27] *Let $\mathcal{F} := (W, \leq, Y_0)$ be a j-frame. Define the relation R_N over W as follows: for all $x, y \in W$, $x R_N y$ if and only if $\exists z \in W(x \leq z \ \& \ y \leq z \ \& \ z \notin Y_0)$. Then we have the following.*

(1) $Y_0 := \{z \in W \mid \exists x, y \in W(x \leq z \ \& \ y \leq z \ \& \ x \acute{R}_N y)\}$.
(2) $(W, \leq, R_N)$ *is a strictly condensed J-frame. Denote this frame as $\Phi(\mathcal{F})$.*
(3) *If v is a valuation on $\mathcal{F}$, then v is a valuation on $\Phi(\mathcal{F})$ such that for all ML-formulas ϕ and $x \in W$, $(\mathcal{F}, v), x \vDash \phi$ if and only if $(\Phi(\mathcal{F}), v), x \vDash \phi$.*

Theorem 8. [27] *Consider a strictly condensed J-frame $\mathcal{G} := (W, \leq, R_N)$, where $\leq$ is a partial order. Define $Y_0 \subseteq W$ as follows: $Y_0 := \{z \in W \mid \exists x, y \in W(x \leq z \ \& \ y \leq z \ \& \ x \acute{R}_N y)\}$. Then we have the following.*

(1) $x R_N y$ *if and only if $\exists z \in W(x \leq z \ \& \ y \leq z \ \& \ z \notin Y_0)$.*
(2) $(W, \leq, Y_0)$ *is a j-frame. Denote this frame by $\Psi(\mathcal{G})$.*
(3) *If v is a valuation on $\mathcal{G}$, then v is a valuation on $\Psi(\mathcal{G})$ such that for all ML-formulas ϕ and $x \in W$, $(\mathcal{G}, v), x \vDash \phi$ if and only if $(\Psi(\mathcal{G}), v), x \vDash \phi$.*

Theorems 7 and 8 can be extended to the case of normal and strictly condensed H-frames, thus giving an inter-translation in case of IL also.

In Theorem 1, we observed duality between topological spaces and algebras. Connections between frames and algebras have also been studied (cf. [50]). Kripke [54] has shown the connection between normal frames and pBa (cf. [11,17]): every pBa H can be 'embedded' into a pBa of upsets of W, where $(W, \leq)$ is a normal frame constructed using H. Embeddings between two normal frames are just embeddings between posets (Definition 4), since any normal frame is a poset.

Proposition 26 (Complex algebra of a normal frame). *Consider a normal frame $(W, \leq)$. Let $Up(W)$ denote the set of upsets of W. Define the following operators on $Up(W)$. For $U, V \in Up(W)$,*

(1) $U \to V := \{w \in W \mid \forall v \in W(w \leq v \Rightarrow (v \in U \Rightarrow v \in V))\}$

(2) $\neg U := U \to \emptyset = \{w \in W \mid \forall v \in W(w \le v \Rightarrow v \notin U)\}$.
The algebra $(Up(W), W, \emptyset, \cap, \cup, \to, \neg)$ is called the complex algebra of the frame $(W, \le)$.

Proposition 27. $(Up(W), W, \emptyset, \cap, \cup, \to, \neg)$ forms a pBa.

Recall the concepts of filters and prime filters in any distributive lattice (Definition 11). Since any pBa is a distributive lattice, we can define filters in pBa.

Definition 75 (Canonical frame of a pBa).
For a pBa $\mathcal{A} := (A, 1, 0, \wedge, \vee, \to, \neg)$, let X_A be the set of all prime filters in $\mathcal{A}$. Then $(X_A, \subseteq)$ is called the canonical frame of $\mathcal{A}$.

We have the following connection between pBa and normal frames.

Theorem 9.

1. *Every pBa* $\mathcal{A} := (A, 1, 0, \wedge, \vee, \to, \neg)$ *can be embedded into the complex algebra* $Up(\mathcal{A}) := (Up(X_A), X_A, \emptyset, \cap, \cup, \to, \neg)$ *of the canonical frame* $(X_A, \subseteq)$ *of* $\mathcal{A}$. *The monomorphism* $h : A \to Up(X_A)$ *is defined as follows, for any* a *in* A:

$$h(a) := \{P \in X_A \mid a \in P\}.$$

2. *Every normal frame* $(W, \le)$ *can be embedded into the canonical frame* $(X_A, \subseteq)$ *of the complex algebra* $(Up(W), W, \emptyset, \cap, \cup, \to, \neg)$ *of the frame* $(W, \le)$. *The embedding* $g : W \to X_A$ *is defined as follows, for any* $w \in W$:

$$g(w) := \{U \in Up(W) \mid w \in U\}.$$

2.6   Bounded Distributive Lattice Logic

In the language of IL or ML, the basic logical connectives are conjunction, disjunction, implication and negation, other than the propositional constants 'top' and 'bottom'. In the logical systems given by Došen [27], the alphabet of the language has the connectives $\to$ (implication), $\vee$ (disjunction), $\wedge$ (conjunction) and $\sim$ (negation). A question that arises is - what kind of properties of negation can be studied that do not involve all the other connectives in the alphabet? We have earlier referred to the work of Vakarelov [85] in connection with the study of negation as modal impossibility operators. A basic difference between his work and that of Došen's is that the logic in the former does not have implication ('$\to$'). The logical consequence is defined as a pair of formulas (ϕ, ψ), written as $\phi \vdash \psi$, expressing 'if ϕ then ψ'. $\phi \vdash \psi$ is called a *sequent*. The rule expressing 'if $\alpha \vdash \beta$ then $\gamma \vdash \delta$' is written as $\alpha \vdash \beta / \gamma \vdash \delta$. As a consequence, there are no theorems or axioms involving implication, such as $\phi \to \phi$ - instead it is replaced by '$\phi \vdash \phi$' in the logical system. In fact, Dunn [30,32] has studied negation independent of any other connectives. He has also defined the corresponding algebraic models of such logic. So what are the different properties of negation considered? In [30], Dunn has mentioned several conditions on negations. Some of them are as follows.

(1) Constructive double negation: $\alpha \vdash {\sim}{\sim}\alpha$.
(2) Classical double negation: ${\sim}{\sim}\alpha \vdash \alpha$.
(3) Contraposition: $\alpha \vdash \beta/{\sim}\beta \vdash {\sim}\alpha$.
(4) Absurdity: $\alpha \vdash \beta$, $\alpha \vdash {\sim}\beta/\alpha \vdash {\sim}\gamma$.
(5) Antilogism: $\alpha \wedge \beta \vdash \gamma/\alpha \wedge {\sim}\gamma \vdash {\sim}\beta$ (involves conjunction).

Combining one or more of them gives 'familiar' negations. For example - (1) and (5) give minimal negation; (1), (4) and (5) give intuitionistic negation. The basic definitions and notations that we use are from Dunn's more recent work on negations [33]. We present below some of the logics that will form the base of the logics in our work. Let us begin with the logic corresponding to distributive lattices.

Definition 76 (DLL: Distributive Lattice Logic).
The alphabet of DLL consists of propositional variables, binary connectives $\vee$ and $\wedge$. The axioms and rules are as follows:

> A1. $\alpha \vdash \alpha$
> A2. $\alpha \vdash \beta$, $\beta \vdash \gamma$ / $\alpha \vdash \gamma$
> A3. $\alpha \wedge \beta \vdash \alpha$; $\alpha \wedge \beta \vdash \beta$
> A4. $\alpha \vdash \beta$, $\alpha \vdash \gamma$ / $\alpha \vdash \beta \wedge \gamma$
> A5. $\alpha \vdash \gamma$, $\beta \vdash \gamma$ / $\alpha \vee \beta \vdash \gamma$
> A6. $\alpha \vdash \alpha \vee \beta$; $\beta \vdash \alpha \vee \beta$
> A7. $\alpha \wedge (\beta \vee \gamma) \vdash (\alpha \wedge \beta) \vee (\alpha \wedge \beta)$

For any two formulas α and β for a given logic L, we write $\alpha \vdash_L \beta$, if the sequent $\alpha \vdash \beta$ is derivable using the axioms and rules of L.

Now consider a language with the same alphabet as that of *DLL*, along with the unary connective 'negation' $\sim$ and constants $\top$ and $\bot$. We define K_i, the logic with preminimal negation [33], that forms the base of the logics in our work.

Definition 77 (K_i).
The logic K_i has axioms and rules as follows:

> A1-A7 of logic DLL
> A8. $\alpha \vdash \top$ *(Top)*
> A9. $\bot \vdash \alpha$ *(Bottom)*
> A10. $\alpha \vdash \beta$ / ${\sim}\beta \vdash {\sim}\alpha$ *($\sim$-contraposition)*
> A11. ${\sim}\alpha \wedge {\sim}\beta \vdash {\sim}(\alpha \vee \beta)$ *($\sim$-$\vee$-linearity)*
> A12. $\top \vdash {\sim}\bot$ *($\sim$-Nor)*

A negation satisfying the above axioms and rules is called 'preminimal negation'.

Remark 7. As mentioned earlier, various properties of negation can be included in the logic K_i to obtain other logics with negation. These are as follows:

(1) Quasi-minimal negation: Preminimal negation $+ \alpha \vdash \sim\sim\alpha$.
(2) Minimal negation: Quasi-minimal negation $+ \alpha \wedge \beta \vdash \gamma/\alpha \wedge \sim\gamma \vdash \sim\beta$.
(3) Intuitionistic negation: Minimal negation $+ \alpha \wedge \sim\alpha \vdash \beta$.
(4) De-Morgan negation: Quasi-minimal negation $+ \sim\sim\alpha \vdash \alpha$.
(5) Ortho-negation: Intuitionistic negation $+$ De-Morgan negation.

The above characterizations result in Dunn's Kite diagram (Fig. 23), representing various negations as nodes in a kite-like figure. Here, the arrows represent containment: for example, an arrow from preminimal to quasi-minimal represents the fact that a quasi-minimal negation is also a preminimal negation.

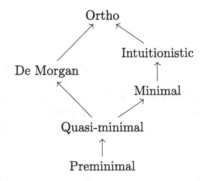

Fig. 23. Dunn's Kite representing various negations

Remark 8. The logic K_i^m with minimal negation is different from minimal logic (ML) mentioned in the previous sections: the alphabets of the languages of the two are different, and the consequence relation '$\vdash$' here differs from the '$\vdash$' defined for ML.

For ML and IL, we saw both algebraic and relational semantics. To define algebraic semantics for an extension of K_i, we need the notion of validity with respect to a 'corresponding' class $\mathcal{A}$ of algebras. As defined in Definition 52, here also we shall use the same terminology 'L-algebras' - for the corresponding class of algebras with respect to the logic L. Any L-algebra here will have a partial order. The definition of such an algebra otherwise remains the same as Definition 52 with just the conditions (1) and (2) (as the systems here do not have '$\rightarrow$'). As before, a valuation is a map $v : PV \rightarrow A$. A derivable sequent $\alpha \vdash_L \beta$ is valid in an algebra $\mathcal{A}$ (notation: $\alpha \vDash_{\mathcal{A}} \beta$), if for every valuation v on $\mathcal{A}$, $v(\alpha) \le v(\beta)$. The Lindenbaum-Tarski-style algebra $\mathcal{U}(L)$ for L can be defined here also. It is based on a relation $\simeq$ on the set F of all the formulas of L: $\alpha \simeq \beta$ if and only if $\alpha \vdash \beta$ and $\beta \vdash \alpha$. $\simeq$ is an equivalence relation on F and the domain of $\mathcal{U}(L)$ is the quotient set $F/\simeq$. Operations on $\mathcal{F}/\simeq$ are induced by the connectives in L. $\mathcal{C}$ is said to be the class of algebras *corresponding* to the logic L, if (i) (soundness) all derivable sequents are valid in every algebra belonging to $\mathcal{C}$, and (ii) (completeness) all sequents that are valid in every algebra of $\mathcal{C}$, are derivable in L.

As the name suggests, the algebraic class corresponding to DLL is the class of distributive lattices. Consider the following algebra.

Definition 78 (Distributive lattice with preminimal negation).
A distributive lattice with preminimal negation $\mathcal{A}$ is of the form $(A, 1, 0, \vee, \wedge, \sim)$, where the reduct $(A, 1, 0, \vee, \wedge)$ is a bounded distributive lattice, satisfying the following properties: For all $a, b \in A$,

(1) $a \leq b \Rightarrow \sim b \leq \sim a$,
(2) $\sim a \wedge \sim b \leq \sim(a \vee b)$, and
(3) $1 = \sim 0$.

If a distributive lattice with preminimal negation $\mathcal{A} := (A, 1, 0, \vee, \wedge, \sim)$ satisfies (1) $a \leq \sim\sim a$ and (2) $a \wedge b \leq c \Rightarrow a \wedge \sim c \leq \sim b$, for all $a, b, c \in A$, then it is called a distributive lattice with minimal negation. Moreover, if a distributive lattice $\mathcal{A}$ with minimal negation satisfies $a \wedge \sim a \leq b$ for all $a, b \in A$, then it is called a distributive lattice with intuitionistic negation.

Theorem 10.

(1) K_i corresponds to the class of distributive lattices with preminimal negation.
(2) The logic K_i^m with minimal negation corresponds to the class of distributive lattices with minimal negation.
(3) The logic K_i^i with intuitionistic negation corresponds to the class of distributive lattices with intuitionistic negation.

We next turn to relational semantics for the logics. One of the motivations of Dunn was to study negation as a modal impossibility operator in 'compatibility' frames, in line with the work of negation and consistency by Vakarelov [85]. Let us first see what 'compatibility' frames are, and define the semantics over such frames for the logic K_i.

Definition 79 (Compatibility frames [33]).
Consider a tuple $\mathcal{F} := (W, C, \leq)$, where W is a non-empty set, $\leq$ is a partial order on the set W, and C is a binary relation on W satisfying the following condition: for all $x, x', y, y' \in W$,

$$\text{If } x' \leq x, \ y' \leq y \text{ and } xCy \text{ then } x'Cy' \quad (C)$$

$\mathcal{F}$ is called a compatibility frame.

Definition 80 (Valuation on a compatibility frame).
A valuation $v : PV \to \mathcal{P}(W)$ on a compatibility frame $\mathcal{F} := (W, C, \leq)$ is defined as a mapping from the set of propositional variables in the language of K_i to the power set of W such that $v(p)$ is an upset. We use the notation $x \vDash p$, whenever $x \in v(p)$.
The mapping v can be extended to all the formulas of K_i inductively as follows:

(1) $x \vDash \alpha \wedge \beta$ *if and only if* $x \vDash \alpha$ *and* $x \vDash \beta$.
(2) $x \vDash \alpha \vee \beta$ *if and only if* $x \vDash \alpha$ *or* $x \vDash \beta$.
(3) $x \vDash \top$.
(4) $x \nvDash \bot$.
(5) $x \vDash \neg\alpha$ *if and only if* $\forall y \in W$, $xCy \Rightarrow y \nvDash \alpha$.

For the compatibility frame $\mathcal{F} := (W, C, \leq)$, the pair $\mathcal{M} := (F, \vDash)$ is called a model of K_i.
We say a consequence pair (α, β) is valid *in the model $\mathcal{M}$, denoted by $\alpha \vDash_{\mathcal{M}} \beta$, if for all $x \in W$, $x \vDash \alpha \Rightarrow x \vDash \beta$.*
The validity *of a consequence pair (α, β) in a frame $\mathcal{F}$, denoted as $\alpha \vDash_{\mathcal{F}} \beta$, is defined as $\alpha \vDash_{\mathcal{M}} \beta$ for every model $\mathcal{M}$ on the frame $\mathcal{F}$.*
For a class $\mathfrak{F}$ of frames, if $\alpha \vDash_{\mathcal{F}} \beta$ for every $\mathcal{F} \in \mathfrak{F}$, we write $\alpha \vDash_{\mathfrak{F}} \beta$.

It can be checked using induction on the number of connectives in a formula, that the following condition, called as '*hereditary condition*', holds for any formula α in K_i, that is,

$$\text{for all } x, y \in W, x \leq y \text{ and } x \vDash \alpha \text{ imply } y \vDash \alpha$$

We have the following soundness and completeness results for the logic K_i.

Theorem 11. [33] *For any two formulas α and β in the language of K_i, the following are equivalent:*

(1) $\alpha \vdash_{K_i} \beta$.
(2) $\alpha \vDash_{\mathcal{F}} \beta$ for any compatibility frame $\mathcal{F} := (W, C, \leq)$.

3 Categories of Rough Sets

Our interest in this section lies primarily in the category-theoretic studies of rough sets. In [4] the category *ROUGH* and ξ-*ROUGH* were proposed. Interesting category-theoretic properties immediately came up, substantiating the non-classical nature of rough sets, as contrasted to classical sets. The main observation was that *ROUGH* is *not a topos*. One could also show that the category 'lies between' two equivalent toposes.

Subsequently, Li and Yuan defined another category RSC of rough sets in [57], apparently unaware of the existing work on $ROUGH$. It turns out that RSC is already mentioned as an example of a topological quasitopos in [88]. The quasitopos of L-fuzzy sets [41] for a complete lattice or a pseudo-Boolean algebra L has been well-studied [45,88], and RSC is a special case when $L := \{0,1\}$.

We observe in this section that RSC is, in fact, equivalent to $ROUGH$ and hence, effectively, the two share all category-theoretic properties. So, for instance, $ROUGH$ is a weak topos as proved for RSC in [57]. Further we show in this section that the construction of both $ROUGH$ and ξ-$ROUGH$ can be lifted to a more general set-up, with an arbitrary topos $\mathscr{C}$ serving as a 'base', just as sets constitute a base for the definitions of rough sets. The first generalization, $RSC(\mathscr{C})$, yields a quasitopos, while the second, ξ-$RSC(\mathscr{C})$, gives a topos when the base $\mathscr{C}$ is a Boolean topos. In the special case when we take the topos SET of sets, $RSC(SET)$ is just the category RSC. Hence, RSC and $ROUGH$ are quasitopos. This development assumes significance from the perspective of foundations of rough sets. Just as a topos is a category-theoretic abstraction of SET, the quasitopos $RSC(\mathscr{C})$ becomes an abstraction of the category $ROUGH$ of rough sets.

Next, looking for instances of $RSC(\mathscr{C})$ where $\mathscr{C}$ is not SET, one focuses on the category $RSC(\textbf{M-Set})$ for any monoid $\textbf{M}$. We observe that the objects of this category give rise to *monoid actions on rough sets*. This opens up a direction of work worthy of further investigation (Sect. 4), both in terms of algebraic studies and in the direction of applications.

As RSC and $ROUGH$ are equivalent and the categorial constructions can be presented in a simpler manner in RSC, we adopt the latter's notations in the paper. In the next section, we present the definitions of the two categories $ROUGH$ and RSC, and show their equivalence. In Sect. 3.2, we define the category-theoretic generalizations $RSC(\mathscr{C})$ and ξ-$RSC(\mathscr{C})$, give the example of $RSC(\textbf{M-Set})$, and observe the topos-related nature of ξ-$RSC(\mathscr{C})$. Some fundamental properties of $RSC(\mathscr{C})$ are presented in Sect. 3.3, culminating in the result that it is a quasitopos. In Sect. 3.4, we define some more categories of rough sets, and compare them with existing categories of rough sets. We shall end this section in Sect. 3.5.

3.1   Categories - RSC and $ROUGH$

We have seen the two categories of rough sets $ROUGH$ and RSC (Definition 32 and 33). Recall the definition of rough universes used in Definition 34.

Observation 12.

(1) Given any set C, a pair of sets (A_1, A_2), where $A_1 \subseteq A_2 \subseteq C$, can form an I-rough set of the rough universe $(C, \textbf{B})$, where $\textbf{B}$ is the Boolean algebra $(\mathcal{P}(C), C, \emptyset, \cap, \cup, \rightarrow, \neg)$. In fact, taking C to be A_2 itself, any pair of the form (A_1, A_2), where $A_1 \subseteq A_2$, can be considered as an RSC-object.

(2) In the definition of an *RSC* object (A_1, A_2), a universe U is not explicitly mentioned, whereas in a *ROUGH*-object (U, R, X), the universe U is included in the definition of the object. However, note that the arrows in *RSC* as well as *ROUGH* are confined only to the upper approximation region of rough sets. It may be expected that the universe U would come into play explicitly, when the negative region features in the definition of arrows.

The first result that we start our work with is that both the categories are same upto equivalence of categories (Definition 17).

Theorem 12. $RSC \simeq ROUGH$.

Proof. Define two maps, both denoted by G, one from $Obj(RSC)$ to $Obj(ROUGH)$, and another from $Arr(RSC)$ to $Arr(ROUGH)$ as follows.
(1). G maps an *RSC*-object (X_1, X_2) to the *ROUGH*-object (U, R, X), where
$U := X_2 + (X_2 \backslash X_1) = \{(a, 1) \mid a \in X_2\} \cup \{(b, 2) \mid b \in X_2 \backslash X_1\}$,
$R := \{((x, 1), (x, 1)) \mid x \in X_1\} \cup$
$\{((x, 1), (x, 1)), ((x, 1), (x, 2)), ((x, 2), (x, 1)), ((x, 2), (x, 2)) \mid x \in X_2 \backslash X_1\}$, and
$X := X_2 + \emptyset = \{(x, 1) \mid x \in X_2\}$.
It can be easily observed here that R is an equivalence relation. Note that $\overline{X}_R \cong X_2$ ($\cong$ represents set bijection) under iso *SET*-arrow $m_X : X_2 \to \overline{X}_R$, mapping x to the equivalence class $[(x, 1)]$ and $\underline{X}_R \cong X_1$ (restriction of m_X to X_1 is also an iso).
(2). Consider *RSC*-arrow $f : (X_1, X_2) \to (Y_1, Y_2)$ and *ROUGH*-arrow
$\overline{f} : (U, R, X) \to (V, S, Y)$, where $(U, R, X) = G((X_1, X_2))$ and $(V, S, Y) = G((Y_1, Y_2))$ are *ROUGH*-objects. Then G maps f to $\overline{f}$ such that $\overline{f} \circ m_X = n_Y \circ f$ in *SET* where $m_X : X_2 \to \overline{X}_R$ and $n_Y : Y_2 \to \overline{Y}_S$. So, $\overline{f} = n_Y \circ f \circ m_X^{-1}$, where m_X^{-1} is the inverse function of m_X.
G is a functor. Let us check that G is a full and faithful functor, i.e. the following function (Definition 16) is one-one and onto.

$$\Phi : Hom_{RSC}((X_1, X_2), (Y_1, Y_2)) \to Hom_{ROUGH}((U, R, X), (V, S, Y)), \text{ where}$$

$$\Phi(f) := \overline{f}.$$

Φ is one-one: For $f, g \in Hom_{RSC}((X_1, X_2), (Y_1, Y_2))$, $\overline{f} = \overline{g} \Rightarrow \overline{f} \circ m_X = \overline{g} \circ m_X \Rightarrow n_Y \circ f = n_Y \circ g \Rightarrow f = g$, as n_Y is iso and thus a monic (Definition 14).
Φ is onto: For $h \in Hom_{ROUGH}((U, R, X), (V, S, Y))$, define a function $f = n_Y^{-1} \circ h \circ m_X : X_2 \to Y_2$. We have $f \in Hom_{RSC}((X_1, X_2), (Y_1, Y_2))$, because $f(X_1) \subseteq Y_1$. Further $\overline{f} \circ m_X = n_Y \circ f = n_Y \circ n_Y^{-1} \circ h \circ m_X = h \circ m_X$. Since, m_X is an iso, we have $\overline{f} = h$.
G is essentially surjective: Consider an object (U, R, X) in *ROUGH*. We have $(\underline{X}, \overline{X})$ as an object in *RSC* (cf. Remark 12) and by construction of G, $G((\underline{X}, \overline{X})) \cong (U, R, X)$.

Remark 9. We observe the following from the proof.

(i) The above equivalence *will not work* if all *ROUGH* objects and arrows are defined for a fixed approximation space (U, R).
(ii) Any *RSC*-object (A_1, A_2) can be expressed, upto set isomorphism, as $(\underline{\mathcal{X}}, \overline{\mathcal{X}})$ for some *ROUGH*-object (U, R, X).

In [4], it was shown that *ROUGH* is finitely complete and is not a topos. [57] showed that *RSC* has limits, exponents and is not a topos, but a weak topos. Due to Theorem 12 therefore, *ROUGH* and *RSC* share all these properties.

We have already seen another category of rough sets (Definition 33) based on the idea that during 'communications' (being represented by arrows) between rough sets, the boundary region may be another invariant. One can define ξ-*RSC* in a similar way.

Definition 81. *(ξ-RSC category)*
Objects of ξ-RSC are all I-rough sets of all rough universes. An arrow in ξ-RSC with domain (X_1, X_2) and codomain (Y_1, Y_2) is a map $f : X_2 \to Y_2$ such that $f(X_1) \subseteq Y_1$ and $f(X_2 \backslash X_1) \subseteq Y_2 \backslash Y_1$.

These two categories can be shown to be equivalent, using the same functor as defined in Theorem 12. In fact, one can show more. Let us consider SET^2 (Example 3).

Proposition 28. *ξ-RSC is equivalent to ξ-ROUGH, and also equivalent to SET^2.*

Proof. Define two maps, both denoted by H, one from $Obj(\xi$-$RSC)$ to $Obj(SET^2)$ and another from $Arr(\xi$-$RSC)$ to $Arr(SET^2)$ as follows:
(1). H maps ξ-RSC-object (X_1, X_2) to SET^2-object $(X_1, X_2 \backslash X_1)$, and
(2). H maps ξ-RSC-arrow $f : (X_1, X_2) \to (Y_1, Y_2)$ to SET^2-arrow (f_1, f_2) : $(X_1, X_2 \backslash X_1) \to (Y_1, Y_2 \backslash Y_1)$, where $f_1 := f|_{X_1}$ (restriction of f on X_1) and $f_2 := f|_{X_2 \backslash X_1}$ (restriction of f on $X_2 \backslash X_1$).
H is a functor. Let us now check that H is full and faithful. Consider

$$\Phi : Hom_{\xi\text{-}RSC}((X_1, X_2), (Y_1, Y_2)) \to Hom_{SET^2}((X_1, X_2 \backslash X_1), (Y_1, Y_2 \backslash Y_1)),$$

where

$$\Phi(f) := (f_1, f_2).$$

For $f, g \in Hom_{\xi\text{-}RSC}((X_1, X_2), (Y_1, Y_2))$, $(f_1, f_2) = (g_1, g_2) \Rightarrow f_1 = g_1$ and $f_2 = g_2$. Now, for $x \in X_1$, $f(x) = f_1(x) = g_1(x) = g(x)$ and for $x \in X_2 \backslash X_1$, $f(x) = f_2(x) = g_2(x) = g(x)$. Therefore, $f = g$ and Φ is one-one.
For $(f_1, f_2) \in Hom_{SET^2}((X_1, X_2 \backslash X_1), (Y_1, Y_2 \backslash Y_1))$, define a function $h : X_2 \to Y_2$ as follows: for $x \in X_1$, $h(x) = f_1(x)$ and for $x \in X_2 \backslash X_1$, $h(x) = f_2(x)$. Therefore, we have $H(h) = (f_1, f_2)$ and Φ is onto.
Now given any object (X, Y) in SET^2, we have the object $(X + \emptyset, X + Y)$ in ξ-RSC (where $\emptyset$ is the initial object in SET) such that $H((X + \emptyset, X + Y)) \cong (X, Y)$. Thus, H is essentially surjective, and we have ξ-$RSC \simeq SET^2$.

As noted in Example 4, SET^2 is a topos. Therefore, we have the following.

Corollary 1. *ξ-ROUGH ($\simeq \xi$-RSC) is a topos.*

We shall see in the sequel that RSC forms a quasitopos (Definition 28).

3.2   Generalizations of RSC and ξ-RSC

The rough set categories RSC and ξ-RSC have objects as I-rough sets (A_1, A_2), where A_1 and A_2 are objects of SET, which is a topos (Example 4). In this section, we propose a natural generalization of these categories, with base as an arbitrary topos instead of SET. Let $\mathscr{C}$ be an arbitrary non-degenerate topos (Remark 3).

3.2.1 $RSC(\mathscr{C})$

The category RSC has 'base' as sets: an RSC-object (A, B) has two underlying base sets A and B such that $A \subseteq B$. An RSC-arrow is a set function satisfying certain properties (Definition 34). Let us note the following.

Observation 13.

(1) For an RSC-object (X_1, X_2), the property $X_1 \subseteq X_2$ implies that there exists an inclusion function $m : X_1 \to X_2$. Further, in SET, any inclusion function is a monic arrow.

(2) Consider an RSC-arrow f from (X_1, X_2) to (Y_1, Y_2), i.e. $f : X_2 \to Y_2$ is a map such that $f(X_1) \subseteq Y_1$. This implies that there exist two functions $f : X_2 \to Y_2$ and $f' : X_1 \to Y_1$ such that $f'(x) = f(x)$ for all $x \in X_1$, i.e. $f(m(x)) = f'(x)$ for all $x \in X_1$, i.e. $f \circ m = f' \circ m'$, where $m : X_1 \to X_2$ and $m' : Y_1 \to Y_2$ are inclusion functions.

In the category-theoretic terminology, this property can be expressed as follows. There exists a SET-arrow $f : X_2 \to Y_2$ such that $m' \circ f' = f \circ m$, where $m : X_1 \to X_2$ and $m' : Y_1 \to Y_2$ are monic arrows.

Taking a cue from the above observation, we propose the following generalization of the category RSC.

Definition 82 ($RSC(\mathscr{C})$ category).
Objects of $RSC(\mathscr{C})$ are triples of the form (X_1, X_2, m), where X_1 and X_2 are $\mathscr{C}$-objects and $m : X_1 \to X_2$ is a monic arrow in $\mathscr{C}$.
An arrow in $RSC(\mathscr{C})$ with domain (X_1, X_2, m) and codomain (Y_1, Y_2, m') is a pair of arrows (f', f) where $f' : X_1 \to Y_1$ and $f : X_2 \to Y_2$ in $\mathscr{C}$ such that the diagram in Fig. 24 commutes in $\mathscr{C}$, i.e. $f \circ m = m' \circ f'$.
Composition of arrow is component-wise. Identity arrow on object (X_1, X_2, m) is $(1_{X_1}, 1_{X_2})$.

Observation 14.

(1) In place of (X_1, X_2, m_1), we may sometimes simply refer to the monic m_1 as an $RSC(\mathscr{C})$-object.

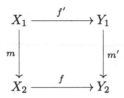

Fig. 24. An $RSC(\mathscr{C})$-arrow $f : (X_1, X_2) \to (Y_1, Y_2)$

(2) The category $RSC(\mathscr{C})$ is a full subcategory (Definition 13) of the arrow category $\mathscr{C}^{\to}$ (Example 3) consisting of all the monics in $\mathscr{C}$ as its objects. In [52], a general result is obtained about quasitoposes. Given two quasitoposes $\mathscr{C}$, $\mathscr{D}$ and a functor $T : \mathscr{D} \to \mathscr{C}$, a new category $\mathscr{C}/T$ is defined using a notion of 'Artin glueing'. Then it is shown that a particular full subcategory of the category $\mathscr{C}/T$, denoted $\mathscr{C}//T$, forms a quasitopos. As a special case when $\mathscr{C} = \mathscr{D}$ and T is the identity functor, $\mathscr{C}/T = \mathscr{C}^{\to}$ and $\mathscr{C}//T = RSC(\mathscr{C})$. Thereby one can conclude that $RSC(\mathscr{C})$ is a quasitopos.

In the following, we shall derive explicitly all the constructs that make $RSC(\mathscr{C})$ a quasitopos.

What happens in the special case when $\mathscr{C}$ is the topos SET? Let us define a notation first.

Notation 4.

(1) For any two sets X_1 and X_2 such that $X_1 \subseteq X_2$, the inclusion function of X_1 into X_2 is denoted by $i_{X_1} : X_1 \to X_2$. Note that i_{X_1} is a monic arrow in SET.

(2) Let $X_1 \subseteq X_2$ and $Y_1 \subseteq Y_2$. For a function $f : X_2 \to Y_2$ such that $f(X_1) \subseteq Y_1$, define $f' : X_1 \to Y_1$ by $f'(x) := f(x)$ for all $x \in X_1$. The 'restriction' function f' is denoted by $f|_{X_1}$.

Notice that any RSC-object (X_1, X_2) can be re-expressed as (X_1, X_2, i_{X_1}). Since any inclusion is a monic arrow in SET, (X_1, X_2, i_{X_1}) is an $RSC(SET)$-object. For any $RSC(SET)$-arrow $(f', f) : (X_1, X_2, i_{X_1}) \to (Y_1, Y_2, i_{Y_1})$, we have $f \circ i_{X_1} = i_{Y_1} \circ f'$, and thus f' is just $f|_{X_1}$ (defined in Notation 4). Moreover, for any RSC-arrow $f : (X_1, X_2) \to (Y_1, Y_2)$, we have the $RSC(SET)$-arrow $(f|_{X_1}, f) : (X_1, X_2, i_{X_1}) \to (Y_1, Y_2, i_{X_1})$. Thus, we can say that RSC is isomorphic to a full subcategory of $RSC(SET)$ with objects of the form (X_1, X_2, i_{X_1}). Further, we have the following.

Proposition 29. $RSC \simeq RSC(SET)$.

Proof. Define two maps, both denoted by F, one from $Obj(RSC)$ to $Obj(RSC(SET))$, and another from $Arr(RSC)$ to $Arr(RSC(SET))$ as follows.
(1). F maps an RSC-object (X_1, X_2) to the $RSC(SET)$-object (X_1, X_2, i_{X_1}), where $i_{X_1} : X_1 \to X_2$ is the inclusion of X_1 into X_2.

(2). F maps an RSC-arrow $f : (X_1, X_2) \rightarrow (Y_1, Y_2)$ to $RSC(SET)$-arrow $(f|_{X_1}, f) : (X_1, X_2, i_{X_1}) \rightarrow (Y_1, Y_2, i_{Y_1})$.

F is full and faithful.

Now, consider an $RSC(SET)$-object (X_1, X_2, m). Then $(m(X_1), X_2)$ is an RSC-object, and $(m(X_1), X_2, i_{m(X_1)})$ is an $RSC(SET)$-object. Define a map $g : m(X_1) \rightarrow X_1$ mapping $g(m(x)) = x$ for all $m(x) \in m(X_1)$. Since m is monic, g is well-defined. Consider another function $m|_{X_1} : X_1 \rightarrow m(X_1)$. Observe that $(g, 1_{X_2}) : (m(X_1), X_2, i_{m(X_1)}) \rightarrow (X_1, X_2, m)$ and $(m|_{X_1}, 1_{X_2}) : (X_1, X_2, m) \rightarrow (m(X_1), X_2, i_{m(X_1)})$ are $RSC(SET)$-arrows. Moreover, they are isos because $(g, 1_{X_2}) \circ (m|_{X_1}, 1_{X_2}) = (1_{X_1}, 1_{X_2})$ and $(m|_{X_1}, 1_{X_2}) \circ (g, 1_{X_2}) = (1_{m(X_1)}, 1_{X_2})$. Now, $F((m(X_1), X_2)) = (m(X_1), X_2, i_{m(X_1)}) \cong (X_1, X_2, m)$. Therefore, F is essentially surjective and $RSC \simeq RSC(SET)$.

Observation 15.

(1) In view of the above proposition, we can say that $RSC(\mathscr{C})$ is a generalization of the category RSC.

(2) $RSC(\mathscr{C})$ is a generalization of RSC and not $ROUGH$, as in defining the $RSC(\mathscr{C})$-objects (X_1, X_2, m), we are not making any reference to a $\mathscr{C}$-object 'U' of which both A_1 and A_2 are subobjects, and also no reference to any equivalence relation on U has been made.

It is natural to wonder what $RSC(\mathscr{C})$ actually looks like, for a topos $\mathscr{C}$ that is not SET. Let us consider the topos **M-Set**, where **M** is a monoid (Example 4). When $\mathbf{M} := \mathbf{1} = \{0\}$, we have nothing new, as in this case, **M-Set** is isomorphic to SET. So, let us take the next simplest monoid $\mathbf{M} := \mathbf{2} = \{0, 1\}$ with multiplication $\cdot$, where 1 is the identity element, and $0 \cdot 0 = 0 \cdot 1 = 1 \cdot 0 = 0$. The objects of **2-Set** can be identified with the pairs (X, μ), where X is a set and $\mu : X \rightarrow X$ is a set function such that $\mu^2 = \mu$. Arrows of **2-Set** are of the form $f : (X_1, \mu) \rightarrow (X_2, \lambda)$ such that $f : X_1 \rightarrow X_2$ is a function satisfying $\lambda \circ f = f \circ \mu : X_1 \rightarrow X_2$.

In the category $RSC(\mathbf{2\text{-}Set})$, an object is the triple $((X_1, \mu), (X_2, \lambda), m)$, where $m : (X_1, \mu) \rightarrow (X_2, \lambda)$ is a monic arrow in **2-Set**, i.e. $m : X_1 \rightarrow X_2$ is an injective function satisfying $m \circ \mu = \lambda \circ m$ (Fig. 25a). An $RSC(\mathbf{2\text{-}Set})$-arrow $(f', f) : ((X_1, \mu), (X_2, \lambda), m_1) \rightarrow ((Y_1, \nu), (Y_2, \delta), m_2)$ is a pair of functions $f' : X_1 \rightarrow Y_1$ and $f : X_2 \rightarrow Y_2$ such that the diagram in Fig. 25b commutes.

Now, consider the full subcategory of $RSC(\mathbf{2-Set})$, with objects as $i_{X_1} : (X_1, \mu) \rightarrow (X_2, \lambda)$, i_{X_1} being the inclusion of X_1 into X_2 and $\mu = \lambda|_{X_1}$. Then we can re-present the object $((X_1, \lambda|_{X_1}), (X_2, \lambda), i_{X_1})$ as simply (X_1, X_2, λ), where $X_1 \subseteq X_2$ and $\lambda : X_2 \rightarrow X_2$ is such that $\lambda^2 = \lambda$ and $\lambda(X_1) \subseteq X_1$. An $RSC(\mathbf{2-Set})$ arrow $(f|_{X_1}, f) : (X_1, X_2, \lambda) \rightarrow (Y_1, Y_2, \mu)$ can then also be simply re-presented as the function $f : X_2 \rightarrow Y_2$ such that $f(X_1) \subseteq Y_1$ and $\mu \circ f = f \circ \lambda$. The complete definition of this category is as follows.

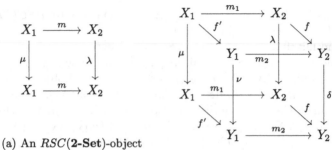

(a) An $RSC(\mathbf{2\text{-}Set})$-object

(b) An $RSC(\mathbf{2\text{-}Set})$-arrow

Fig. 25. The category $RSC(2-\mathbf{Set})$

Definition 83 ($RSC_{2\text{-}Set}$).
Objects of $RSC_{2\text{-}Set}$ are the triples (X_1, X_2, λ), where $X_1 \subseteq X_2$ and $\lambda : X_2 \to X_2$ is a set function such that $\lambda^2 = \lambda$, and $\lambda(X_1) \subseteq X_1$.
An $RSC_{2\text{-}Set}$-arrow $f : (X_1, X_2, \lambda) \to (Y_1, Y_2, \mu)$ is the set function $f : X_2 \to Y_2$ with $f(X_1) \subseteq Y_1$ such that the diagram in Fig. 26 commutes.
Composition of arrows is the usual function composition. Identity arrow for $RSC_{2\text{-}Set}$-object (X_1, X_2, μ) is 1_{X_2}.

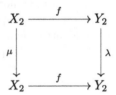

Fig. 26. An $RSC_{2\text{-}Set}$-arrow $f : (X_1, X_2, \mu) \to (Y_1, Y_2, \lambda)$

Thus, we can say that $RSC_{2\text{-}Set}$ is isomorphic to a full subcategory of $RSC(\mathbf{2\text{-}Set})$. Moreover, as observed for RSC and $RSC(SET)$, we have the following result for $RSC_{2\text{-}Set}$ and $RSC(\mathbf{2\text{-}Set})$.

Proposition 30. $RSC_{2\text{-}Set} \simeq RSC(2-\mathbf{Set})$.

Proof. Define two maps, both denoted by F,-
one from $Obj(RSC_{2\text{-}Set})$ to $Obj(RSC(\mathbf{2\text{-}Set}))$, and
another from $Arr(RSC_{2\text{-}Set})$ to $Arr(RSC(\mathbf{2\text{-}Set}))$ as follows:
(1). F maps an $RSC_{2\text{-}Set}$-object (X_1, X_2, μ) to the $RSC(\mathbf{2\text{-}Set})$-object $((X_1, \mu|_{X_1}), (X_2, \mu), i_{X_1})$.

(2). F maps an $RSC_{2\text{-Set}}$-arrow $f : (X_1, X_2, \mu) \to (Y_1, Y_2, \lambda)$ to $RSC(\mathbf{2\text{-Set}})$-arrow $(f|_{X_1}, f) : ((X_1, \mu|_{X_1}), (X_2, \mu), i_{X_1}) \to ((Y_1, \lambda|_{Y_1}), (Y_2, \lambda), i_{Y_1})$.
F is full and faithful.
Now, consider a $RSC(\mathbf{2\text{-Set}})$-object $((X_1, \mu), (X_2, \lambda), m)$.
Here, λ restricts $m(X_1)$ to $m(X_1)$ (Fig. 25a). Therefore, we can obtain
an $RSC_{2\text{-Set}}$-object $(m(X_1), X_2, \lambda)$ and
an $RSC(\mathbf{2\text{-Set}})$-object $((m(X_1), \lambda|_{m(X_1)}), (X_2, \lambda), i_{m(X_1)})$.
As in Proposition 29, define a map $g : m(X_1) \to X_1$ such that $g(m(x)) = x$ for
all $m(x) \in m(X_1)$, which is well-defined as m is monic. Here,

$$(g, 1_{X_2}) : ((m(X_1), \lambda|_{m(X_1)}), (X_2, \lambda), i_{m(X_1)}) \to ((X_1, \mu), (X_2, \lambda), m) \text{ and}$$

$$(m|_{X_1}, 1_{X_2}) : ((X_1, \mu), (X_2, \lambda), m) \to ((m(X_1), \lambda|_{m(X_1)}), (X_2, \lambda), i_{m(X_1)})$$

are arrows in $RSC(\mathbf{2\text{-Set}})$, and $(g, 1_{X_2})$ is an iso.
Therefore,
$F((m(X_1), X_2, \lambda)) \qquad\qquad\qquad\qquad\qquad\qquad =$
$((m(X_1), \lambda|_{m(X_1)}), (X_2, \lambda), i_{m(X_1)}) \cong ((X_1, \mu), (X_2, \lambda), m))$. So, F is essentially surjective and $RSC_{2\text{-Set}} \simeq RSC(\mathbf{2 - Set})$.

In an $RSC_{2\text{-Set}}$-object (X_1, X_2, μ), the monoid action μ preserves the rough set lower approximation (as $\mu|_{X_1} : X_1 \to X_1$). This signifies that an object (X_1, X_2, μ) of $RSC_{2\text{-Set}}$ gives rise to an action of the monoid $\mathbf{2}$ on the rough set (X_1, X_2). The arrow f preserves these monoid actions (through the condition $\lambda \circ f = f \circ \mu$ in the diagram), apart from preserving information in the lower approximation (given by the RSC arrow condition $f(X_1) \subseteq Y_1$).

In fact, we can look at the above constructions in the more general framework of $RSC(\mathbf{M\text{-Set}})$ and $RSC_{\mathbf{M\text{-Set}}}$, where $\mathbf{M}$ is any monoid, and obtain the following definition.

Definition 84 (Monoid Actions on Rough sets).

A monoid action on a rough set (X_1, X_2) is a triple (X_1, X_2, μ) such that $\mu : \mathbf{M} \times X_2 \to X_2$ is a monoid action of $\mathbf{M}$ on the set X_2, with the condition that $\mu|_{X_1}$ is a monoid action of $\mathbf{M}$ on X_1.

This notion appears to hold promise for future investigation, algebraically, as well as in the direction of applications. We shall study this in some detail in Sect. 4.

3.2.2 ξ-$RSC(\mathscr{C})$

Recall the definition of $\neg$ operator on $Sub(D)$ for an object D in the topos $\mathscr{C}$ (Definition 30(4)). As observed in case of RSC (Observation 13), let us first make a similar observation for ξ-RSC.

Observation 16.

(1) Objects of ξ-RSC are same as objects of RSC.
(2) Consider a ξ-RSC-arrow f from (X_1, X_2) to (Y_1, Y_2), i.e. $f : X_2 \to Y_2$ is a map such that $f(X_1) \subseteq Y_1$ and $f(X_2 \backslash X_1) = (Y_2 \backslash X_1)$. This implies that there exist two functions $f' : X_1 \to Y_1$ and $f'' : X_2 \backslash X_1 \to Y_2 \backslash Y_1$ such that the following hold.

(a) $f'(x) = f(x)$ for all $x \in X_1$, i.e. $f(m(x)) = f'(x)$ for all $x \in X_1$, i.e. $f \circ m = f' \circ m'$, where $m : X_1 \to X_2$ and $m' : Y_1 \to Y_2$ are inclusion functions. In category-theoretic terminology, there exists a SET-arrow $f : X_2 \to Y_2$ such that $m' \circ f' = f \circ m$, where $m : X_1 \to X_2$ and $m' : Y_1 \to Y_2$ are monic arrows.

(b) $f''(x) = f(x)$ for all $x \in X_2\backslash X_1$, i.e. $f(\neg m(x)) = f''(x)$ for all $x \in X_2\backslash X_1$, i.e. $f \circ \neg m = \neg m' \circ f''$, where $\neg m : X_2\backslash X_1 \to X_2$ and $\neg m' : Y_2\backslash Y_1 \to Y_2$ are inclusion functions. In category-theoretic terminology, there exists a SET-arrow $f : X_2 \to Y_2$ such that $\neg m' \circ f'' = f \circ \neg m$, where $\neg m : \neg X_1 \to X_2$ and $\neg m' : \neg Y_1 \to Y_2$ are monics as defined in Definition 30(4), and m and m' are as defined above in (a). Note that here $\neg X_1 = X_2\backslash X_1$ and $\neg Y_1 = Y_2\backslash Y_1$.

Based on the above observation, we define the generalization of the category $\xi\text{-}RSC(\mathscr{C})$.

Definition 85 ($\xi\text{-}RSC(\mathscr{C})$ category).
Objects of $\xi\text{-}RSC(\mathscr{C})$ are same as objects of $RSC(\mathscr{C})$. An arrow (f', f) : $(X_1, X_2.m) \to (Y_1, Y_2, m')$ of $RSC(\mathscr{C})$, where $f' : X_1 \to Y_1$ and $f : X_2 \to Y_2$, is an arrow in $\xi\text{-}RSC(\mathscr{C})$ if there exists an arrow $f'' : \neg X_1 \to \neg Y_1$ such that the outer square in Fig. 27 commutes. An arrow in $\xi\text{-}RSC(\mathscr{C})$ is represented as a triple (f'', f', f).
Composition is component-wise.
Identity arrow on object (X_1, X_2, m) is $(1_{\neg X_1}, 1_{X_1}, 1_{X_2})$.

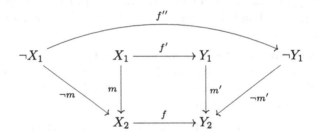

Fig. 27. A $\xi\text{-}RSC(\mathscr{C})$-arrow $(f'', f', f) : (X_1, X_2) \to (Y_1, Y_2)$

Observation 17. As in the case of $RSC(\mathscr{C})$, we observe that in the special case when $\mathscr{C}$ is the topos SET, $\xi\text{-}RSC$ is just the full subcategory of $\xi\text{-}RSC(SET)$, and equivalent to it.

Similar to the question addressed in Proposition 28 for $\xi\text{-}RSC$, it is natural to enquire whether $\xi\text{-}RSC(\mathscr{C})$ is also a topos, and whether it is isomorphic to $\mathscr{C}^2$ (Example 3). For an arbitrary topos $\mathscr{C}$, we are able to show that $\xi\text{-}RSC(\mathscr{C})$ is not always a topos (or a quasitopos), and that depends on whether the base topos $\mathscr{C}$ has the Boolean property (Definition 31).

Theorem 13. *ξ-RSC($\mathscr{C}$) is a topos if and only if $\mathscr{C}$ is a Boolean topos.*

Proof. ($\Leftarrow$) Let $\mathscr{C}$ be a Boolean topos. We shall show that ξ-RSC($\mathscr{C}$) is equivalent to $\mathscr{C}^2$ and thus, a topos (and quasitopos) by Example 4 and Proposition 11. Define $G : \xi$-RSC($\mathscr{C}$) $\to \mathscr{C}^2$, mapping ξ-RSC($\mathscr{C}$)-object (X_1, X_2, m) to $\mathscr{C}^2$-object $(X_1, \neg X_1)$ and ξ-RSC($\mathscr{C}$)-arrow $(f'', f', f) : (X_1, X_2, m_1) \to (Y_1, Y_2, m_2)$ to $\mathscr{C}^2$-arrow $(f', f'') : (X_1, \neg X_1) \to (Y_1, Y_2)$, where $f : X_2 \to Y_2$, $f' : X_1 \to Y_1$ and $f'' : \neg X_1 \to \neg Y_1$ are $\mathscr{C}$-arrows. We shall show that G is full, faithful and essentially surjective.

(1). G is full: let $(h, k) : (X_1, \neg X_1) \to (Y_1, \neg Y_1)$ in $\mathscr{C}^2$, where $h : X_1 \to Y_1$ and $k : \neg X_1 \to \neg Y_1$ are $\mathscr{C}$-arrows. We have to find $(f'', f', f) : (X_1, X_2, m_1) \to (Y_1, Y_2, m_2)$ such that $f' = h$ and $f'' = k$. Since $\mathscr{C}$ is a topos, $\mathcal{H} := (Sub(X_2), 1_{X_2}, i_{X_2}, \cup, \cap, \to, \neg)$ is a pBa (Proposition 13). Using Proposition 12, we have the following diagram, where inner square is a pushout and outer square is a pullback.

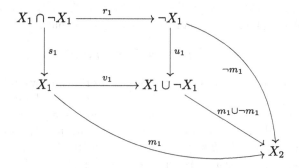

In the *pBa* $\mathcal{H}$, $m_1 \cap \neg m_1 \simeq !_{X_2}$. We have $X_1 \cap \neg X_1 \cong 0$, and the above square can be re-expressed as follows such that inner square is a pushout and outer square is a pullback.

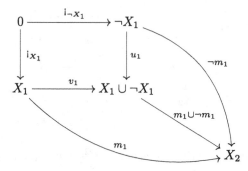

Further, since $\mathscr{C}$ is a Boolean topos, $\mathcal{H}$ is also a Boolean algebra (Definition 31), implying $m_1 \cup \neg m_1 \simeq 1_{X_2}$ and $X_1 \cup \neg X_1 \cong X_2$. Identity arrows 1_{X_2} and 1_{Y_2} are always iso, implying the following arrow equivalences in $\mathscr{C}$: $v_1 \simeq m_1$ and $u_1 \simeq \neg m_1$. Therefore, the above square can be re-expressed as the left hand side square below, which is a pullback as well as pushout.

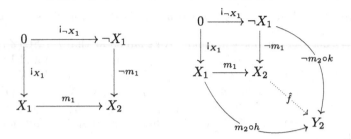

Now, in the right hand side diagram above, we have $\neg m_2 \circ k \circ i_{\neg X_1} = m_2 \circ h \circ i_{X_1}$, because there is a unique arrow from 0 to Y_2 (universal property of the initial object 0). Using the universal property of the pushout square, there exists a unique arrow $\hat{f} : X_2 \to Y_2$ such that $\hat{f} \circ m_1 = m_2 \circ h$ and $\hat{f} \circ \neg m_1 = \neg m_2 \circ k$ (Right hand side square above). Therefore, $(k, h, \hat{f}) : (X_1, X_2, m_1) \to (Y_1, Y_2, m_2)$ is the required ξ-$RSC(\mathscr{C})$-arrow such that $G((k, h, \hat{f})) = (h, k)$.

(2). G is faithful: consider two ξ-$RSC(\mathscr{C})$-arrows (f'', f', f) and (g'', g', g) with domain (X_1, X_2, m_1) and codomain (Y_1, Y_2, m_2) such that

$$G((f'', f', f)) = G((g'', g', g)),$$

implying $f'' = g''$ and $f' = g'$. Using the property of the arrows in ξ-$RSC(\mathscr{C})$, we have $f \circ m_1 = m_2 \circ f'$, $f \circ \neg m_1 = \neg m_2 \circ f''$, $g \circ m_1 = m_2 \circ g'$ and $g \circ \neg m_1 = \neg m_2 \circ g''$. As argued above, the left hand side diagram below is again a pullback and a pushout. In the right hand side diagram below, taking $\hat{f} = f$, both the triangles commute when (i). $h = f'$, $k = f''$, and also when (ii). $h = g'$, $k = g''$. So there is a unique $\mathscr{C}$-arrow $\hat{f_1} : X_2 \to Y_2$ such that $\hat{f_1} \circ m_1 = m_2 \circ f'$ and $\hat{f_1} \circ \neg m_1 = \neg m_2 \circ f''$, and a unique $\mathscr{C}$-arrow $\hat{f_2} : X_2 \to Y_2$ such that $\hat{f_2} \circ m_1 = m_2 \circ g'$ and $\hat{f_2} \circ \neg m_1 = \neg m_2 \circ g''$. This implies that $\hat{f_1} = f$ and $\hat{f_2} = g$. Finally, since $f' = g'$ and $f'' = g''$, we get $f = g$.

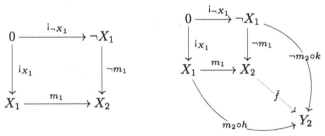

(3). G is essentially surjective: consider a $\mathscr{C}^2$-object (X_1, X_1'). In any topos, the coproduct arrows $X_1 \xrightarrow{i_1} X_1 + X_1' \xleftarrow{i_1'} X_1'$ are monics and the left hand side square below is a pullback and a pushout (Proposition 9). Further using the fact that $(Sub(X_1 + X_1'), 1_{X_1 + X_1'}, i_{X_1 + X_1'}, \cup, \cap, \to, \neg)$ is a Boolean algebra, we have $i_{X_1+X_1'} \simeq i_1 \cap \neg i_1$ and $1_{X_1+X_1'} \simeq i_1 \cup \neg i_1$, i.e. $0 \cong (X_1 \cap \neg X_1)$ and $(X_1+X_1') \cong (X_1 \cup \neg X_1)$. Using Proposition 12, the right hand side square is also a pushout and pullback. Comparing both the pushout squares, we have $i_1' \simeq \neg i_1$ and $X_1' \cong \neg X_1$. Therefore, $G((X_1, X_1 + X_1', i_1)) \cong (X_1, X_1')$.

$$
\begin{array}{ccc}
0 & \xrightarrow{\ i_{X_1'}\ } & X_1' \\
{\scriptstyle i_{X_1}}\big\downarrow & & \big\downarrow{\scriptstyle i_1'} \\
X_1 & \xrightarrow{\ i_1\ } & (X_1 + X_1')
\end{array}
\qquad
\begin{array}{ccc}
0 & \xrightarrow{\ i_{X_1}\ } & \neg X_1 \\
{\scriptstyle i_{X_1}}\big\downarrow & & \big\downarrow{\scriptstyle \neg i_1} \\
X_1 & \xrightarrow{\ i_1\ } & (X_1 + X_1')
\end{array}
$$

($\Rightarrow$) Assume $\mathscr{C}$ to be a non-Boolean topos, that is, there exists a monic arrow $m : A \to B$ such that $(m \cup \neg m) \not\simeq 1_B$ in the pBa $(Sub(B), 1_B, i_B, \cup, \cap, \to, \neg)$. Note the following.

(1). $\neg(m \cup \neg m) \simeq i_B$ (Proposition 2(11)) and $\neg(A \cup \neg A) \cong 0$.

(2). $(m \cup \neg m) : (A \cup \neg A) \to B$ is not an iso, because if $(m \cup \neg m)$ is an iso, then $A \cup \neg A \cong B$ and $(m \cup \neg m) \simeq 1_B$, which is not true, by assumption.

Now, consider the $\xi\text{-}RSC(\mathscr{C})$-arrow $(1_0, m \cup \neg m, 1_B) : (A \cup \neg A, B, m \cup \neg m) \to (B, B, 1_B)$. Let us show that this arrow is a monic and epic, but not an iso.

$(1_0, m \cup \neg m, 1_B)$ is monic: consider the $\xi\text{-}RSC(\mathscr{C})$-arrows

$$(f'', f', f), (g'', g', g) : (C_1, C_2, m_3) \to (A \cup \neg A, B, m \cup \neg m),$$

such that $(1_0, m \cup \neg m, 1_B) \circ (f'', f', f) = (1_0, m \cup \neg m, 1_B) \circ (g'', g', g)$. We have $(1_0 \circ f'', (m \cup \neg m) \circ f', 1_B \circ f) = (1_0 \circ g'', (m \cup \neg m) \circ g', 1_B \circ g)$. So $f = g$.
$\Rightarrow 1_0 \circ f'' = 1_0 \circ g''$, $(m \cup \neg m) \circ f' = (m \cup \neg m) \circ g'$ and $1_B \circ f = 1_B \circ g$. Thus $f'' = g''$.
Since $(m \cup \neg m)$ is a monic in $\mathscr{C}$, $(m \cup \neg m) \circ f' = (m \cup \neg m) \circ g' \Rightarrow f' = g'$.

$(1_0, m \cup \neg m, 1_B)$ is epic: consider the $\xi\text{-}RSC(\mathscr{C})$-arrows $(f'', f', f), (g'', g', g) : (B, B, 1_B) \to (C_1, C_2, m_3)$, such that $(f'', f', f) \circ (1_0, m \cup \neg m, 1_B) = (g'', g', g) \circ (1_0, m \cup \neg m, 1_B)$. We have $(f'' \circ 1_0, f' \circ (m \cup \neg m), f \circ 1_B) = (g'' \circ 1_0, g' \circ (m \cup \neg m), g \circ 1_B)$
$\Rightarrow f'' \circ 1_0 = g'' \circ 1_0$, $f' \circ (m \cup \neg m) = g' \circ (m \cup \neg m)$ and $f \circ 1_B = g \circ 1_B$.
So, $f'' = g''$. Since (f'', f', f) and (g'', g', g) are $\xi\text{-}RSC(\mathscr{C})$-arrows, we have $m_3 \circ f' = f \circ 1_B$ and $m_3 \circ g' = g \circ 1_B$. Now, $f = g \Rightarrow m_3 \circ f' = m_3 \circ g'$. As m_3 is monic, $f' = g'$.

$(1_0, m \cup \neg m, 1_B)$ is not iso: suppose to the contrary, $(1_0, m \cup \neg m, 1_B)$ is an iso in $\xi\text{-}RSC(\mathscr{C})$, i.e. there exists a $\xi\text{-}RSC(\mathscr{C})$-arrow $(g'', g', g) : (B, B, 1_B) \to (A \cup \neg A, B, m \cup \neg m)$ such that $(g'', g', g) \circ (1_0, m \cup \neg m, 1_B) = 1_{(A \cup \neg A, B)}$ and $(1_0, m \cup \neg m, 1_B) \circ (g'', g', g) = 1_{(B,B)}$.
$\Rightarrow g' \circ (m \cup \neg m) = 1_{A \cup \neg A}$, and $(m \cup \neg m) \circ g' = 1_B$.
$\Rightarrow m \cup \neg m$ is an iso, a contradiction.
Therefore using Proposition 9, $\xi\text{-}RSC(\mathscr{C})$ is not a topos.

Recall the category $RSC(\mathbf{2\text{-}Set})$, where $\mathbf{2\text{-}Set}$ is the category of monoid actions of $\mathbf{2} := \{0, 1\}$ on sets. Since, $\mathbf{2\text{-}Set}$ is a topos but not a Boolean topos [43], $\xi\text{-}RSC(\mathbf{2\text{-}Set})$ does not form a topos. In general, any topos $\mathbf{M\text{-}Set}$, where $\mathbf{M}$ is a monoid but not a group, is non-Boolean [43]. Therefore, we have the following.

Corollary 2.

(1) ξ-RSC($\mathbf{M}$-$\mathbf{Set}$) is not a topos, where $\mathbf{M}$ is any monoid that is not a group.
(2) ξ-RSC(SET)($\simeq SET^2$) is a topos.

3.3   Properties of $RSC(\mathscr{C})$

As checked for RSC in [57,88], various category-theoretic properties can be
established for $RSC(\mathscr{C})$, leading to the result that it is a quasitopos.

Proposition 31. $RSC(\mathscr{C})$ *is finitely complete and cocomplete.*

Proof.
Products: Given objects (X_1, X_2, m_1) and (Y_1, Y_2, m_2) in $RSC(\mathscr{C})$, $((X_1 \times Y_1, X_2 \times Y_2, m_1 \times m_2), (p_1, p_3), (p_2, p_4))$ is the product, where $m_1 \times m_2$ is as in
Remark 2 and $(X_1 \times Y_1, p_1, p_2)$ is the product of X_1 and Y_1, $(X_2 \times Y_2, p_3, p_4)$ is
the product of X_2 and Y_2 in $\mathscr{C}$. This is established as follows (Fig. 28).
Claim (1) - $(X_1 \times Y_1, X_2 \times Y_2, m_1 \times m_2)$ is an $RSC(\mathscr{C})$-object, i.e. $m_1 \times m_2$ is
monic in $\mathscr{C}$.
Suppose $f, g : Z \to (X_1 \times Y_1)$ is a $\mathscr{C}$-arrow such that $(m_1 \times m_2) \circ f = (m_1 \times m_2) \circ g$.
This implies $p_3 \circ (m_1 \times m_2) \circ f = p_3 \circ (m_1 \times m_2) \circ g \Rightarrow m_1 \circ p_1 \circ f = m_1 \circ p_1 \circ g \Rightarrow p_1 \circ f = p_1 \circ g : Z \to X_1$ (as m_1 is monic). Similarly,
$p_4 \circ (m_1 \times m_2) \circ f = p_4 \circ (m_1 \times m_2) \circ g \Rightarrow p_2 \circ f = p_2 \circ g : Z \to Y_1$. Using
universal property of the product $X_1 \times Y_1$, we have $f = g$.
By definition of the arrow $m_1 \times m_2$, $(p_1, p_3) : (X_1 \times Y_1, X_2 \times Y_2, m_1 \times m_2) \to (X_1, X_2, m_1)$ and $(p_2, p_4) : (X_1 \times Y_1, X_2 \times Y_2, m_1 \times m_2) \to (Y_1, Y_2, m_2)$ are
$RSC(\mathscr{C})$-arrows.
Claim (2) - $((X_1 \times Y_1, X_2 \times Y_2, m_1 \times m_2), (p_1, p_3), (p_2, p_4))$ is the product of
(X_1, X_2, m_1) and (Y_1, Y_2, m_2) in $RSC(\mathscr{C})$.
Universal property: consider an $RSC(\mathscr{C})$-object (Z_1, Z_2, m_3) and $RSC(\mathscr{C})$-
arrows

$$(u', u) : (Z_1, Z_2, m_3) \to (X_1, X_2, m_1) \text{ and } (v', v) : (Z_1, Z_2, m_3) \to (Y_1, Y_2, m_2).$$

We need to find an $RSC(\mathscr{C})$-arrow $\langle(u', u), (v', v)\rangle : (Z_1, Z_2) \to (X_1 \times Y_1, X_2 \times Y_2)$
satisfying $(p_1, p_3) \circ \langle(u', u), (v', v)\rangle = (u', u)$ and $(p_2, p_4) \circ \langle(u', u), (v', v)\rangle = (v', v)$.
Using universal property of the products $(X_1 \times Y_1, p_1, p_2)$ and $(X_2 \times Y_2, p_3, p_4)$,
there exist unique arrows $\mathscr{C}$-arrows $\langle u', v'\rangle : Z_1 \to (X_1 \times Y_1)$ and $\langle u, v\rangle : Z_2 \to (X_2 \times Y_2)$ such that $p_1 \circ \langle u', v'\rangle = u'$, $p_2 \circ \langle u', v'\rangle = v'$, $p_3 \circ \langle u, v\rangle = u$ and
$p_4 \circ \langle u, v\rangle = v$. Since (u', u) and (v', v) are $RSC(\mathscr{C})$-arrows, we have $u \circ m_3 = m_1 \circ u'$ and $v \circ m_3 = m_2 \circ v'$. We make the following claims.
(A). $(\langle u', v'\rangle, \langle u, v\rangle) : (Z_1, Z_2, m_3) \to (X_1 \times Y_1, X_2 \times Y_2, m_1 \times m_2)$ is an $RSC(\mathscr{C})$-
arrow, i.e. $(m_1 \times m_2) \circ \langle u', v'\rangle = \langle u, v\rangle \circ m_3$. We have $p_3 \circ (m_1 \times m_2) \circ \langle u', v'\rangle = m_1 \circ p_1 \circ \langle u', v'\rangle = m_1 \circ u' = u \circ m_3 = p_3 \langle u, v\rangle \circ m_3$, and $p_4 \circ (m_1 \times m_2) \circ \langle u', v'\rangle = m_2 \circ p_2 \circ \langle u', v'\rangle = m_2 \circ v' = v \circ m_3 = p_4 \circ \langle u, v\rangle \circ m_3$. Using universal property
of the product $(X_2 \times Y_2, p_3, p_4)$, we have $(m_1 \times m_2) \circ \langle u', v'\rangle = \langle u, v\rangle \circ m_3$.

(B). $(p_1, p_3) \circ (\langle u', v' \rangle, \langle u, v \rangle) = (u', u)$ and $(p_2, p_4) \circ (\langle u', v' \rangle, \langle u, v \rangle) = (v', v)$, i.e. $p_1 \circ \langle u', v' \rangle = u'$, $p_3 \circ \langle u, v \rangle = u$, $p_2 \circ \langle u', v' \rangle = v'$ and $p_4 \circ \langle u, v \rangle = v$, all of which hold in $\mathscr{C}$.

(C). $(\langle u', v' \rangle, \langle u, v \rangle)$ is the unique arrow satisfying (B). Uniqueness of $\langle u', v' \rangle$ and $\langle u, v \rangle$ entails the uniqueness of $(\langle u', v' \rangle, \langle u, v \rangle)$.

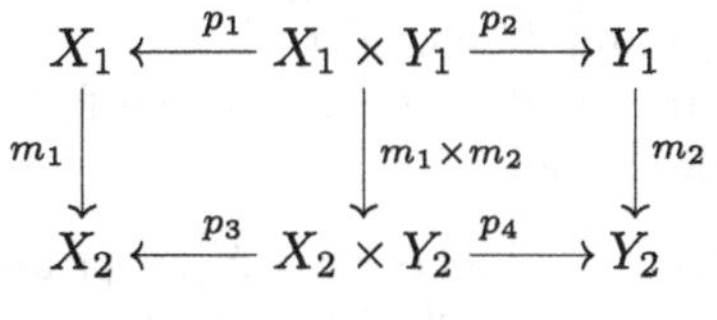

$$
\begin{array}{ccccc}
X_1 & \xleftarrow{\ p_1\ } & X_1 \times Y_1 & \xrightarrow{\ p_2\ } & Y_1 \\
{\scriptstyle m_1}\downarrow & & {\scriptstyle m_1 \times m_2}\downarrow & & \downarrow{\scriptstyle m_2} \\
X_2 & \xleftarrow{\ p_3\ } & X_2 \times Y_2 & \xrightarrow{\ p_4\ } & Y_2
\end{array}
$$

Fig. 28. Products in $RSC(\mathscr{C})$

Equalizers: Given $RSC(\mathscr{C})$-arrows $(u', u), (v', v) : (X_1, X_2, m_1) \to (Y_1, Y_2, m_2)$, the equalizer is the $RSC(\mathscr{C})$-arrow $(e', e) : (E', E, (\widetilde{m_1 \circ e'})) \to (X_1, X_2, m_1)$, where $e' : E' \to X_1$ and $e : E \to X_2$ are equalizers of the pairs u', v' and u, v respectively in $\mathscr{C}$. This is shown as follows.

We have $u \circ m_1 \circ e' = m_2 \circ u' \circ e' = m_2 \circ v' \circ e' = v \circ m_1 \circ e'$. Thus, there exists a unique $\mathscr{C}$-arrow $(\widetilde{m_1 \circ e'}) : E' \to E$ obtained through the universal property of the equalizer e (Fig. 29) such that $e \circ (\widetilde{m_1 \circ e'}) = m_1 \circ e'$. We claim that this arrow is monic in $\mathscr{C}$. Let $f, g : Z \to E'$ such that $(\widetilde{m_1 \circ e'}) \circ f = (\widetilde{m_1 \circ e'}) \circ g$. We have $e \circ (\widetilde{m_1 \circ e'}) \circ f = e \circ (\widetilde{m_1 \circ e'}) \circ g \Rightarrow m_1 \circ e' \circ f = m_1 \circ e' \circ g \Rightarrow e' \circ f = e' \circ g$ (as m_1 is monic). Further, in any topos, any equalizer is always monic (Proposition 9), therefore e' is a monic, implying $f = g$. So, $(E', E, (\widetilde{m_1 \circ e'}))$ is an $RSC(\mathscr{C})$-object.

We have to show that (e', e) is indeed an equalizer of (u', u) and (v', v). $(u', u) \circ (e', e) = (v', v) \circ (e', e)$, as e, e' are equalizers in $\mathscr{C}$ of u', v' and u, v respectively.

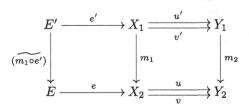

Fig. 29. Equalizers in $RSC(\mathscr{C})$

Universal property:
Consider an $RSC(\mathscr{C})$-arrow $(r', r) : (Z_1, Z_2, m_3) \to (X_1, X_2, m_1)$ such that $(u', u) \circ (r', r) = (v', v) \circ (r', r)$, i.e. we have $u' \circ r' = v' \circ r'$ and $u \circ r = v \circ r$. Since

(r', r) is an $RSC(\mathscr{C})$-arrow, we have $m_1 \circ r' = r \circ m_3$. Using universal properties of the equalizers e and e', there exist $\mathscr{C}$-arrows $\tilde{e} : Z_2 \to E$ and $\tilde{e}' : Z_1 \to E'$ such that $r' = e' \circ \tilde{e}'$ and $r = e \circ \tilde{e}$ respectively. We have $e \circ \tilde{e} \circ m_3 = r \circ m_3 = m_1 \circ r' = m_1 \circ e' \circ \tilde{e}' = e \circ \widetilde{(m_1 \circ e')} \circ \tilde{e}'$. Since e is an equalizer, and thus a monic, $\tilde{e} \circ m_3 = \widetilde{(m_1 \circ e')} \circ \tilde{e}'$. Thus, $(\tilde{e}', \tilde{e}) : (Z_1, Z_2, m_3) \to (E', E, \widetilde{(m_1 \circ e')})$ is an $RSC(\mathscr{C})$-arrow. Moreover, $(e', e) \circ (\tilde{e}', \tilde{e}) = (e' \circ \tilde{e}', e \circ \tilde{e}) = (r', r)$.
Uniqueness of $(\tilde{e}', \tilde{e})$ follows from the uniqueness of $\tilde{e}'$ and $\tilde{e}$. Therefore, using Proposition 6, $RSC(\mathscr{C})$ is finitely complete.
To show finite cocompleteness of $RSC(\mathscr{C})$, image factorization (Definition 29) has to be used as we show below.

Coproducts: Given $RSC(\mathscr{C})$-objects (X_1, X_2, m_1) and (Y_1, Y_2, m_2), take the image factorization of $(m_1 + m_2) : (X_1 + Y_1) \to (X_2 + Y_2)$ (Remark 2 and Definition 29). $((m_1 + m_2)(X_1 + Y_1), (X_2 + Y_2), Im(m_1 + m_2))$ is the coproduct object of (X_1, X_2, m_1) and (Y_1, Y_2, m_2) in $RSC(\mathscr{C})$ (Fig. 30), where $(X_1 + Y_1, i_1, i_2)$ is the coproduct of X_1 and Y_1, and $(X_2 + Y_2, i_3, i_4)$ is the coproduct of X_2 and Y_2. Let us recall how $Im(m_1 + m_2)$ is obtained. $Im(m_1 + m_2)$ is the equalizer of the $\mathscr{C}$-arrows $p : (X_2 + Y_2) \to D''$ and $q : (X_1 + Y_2) \to D''$, where p and q are pushout of $(m_1 + m_2)$ with $(m_1 + m_2)$ (diagram below). Therefore $q \circ (m_1 + m_2) = p \circ (m_1 + m_2)$, $q \circ Im(m_1 + m_2) = p \circ Im(m_1 + m_2)$, and there exists a unique arrow $(m_1 + m_2)^* : X_1 + Y_1 \to ((m_1 + m_2)(X_1 + Y_1))$ such that $Im(m_1 + m_2) \circ (m_1 + m_2)^* = (m_1 + m_2)$.

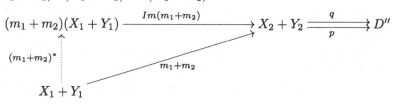

$((m_1 + m_2)^* \circ i_1, i_3) : (X_1, X_2, m_1) \to ((m_1 + m_2)(X_1 + Y_1), (X_2 + Y_2), Im(m_1 + m_2))$ and $((m_1 + m_2)^* \circ i_2, i_4) : (Y_1, Y_2, m_2) \to ((m_1 + m_2)(X_1 + Y_1), (X_2 + Y_2), Im(m_1 + m_2))$ are $RSC(\mathscr{C})$-arrows, because $Im(m_1 + m_2) \circ (m_1 + m_2)^* \circ i_1 = (m_1 + m_2) \circ i_1 = i_3 \circ m_1$ and $Im(m_1 + m_2) \circ (m_1 + m_2)^* \circ i_2 = (m_1 + m_2) \circ i_2 = i_4 \circ m_2$ (last step is using Remark 2).

$$X_1 \xrightarrow{(m_1 + m_2)^* \circ i_1} (m_1 + m_2)(X_1 + Y_1) \xleftarrow{(m_1 + m_2)^* \circ i_2} Y_1$$

with vertical arrows m_1, $Im(m_1 + m_2)$, m_2 and bottom row

$$X_2 \xrightarrow{i_3} X_2 + Y_2 \xleftarrow{i_4} Y_2$$

Claim: $(((m_1 + m_2)(X_1 + Y_1), (X_2 + Y_2), Im(m_1 + m_2)), ((m_1 + m_2)^* \circ i_1, i_3), ((m_1 + m_2)^* \circ i_2, i_4))$ is the coproduct of (X_1, X_2, m_1) and (Y_1, Y_2, m_2) in $RSC(\mathscr{C})$ (Fig. 30).

Universal property:
Suppose there exists an $RSC(\mathscr{C})$-object (Z_1, Z_2, m_3), and two $RSC(\mathscr{C})$-arrows
$(u', u) : (X_1, X_2, m_1) \rightarrow (Z_1, Z_2, m_3)$ and $(v', v) : (Y_1, Y_2, m_2) \rightarrow (Z_1, Z_2, m_3)$.
Since (u', u) and (v', v) are $RSC(\mathscr{C})$-arrows, $m_3 \circ u' = u \circ m_1$ and $m_3 \circ v' = v \circ m_2$.
Using universal property of the coproducts $(X_1 + Y_1, i_1, i_2)$ and $(X_2 + Y_2, i_3, i_4)$,
there exist unique $\mathscr{C}$-arrows $[u', v'] : (X_1 + Y_1) \rightarrow Z_1$ and $[u, v] : (X_2 + Y_2) \rightarrow Z_2$
such that $[u', v'] \circ i_1 = u'$, $[u', v'] \circ i_2 = v'$, $[u, v] \circ i_3 = u$ and $[u, v] \circ i_4 = v$. We
have to find a unique $RSC(\mathscr{C})$-arrow

$$(l', l) : ((m_1 + m_2)(X_1 + Y_1), (X_2 + Y_2), Im(m_1 + m_2)) \rightarrow (Z_1, Z_2, m_3)$$

such that $(l', l) \circ ((m_1 + m_2)^* \circ i_1, i_3) = (u', u)$ and $(l', l) \circ ((m_1 + m_2)^* \circ i_2, i_4) = (v', v)$.
We have to find a unique $RSC(\mathscr{C})$-arrow $(l', l) : ((m_1 + m_2)(X_1 + X_2), (X_2 + Y_2), Im(m_1 + m_2)) \rightarrow (Z_1, Z_2, m_3)$ such that $(l', l) \circ ((m_1 + m_2)^* \circ i_1, i_3) = (u', u)$
and $(l', l) \circ ((m_1 + m_2)^* \circ i_2, i_4) = (v', v)$.

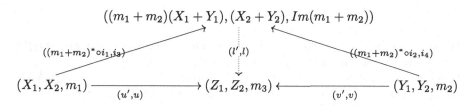

$$((m_1 + m_2)(X_1 + Y_1), (X_2 + Y_2), Im(m_1 + m_2))$$

$((m_1+m_2)^* \circ i_1, i_3)$ (l', l) $((m_1+m_2)^* \circ i_2, i_4)$

$(X_1, X_2, m_1) \xrightarrow{(u', u)} (Z_1, Z_2, m_3) \xleftarrow{(v', v)} (Y_1, Y_2, m_2)$

Fig. 30. Coproducts in $RSC(\mathscr{C})$

We have $[u, v] \circ (m_1 + m_2) \circ i_1 = [u, v] \circ i_3 \circ m_1 = u \circ m_1 = m_3 \circ u' = m_3 \circ [u', v'] \circ i_1$, and $[u, v] \circ (m_1 + m_2) \circ i_2 = [u, v] \circ i_4 \circ m_2 = v \circ m_2 = m_3 \circ v' = m_3 \circ [u', v'] \circ i_2$.
Using the universal property of the coproduct $(X_1 + Y_1, i_1, i_2)$,

$$m_3 \circ [u', v'] = [u, v] \circ (m_1 + m_2).$$

Now, since m_3 is monic, and in a topos, any monic is an equalizer (Proposition
9), we have m_3 is the equalizer of some pair of $\mathscr{C}$-arrows $r, s : Z_2 \rightarrow D'''$ such
that $r \circ m_3 = s \circ m_3$.
We have $s \circ [u, v] \circ (m_1 + m_2) = s \circ m_3 \circ [u', v'] = r \circ m_3 \circ [u', v'] = r \circ [u, v] \circ (m_1 + m_2)$.
Using the universal property of the pushout arrows p and q, there exists $\alpha :$
$D'' \rightarrow D'''$ such that $\alpha \circ p = r \circ [u, v]$ and $\alpha \circ q = s \circ [u, v]$. Now, consider the
diagram below. We have $r \circ [u, v] \circ Im(m_1 + m_2) = \alpha \circ p \circ Im(m_1 + m_2) = \alpha \circ q \circ Im(m_1 + m_2) = s \circ [u, v] \circ Im(m_1 + m_2)$. Using the equalizer property of
m_3, there exists t such that the left hand square in the diagram below commutes.

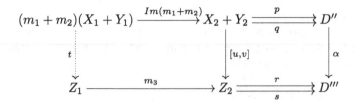

Define $(l', l) := (t, [u, v]) : ((m_1 + m_2)(X_1 + Y_1), (X_2 + Y_2), Im(m_1 + m_2)) \to$
(Z_1, Z_2, m_3) in the diagram in Fig. 30. This is an arrow in $RSC(\mathscr{C})$ because
$m_3 \circ t = [u, v] \circ Im(m_1 + m_2)$.
We have $m_3 \circ t \circ (m_1 + m_2)^* \circ i_1 = [u, v] \circ Im(m_1 + m_2) \circ (m_1 + m_2)^* \circ i_1 =$
$[u, v] \circ (m_1 + m_2) \circ i_1 = [u, v] \circ i_3 \circ m_1 = u \circ m_1 = m_3 \circ u'$. Since m_3 is monic,

$$t \circ (m_1 + m_2)^* \circ i_1 = u'.$$

Again, $m_3 \circ t \circ (m_1 + m_2)^* \circ i_2 = [u, v] \circ Im(m_1 + m_2) \circ (m_1 + m_2)^* \circ i_2 =$
$[u, v] \circ (m_1 + m_2) \circ i_2 = [u, v] \circ i_4 \circ m_2 = v \circ m_2 = m_3 \circ v'$ and using m_3 monic,

$$t \circ (m_1 + m_2)^* \circ i_2 = v'.$$

We also have $[u, v] \circ i_3 = u$ and $[u, v] \circ i_4 = v$. Therefore, $(t, [u, v]) \circ ((m_1 + m_2)^* \circ$
$i_1, i_3) = (u', u)$ and $(t, [u, v]) \circ ((m_1 + m_2)^* \circ i_2, i_4) = (v', v)$.
Uniqueness of $(t, [u, v])$: take an $RSC(\mathscr{C})$-arrow $(l', l) : ((m_1 + m_2)(X_1 +$
$X_2), (X_2 + Y_2), Im(m_1 + m_2)) \to (Z_1, Z_2, m_3)$ such that $(l', l) \circ ((m_1 + m_2)^* \circ$
$i_1, i_3) = (u', u)$ and $(l', l) \circ ((m_1 + m_2)^* \circ i_2, i_4) = (v', v)$. Then $l \circ i_3 = u$ and
$l \circ i_4 = v$ imply $l = [u, v]$, by uniqueness of $[u, v]$. Now, since (l', l) is an $RSC(\mathscr{C})$-
arrow, $m_3 \circ l' = l \circ Im(m_1 + m_2)$.
So $m_3 \circ l' = l \circ Im(m_1 + m_2) = [u, v] \circ Im(m_1 + m_2) = m_3 \circ t$. Since m_3 is a
monic, $l' = t$.

Coequalizer: Consider $RSC(\mathscr{C})$-arrows $(u', u) : (X_1, X_2, m_1) \to (Y_1, Y_2, m_2)$
and $(v', v) : (X_1, X_2, m_1) \to (Y_1, Y_2, m_2)$, i.e. $m_2 \circ u' = u \circ m_1$ and $m_2 \circ v' = v \circ m_1$.
Further, let $f' : Y_1 \to C'$ and $f : Y_2 \to C$ be coequalizers of the pairs u', v' and
u, v respectively in $\mathscr{C}$, i.e. $f' \circ u' = f' \circ v'$ and $f \circ u = f \circ v$.
Now, we have $f \circ m_2 \circ u' = f \circ u \circ m_1 = f \circ v \circ m_1 = f \circ m_2 \circ v'$. Using the
universal property of f', there exists a unique $\mathscr{C}$-arrow $(\widetilde{f \circ m_2}) : C' \to C$ such
that

$$(\widetilde{f \circ m_2}) \circ f' = f \circ m_2.$$

Now, take the image factorization of $(\widetilde{f \circ m_2})$ as follows.
$Im((\widetilde{f \circ m_2})) : ((\widetilde{f \circ m_2}))(C') \to C$ is the equalizer of the $\mathscr{C}$-arrows $p : C \to D''$
and $q : C \to D''$, where p and q are pushout of $(\widetilde{f \circ m_2})$ with $(\widetilde{f \circ m_2})$ (Diagram
below). Here, $p \circ (\widetilde{f \circ m_2}) = q \circ (\widetilde{f \circ m_2})$, $p \circ Im((\widetilde{f \circ m_2})) = q \circ Im((\widetilde{f \circ m_2}))$
and $Im((\widetilde{f \circ m_2})) \circ (\widetilde{f \circ m_2})^* = ((\widetilde{f \circ m_2}))$.

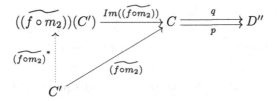

We claim that

$$((\widetilde{f \circ m_2})^* \circ f', f) : (Y_1, Y_2, m_2) \to (((\widetilde{f \circ m_2}))(C'), C, Im((\widetilde{f \circ m_2})))$$

is an $RSC(\mathscr{C})$-arrow, and a coequalizer of (u', u) and (v', v) (Fig. 31). Now,

$$((\widetilde{f \circ m_2})^* \circ f', f) \circ (u', u) = ((\widetilde{f \circ m_2})^* \circ f' \circ u', f \circ u) =$$
$$((\widetilde{f \circ m_2})^* \circ f' \circ v', f \circ v) = (\widetilde{f \circ m_2})^* \circ f', f) \circ (v', v).$$

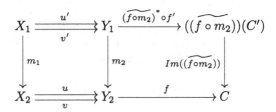

Fig. 31. Coequalizers in $RSC(\mathscr{C})$

Universal property: suppose there exists an $RSC(\mathscr{C})$-object (Z_1, Z_2, m_3) and an $RSC(\mathscr{C})$-arrow $(k', k) : (Y_1, Y_2, m_2) \to (Z_1, Z_2, m_3)$ such that $(k', k) \circ (u', u) = (k', k) \circ (v', v)$, i.e. $k' \circ u' = k' \circ v'$ and $k \circ u = k \circ v$. Using universal property of the coequalizers f' and f, there exist unique arrows $\mathscr{C}$-arrows $\tilde{k}' : C' \to Z_1$ and $\tilde{k} : C \to Z_2$ such that $\tilde{k} \circ f = k$ and $\tilde{k}' \circ f' = k'$. Since (k', k) is an $RSC(\mathscr{C})$-arrow, we have $m_3 \circ k' = k \circ m_2$. We have to find a unique $RSC(\mathscr{C})$-arrow $(l', l) : (((\widetilde{f \circ m_2}))(C'), C, Im((\widetilde{f \circ m_2}))) \to (Z_1, Z_2, m_3)$ such that $(l', l) \circ ((\widetilde{f \circ m_2})^* \circ f', f) = (k', k)$.

We have $\tilde{k} \circ (\widetilde{f \circ m_2}) \circ f' = \tilde{k} \circ f \circ m_2 = k \circ m_2 = m_3 \circ k' = m_3 \circ \tilde{k}' \circ f'$. Since, any coequalizer is always an epic in any category [43], so f' is an epic and

$$\tilde{k} \circ (\widetilde{f \circ m_2}) = m_3 \circ \tilde{k}'.$$

Now, since m_3 is monic, and in a topos, any monic is an equalizer (Proposition 9), we have m_3 as the equalizer of some pair of $\mathscr{C}$-arrows $r, s : Z_2 \to D'''$ such that $r \circ m_3 = s \circ m_3$.
$r \circ \tilde{k} \circ Im((\widetilde{f \circ m_2})) \circ (\widetilde{f \circ m_2})^* = r \circ \tilde{k} \circ (\widetilde{f \circ m_2}) = r \circ m_3 \circ \tilde{k}' = s \circ m_3 \circ \tilde{k}' = s \circ \tilde{k} \circ (\widetilde{f \circ m_2}) = s \circ \tilde{k} \circ Im((\widetilde{f \circ m_2})) \circ (\widetilde{f \circ m_2})^*$. Since $(\widetilde{f \circ m_2})^*$ is epic (Definition 29),

$$r \circ \tilde{k} \circ Im((\widetilde{f \circ m_2})) = s \circ \tilde{k} \circ Im((\widetilde{f \circ m_2})).$$

Using universal property of the pushout arrows p and q, there exists $\alpha : D'' \to D'''$ such that $\alpha \circ p = r \circ \tilde{k}$ and $\alpha \circ q = s \circ \tilde{k}$. Now, consider the diagram below. We have $r \circ \tilde{k} \circ Im((\widetilde{f \circ m_2})) = \alpha \circ p \circ Im((\widetilde{f \circ m_2})) = \alpha \circ q \circ Im((\widetilde{f \circ m_2})) = s \circ \tilde{k} \circ Im((\widetilde{f \circ m_2}))(C')$. Using the equalizer property of m_3, there exists $t : ((\widetilde{f \circ m_2}))(C') \to Z_1$ such that the left hand square in the diagram below commutes.

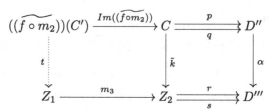

Define $(l', l) := (t, \tilde{k}) : (((\widetilde{f \circ m_2}))(C'), C, Im((\widetilde{f \circ m_2}))) \to (Z_1, Z_2, m_3)$. This is an arrow in $RSC(\mathscr{C})$ because $m_3 \circ t = \tilde{k} \circ Im((\widetilde{f \circ m_2}))$. Moreover, $m_3 \circ t \circ ((\widetilde{f \circ m_2}))^* \circ f' = \tilde{k} \circ Im(((\widetilde{f \circ m_2}))) \circ ((\widetilde{f \circ m_2}))^* \circ f' = \tilde{k} \circ ((\widetilde{f \circ m_2})) \circ f' = \tilde{k} \circ f \circ m_2 = k \circ m_2 = m_3 \circ k'$. Since m_3 is monic,

$$t \circ ((\widetilde{f \circ m_2}))^* \circ f' = k'.$$

Also $\tilde{k} \circ f = k$. So $(t, \tilde{k}) \circ ((\widetilde{f \circ m_2})^* \circ f', f) = (t \circ ((\widetilde{f \circ m_2})^* \circ f', \tilde{k} \circ f) = (k', k)$. Uniqueness of $(t', \tilde{k})$: let $(l', l) : (((\widetilde{f \circ m_2}))(C'), C, Im((\widetilde{f \circ m_2}))) \to (Z_1, Z_2, m_3)$ be an $RSC(\mathscr{C})$-arrow such that $(l', l) \circ ((\widetilde{f \circ m_2})^* \circ f', f) = (k', k)$. Uniqueness of $\tilde{k}$ gives $l = \tilde{k}$. Then the facts that (l', l) is an $RSC(\mathscr{C})$-arrow and m_3 is monic give $l' = t$.

Observation 18. *Consider* $RSC(\mathscr{C})$-*arrows* $(f', f) : (X_1, X_2, m_1) \to (Y_1, Y_2, m_2)$ *and* $(g', g) : (D_1, D_2, m_3) \to (E_1, E_2, m_4)$. *Then* $(f', f) \times (g', g) = (f' \times g', f \times g)$.

Proof. Let $(X_1 \times D_1, p_1, p_2)$, $(X_2 \times D_2, p_3, p_4)$, $(Y_1 \times E_1, p_5, p_6)$ and $(Y_2 \times E_2, p_7, p_8)$ be products in $\mathscr{C}$. The above proposition gives that $((X_1 \times D_1, X_2 \times D_2, m_1 \times m_3), (p_1, p_3), (p_2, p_4))$ is a product in $RSC(\mathscr{C})$. Then , using Remark 2,
$(f', f) \times (g', g) = \langle (f', f) \circ (p_1, p_3), (g', g) \circ (p_2, p_4) \rangle = \langle (f' \circ p_1, f \circ p_3), (g' \circ p_2, g \circ p_4) \rangle = (\langle f' \circ p_1, g' \circ p_2 \rangle, \langle f \circ p_3, g \circ p_4 \rangle) = (f' \times g', f \times g)$.

Proposition 32. *Consider the diagram 32a in* $RSC(\mathscr{C})$. *This diagram is a pullback if and only if both the squares in the diagram 32b are pullbacks in* $\mathscr{C}$.

Proof. Here, $f \circ m_1 = m_4 \circ f'$, $g \circ m_2 = m_4 \circ g'$, $s \circ m_3 = m_2 \circ s'$ and $r \circ m_3 = m_1 \circ r'$.
($\Rightarrow$) Let the diagram in Fig. 32a be a pullback. Consider a $\mathscr{C}$-object E such that $f \circ n = g \circ m$, where $n : E \to X_2$ and $m : E \to Y_2$. Consider the $RSC(\mathscr{C})$-object

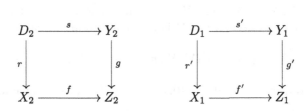

(a) Pullbacks in $RSC(\mathscr{C})$

$$D_2 \xrightarrow{\quad s \quad} Y_2 \qquad D_1 \xrightarrow{\quad s' \quad} Y_1$$

$$\begin{array}{ccc} r \downarrow & & \downarrow g \\ X_2 \xrightarrow{\quad f \quad} Z_2 \end{array} \qquad \begin{array}{ccc} r' \downarrow & & \downarrow g' \\ X_1 \xrightarrow{\quad f' \quad} Z_1 \end{array}$$

(b) Pullbacks in $\mathscr{C}$

Fig. 32. Pullbacks in $RSC(\mathscr{C})$

$(0, E, i_E)$. Since, $n \circ i_E = m_1 \circ i_{X_1}$ and $m \circ i_E = m_2 \circ i_{Y_1}$, we have $RSC(\mathscr{C})$-arrows $(i_{X_1}, n) : (0, E, i_E) \to (X_1, X_2, m_1)$ and $(i_{Y_1}, m) : (0, E, i_E) \to (Y_1, Y_2, m_2)$. Moreover, using $f \circ n = g \circ m$ and $f' \circ i_{X_1} = g' \circ i_{Y_1}$, we have $(f', f) \circ (i_{X_1}, n) = (g', g) \circ (i_{Y_1}, m)$ in $RSC(\mathscr{C})$. Using the given condition that diagram in Fig. 32a is a pullback, there exists a unique arrow $(i_{D_1}, h) : (0, E, i_E) \to (D_1, D_2, m_3)$ such that $(i_{X_1}, n) = (r', r) \circ (i_{D_1}, h)$ and $(i_{Y_1}, m) = (s', s) \circ (i_{D_1}, h)$. This implies that $h : E \to D_2$ is such that $n = r \circ h$ and $m = s \circ h$. Therefore, h is the required $\mathscr{C}$-arrow for left hand side diagram in Fig. 32b to be a pullback. Uniqueness of h follows from the uniqueness of (i_{D_1}, h)

To show that the right hand side diagram in Fig. 32b is a pullback. Consider a $\mathscr{C}$-object E such that $f' \circ n' = g' \circ m'$, where $m' : E \to Y_1$ and $n' : E \to X_1$. Consider the $RSC(\mathscr{C})$-object $(E, E, 1_E)$. Then $(f', f) \circ (n', m_1 \circ n') = (g', g) \circ (m', m_2 \circ m')$ in $RSC(\mathscr{C})$. Using the condition that the diagram in Fig. 32a is a pullback, there exists a unique $RSC(\mathscr{C})$-arrow $(h', h) : (E, E, 1_E) \to (D_1, D_2, m_3)$ such that $(r', r) \circ (h', h) = (n', m_1 \circ n')$ and $(s', s) \circ (h', h) = (m', m_2 \circ m')$. This implies $r' \circ h' = n'$ and $s' \circ h' = m'$. Therefore, h' is the required $\mathscr{C}$-arrow for right hand side diagram in Fig. 32b to be a pullback. Uniqueness of h' follows from the uniqueness of (h', h)

$(\Leftarrow)$ Let (E_1, E_2, m_5) be an $RSC(\mathscr{C})$-object such that $(f', f) \circ (n', n) = (g', g) \circ (m', m)$ in $RSC(\mathscr{C})$, where $(n', n) : (E_1, E_2, m_5) \to (X_1, X_2, m_1)$ and $(m', m) : (E_1, E_2, m_5) \to (Y_1, Y_2, m_2)$. This implies $f' \circ n' = g' \circ m'$, $f \circ n = g \circ m$. Then using the pullback property of the diagrams in Fig. 32b, we can obtain the unique $\mathscr{C}$-arrows $h' : E_1 \to D_1$ and $h : E_2 \to D_2$ such that $r' \circ h' = n'$, $r \circ h = n$, $s' \circ h' = m'$ and $s \circ h = m$. Note that $f \circ r \circ m_3 \circ h' = g \circ s \circ m_3 \circ h'$, $r \circ m_3 \circ h' = r \circ h \circ m_5$ and $s \circ m_3 \circ h' = s \circ h \circ m_5$. Using pullback property of the left hand square in Fig. 32b, we have $h \circ m_5 = m_3 \circ h'$. Therefore, (h', h) is an $RSC(\mathscr{C})$-arrow. Uniqueness of (h', h) follows from the uniqueness of h' and h in Fig. 32b.

Proposition 33. The following can be observed in $RSC(\mathscr{C})$.
Let $(f', f) : (X_1, X_2, m_1) \to (Y_1, Y_2, m_2)$ be an $RSC(\mathscr{C})$-arrow.

(1) $(1, 1)$ and $(0, 0)$ are the terminal and initial objects of $RSC(\mathscr{C})$ respectively, where 1 is the terminal object and 0 is the initial object of $\mathscr{C}$.
(2) $(f', f) : (X_1, X_2, m_1) \to (Y_1, Y_2, m_2)$ is monic if and only if $f : X_2 \to Y_2$ is a monic in $\mathscr{C}$. Moreover, (f', f) is monic also implies that f' is a monic in $\mathscr{C}$.
(3) If $(f', f) : (X_1, X_2, m_1) \to (Y_1, Y_2, m_2)$ is epic then $f : X_2 \to Y_2$ is epic in $\mathscr{C}$.
(4) For a $\mathscr{C}$-object Z, $(i_Z, 1_Z) : (0, Z, i_Z) \to (Z, Z, 1_Z)$ is epic and monic in $RSC(\mathscr{C})$.
(5) $(f', f) : (X_1, X_2, m_1) \to (Y_1, Y_2, m_2)$ is an iso in $RSC(\mathscr{C})$ if and only if $f : X_2 \to Y_2$ and $f' : X_1 \to Y_1$ are isos in $\mathscr{C}$.

Proof.
(1). Consider any $RSC(\mathscr{C})$-object (X_1, X_2, m_1).
(A). We have an $RSC(\mathscr{C})$-arrow $(!_{X_1}, !_{X_2}) : (X_1, X_2, m_1) \to (1, 1, 1_1)$. Now, if there exists any other $RSC(\mathscr{C})$-arrow $(f', f) : (X_1, X_2, m_1) \to (1, 1, 1_1)$, then $f' = !_{X_1}$ and $f = !_{X_2}$ because of the universal property of the terminal object 1 in $\mathscr{C}$.
(B). We have an $RSC(\mathscr{C})$-arrow $(i_{X_1}, i_{X_2}) : (0, 0, 1_0) \to (X_1, X_2, m_1)$. Now, if there exists any other $RSC(\mathscr{C})$-arrow $(f', f) : (0, 0, 1_0) \to (X_1, X_2, m_1)$, then $f' = i_{X_1}$ and $f = i_{X_2}$ because of the universal property of the initial object 1 in $\mathscr{C}$.
For (2) and (3): we have $m_2 \circ f' = f \circ m_1$.
(2). ($\Rightarrow$) Let $\mathscr{C}$-arrows $u, v : Z \to X_2$ be such that $f \circ u = f \circ v$. Consider the following diagram.

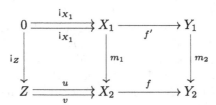

As $\mathscr{C}$ is cartesian closed, i_Z is monic. Thus, $(0, Z, i_Z)$ is an $RSC(\mathscr{C})$-object. Moreover, (i_{X_1}, u) and (i_{X_1}, v) are $RSC(\mathscr{C})$-arrows, and $(f', f) \circ (i_{X_1}, u) = (f', f) \circ (i_{X_1}, v)$. Since (f', f) is monic, we have $(i_{X_1}, u) = (i_{X_1}, v)$, implying $u = v$.
($\Leftarrow$) Let $RSC(\mathscr{C})$-arrows $(u', u), (v', v) : (Z_1, Z_2, m_3) \to (X_1, X_2, m_1)$ such that $(f', f) \circ (u', u) = (f', f) \circ (v', v)$. Therefore, $m_1 \circ u' = u \circ m_3$ and $m_1 \circ v' = v \circ m_3$. We have $f \circ u = f \circ v$ implies $u = v$ (as f is monic). Further, $m_1 \circ u' = u \circ m_3 =$

$v \circ m_3 = m_1 \circ v'$. Since m_1 is monic, we have $u' = v'$. Therefore, $(u', u) = (v', v)$ and (f', f) is monic.

Now, let (f', f) be monic. This implies that f is monic. Let $\mathscr{C}$-arrows $g, h : Z \to X_1$ be such that $f' \circ g = f' \circ h$. We have $m_2 \circ f' \circ g = m_2 \circ f' \circ h \Rightarrow f \circ m_1 \circ g = f \circ m_1 \circ h$. Since, f and m_1 are monics, we have $g = h$, and thus f' is monic.

(3). Let $\mathscr{C}$-arrows $u, v : X_2 \to Z$ be such that $u \circ f = v \circ f$. Consider the following diagram.

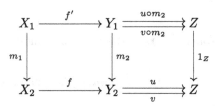

Now, $(u \circ m_2, u)$ and $(v \circ m_2, v)$ are $RSC(\mathscr{C})$-arrows from (Y_1, Y_2, m_2) to $(Z, Z, 1_Z)$ such that $(u \circ m_2, u) \circ (f', f) = (v \circ m_2, v) \circ (f', f)$. Since (f', f) is epic, we have $(u \circ m_2, u) = (v \circ m_2, v)$, implying $u = v$.

(4). Consider $RSC(\mathscr{C})$-arrows (f', f) and (g', g) with domain $(Z, Z, 1_Z)$ and codomain (X_1, X_2, m_1), such that $(f', f) \circ (i_Z, 1_Z) = (g', g) \circ (i_Z, 1_Z)$. This implies $f = g$. Since (f', f) and (g', g) are $RSC(\mathscr{C})$-arrows, we have $m_1 \circ f' = f \circ 1_Z = f$ and $m_1 \circ g' = g \circ 1_Z = g$. Then $f = g$ implies $m_1 \circ f' = m_1 \circ g'$. m_1 monic gives $f' = g'$.

(5). ($\Rightarrow$) There exists an $RSC(\mathscr{C})$-arrow $(g', g) : (Y_1, Y_2, m_2) \to (X_1, X_2, m_1)$ such that $(g', g) \circ (f', f) = 1_{(X_1, X_2, m_1)}$ and $(f', f) \circ (g', g) = 1_{(Y_1, Y_2, m_2)}$. This implies $g' \circ f' = 1_{X_1}$, $g \circ f = 1_{X_2}$, $f' \circ g' = 1_{Y_1}$ and $f \circ g = 1_{Y_2}$. Therefore, f and f' are iso arrows in $\mathscr{C}$.

($\Leftarrow$) Let f' and f be isos in $\mathscr{C}$, i.e. there exist $\mathscr{C}$-arrows $g' : Y_1 \to X_1$ and $g : Y_2 \to X_2$ such that $f \circ g = 1_{Y_2}$, $g \circ f = 1_{X_2}$, $f' \circ g' = 1_{Y_1}$, and $g' \circ f' = 1_{X_1}$. Therefore, $(g', g) \circ (f', f) = 1_{(X_1, X_2, m_1)}$ and $(f', f) \circ (g', g) = 1_{(Y_1, Y_2, m_2)}$; and $(f', f) : (X_1, X_2, m_1) \to (Y_1, Y_2, m_2)$ is an iso in $RSC(\mathscr{C})$.

Observation 19. *In the above proposition, (4) gives an example of an epic $(i_Z, 1_Z) : (0, Z, i_Z) \to (Z, Z, 1_Z)$ in $RSC(\mathscr{C})$, where Z is a $\mathscr{C}$-object. Note here that 1_Z is epic in $\mathscr{C}$, however i_Z need not be epic in $\mathscr{C}$. For example, in SET, $i : \emptyset \to \{1\}$ is not a surjective function, and thus not an epic.*

Proposition 34. *$RSC(\mathscr{C})$ is cartesian closed.*

Proof. Consider objects (X_1, X_2, m_1) and (Y_1, Y_2, m_2) in $RSC(\mathscr{C})$. Let $ev' : Y_2^{X_1} \times X_1 \to Y_2$ in $\mathscr{C}$, $ev'' : Y_1^{X_1} \times X_1 \to Y_1$ and $ev : Y_2^{X_2} \times X_2 \to Y_2$ be the evaluation arrows of the exponents $Y_2^{X_1}$, $Y_1^{X_1}$ and $Y_2^{X_2}$ in $\mathscr{C}$ respectively. Consider the $\mathscr{C}$-arrow $g := m_2 \circ ev'' : Y_1^{X_1} \times X_1 \to Y_2$. Using exponent property of $Y_2^{X_1}$, there exists a unique arrow $\hat{g} : Y_1^{X_1} \to Y_2^{X_1}$ such that $ev' \circ (\hat{g} \times 1_{X_1}) = m_2 \times ev''$ (left hand side diagram below).

$$Y_1^{X_1} \times X_1 \xrightarrow{\ ev''\ } Y_1$$

$$Y_2^{X_2} \times X_1 \xrightarrow{\ (1_{Y_2^{X_2}} \times m_1)\ } Y_2^{X_2} \times X_2 \xrightarrow{\ ev\ } Y_2$$

$$\hat{g} \times 1_{X_1} \qquad m_2 \qquad \hat{h} \times 1_{X_1} \qquad 1_{Y_2}$$

$$Y_2^{X_1} \times X_1 \xrightarrow{\ ev'\ } Y_2 \qquad\qquad Y_2^{X_1} \times X_1 \xrightarrow{\qquad\qquad ev' \qquad\qquad} Y_2$$

Again, consider the $\mathscr{C}$-arrow $h := 1_{Y_2} \circ ev \circ (1_{Y_2^{X_2}} \times m_1) : Y_2^{X_2} \times X_1 \to Y_2$. Using the exponent property of $Y_2^{X_1}$, there exists a unique arrow $\hat{h} : Y_2^{X_2} \to Y_2^{X_1}$ such that

$$ev' \circ (\hat{h} \times 1_{X_1}) = 1_{Y_2} \circ ev \circ (1_{Y_2}^{X_2} \times m_1) \text{ (right hand diagram above).}$$

Now, let $r : D \to Y_1^{X_1}$ and $s : D \to Y_2^{X_2}$ be the pullback of $\hat{h}$ with $\hat{g}$ in $\mathscr{C}$ (diagram below).

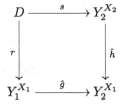

We show that the exponent object is given by $(D, Y_2^{X_2}, s)$ and the evaluation arrow is given by $(ev'' \circ (r \times 1_{X_1}), ev) : (D, Y_2^{X_2}, s) \times (X_1, X_2, m_1) \to (Y_1, Y_2, m_2)$ (Fig. 33). Note that $(D, Y_2^{X_2}, s) \times (X_1, X_2, m_1) = (D \times X_1, Y_2^{X_2} \times X_2, s \times m_1)$.

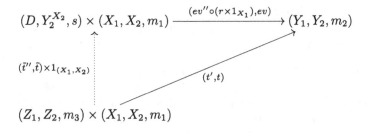

Fig. 33. Exponents in $RSC(\mathscr{C})$

Claim 1: $(D, Y_2^{X_2}, s)$ is an $RSC(\mathscr{C})$-object, i.e. s is a monic. For that, we first show that $\hat{g}$ is monic: let $p, q : Z \to Y_1^{X_1}$ such that $\hat{g} \circ p = \hat{g} \circ q$. Note that using the product properties in $\mathscr{C}$, we have the following:

$$(\hat{g} \times 1_{X_1}) \circ (p \times 1_{X_1}) = (\hat{g} \circ p) \times 1_{X_1} \text{ and } (\hat{g} \times 1_{X_1}) \circ (q \times 1_{X_1}) = (\hat{g} \circ q) \times 1_{X_1}.$$

Since $(\hat{g} \circ q) = (\hat{g} \circ p)$, we have $m_2 \circ ev'' \circ (p \times 1_{X_1}) = ev' \circ (\hat{g} \times 1_{X_1}) \circ (p \times 1_{X_1}) = ev' \circ ((\hat{g} \circ p) \times 1_{X_1}) = ev' \circ ((\hat{g} \circ q) \times 1_{X_1}) = ev' \circ (\hat{g} \times 1_{X_1}) \circ (q \times 1_{X_1}) = m_2 \circ ev'' \circ (q \times 1_{X_1})$.

Since m_2 is monic, we have $ev'' \circ (p \times 1_{X_1}) = ev'' \circ (q \times 1_{X_1})$. Using the property of the exponent $Y_1^{X_1}$, $p = q$.

Now, using Proposition 5, $\hat{g}$ is monic implies that s is a monic.

Claim 2: $(ev'' \circ (r \times 1_{X_1}), ev)$ is an $RSC(\mathscr{C})$-arrow, i.e. $m_2 \circ ev'' \circ (r \times 1_{X_1}) = ev \circ (s \times m_1)$: Using product properties, we have $(1_{Y_2^{X_2}} \times m_1) \circ (s \times 1_{X_1}) = s \times m_1$.

Again using product properties and $\hat{g} \circ r = \hat{h} \circ s$, we obtain $(\hat{g} \times 1_{X_1}) \circ (r \times 1_{X_1}) = (\hat{h} \times 1_{X_1}) \circ (s \times 1_{X_1})$ implying $(\hat{g} \circ r) \times 1_{X_1} = (\hat{h} \circ s) \times 1_{X_1}$. Now,

$$
\begin{aligned}
m_2 \circ ev'' \circ (r \times 1_{X_1}) &= ev' \circ (\hat{g} \times 1_{X_1}) \circ (r \times 1_{X_1}) \\
&= ev' \circ (\hat{h} \times 1_{X_1}) \circ (s \times 1_{X_1}) \\
&= ev \circ (1_{Y_2^{X_2}} \times m_1) \circ (s \times 1_{X_1}) \\
&= ev \circ (s \times m_1)
\end{aligned}
$$

Consider an $RSC(\mathscr{C})$-object (Z_1, Z_2, m_3) and an $RSC(\mathscr{C})$-arrow $(t', t) : (Z_1, Z_2, m_3) \times (X_1, X_2, m_1) \to (Y_1, Y_2, m_2)$. We have the following:

$$
t \circ (m_3 \times m_1) = m_2 \times t'.
$$

Claim 3: There exists a unique $RSC(\mathscr{C})$-arrow $(l', l) : (Z_1, Z_2, m_3) \to (D, Y_2^{X_2}, s)$ such that $(ev'' \circ (r \times 1_{X_1}), ev) \circ (l' \times 1_{X_1}, l \times 1_{X_2}) = (t', t)$. Using the property of the exponent $Y_2^{X_2}$, there exists a unique $\tilde{t} : Z_2 \to Y_2^{X_2}$ such that $ev \circ (\tilde{t} \times 1_{X_2}) = t$. Further, using the property of the exponent $Y_1^{X_1}$, there exists a unique $\tilde{t}' : Z_1 \to Y_1^{X_1}$ such that $ev'' \circ (\tilde{t}' \times 1_{X_1}) = t'$. The following diagram commutes.

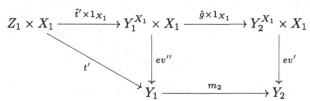

Here, $(\hat{g} \times 1_{X_1}) \circ (\tilde{t}' \times 1_{X_1}) = (\hat{g} \circ \tilde{t}') \times 1_{X_1}$. Consider another diagram as follows:

Using product properties, we have
$ev' \circ ((\hat{h} \circ \tilde{t} \circ m_3) \times 1_{X_1}) = ev' \circ (\hat{h} \times 1_{X_1}) \circ (\tilde{t} \times 1_{X_1}) \circ (m_3 \times 1_{X_1}) = ev \circ (1_{Y_2^{X_2}} \times m_1) \circ (\tilde{t} \times 1_{X_1}) \circ (m_3 \times 1_{X_1}) = ev \circ (\tilde{t} \times m_1) \circ (m_3 \times 1_{X_1}) = ev \circ (\tilde{t} \times 1_{X_2}) \circ (1_{Z_2} \times m_1) \circ (m_3 \times 1_{X_1}) = ev \circ (\tilde{t} \times 1_{X_2}) \circ (m_3 \times m_1)$. Therefore, the above diagram commutes.

We also have $m_2 \circ t' = t \circ (m_3 \times m_1) = ev \circ (\tilde{t} \times 1_{X_2}) \times (m_3 \times m_1)$. Comparing both the diagrams above, and the fact that $Y_2^{X_1}$ is an exponent, we obtain:

$$(\hat{h} \circ \tilde{t} \circ m_3) = \hat{g} \circ \tilde{t}'.$$

Therefore, the outer square in the following diagram commutes.

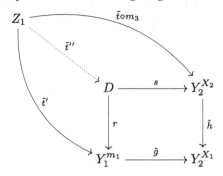

Using the pullback property of the inner square there exists a unique $\mathscr{C}$-arrow $\tilde{t}'' : Z_1 \to D$ such that $r \circ \tilde{t}'' = \tilde{t}'$ and $s \circ \tilde{t}'' = \tilde{t} \circ m_3$. Here, we have $(ev'' \circ (r \times 1_{X_1}), ev) \circ (\tilde{t}'' \times 1_{X_1}, \tilde{t} \times 1_{X_2}) = (ev'' \circ (r \times 1_{X_1}) \circ (\tilde{t}'' \times 1_{X_1}), ev \circ (\tilde{t} \times 1_{X_2})) = (ev'' \circ ((r \circ \tilde{t}'') \times 1_{X_1}), t) = (ev'' \circ (\tilde{t}' \times 1_{X_1}), t) = (t', t)$. Therefore, take $(l', l) = (\tilde{t}'', \tilde{t})$. This is an $RSC(\mathscr{C})$ arrow from (Z_1, Z_2, m_3) to $(D, Y_2^{X_2}, s)$ because $s \circ \tilde{t}'' = \tilde{t} \circ m_3$.

Claim 4: Uniqueness of $(\tilde{t}'', \tilde{t})$. Let there exist another $RSC(\mathscr{C})$-arrow (l', l) : $(Z_1, Z_2, m_3) \to (D, Y_2^{X_2}, s)$ such that $(ev'' \circ (r \times 1_{X_1}), ev) \circ (l' \times 1_{X_1}, l \times 1_{X_2}) = (t', t)$. This implies $ev \circ (l \times 1_{X_2}) = t$. Since $\tilde{t}$ is the unique arrow satisfying $ev \circ (\tilde{t} \times 1_{X_2}) = t$, we have $l = \tilde{t}$. Moreover, since (l', l) is an $RSC(\mathscr{C})$-arrow, $s \circ l' = l \circ m_3 = \tilde{t} \circ m_3 = s \circ \tilde{t}''$. s is monic implies $l' = \tilde{t}''$.

Observation 20. $RSC(\mathscr{C})$ does not have a subobject classifier for the class of monics. This can be established through a counterexample, in the same way as shown in [57] for RSC. Suppose to the contrary, for the class of monics, $RSC(\mathscr{C})$ has a subobject classifier $(\Omega', \Omega, m_\Omega)$ with *true* arrow (t_1, t_2) : $(1, 1, 1_1) \to (\Omega', \Omega, m_\Omega)$. Consider the monic arrow $(!, 1_1) : (0, 1, !) \to (1, 1, 1_1)$ in $RSC(\mathscr{C})$ (Proposition 33) and its characteristic arrow (χ_1, χ_2). Then we have the diagram in Fig. 34 in $RSC(\mathscr{C})$. By the commutativity of the inner square, we have $\chi_2 = t_2$. Since (χ_1, χ_2) and (t_1, t_2) are $RSC(\mathscr{C})$-arrows, $\chi_2 = m_\Omega \circ \chi_1$ and $t_2 = m_\Omega \circ t_1$. $\chi_2 = t_2$ implies $m_\Omega \circ \chi_1 = m_\Omega \circ t_1$. Since m_Ω is monic, $t_1 = \chi_1$. Therefore $(\chi_1, \chi_2) = (t_1, t_2)$ and $(\chi_1, \chi_2) \circ (1_1, 1_1) = (t_1, t_2) \circ (1_1, 1_1)$. By the universal property of the pullbacks in $RSC(\mathscr{C})$, we get a unique $RSC(\mathscr{C})$-arrow $(h', h) : (1, 1, 1_1) \to (0, 1, !)$ such that $(!, 1_1) \circ (h', h) = (1_1, 1)$. The existence of $h' : 1 \to 0$ implies that $\mathscr{C}$ is a degenerate topos (Remark 3), contrary to our assumption.

However, $RSC(\mathscr{C})$ has a subobject classifier for a smaller class of arrows, namely that of *strong monics* (Definition 14). We shall see this in Proposition

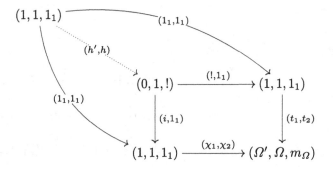

Fig. 34. $RSC(\mathscr{C})$ is not a topos

38 at the end of this section. Let us give a characterization of strong monics in $RSC(\mathscr{C})$ first.

Proposition 35. *Consider an $RSC(\mathscr{C})$-arrow $(f', f) : (X_1, X_2, m_1) \to (Y_1, Y_2, m_2)$. (f', f) is a strong monic if and only if $f : X_2 \to Y_2$ is a monic in $\mathscr{C}$ and the diagram in Fig. 35 is a pullback.*

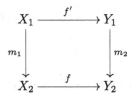

Fig. 35. An arrow (f', f) in $RSC(\mathscr{C})$

Proof. Let (f', f) be a strong monic in $RSC(\mathscr{C})$. Then, by definition of strong monic, (f', f) is monic in $RSC(\mathscr{C})$, which further implies f and f' both to be monic in $\mathscr{C}$ (Proposition 33). Consider $\mathscr{C}$-arrows $r : Z \to Y_1$ and $s : Z \to X_2$ such that $f \circ s = m_2 \circ r$. For the diagram in Fig. 35 to be a pullback, we must obtain a unique arrow h such that $r = m_1 \circ h$ and $s = f' \circ h$. We have $(r, m_2 \circ r) \circ (i_Z, 1_Z) = (f', f) \circ (i_{X_1}, s)$ in $RSC(\mathscr{C})$. Here, $(i_Z, 1_Z)$ is an epic arrow (Proposition 33).

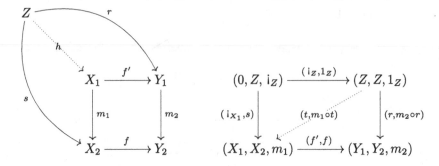

Since (f', f) is a strong monic, there exists a unique 'diagonal' arrow

$$(t, m_1 \circ t) : (Z, Z, 1_Z) \to (X_1, X_2, m_1)$$

such that the both the inner triangles commute. Taking $h := t$, we get the required unique arrow. Therefore, the diagram in Fig. 35 is a pullback.

For the converse, consider the following commutative square in $RSC(\mathscr{C})$ with (e', e) an epic arrow. To show that (f', f) is a strong monic, we have to show the existence of a unique 'diagonal' $RSC(\mathscr{C})$-arrow (t', t), such that the inner triangles commute.

$$
\begin{array}{ccc}
(Z_1, Z_2, m_3) & \xrightarrow{\;(e',e)\;} & (E_1, E_2, m_4) \\
{\scriptstyle(r',r)}\downarrow & {\scriptstyle(t',t)} & \downarrow{\scriptstyle(s',s)} \\
(X_1, X_2, m_1) & \xrightarrow{\;(f',f)\;} & (Y_1, Y_2, m_2)
\end{array}
$$

In the topos $\mathscr{C}$, any monic is also a strong monic. We also have the fact that f is monic and e is epic in $\mathscr{C}$ because (f', f) is monic and (e', e) is epic in $RSC(\mathscr{C})$ (Proposition 33). Therefore, the commutative square $s \circ e = f \circ r$ has a unique diagonal $t : E_2 \to X_2$ such that $f \circ t = s$ and $t \circ e = r$ in $\mathscr{C}$.

Moreover, $f \circ t \circ m_4 = m_2 \circ s'$, because $f \circ t = s$ and $s \circ m_4 = m_2 \circ s'$, in $\mathscr{C}$. Using the assumption that the square in Fig. 35 is a pullback, there exists a unique arrow $t' : E_1 \to X_1$ such that $f' \circ t' = s'$ and $m_1 \circ t' = t \circ m_4$.

Claim 1: (t', t) is an $RSC(\mathscr{C})$-arrow from (E_1, E_2) to (X_1, X_2), i.e. $m_1 \circ t' = t \circ m_4$. Since (f', f) and (s', s) are $RSC(\mathscr{C})$-arrows, we have $f \circ m_1 = m_2 \circ f'$ and $s \circ m_4 = m_2 \circ s'$. Now, $f \circ m_1 \circ t' = m_2 \circ f' \circ t' = m_2 \circ s' = s \circ m_4 = f \circ t \circ m_4$. f is monic implies $m_1 \circ t' = t \circ m_4$.

Claim 2: $t' \circ e' = r'$. We have $m_1 \circ t' \circ e' = t \circ m_4 \circ e' = t \circ e \circ m_3 = r \circ m_3 = m_4 \circ r'$. m_1 is monic implies $t' \circ e' = r'$.

Claim 3: (t', t) is the required diagonal. We have $(f', f) \circ (t', t) = (f' \circ t', f \circ t) = (s', s)$ and $(t', t) \circ (e', e) = (t' \circ e', t \circ e) = (r', r)$.

Claim 4: Uniqueness of (t', t). Let there exist another $RSC(\mathscr{C})$-arrow (l', l) : $(E_1, E_2, m_4) \to (X_1, X_2, m_1)$ such that $(f', f) \circ (l', l) = (s', s)$ and $(l', l) \circ (e', e) = (r', r)$. This implies $f \circ l = s$ and $l \circ e = r$. Uniqueness of t implies $t = l$. Now, since (l', l) is an $RSC(\mathscr{C})$-arrow, $m_1 \circ l' = l \circ m_4 = t \circ m_4 = m_1 \circ t'$. m_1 is monic implies $l' = t'$.

Note that in general the class of strong monics is not the same as the class of monics in $RSC(\mathscr{C})$. This can be seen through the following example in RSC.

Example 8. Consider the RSC-arrow $(!, 1_{\{1\}}) : (\emptyset, \{1\}) \rightarrow (\{1\}, \{1\})$. $1_{\{1\}}$ is monic in SET implies $(!, 1_{\{1\}})$ is a monic in RSC (Proposition 33). Consider the following square.

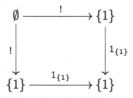

Suppose the square is a pullback. Consider two SET-arrows $r = s = 1_{\{1\}} :$ $\{1\} \rightarrow \{1\}$. We have $1_{\{1\}} \circ r = 1_{\{1\}} \circ s$. Using pullback property, there exists a unique function $h : \{1\} \rightarrow \emptyset$ such that $! \circ h = r = s$. However, no such h exists, because there does not exist any function from $\{1\}$ to $\emptyset$, a contradiction. Therefore, the above square is not a pullback in SET. Using Proposition 35, $(!, 1_{\{1\}})$ is not a strong monic in RSC.

Observation 21. An RSC-arrow $m : (X_1, X_2) \rightarrow (Y_1, Y_2)$ is a monic implies $m : X_2 \rightarrow Y_2$ is an injective function.

Proposition 36. *For a monic arrow $m : (X_1, X_2) \rightarrow (Y_1, Y_2)$ in RSC such that m is an inclusion of X_2 into Y_2, the following are equivalent.*

(1) *m is a strong monic.*
(2) *m is an equalizer.*
(3) *$X_1 = X_2 \cap Y_1$.*
(4) *$X_2 \backslash X_1 \subseteq Y_2 \backslash Y_1$.*

Proof. Observe that $X_1 \subseteq Y_1 \cap X_2$.
(1) $\Rightarrow$ (2) : Consider the left hand side triangle in the diagram below. By the image factorization of m, we obtain an epic arrow e and an equalizer n in RSC (Definition 29).

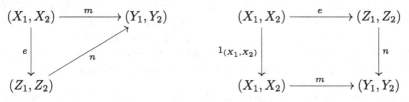

Now, consider the right hand side square, where m is strong monic and e is epic. Therefore, there exists a diagonal arrow $t : (Z_1, Z_2) \rightarrow (X_1, X_2)$ such that

$t \circ e = 1_{(X_1, X_2)}$. Thus, we have $e \circ t \circ e = e \Rightarrow e \circ t = 1_{(Z_1, Z_2)}$, so t is an iso arrow and $m \simeq n$. Since n is an equalizer, therefore m is also an equalizer.

$(2) \Rightarrow (3)$: Let $m : (X_1, X_2) \to (Y_1, Y_2)$ be an equalizer in RSC of RSC-arrows $f, g : (Y_1, Y_2) \to (Z_1, Z_2)$. We need to show that $X_2 \cap Y_1 \subseteq X_1$. Now, $m : (X_2 \cap Y_1, X_2) \to (Y_1, Y_2)$ is also an RSC-arrow. Therefore, there exists a unique RSC-arrow $i : (X_2 \cap Y_1, X_2) \to (X_1, X_2)$ such that $m \circ i = m$ (diagram below). So for all $x \in X_2$, $m(i(x)) = m(x)$, i.e. $i(x) = x$, implying $X_2 \cap Y_1 \subseteq X_1$.

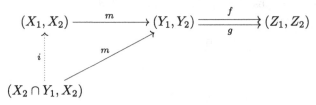

$(3) \Rightarrow (1)$: Let $X_1 = X_2 \cap Y_1$. Consider the diagram below in SET.

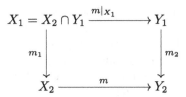

Here m_1 and m_2 are inclusion functions of X_1 and Y_1 into X_2 and Y_2 respectively. Since m is an RSC-arrow, the square is commutative.

Claim: The square is a pullback. Consider the functions $r : Z \to X_2$ and $s : Z \to Y_1$ such that $m_2 \circ s = m \circ r$. Since both the functions m and m_2 are inclusions, we have $r(z) = s(z)$ for all $z \in Z$. So, $s(z) \in Y_1 \cap X_2 = X_1$. Define $h : Z \to X_1$ as $h(z) := s(z)$ for all $z \in Z$. Since m_1 and $m|_{X_1}$ are inclusions, we have $m_1 \circ h = r$ and $m|_{X_1} \circ h = s$. This implies that the diagram above is a pullback.

Finally using Proposition 35, m is a strong monic in RSC.

$(3) \Rightarrow (4)$: $X_2 \backslash X_1 = X_2 \backslash (X_2 \cap Y_1) = X_2 \backslash Y_1 \subseteq Y_2 \backslash Y_1$.

$(4) \Rightarrow (3)$: Since $m|_{X_1} : X_1 \to Y_1$ is an inclusion, $X_1 \subseteq X_2 \cap Y_1$. Let $x \in X_2 \cap Y_1$. This implies $x \notin Y_2 \backslash Y_1 \Rightarrow x \notin X_2 \backslash X_1 \Rightarrow x \in X_1$.

Consider a strong monic $m : (X_1, X_2) \to (Y_1, Y_2)$ in RSC. Then m is a monic. Here, $m : X_2 \to Y_2$ and $m|_{X_1} : X_1 \to Y_1$ are monics in SET (Proposition 33), i.e. m is an injective function. This implies that $X_2 \cong m(X_2) \subseteq Y_2$ and $X_1 \cong m(X_1) \subseteq Y_1$. Therefore, $(X_1, X_2) \cong (m(X_1), m(X_2))$. Consider the inclusion function $i : m(X_2) \to Y_2$, then $i : (m(X_1), m(X_2)) \to (Y_1, Y_2)$ is an RSC-arrow, and $i \simeq m$ in RSC. Since, m is a strong monic, therefore i is also a strong monic in RSC. Using the Proposition above, we have the following corollary.

Corollary 3. *For a monic arrow $m : (X_1, X_2) \to (Y_1, Y_2)$ in RSC, the following are equivalent.*

(1) m *is a strong monic.*
(2) m *is an equalizer.*
(3) $m(X_1) = m(X_2) \cap Y_1$.
(4) $m(X_2) \backslash m(X_1) \subseteq Y_2 \backslash Y_1$.

Let us see what are the monics in ξ-RSC.

Proposition 37. *Consider a ξ-RSC-arrow $m : (X_1, X_2) \rightarrow (Y_1, Y_2)$. m is monic in ξ-RSC if and only if m is an injective function and $m(X_2) \backslash m(X_1) \subseteq Y_2 \backslash Y_1$.*

Proof. Since m is a ξ-RSC-arrow, $m(X_1) \subseteq Y_1$ and $m(X_2 \backslash X_1) \subseteq Y_2 \backslash Y_1$.
($\Rightarrow$) Let m be monic in ξ-RSC. Let $f, g : Z \rightarrow X_2$ such that $m \circ f = m \circ g$. Define
a set $Z_1 := \{z \in Z \mid f(z) \in X_1\}$. Claim: (1) $f(Z_1) \subseteq X_1$, (2). $f(Z \backslash Z_1) \subseteq X_2 \backslash X_1$,
(3) $g(Z_1) \subseteq X_1$, and (4). $g(Z \backslash Z_1) \subseteq X_2 \backslash X_1$.
(1) and (2) are direct by the definition of the set Z_1.
For (3), let $z \in Z_1$ such that $g(z) \notin X_1$. $z \in Z_1 \Rightarrow f(z) \in X_1 \Rightarrow m(f(z)) \in m(X_1) \subseteq Y_1 \Rightarrow m(g(z)) \in Y_1$, because $m(f(z)) = m(g(z))$. $g(z) \notin X_1 \Rightarrow g(z) \in X_2 \backslash X_1 \Rightarrow m(g(z)) \in m(X_2 \backslash X_1) \subseteq Y_2 \backslash Y_1 \Rightarrow m(g(z)) \notin Y_1$, a contradiction. So, $g(Z_1) \subseteq X_1$.
For (4), let $z \in Z \backslash Z_1$ such that $g(z) \notin X_2 \backslash X_1$. $z \in Z \backslash Z_1 \Rightarrow z \notin Z_1 \Rightarrow f(z) \notin X_1 \Rightarrow f(z) \in X_2 \backslash X_1 \Rightarrow m(f(z)) \in Y_2 \backslash Y_1 \Rightarrow m(g(z)) \in Y_2 \backslash Y_1$. $g(z) \notin X_2 \backslash X_1 \Rightarrow g(z) \in X_1 \Rightarrow m(g(z)) \in Y_1 \Rightarrow m(g(z)) \notin Y_2 \backslash Y_1$, a contradiction. So, we have
(4). These four properties define two ξ-RSC-arrows $f, g : (Z_1, Z) \rightarrow (X_1, X_2)$ such that $m \circ f = m \circ g$. Since m is monic in ξ-RSC, we have $f = g$. Therefore, m is monic in category SET, and hence it is an injective function. Moreover, m injective implies $m(X_2 \backslash X_1) = m(X_2) \backslash m(X_1)$. So, $m(X_2) \backslash m(X_1) \subseteq Y_2 \backslash Y_1$.
($\Leftarrow$) Let m be an injective function and $m(X_2) \backslash m(X_1) \subseteq Y_2 \backslash Y_1$. Consider two ξ-RSC-arrows $f, g : (Z_1, Z_2) \rightarrow (X_1, X_2)$ such that $m \circ f = m \circ g$. Since $f, g : Z_2 \rightarrow X_2$ are functions and m is injective, we have $f = g$. Therefore m is a monic in ξ-RSC. ∎

Proposition 37 and Corollary 3 imply the following observation.

Observation 22. *The class of monics in ξ-RSC is the same as the class of strong monics in RSC.*

Recall the definition of the representations of partial morphisms (Definition 27) and quasitoposes (Definition 28). We have the following result.

Theorem 14. $RSC(\mathscr{C})$ *is a quasitopos.*

Proof. Propositions 34 and 31 gives that $RSC(\mathscr{C})$ is finitely cocomplete and cartesian closed. We have to only show that partial morphisms in $RSC(\mathscr{C})$ are represented. Consider any $RSC(\mathscr{C})$-object (Y_1, Y_2, m_2). We need to find an $RSC(\mathscr{C})$-object $(\tilde{Y}_1, \tilde{Y}_2, m)$ and a strong monic $\nu_{(Y_1, Y_2, m_2)} : (Y_1, Y_2, m_2) \rightarrow (\tilde{Y}_1, \tilde{Y}_2, m)$ satisfying the following property.
Condition (1): For every partial morphism $((u', u), (v', v)) : (X_1, X_2, m_1) \rightarrow$

(Y_1, Y_2, m_2) with codomain (Y_1, Y_2), there exists a unique $RSC(\mathscr{C})$-arrow (l', l) : $(X_1, X_2, m_1) \to (\tilde{Y}_1, \tilde{Y}_2, m)$ such that the following diagram is a pullback in $RSC(\mathscr{C})$.

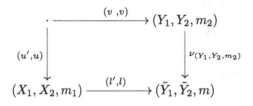

Since $\mathscr{C}$ is a topos, and partial morphisms are represented in any topos [43], we have the following: for $\mathscr{C}$-object Y_2, there exist a $\mathscr{C}$-object $\tilde{Y}_2$ and a strong monic $\nu_{Y_2} : Y_2 \to \tilde{Y}_2$ satisfying the following property.
Condition (2): For every partial morphism $(u, v) : X_2 \to Y_2$ with codomain Y_2, there exists a unique $\mathscr{C}$-arrow $\overline{v} : X_2 \to \tilde{Y}_2$ such that the following diagram is a pullback in $\mathscr{C}$.

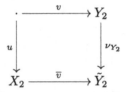

Consider the $RSC(\mathscr{C})$-object $(\tilde{Y}_2, \tilde{Y}_2, 1_{\tilde{Y}_2})$. We have $\nu_{Y_2} \circ m_2 = 1_{\tilde{Y}_2} \circ \nu_{Y_2} \circ m_2$: $Y_1 \to \tilde{Y}_2$. Therefore, $(\nu_{Y_2} \circ m_2, \nu_{Y_2})$ is an $RSC(\mathscr{C})$-arrow from (Y_1, Y_2) to $(\tilde{Y}_2, \tilde{Y}_2)$. We prove Condition (1), taking $(\tilde{Y}_1, \tilde{Y}_2, m) := (\tilde{Y}_2, \tilde{Y}_2, 1_{\tilde{Y}_2})$, $\nu_{(Y_1, Y_2, m_2)} := (\nu_{Y_2} \circ m_2, \nu_{Y_2})$ and $(l', l) := (\overline{v} \circ m_1, \overline{v})$. Note that $(\overline{v} \circ m_1, \overline{v}) : (X_1, X_2, m_1) \to (\tilde{Y}_2, \tilde{Y}_2, 1_{\tilde{Y}_2})$ is an $RSC(\mathscr{C})$-arrow.
Let $((u', u), (v', v)) : (X_1, X_2, m_1) \to (Y_1, Y_2, m_2)$ be a partial morphism, i.e. $(u', u) : (Z_1, Z_2, m_3) \to (X_1, X_2, m_1)$ is a strong monic in $RSC(\mathscr{C})$, where (Z_1, Z_2, m_3) is the domain of (u', u), as well as domain of (v', v), in the partial morphism $((u', u), (v', v))$. Further, we have $m_2 \circ v' = v \circ m_3$ and $u \circ m_3 = m_1 \circ u'$. Proposition 35 implies that u is a monic and the square in Fig. 36a is a pullback in $\mathscr{C}$.
Now, u monic implies that $(u, v) : X_2 \to Y_2$ is a partial morphism in $\mathscr{C}$. Using condition (2), there exists a $\mathscr{C}$-arrow $\overline{v} : X_2 \to \tilde{Y}_2$ such that the square in Fig. 36b is a pullback in $\mathscr{C}$.

Claim: The diagram in Fig. 36d is also a pullback in $\mathscr{C}$.
Note that if we show this diagram is a pullback, then using Proposition 32 and that the diagram in Fig. 36b is a pullback, we get that diagram in Fig. 37 is a pullback.
Indeed, we have $\nu_{Y_2} \circ m_2 \circ v' = \nu_{Y_2} \circ v \circ m_3 = \overline{v} \circ u \circ m_3 = \overline{v} \circ m_1 \circ u'$, and thus the square in Fig. 36d commutes. Using Proposition 4 and that the diagrams in Fig. 36a and Fig. 36b are pullbacks, the outer square in Fig. 36c is a pullback.

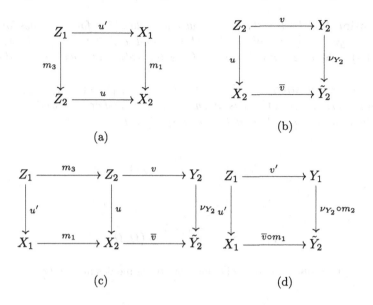

Fig. 36. Some pullbacks in $\mathscr{C}$

Now, let $s : E \to X_1$ and $r : E \to Y_1$ such that $\nu_{Y_2} \circ m_2 \circ r = \bar{v} \circ m_1 \circ s$. Define $p := m_2 \circ r : E \to Y_2$ and $q := s$, then in the outer pullback in Fig. 36c, there exists a unique arrow $h : E \to Z_1$ such that $v \circ m_3 \circ h = m_2 \circ r$ and $u' \circ h = s$. As $v \circ m_3 = m_2 \circ v'$, so $m_2 \circ v' \circ h = m_2 \circ r$. m_2 is monic implies $r = v' \circ h$. Thus, the claim.

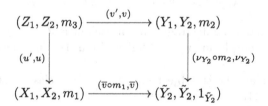

Fig. 37. Partial morphisms are represented in $RSC(\mathscr{C})$

The uniqueness of $\bar{v}$ implies the uniqueness of $(\bar{v} \circ m_1, \bar{v})$ in Fig. 37.

Corollary 4. *ROUGH and RSC form quasitoposes.*

Proposition 10 gives us the fact that $RSC(\mathscr{C})$ has a subobject classifier for the class of strong monics. Explicitly, we can derive the subobject classifier in $RSC(\mathscr{C})$ using the subobject classifier in $\mathscr{C}$, Proposition 32 and Observation 33(2).

Proposition 38. *In* $RSC(\mathscr{C})$, *the subobject classifier for the class of strong monics is given by the object* $(\Omega, \Omega, 1_\Omega)$ *and arrow* $(\top, \top) : (1, 1, 1_1) \to$ $(\Omega, \Omega, 1_\Omega)$, *where* Ω *with* $\top : 1 \to \Omega$ *is the subobject classifier in the topos* $\mathscr{C}$.

Further, given any strong monic $(f', f) : (X_1, X_2, m_1) \to (Y_1, Y_2, m_2)$, *the characteristic morphism of* (f', f) *is given by* $(\chi_{f'}, \chi_f)$, *where* $\chi_{f'}$ *and* χ_f *are characteristic morphisms of* f' *and* f *in* $\mathscr{C}$ *respectively.*

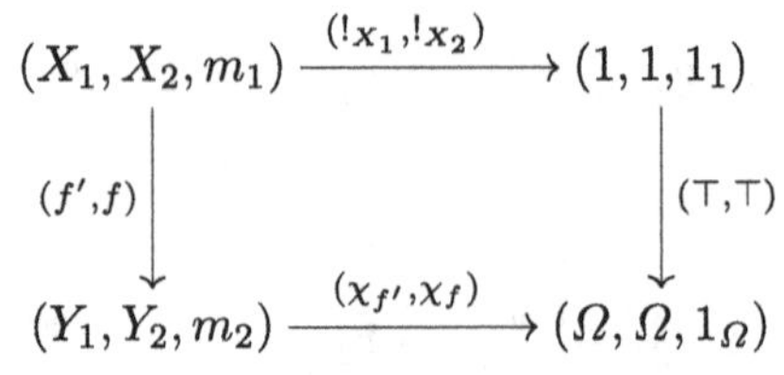

$$
\begin{array}{ccc}
(X_1, X_2, m_1) & \xrightarrow{(!_{x_1}, !_{x_2})} & (1, 1, 1_1) \\
{\scriptstyle (f', f)} \downarrow & & \downarrow {\scriptstyle (\top, \top)} \\
(Y_1, Y_2, m_2) & \xrightarrow{(\chi_{f'}, \chi_f)} & (\Omega, \Omega, 1_\Omega)
\end{array}
$$

Fig. 38. Subobject classifier for strong monics in $RSC(\mathscr{C})$

Proof. Consider a strong monic $(f', f) : (X_1, X_2, m_1) \to (Y_1, Y_2, m_2)$. Using Proposition 35, f is monic in $\mathscr{C}$. Using Observation 33(2), f' is also a monic in $\mathscr{C}$. The characteristic morphisms for f and f' in $\mathscr{C}$ are χ_f and $\chi_{f'}$ respectively, with the appropriate pullback diagrams. Consider the diagram in Fig. 38. (f', f), $(!_{x_1}, !_{x_2})$ and $(\top, \top)$ are arrows in $RSC(\mathscr{C})$, and thus $m_2 \circ f' = f \circ m_1$, $!_{x_2} \circ m_1 = 1_1 \circ !_{x_1}$ and $\top \circ 1_1 = 1_\Omega \circ \top$. Observe the following:
$\chi_f \circ m_2 \circ f' = \chi_f \circ f \circ m_1 = \top \circ !_{x_2} \circ m_1 = \top \circ 1_1 \circ !_{x_1} = \top \circ !_{x_1}$.
Since $\chi_{f'}$ is a unique arrow satisfying $\chi_{f'} \circ f' = \top \circ !_{x_1}$ (Subobject classifier property), we have $\chi_{f'} = \chi_f \circ m_2$. Therefore, $(\chi_{f'}, \chi_f) : (Y_1, Y_2, m_2) \to (\Omega, \Omega, 1_\Omega)$ is an arrow in $RSC(\mathscr{C})$.
Using Proposition 32, the diagram is a pullback in $RSC(\mathscr{C})$. The uniqueness of the characteristic morphism $(\chi_{f'}, \chi_f)$ follows from the uniqueness of $\chi_{f'}$ and χ_f. ∎

Note: In particular, the subobject classifier for the class of strong monics in RSC is $(\top, \top) : (1, 1) \to (2, 2)$, where $2 := \{0, 1\}$, $1 := \{0\}$ and $\top : 1 \to 2$ mapping 0 to 1 is the subobject classifier in SET.

3.4 Some More Categories of Rough Sets

In the previous sections, we dealt with some categories of rough sets, namely RSC, $ROUGH$, ξ-$ROUGH$ and ξ-RSC. However, depending on the different definitions of rough sets and the mappings between two rough sets, we can define other categories of rough sets. We shall explore some of these in this section.

Observe that while defining RSC, the domain U is not explicitly mentioned in the RSC-object. One basic possibility is to define the objects with the appropriate domain attached. Recall the definition of rough universes used in Definition 34.

Definition 86 (RSC^* and ξ-RSC^* Categories).

(1) *Objects of RSC^* are of the form (U, X_1, X_2) such that $X_1 \subseteq X_2 \subseteq U$, where X_1, $X_2 \in \mathbf{B}$ for the rough universe $(U, \mathbf{B})$. An arrow in RSC^* with domain (U, X_1, X_2) and codomain (V, Y_1, Y_2) is a map $f : X_2 \to Y_2$ such that $f(X_1) \subseteq Y_1$.*
Composition of two arrows is defined as function composition.
Identity arrow on (U, X_1, X_2) is the identity function 1_{X_2} on X_2.

(2) *Objects of ξ-RSC^* are same as objects of RSC^*. An arrow in ξ-RSC^* with domain (U, X_1, X_2) and codomain (V, Y_1, Y_2) is a map $f : X_2 \to Y_2$ such that $f(X_1) \subseteq Y_1$ and $f(X_2 \backslash X_1) \subseteq Y_2 \backslash Y_1$.*
Composition of two arrows and identity arrows is defined as in RSC^*.

Proposition 39. *1. In RSC^*, $(U, X_1, X_2) \cong (X_2, X_1, X_2)$.*
2. $RSC \simeq RSC^* \simeq ROUGH$.
3. ξ-$RSC^* \simeq \xi$-RSC.
4. RSC^* is a quasitopos, and ξ-RSC^* is a topos.

Proof.
(1). Consider the identity function 1_{X_2} on X_2. $1_{X_2}(X_1) = X_1$. Therefore, $f := 1_{X_2}$ is an arrow in RSC^* from (U, X_1, X_2) to (X_2, X_1, X_2).
Note that $g := 1_{X_2}$ can also be considered as an arrow from (X_2, X_1, X_2) to (U, X_1, X_2). We have $f \circ g$ is identity on (X_2, X_1, X_2), and $g \circ f$ is identity on (U, X_1, X_2) in RSC^*. Therefore, f is iso and $(U, X_1, X_2) \cong (X_2, X_1, X_2)$.
(2). Define a mapping G such that it maps arrows of RSC to arrows of RSC^*, and objects of RSC to objects of RSC^* as follows: For RSC-object (X_1, X_2), G maps (X_1, X_2) to (X_2, X_1, X_2), and for RSC-arrow $f : (X_1, X_2) \to (Y_1, Y_2)$ to $f : (X_2, X_1, X_2) \to (Y_2, Y_1, Y_2)$. G is a functor.
G if full: For $f : (X_2, X_1, X_2) \to (Y_2, Y_1, Y_2)$, f is an arrow in RSC such that $G(f) = f$.
G is faithful: For f and g from (X_1, X_2) to (Y_1, Y_2), $G(f) = G(g)$ implies $f = g$.
G is essentially surjective: Given a RSC^*-object (U, X_1, X_2), consider the RSC-object (X_1, X_2). $G((X_1, X_2)) = (X_2, X_1, X_2)$.
Using (1), $(U, X_1, X_2) \cong G((X_1, X_2))$.
(3). The same map G as above gives ξ-$RSC^* \simeq \xi$-RSC.
(4). Since RSC is a quasitopos (Theorem 14), RSC^* is also quasitopos using (2). Since ξ-RSC is a topos (Proposition 28), RSC^* is also a topos using (3).

Observation 23.

(1) The equivalence tells us that the rough universe does not play any role in defining the categorial properties of RSC^* or RSC or $ROUGH$. This is because for an object (U, R, X) of $ROUGH$, the negative region '$(U/R)\backslash \overline{\mathcal{X}}_R$' does not picture in the definition of the arrows in $ROUGH$.

(2) We can use the representations of RSC^* instead of RSC to study the categories of rough sets in the previous section. However, as we are interested in category-theoretic aspects and the categories are equivalent, for simplicity, we do not bring the universe U inside the notations used in the category RSC.

Before proceeding further, let us recall from Sect. 2.3, the definition of a rough set and the four definable regions in the domain, namely, the 'definite' $\underline{X}$ (d), 'possible' $\overline{X}$ (p), 'boundary' $\overline{X}\backslash\underline{X}$ (b), and 'negative' $U\backslash\overline{X}$ (n) regions. We have already observed that an arrow of RSC preserves information in the possible and definite regions of a rough set. Similarly, an arrow of ξ-RSC preserves information in the possible, definite, as well as the boundary regions. Depending on the choice of the region (amongst definite, possible, boundary and negative) with respect to which one wants information to be preserved, the definition of an arrow can change, whereby the definition of the category itself can change. We shall see these categories in the next definition.

Definition 87.
Objects in each of these cases is (U, X_1, X_2) - same as objects of RSC^*. An arrow from (U, X_1, X_2) to (U, Y_1, Y_2) is defined depending on the regions that are required to be preserved.
Composition of arrows is defined as component-wise (in case there is more than one function associated with an arrow) function composition.
Identity arrow is the identity function (or tuples of identity functions).

1. *'d' and 'p' preserved: Arrow is a map $f : X_2 \to Y_2$ such that $f(X_1) \subseteq Y_1$. This is RSC^*, equivalent to RSC and $ROUGH$ (Proposition 39).*
2. *'d', 'p' and 'b' preserved: Arrow is a map $f : X_2 \to Y_2$ such that $f(X_1) \subseteq Y_1$ and $f(X_2\backslash X_1) \subseteq (Y_2\backslash Y_1)$. This is ξ-RSC^*, equivalent to ξ-RSC and ξ-$ROUGH$ (Proposition 39).*
3. *'d', 'p' and 'n' preserved: Arrow is a pair of maps (f, f') such that $f : X_2 \to Y_2$ and $f' : (U\backslash X_2) \to (V\backslash Y_2)$ such that $f(X_1) \subseteq Y_1$ (Say RSC_1).*
4. *'d', 'p', 'b' and 'n' preserved: Arrow is a pair of maps (f, f') such that $f : X_2 \to Y_2$ and $f' : (U\backslash X_2) \to (V\backslash Y_2)$ such that $f(X_1) \subseteq Y_1$ and $f(X_2\backslash X_1) \subseteq (Y_2\backslash Y_1)$ (Say ξ-RSC_1).*

Observation 24. *In RSC_1 and ξ-RSC_1, for an arrow $(f, f') : (U, X_1, X_2) \to (V, Y_1, Y_2)$, we can obtain a function $g : U \to V$ such that $g|_{X_2} = f$ and $g|_{(U\backslash X_2)} = f'$.*

Let us see some equivalences between the defined categories.

Proposition 40.

(1) $RSC_1 \simeq RSC \times SET$.
(2) ξ-$RSC_1 \simeq \xi$-$RSC \times SET \simeq SET^3$ (because ξ-$RSC \simeq SET^2$).

Proof.
(1). Define F from RSC_1 to $RSC \times SET$ such that F maps RSC_1-object (U, X_1, X_2) to $(RSC \times SET)$-object $((X_1, X_2), (U\backslash X_2))$. Note that an RSC_1-arrow $(f, f') : (U, X_1, X_2) \to (V, Y_1, Y_2)$ is also an $(RSC \times SET)$-arrow (f, f'). Let F act as identity on the set of RSC_1-arrows.
It is easy to observe that F is a full and faithful functor. Let us show that F is essentially surjective.
Consider an $(RSC \times SET)$-object $((X_1, X_2), Y)$. Then $(X_2 + Y, X_1 + \emptyset, X_2 + \emptyset)$ is an RSC_1-object, and $F((X_2 + Y, X_1 + \emptyset, X_2 + \emptyset)) = ((X_1 + \emptyset, X_2 + \emptyset), (X_2 + $

$Y)\backslash(X_2+\emptyset))$. In SET, $(X_2+Y)\backslash(X_2+\emptyset) \cong Y$, and in RSC, $(X_1+\emptyset, X_2+\emptyset) \cong (X_1, X_2)$. Therefore, $((X_1+\emptyset, X_2+\emptyset), (X_2+Y)\backslash(X_2+\emptyset)) \cong ((X_1, X_2), Y)$ in $RSC \times SET$.

(2). The same mapping as above gives the first equivalence in this case. Using $\xi\text{-}RSC \simeq SET^2$ (Proposition 28), we have $\xi\text{-}RSC \times SET \simeq SET^3$.

This gives an insight into what RSC_1 is, in category-theoretic terms. Any object (U, X_1, X_2) can be looked upon as having two components - an RSC-object (or a $\xi\text{-}RSC$-object) (X_1, X_2) and a set $(U\backslash X_2)$. Since, RSC is a quasitopos and $\xi\text{-}RSC$ is a topos, using the above equivalence conditions and Proposition 11, we have the following result.

Proposition 41. *RSC_1 is a quasitopos, and $\xi\text{-}RSC_1$ is a topos.*

All the definitions of categories on rough sets we have presented so far, are based on the Pawlak's definition of rough sets. But there are other definitions of rough sets (cf. [7]). For instance by Pagliani's definition [72], in an approximation space (U, R), for $X \subseteq U$, a rough set is the pair $(\underline{X}, \overline{X}^c)$, where $\overline{X}^c := U\backslash\overline{X}$. It may be possible to define categories on rough sets different than the ones above, using these alternate definitions of rough sets.

3.4.1   Generalization of RSC_1

First let us see what is $RSC_1(\mathscr{C})$, the generalization of RSC_1 over an arbitrary non-degenerate topos $\mathscr{C}$.

Definition 88 ($RSC_1(\mathscr{C})$ category).
Objects of $RSC_1(\mathscr{C})$ are represented by the tuple $(U, X_1, X_2, m_{X_1}, m_{X_2})$, where X_1, X_2 and U are $\mathscr{C}$-objects and $m_{X_1} : X_1 \rightarrow X_2$ and $m_{X_2} : X_2 \rightarrow U$ are monics in $\mathscr{C}$.

An $RSC_1(\mathscr{C})$-arrow with domain $(U, X_1, X_2, m_{X_1}, m_{X_2})$ and codomain $(V, Y_1, Y_2, m_{Y_1}, m_{Y_2})$ is a triplet (f', f, f''), where $f' : X_1 \rightarrow Y_1$, $f : X_2 \rightarrow Y_2$ and $f'' : \neg X_2 \rightarrow \neg Y_2$ are arrows in $\mathscr{C}$ and $\neg X_2$, $\neg Y_2$ are the domains of $\neg m_{X_2} : \neg X_2 \rightarrow U$ and $\neg m_{Y_2} : \neg Y_2 \rightarrow V$ respectively, such that the following diagram commutes in $\mathscr{C}$.

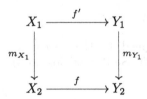

The composition of arrows is defined using the composition of arrows in $\mathscr{C}$. The identity arrow on $(U, X_1, X_2, m_{X_1}, m_{X_2})$ is $(1_{X_1}, 1_{X_2}, 1_{\neg X_2})$.

Observation 25. For any $RSC_1(\mathscr{C})$-object $(U, X_1, X_2, m_{X_1}, m_{X_2})$, (X_1, X_2, m_{X_1}) is an $RSC(\mathscr{C})$-object; and for any $RSC_1(\mathscr{C})$-arrow

$$(f', f, f'') : (U, X_1, X_2, m_{X_1}, m_{X_2}) \to (V, Y_1, Y_2, m_{Y_1}, m_{Y_2}),$$

$(f', f) : (X_1, X_2, m_{X_1}) \to (Y_1, Y_2, m_{Y_1})$ is an $RSC(\mathscr{C})$-arrow.

What happens when $\mathscr{C}$ is SET? Let us first observe the following result.

Proposition 42. If $\mathscr{C}$ is a Boolean topos, $RSC_1(\mathscr{C}) \simeq RSC(\mathscr{C}) \times \mathscr{C}$, and forms a quasitopos.

Proof. Define F from $RSC_1(\mathscr{C})$ to $RSC(\mathscr{C}) \times \mathscr{C}$ as follows:
(1). F maps $RSC_1(\mathscr{C})$-object $(U, X_1, X_2, m_{X_1}, m_{X_2})$ to $(RSC(\mathscr{C}) \times \mathscr{C})$-object $((X_1, X_2, m_{X_1}), \neg X_2)$;
(2). F maps $RSC_1(\mathscr{C})$-arrow

$$(f', f, f'') : (U, X_1, X_2, m_{X_1}, m_{X_2}) \to (V, Y_1, Y_2, m_{Y_1}, m_{Y_2})$$

to $(RSC(\mathscr{C}) \times \mathscr{C})$-arrow

$$((f', f), f'') : ((X_1, X_2, m_{X_1}), \neg X_2) \to ((Y_1, Y_2, m_{Y_1}), \neg Y_2).$$

Clearly, F is a full and faithful functor.
Claim: F is essentially surjective: let $((X_1, X_2, m_{X_1}), X_3)$ be an $(RSC(\mathscr{C}) \times \mathscr{C})$-object. Consider the $RSC_1(\mathscr{C})$-object $(X_2 + X_3, X_1, X_2, m_{X_1}, i_1)$, where $i_1 : X_2 \to X_2 + X_3$ is a $\mathscr{C}$-arrow such that $(X_2 + X_3, i_1, i_2)$ is a coproduct in the topos $\mathscr{C}$. Note that $i_1 : X_2 \to X_2 + X_3$ and $i_2 : X_3 \to X_2 + X_3$ are monics in any topos $\mathscr{C}$ (Proposition 9(3)). Moreover, as observed in the proof of the essential surjectivity of the functor G in Theorem 13, using the fact that $\mathscr{C}$ is a Boolean topos, we get $\neg i_1 \simeq i_2$ and $\neg X_2 \cong X_3$. Therefore, $F((X_2 + X_3, X_1, X_2, m_{X_1}, i_1)) = ((X_1, X_2, m_{X_1}), \neg X_2) \cong ((X_1, X_2, m_{X_1}), X_3)$, and F is essentially surjective.
$RSC(\mathscr{C})$ is a quasitopos. $\mathscr{C}$ is a topos, and thus a quasitopos. Therefore, by Proposition 11, $RSC(\mathscr{C}) \times \mathscr{C}$ is a quasitopos.

In Proposition 29, we observed that $RSC(SET)$ is equivalent to RSC. Similar to that, we obtain

Corollary 5. $RSC_1(SET) \simeq RSC_1$.

Proof. Proposition 42 applied to the special case when $\mathscr{C}$ is SET gives $RSC_1(SET) \simeq RSC \times SET$. So Proposition 40(3) gives the result.

3.5   Conclusions

In this section, we defined and studied various categories of rough sets and their properties. In particular in the previous section, we saw some new categories on rough sets. There are still other possibilities that may be worth exploring. For instance, we can get another category based on the following observation.

For RSC_1, given an arrow $(f, f') : (U, X_1, X_2) \to (V, Y_1, Y_2)$, we can obtain a function $g : U \to V$ such that $g|_{X_2} = f$ and $g|_{(U \backslash X_2)} = f'$. Thus, we can define a new category RSC_2 as follows: objects of RSC_2 are same as objects of RSC_1. An arrow of RSC_2 is a map $f : U \to V$ such that $f(X_2) \subseteq Y_2$, $f(U \backslash X_2) \subseteq (V \backslash Y_2)$ and $f(X_1) \subseteq Y_1$.

Can this be obtained in the generalized scenario of $RSC_1(\mathscr{C})$? Consider an arrow $RSC_1(\mathscr{C})$, $((f', f), f'') : (U, X_1, X_2, m_{X_1}, m_{X_2}) \to (V, Y_1, Y_2, m_{Y_1}, m_{Y_2})$. In the topos $\mathscr{C}$, using Proposition 12, we can obtain the arrow $(\neg m_{Y_2} \circ f'') \cup (m_{Y_2} \circ f) : X_2 \cup \neg X_2 \to Y_2 \cup \neg Y_2$ in $\mathscr{C}$ (diagram below).

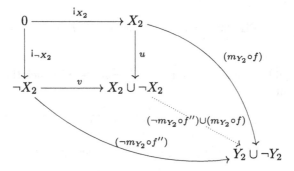

However, in a general non-Boolean topos $\mathscr{C}$, we may not always have $(X_2 \cup \neg X_2) \cong U$ and $(Y_2 \cup \neg Y_2) \cong V$, and thus the above construction may not give a unique arrow from U to V. So a possible way to overcome this issue is by starting with a 'fixed' arrow $g : U \to V$ and replacing (i) objects $X_2 \cup \neg X_2$ by U, $Y_2 \cup \neg Y_2$ by V, (ii) arrows $(\neg m_{Y_2} \circ f'') \cup (m_{Y_2} \circ f)$ by g, u by m_{X_2}, and v by $\neg m_{X_2}$ such that both the triangles commute in the diagram. This could then be a generalization of RSC_2. The exact definition is as follows.

Definition 89 ($RSC_2(\mathscr{C})$ category).
Objects of $RSC_2(\mathscr{C})$ are same as objects of $RSC_1(\mathscr{C})$.
An arrow in $RSC_2(\mathscr{C})$ with domain $(U, X_1, X_2, m_{X_1}, m_{X_2})$ and codomain $(V, Y_1, Y_2, m_{Y_1}, m_{Y_2})$ is a quadruple (f'', f', f, g) where $g : U \to V$, $f' : X_1 \to Y_1$, $f : X_2 \to Y_2$ and $f'' : \neg X_2 \to \neg Y_2$ are arrows in $\mathscr{C}$ such that the diagrams in Fig. 39 commute in $\mathscr{C}$, where $\neg X_2$ and $\neg Y_2$ are domains of $\neg m_{X_2} : \neg X_2 \to U$ and $\neg m_{Y_2} : \neg Y_2 \to V$ respectively. The composition of arrows is defined using the composition of arrows in $\mathscr{C}$.
The identity arrow on $(U, X_1, X_2, m_{X_1}, m_{X_2})$ is $(1_{\neg X_2}, 1_{X_1}, 1_{X_2}, 1_U)$.

Just as ξ-$RSC(\mathscr{C})$ (Definition 85) is defined for the case when the boundary is preserved, we can define ξ-$RSC_1(\mathscr{C})$ and ξ-$RSC_2(\mathscr{C})$ by adding the following condition: an arrow with domain $(U, X_1, X_2, m_{X_1}, m_{X_2})$ and codomain $(V, Y_1, Y_2, m_{Y_1}, m_{Y_2})$ in $RSC_1(\mathscr{C})$ ($RSC_2(\mathscr{C})$) is an arrow in $\xi - RSC_1(\mathscr{C})$

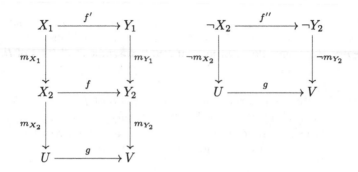

Fig. 39. An $RSC_2(\mathscr{C})$-arrow (f'', f', f, g)

$(\xi - RSC_2(\mathscr{C}))$, if there exists a $\mathscr{C}$-arrow $f''' : \neg X_1 \to \neg Y_1$ such that the following diagram commutes.

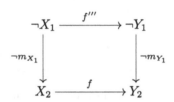

where m_{X_1}, m_{Y_1} and f are arrows as defined for $RSC_1(\mathscr{C})$ in Definition 88 (for $RSC_2(\mathscr{C})$ in Definition 89).

The most recent work that we find on categories based on rough sets is in [13]. Two categories $AprS$ and $\overline{Apr}S$ are defined, objects in each being approximation spaces (U, R). That is in contrast to $ROUGH$ (or equivalently, RSC), where objects are rough sets (U, R, X). An arrow ϕ in $\underline{Apr}S$ from (U, R) to (V, S) is a function $\phi : U \to V$ such that for every $X \subseteq U$, $\phi(\underline{X}) = \phi(X)$. In $\overline{Apr}S$, an arrow ϕ from (U, R) to (V, S) is again a function $\phi : U \to V$ such that for every $X \subseteq U$, $\phi(\overline{X}) = \phi(X)$. A third category Apr has the same objects, and arrows that satisfy both the properties $\phi(\underline{X}) = \phi(X)$ and $\phi(\overline{X}) = \phi(X)$ for every $X \subseteq U$. So the set $X \subseteq U$ is explicitly brought into the picture in all the three categories, whereas the categories that we have studied deal only with equivalence classes in $\underline{\mathcal{X}}$ or $\overline{\mathcal{X}}$. In $ROUGH$, the idea of defining an arrow on $\overline{\mathcal{X}}$, and not X, is that a rough set is recognized by its lower and upper approximations only; then we address the point of preserving information in the respective regions.

However, one may study new categories that imbibe some features of the arrow properties of the categories defined above and also those of $ROUGH$. Such a category may have objects of the form (U, R, X). A possible way to define an arrow from (U, R, X) to (V, S, Y) could be to consider a function $\phi : U \to V$ such that (i) $\phi(X) \subseteq Y$, (ii) $\underline{\phi(X)} = \phi(\underline{X})$, and (iii) $\overline{\phi(X)} = \phi(\overline{X})$. Observe that $\phi([x])$, $[x] \in \overline{\mathcal{X}}$, would always be a definable set contained in $\overline{Y}$ in the approximation space (V, S).

4 Transformation Semigroups and Semiautomata for Rough Sets

In the previous section we studied categories of rough sets and generalizations. An instance of the generalizations is found to be the special class of categories $RSC(\mathbf{M\text{-}Set})$ for monoids $\mathbf{M}$, which yields the definition of *monoid actions on rough sets* [63]. We observe that the objects of the category $RSC(\mathbf{M\text{-}Set})$ may be interpreted as transformations for rough sets. By taking the more general structure of semigroups instead of monoids, we obtain here a natural definition of a *transformation semigroup for a rough set*. The algebra of these transformation semigroups is developed in this section, by defining basic constructions of *ts* theory such as resets, coverings, (direct and wreath) products and admissible partitions for the structures (cf. Sects. 4.1, 4.2). Our main goal is to look for a decomposition result (cf. [2]) for transformation semigroups for rough sets (cf. Sect. 4.3).

Moreover, there is a natural connection of classical transformation semigroup with automata theory [46]. We shall also study this connection in case of rough sets, by defining a *semiautomaton for a rough set* (cf. Sect. 4.4). In Sect. 4.5, some future directions of work related to transformation semigroups for rough sets shall be discussed.

We shall follow the notations and terminologies of Sect. 2.4.1 throughout this section.

4.1 Transformation Semigroups for Rough Sets

Instances of $RSC(\mathscr{C})$ are provided by the class of categories $RSC(\mathbf{M\text{-}Set})$ for monoids $\mathbf{M}$; the properties of objects and morphisms therein yield the definition of *monoid actions on rough sets* (Definition 84). We re-present the definition in case of the more general structure of semigroups, and $ROUGH$-objects (U, R, X) with $\overline{X}$ finite.

Note: In this section, we consider the triple (U, R, X) as a rough set (cf. Sect. 1).

Definition 90 (Semigroup action on rough sets).
A semigroup action on a rough set (U, R, X) *with* $\overline{X}$ *finite, is a quadruple* (U, R, X, δ) *where* $\delta : \overline{X} \times S \to \overline{X}$ *is an action of a semigroup* S *on* $\overline{X}$ *such that the restriction* $\delta|_{\underline{X}} : \underline{X} \times S \to \underline{X}$ *is an action of* S *on* $\underline{X}$. *Note that* $\delta|_{\underline{X}}((q, s)) := \delta((q, s))$, *for all* $q \in \underline{X}, s \in S$.

Using Observation 5, we get a *ts* $(\overline{X}, S/\sim_2)$ for δ, where $s_1 \sim_2 s_2$ if and only if $qs_1 = qs_2$ for all $q \in \overline{X}$. As $\overline{X}$ is finite, $S/\sim_2$ is also finite. Now consider the action $\delta' : \overline{X} \times S/\sim_2 \to \overline{X}$ associated with the *ts* $(\overline{X}, S/\sim_2)$. By Observation 4, we can identify $\{\delta'_{[s]\sim_2}\}_{s \in S}$ with $S/\sim_2$ which is a semigroup of transformations of $\overline{X}$. These transformations also restrict $\underline{X}$ to $\underline{X}$. Thus, $(\underline{X}, S/\sim_1)$ forms another *ts*, where $s_1 \sim_1 s_2$ if and only if $qs_1 = qs_2$ for all $q \in \underline{X}$. We now arrive at the definition of a *ts* for a rough set. Henceforth, when the contexts are clear, we shall drop suffixes and simply write $\sim$.

Definition 91 (Transformation semigroups for rough sets).
A transformation semigroup for a rough set $\mathcal{U} := (U, R, X)$ *is* $\mathcal{A} := (U, R, X, S)$, *denoted as the pair* $(\mathcal{U}, S)$, *where* $(\overline{\mathcal{X}}, S)$ *is a ts such that* $\underline{\mathcal{X}}s \subseteq \underline{\mathcal{X}}$ *for all* $s \in S$. $(\overline{\mathcal{X}}, S)$ *is called the* upper ts *and* $(\underline{\mathcal{X}}, S/\sim)$ *the* lower ts *for* $\mathcal{A}$, *where* $s_1 \sim s_2$ *if and only if* $qs_1 = qs_2$ *for all* $q \in \underline{\mathcal{X}}$.
If S *contains the identity transformation,* $\mathcal{A}$ *is a* transformation monoid *for the rough set* $\mathcal{U}$. *If* $Q \neq \emptyset$ *and* S *is a group such that* S *contains the identity transformation,* $\mathcal{A}$ *is a* transformation group *for the rough set* $\mathcal{U}$.

Note: A transformation semigroup $\mathcal{A} := (\mathcal{U}, S)$ for rough set $\mathcal{U}$ shall be abbreviated as '*ts for rough set* $\mathcal{U}$'. Similarly, '*tm for a rough set*' and '*tg for a rough set*' will be used for a transformation monoid for a rough set and a transformation group for a rough set respectively. Plurals of each of them shall also be denoted by the same abbreviation.

Remark 10. A notion of *rough transformation semigroup* was defined in [82], and was motivated by the *rough finite semi-automaton* defined by Basu [9]. We shall make a comparison of these structures in Sect. 4.4.

Notation 5. *Since the elements of the set* $\overline{\mathcal{X}}$ *are equivalence classes, wherever there is no confusion, we shall use the variables* $q, q_1, q_2, \ldots$ *to represent the equivalence classes in* $\overline{\mathcal{X}}$ *(as done in the above definition).*

Recall Remark 5.

Definition 92 (Transformation monoid for a ts for rough set).
For a ts $\mathcal{A} := (\mathcal{U}, S)$ *for the rough set* $\mathcal{U} := (U, R, X)$, *we get* $S^{\cdot} := S \cup \{1_{\overline{\mathcal{X}}}\}$, *where* $1_{\overline{\mathcal{X}}}$ *is the identity transformation on* $\overline{\mathcal{X}}$. *The tm* $\mathcal{A}^{\cdot} := (\mathcal{U}, S^{\cdot})$, *where the identity transformation of* $\overline{\mathcal{X}}$ *is added to* S, *is called a tm for the ts* $\mathcal{A}$.

We note that given a semigroup action (U, R, X, δ) on rough set $\mathcal{U} := (U, R, X)$, we can obtain a ts $(\mathcal{U}, S/\sim_2)$ for the same rough set $\mathcal{U}$. Conversely, a ts $(\mathcal{U}, S)$ for a rough set $\mathcal{U}$ gives a semigroup action (U, R, X, δ) on $\mathcal{U}$, where δ is the action associated with ts $(\overline{\mathcal{X}}, S)$ (cf. Definition 35).

Observation 26.

1. Relating *ts* and *ts* for rough sets (Definitions 35 and 91):
   (A). Consider a rough set $\mathcal{U} := (U, R, X)$ such that X is definable, i.e. $\underline{\mathcal{X}} = \overline{\mathcal{X}}$ (Definition 1), and a ts $\mathcal{A} := (\overline{\mathcal{X}}, S)$. Here, $\underline{\mathcal{X}}s \subseteq \underline{\mathcal{X}}$ for all $s \in S$. Therefore, $(\mathcal{U}, S)$ is a ts for rough set $\mathcal{U}$ for which, trivially, $\mathcal{A}$ is the upper ts, and also the lower ts up to ts isomorphism. In this case, $(\mathcal{U}, S)$ shall also be denoted as $(\overline{\mathcal{X}}, S)$, by abuse of notation.
   (B). Now, consider a rough set $\mathcal{U} := (U, R, X)$ such that $\underline{\mathcal{X}} = \emptyset$ and a ts $(\overline{\mathcal{X}}, S)$. $(\mathcal{U}, S)$ is then a ts for rough set $\mathcal{U}$, for which $(\overline{\mathcal{X}}, S)$ is the upper ts and $(\emptyset, S/\sim)$ is the lower ts, where $S/\sim$ is a 1-element semigroup.
2. Recall Observation 6 and Proposition 14. Now, for ts $(\mathcal{U}, S)$ for a rough set $\mathcal{U} := (U, R, X)$, since $\underline{\mathcal{X}}s \subseteq \underline{\mathcal{X}}$ for all $s \in S$, we have $(\overline{\mathcal{X}}, S)_{\underline{\mathcal{X}}} = (\overline{\mathcal{X}}, S)|\underline{\mathcal{X}}$. Moreover by definition, $(\overline{\mathcal{X}}, S)_{\underline{\mathcal{X}}}$ is just $(\underline{\mathcal{X}}, S/\sim)$.

Example 9. Consider a rough set $\mathcal{U} := (U, R, X)$ such that $\overline{\mathcal{X}}$ is 7-element set $\{1, 2, 3, 4, 5, 6, 7\}$, and $\underline{\mathcal{X}} := \{1, 5, 6, 7\}$. Note that each $i \in \overline{\mathcal{X}}$ $(1 \le i \le 7)$ represents an equivalence class in U.

Now, consider the ts $(\overline{\mathcal{X}}, S)$ from [59] where $S := \{s_i \mid 1 \le i \le 7\}$ is the semigroup with $s_1 := 1_{\overline{\mathcal{X}}}$ and

$$s_2 := \begin{pmatrix} 1\,2\,3\,4\,5\,6\,7 \\ 1\,2\,4\,3\,6\,5\,7 \end{pmatrix} \quad s_3 := \begin{pmatrix} 1\,2\,3\,4\,5\,6\,7 \\ 1\,5\,6\,7\,1\,1\,1 \end{pmatrix} \quad s_4 := \begin{pmatrix} 1\,2\,3\,4\,5\,6\,7 \\ 1\,5\,7\,6\,1\,1\,1 \end{pmatrix}$$

$$s_5 := \begin{pmatrix} 1\,2\,3\,4\,5\,6\,7 \\ 1\,6\,5\,7\,1\,1\,1 \end{pmatrix} \quad s_6 := \begin{pmatrix} 1\,2\,3\,4\,5\,6\,7 \\ 1\,6\,7\,5\,1\,1\,1 \end{pmatrix} \quad s_7 := \begin{pmatrix} 1\,2\,3\,4\,5\,6\,7 \\ 1\,1\,1\,1\,1\,1\,1 \end{pmatrix}$$

Since $\underline{\mathcal{X}} s_i \subseteq \underline{\mathcal{X}}$ for all $s_i \in S$, $(\mathcal{U}, S)$ forms a ts for rough set $\mathcal{U}$. The upper ts is $(\overline{\mathcal{X}}, S)$ and the lower ts is $(\underline{\mathcal{X}}, S/\sim)$, where $S/\sim = \{\{s_1\}, \{s_2\}, \{s_3, s_4, s_5, s_6, s_7\}\}$.

4.1.1   Resets and Closures

What could be an appropriate definition for a reset ts here?

Definition 93 (Reset ts for rough sets).
The tuple (U, R, X, S) is a reset if the upper ts $(\overline{\mathcal{X}}, S)$ is a reset, that is, $|\overline{\mathcal{X}} s| \le 1$ for all $s \in S$.

Observe that the lower ts $(\underline{\mathcal{X}}, S/\sim)$ is also a reset in the definition above, because $|\underline{\mathcal{X}} s| \le |\overline{\mathcal{X}} s| \le 1$ for all $s \in S$.

Example 10. Recall Example 6. A trivial reset ts for a rough set $\mathcal{U} := (U, R, X)$ is $\mathcal{A} := (\mathcal{U}, \emptyset)$. The reset ts $(\underline{\mathcal{X}}, \emptyset)$ and $(\overline{\mathcal{X}}, \emptyset)$ are respectively the lower ts and the upper ts. If $|\underline{\mathcal{X}}| = m \le n = |\overline{\mathcal{X}}|$, $\mathcal{A}$ shall be denoted by $(\mathbf{m}, \mathbf{n})$.

Definition 94 (Permutation groups for rough sets).
A ts $\mathcal{A} := (\mathcal{U}, S)$ for the rough set $\mathcal{U} := (U, R, X)$ is called a permutation group for $\mathcal{U}$ if $(\overline{\mathcal{X}}, S)$ is a permutation group, that is $\overline{\mathcal{X}} s = \overline{\mathcal{X}}$, for any $s \in S$.

Example 11. Consider the rough set $\mathcal{U} := (U, R, X)$ and the permutation group $(\overline{\mathcal{X}}, S)$, where $\overline{\mathcal{X}} := \{1, 2, 3, 4\}$ and the semigroup $S := \{s_i \mid 1 \le i \le 4\}$ such that $s_1 := 1_{\overline{\mathcal{X}}}$ and

$$s_2 := \begin{pmatrix} 1\,2\,3\,4 \\ 2\,1\,3\,4 \end{pmatrix} \quad s_3 := \begin{pmatrix} 1\,2\,3\,4 \\ 1\,2\,4\,3 \end{pmatrix} \quad s_4 := \begin{pmatrix} 1\,2\,3\,4 \\ 2\,1\,4\,3 \end{pmatrix}$$

Take $\underline{\mathcal{X}} := \{1, 2\}$. Since $\underline{\mathcal{X}} s_i \subseteq \underline{\mathcal{X}}$ for all $s_i \in S$, $(\mathcal{U}, S)$ is a tg for the rough set $\mathcal{U}$ and hence a permutation group for $\mathcal{U}$. The upper ts is $(\overline{\mathcal{X}}, S)$ and the lower ts is $(\underline{\mathcal{X}}, S/\sim)$ where $S/\sim = \{\{s_1, s_3\}, \{s_2, s_4\}\}$.

In the above example the lower ts $(\underline{\mathcal{X}}, S/\sim)$ is also a permutation group. Is this always the case with permutation groups for rough sets?

Proposition 43. *For a permutation group $(\mathcal{U}, G)$ for rough set $\mathcal{U} := (U, R, X)$, $(\underline{\mathcal{X}}, G/\sim)$ is a permutation group.*

Proof. Since $(\overline{\mathcal{X}}, G)$ is a permutation group, any $g \in G$ is a permutation of $\overline{\mathcal{X}}$. Let $|\overline{\mathcal{X}}| = n$. As $(\mathcal{U}, G)$ is a ts for rough set $\mathcal{U}$, $\underline{\mathcal{X}}g \subseteq \underline{\mathcal{X}}$, for all $g \in G$. Now, suppose to the contrary, $(\underline{\mathcal{X}}, G/\sim)$ is not a permutation group of $\underline{\mathcal{X}}$, i.e. there exist $g \in G$ and $q_1 \in \underline{\mathcal{X}}$ such that $q_1 \notin \underline{\mathcal{X}}g$. Then there exists a $q_2 \in \overline{\mathcal{X}} \backslash \underline{\mathcal{X}}$ such that $q_2g = q_1$; for q_2, there exists $q_3 \in \overline{\mathcal{X}} \backslash \underline{\mathcal{X}}$ such that $q_3g = q_2$, and so on; there exists $q_n \in \overline{\mathcal{X}} \backslash \underline{\mathcal{X}}$ such that $q_ng = q_{n-1}$. Since g is a permutation, $g^n = 1_{\overline{\mathcal{X}}}$ and $q_n = q_ng^n = q_1g$. Now, $q_n \in \overline{\mathcal{X}} \backslash \underline{\mathcal{X}}$ implies $q_1 \in \overline{\mathcal{X}} \backslash \underline{\mathcal{X}}$, a contradiction.

Example 12. Recall Example 7, we have defined the ts $(S^\bullet, S)$ for a semigroup S and the tg (G, G) for a group G, denoted by S and G respectively.
Consider a semigroup S and the rough set $\mathcal{U} := (S^\bullet, Id_{S^\bullet}, S^\bullet)$, $Id_{S^\bullet}$ being the identity relation on $S^\bullet$. $(\mathcal{U}, S)$ is a ts for rough set $\mathcal{U}$ (Observation 26(1A)). This shall also be denoted as S.
If S is replaced by a monoid M or a group G, then $G^\bullet = G$ and $M^\bullet = M$. In this case, $(\mathcal{U}, G)$ and $(\mathcal{U}, M)$ are ts for the rough set $\mathcal{U}$, denoted by G and M respectively.

Consider a rough set $\mathcal{U} := (U, R, X)$ and ts $\mathcal{A} := (\mathcal{U}, S)$ for rough set $\mathcal{U}$. What can be a suitable extension of the 'closure' $\overline{\mathcal{A}}$ of the ts $\mathcal{A}$ for rough set $\mathcal{U}$? Note that for $q \in \overline{\mathcal{X}} \backslash \underline{\mathcal{X}}$, the constant functions $\overline{q}$ on $\overline{\mathcal{X}}$ do not restrict $\overline{\mathcal{X}}$ into $\underline{\mathcal{X}}$. We have the following.

Definition 95 (Closure of a ts for a rough set).
The closure of $\mathcal{A} := (\mathcal{U}, S)$ for a rough set $\mathcal{U} := (U, R, X)$ is defined as

$$\overline{\mathcal{A}} := (\mathcal{U}, S') \text{ with } S' := \langle S \cup \{\overline{q} \mid q \in \underline{\mathcal{X}}\} \cup \{\widetilde{q} \mid q \in \overline{\mathcal{X}} \backslash \underline{\mathcal{X}}\}\rangle, \text{ where}$$

for each $q \in \overline{\mathcal{X}}$, $\overline{q}$ is the constant function on $\overline{\mathcal{X}}$ such that $\overline{q}(x) = q$, for any $x \in \overline{\mathcal{X}}$, and

$$\widetilde{q}(x) := q \quad \text{if } x \in \overline{\mathcal{X}} \backslash \underline{\mathcal{X}}, \text{ and}$$
$$\text{undefined otherwise.}$$

Observation 27.

(1) If $\underline{\mathcal{X}} = \overline{\mathcal{X}}$ (i.e. X is definable) or $\underline{\mathcal{X}} = \emptyset$ then the semigroup S' in $\overline{\mathcal{A}}$ (the same as the semigroup in $(\overline{\mathcal{X}}, S)$, as expected) contains all functions $\overline{q}$ for $q \in \overline{\mathcal{X}}$.
(2) Let $\emptyset_{\overline{\mathcal{X}}}$ denote the empty partial function, i.e. $\emptyset_{\overline{\mathcal{X}}}(q)$ is not defined for any $q \in \overline{\mathcal{X}}$. If $\underline{\mathcal{X}} \neq \emptyset$ and $\underline{\mathcal{X}} \neq \overline{\mathcal{X}}$, then S' contains the following:
   (a) all $\widetilde{q}$ for $q \in \overline{\mathcal{X}}$, since if $q \in \underline{\mathcal{X}}$ then $\widetilde{q} = \widetilde{q'q} \in S'$ for any $q' \in \overline{\mathcal{X}} \backslash \underline{\mathcal{X}}$,
   (b) $\emptyset_{\overline{\mathcal{X}}}$, because $\emptyset_{\overline{\mathcal{X}}} = \overline{q'}\widetilde{q} \in S'$ for any $q \in \overline{\mathcal{X}} \backslash \underline{\mathcal{X}}$ and $q' \in \underline{\mathcal{X}}$.
(3) Closure is idempotent, i.e. $\overline{\mathcal{A}} = \overline{(\overline{\mathcal{A}})}$.

Example 13. Consider the reset ts $(\mathbf{m}, \mathbf{n})$ for the rough set $\mathcal{U} := (U, R, X)$ of Example 10. For $m \neq n$ and $m \neq 0$,

$$\overline{(\mathbf{m}, \mathbf{n})} := (\mathcal{U}, S'), \text{ where } S' = \{\overline{q} \mid q \in \underline{\mathcal{X}}\} \cup \{\widetilde{q} \mid q \in \overline{\mathcal{X}}\} \cup \{\emptyset_{\overline{\mathcal{X}}}\}.$$

In particular, for the reset ts $(\mathbf{1,2})$ for rough set $\mathcal{U}$ where $\overline{\mathcal{X}} \cong \{0,1\}$ and $\underline{\mathcal{X}} \cong \{0\}$, $S' := \{\overline{0}, \widetilde{1}, \widetilde{0}, \emptyset_{\overline{\mathcal{X}}}\}$. Diagrammatically, the upper ts $(\overline{\mathcal{X}}, S')$ may be represented as:

$$\widetilde{1} \;\circlearrowleft\; 1 \xrightarrow{\;\;\overline{0},\widetilde{0}\;\;} 0 \;\circlearrowleft\; \overline{0}$$

Note that the upper ts $(\overline{\mathcal{X}}, S')$ of the closure $\overline{(1,2)}$ is not isomorphic to the closure $\overline{2}$ of the upper reset ts $\mathbf{2}$ (cf. Example 10). The lower ts $(\underline{\mathcal{X}}, \{\{\overline{0}\}, \{\emptyset_{\overline{\mathcal{X}}}, \widetilde{0}, \widetilde{1}\}\})$ of the closure $\overline{(1,2)}$ is also not isomorphic to the closure $\overline{1}$ of the lower reset ts $\mathbf{1}$. However, the following holds.

Proposition 44. *For ts $\mathcal{A} := (\mathcal{U}, S)$ for a rough set $\mathcal{U} := (U, R, X)$,*

(1) the closure of the upper ts of $\mathcal{A}$ covers the upper ts of the closure of $\mathcal{A}$,
(2) the closure of the lower ts of $\mathcal{A}$ covers the lower ts of the closure of $\mathcal{A}$.

Proof. We refer to $\overline{\mathcal{A}}$ as in Definition 95.
(1). Closure of the upper ts of $\mathcal{A}$ is $(\overline{\mathcal{X}}, \langle S \cup \{\overline{q}\}_{q \in \overline{\mathcal{X}}}\rangle)$ and the upper ts of the closure $\overline{\mathcal{A}}$ is $(\overline{\mathcal{X}}, S')$. The cover map η is $1_{\overline{\mathcal{X}}}$, where (i). for $s \in S$ we take $t_s = s$, (ii). for $s = \overline{q}$ $(q \in \overline{\mathcal{X}})$ we take $t_s = \overline{q}$ and (iii). for $s = \widetilde{q}$ $(q \in \overline{\mathcal{X}} \backslash \underline{\mathcal{X}})$ we take $t_s = \overline{q}$.
(2). Closure of the lower ts $(\underline{\mathcal{X}}, S/\sim)$ of $\mathcal{A}$ is $(\underline{\mathcal{X}}, \langle S/\sim \cup \{[\overline{q}]\}_{q \in \underline{\mathcal{X}}}\rangle)$ and the lower ts of the closure $\overline{\mathcal{A}}$ is $(\underline{\mathcal{X}}, S'/\sim)$, where $\sim$'s are defined appropriately. The cover map η is $1_{\underline{\mathcal{X}}}$, where
(i). for $u = [s] \in S'/\sim$, take $t_u := [s]$;
(ii). for $u = [\overline{q}]$ $(q \in \underline{\mathcal{X}})$, take $t_u := [\overline{q}]$; and
(iii). for $u = [\widetilde{q}]$ $(q \in \overline{\mathcal{X}} \backslash \underline{\mathcal{X}})$, take t_u to be any element in $S/\sim$. Note that $\eta(x)[\widetilde{q}]$ is not defined for any $x \in \underline{\mathcal{X}}$.

4.2   Algebra on Transformation Semigroups for Rough Sets

Consider two ts $\mathcal{A} := (\mathcal{U}, S)$ and $\mathcal{B} := (\mathcal{V}, T)$ for rough sets $\mathcal{U} := (U, R, X)$ and $\mathcal{V} := (V, R', Y)$ respectively, and the ts homomorphism (α, β) from the upper ts $(\overline{\mathcal{X}}, S)$ to the upper ts $(\overline{\mathcal{Y}}, T)$ (Definition 39) satisfying the condition $\alpha(\underline{\mathcal{X}}) \subseteq \underline{\mathcal{Y}}$. Would this imply that the pair $(\alpha|_{\underline{\mathcal{X}}}, \widehat{\beta})$ is a ts homomorphism from $(\underline{\mathcal{X}}, S/\sim)$ to $(\underline{\mathcal{Y}}, T/\sim')$, where $\widehat{\beta} : S/\sim \to T/\sim'$ is defined as $\widehat{\beta}([s]_\sim) := [\beta(s)]_{\sim'}$, $s \in S$? The answer is no, as $\widehat{\beta}$ may not be well-defined: recall the ts $(\mathcal{U}, S)$ for a rough set $\mathcal{U}$ and the upper ts $(\overline{\mathcal{X}}, S)$ of $(\mathcal{U}, S)$ from Example 9. Consider the rough set $\mathcal{U}' := (U, R, \overline{X})$. The lower approximation and upper approximation of $\overline{X}$ are same, and equal to $\overline{X}$. By Observation 26(1), the pair $(\mathcal{U}', S)$ is also a ts for rough set with lower ts identifiable with the upper ts $(\overline{\mathcal{X}}, S)$. $(1_{\overline{\mathcal{X}}}, 1_S) : (\overline{\mathcal{X}}, S) \to (\overline{\mathcal{X}}, S)$ is a ts homomorphism and $1_{\overline{\mathcal{X}}}(\underline{\mathcal{X}}) \subseteq \overline{\mathcal{X}}$. $\widehat{1_S} : S/\sim \to S$ is such that $\widehat{1_S}([s_i]_\sim) := s_i$, $s_i \in S$; however, $[s_3]_\sim = [s_7]_\sim$ but $s_3 \neq s_7$. So we have the following.

Definition 96 (*ts* homomorphisms for rough sets).
The pair (α, β) is a ts homomorphism from $\mathcal{A} := (\mathcal{U}, S)$ to $\mathcal{B} := (\mathcal{V}, T)$, provided

(a) (α, β) is a ts homomorphism between the upper ts $(\overline{\mathcal{X}}, S)$ and $(\overline{\mathcal{Y}}, T)$,
(b) $\alpha(\underline{\mathcal{X}}) \subseteq \underline{\mathcal{Y}}$, and
(c) for any $s, s' \in S$,

$$\text{if } (qs = qs' \text{ for all } q \in \underline{\mathcal{X}}) \text{ then } (p\beta(s) = p\beta(s') \text{ for all } p \in \underline{\mathcal{Y}}). \qquad (4.1)$$

A ts homomorphism (α, β) for a rough set is a ts monomorphism, epimorphism or isomorphism for a rough set, according as it is a ts monomorphism, epimorphism or isomorphism respectively.

Observation 28.

(1) Condition 4.1 in the above definition ensures that the pair $(\alpha|_{\underline{\mathcal{X}}}, \widehat{\beta})$ is a *ts* homomorphism between the lower *ts* $(\underline{\mathcal{X}}, S/\sim)$ and $(\underline{\mathcal{Y}}, T/\sim)$.
(2) If in the above definition, $\alpha|_{\underline{\mathcal{X}}}$ is also a bijection, Condition 4.1 is always true: suppose $qs = qs'$ for all $q \in \underline{\mathcal{X}}$. For $p \in \underline{\mathcal{Y}}$, there exists $q \in \underline{\mathcal{X}}$ such that $\alpha(q) = p$. Then $qs = qs' \Rightarrow \alpha(q)\beta(s) = \alpha(q)\beta(s') \Rightarrow p\beta(s) = p\beta(s')$.

How are *ts* for rough sets and *ts* for sets related? A direct relationship may be observed using category theory. Recall Definition 40 of the category **TS** of transformation semigroups for sets.

Definition 97 (The category RTS of *ts* for rough sets).
Objects are ts for rough sets and arrows are homomorphisms of ts for rough sets. Composition of two homomorphisms of ts for rough sets is defined componentwise.
The identity arrow for a ts for rough sets $(\mathcal{U}, S)$ is given by $(1_{\overline{\mathcal{X}}}, 1_S)$, where $\mathcal{U} := (U, R, X)$, $1_{\overline{\mathcal{X}}}$ is identity function on $\overline{\mathcal{X}}$, and 1_S is identity function on S.

Using Observation 26 (1) and Definition 96 of homomorphisms, we easily obtain

Theorem 15. *The category **TS** is equivalent to each of the following categories:*

1. *the full subcategory of **RTS** with objects of the type $(\mathcal{U}, S)$, where $\mathcal{U} := (U, R, X)$ such that X is definable $(\overline{\mathcal{X}} = \underline{\mathcal{X}})$; and*
2. *the full subcategory of **RTS** with objects of the type $(\mathcal{U}, S)$, where $\mathcal{U} := (U, R, X)$ such that $\underline{\mathcal{X}} = \emptyset$.*

Proof.
(1). Define two maps, both denoted by F, as follows: (i). F maps an **RTS**-object $(\mathcal{U}, S)$, where $\mathcal{U} := (U, R, X)$ such that X is definable, to a **TS**-object $(\overline{\mathcal{X}}, S)$; and (ii). F maps an **RTS**-arrow $(\alpha, \beta) : (\mathcal{U}, S) \to (\mathcal{V}, T)$, where $\mathcal{U} := (U, R, X)$, $\mathcal{V} := (V, R', Y)$, and X and Y are definable, to a **TS**-arrow $(\alpha, \beta) : (\overline{\mathcal{X}}, S) \to (\overline{\mathcal{Y}}, T)$.
F is a functor. Consider two *ts* $\mathcal{A} := (\mathcal{U}, S)$ and $\mathcal{B} := (\mathcal{V}, T)$ for rough sets $\mathcal{U} := (U, R, X)$ and $\mathcal{V} := (V, R', Y)$ respectively such that X and Y are definable sets, i.e. we have $\overline{\mathcal{X}} = \underline{\mathcal{X}}$ and $\overline{\mathcal{Y}} = \underline{\mathcal{Y}}$.

(i). Let **RTS**-arrows (α, β) and (α', β') with domain as $\mathcal{A}$ and codomain as $\mathcal{B}$ be such that $(\alpha, \beta) = (\alpha', \beta')$ in **RTS**. Then $\alpha = \alpha'$ and $\beta = \beta'$. Therefore, F is faithful.

(ii). Let $(\alpha, \beta) : (\overline{\mathcal{X}}, S) \to (\overline{\mathcal{Y}}, T)$ be a **TS**-arrow. Then $\alpha(\underline{\mathcal{X}}) \subseteq \underline{\mathcal{Y}}$. Further, Condition 4.1 also holds because: for $s, s' \in S$, if $qs = qs'$ for all $q \in \underline{\mathcal{X}} = \overline{\mathcal{X}}$, then $s = s'$ and $p\beta(s) = p\beta(s')$ for all $p \in \underline{\mathcal{Y}}$. Therefore, (α, β) is an **RTS**-arrow and F is full.

(iii). Consider a **TS**-object (Q, S). Using the proof of Theorem 12, for the RSC-object (Q, Q), we can obtain an approximation space (U, R) and $X \subseteq U$ such that $(\underline{\mathcal{X}}, \overline{\mathcal{X}}) \cong (Q, Q)$. Here, $Q \cong \overline{\mathcal{X}}$, given by the bijection $\alpha : \overline{\mathcal{X}} \to Q$. Then by Observation 7 $(\overline{\mathcal{X}}, S)$ is a ts such that $(\overline{\mathcal{X}}, S) \cong (Q, S)$. Moreover, $\overline{\mathcal{X}} = \underline{\mathcal{X}}$ implies that X is definable. Therefore, $(\mathcal{U}, S)$ is an **RTS**-object and $F((\mathcal{U}, S)) = (\overline{\mathcal{X}}, S) \cong (Q, S)$, where $\mathcal{U} := (U, R, X)$ and X is definable. This makes F essentially surjective.

(2). Define two maps, both denoted by F, as follows: (i). F maps an **RTS**-object $(\mathcal{U}, S)$, where $\mathcal{U} := (U, R, X)$ such that $\underline{\mathcal{X}} = \emptyset$, to a **TS**-object $(\overline{\mathcal{X}}, S)$; and (ii). F maps an **RTS**-arrow $(\alpha, \beta) : (\mathcal{U}, S) \to (\mathcal{V}, T)$, where $\mathcal{U} := (U, R, X)$, $\mathcal{V} := (V, R', Y)$, and $\underline{\mathcal{X}} = \underline{\mathcal{Y}} = \emptyset$, to a **TS**-arrow $(\alpha, \beta) : (\overline{\mathcal{X}}, S) \to (\overline{\mathcal{Y}}, T)$. F is a full and faithful functor. Let us show that F is essentially surjective. Consider a **TS**-object (Q, S). Using the proof of Theorem 12, for the RSC-object $(\emptyset, \overline{\mathcal{X}})$, we can define an appropriate approximation space (V, R) and $Y \subseteq V$ such that $(\underline{\mathcal{Y}}, \overline{\mathcal{Y}}) \cong (\emptyset, Q)$. Using a similar argument as above, we have $(\mathcal{V}, S)$ as an **RTS**-object, where $\mathcal{V} := (V, S, Y)$. Moreover, $F((\mathcal{V}, S)) = (\overline{\mathcal{Y}}, S) \cong (Q, S)$ such that $\underline{\mathcal{Y}} = \emptyset$.

Consider two rough sets $\mathcal{U} := (U, R, X)$ and $\mathcal{V} := (V, R', Y)$. We can define the approximation space $(U \times V, R \times R')$, where $R \times R'$ is defined as follows: for $(x, y), (x', y') \in U \times V$, $(x, y)(R \times R')(x', y')$ if and only if xRx' and $yR'y'$. Observe that $R \times R'$ is an equivalence relation and $[(x, y)]_{R \times R'} = [x]_R \times [y]_{R'}$ for every $(x, y) \in X \times Y$. Here, the triple $\mathcal{U} \times \mathcal{V} := (U \times V, R \times R', X \times Y)$ is also a rough set, for which $(R \times R')$-upper approximation is denoted by $\overline{\mathcal{X} \times \mathcal{Y}}$ and $(R \times R')$-lower approximation is denoted by $\underline{\mathcal{X} \times \mathcal{Y}}$. Moreover, $\overline{\mathcal{X} \times \mathcal{Y}} = \overline{\mathcal{X}} \times \overline{\mathcal{Y}}$ and $\underline{\mathcal{X} \times \mathcal{Y}} = \underline{\mathcal{X}} \times \underline{\mathcal{Y}}$ [4].

Let us now define direct and wreath products of ts for rough sets. Recall the Definition 41 of products in ts.

Definition 98 (Direct and Wreath products of ts for rough sets).
Consider the ts $\mathcal{A} := (\mathcal{U}, S)$ and $\mathcal{B} := (\mathcal{V}, T)$ for rough sets $\mathcal{U} := (U, R, X)$ and $\mathcal{V} := (V, R', Y)$ respectively, and the rough set $\mathcal{U} \times \mathcal{V} := (U \times V, R \times R', X \times Y)$.
(1). The direct product of $\mathcal{A}$ and $\mathcal{B}$ is

$$\mathcal{A} \times \mathcal{B} := (\mathcal{U} \times \mathcal{V}, S \times T).$$

(2). The wreath product of $\mathcal{A}$ and $\mathcal{B}$ is

$$\mathcal{A} \circ \mathcal{B} := (\mathcal{U} \times \mathcal{V}, S^{\overline{\mathcal{Y}}} \times T).$$

Observation 29. Let us check if the above notions are well-defined. In both the products, $\underline{\mathcal{X}} \times \underline{\mathcal{Y}} \subseteq \overline{\mathcal{X}} \times \overline{\mathcal{Y}}$.

(1) In the direct product of $\mathcal{A}$ and $\mathcal{B}$, for $(q,p) \in \underline{\mathcal{X}} \times \underline{\mathcal{Y}}$, we have $(q,p)(s,t) = (qs,pt) \in \underline{\mathcal{X}} \times \underline{\mathcal{Y}}$. Thus the direct product is a ts for rough set.
(2) For the wreath product $\mathcal{A} \circ \mathcal{B}$ to be a ts for rough set, we have to check whether the semigroup action of $S^{\overline{\mathcal{Y}}} \times T$ on $\overline{\mathcal{X}} \times \overline{\mathcal{Y}}$ restricts $\underline{\mathcal{X}} \times \underline{\mathcal{Y}}$ to $\underline{\mathcal{X}} \times \underline{\mathcal{Y}}$. Let $(q,p) \in \underline{\mathcal{X}} \times \underline{\mathcal{Y}}$ and $(f,t) \in S^{\overline{\mathcal{Y}}} \times T$. Then $(q,p)(f,t) = (qf(p),pt) \in \underline{\mathcal{X}} \times \underline{\mathcal{Y}}$.

For $\mathcal{A} \times \mathcal{B}$, the upper ts is $(\overline{\mathcal{X}} \times \overline{\mathcal{Y}}, S \times T)$ and the lower ts is $(\underline{\mathcal{X}} \times \underline{\mathcal{Y}}, (S \times T)/\sim)$. Similarly, for $\mathcal{A} \circ \mathcal{B}$, the upper ts is $(\overline{\mathcal{X}} \times \overline{\mathcal{Y}}, S^{\overline{\mathcal{Y}}} \times T)$ and the lower ts is $(\underline{\mathcal{X}} \times \underline{\mathcal{Y}}, (S^{\overline{\mathcal{Y}}} \times T)/\sim)$. Let us observe the following.

Observation 30. The upper ts of the direct/wreath products of $\mathcal{A}$ and $\mathcal{B}$ is equal to the direct/wreath products of the upper ts of $\mathcal{A}$ and $\mathcal{B}$.

$$(\overline{\mathcal{X}} \times \overline{\mathcal{Y}}, S \times T) = (\overline{\mathcal{X}}, S) \times (\overline{\mathcal{Y}}, T)$$
$$(\overline{\mathcal{X}} \times \overline{\mathcal{Y}}, S^{\overline{\mathcal{Y}}} \times T) = (\overline{\mathcal{X}}, S) \circ (\overline{\mathcal{Y}}, T)$$

Can we say the same for the lower ts?

Proposition 45. *For two ts $\mathcal{A} := (\mathcal{U}, S)$ and $\mathcal{B} := (\mathcal{V}, T)$ for rough sets $\mathcal{U} := (U, R, X)$ and $\mathcal{V} := (V, R', Y)$ respectively, we have the following.*

1. $(\underline{\mathcal{X}} \times \underline{\mathcal{Y}}, (S \times T)/\sim) \cong (\underline{\mathcal{X}}, S/\sim) \times (\underline{\mathcal{Y}}, T/\sim)$, and
2. $(\underline{\mathcal{X}} \times \underline{\mathcal{Y}}, (S^{\overline{\mathcal{Y}}} \times T)/\sim) \cong (\underline{\mathcal{X}}, S/\sim) \circ (\underline{\mathcal{Y}}, T/\sim)$.

Proof. (1). The right hand side direct product ts is $(\underline{\mathcal{X}} \times \underline{\mathcal{Y}}, S/\sim \times T/\sim)$. Define a function $\phi : (S \times T)/\sim \to S/\sim \times T/\sim$ as follows:

$$\phi([(s,t)]) := ([s],[t])$$

ϕ is a semigroup isomorphism.

(2). The right hand side wreath product ts is $(\underline{\mathcal{X}} \times \underline{\mathcal{Y}}, (S/\sim)^{\underline{\mathcal{Y}}} \times (T/\sim))$. Take α to be the identity map on $\underline{\mathcal{X}} \times \underline{\mathcal{Y}}$. Define β from $((S^{\overline{\mathcal{Y}}} \times T)/\sim)$ to $(S/\sim)^{\underline{\mathcal{Y}}} \times (T/\sim)$ as follows:

$$\beta([(g,t)]) = (ig\nu, [t]),$$

where $g : \overline{\mathcal{Y}} \to S$, $t \in T$, i is the inclusion map from $\underline{\mathcal{Y}}$ to $\overline{\mathcal{Y}}$ and ν is the canonical semigroup homomorphism from S to $S/\sim$ mapping s to $[s]$. We claim that β is a semigroup isomorphism.

Well defined: Let $g_1, g_2 : \underline{\mathcal{Y}} \to T$ and $t_1, t_2 \in T$ such that $(g_1, t_1) \sim (g_2, t_2)$, that is,

$$(q,p)(g_1,t_1) = (q,p)(g_2,t_2) \text{ for all } q \in \underline{\mathcal{X}} \text{ and } p \in \underline{\mathcal{Y}}$$
$$\Leftrightarrow (qg_1(p),pt_1) = (qg_2(p),pt_2) \text{ for all } q \in \underline{\mathcal{X}} \text{ and } p \in \underline{\mathcal{Y}}$$
$$\Leftrightarrow (qg_1(p) = qg_2(p) \text{ for all } q \in \underline{\mathcal{X}} \text{ and } p \in \underline{\mathcal{Y}}) \text{ and } (pt_1 = pt_2 \text{ for all } p \in \underline{\mathcal{Y}})$$
$$\Leftrightarrow (g_1(p) \sim g_2(p) \text{ for all } p \in \underline{\mathcal{Y}}) \text{ and } (p[t_1] = p[t_2] \text{ for all } p \in \underline{\mathcal{Y}})$$
$$\Leftrightarrow ig_1\nu = ig_2\nu \text{ and } [t_1] = [t_2].$$

One-one: Since all the steps in the above argument are equivalent, one-oneness can be obtained by reversing the previous argument.

Onto: Consider the pair $(f, [t]) \in (S/\sim)^{\underline{\mathcal{Y}}} \times (T/\sim)$. Choose and fix $s_o \in S$. For each $p \in \underline{\mathcal{Y}}$, choose $s_p \in S$ such that $s_p \in f(p)$. Now define a function $g_f : \overline{\mathcal{Y}} \to S$ in the following way:

$$g_f(p) = s_p \text{ for } p \in \underline{\mathcal{Y}}, \text{ and}$$

$$= s_o \text{ otherwise.}$$

Then we have $\beta([(g_f, t)]) = (f, [t])$.

Semigroup homomorphism: Recall the semigroup operation defined in the wreath products of two ts (Definition 41). For two ts (A, S_1) and (B, S_2), the semigroup operation in the semigroup $S_1^B \times S_2$ is defined as $(f, t) \cdot (g, t') := (f * g, tt')$, where $f, g \in S_1^B, t, t' \in S_2$ and $f * g$ is a map from B to S_1 given by $(f * g)(p) := f(p)g(pt')$.

Now, consider $[(g_1, t_1)], [(g_2, t_2)] \in ((S^{\overline{\mathcal{Y}}} \times T)/\sim)$. We have to show

$$\beta([(g_1, t_1)][(g_2, t_2)]) = \beta([(g_1, t_1)])\beta([(g_2, t_2)]).$$

We have

$$\beta([(g_1, t_1)][(g_2, t_2)]) = \beta([(g_1 * g_2, t_1 t_2)]) = (i(g_1 * g_2)\nu, [t_1 t_2]). \qquad (4.2)$$

Moreover, for all $p \in \underline{\mathcal{Y}}$,

$$(i(g_1 * g_2)\nu)(p) = \nu(g_1(p)g_2(pt_2))$$
$$= \nu(g_1(p))\nu(g_2(pt_2))$$
$$= (ig_1\nu)(p)(ig_2\nu)(pt_2)$$
$$= ((ig_1\nu) * (ig_2\nu))(p).$$
$$\therefore i(g_1 * g_2)\nu = (ig_1\nu) * (ig_2\nu).$$

Inserting this in Eq. 4.2, $\beta([(g_1, t_1)][(g_2, t_2)]) = ((ig_1\nu) * (ig_2\nu), [t_1][t_2])$ $= (ig_1\nu, [t_1])(ig_2\nu, [t_2]) = \beta([(g_1, t_1)])\beta([(g_2, t_2)])$. Therefore (α, β) is an isomorphism from $(\underline{\mathcal{X}} \times \underline{\mathcal{Y}}, (S^{\overline{\mathcal{Y}}} \times T)/\sim)$ to $(\underline{\mathcal{X}} \times \underline{\mathcal{Y}}, (S/\sim)^{\underline{\mathcal{Y}}} \times (T/\sim))$.

We next move to admissible partitions and quotients. In a ts $\mathcal{A} := (\mathcal{U}, S)$ for rough set $\mathcal{U} := (U, R, X)$, given an admissible partition $\pi_2 := \{H_i\}_{i \in I}$ on $\overline{\mathcal{X}}$ in the upper ts $(\overline{\mathcal{X}}, S)$, one obtains a partition on $\underline{\mathcal{X}}$ in the lower ts $(\underline{\mathcal{X}}, S/\sim)$ as follows:

$$\pi_2 \sqcap \underline{\mathcal{X}} := \{H_i \cap \underline{\mathcal{X}} \mid H_i \in \pi_2 \text{ and } H_i \cap \underline{\mathcal{X}} \neq \emptyset\}.$$

Note that for a non-trivial partition π_2 on $\overline{\mathcal{X}}$, the partition $\pi_2 \sqcap \underline{\mathcal{X}}$ on $\underline{\mathcal{X}}$ can be trivial. Moreover, we have the following result.

Proposition 46. $\pi_2 \sqcap \underline{\mathcal{X}}$ is admissible on $\underline{\mathcal{X}}$ in the lower ts $(\underline{\mathcal{X}}, S/\sim)$.

Proof. Let $H_i \cap \underline{\mathcal{X}} \in \pi_2 \sqcap \underline{\mathcal{X}}$ and $[s] \in S/\sim$ such that $(H_i \cap \underline{\mathcal{X}})[s]$ is non-empty. Consider $q \in H_i \cap \underline{\mathcal{X}}$ such that qs is defined. Note that $qs \in \underline{\mathcal{X}}$ as $q \in \underline{\mathcal{X}}$. As π_2 is admissible, there exists H_j such that $H_i s \subseteq H_j$. Then $qs \in (H_i \cap \underline{\mathcal{X}})s \subseteq H_j \cap \underline{\mathcal{X}}$, that is $H_j \cap \underline{\mathcal{X}} \neq \emptyset$ and $(H_i \cap \underline{\mathcal{X}})[s] \subseteq H_j \cap \underline{\mathcal{X}}$.

However we note that $\pi_2 \sqcap \underline{\mathcal{X}} \not\subseteq \pi_2$, in general. Take the following example.

Example 14. Let us consider the reset $ts\ \overline{A} := \overline{(2,4)}$ (Example 13) for a rough set $\mathcal{U} := (U, R, X)$ such that $\overline{\mathcal{X}} := \{q_1, q_2, q_3, q_4\}$ and $\underline{\mathcal{X}} := \{q_1, q_2\}$. The semigroup associated with the upper ts of $\overline{A}$ is $S := \{\overline{q}_1, \overline{q}_2, \widetilde{q}_1, \widetilde{q}_2, \widetilde{q}_3, \widetilde{q}_4, \emptyset_{\overline{\mathcal{X}}}\}$. Consider the partition $\pi_2 := \{\{q_1, q_3\}, \{q_2\}, \{q_4\}\}$ of $\overline{\mathcal{X}}$. Then π_2 is an admissible partition on $\overline{\mathcal{X}}$ in the ts $(\overline{\mathcal{X}}, S)$. Moreover, $\pi_2 \sqcap \underline{\mathcal{X}} = \{\{q_1\}, \{q_2\}\} \not\subseteq \pi_2$.

In order to define an admissible partition on a rough set $\mathcal{U}$ in $ts\ \mathcal{A} := (\mathcal{U}, S)$, we require a pair (π_1, π_2) such that π_2 is an admissible partition on the upper ts of $\mathcal{A}$, and π_1 $(\subseteq \pi_2)$ gives a partition on the lower ts. We have the following.

Definition 99 (Admissible partitions on a rough set).
Let $\mathcal{A} := (\mathcal{U}, S)$ be a ts for rough set $\mathcal{U} := (U, R, X)$ and $\pi_2 := \{H_i\}_{i \in I}$ be an admissible partition on $\overline{\mathcal{X}}$ in the upper ts $(\overline{\mathcal{X}}, S)$. Consider the quotient ts $(\pi_2, S/\sim)$, and let $\pi_1 := \{H_i \in \pi_2 \mid H_i \cap \underline{\mathcal{X}} \neq \emptyset\}$ be such that π_1 satisfies the condition:

$$\pi_1[s] := \{H_i * [s] \mid H_i \in \pi_1 \text{ and } H_i * [s] \text{ is defined}\} \subseteq \pi_1 \text{ for all } [s] \in S/\sim. \tag{4.3}$$

Then $\pi := (\pi_1, \pi_2)$ is termed an admissible partition on the rough set $\mathcal{U}$ in $\mathcal{A}$. An admissible partition π in $\mathcal{A}$ is non-trivial, if π_2 is non-trivial on $\overline{\mathcal{X}}$ in the upper ts $(\overline{\mathcal{X}}, S)$.

In case of ts, an admissible partition π_2 gives rise to a quotient ts $(\pi_2, S/\sim)$ (Definition 42). Given a ts $\mathcal{A} := (\mathcal{U}, S)$ for $\mathcal{U} := (U, R, X)$ and an admissible partition $\pi := (\pi_1, \pi_2)$ on $\mathcal{U}$ in $\mathcal{A}$, what could be the definition of a quotient ts for rough set with respect to π?
Let us define a collection $\mathcal{C}_\pi$ of subsets of U as:

$$\mathcal{C}_\pi := \{[x]_R \in U/R \mid x \in U \backslash \overline{X}\} \cup \{\bigcup H_i \mid H_i \in \pi_2\}.$$

Then $\mathcal{C}_\pi$ is a partition of U as R is an equivalence relation on U and π_2 is a partition of $\overline{\mathcal{X}}$.
Let R_π be the equivalence relation on U corresponding to the partition $\mathcal{C}_\pi$. One may observe that for any $x \in U$, $[x]_{R_\pi}$ is a union of R-equivalence classes.
Now, consider the rough set $\mathcal{U}_\pi := (U, R_\pi, X_\pi)$, where X_π is defined as:

$$X_\pi := X \cup \left(\bigcup_{H_i \in \pi_1} (\bigcup\{[x]_R \in U/R \mid [x]_R \in H_i\}) \right).$$

Claim: (1) $\pi_2 \cong \overline{\mathcal{X}}_\pi$ and (2) $\pi_1 \cong \underline{\mathcal{X}}_\pi$.
Proof of claim: Define a map $\phi : \overline{\mathcal{X}}_\pi \to \pi_2$ as follows:

$$\text{For all } [x]_{R_\pi} \in \overline{\mathcal{X}}_\pi, \phi([x]_{R_\pi}) := H_i \text{ such that } [x]_R \in H_i.$$

Does such a H_i always exist? Given $[x]_{R_\pi} \in \overline{\mathcal{X}}_\pi$ i.e. $[x]_{R_\pi} \cap X_\pi \neq \emptyset$, there exists $z \in [x]_{R_\pi}$ such that $z \in X_\pi$. Using the definition of R_π and X_π, we have the following cases:

(A). xRz and $z \in X$, then $[x]_R = [z]_R \in \overline{\mathcal{X}}$. Since π_2 is a partition of $\overline{\mathcal{X}}$, there exists a H_i such that $[x]_R \in H_i$.

(B). xRz and $z \in (\bigcup_{H_i \in \pi_1} (\bigcup\{[x]_R \mid [x]_R \in H_i\}))$. This implies that $[x]_R = [z]_R \in (\bigcup_{H_i \in \pi_1} (\bigcup\{[x]_R \mid [x]_R \in H_i\}))$, and therefore $[x]_R \in H_i$ for some $H_i \in \pi_1$.

(C). $[x]_R \neq [z]_R$. Then there exists $H_i \in \pi_2$ such that $[x]_R, [z]_R \in H_i$ and $z \in X$.

(1). ϕ is well-defined follows from the fact that C_π is a partition.

ϕ is one-one: Let $[x]_{R_\pi}, [x']_{R_\pi} \in \overline{\mathcal{X}}_\pi$ such that $\phi([x]_{R_\pi}) = \phi([x']_{R_\pi})$, i.e. $[x]_R, [x']_R \in H_i$ for some $H_i \in \pi_2$. Then using definition of R_π, $[x]_{R_\pi} = [x']_{R_\pi}$.

ϕ is onto: Let $H_i \in \pi_2$. Since H_i is non-empty, there exists $x \in U$ such that $[x]_R \in H_i$. Since $H_i \in \pi_2$ and π_2 is a partition of $\overline{\mathcal{X}}$, we have $[x]_R \in \overline{\mathcal{X}}$. This implies that there exists $z \in X$ such that $z \in [x]_R$. Therefore, $z \in X_\pi$ and $[z]_{R_\pi} \in \overline{\mathcal{X}}_\pi$. Finally, $[z]_R = [x]_R \in H_i$ and $[z]_{R_\pi} \in \overline{\mathcal{X}}_\pi$ imply $\phi([z]_{R_\pi}) = H_i$. Therefore, ϕ is a bijection.

(2). ϕ restricts $\underline{\mathcal{X}}_\pi$ to π_1: let $[x]_{R_\pi} \in \underline{\mathcal{X}}_\pi$ and $\phi([x]_{R_\pi}) = H_i$, i.e. $[x]_{R_\pi} \subseteq X_\pi$ and $[x]_R \in H_i \in \pi_2$. We have to show that $H_i \in \pi_1$.

$[x]_R \subseteq [x]_{R_\pi}$ implies $[x]_R \subseteq X_\pi$. Note that if $[x]_R \cap (\bigcup_{H_i \in \pi_1} (\bigcup\{[x]_R \in U/R \mid [x]_R \in H_i\})) \neq \emptyset$ then $[x]_R \subseteq (\bigcup_{H_i \in \pi_1} (\bigcup\{[x]_R \in U/R \mid [x]_R \in H_i\}))$. Therefore, either $[x]_R \subseteq X$ or $[x]_R \subseteq (\bigcup_{H_i \in \pi_1} (\bigcup\{[x]_R \in U/R \mid [x]_R \in H_i\}))$.

Case 1: $[x]_R \subseteq X$, i.e. $[x]_R \in H_i \cap \underline{\mathcal{X}}$, implying $H_i \in \pi_1$.

Case 2: $[x]_R \subseteq (\bigcup_{H_i \in \pi_1} (\bigcup\{[x]_R \in U/R \mid [x]_R \in H_i\}))$, i.e. there exists $H_t \in \pi_1$ such that $[x]_R \in H_t$. Since π_2 is a partition of $\overline{\mathcal{X}}$, $H_t = H_i$ and thus $H_i \in \pi_1$.

Now, $\pi_2 \cong \overline{\mathcal{X}}_\pi$ and $(\pi_2, S/\sim)$ is a ts, by Observation 7. So, we have a ts $(\overline{\mathcal{X}}_\pi, S/\sim)$ such that $(\overline{\mathcal{X}}_\pi, S/\sim) \cong (\pi_2, S/\sim)$. Condition 4.3 ensures that

$$\underline{\mathcal{X}}_\pi[s] = \{[x]_{R_\pi} * [s] \mid [x]_{R_\pi} \in \underline{\mathcal{X}}_\pi \text{ and } [x]_{R_\pi} * [s] \text{ is defined}\} \subseteq \underline{\mathcal{X}}_\pi \text{ for all } [s] \in S/\sim$$

and $(\underline{\mathcal{X}}_\pi, (S/\sim)/\sim)$ is also a ts. We thus obtain a ts $(\mathcal{U}_\pi, S/\sim)$ for the rough set $\mathcal{U}_\pi$, with upper ts $(\overline{\mathcal{X}}_\pi, S/\sim)$ and lower ts $(\underline{\mathcal{X}}_\pi, (S/\sim)/\sim)$. We give the definition for quotients as follows.

Definition 100 (Quotients).
Let $\mathcal{A} := (\mathcal{U}, S)$ *be a ts for rough set* $\mathcal{U} := (U, R, X)$ *and* $\pi := (\pi_1, \pi_2)$ *be an admissible partition on rough set* $\mathcal{U}$ *in* $\mathcal{A}$. *The quotient of* $\mathcal{A}$ *with respect to* π *is the ts* $\mathcal{A}/\langle \pi \rangle := (\mathcal{U}_\pi, S/\sim)$ *for the rough set* $\mathcal{U}_\pi := (U, R_\pi, X_\pi)$, *where* R_π *is the equivalence relation on* U *corresponding to the partition* C_π *on* U *given as*

$$C_\pi := \{[x]_R \in U/R \mid x \in U \backslash \overline{X}\} \cup \{\bigcup H_i \mid H_i \in \pi_2\},$$

and $X_\pi := X \cup (\bigcup_{H_i \in \pi_1} (\bigcup\{[x]_R \in U/R \mid [x]_R \in H_i\})).$

Note:

(1). Based on the above definition and the discussion preceding it, we shall iden-
tify the sets $\overline{\mathcal{X}_\pi}$ and $\underline{\mathcal{X}_\pi}$ with π_2 and π_1 respectively, and the upper ts $(\overline{\mathcal{X}_\pi}, S/\sim)$
and the lower ts $(\underline{\mathcal{X}_\pi}, (S/\sim)/\sim)$ in $\mathcal{A}/\langle\pi\rangle$ with $(\pi_2, S/\sim)$ and $(\pi_1, (S/\sim)/\sim)$
respectively.

(2). π_1 need not be a partition on $\underline{\mathcal{X}}$: take Example 14 for $\mathcal{U}$ and the reset ts $\mathcal{A} :=$
$(\mathcal{U}, S)$ $(=(\mathbf{2}, \mathbf{4}))$ such that $S = \emptyset$. $\pi_2 := \{\{q_1, q_3\}, \{q_2\}, \{q_4\}\}$ is still an admissible partition on $\overline{\mathcal{X}}$ in the upper ts of $\mathcal{A}$. Observe that $\pi_1 := \{\{q_1, q_3\}, \{q_2\}\}$ is
not a partition of $\underline{\mathcal{X}}$ and π_1 satisfies Condition 4.3 vacuously.
However, π_1 induces an admissible partition $\pi_1 \sqcap \underline{\mathcal{X}}$ on $\underline{\mathcal{X}}$ in the lower ts $(\underline{\mathcal{X}}, S/\sim)$
of $\mathcal{A}$, as shown in Proposition 46.

Definition 101 (Orthogonal partitions on a rough set).
*Consider a ts $\mathcal{A} := (\mathcal{U}, S)$ for a rough set $\mathcal{U} := (U, R, X)$. A non-trivial admis-
sible partition $\pi := (\pi_1, \pi_2)$ on $\mathcal{U}$ in $\mathcal{A}$ is called* orthogonal *if there exists a
non-trivial admissible partition $\tau := (\tau_1, \tau_2)$ on $\mathcal{U}$ in $\mathcal{A}$ such that $\pi_2 \cap \tau_2 = 1_{\overline{\mathcal{X}}}$
and $\pi_1 \cap \tau_1 = 1_{\underline{\mathcal{X}}}$.*

It is clear that τ is also orthogonal.

Example 15. Consider the ts $\mathcal{A}$ for rough set of Example 9. Define a partition
π_2 on $\overline{\mathcal{X}}$ as $\pi_2 := \{\{1\}, \{2, 3, 4\}, \{5, 6, 7\}\}$. Let $H_1 := \{1\}$, $H_2 := \{2, 3, 4\}$ and
$H_3 := \{5, 6, 7\}$. Here,

(1) $H_1 s_i \subseteq H_1$ for $i = 1, \ldots, 7$.
(2) $H_2 s_i \subseteq H_2$ for $i = 1, 2$, $H_2 s_i \subseteq H_3$ for $i = 3, \ldots, 6$, and $H_2 s_7 \subseteq H_1$.
(3) $H_3 s_i \subseteq H_3$ for $i = 1, 2$, $H_3 s_i \subseteq H_1$ for $i = 3, \ldots, 7$.

So, π_2 is an admissible partition on $\overline{\mathcal{X}}$ in the ts $(\overline{\mathcal{X}}, S)$. In the quotient ts
$(\pi_2, S/\sim)$, the semigroup $S/\sim := \{\{s_1, s_2\}, \{s_3, s_4, s_5, s_6\}, \{s_7\}\}$ and $\pi_1 :=$
$\{H_i \in \pi_2 \mid H_i \cap \underline{\mathcal{X}} \neq \emptyset\} = \{\{1\}, \{5, 6, 7\}\}$. For $[s_1] \in S/\sim$, $\pi_1[s_1] = \pi_1$, while
$\pi_1[s_i] = \{\{1\}\} \subseteq \pi_1$ for $i = 3, 7$. Therefore $\pi := (\pi_1, \pi_2)$ is an admissible parti-
tion on $\mathcal{U}$ in $\mathcal{A}$.

Consider another partition τ_2 on $\overline{\mathcal{X}}$ as $\tau_2 := \{\{1, 7\}, \{2\}, \{3\}, \{4\}, \{5\}, \{6\}\}$. Let
$H_1 := \{1, 7\}$, $H_2 := \{2\}$, $H_3 := \{3\}$, $H_4 := \{4\}$, $H_5 := \{5\}$, and $H_6 := \{6\}$.
Here,

(1) $H_1 s_i \subseteq H_1$ for $i = 1, \ldots, 7$.
(2) $H_2 s_i \subseteq H_2$ for $i = 1, 2$, $H_2 s_i \subseteq H_5$ for $i = 3, 4$, $H_2 s_i \subseteq H_6$ for $i = 5, 6$, and
$H_2 s_7 \subseteq H_1$.
(3) $H_3 s_1 \subseteq H_3$, $H_3 s_2 \subseteq H_4$, $H_3 s_3 \subseteq H_6$, $H_3 s_4 \subseteq H_1$, $H_3 s_5 \subseteq H_5$, and $H_3 s_i \subseteq$
H_1 for $i = 6, 7$.
(4) $H_4 s_1 \subseteq H_4$, $H_4 s_2 \subseteq H_3$, $H_4 s_3 \subseteq H_1$, $H_4 s_4 \subseteq H_6$, $H_4 s_5 \subseteq H_1$, $H_4 s_6 \subseteq H_5$,
and $H_4 s_7 \subseteq H_1$.
(5) $H_5 s_1 \subseteq H_5$, $H_5 s_2 \subseteq H_6$, and $H_5 s_i \subseteq H_1$ for $i = 3, \ldots, 7$.
(6) $H_6 s_1 \subseteq H_6$, $H_6 s_2 \subseteq H_5$, and $H_5 s_i \subseteq H_1$ for $i = 3, \ldots, 7$.

So, τ_2 is an admissible partition on $\overline{\mathcal{X}}$ in the *ts* $(\overline{\mathcal{X}}, S)$. Moreover, for every $s_i, s_j \in S$ $(i \neq j)$, there exists a H_k in τ_2 such that $H_k * s_i \neq H_k * s_j$. So, in the quotient *ts* $(\tau_2, S/\sim)$, the semigroup $S/\sim \cong S$. Now, $\tau_1 := \{H_i \in \tau_2 \mid H_i \cap \underline{\mathcal{X}} \neq \emptyset\} = \{\{1,7\}, \{5\}, \{6\}\}$. For $[s_1] \in S/\sim$, $\tau_1[s_i] = \tau_1$ for $i = 1, 2$, and $\tau_1[s_i] = \{\{1,7\}\} \in \tau_1$ for $i = 3, \ldots, 7$. Therefore $\tau := (\tau_1, \tau_2)$ is an admissible partition on $\mathcal{U}$ in $\mathcal{A}$.

Moreover, $\pi_2 \cap \tau_2 = 1_{\overline{\mathcal{X}}}$ and $\pi_1 \cap \tau_1 = 1_{\underline{\mathcal{X}}}$. Therefore π and τ are orthogonal partitions on $(\underline{\mathcal{X}}, \overline{\mathcal{X}})$ in $\mathcal{A}$.

We now come to the last definition in this section. If $\mathcal{A}$ and $\mathcal{B}$ are *ts* for rough sets, a covering of $\mathcal{A}$ by $\mathcal{B}$ should result in two coverings (cf. Definition 44): one of the upper *ts* of $\mathcal{A}$ by the upper *ts* of $\mathcal{B}$ and another of the lower *ts* of $\mathcal{A}$ by the lower *ts* of $\mathcal{B}$.

Definition 102 (Coverings).
A ts $\mathcal{A} := (\mathcal{U}, S)$ for rough set $\mathcal{U} := (U, R, X)$ is covered by ts $\mathcal{B} := (\mathcal{V}, T)$ for rough set $\mathcal{V} := (V, R', Y)$, written as $\mathcal{A} \preccurlyeq \mathcal{B}$, if there exists a surjective partial function $\eta : \overline{\mathcal{Y}} \rightarrow \overline{\mathcal{X}}$ such that

(a) η restricts $\underline{\mathcal{Y}}$ onto $\underline{\mathcal{X}}$, that is $\eta(\underline{\mathcal{Y}}) = \underline{\mathcal{X}}$, and
(b) η is a covering of $(\overline{\mathcal{X}}, S)$ by $(\overline{\mathcal{Y}}, T)$.

η is called a covering of $\mathcal{A}$ by $\mathcal{B}$, or $\mathcal{B}$ is said to cover $\mathcal{A}$ by η. t_s is said to cover s.

Proposition 47. *If η is a covering of $\mathcal{A} := (\mathcal{U}, S)$ by $\mathcal{B} := (\mathcal{V}, T)$, where $\mathcal{U} := (U, R, X)$ and $\mathcal{V} := (V, R', Y)$, $\eta|_{\underline{\mathcal{Y}}}$ is a covering of the lower ts $(\underline{\mathcal{X}}, S/\sim)$ of $\mathcal{A}$ by the lower ts $(\underline{\mathcal{Y}}, T/\sim)$ of $\mathcal{B}$.*

Proof. By Definition 102, $\eta|_{\underline{\mathcal{Y}}} : \underline{\mathcal{Y}} \rightarrow \underline{\mathcal{X}}$ is a surjective partial function. Let $[s] \in S/\sim$ such that $\eta|_{\underline{\mathcal{Y}}}(p)[s]$ is defined for $p \in \underline{\mathcal{Y}}$. Since η is a covering of $(\overline{\mathcal{X}}, S)$ by $(\overline{\mathcal{Y}}, T)$ and $p \in \overline{\mathcal{Y}}$, there exists $t_s \in T$ such that $\eta(p)s = \eta(pt_s)$. Consider the equivalence class $[t_s] \in T/\sim$ in *ts* $(\underline{\mathcal{Y}}, T/\sim)$. For $p \in \underline{\mathcal{Y}}$, $\eta|_{\underline{\mathcal{Y}}}(p)[s]$ is defined $\Rightarrow \eta(p)s$ is defined and $\eta|_{\underline{\mathcal{Y}}}(p)[s] = \eta(p)[s] = \eta(p)s = \eta(pt_s) = \eta(p[t_s]) = \eta|_{\underline{\mathcal{Y}}}(p[t_s])$.

We now prove the following results that will be helpful in the study of decomposition theorems for *ts* for rough sets. Recall Definition 92 of *tm* $\mathcal{A}^{\cdot} := (\mathcal{U}, S^{\cdot})$.

Proposition 48. *Let $\mathcal{A}, \mathcal{B}, \mathcal{C}, \mathcal{D}$ be ts for rough sets.*

1. *$\mathcal{A} \preccurlyeq \overline{\mathcal{A}} = \overline{(\mathcal{A})}$.*
2. *For $\mathcal{A} \preccurlyeq \mathcal{B}$, we have $\mathcal{A}^{\cdot} \preccurlyeq \mathcal{B}^{\cdot}$.*
3. *If $\mathcal{A} \preccurlyeq \mathcal{B}$ then $\overline{\mathcal{A}} \preccurlyeq \overline{\mathcal{B}}$.*
4. *If $\mathcal{A} \preccurlyeq \mathcal{C}$ and $\mathcal{B} \preccurlyeq \mathcal{D}$ then $\mathcal{A} \times \mathcal{B} \preccurlyeq \mathcal{C} \times \mathcal{D}$.*
5. *For $\mathcal{A} \preccurlyeq \mathcal{C}$ and $\mathcal{B} \preccurlyeq \mathcal{D}$, we have $\mathcal{A} \circ \mathcal{B} \preccurlyeq \mathcal{C} \circ \mathcal{D}$.*
6. *If $\mathcal{A} \preccurlyeq \mathcal{B}$ and $\mathcal{B} \preccurlyeq \mathcal{C}$ then $\mathcal{A} \preccurlyeq \mathcal{C}$.*
7. *$\mathcal{A} \times \mathcal{B} \preccurlyeq \mathcal{A} \circ \mathcal{B}$.*

8. $A \preccurlyeq A^{\bullet}$.

9. $(A \circ B)^{\bullet} \preccurlyeq A^{\bullet} \circ B^{\bullet}$.

Proof. Let $A := (\mathcal{U}_a, S_a)$, $B := (\mathcal{U}_b, S_b)$, $C := (\mathcal{U}_c, S_c)$ and $D := (\mathcal{U}_d, S_d)$, where $\mathcal{U}_i := (U_i, R_i, X_i)$ for $i = a, b, c, d$.

(1). The cover map is $\eta = 1_{\overline{\mathcal{X}}_a}$ and for any $s \in S_a$, take $t_s := s$. η restricts $\underline{\mathcal{X}}_a$ to $\underline{\mathcal{X}}_a$ and t_s covers s.

(2). Let η be the covering of A by B. Then η is a covering of $A^{\bullet}$ by $B^{\bullet}$ also. $1_{\overline{\mathcal{X}}_a} \in S_a$ is covered by $1_{\overline{\mathcal{X}}_b} \in S_b$.

(3). Since $A \preccurlyeq B$, there is a covering η of A by B and for any $s \in S_a$, there exists a $t_s \in S_b$ such that t_s covers s. We claim that η is also a covering of $\overline{A}$ by $\overline{B}$. The semigroups associated with $\overline{A}$ and $\overline{B}$ are

$$S_a' := \langle S_a \cup \{\overline{q} \mid q \in \underline{\mathcal{X}}_a\} \cup \{\widetilde{q} \mid q \in \overline{\mathcal{X}}_a \backslash \underline{\mathcal{X}}_a\}\rangle, \text{ and}$$
$$S_b' := \langle S_b \cup \{\overline{q} \mid q \in \underline{\mathcal{X}}_b\} \cup \{\widetilde{q} \mid q \in \overline{\mathcal{X}}_b \backslash \underline{\mathcal{X}}_b\}\rangle \text{ respectively.}$$

Note that $\eta : \overline{\mathcal{X}}_b \to \overline{\mathcal{X}}_a$ is surjective such that $\eta(\underline{\mathcal{X}}_b) = \underline{\mathcal{X}}_a$.

For $s = \overline{q} \in S_a'$ (so that $q \in \underline{\mathcal{X}}_a$), choose $q' \in \eta^{-1}(q)$. $q' \in \underline{\mathcal{X}}_b$. So take $t_s := \overline{q'} \in S_b'$.

For $s = \widetilde{q} \in S_a'$ (so that $q \in \overline{\mathcal{X}}_a \backslash \underline{\mathcal{X}}_a$), choose $q' \in \eta^{-1}(q)$.

(a) If $q' \in \underline{\mathcal{X}}_b$, take $t_s := \overline{q'} \in S_b'$; (b) If $q' \in \overline{\mathcal{X}}_b \backslash \underline{\mathcal{X}}_b$, take $t_s := \widetilde{q'} \in S_b'$.

(4). Let η and ν be the coverings that give $A \preccurlyeq C$ and $B \preccurlyeq D$ respectively. Consider the map $(\eta, \nu) : \overline{\mathcal{X}}_c \times \overline{\mathcal{X}}_d \to \overline{\mathcal{X}}_a \times \overline{\mathcal{X}}_b$ defined as $(\eta, \nu)(a, b) := (\eta(a), \nu(b))$ for any $(a, b) \in \overline{\mathcal{X}}_c \times \overline{\mathcal{X}}_d$. It is a surjective partial map and restricts $\underline{\mathcal{X}}_c \times \underline{\mathcal{X}}_d$ to $\underline{\mathcal{X}}_a \times \underline{\mathcal{X}}_b$. Further, any $(s, s') \in S_a \times S_b$ is covered by $t_{(s,s')} := (t_s, t_{s'}) \in S_c \times S_d$. So (η, ν) is a covering of $\overline{\mathcal{X}}_a \times \overline{\mathcal{X}}_b$ by $\overline{\mathcal{X}}_c \times \overline{\mathcal{X}}_d$.

(5). The same map (η, ν) as above is the covering that gives $A \circ B \preccurlyeq C \circ D$: any $(f, s) \in (S_a)^{\overline{\mathcal{X}}_b} \times S_b$ is covered by $t_{(s,s')} := (f', t_s) \in (S_c)^{\overline{\mathcal{X}}_d} \times S_d$, where $f' : \overline{\mathcal{X}}_d \to S_c$ is defined as $f'(q') = t_{f(\nu(q'))}$, for all $q' \in \overline{\mathcal{X}}_d$.

Now, for $(a, b) \in \overline{\mathcal{X}}_c \times \overline{\mathcal{X}}_d$, whenever $(\eta, \nu)((a, b))(f, s)$ is defined,

$$
\begin{aligned}
(\eta, \nu)((a, b)(f', t_s)) &= (\eta, \nu)((a f'(b), b t_s)) \\
&= (\eta, \nu)((a t_{f(\nu(b))}, b t_s)) \\
&= (\eta(a t_{f(\nu(b))}), \nu(b t_s)) \\
&= (\eta(a) f(\nu(b)), \nu(b) s) \\
&= (\eta(a), \nu(b))(f, s) \\
&= (\eta, \nu)((a, b))(f, s).
\end{aligned}
$$

(6). If η and ν are the coverings that give $A \preccurlyeq B$ and $B \preccurlyeq C$ respectively, then $\eta\nu$ is the covering of A by C. For $q \in \underline{\mathcal{X}}_a$, there exists $q' \in \underline{\mathcal{X}}_b$ such that $\eta(q') = q$. For $q' \in \underline{\mathcal{X}}_b$, there exists $q'' \in \underline{\mathcal{X}}_c$ such that $\nu(q'') = q'$. So, $\eta(\nu(q'')) = q$.

For $s \in S_a$, take $t_s := r$, where r is the cover of r' with respect to ν and r' is the cover of s with respect to η. Now, for $q \in \overline{\mathcal{X}}_c$, $\nu(\eta(qr)) = \eta(\nu(q)r') = \eta(\nu(q))s$.

(7). The covering is $1_{\overline{\mathcal{X}}_a \times \overline{\mathcal{X}}_b}$. Any $(s, s') \in (S_a, S_b)$ is covered by $t_{(s,s')} := (f, s') \in ((S_a)^{\overline{\mathcal{X}}_b}, S_b)$, where $f : \overline{\mathcal{X}}_b \to S_a$ is the constant map defined as $f(q') = s$ for all $q' \in \overline{\mathcal{X}}_b$. For $(a,b) \in \overline{\mathcal{X}}_a \times \overline{\mathcal{X}}_b$, whenever $(a,b)(s,s')$ is defined, $(a,b)(f,s') = (af(b), bs') = (as, bs') = (a,b)(s,s')$.

(8). The covering is $1_{\overline{\mathcal{X}}_a}$ and for any $s \in S_a$, take $t_s := s$.

(9) The semigroup in $(\mathcal{A} \circ \mathcal{B})^{\bullet}$ and $\mathcal{A}^{\bullet} \circ \mathcal{B}^{\bullet}$ are S' and $(S'_a)^{\overline{\mathcal{X}}_b} \times S'_b$ respectively, where $S' := ((S_a)^{\overline{\mathcal{X}}_b} \times S_b) \cup \{1_{\overline{\mathcal{X}}_a \times \overline{\mathcal{X}}_b}\}$, $S'_a := S_a \cup \{1_{\overline{\mathcal{X}}_a}\}$ and $S'_b := S_b \cup \{1_{\overline{\mathcal{X}}_b}\}$. We show that the identity map $1_{\overline{\mathcal{X}}_a \times \overline{\mathcal{X}}_b}$ is a covering. For every $\alpha \in S'$, we have to find $t_\alpha \in (S'_a)^{\overline{\mathcal{X}}_b} \times S'_b$ such that for every $(q',p') \in (\overline{\mathcal{X}}_a \times \overline{\mathcal{X}}_b)$, if $(q',p')\alpha$ is defined then $(q',p')\alpha = (q',p')t_\alpha$. We have two cases.

(a). $\alpha := (f,s) \in (S_a)^{\overline{\mathcal{X}}_b} \times S_b$. Take $t_\alpha := (if, s)$, where $i : S_a \to S'_a$ is the inclusion map. For $(q',p') \in (\overline{\mathcal{X}}_a \times \overline{\mathcal{X}}_b)$, we have $(q',p')t_\alpha = (q',p')(if, s) = (q'i(f(p')), p's') = (q'f(p'), p's') = (q',p')(f,s) = (q',p')\alpha$.

(b). $\alpha := 1_{\overline{\mathcal{X}}_a \times \overline{\mathcal{X}}_b}$. Take $t_\alpha := (h, 1_{\overline{\mathcal{X}}_b})$, where $h : \overline{\mathcal{X}}_b \to S'_a$ is such that $h(p) := 1_{\overline{\mathcal{X}}_a}$ for all $p \in \overline{\mathcal{X}}_b$. For $(q',p') \in (\overline{\mathcal{X}}_a \times \overline{\mathcal{X}}_b)$, we have $(q',p')t_\alpha = (q',p')(h, 1_{\overline{\mathcal{X}}_b}) = (q'h(p'), p') = (q',p') = (q',p')1_{\overline{\mathcal{X}}_a \times \overline{\mathcal{X}}_b} = (q',p')\alpha$.

Remark 11. One of the important results of ts theory that fails in the ts theory for rough sets is that the closure may not distribute over products, i.e. one can obtain two ts for rough sets $\mathcal{A}$ and $\mathcal{B}$ such that $\overline{\mathcal{A} \times \mathcal{B}} \not\preceq \overline{\mathcal{A}} \times \overline{\mathcal{B}}$ or $\overline{\mathcal{A} \circ \mathcal{B}} \not\preceq \overline{\mathcal{A}} \circ \overline{\mathcal{B}}$. Consider ts $\mathcal{A} = \mathcal{B} := (\mathbf{1}, \mathbf{2})$ for a rough set $\mathcal{U} := (U, R, X)$, where $\overline{\mathcal{X}} := \{q_1, q_2\}$ and $\underline{\mathcal{X}} := \{q_1\}$. $\overline{\mathcal{X}} \times \overline{\mathcal{X}} = \{(q_1,q_1), (q_1,q_2), (q_2,q_1), (q_2,q_2)\}$ and $\underline{\mathcal{X}} \times \underline{\mathcal{X}} = \{(q_1,q_1)\}$. The semigroup associated with the ts $\overline{\mathcal{A} \times \mathcal{B}}$ is

$$S' := \{\widetilde{(q_1,q_1)}, \widetilde{(q_1,q_1)}, \widetilde{(q_1,q_2)}, \widetilde{(q_2,q_1)}, \widetilde{(q_2,q_2)}, \emptyset_{(\overline{\mathcal{X}} \times \overline{\mathcal{X}})}\},$$

and the semigroup associated with the ts $\overline{\mathcal{A}} \times \overline{\mathcal{B}}$ is $S_a \times S_b$, where $S_a = S_b = \{\overline{q}_1, \widetilde{q}_1, \widetilde{q}_2, \emptyset_{\overline{\mathcal{X}}}\}$.

If possible, suppose there exists a surjective partial function $\eta : \overline{\mathcal{X}} \times \overline{\mathcal{X}} \to \overline{\mathcal{X}} \times \overline{\mathcal{X}}$ such that η is a covering of $\overline{\mathcal{A} \times \mathcal{B}}$ by $\overline{\mathcal{A}} \times \overline{\mathcal{B}}$. Note the following two points.

(1). η restricts $\underline{\mathcal{X}} \times \underline{\mathcal{X}}$ to itself. So, it maps (q_1, q_1) to (q_1, q_1).

(2). η is a bijection.

Now, since η is a covering, for every $s \in S'$, there exists a $t_s \in S_a \times S_b$ such that for any $(x,y) \in \overline{\mathcal{X}} \times \overline{\mathcal{X}}$, if $\eta((x,y))s$ is defined then $\eta((x,y))s = \eta((x,y)t_s)$. Let $s = (q_1, q_2)$, then $\eta((q_1,q_2))s = \eta((q_2,q_1))s = \eta((q_2,q_2))s = (q_1, q_2)$. Therefore, there exists $t_s = (x', y') \in S_a \times S_b$ such that $(q_1, q_2)t_s$, $(q_2, q_1)t_s$ and $(q_2, q_2)t_s$ are defined. Since t_s acts on $\overline{\mathcal{X}} \times \overline{\mathcal{X}}$ component-wise, each of q_1x', q_2x', q_1y', q_2y' should be defined. So, the only choice for x' and y' both is $\overline{q}_1$, i.e. $t_s := (\overline{q}_1, \overline{q}_1)$. However, $\eta((q_1,q_2)t_s) = \eta(q_1\overline{q}_1, q_2\overline{q}_1) = \eta((q_1,q_1)) = (q_1, q_1) \neq (q_1, q_2) = \eta((q_1,q_2)s)$. Therefore, no such covering η can exist. For the same $\mathcal{A}$ and $\mathcal{B}$, we can show that $\overline{\mathcal{A} \circ \mathcal{B}} \not\preceq \overline{\mathcal{A}} \circ \overline{\mathcal{B}}$.

Proposition 49. *A reset ts $\mathcal{A} := (\mathcal{U}, S)$ is covered by the reset ts $\overline{(\mathbf{m}, \mathbf{n})}$, where $\mathcal{U} := (U, R, X)$ and $|\underline{\mathcal{X}}| = m \leq n = |\overline{\mathcal{X}}|$.*

Proof. The covering η will be $1_{\overline{\mathcal{X}}}$, and then we argue for the two cases obtained by Observation 27: (1) $\underline{\mathcal{X}} = \emptyset$ or $\underline{\mathcal{X}} = \overline{\mathcal{X}}$, and (2) $\underline{\mathcal{X}} \neq \emptyset$, $\underline{\mathcal{X}} \neq \overline{\mathcal{X}}$.

Case 1: $\underline{\mathcal{X}} = \emptyset$ or $\underline{\mathcal{X}} = \overline{\mathcal{X}}$. The semigroup in $\overline{(\mathbf{m}, \mathbf{n})}$ is $S' = \{\overline{q} \mid q \in \overline{\mathcal{X}}\}$. For any

$s \in S$, (a) if $\overline{\mathcal{X}}s = \emptyset$, choose any $t_0 \in S'$ and take $t_s := t_0$; and (b) if $\overline{\mathcal{X}}s = \{q\}$, take $t_s := \overline{q} \in S'$.

Case 2: $\underline{\mathcal{X}} \neq \emptyset$ and $\underline{\mathcal{X}} \neq \overline{\mathcal{X}}$. As seen in Example 13, the semigroup in $\overline{(\mathbf{m}, \mathbf{n})}$ is $S' = \{\overline{q} \mid q \in \underline{\mathcal{X}}\} \cup \{\widetilde{q} \mid q \in \overline{\mathcal{X}}\} \cup \{\emptyset_{\overline{\mathcal{X}}}\}$. For any $s \in S$, if $\overline{\mathcal{X}}s = \emptyset$, take $t_s := \emptyset_{\overline{\mathcal{X}}} \in S'$. Now suppose $\overline{\mathcal{X}}s = \{q\}$. If $q \in \underline{\mathcal{X}}$, take $t_s := \overline{q} \in S'$; and if $q \in \overline{\mathcal{X}} \backslash \underline{\mathcal{X}}$, take $t_s := \widetilde{q} \in S'$. Observe that $\underline{\mathcal{X}}s \subseteq \underline{\mathcal{X}}$ implies $\underline{\mathcal{X}}s = \emptyset$.

4.3   Decomposition Theorems

We have mentioned decomposition results of ts in Proposition 16 and Theorem 5. Our goal here is to study decomposition results of the above kind in the case of a ts for rough sets. So we shall look for coverings of $\mathcal{A} := (\mathcal{U}, S)$ for rough set $\mathcal{U} := (U, R, X)$ by products of the closure of non-decomposable and smaller ts $\mathcal{A}_i := (\mathcal{U}_i, S_i)$ for rough sets $\mathcal{U}_i := (U_i, R_i, X_i)$, $i \in I$. In ts for rough sets, we say $(\mathcal{U}_1, S_1)$ is 'smaller' than $(\mathcal{U}_2, S_2)$, where $\mathcal{U}_i := (U, R, X)$, $i = 1, 2$, when $(|S_1|, |\overline{\mathcal{X}}_1|, |\underline{\mathcal{X}}_1|) < (|S_2|, |\overline{\mathcal{X}}_2|, |\underline{\mathcal{X}}_2|)$, where the order '$<$' is lexicographical. We present our first result in this direction here, for the special case of reset ts for rough sets. Any reset ts $\mathcal{A} := (\mathcal{U}, S)$ for rough set $\mathcal{U} := (U, R, X)$ with $\overline{\mathcal{X}} \neq \emptyset$, can be covered by the direct product of either the resets $\overline{(\mathbf{0}, \mathbf{2})}$, or $\overline{(\mathbf{2}, \mathbf{2})}$ and $\overline{(\mathbf{1}, \mathbf{n})}$ for some n. Formally, we have

Theorem 16. *For a reset ts $\mathcal{A} := (\mathcal{U}, S)$ for rough set $\mathcal{U} := (U, R, X)$ with $|\overline{\mathcal{X}}| = n$ ($\neq 0$) and $|\underline{\mathcal{X}}| = m$,*

$$(1).\ \text{for } m \geq 1, (\mathcal{U}, S) \preccurlyeq \prod^{m-1} \overline{(\mathbf{2}, \mathbf{2})} \times \overline{(\mathbf{1}, \mathbf{n} - \mathbf{m} + \mathbf{1})}.$$

$$(2).\ \text{for } m = 0, (\mathcal{U}, S) \preccurlyeq \prod^{n-1} \overline{(\mathbf{0}, \mathbf{2})},$$

where $\prod^k \mathcal{A}_i$ denotes the direct product of ts $\mathcal{A}_i$ for rough sets, $i = 1, \ldots, k$.

Proof. Since $|\overline{\mathcal{X}}| = n$ is finite, let us enumerate the elements $\{q_i\}_{i=1}^n$ of $\overline{\mathcal{X}}$ such that the first m elements belong to $\underline{\mathcal{X}}$, i.e. $\underline{\mathcal{X}} = \{q_i\}_{i=1}^m$. Using Proposition 49,

$$\mathcal{A} \preccurlyeq \mathcal{B} := \overline{(\mathbf{m}, \mathbf{n})}. \tag{4.4}$$

We shall focus on the reset $\mathcal{B}$. By Example 13, the semigroup associated with the ts $\mathcal{B}$ is as follows:

$$S' = \{\overline{q_i} \mid q_i \in \underline{\mathcal{X}}\} \cup \{\widetilde{q_i} \mid q_i \in \overline{\mathcal{X}}\} \cup \{\emptyset_{\overline{\mathcal{X}}}\}.$$

<u>Case 1</u>: $|\underline{\mathcal{X}}| = 2$, i.e. $\underline{\mathcal{X}} = \{q_1, q_2\}$, and $|\overline{\mathcal{X}}| > 2$. Define the following partitions on $\overline{\mathcal{X}}$.

$$\pi_2 := \{\ \{q_1\}, \overline{\mathcal{X}} \backslash \{q_1\}\ \}$$
$$\tau_2 := \{\ \{q_1, q_2\}, \{q_3\}, \{q_4\}, \{q_5\}, \ldots, \{q_n\}\ \}$$

Then $\pi := (\pi_1, \pi_2)$, where $\pi_1 = \pi_2 = \{H_i \in \pi_2 \mid H_i \cap \underline{\mathcal{X}} \neq \emptyset\}$, is an admissible partition on $\mathcal{U}$ in $\mathcal{A}$. Recall Definition 100 for quotient ts for rough sets. In the quotient ts $\mathcal{B}/\langle \pi \rangle := (\mathcal{U}_\pi, S'/\sim_{\pi_2})$ for the rough set $\mathcal{U}_\pi := (U, R_\pi, X_\pi)$, the semigroup

$$S'/\sim_{\pi_2} = \{ \{\widetilde{q_1}\}, \{\widetilde{q_2}\}, \{\widetilde{q_1}\}, \{\widetilde{q_i} \mid 2 \leq i \leq n\}, \{\emptyset_{\overline{\mathcal{X}}}\} \}.$$

Here, $(\pi_2, S'/\sim_{\pi_2})$ is a reset ts. So, $\mathcal{B}/\langle \pi \rangle$ is a reset ts. Proposition 49 implies

$$\mathcal{B}/\langle \pi \rangle \preccurlyeq \overline{(\mathbf{2}, \mathbf{2})}. \tag{4.5}$$

Now, $\tau_1 := \{K_i \in \pi_2 \mid K_i \cap \underline{\mathcal{X}} \neq \emptyset\} = \{\{q_1, q_2\}\}$ and $\tau_1[s] \subseteq \tau_1$ for all $[s] \in S'/\sim_{\pi_2}$. Thus, $\tau := (\tau_1, \tau_2)$ is also an admissible partition on $\mathcal{U}$. $\mathcal{B}/\langle \tau \rangle :=$ $(\mathcal{U}_\tau, S'/\sim_{\tau_2})$ for the rough set $\mathcal{U}_\tau := (U, R_\tau, X_\tau)$ is the quotient ts of $\mathcal{B}$ with respect to τ, where the semigroup is

$$S'/\sim_{\tau_2} = \{ \{\overline{q_1}, \overline{q_2}\}, \{\widetilde{q_1}, \widetilde{q_2}\}, \{\widetilde{q_i}\}_{i=3}^n, \{\emptyset_{\overline{\mathcal{X}}}\}\}.$$

In fact, π and τ are orthogonal admissible partitions on $\mathcal{U}$ because $\pi_2 \cap \tau_2 = 1_{\overline{\mathcal{X}}}$, and $\pi_1 \cap \tau_1 = 1_{\underline{\mathcal{X}}}$.

We claim that $\mathcal{B} \preccurlyeq \mathcal{B}/\langle \pi \rangle \times \mathcal{B}/\langle \tau \rangle$. Let us first show the following covering:

$$(\overline{\mathcal{X}}, S') \preccurlyeq (\pi_2, S'/\sim_{\pi_2}) \times (\tau_2, S'/\sim_{\tau_2}).$$

Define the map $\eta : \pi_2 \times \tau_2 \to \overline{\mathcal{X}}$ as follows: for $H_i \in \pi_2$ and $K_j \in \tau_2$,

$$\eta(H_i, K_j) := q_k \text{ if } H_i \cap K_j = \{q_k\}, \text{ and}$$
$$\eta(H_i, K_j) \text{ is not defined if } H_i \cap K_j = \emptyset.$$

- η is well-defined and onto because τ_2 is orthogonal to π_2.
- η is a covering of $(\overline{\mathcal{X}}, S')$ by $(\pi_2, S'/\sim_{\pi_2}) \times (\tau_2, S'/\sim_{\tau_2})$, where
  - $\overline{q_1} \in S'$ is covered by $([\overline{q_1}]_{\pi_2}, [\overline{q_1}]_{\tau_2}) \in S'/\sim_{\pi_2} \times S'/\sim_{\tau_2}$,
  - $\overline{q_2} \in S'$ is covered by $([\overline{q_2}]_{\pi_2}, [\overline{q_1}]_{\tau_2}) \in S'/\sim_{\pi_2} \times S'/\sim_{\tau_2}$, and
  - $\widetilde{q_i} \in S'$ is covered by $([\overline{q_2}]_{\pi_2}, [\widetilde{q_i}]_{\tau_2}) \in S'/\sim_{\pi_2} \times S'/\sim_{\tau_2}$ for all $3 \leq i \leq n$.

Moreover, η restricts $\pi_1 \times \tau_1$ to $\underline{\mathcal{X}}$. So η is a covering of $\mathcal{B}$ by $\mathcal{B}/\langle \pi \rangle \times \mathcal{B}/\langle \tau \rangle$, i.e.

$$\mathcal{B} \preccurlyeq \mathcal{B}/\langle \pi \rangle \times \mathcal{B}/\langle \tau \rangle. \tag{4.6}$$

Observe the following for the quotient ts $\mathcal{B}/\langle \tau \rangle$ for rough set $\mathcal{U}_\tau$.

1. $|\tau_1| = 1$, $|\tau_2| = |\overline{\mathcal{X}}| - 1$, and
2. $\mathcal{B}/\langle \tau \rangle$ is again a reset ts for rough sets, and

$$\mathcal{B}/\langle \tau \rangle \preccurlyeq \overline{(\mathbf{1}, \mathbf{n} - \mathbf{1})}. \tag{4.7}$$

Thus we have, using Propositions 48 (4) and (6) on Eqs. 4.4, 4.5, 4.6 and 4.7,

$$\mathcal{A} \preccurlyeq \mathcal{B} \preccurlyeq \mathcal{B}/\langle \pi \rangle \times \mathcal{B}/\langle \tau \rangle \preccurlyeq \overline{(\mathbf{2}, \mathbf{2})} \times \overline{(\mathbf{1}, \mathbf{n} - \mathbf{1})}. \tag{4.8}$$

<u>Case 2</u>: $|\underline{\mathcal{X}}| > 2$. $\{q_1, q_2, q_3\} \subseteq \underline{\mathcal{X}}$. We consider the following partitions on $\overline{\mathcal{X}}$.

$$\pi_2 := \{\ \{q_1, q_2\}, \overline{\mathcal{X}} \backslash \{q_1, q_2\}\ \},$$
$$\tau_2 := \{\ \{q_1, q_3\}, \{q_2\}, \{q_4\}, \{q_5\}, \ldots, \{q_n\}\ \}.$$

We have $\pi_1 = \{H_i \in \pi_2 \mid H_i \cap \underline{\mathcal{X}} \neq \emptyset\} = \pi_2$ and $\pi := (\pi_1, \pi_2)$ is an admissible partition on $(\underline{\mathcal{X}}, \overline{\mathcal{X}})$. The semigroup in the upper ts of the quotient $\mathcal{B}/\langle\pi\rangle$ is

$$S'/{\sim}_{\pi_2} = \{\ \{\overline{q}_1, \overline{q}_2\},\ \{\widetilde{q}_1, \widetilde{q}_2\},\ \{\widetilde{q}_i \mid 3 \leq i \leq n\},\ \{\overline{q}_i \mid 3 \leq i \leq m\},\ \{\emptyset_{\overline{\mathcal{X}}}\}\},$$

and as $\mathcal{B}/\langle\pi\rangle$ is a reset ts, using Proposition 49,

$$\mathcal{B}/\langle\pi\rangle \preccurlyeq \overline{(\mathbf{2}, \mathbf{2})}. \tag{4.9}$$

The semigroup in the upper ts of the quotient $\mathcal{B}/\langle\tau\rangle$ is

$$S'/{\sim}_{\tau_2} = \{\ \{\overline{q}_1, \overline{q}_3\}, \{\overline{q}_2\}, \{\overline{q}_i\}_{i=4}^m, \{\widetilde{q}_1, \widetilde{q}_3\}, \{\widetilde{q}_2\}, \{\widetilde{q}_i\}_{i=4}^n, \{\emptyset_{\overline{\mathcal{X}}}\}\}.$$

Further,

$$\tau_1 = \{K_i \in \tau_2 \mid K_i \cap \underline{\mathcal{X}} \neq \emptyset\} = \{\ \{q_1, q_3\}, \{q_2\}, \{q_4\}, \{q_5\}, \ldots, \{q_m\}\ \},\ \text{ and}$$
$$\pi_2 \cap \tau_2 = 1_{\overline{\mathcal{X}}},\ \pi_1 \cap \tau_1 = 1_{\underline{\mathcal{X}}}\ \text{and}\ \tau_1[s] \subseteq \tau_1\ \text{for all}\ [s] \in S'/{\sim}_{\pi_2}.$$

This results in orthogonal admissible partitions $\pi := (\pi_1, \pi_2)$ and $\tau := (\tau_1, \tau_2)$ on $\mathcal{U}$.

We claim that $\mathcal{B}$ is covered by the direct product of the quotient $\mathcal{B}/\langle\pi\rangle$ and $\mathcal{B}/\langle\tau\rangle$. Define the map $\eta : \pi_2 \times \tau_2 \to \overline{\mathcal{X}}$ as follows:

$$\eta(H_i, K_j) = q_k \text{ for non-empty } H_i \cap K_j = \{q_k\}, \text{ and}$$
$$\eta(H_i, K_j) \text{ is not defined if } H_i \cap K_j \text{ is empty,}$$

for $H_i \in \pi_2$ and $K_j \in \tau_2$.

- η is well-defined and onto because τ_2 is orthogonal to π_2.
- η is a covering of $(\overline{\mathcal{X}}, S')$ by $(\pi_2, S'/{\sim}_{\pi_2}) \times (\tau_2, S'/{\sim}_{\tau_2})$, where for any $s \in S'$, t_s is $([s]_{\pi_2}, [s]_{\tau_2})$.
- η restricts $\pi_1 \times \tau_1$ to $\underline{\mathcal{X}}$.

$$\therefore\ \mathcal{B} \preccurlyeq \mathcal{B}/\langle\pi\rangle \times \mathcal{B}/\langle\tau\rangle. \tag{4.10}$$

Observe the following for the quotient ts $\mathcal{B}/\langle\tau\rangle$:

1. $|\tau_1| = |\underline{\mathcal{X}}| - 1$, $|\tau_2| = |\overline{\mathcal{X}}| - 1$, and
2. $\mathcal{B}/\langle\tau\rangle$ is again a reset ts for rough set, and

$$\mathcal{B}/\langle\tau\rangle \preccurlyeq \overline{(\mathbf{m-1,n-1})} \qquad (4.11)$$

Thus as in Case 1, using Proposition 48 (4) and (6) on Eqs. 4.4, 4.9, 4.10 and 4.11, we get

$$\mathcal{A} \preccurlyeq \mathcal{B} \preccurlyeq \mathcal{B}/\langle\pi\rangle \times \mathcal{B}/\langle\tau\rangle \preccurlyeq \overline{(\mathbf{2,2})} \times \overline{(\mathbf{m-1,n-1})}.$$

By repeating the process in Case 2 above $m-2$ times, we obtain the following decomposition:

$$\mathcal{A} \preccurlyeq \prod^{m-2} \overline{(\mathbf{2,2})} \times \overline{(\mathbf{2,n-m+2})}.$$

Applying Case 1 on $\overline{(\mathbf{2,n-m+2})}$, we have

$$\mathcal{A} \preccurlyeq \prod^{m-1} \overline{(\mathbf{2,2})} \times \overline{(\mathbf{1,n-m+1})}. \qquad (4.12)$$

In case $|\underline{\mathcal{X}}| = |\overline{\mathcal{X}}| = m = n$, Eq. 4.12 reduces to

$$\mathcal{A} \preccurlyeq \prod^{n-1} \overline{(\mathbf{2,2})} \times \overline{(\mathbf{1,1})} \cong \prod^{n-1} \overline{(\mathbf{2,2})}. \qquad (4.13)$$

<u>Case 3</u>: $|\underline{\mathcal{X}}| = |\overline{\mathcal{X}}| = 2$. $\mathcal{A} \preccurlyeq \overline{(\mathbf{2,2})}$, by Proposition 49.

<u>Case 4</u>: $|\underline{\mathcal{X}}| = 0$. For the reset ts $(\mathbf{0,n})$ where $n > 2$, the proof of decomposition is again similar as that of Case 2, with the change that $\pi_1 = \tau_1 = \emptyset$. We get in this case

$$\mathcal{A} \preccurlyeq \overline{(\mathbf{0,n})} \preccurlyeq \prod^{n-1} \overline{(\mathbf{0,2})}. \qquad (4.14)$$

We now present some more decomposition results for ts for rough sets analogous to the ones for ts given in Proposition 17. The first one gives the decomposition in case of tm for rough sets.

Proposition 50. *Let* $\mathcal{A} := (\mathcal{U}, M)$ *be a tm for the rough set* $\mathcal{U} := (U, R, X)$ *and* G *be the maximal subgroup of* M, *then* $\mathcal{A} \preccurlyeq (\mathcal{U}, M\backslash G)^{\cdot} \circ G$.

Proof. For the group G, let the associated rough set be $\mathcal{V} := (V, R', Y)$ (Example 12), where $\underline{\mathcal{Y}} = \overline{\mathcal{Y}} \cong G$. We have to show the following:

$$\mathcal{A} \preccurlyeq (\mathcal{U} \times \mathcal{V}, ((M\backslash G) \cup \{1_{\overline{\mathcal{X}}}\})^{\overline{\mathcal{Y}}} \times G).$$

Let $\mathcal{B} := (\mathcal{U} \times \mathcal{V}, ((M\backslash G) \cup \{1_{\overline{\mathcal{X}}}\})^{\overline{\mathcal{Y}}} \times G)$. Since the upper ts $(\overline{\mathcal{X}}, M)$ of $\mathcal{A}$ is a tm, by Proposition 17(1), there is a covering $\eta : \overline{\mathcal{X}} \times G \to \overline{\mathcal{X}}$ that gives

$$(\overline{\mathcal{X}}, M) \preccurlyeq ((\overline{\mathcal{X}} \times G), (((M\backslash G) \cup \{1_{\overline{\mathcal{X}}}\})^{G} \times G)).$$

In [34], η is defined as $\eta((q,g)) := qg$ for $q \in \overline{\mathcal{X}}$ and $g \in G$. Observe that η restricts $\underline{\mathcal{X}} \times G$ onto $\underline{\mathcal{X}}$: (i). any $q \in \overline{\mathcal{X}}$ is expressed as $q1_{\overline{\mathcal{X}}}$, $1_{\overline{\mathcal{X}}}$ being the identity in G. (ii). As $\mathcal{A}$ is a ts for rough sets, if $q \in \underline{\mathcal{X}}$ then $qg \in \underline{\mathcal{X}}$, for all $g \in G$. Therefore, $\mathcal{A} \preccurlyeq (\mathcal{U}, M\backslash G)^{\cdot} \circ G$.

A corollary of the above proposition gives the decomposition in case of tg for rough sets. It says that any tg can be decomposed into two parts - (1) reset ts for rough sets, and (2) groups.

Corollary 6. *Given a tg* $\mathcal{A} := (\mathcal{U}, G)$ *for the rough set* $\mathcal{U} := (U, R, X)$ *such that* $|\underline{\mathcal{X}}| = m$ *and* $|\overline{\mathcal{X}}| = n$, *we have*

$$\overline{\mathcal{A}} \preccurlyeq \overline{(\mathbf{m}, \mathbf{n})}^{\cdot} \circ G$$

Proof. When $|\overline{\mathcal{X}}| = 1$, G is the 1-element trivial group. $\overline{\mathcal{A}}$ is a reset and Proposition 49 implies $\overline{\mathcal{A}} \preccurlyeq \overline{(\mathbf{m}, \mathbf{1})}$, where $|\underline{\mathcal{X}}| = m$. Since G is a 1-element group, for any ts $(\mathcal{V}, T)$ for the rough set $\mathcal{V} := (V, S', Y)$, we have $(\mathcal{V}, T) \cong (\mathcal{V}, T) \circ G$, because $\overline{\mathcal{Y}} \times G \cong \overline{\mathcal{Y}}$, $\underline{\mathcal{Y}} \times G \cong \underline{\mathcal{Y}}$, and $T^G \times G \cong T$. In particular, for the ts $\overline{(\mathbf{m}, \mathbf{1})}^{\cdot}$, $\overline{(\mathbf{m}, \mathbf{1})}^{\cdot} \cong \overline{(\mathbf{m}, \mathbf{1})}^{\cdot} \circ G$. Finally,

$$\overline{\mathcal{A}} \preccurlyeq \overline{(\mathbf{m}, \mathbf{1})} \preccurlyeq \overline{(\mathbf{m}, \mathbf{1})}^{\cdot} \preccurlyeq \overline{(\mathbf{m}, \mathbf{1})}^{\cdot} \circ G.$$

General case: Observe that $\overline{\mathcal{A}}$ is a tm for rough set $\mathcal{U}$ with the associated monoid as $S' = \langle G \cup \{\overline{q} \mid q \in \underline{\mathcal{X}}\} \cup \{\widetilde{q} \mid q \in \overline{\mathcal{X}} \backslash \underline{\mathcal{X}}\}\rangle$. The elements of $S' \backslash G$ are constant functions or partial constant functions or empty function, which are not invertible for $|\overline{\mathcal{X}}| \neq 1$. Thus, we can conclude the following:
(1). $(\mathcal{U}, S' \backslash G)$ is a reset.
(2). The maximal subgroup of S' is equal to G because elements of G in S' are the only invertible elements in S', and thus form a group.
Applying Proposition 50 on $\overline{\mathcal{A}}$, we obtain $\overline{\mathcal{A}} \preccurlyeq (\mathcal{U}, S' \backslash G)^{\cdot} \circ G$. Since $(\mathcal{U}, S' \backslash G)$ is a reset, Proposition 49 implies $\overline{(\mathcal{U}, S' \backslash G)} \preccurlyeq \overline{(\mathbf{m}, \mathbf{n})}$. Finally, Proposition 48(2), (5) and (6) imply $\overline{\mathcal{A}} \preccurlyeq \overline{(\mathbf{m}, \mathbf{n})}^{\cdot} \circ G$.

In Theorem 16, we observed a decomposition for resets. Moreover, any group can be decomposed into non-trivial simple groups [34]. Thus, we can obtain a complete decomposition in case of tg for rough sets.

Another straight-forward extension is the decomposition result of ts theory given in Proposition 17(3).

Proposition 51. *Let* $\mathcal{A} := (\mathcal{U}, S)$ *be a ts for the rough set* $\mathcal{U} := (U, R, X)$, L *a left ideal in* S, *and* T *a subsemigroup of* S *such that* $L \cup T = S$. *Then* $\mathcal{A} \preccurlyeq (\mathcal{U}, L)^{\cdot} \circ \overline{\mathcal{B}}$, *where* $\mathcal{B} = (\mathcal{V}, T)$ *is a ts for the rough set* $\mathcal{V} := (V, R', Y)$ *such that* $\overline{\mathcal{Y}} \cong \underline{\mathcal{Y}} \cong T \cup \{1_{\overline{\mathcal{X}}}\}$.

Proof. Consider the ts $(\overline{\mathcal{X}}, S)$. From Proposition 17(3), we obtain the following covering: $(\overline{\mathcal{X}}, S) \preccurlyeq (\overline{\mathcal{X}}, L)^{\cdot} \circ (T \cup \{1_{\overline{\mathcal{X}}}\}, T)$, i.e.

$$(\overline{\mathcal{X}}, S) \preccurlyeq (\overline{\mathcal{X}} \times (T \cup \{1_{\overline{\mathcal{X}}}\}), (L \cup \{1_{\overline{\mathcal{X}}}\})^{(T \cup \{1_{\overline{\mathcal{X}}}\})} \times T'),$$

where $T' := \langle T \cup \{\overline{q} \mid q \in T \cup \{1_{\overline{\mathcal{X}}}\}\}\rangle$. In [34], the covering $\eta : \overline{\mathcal{X}} \times (T \cup \{1_{\overline{\mathcal{X}}}\}) \to \overline{\mathcal{X}}$ is defined as $\eta((q, t)) := qt$ and $\eta((q, 1_{\overline{\mathcal{X}}})) := q$ for all $q \in \overline{\mathcal{X}}$ and $t \in T$.
Define a rough set $\mathcal{V} := (V, R', Y)$ and a ts $\mathcal{B} := (\mathcal{V}, T)$ for the rough set $\mathcal{V}$ as done in Example 12, such that $\overline{\mathcal{Y}} = \underline{\mathcal{Y}} \cong T \cup \{1_{\overline{\mathcal{X}}}\}$. Since $\overline{\mathcal{Y}} = \underline{\mathcal{Y}}$, we observe

that the semigroup in the closure $\overline{\mathcal{B}}$ is isomorphic to T'.
Then we get the following covering:

$$(\overline{\mathcal{X}}, S) \preccurlyeq (\overline{\mathcal{X}}, L)^{\bullet} \circ \overline{(\overline{\mathcal{Y}}, T)}, \text{ i.e. } (\overline{\mathcal{X}}, S) \preccurlyeq (\overline{\mathcal{X}} \times \overline{\mathcal{Y}}, (L \cup \{1_{\overline{\mathcal{X}}}\})^{\overline{\mathcal{Y}}} \times T').$$

Moreover, the restriction of η on $\underline{\mathcal{X}} \times \overline{\mathcal{Y}}$ maps onto $\underline{\mathcal{X}}$, as for $q \in \underline{\mathcal{X}}$ and $s \in S$, $qs \in \underline{\mathcal{X}}$. Therefore, we finally get a covering $\mathcal{A} \preccurlyeq (\mathcal{U}, L)^{\bullet} \circ \overline{\mathcal{B}}$.

The following Proposition gives a decomposition result applying specifically to the theory of ts for rough sets. It is motivated by the decomposition of ts given in Proposition 17(2).

Proposition 52. *Let $\mathcal{U} := (U, R, X)$ be a rough set and $p \in \overline{\mathcal{X}}$ be such that $\underline{\mathcal{X}} \backslash \{p\} \neq \emptyset$ as well as $(\overline{\mathcal{X}} \backslash \underline{\mathcal{X}}) \backslash \{p\} \neq \emptyset$. If $\mathcal{A} := (\mathcal{U}, S)$ is a ts for $\mathcal{U}$ such that $\overline{\mathcal{X}} s \subseteq \overline{\mathcal{X}} \backslash \{p\}$ for all $s \in S$, we have the following decomposition.*

(1) If $p \in \underline{\mathcal{X}}$ then $(\mathcal{U}, S) \preccurlyeq \overline{(\mathcal{V}', S')} \circ \overline{(2, 2)}$, where $\mathcal{V}' := (V', R', Y')$, $\underline{\mathcal{Y}}' = \underline{\mathcal{X}} \backslash \{p\}$ and $\overline{\mathcal{Y}}' = \overline{\mathcal{X}} \backslash \{p\}$.
(2) If $p \in \overline{\mathcal{X}} \backslash \underline{\mathcal{X}}$, then $(\mathcal{U}, S) \preccurlyeq \overline{(\mathcal{V}'', S'')} \circ \overline{(1, 2)}$, where $\mathcal{V}'' := (V'', R'', Y'')$, $\underline{\mathcal{Y}}'' = \underline{\mathcal{X}}$ and $\overline{\mathcal{Y}}'' = \overline{\mathcal{X}} \backslash \{p\}$.

Here, S' and S'' are semigroups such that $|S'| \leq |S|$ and $|S''| \leq |S|$.

Proof. $\underline{\mathcal{X}} \backslash \{p\} \neq \emptyset$ implies $\overline{\mathcal{X}} \backslash \{p\} \neq \emptyset$. Observe that if $\overline{\mathcal{X}} \backslash \{p\} = \emptyset$, i.e. $\overline{\mathcal{X}} = \{p\}$, by Proposition 49, $\mathcal{A} \preccurlyeq \overline{(0, 1)}$ or $\mathcal{A} \preccurlyeq \overline{(1, 1)}$.

(1). $p \in \underline{\mathcal{X}}$.
Define the rough set $\mathcal{V}' := (V', R', Y')$ as follows:

$$V' := U \backslash p, R' := R \backslash \{(x, y) \mid x, y \in p\}, \text{ and } Y' := X \backslash p.$$

Then
(a). $\underline{\mathcal{Y}}' = \underline{\mathcal{X}} \backslash \{p\}$ and $\overline{\mathcal{Y}}' = \overline{\mathcal{X}} \backslash \{p\}$.
(b). $(\mathcal{V}', S/\sim)$ is a ts for the rough set $\mathcal{V}'$.
Now, define the following partition on $\overline{\mathcal{X}}$: $\tau_2 := \{\{p\}, \overline{\mathcal{X}} \backslash \{p\}\}$. Observe that τ_2 is an admissible partition of $(\overline{\mathcal{X}}, S)$, and therefore we have the quotient ts $(\tau_2, S/\sim_{\tau_2})$. Here, $S/\sim_{\tau_2}$ is a 1-element semigroup, as for any $s \in S$, $\{p\}s \subseteq \overline{\mathcal{X}} \backslash \{p\}$ and $(\overline{\mathcal{X}} \backslash \{p\})s \subseteq \overline{\mathcal{X}} \backslash \{p\}$. Define $\tau_1 := \{K_i \in \tau_2 \mid K_i \cap \underline{\mathcal{X}} \neq \emptyset\}$. Since $\{p\} \in \underline{\mathcal{X}}$ and $\underline{\mathcal{X}} \backslash \{p\} \neq \emptyset$, $\tau_1 = \tau_2$. Thus, $\tau := (\tau_1, \tau_2)$ is an admissible partition on $\mathcal{U}$ (Definition 99). Further, using Definition 100, we have the following:
(a). $\mathcal{A}/\langle \tau \rangle := (\mathcal{U}_\tau, S/\sim_{\tau_2})$ for the rough set $\mathcal{U}_\tau := (U, R_\tau, X_\tau)$ is the quotient ts of $\mathcal{A}$ with respect to τ. $(\tau_2, S/\sim_{\tau_2})$ is identifiable to both the upper and lower ts of $\mathcal{A}/\langle \tau \rangle$.
(b). $(\tau_2, S/\sim_{\tau_2}) \cong (\{0, 1\}, \{\overline{0}\})$, because $|\tau_2| = 2$ and $\tau_2[s] \subseteq \{\overline{\mathcal{X}} \backslash \{p\}\}$ for $[s] \in S/\sim_{\tau_2}$.
Identify the upper and lower ts of $\mathcal{A}/\langle \tau \rangle$ with $(\{0, 1\}, \{\overline{0}\})$.

We claim that $\mathcal{A} \preccurlyeq \overline{(\mathcal{V}', S/\sim)} \circ \mathcal{A}/\langle \tau \rangle$. For this, we have to find a covering $\eta : (\overline{\mathcal{X}} \backslash \{p\}) \times \{0,1\} \to \overline{\mathcal{X}}$ that gives

(i). $(\overline{\mathcal{X}}, S) \preccurlyeq ((\overline{\mathcal{X}} \backslash \{p\}), S/\sim) \circ (\{0,1\}, \{\overline{0}\})$, i.e.

$$(\overline{\mathcal{X}}, S) \preccurlyeq ((\overline{\mathcal{X}} \backslash \{p\}) \times \{0,1\}, \ (T')^{\{0,1\}} \times \{\overline{0}\}) \tag{4.15}$$

where $T' := \langle\ (S/\sim) \cup \{\overline{q} \mid q \in \underline{\mathcal{X}} \backslash \{p\}\} \cup \{\widetilde{q} \mid q \in (\overline{\mathcal{X}} \backslash \underline{\mathcal{X}})\}\ \rangle$. Further,

(ii). η restricts $(\underline{\mathcal{X}} \backslash \{p\}) \times \{0,1\}$ to $\underline{\mathcal{X}}$.

Define η as in [34]:
$\eta((q,0)) := q$ and $\eta((q,1)) := p$ for all $q \in \overline{\mathcal{X}} \backslash \{p\}$. For any $s \in S$, define $t_s := (f, \overline{0})$ where $f : \{0,1\} \to T'$ as follows:

If ps is defined then $f(0) = [s]$ and $f(1) = \overline{ps}$,

If ps is not defined then $f(0) = [s]$ and $f(1) = [s_0]$,

where $s_0 \in S$ is an arbitrary (fixed) element. When ps is defined, $p \in \underline{\mathcal{X}}$, $ps \neq p$ (as $ps \in \overline{\mathcal{X}} \backslash \{p\}$) and $\mathcal{A}$ is a ts for rough sets together imply $ps \in \underline{\mathcal{X}} \backslash \{p\}$. Therefore, $\overline{ps} \in T'$, and f is well-defined.

For (i), we show that given any $s \in S$, for all $q \in \overline{\mathcal{X}} \backslash \{p\}$ and $i = 0,1$, if $\eta((q,i))s$ is defined, then $\eta((q,i))s = \eta((q,i)t_s)$.

Let $\eta((q,0))s$ be defined, i.e. qs is defined. Then $\eta((q,0)t_s) = \eta((q,0)(f,\overline{0})) = \eta((qf(0),0)) = \eta((q[s],0)) = q[s] = qs = \eta((q,0))s$.

Let $\eta((q,1))s$ be defined, i.e. ps is defined. Then $\eta((q,1)t_s) = \eta((q,1)(f,\overline{0})) = \eta((qf(1),0)) = \eta((q(\overline{ps}),0)) = q\overline{ps} = ps = \eta((q,1))s$.

Let us now check that η is onto. Let $q \in \overline{\mathcal{X}}$. Then, $\eta((q,0)) = q$ for any $q \in \overline{\mathcal{X}} \backslash \{p\}$. Moreover, since $\overline{\mathcal{X}} \backslash \{p\}$ is non-empty, there exists a $q_0 \in \overline{\mathcal{X}} \backslash \{p\}$ such that $\eta((q_0,1)) = p$. So, η is onto. Therefore, η is a covering for Eq. 4.15.

(ii). η restricts $(\underline{\mathcal{X}} \backslash \{p\}) \times \{0,1\}$ to $\underline{\mathcal{X}}$, by definition. Let us check that the restriction is onto: for $q \in \underline{\mathcal{X}} \backslash \{p\}$, $\eta((q,0)) = q$. Moreover, since $\underline{\mathcal{X}} \backslash \{p\}$ is non-empty, there exists a $q_0 \in \underline{\mathcal{X}} \backslash \{p\}$ such that $\eta((q_0,1)) = p$.

Thus η is the required covering that gives $\mathcal{A} \preccurlyeq \overline{(\mathcal{V}', S/\sim)} \circ \mathcal{A}/\langle \tau \rangle$.

Finally, since $\mathcal{A}/\langle \tau \rangle$ is a reset, $\mathcal{A}/\langle \tau \rangle \preccurlyeq \overline{(\mathbf{2,2})}$, and we have $\mathcal{A} \preccurlyeq \overline{(\mathcal{V}'', S/\sim)} \circ \overline{(\mathbf{2,2})}$. Note that $|S/\sim| \leq |S|$.

(2). $p \in \overline{\mathcal{X}} \backslash \underline{\mathcal{X}}$. Note that $\underline{\mathcal{X}} \backslash \{p\} \neq \emptyset$ implies $\underline{\mathcal{X}} \neq \emptyset$.

Define the rough set $\mathcal{V}'' := (V'', R'', Y'')$ as follows: (1). $V'' := U \backslash p$, (2). $R'' := R \backslash \{(x,y) \mid x,y \in p\}$, and (3). $Y'' := X \backslash p$. Observe that (a). $\underline{y}'' = \underline{\mathcal{X}}$, and (b). $\overline{y}'' = \overline{\mathcal{X}} \backslash \{p\}$. Then, $(\mathcal{V}'', S/\sim)$ is a ts for the rough set $\mathcal{V}''$.

We have observed in (1) that $\tau_2 := \{\{p\}, \overline{\mathcal{X}} \backslash \{p\}\}$ is an admissible partition on $\overline{\mathcal{X}}$, and therefore we have the quotient ts $(\tau_2, S/\sim_{\tau_2})$, where $S/\sim_{\tau_2}$ is a 1-element semigroup. Define $\tau_1 := \{K_i \in \tau_2 \mid K_i \cap \underline{\mathcal{X}} \neq \emptyset\}$. Since, $\underline{\mathcal{X}}$ is non-empty, $\tau_1 = \{\overline{\mathcal{X}} \backslash \{p\}\}$. We also have $\tau_1[s] \subseteq \tau_1$ for $[s] \in S/\sim_{\tau_2}$. Thus, $\tau := (\tau_1, \tau_2)$ is an admissible partition on $\mathcal{U}$ (Definition 99). Further, using Definition 100, we have the following:

(A). $\mathcal{A}/\langle \tau \rangle := (\mathcal{U}_\tau, S/\sim_{\tau_2})$ for the rough set $\mathcal{U}_\tau := (U, R_\tau, X_\tau)$ is the quotient ts of $\mathcal{A}$ with respect to τ.

(B). $(\tau_2, S/\!\sim_{\tau_2}) \cong (\{0,1\}, \{\overline{0}\})$, because $|\tau_2| = 2$ and $\tau_2[s] \subseteq \{\overline{\mathcal{X}}\backslash\{p\}\}$ for $[s] \in S/\!\sim_{\tau_2}$. Identify the upper ts of $\mathcal{A}/\langle\tau\rangle$ with $(\{0,1\}, \{\overline{0}\})$, the lower ts of $\mathcal{A}/\langle\tau\rangle$ with $(\{0\}, \{\overline{0}\})$.

We claim that $\mathcal{A} \preccurlyeq \overline{(\mathcal{V}'', S/\!\sim)} \circ \mathcal{A}/\langle\tau\rangle$. For this, we have to find a covering
$\eta : (\overline{\mathcal{X}}\backslash\{p\}) \times \{0,1\} \to \overline{\mathcal{X}}$ that gives

(i). $(\overline{\mathcal{X}}, S) \preccurlyeq (\overline{\mathcal{Y}}'', T'') \circ (\overline{\mathcal{X}_\tau}, S/\!\sim_{\tau_2})$, i.e.

$$(\overline{\mathcal{X}}, S) \preccurlyeq (\overline{\mathcal{X}}\backslash\{p\} \times \{0,1\}, (T'')^{\{0,1\}} \times \{\overline{0}\}), \tag{4.16}$$

where $T'' := \langle\, (S/\!\sim) \cup \{\overline{q} \mid q \in \underline{\mathcal{X}}\} \cup \{\widetilde{q} \mid q \in (\overline{\mathcal{X}}\backslash\underline{\mathcal{X}})\backslash\{p\}\} \,\rangle$; and
(ii). restricts $(\underline{\mathcal{X}} \times \{0\})$ onto $\underline{\mathcal{X}}$.

Define η as follows:

$\eta((q,0)) := q$ for all $q \in \overline{\mathcal{X}}\backslash\{p\}$.
$\eta((q,1)) := p$ for all $q \in (\overline{\mathcal{X}}\backslash\underline{\mathcal{X}})\backslash\{p\}$.
$\eta((q,1))$ is not defined for any $q \in \underline{\mathcal{X}}$.

Note that $ps \neq p$. For $s \in S$, define $t_s := (g, \overline{0})$, where $g : \{0,1\} \to T''$, as follows:

If ps is defined and $ps \in \underline{\mathcal{X}}$ then $g(0) = [s]$ and $g(1) = \overline{ps}$.
If ps is defined and $ps \in (\overline{\mathcal{X}}\backslash\underline{\mathcal{X}})\backslash\{p\}$ then $g(0) = [s]$ and $g(1) = \widetilde{ps}$.
If ps is not defined then $g(0) = [s]$ and $g(1) = [s_0]$, where $s_0 \in S$ is any fixed element.

g is well defined.
For (i), we show that given any $s \in S$, for any $q \in \overline{\mathcal{X}}\backslash\{p\}$ and $i = 0, 1$, if $\eta((q,i))s$ is defined, then $\eta((q,i))s = \eta((q,i)t_s)$.

– Let $(\eta((q,0))s$ be defined, i.e. qs is defined. Then $\eta((q,0)t_s) = \eta((q,0)(g,\overline{0})) = \eta((qg(0),0)) = \eta((q[s],0)) = \eta((qs,0)) = qs = (\eta((q,0))s$.
– Let $(\eta((q,1))s$ be defined, i.e. $q \in (\overline{\mathcal{X}}\backslash\underline{\mathcal{X}})\backslash\{p\}$, and ps is defined. Then $\eta((q,1)t_s) = \eta((q,1)(g,\overline{0})) = \eta((qg(1),0))$.
If $ps \in (\overline{\mathcal{X}}\backslash\underline{\mathcal{X}})\backslash\{p\}$, then $\eta((qg(1),0)) = \eta((q\widetilde{ps},0)) = \eta((ps,0)) = ps = \eta((q,1))s$.
If $ps \in \underline{\mathcal{X}}$, then $\eta((qg(1),0)) = \eta((q\overline{ps},0)) = \eta((ps,0)) = ps = \eta((q,1))s$.

Let us now check that η is onto: for $q \in \overline{\mathcal{X}}$, $\eta((q,0)) = q$, if $q \in \overline{\mathcal{X}}\backslash\{p\}$. Moreover, since $(\overline{\mathcal{X}}\backslash\underline{\mathcal{X}})\backslash\{p\}$ is non-empty, there exists a $q_0 \in (\overline{\mathcal{X}}\backslash\underline{\mathcal{X}})\backslash\{p\}$ such that $\eta((q_0,1)) = p$.
For (ii), note that η restricts $\underline{\mathcal{X}} \times \{0\}$ onto $\underline{\mathcal{X}}$, because $\eta((q,0)) = q$ for all $\underline{\mathcal{X}}$.
Therefore, η is the required covering that gives $\mathcal{A} \preccurlyeq \overline{(\mathcal{V}'', S/\!\sim)} \circ \mathcal{A}/\langle\tau\rangle$.
Finally, since $\mathcal{A}/\langle\tau\rangle$ is a reset, $\mathcal{A}/\langle\tau\rangle \preccurlyeq (\mathbf{1}, \mathbf{2})$, and we have $\mathcal{A} \preccurlyeq \overline{(\mathcal{V}'', S/\!\sim)} \circ (\mathbf{1}, \mathbf{2})$.
Again, $|S/\!\sim| \leq |S|$.

The assumption $Q\backslash\{p\} \neq \emptyset$ of Proposition 17(2) translated here would give only $\overline{\mathcal{X}}\backslash\{p\} \neq \emptyset$, i.e. the possible region of the rough set has at least 2 equivalence classes. To get the above decomposition in case of ts for rough sets, we separately assumed the same for the lower and the boundary regions as well, i.e. $\underline{\mathcal{X}}\backslash\{p\} \neq \emptyset$ and $(\overline{\mathcal{X}}\backslash\underline{\mathcal{X}})\backslash\{p\} \neq \emptyset$.

4.4 Rough Sets and Automata Theory

We now focus on connections of transformation semigroups with *semiautomata*, and how these could apply to the study here in the context of rough sets. Our aim is to see how the *ts* for rough sets that we study in this section, are related to an appropriate notion of semiautomaton that may be defined in the context of rough sets. We have already defined the concept of 'subautomaton' for this purpose in Definition 47. This definition suits us here, as a natural relation with *ts* for rough sets in the lines described in Remark 6 (for *ts* and semiautomata) may be arrived at, if the input states constitute a rough set and Σ is fixed.

Definition 103 (Semiautomaton for a rough set).
A semiautomaton for a rough set $\mathcal{U} := (U, R, X)$ is a triple $\mathcal{M} := (\mathcal{U}, \Sigma, \delta)$, where $(\overline{\mathcal{X}}, \Sigma, \delta)$ is a semiautomaton ($\delta : \overline{\mathcal{X}} \times \Sigma \to \overline{\mathcal{X}}$ is a partial function), $\underline{\mathcal{X}} \subseteq \overline{\mathcal{X}}$, and $(\underline{\mathcal{X}}, \Sigma, \delta \mid_{\underline{\mathcal{X}}})$ is a subautomaton of $(\overline{\mathcal{X}}, \Sigma, \delta)$.

Remark 12. Let us compare semiautomaton for rough sets with Basu's definition [9,79] of *rough semi-automaton*, and also compare *rough transformation semigroups* defined in [82] with the *ts* for rough sets considered in this section.

1. A rough semi-automaton generalizes the concept of a non-deterministic automaton, in which the transition function maps an input state to a set of input states. For the definition in [9], the set U of input states has a partition R yielding an approximation space on U. For a given state and an input symbol, the transition function gives an output that is a rough set on the approximation space (U, R).
   In our case also, there is an underlying partition R on a set U of states; for any subset X of states in the approximation space (U, R), the triple (U, R, X) is a rough set. However, on any given input symbol, the transition function is a (partial) map between *equivalence classes of states* in $\overline{\mathcal{X}}$ such that classes in $\underline{\mathcal{X}}$ are mapped to those in $\underline{\mathcal{X}}$ only.
2. Rough transformation semigroups are derived from rough semi-automata [9], and thus involve transformations of the set U into the collection of rough sets on the approximation space on U. In contrast, if we consider the interpretation given above for *ts* for rough sets defined here, these structures are semigroups of transformations of the set $\overline{\mathcal{X}}$ of equivalence classes to itself that also restrict the set $\underline{\mathcal{X}}$ to itself.

Now to get the exact connection with *ts* for rough sets, we define homomorphisms. Recall Definition 46.

Definition 104 (Homomorphisms).
Let $\mathcal{M} := (\mathcal{U}, \Sigma, \delta)$ and $\mathcal{N} := (\mathcal{V}, \Lambda, \gamma)$ be two semiautomata for rough sets $\mathcal{U} := (U, R, X)$ and $\mathcal{V} := (V, S', Y)$ respectively. The semiautomaton homomorphism (α, β) from $(\overline{\mathcal{X}}, \Sigma, \delta)$ to $(\overline{\mathcal{Y}}, \Lambda, \gamma)$ such that $\alpha(\underline{\mathcal{X}}) \subseteq \underline{\mathcal{Y}}$ is called a semiautomaton homomorphism for rough sets.

Let us now consider a ts $\mathcal{A} := (\mathcal{U}, S)$ for a rough set $\mathcal{U} := (U, R, X)$. The tuple $(\mathcal{U}, S, \delta)$ is a semiautomaton for rough set, denoted by $RSM(\mathcal{A})$, where δ is the partial semigroup action associated with the upper ts of $\mathcal{A}$. On the other hand, starting from a semiautomaton $\mathcal{M} := (\mathcal{U}, \Sigma, \delta)$ for a rough set $\mathcal{U} := (U, R, X)$, we have $TS((\overline{\mathcal{X}}, \Sigma, \delta)) := (\overline{\mathcal{X}}, \Sigma^*/\sim)$ as a ts. As $\underline{\mathcal{X}} \subseteq \overline{\mathcal{X}}$ and $q[s] \in \underline{\mathcal{X}}$ for all $q \in \underline{\mathcal{X}}$, $s \in \Sigma^*$, $(\mathcal{U}, \Sigma^*/\sim)$ forms a ts for rough set – it is denoted as $RTS(\mathcal{M})$.

Theorem 17. *Consider a ts $\mathcal{A} := (\mathcal{U}, S)$ for a rough set $\mathcal{U} := (U, R, X)$ and a semiautomaton $\mathcal{M} := (\mathcal{V}, \Sigma, \delta)$ for a rough set $\mathcal{V} := (V, S', Y)$. The following results hold.*

(a) *$RTS(RSM(\mathcal{A})) \cong \mathcal{A}$, and*
(b) *there exists a homomorphism from $\mathcal{M}$ to $RSM(RTS(\mathcal{M}))$.*

Proof.
(a). $RTS(RSM(\mathcal{A})) = RTS(\mathcal{U}, S, \delta) = (\mathcal{U}, S^*/\sim)$. The semigroup $S^*/\sim$ is isomorphic to S, because S itself being a semigroup, $S^* \cong S$ and the congruence relation $\sim$ is the identity. Thus $RTS(RSM(\mathcal{A})) \cong \mathcal{A}$.
(b). $RTS(\mathcal{M}) := (\mathcal{V}, \Sigma^*/\sim)$ and $RSM(RTS(\mathcal{M})) = (\mathcal{V}, \Sigma^*/\sim, \widetilde{\delta})$, where $\widetilde{\delta}(p, [s_1 \cdots s_n]) = \delta(\cdots \delta((\delta(p, s_1), s_2), \cdots), s_n)$ for all $p \in \overline{\mathcal{Y}}$ and $s_i \in \Sigma$. Define the semiautomaton homomorphism $(1_{\overline{\mathcal{Y}}}, \beta)$ from $(\overline{\mathcal{Y}}, \Sigma, \delta)$ to $(\overline{\mathcal{Y}}, \Sigma^*/\sim, \widetilde{\delta})$, where $\beta : \Sigma \to \Sigma^*/\sim$ maps s to $[s]$ for all $s \in \Sigma$. Since $1_{\overline{\mathcal{Y}}}(\underline{\mathcal{Y}}) \subseteq \underline{\mathcal{Y}}$, we have the required semiautomaton homomorphism from $\mathcal{M}$ to $RSM(RTS(\mathcal{M}))$.

Due to Theorem 17, studying any one of semiautomata for rough sets or transformation semigroups for rough sets is enough to get similar results for the other. In particular, all the concepts defined in our work on ts for rough sets can be carried over to semiautomata for rough sets.

4.5 Conclusions

The theory of transformation semigroups has two strong motivations – one from semigroup theory, another from automata theory. This work marks the beginning of a study of transformation semigroups for rough sets, that is distinct from the notion of rough transformation semigroups defined earlier by [82]. A goal is to obtain decomposition results; the work introduces some basic notions for the purpose, culminating in a decomposition theorem for reset ts for rough sets.

Let us make some comments about the ts for rough sets, and the results obtained in Sect. 4.1. Any rough set (U, R, X) gives a pair of objects $(\underline{\mathcal{X}}, \overline{\mathcal{X}})$, such that $\underline{\mathcal{X}} \subseteq \overline{\mathcal{X}}$. Further, for any pair of objects (Q_1, Q_2), we can obtain a rough set (U, R, X) such that $\overline{\mathcal{X}} \cong Q_2$ and $\underline{\mathcal{X}} \cong Q_1$ (cf. proof of Proposition 12). In Definitions 91 and 103 of ts for rough sets or semiautomaton for rough sets, the semigroup action is defined on the set $\overline{\mathcal{X}}$ for the rough set (U, R, X). The approximation space (U, R) and the subset $X \subseteq U$ do not explicitly feature in the definition of semigroup action. This direction has been explored in [65], where in place of the rough set (U, R, X), a pair of sets (Q_1, Q_2) with the condition that $Q_1 \subseteq Q_2$ is considered.

There are some basic points of differences between the theory of ts and the theory of ts for rough sets. Let us observe them.

1. The first result that differs in the theories is mentioned in Remark 11. The closure of the (direct or wreath) products of two *ts* for rough sets may not be covered by the products of the their closures. This is established by a counter-example.

2. Let us recall decomposition results for the reset *ts* in both the theories, i.e. Proposition 16 and Theorem 16. Any reset *ts* can be covered by $\prod^k \overline{\mathbf{2}}$. For reset *ts* for rough sets, the decomposition has the basic entities as $\overline{(\mathbf{0}, \mathbf{2})}$, and $\overline{(\mathbf{1}, \mathbf{n})}$, other than $\overline{(\mathbf{2}, \mathbf{2})}$. We conjecture that the *ts* $\overline{(\mathbf{1}, \mathbf{n})}$ for some rough set may not be decomposable.

Conjecture 1. For any reset *ts* $\mathcal{A} := (\mathcal{U}, S)$ for rough set $\mathcal{U} := (U, R, X)$ with $|\overline{\mathcal{X}}| = n$ and $|\underline{\mathcal{X}}| = 1$, the closure $\overline{\mathcal{A}}$ of $\mathcal{A}$ cannot be decomposed into smaller *ts* for rough sets.

3. In Propositions 50, 52 and 51, we obtained some decomposition results for *ts* for rough sets, based on the decomposition results for *ts* in Proposition 17. Among these three, the first and second ones are direct 'extensions' of those for *ts*. However, Proposition 52 becomes different, giving two distinct decompositions that are based on a condition on $\underline{\mathcal{X}}$ in the rough set (U, R, X).

4. The main aim in [34] for Proposition 17 is to obtain Krohn-Rhodes decomposition Theorem 5 for *ts*. However, the proof of Theorem 5 cannot be extended to *ts* for rough sets, because of the failure of the closures to distribute over products (Point (1) above). As shown in Theorem 16, $\overline{(\mathbf{2}, \mathbf{2})}^{\cdot}$, $\overline{(\mathbf{0}, \mathbf{2})}^{\cdot}$ and $\overline{(\mathbf{1}, \mathbf{r})}^{\cdot}$ are the basic entities that reset *ts* for rough sets are decomposed into. So following the lines of Krohn-Rhodes decomposition, we may conjecture the following for a general *ts* for rough set.

Conjecture 2. Every *ts* $\mathcal{A} := (\mathcal{U}, S)$ for rough set $\mathcal{U} := (U, R, X)$ admits a decomposition in the following way:

$$\mathcal{A} \preccurlyeq \mathcal{A}_1 \circ \ldots \circ \mathcal{A}_n,$$

where $\mathcal{A}_i$ is either $\overline{(\mathbf{2}, \mathbf{2})}^{\cdot}$ or $\overline{(\mathbf{0}, \mathbf{2})}^{\cdot}$ or $\overline{(\mathbf{1}, \mathbf{r})}^{\cdot}$, or $\mathcal{A}_i$ is a simple group satisfying $\mathcal{A}_i \preccurlyeq S$ for all $1 \leq i \leq n$.

5   Contrapositionally Complemented Pseudo-Boolean Algebras

In this Section, we study the algebra of strong subobjects of an $RSC(\mathscr{C})$-object. In particular, we study it for the special case when $\mathscr{C}$ is SET. It is observed that if one follows the standard categorial constructions of the operations of intersection, union, complement and implication on the collection of strong subobjects of any $ROUGH$-object, one arrives at a Boolean algebra. Specifically, we see that the negation does not capture *relative rough complementation*, which is really what is required, as we are studying subobjects of a $ROUGH$-object. On considering a different definition of negation that, in fact, is in line with Iwiński's *rough difference* operator, one finds a complete change in the picture. We show that strong *ROUGH*-subobjects of an object form *contrapositionally complemented lattices* [75] with the excluded middle property, called '*c*∨*c* lattices' [39,68].

For the general case of an $RSC(\mathscr{C})$-object and its strong subobjects, we know that the algebra these form is a *pBa* (as $RSC(\mathscr{C})$ is a quasitopos). So we start with an arbitrary *pBa*, and implement Iwiński's *rough difference* operator in that scenario. We find that the strong subobjects of an $RSC(\mathscr{C})$-object form a new algebraic structure with two negations which we call 'contrapositionally complemented pseudo-Boolean algebra'. In this section, we present a detailed study of this class of algebras and a subclass, the algebras in the latter having the excluded middle property.

In the next section, algebras of strong subobjects in RSC and $RSC(\mathscr{C})$ are discussed, culminating in the definitions of the two new algebraic structures contrapositionally complemented pseudo-Boolean algebras and contrapositionally ∨ complemented pseudo-Boolean algebras. In Sect. 5.2, we observe properties and examples of the contrapositionally complemented pseudo-Boolean algebras. We compare these algebraic structures with known algebras in Sect. 5.3. Some representation theorems for these algebras with respect to topological spaces are obtained in Sect. 5.4. We end this section by giving some future directions of work in Sect. 5.5.

5.1   Algebras of Strong Subobjects in RSC and $RSC(\mathscr{C})$

As mentioned in the discussion preceding Proposition 13, in any topos (quasitopos), the subobject classifier for the class of monics (strong monics) defines operators ∩, ∪, → and ¬ over the class of monics (strong monics) with fixed codomain. These operators give an algebra (a *pBa*) over the strong subobjects of any object in a quasitopos (Proposition 13). We now study the constructions of these operators in this algebra for the quasitopos $RSC(\mathscr{C})$.

Consider an $RSC(\mathscr{C})$-object (U_1, U_2, m). Let $\mathcal{M}((U_1, U_2, m))$ denote the set of all strong subobjects in $RSC(\mathscr{C})$ with codomain (U_1, U_2, m). Note that since $\mathscr{C}$ is a topos, there exists a subobject classifier Ω with $\top : 1 \to \Omega$ in $\mathscr{C}$ for the class of monics.

In Proposition 38, we have obtained the subobject classifier for the class of strong monics in $RSC(\mathscr{C})$, given by the $RSC(\mathscr{C})$-object $(\Omega, \Omega, 1_\Omega)$ with arrow $(\top, \top) : (1, 1, 1_1) \to (\Omega, \Omega, 1_\Omega)$. Further, given any strong monic $(f', f) : (X_1, X_2, m_1) \to (U_1, U_2, m)$, the characteristic morphism of (f', f) is given by $(\chi_{f'}, \chi_f)$, where $\chi_{f'}$ and χ_f are characteristic morphisms of f' and f in $\mathscr{C}$ respectively. Let us characterize the operators on $\mathcal{M}((U_1, U_2, m))$. Recall the operators defined for subobjects in the topos $\mathscr{C}$ (Definition 30) - denoted by ∩, ∪, → and ¬.

Intersection: First let us see what is the arrow

$$\cap_{RSC(\mathscr{C})} : (\Omega, \Omega, 1_\Omega) \times (\Omega, \Omega, 1_\Omega) \to (\Omega, \Omega, 1_\Omega),$$

which is defined as the characteristic morphism of $\langle(\top,\top),(\top,\top)\rangle$.

Using the universal property of products in $RSC(\mathscr{C})$ (Proposition 31), $(\Omega,\Omega,1_\Omega) \times (\Omega,\Omega,1_\Omega) = (\Omega \times \Omega, \Omega \times \Omega, 1_\Omega \times 1_\Omega) = (\Omega \times \Omega, \Omega \times \Omega, 1_{\Omega\times\Omega})$ and $\langle(\top,\top),(\top,\top)\rangle = (\langle\top,\top\rangle,\langle\top,\top\rangle)$. The following diagram is a pullback in $\mathscr{C}$ (Fig. 15).

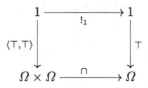

Claim: $\cap_{RSC(\mathscr{C})} = (\cap,\cap)$.

Consider the right hand side square in Fig. 40. Using Proposition 32 and the above diagram as a pullback, we have the right hand side square as a pullback in $RSC(\mathscr{C})$. Therefore, $\cap_{RSC(\mathscr{C})} = (\cap,\cap)$.

Now, let $(f',f),(g',g) \in \mathcal{M}((U_1,U_2,m))$ such that $(f',f) : (X_1,X_2,m_1) \to (U_1,U_2,m)$ and $(g',g) : (Y_1,Y_2,m_2) \to (U_1,U_2,m)$.

Using Proposition 38, $\chi_{(f',f)} = (\chi_{f'},\chi_f) : (U_1,U_2,m) \to (\Omega,\Omega,1_\Omega)$ and $\chi_{(g',g)} = (\chi_{g'},\chi_g) : (U_1,U_2,m) \to (\Omega,\Omega,1_\Omega)$. Therefore,

$$\langle\chi_{(f',f)},\chi_{(g',g)}\rangle = \langle(\chi_{f'},\chi_f),(\chi_{g'},\chi_g)\rangle.$$

Using the property of products in $RSC(\mathscr{C})$ (Proposition 31), we have

$$\Rightarrow \langle\chi_{(f',f)},\chi_{(g',g)}\rangle = (\langle\chi_{f'},\chi_{g'}\rangle,\langle\chi_f,\chi_g\rangle).$$

This also implies that $(\langle\chi_{f'},\chi_{g'}\rangle,\langle\chi_f,\chi_g\rangle) : (U_1,U_2,m) \to (\Omega\times\Omega,\Omega\times\Omega,1_{\Omega\times\Omega})$ is an $RSC(\mathscr{C})$-arrow. So, $\langle\chi_{f'},\chi_{g'}\rangle = \langle\chi_f,\chi_g\rangle \circ m$. From diagram in Fig. 15, the following two diagrams are pullbacks in $\mathscr{C}$.

$$
\begin{array}{ccc}
X_1 \cap Y_1 & \xrightarrow{!_{X_1\cap Y_1}} & 1 \\
\downarrow{\scriptstyle f'\cap g'} & & \downarrow{\scriptstyle \langle\top,\top\rangle} \\
U_1 & \xrightarrow{\langle\chi'_f,\chi'_g\rangle} & \Omega\times\Omega
\end{array}
\qquad
\begin{array}{ccc}
X_2 \cap Y_2 & \xrightarrow{!_{X_2\cap Y_2}} & 1 \\
\downarrow{\scriptstyle f\cap g} & & \downarrow{\scriptstyle \langle\top,\top\rangle} \\
U_2 & \xrightarrow{\langle\chi_f,\chi_g\rangle} & \Omega\times\Omega
\end{array}
$$

Observe the following: $\langle\chi_f,\chi_g\rangle \circ m \circ (f'\cap g') = 1_{\Omega\times\Omega}\circ\langle\chi_{f'},\chi_{g'}\rangle\circ(f'\cap g') = 1_{\Omega\times\Omega}\circ\langle\top,\top\rangle\circ!_{X_1\cap Y_1} = \langle\top,\top\rangle\circ 1_1\circ!_{X_1\cap Y_1}$. So, using the pullback property of the right hand side diagram above, there exists a unique arrow $h : X_1\cap Y_1 \to X_1 \cap Y_2$ such that $(f\cap g)\circ h = m\circ(f'\cap g')$ and $!_{X_2\cap Y_2}\circ h = 1_1 \circ !_{X_1\cap Y_1}$.

Claim: h is monic in $\mathscr{C}$. Let $r,s : D \to X_1\cap Y_1$ such that $h\circ r = h\circ s$. Then $(f\cap g)\circ h\circ r = (f\cap g)\circ h\circ s \Rightarrow m\circ(f'\cap g')\circ r = m\circ(f'\cap g')\circ s \Rightarrow r = s$, because m and $f'\cap g'$ are monics in $\mathscr{C}$. Therefore, $(X_1\cap Y_1, X_2\cap Y_2, h)$ is an $RSC(\mathscr{C})$-object.

Moreover, $(f\cap g)\circ h = m\circ(f'\cap g')$ implies that $(f'\cap g', f\cap g) : (X_1\cap Y_1, X_2\cap Y_2, h) \to (U_1,U_2,m)$ is an $RSC(\mathscr{C})$-arrow. Now, consider the left hand side diagram in Fig. 40. This is a diagram in $RSC(\mathscr{C})$ and using Proposition 32, the square is a pullback in $RSC(\mathscr{C})$. Finally, using Proposition 4, the outer square is a pullback in $RSC(\mathscr{C})$. Thus, $(f' \cap g', f \cap g)$ is the the characteristic morphism of $\cap_{RSC(\mathscr{C})} \circ (\langle \chi_{f'}, \chi_{g'} \rangle, \langle \chi_f, \chi_g \rangle)$, and $(f', f) \cap (g', g) = (f' \cap g', f \cap g)$.

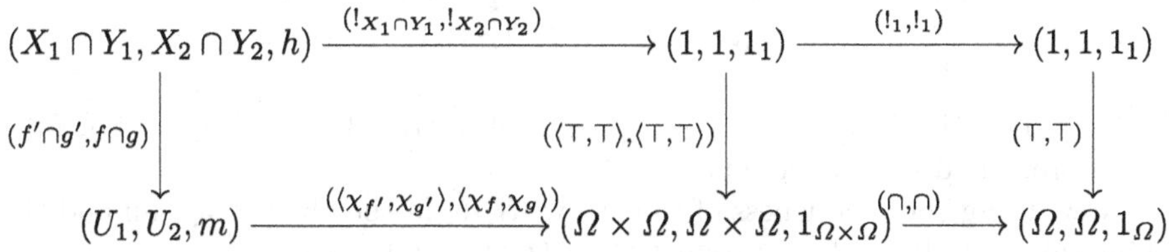

$$
\begin{array}{ccccc}
(X_1 \cap Y_1, X_2 \cap Y_2, h) & \xrightarrow{(!_{x_1 \cap Y_1}, !_{x_2 \cap Y_2})} & (1,1,1_1) & \xrightarrow{(!_1,!_1)} & (1,1,1_1) \\
{\scriptstyle (f' \cap g', f \cap g)} \downarrow & & \downarrow {\scriptstyle (\langle \top, \top \rangle, \langle \top, \top \rangle)} & & \downarrow {\scriptstyle (\top, \top)} \\
(U_1, U_2, m) & \xrightarrow{(\langle \chi_{f'}, \chi_{g'} \rangle, \langle \chi_f, \chi_g \rangle)} & (\Omega \times \Omega, \Omega \times \Omega, 1_{\Omega \times \Omega}) & \xrightarrow{(\cap, \cap)} & (\Omega, \Omega, 1_\Omega)
\end{array}
$$

Fig. 40. Intersection of (f', f) and (g', g) in $RSC(\mathscr{C})$

Union: $\cup_{RSC(\mathscr{C})} : (\Omega, \Omega, 1_\Omega) \times (\Omega, \Omega, 1_\Omega) \to (\Omega, \Omega, 1_\Omega)$ (Definition 30) is defined as the characteristic morphism of

$$[\langle (\top_\Omega, \top_\Omega), (1_\Omega, 1_\Omega) \rangle, \langle (1_\Omega, 1_\Omega), (\top_\Omega, \top_\Omega) \rangle] :$$
$$(\Omega, \Omega, 1_\Omega) + (\Omega, \Omega, 1_\Omega) \to (\Omega, \Omega) \times (\Omega, \Omega),$$

where $\top_\Omega = \top \circ !_\Omega$ in $\mathscr{C}$.
Claim 1: $(\Omega, \Omega, 1_\Omega) + (\Omega, \Omega, 1_\Omega) = (\Omega + \Omega, \Omega + \Omega, 1_{\Omega + \Omega})$.
Using the property of coproducts in $RSC(\mathscr{C})$ (Proposition 31), $(\Omega, \Omega, 1_\Omega) + (\Omega, \Omega, 1_\Omega)$ is the domain of the image of $1_\Omega + 1_\Omega$. Consider the diagram below, where $(\Omega + \Omega, i_1, i_2)$ is the coproduct in $\mathscr{C}$. Both the squares commute trivially. Using the universal property of $\Omega + \Omega$ in $\mathscr{C}$, $1_\Omega + 1_\Omega = [i_1, i_2] = 1_{\Omega + \Omega}$, which is a monic. Therefore, image of $1_\Omega + 1_\Omega$ is $1_\Omega + 1_\Omega$ itself, and $(\Omega, \Omega, 1_\Omega) + (\Omega, \Omega, 1_\Omega) = (\Omega + \Omega, \Omega + \Omega, 1_{\Omega + \Omega})$.

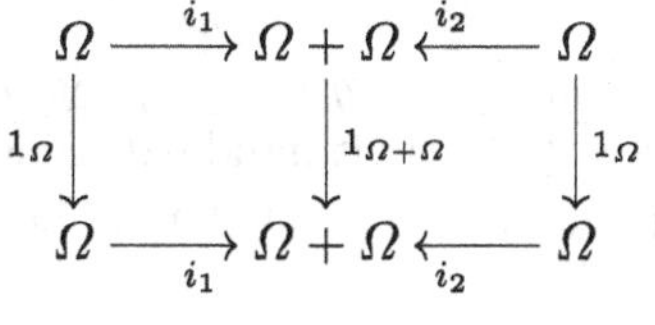

$$
\begin{array}{ccccc}
\Omega & \xrightarrow{i_1} & \Omega + \Omega & \xleftarrow{i_2} & \Omega \\
{\scriptstyle 1_\Omega} \downarrow & & \downarrow {\scriptstyle 1_{\Omega + \Omega}} & & \downarrow {\scriptstyle 1_\Omega} \\
\Omega & \xrightarrow{i_1} & \Omega + \Omega & \xleftarrow{i_2} & \Omega
\end{array}
$$

As we have already seen, $(\Omega, \Omega, 1_\Omega) \times (\Omega, \Omega, 1_\Omega) = (\Omega \times \Omega, \Omega \times \Omega, 1_{\Omega \times \Omega})$. Further,

$$\langle (\top_\Omega, \top_\Omega), (1_\Omega, 1_\Omega) \rangle = (\langle \top_\Omega, 1_\Omega \rangle, \langle \top_\Omega, 1_\Omega \rangle) \text{ and}$$
$$\langle (1_\Omega, 1_\Omega), (\top_\Omega, \top_\Omega) \rangle = (\langle 1_\Omega, \top_\Omega \rangle, \langle 1_\Omega, \top_\Omega \rangle).$$

Both the arrows have domain $(\Omega, \Omega, 1_\Omega)$ and codomain as $(\Omega \times \Omega, \Omega \times \Omega, 1_{\Omega \times \Omega})$.
Claim 2: The diagram below commutes in $RSC(\mathscr{C})$, where (α, α) is as follows:

$$(\alpha, \alpha) := ([\langle \top_\Omega, 1_\Omega \rangle, \langle 1_\Omega, \top_\Omega \rangle], [\langle \top_\Omega, 1_\Omega \rangle, \langle 1_\Omega, \top_\Omega \rangle]),$$

where $[\langle \top_\Omega, 1_\Omega \rangle, \langle 1_\Omega, \top_\Omega \rangle] : \Omega + \Omega \to \Omega \times \Omega.$

$$(\Omega, \Omega, 1_\Omega) \xrightarrow{(i_1, i_1)} (\Omega + \Omega, \Omega + \Omega, 1_{\Omega+\Omega}) \xleftarrow{(i_2, i_2)} (\Omega, \Omega, 1_\Omega)$$

with $\langle \langle \top_\Omega, 1_\Omega \rangle, \langle \top_\Omega, 1_\Omega \rangle \rangle$, (α, α), $\langle \langle 1_\Omega, \top_\Omega \rangle, \langle 1_\Omega, \top_\Omega \rangle \rangle$ mapping to

$$(\Omega \times \Omega, \Omega \times \Omega, 1_{\Omega \times \Omega})$$

$[\langle \top_\Omega, 1_\Omega \rangle, \langle 1_\Omega, \top_\Omega \rangle] \circ i_1 = \langle \top_\Omega, 1_\Omega \rangle$ and $[\langle \top_\Omega, 1_\Omega \rangle, \langle 1_\Omega, \top_\Omega \rangle] \circ i_2 = \langle 1_\Omega, \top_\Omega \rangle.$ Therefore, the diagram commutes.

Now, using the uniqueness of the arrow (α, α) in the above diagram and the universal property of the coproduct $(\Omega + \Omega, \Omega + \Omega, 1_{\Omega+\Omega})$,

$$[\langle (\top_\Omega, \top_\Omega), (1_\Omega, 1_\Omega) \rangle, \langle (1_\Omega, 1_\Omega), (\top_\Omega, \top_\Omega) \rangle] = (\alpha, \alpha).$$

Finally, using Proposition 32, we have the characteristic morphism of the above arrow as $(\cup, \cup)$, where $\cup : \Omega \times \Omega \to \Omega$ denotes the union arrow in $\mathscr{C}$ (Fig. 16). So, we obtain $\cup_{RSC(\mathscr{C})} = (\cup, \cup)$.

Consider the diagram in Fig. 41. Right hand side square commutes and is a pullback (Proposition 32). Let $(f', f), (g', g) \in \mathcal{M}((U_1, U_2, m))$ such that $(f', f) : (X_1, X_2, m_1) \to (U_1, U_2, m)$ and $(g', g) : (Y_1, Y_2, m_2) \to (U_1, U_2, m)$. As observed above in case of intersection, $\langle \chi_{(f',f)}, \chi_{(g',g)} \rangle = (\langle \chi_{f'}, \chi_{g'} \rangle, \langle \chi_f, \chi_g \rangle) : (U_1, U_2, m) \to (\Omega \times \Omega, \Omega \times \Omega, 1_{\Omega \times \Omega})$ is an $RSC(\mathscr{C})$-arrow, i.e. $\langle \chi_{f'}, \chi_{g'} \rangle = \langle \chi_f, \chi_g \rangle \circ m.$

From diagram in Fig. 16, the following two diagrams are pullbacks in $\mathscr{C}$.

$$\begin{array}{ccc} X_1 \cup Y_1 & \xrightarrow{t_1} & \Omega + \Omega \\ {\scriptstyle f' \cup g'} \downarrow & & \downarrow {\scriptstyle \alpha} \\ U_1 & \xrightarrow{\langle \chi_{f'}, \chi_{g'} \rangle} & \Omega \times \Omega \end{array} \qquad \begin{array}{ccc} X_2 \cup Y_2 & \xrightarrow{t_2} & \Omega + \Omega \\ {\scriptstyle f \cup g} \downarrow & & \downarrow {\scriptstyle \alpha} \\ U_2 & \xrightarrow{\langle \chi_f, \chi_g \rangle} & \Omega \times \Omega \end{array}$$

Observe that $\langle \chi_f, \chi_g \rangle \circ m \circ (f' \cup g') = \langle \chi_{f'}, \chi_{g'} \rangle \circ (f' \cup g') = \alpha \circ t_1.$ So, using the pullback property of the right hand side diagram above, there exists a unique arrow $h : X_1 \cup Y_1 \to X_1 \cup Y_2$ such that $(f \cup g) \circ h = m \circ (f' \cup g')$ and $t_2 \circ h = 1_{\Omega+\Omega} \circ t_1.$

Claim: h is monic in $\mathscr{C}$. Let $r, s : D \to X_1 \cup Y_1$ such that $h \circ r = h \circ s.$ Then $(f \cup g) \circ h \circ r = (f \cup g) \circ h \circ s \Rightarrow m \circ (f' \cup g') \circ r = m \circ (f' \cup g') \circ s \Rightarrow r = s,$ because m and $f' \cup g'$ are monics in $\mathscr{C}$. Therefore, $(X_1 \cup Y_1, X_2 \cup Y_2, h)$ is an $RSC(\mathscr{C})$-object.

Moreover, $(f \cup g) \circ h = m \circ (f' \cup g')$ implies that $(f' \cup g', f \cup g)$ is an $RSC(\mathscr{C})$-arrow.

So, in Fig. 41, the left hand side square is a diagram in $RSC(\mathscr{C})$. Using Proposition 32, this square is a pullback in $RSC(\mathscr{C})$. Finally, using Proposition 4, the outer square is a pullback in $RSC(\mathscr{C})$, and thus, $(f' \cup g', f \cup g)$ is the the characteristic morphism of $\cup_{RSC(\mathscr{C})} \circ (\langle \chi_{f'}, \chi_{g'} \rangle, \langle \chi_f, \chi_g \rangle)$, and $(f', f) \cup (g', g) = (f' \cup g', f \cup g).$

$$(X_1 \cup Y_1, X_2 \cup Y_2, h) \xrightarrow{(t_1, t_2)} (\Omega + \Omega, \Omega + \Omega, 1_{\Omega + \Omega}) \xrightarrow{(!_{\Omega + \Omega}, !_{\Omega + \Omega})} (1, 1, 1_1)$$

$$\downarrow (f' \cup g', f \cup g) \qquad\qquad \downarrow (\alpha, \alpha) \qquad\qquad\qquad \downarrow (\top, \top)$$

$$(U_1, U_2, m) \xrightarrow{(\langle \chi_{f'}, \chi_{g'} \rangle, \langle \chi_f, \chi_g \rangle)} (\Omega \times \Omega, \Omega \times \Omega, 1_{\Omega \times \Omega}) \xrightarrow{(\cup, \cup)} (\Omega, \Omega, 1_\Omega)$$

Fig. 41. Union of (f', f) and (g', g) in $RSC(\mathscr{C})$

Negation: Using Proposition 33, $(0, 0, 1_0)$ and $(1, 1, 1_1)$ are the initial and terminal objects in $RSC(\mathscr{C})$ respectively. Consider the diagram below.

$$(0, 0, 1_0) \xrightarrow{(!,!)} (1, 1, 1_1)$$

$$\downarrow (!,!) \qquad\qquad \downarrow (\top, \top)$$

$$(1, 1, 1_1) \xrightarrow{(\bot, \bot)} (\Omega, \Omega, 1_\Omega)$$

The $\mathscr{C}$-arrow $\bot : 1 \to \Omega$ is the characteristic morphism of $! : 0 \to 1$ in $\mathscr{C}$. Using Proposition 32, the above diagram is a pullback in $RSC(\mathscr{C})$. Therefore, $(\bot, \bot)$ is the characteristic morphism of $(!, !)$.

The arrow $\neg_{RSC(\mathscr{C})} : (\Omega, \Omega, 1_\Omega) \to (\Omega, \Omega, 1_\Omega)$ (Definition 30) is defined as the characteristic morphism of $(\bot, \bot)$. Since $\neg : \Omega \to \Omega$ is the characteristic morphism of $\bot$, again using Proposition 32, the right hand side diagram in Fig. 42 is a pullback, and $\neg_{RSC(\mathscr{C})} = (\neg, \neg)$.

Let $(f', f) \in \mathcal{M}((U_1, U_2, m))$ such that $(f', f) : (X_1, X_2, m_1) \to (U_1, U_2, m)$. The following diagrams are pullbacks in $\mathscr{C}$ (Fig. 18).

$$\neg X_2 \xrightarrow{!_{\neg X_2}} 1 \qquad\qquad \neg X_1 \xrightarrow{!_{\neg X_1}} 1$$

$$\downarrow \neg f \qquad\qquad \downarrow \bot \qquad\qquad \downarrow \neg f' \qquad\qquad \downarrow \bot$$

$$U_2 \xrightarrow{\chi_f} \Omega \qquad\qquad U_1 \xrightarrow{\chi_{f'}} \Omega$$

In Proposition 38, we have observed that $(\chi_{f'}, \chi_f)$ is an $RSC(\mathscr{C})$-arrow. So, $\chi_f \circ m = \chi_{f'}$. Now, $\chi_f \circ m \circ \neg f' = 1_\Omega \circ \chi_{f'} \circ \neg f' = 1_\Omega \circ \bot \circ !_{\neg X_1}$. So, using the pullback property of the left hand side diagram above, there exists a unique arrow $h : \neg X_1 \to \neg X_2$ such that $\neg f \circ h = m \circ \neg f'$ and $!_{\neg X_2} \circ h = !_{\neg X_1}$.

Claim: h is monic in $\mathscr{C}$. Let $r, s : D \to \neg X_1$ such that $h \circ r = h \circ s$. Then $\neg f \circ h \circ r = \neg f \circ h \circ s \Rightarrow m \circ \neg f' \circ r = m \circ \neg f' \circ s \Rightarrow r = s$, because m and $\neg f'$ are monics in $\mathscr{C}$. Therefore, $(\neg X_1, \neg X_2, h)$ is an $RSC(\mathscr{C})$-object.

Moreover, $\neg f \circ h = m \circ \neg f'$ implies that $(\neg f', \neg f)$ is an $RSC(\mathscr{C})$-arrow.

So, the left hand side square in Fig. 42 is a diagram in $RSC(\mathscr{C})$. Using Proposition 32 in $RSC(\mathscr{C})$, this square is a pullback in $RSC(\mathscr{C})$. Finally, using Proposition 4, the outer square is a pullback in $RSC(\mathscr{C})$, and and thus, $\neg(f', f)$ is the the characteristic morphism of $\neg_{RSC(\mathscr{C})} \circ (\chi_{f'}, \chi_f)$, and $\neg(f', f) = (\neg f', \neg f)$.

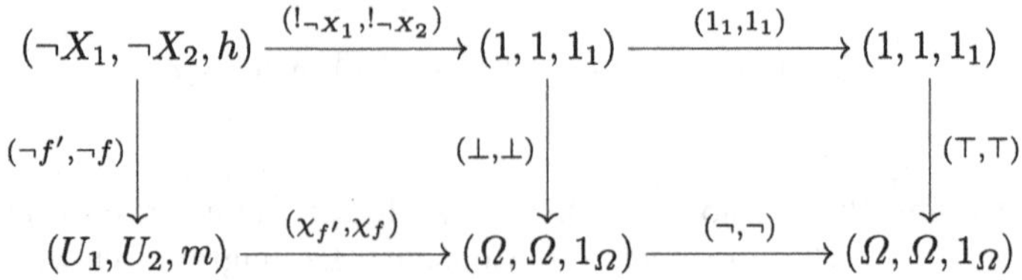

$$
\begin{array}{ccccc}
(\neg X_1, \neg X_2, h) & \xrightarrow{(!\neg x_1, !\neg x_2)} & (1, 1, 1_1) & \xrightarrow{(1_1, 1_1)} & (1, 1, 1_1) \\
\downarrow{\scriptstyle (\neg f', \neg f)} & & \downarrow{\scriptstyle (\bot, \bot)} & & \downarrow{\scriptstyle (\top, \top)} \\
(U_1, U_2, m) & \xrightarrow{(\chi_{f'}, \chi_f)} & (\Omega, \Omega, 1_\Omega) & \xrightarrow{(\neg, \neg)} & (\Omega, \Omega, 1_\Omega)
\end{array}
$$

Fig. 42. Negation of (f', f) in $RSC(\mathscr{C})$

Implication: Let $e : E \to \Omega \times \Omega$ be the equalizer of $\cap : \Omega \times \Omega \to \Omega$ and $p_1 : \Omega \times \Omega \to \Omega$ in $\mathscr{C}$, where $(\Omega \times \Omega, p_1, p_2)$ is the product. We have already observed that $\cap_{RSC(\mathscr{C})} = (\cap, \cap)$. From the property of products in Proposition 31, $((\Omega \times \Omega, \Omega \times \Omega, 1_{\Omega \times \Omega}), (p_1, p_1), (p_2, p_2))$ is a product in $RSC(\mathscr{C})$. So Proposition 31 also implies that the equalizer of $(\cap, \cap)$ and (p_1, p_1) is $(e, e) : (E, E, 1_E) \to (\Omega \times \Omega, \Omega \times \Omega, 1_{\Omega \times \Omega})$.

Since $\to$ is the characteristic morphism of the equalizer e in $\mathscr{C}$, using Proposition 32, we have $(\to, \to)$ as the characteristic morphism of (e, e) in $RSC(\mathscr{C})$. Therefore, we have $\to_{RSC(\mathscr{C})} = (\to, \to)$.

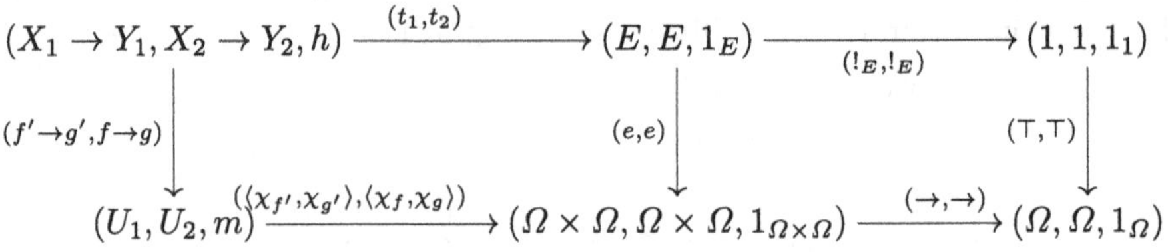

$$
\begin{array}{ccccc}
(X_1 \to Y_1, X_2 \to Y_2, h) & \xrightarrow{(t_1, t_2)} & (E, E, 1_E) & \xrightarrow{(!_E, !_E)} & (1, 1, 1_1) \\
\downarrow{\scriptstyle (f' \to g', f \to g)} & & \downarrow{\scriptstyle (e, e)} & & \downarrow{\scriptstyle (\top, \top)} \\
(U_1, U_2, m) & \xrightarrow{(\langle \chi_{f'}, \chi_{g'} \rangle, \langle \chi_f, \chi_g \rangle)} & (\Omega \times \Omega, \Omega \times \Omega, 1_{\Omega \times \Omega}) & \xrightarrow{(\to, \to)} & (\Omega, \Omega, 1_\Omega)
\end{array}
$$

Fig. 43. Implication of (f', f) and (g', g) in $RSC(\mathscr{C})$

As observed above for intersections $(\cap)$ and union $(\cup)$ in $RSC(\mathscr{C})$, for (f', f), $(g', g) \in \mathcal{M}((U_1, U_2, m))$ such that $(f', f) : (X_1, X_2, m_1) \to (U_1, U_2, m)$ and $(g', g) : (Y_1, Y_2, m_2) \to (U_1, U_2, m)$, $\langle \chi_{(f', f)}, \chi_{(g', g)} \rangle = (\langle \chi_{f'}, \chi_{g'} \rangle, \langle \chi_f, \chi_g \rangle)$. Since pullback of e with $\langle \chi_f, \chi_g \rangle$ gives $f \to g$, using Proposition 32 again and similar arguments as for the previous operators of $\cap$ and $\cup$ we have $(f', f) \to (g', g) = (f' \to g', f \to g)$ (Fig. 43).

In the subobject lattice $\mathcal{M}((U_1, U_2, m))$, the top element is $(1_{U_1}, 1_{U_2})$ with domain $(1, 1, 1_1)$, and the bottom element is (i_{U_1}, i_{U_2}) with domain $(0, 0, 1_0)$. Using Proposition 13, $\mathcal{M}((U_1, U_2, m))$ with the above operators forms a pBa. Let us summarize these results.

Proposition 53. *The operators on $\mathcal{M}((U_1, U_2, m))$ obtained by taking the pull-backs of the respective characteristic morphisms along the $RSC(\mathscr{C})$-subobject classifier $(\top, \top) : (1, 1, 1_1) \to (\Omega, \Omega, 1_\Omega)$ are as follows:*

$$\cap:\ (f', f) \cap (g', g) = (f' \cap g', f \cap g),$$
$$\cup:\ (f', f) \cup (g', g) = (f' \cup g', f \cup g),$$
$$\neg:\ \neg(f', f) = (\neg f', \neg f),$$
$$\to:\ (f', f) \to (g', g) = (f' \to g', f \to g),$$

where (f', f) and (g', g) are strong monics with codomain (U_1, U_2, m), and $\top : 1 \to \Omega$ is the subobject classifier of the topos $\mathscr{C}$. The operators on f', g' (and f, g) used above are those of the algebra of subobjects of U_1 (and U_2) in the topos $\mathscr{C}$. For any $RSC(\mathscr{C})$-object, following is a pBa.

$$(U_1, U_2, m), (\mathcal{M}((U_1, U_2, m)), (1_{U_1}, 1_{U_2}), (i_{U_1}, i_{U_2}), \cap, \cup, \to, \neg)$$

Before proceeding any further in the generalized scenario of $RSC(\mathscr{C})$, let us first study the above algebra when $\mathscr{C}$ is *SET*. Recall Definition 2 of rough sets. Let $(\mathbb{U}, R)$ be an approximation space and $U \subseteq \mathbb{U}$. Consider the object $(\underline{\mathcal{U}}, \overline{\mathcal{U}})$ of RSC, the pair of collections of equivalence classes in $\mathbb{U}$ contained in the R-lower and R-upper approximations of U respectively.

Remark 13.

1. Let $\mathcal{M}(U)$ denote the set of all strong subobjects of $(\underline{\mathcal{U}}, \overline{\mathcal{U}})$. In other words, $\mathcal{M}(U)$ consists of all RSC-objects (X_1, X_2) such that $X_1 \subseteq X_2 \subseteq \overline{\mathcal{U}}$ and $X_1 = X_2 \cap \underline{\mathcal{U}}$, that is $X_2 \backslash X_1 \subseteq \overline{\mathcal{U}} \backslash \underline{\mathcal{U}}$ (cf. Proposition 36). We show below in Proposition 55 that all objects in $\mathcal{M}(U)$ are, in fact, of the form $(\underline{A}, \overline{A})$, for some $A \subseteq U$.

2. It must also be noted here that, as observed in Remark 9(ii), any RSC-object (U_1, U_2) can be expressed, upto set isomorphism, as $(\underline{\mathcal{U}}, \overline{\mathcal{U}})$ for some $ROUGH$-object $(\mathbb{U}, R, U)$. So the set of all strong subobjects of the RSC-object (U_1, U_2) can be identified with $\mathcal{M}(U)$. Thus, by studying the algebra formed by $\mathcal{M}(U)$ for U in any approximation space $(\mathbb{U}, R)$, we shall cover the study of the algebra formed by the set of strong subobjects of any RSC-object (U_1, U_2).

3. $Sub(A) \cong (\mathcal{P}(A), A, \emptyset, \cup, \cap, \to, \neg)$ for any set A. Thus, $(\mathcal{P}(\overline{\mathcal{U}}), \overline{\mathcal{U}}, \emptyset, \cup, \cap, \to, \neg)$ and $(\mathcal{P}(\underline{\mathcal{U}}), \underline{\mathcal{U}}, \emptyset, \cup, \cap, \to, \neg)$ both form Boolean algebras. Here, $\cup$ and $\cap$ are the usual set intersections and unions; $\neg$ is the relative complement with respect to $\overline{\mathcal{U}}$ and $\underline{\mathcal{U}}$, that is, for $B \subseteq \overline{\mathcal{U}} \ (\underline{\mathcal{U}})$, $\neg B = \overline{\mathcal{U}} \backslash B \ (\underline{\mathcal{U}} \backslash B)$; and $\to$ is defined using Proposition 2(12), that is, for $B, C \subseteq \overline{\mathcal{U}} \ (\underline{\mathcal{U}})$, $B \to C = (\overline{\mathcal{U}} \backslash B) \cup C \ ((\underline{\mathcal{U}} \backslash B) \cup C)$.

Let us see exactly what form the operators on $\mathcal{M}(U)$ take in RSC. As mentioned in Notation 2, in a topos or a quasitopos $\mathscr{C}$, for a $\mathscr{C}$-object D, we can identify a subobject f in $Sub(D)$ with its domain. Then Proposition 53 and Remark 13(3) give operators on $\mathcal{M}(U)$ as follows.

Proposition 54. *The following operators on $\mathcal{M}(U)$ are obtained using pullbacks of the respective characteristic morphisms along the RSC subobject classifier* $(\top, \top) : (1,1) \to (2,2)$:

$$\cap : (A_1, A_2) \cap (B_1, B_2) = (A_1 \cap B_1, A_2 \cap B_2),$$
$$\cup : (A_1, A_2) \cup (B_1, B_2) = (A_1 \cup B_1, A_2 \cup B_2),$$
$$\neg : \neg(A_1, A_2) = (\neg A_1, \neg A_2)$$
$$= (\underline{\mathcal{U}}\backslash A_1, \overline{\mathcal{U}}\backslash A_2), \ and$$
$$\to : (A_1, A_2) \to (B_1, B_2) = (A_1 \to B_1, A_2 \to B_2)$$
$$= (\underline{\mathcal{U}}\backslash A_1, \overline{\mathcal{U}}\backslash A_2) \cup (B_1, B_2).$$

$(\mathcal{M}(U), (\underline{\mathcal{U}}, \overline{\mathcal{U}}), (\emptyset, \emptyset), \cup, \cap, \neg, \to)$ *forms a Boolean algebra, where the top and bottom elements are $(\underline{\mathcal{U}}, \overline{\mathcal{U}})$ and $(\emptyset, \emptyset)$ respectively.*

Remark 14. As observed earlier (cf. Observation 22), the class of monics in ξ-*RSC* and the class of strong monics in *RSC* is the same, that is, it is $\mathcal{M}(U)$. The operators $\cup, \cap, \neg, \to$ on $\mathcal{M}(U)$ in ξ-*RSC* are also the same.

5.1.1   The Relative Rough Complement

From Proposition 54 we conclude that in the quasitopos *RSC*, the algebra obtained over the class of strong subobjects of any *RSC*-object (U_1, U_2) is a Boolean algebra. This makes *RSC* a Boolean quasitopos [62]. However, as is well-known, algebraic structures formed from rough sets (e.g. cf. [5]) are non-Boolean. The prime reason for the classical behaviour here, we find, lies in the definition of negation $\neg$ of the *RSC*-object (A_1, A_2). So, we take a re-look at this operator. One immediately notes that we require a *relative* rough complement here, as the negation of (A_1, A_2) is *with respect to* the object $(\underline{\mathcal{U}}, \overline{\mathcal{U}})$. We refer to the only proposal of relative complement in rough set literature, namely Iwiński's *rough difference* operator [48], and re-define negation on strong monics in *RSC* as follows.

Definition 105.
On the set $\mathcal{M}(U)$, define negation as:

$$\sim : \ \sim(A_1, A_2) := (\underline{\mathcal{U}}\backslash A_2, \overline{\mathcal{U}}\backslash A_1).$$

All other operators $\cap, \cup$ and $\to$ remain the same.

Observation 31. *The operator $\sim$ is well-defined, because* $(\overline{\mathcal{U}}\backslash A_1) \cap \underline{\mathcal{U}} = (\overline{\mathcal{U}}\backslash(A_2 \cap \underline{\mathcal{U}})) \cap \underline{\mathcal{U}} = ((\overline{\mathcal{U}}\backslash A_2) \cup (\overline{\mathcal{U}}\backslash \underline{\mathcal{U}})) \cap \underline{\mathcal{U}} = ((\overline{\mathcal{U}}\backslash A_2) \cap \underline{\mathcal{U}}) \cup ((\overline{\mathcal{U}}\backslash \underline{\mathcal{U}}) \cap \underline{\mathcal{U}}) = (\underline{\mathcal{U}}\backslash A_2) \cup \emptyset = \underline{\mathcal{U}}\backslash A_2$. *Moreover, we also have $\underline{\mathcal{U}}\backslash A_2 = \underline{\mathcal{U}}\backslash A_1$. Therefore, $\sim$ can also be defined as:*

$$\sim : \ \sim(A_1, A_2) = (\underline{\mathcal{U}}\backslash A_1, \overline{\mathcal{U}}\backslash A_1).$$

Remark 15. What does the $\sim$ signify? Consider Fig. 44 representing both the negations $\neg(A_1, A_2)$ and $\sim(A_1, A_2)$ for some $(A_1, A_2) \in \mathcal{M}(U)$. $\overline{U}$ is denoted by the complete outer square, and $\underline{U}$ is denoted by the blue shaded inner square. A_1 is the inner most square, and A_2 is the dark grey shaded square including A_1. The first component of the negation is the area shaded with slanted lines, and the second component of the negation is area shaded with horizontal lines. Note the following points for negations.

(1) The first component $\underline{U} \backslash A_1 = \underline{U} \backslash A_2$ for both the negations are same.
(2) The region $A_2 \backslash A_1$ is not included in the second component $\overline{U} \backslash A_2$ of $\neg(A_1, A_2)$, but is included in the second component of $\sim(A_1, A_2)$.

The set $A_2 \backslash A_1$ is a subset of the boundary region $\overline{U} \backslash \underline{U}$, which contains exactly those equivalence classes which are partially, but not completely, included in the set U. We note that the possible region $\overline{U} \backslash A_2$ of $\neg(A_1, A_2)$ $(= (\underline{U} \backslash A_1, \overline{U} \backslash A_2))$ does not include $A_2 \backslash A_1$ which is part of the 'non-definite' region of (A_1, A_2). Therefore, we would want to add $A_2 \backslash A_1$ to the boundary region of the negation, thereby giving $(\overline{U} \backslash A_2) \cup (A_2 \backslash A_1) = \overline{U} \backslash A_1$ as the new possible region. This ensures that $\sim$ captures negation of a rough set in a better way than $\neg$.

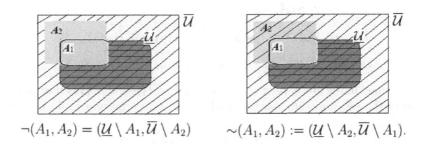

$$\neg(A_1, A_2) = (\underline{U} \backslash A_1, \overline{U} \backslash A_2) \qquad \sim(A_1, A_2) := (\underline{U} \backslash A_2, \overline{U} \backslash A_1).$$

Fig. 44. Relative negations in rough sets

Let us see an example of strong monics and its negation in a rough set.

Example 16. One of the applications of rough sets is in medical science. Consider Table 2, containing some information about patients, and their diagnosis. This table is a modified form of the one given in [83]. Take $\mathbb{U} := \{p_i \mid 1 \leq i \leq 10\}$, where each p_i represents a patient. Define a relation '$\sim$' on $\mathbb{U}$ as follows: $p_i \sim p_j$ if and only if the diagnosis of patients p_i and p_j is the same, i.e. all the attribute values for the attributes cough, vomiting and shivering are the same for p_i and p_j. '$\sim$' is clearly an equivalence relation on $\mathbb{U}$. Consider the approximation space $(\mathbb{U}, \sim)$.

$$\mathbb{U}/\sim = \{\{p_1, p_4\}, \{p_2, p_8, p_7\}, \{p_3, p_5\}, \{p_6\}, \{p_9\}, \{p_{10}\}\}$$

Let $U := \{p_i \mid 1 \leq i \leq 7\}$. Then $(\mathbb{U}, \sim, U)$ is an object of *ROUGH*. Here,

$$\overline{U} := \{\{p_1, p_4\}, \{p_2, p_8, p_7\}, \{p_3, p_5\}, \{p_6\}\} \text{ and } \underline{U} := \{\{p_1, p_4\}, \{p_3, p_5\}, \{p_6\}\}.$$

Table 2. An example of a medical information system

Patient	Cough	Vomiting	Shivering
p_1	Always	Seldom	Seldom
p_2	Seldom	Never	Always
p_3	Never	Always	Seldom
p_4	Always	Seldom	Seldom
p_5	Never	Always	Seldom
p_6	Never	Seldom	Always
p_7	Seldom	Never	Always
p_8	Seldom	Never	Always
p_9	Always	Always	Seldom
p_{10}	Seldom	Seldom	Seldom

Consider the set $\mathcal{M}(U)$ of all the strong subobjects of the RSC-object $(\underline{\mathcal{U}}, \overline{\mathcal{U}})$. Take a set $A := \{p_1, p_2, p_4\}$. Then $\overline{A} := \{\{p_1, p_4\}, \{p_2, p_8, p_7\}\}$ and $\underline{A} := \{\{p_1, p_4\}\}$. Observe that $\underline{A} = \overline{A} \cap \underline{\mathcal{U}}$ and $\underline{A} \subseteq \overline{A}$. Therefore, $(\underline{A}, \overline{A}) \in \mathcal{M}(U)$. What are $\neg(\underline{A}, \overline{A})$ and $\sim(\underline{A}, \overline{A})$?

$$\neg(\underline{A}, \overline{A}) = (\{\{p_3, p_5\}, \{p_6\}\}, \{\{p_3, p_5\}, \{p_6\}\})$$

$$\sim(\underline{A}, \overline{A}) = (\{\{p_3, p_5\}, \{p_6\}\}, \{\{p_2, p_8, p_7\}, \{p_3, p_5\}, \{p_6\}\})$$

As mentioned in the above remark, the set $\overline{A} \backslash \underline{A} := \{\{p_2, p_8, p_7\}\}$ is the boundary region of the set, and thus must be included in the negation of $(\underline{A}, \overline{A})$. This is true for the negation $\sim$, and not for $\neg$.

Now, consider the set $C := U \backslash \underline{A} := \{p_2, p_3, p_5, p_6, p_7\}$. Here,

$$(\underline{C}, \overline{C}) = (\{\{p_3, p_5\}, \{p_6\}\}, \{\{p_2, p_8, p_7\}, \{p_3, p_5\}, \{p_6\}\}).$$

Comparing this with both the negations, we have $(\underline{C}, \overline{C}) = \sim(\underline{A}, \overline{A})$. We shall show in the sequel that given any arbitrary $(A_1, A_2) \in \mathcal{M}(U)$, for a suitable A such that $(\underline{A}, \overline{A}) = (A_1, A_2)$, we can always obtain $(\underline{C}, \overline{C}) = \sim(\underline{A}, \overline{A})$.

Let us now show that any strong subobject of $(\underline{\mathcal{U}}, \overline{\mathcal{U}})$ can be expressed by the collections of equivalence classes in the lower and upper approximations of some subset of U.

Proposition 55. *Let* $(A_1, A_2) \in \mathcal{M}(U)$. *Then there exists* $A \subseteq U$, *given by* $A := U \cap (\bigcup A_2)$, *such that* $(A_1, A_2) = (\underline{A}, \overline{A})$, *that is* $A_1 = \underline{A}$ *and* $A_2 = \overline{A}$.

Proof. Let $[x] \in A_2$, that is, $[x] \subseteq \bigcup A_2 \Rightarrow [x] \cap \bigcup A_2 = [x]$. We also have, $A_2 \subseteq \overline{\mathcal{U}}$, that is, $[x] \cap U \neq \emptyset$. Taking intersection of these two, we have $[x] \cap (U \cap \bigcup A_2) \neq \emptyset$, that is, $[x] \cap A \neq \emptyset \Rightarrow [x] \in \overline{A}$.

The converse can be obtained by following the above argument in the opposite direction. Note that if $[x] \cap \bigcup A_2 \neq \emptyset \Rightarrow [x] \subseteq \bigcup A_2$, because A_2 is a set of equivalence classes. Therefore, we have $A_2 = \overline{A}$.

Now, using the fact that $\underline{X \cap Y} = \underline{X} \cap \underline{Y}$, we have $\underline{A} = \underline{U} \cap (\underline{\bigcup A_2})$, that is $\underline{A} = \underline{U} \cap A_2 = A_1$.

In other words, all objects in $\mathcal{M}(U)$ are of the form $(\underline{A}, \overline{A})$. Moreover, we get closure with respect to all the previously defined operators on $\mathcal{M}(U)$. Before that let us make a useful observation.

Observation 32. Consider $(A_1, A_2) \in \mathcal{M}(U)$. Let $C \subseteq U$ such that $\overline{C} = A_2$ and $\underline{C} = \overline{C} \cap \mathcal{U}$. Then $\overline{C} = A_2 \Rightarrow \overline{C} \cap \mathcal{U} = A_2 \cap \mathcal{U} \Rightarrow \underline{C} = A_1$. Therefore, $(\underline{C}, \overline{C}) = (A_1, A_2)$.

Proposition 56. *Given (A_1, A_2) and (B_1, B_2) as strong subobjects of $(\underline{\mathcal{U}}, \overline{\mathcal{U}})$, we have the following. Let $A := U \cap (\bigcup A_2)$ and $B := U \cap (\bigcup B_2)$.*

1. $\sim(A_1, A_2) = (\underline{C}, \overline{C})$, *where* $C := U \backslash A$.
2. $(A_1, A_2) \cup (B_1, B_2) = (\underline{C}, \overline{C})$, *where* $C := A \cup B$.
3. $(A_1, A_2) \cap (B_1, B_2) = (\underline{C}, \overline{C})$, *where* $C := A \cap B$.
4. $(A_1, A_2) \rightarrow (B_1, B_2) = (\underline{C}, \overline{C})$, *where* $C := (U \backslash A) \cup B$.

Proof. From Proposition 55, $(A_1, A_2) = (\underline{A}, \overline{A})$ and, so

$$\sim(A_1, A_2) = \sim(\underline{A}, \overline{A}) = (\underline{\mathcal{U}} \backslash \underline{A}, \overline{\mathcal{U}} \backslash \underline{A}).$$

(1) : ($\Rightarrow$) Let $[x] \in \underline{\mathcal{U}} \backslash \underline{A}$, i.e. $[x] \in \underline{\mathcal{U}}$ and $[x] \notin \underline{A}$. Since $\underline{A} = \overline{A} \cap \underline{\mathcal{U}}$, $[x] \notin \overline{A} \cap \underline{\mathcal{U}} \Rightarrow [x] \notin \overline{A} \Rightarrow [x] \cap A = \emptyset$. $\underline{A} \subseteq A$ and $[x] \cap A = \emptyset$ implies $[x] \cap \underline{A} = \emptyset$. Therefore, $[x] \subseteq U \backslash \underline{A}$, and $[x] \in \underline{C}$.
Let $[x] \in \overline{\mathcal{U}} \backslash \underline{A}$, i.e. $[x] \cap U \neq \emptyset$ and $[x] \cap \underline{A} = \emptyset$. $[x] \cap U \neq \emptyset$ implies that there exists $y \in [x]$ such that $y \in U$. $[x] \cap \underline{A} = \emptyset$ further implies that $y \notin \underline{A}$. Therefore, $[x] \cap (U \backslash \underline{A}) \neq \emptyset$, implying $[x] \in \overline{C}$.
($\Leftarrow$) Let $[x] \in \underline{C}$, i.e. $[x] \subseteq U \backslash \underline{A}$, i.e. $[x] \subseteq U$ and $[x] \not\subseteq A$. $[x] \subseteq U$ implies $[x] \in \underline{\mathcal{U}}$. $[x] \not\subseteq A$ implies $[x] \notin \underline{A}$. Therefore, $[x] \in \underline{\mathcal{U}} \backslash \underline{A}$.
Let $[x] \in \overline{C}$, i.e. $[x] \cap (U \backslash \underline{A}) \neq \emptyset$, i.e. there exists $y \in [x]$ such that $y \in U \backslash \underline{A}$. This implies $[x] \cap U \neq \emptyset$ and $[x] \not\subseteq A$. Therefore, $[x] \in \overline{\mathcal{U}} \backslash \underline{A}$.
(2) : $\overline{C} = \overline{A \cup B} = \overline{A} \cup \overline{B}$, i.e. $\overline{C} = A_2 \cup B_2$.
$\underline{C} = \underline{A \cup B} = \underline{U \cap ((\bigcup A_2) \cup (\bigcup B_2))} = \underline{U} \cap \underline{A \cup B} = \underline{U} \cap (\underline{A} \cup \underline{B}) = \underline{A} \cup \underline{B} = \bigcup(A_1 \cup B_1)$.
(3) : $\overline{C} = \overline{A \cap B} = \overline{U \cap (\bigcup A_2) \cap (\bigcup B_2)} = \overline{U \cap \overline{A} \cap \overline{B}} = \overline{U} \cap \overline{A} \cap \overline{B} = \overline{A} \cap \overline{B}$.
$\underline{C} = \underline{A \cap B} = \underline{A} \cap \underline{B}$, i.e. $\underline{C} = A_2 \cup B_2$.
(4) : $\overline{C} = \overline{(U \backslash \overline{A}) \cup B} = \overline{(U \backslash \overline{A})} \cup \overline{B} = (\overline{U} \backslash \overline{A}) \cup \overline{B}$.
We have to show that $\underline{C} = (\underline{\mathcal{U}} \backslash \underline{A}) \cup \underline{B}$.
($\subseteq$): Let $[x] \in \underline{C}$, i.e. $[x] \subseteq (U \backslash \overline{A}) \cup B$. Suppose $[x] \notin (\underline{\mathcal{U}} \backslash \underline{A}) \cup \underline{B}$, i.e. $[x] \notin (\underline{\mathcal{U}} \backslash \underline{A})$ and $[x] \notin \underline{B}$. $[x] \notin (\underline{\mathcal{U}} \backslash \underline{A})$ implies that either $[x] \notin \underline{\mathcal{U}}$ or $[x] \in \underline{A}$.
Case 1: $[x] \notin \underline{B}$ and $[x] \notin \underline{\mathcal{U}}$: There exists $z \in [x]$ such that $z \notin U$. $B = U \cap (\bigcup B_2)$ implies $z \notin B$. Therefore, $z \notin (U \backslash \overline{A}) \cup B$, contradicting the fact that $[x] \in \underline{C}$.

Case 2: $[x] \notin \underline{\mathcal{B}}$ and $[x] \in \underline{\mathcal{A}}$: We have $[x] \subseteq \overline{A}$, i.e. for all $y \in [x]$, $y \notin (U \backslash \overline{A})$. Since $[x] \notin \underline{\mathcal{B}}$, there exists $z' \in [x]$ such that $z' \notin B$. Therefore, $z' \notin (U \backslash \overline{A}) \cup B$, again contradicting the assumption that $[x] \in \underline{\mathcal{C}}$.

$(\supseteq)$: Let $[x] \in (\underline{\mathcal{U}} \backslash \underline{\mathcal{A}}) \cup \underline{\mathcal{B}}$, i.e. either $[x] \in (\underline{\mathcal{U}} \backslash \underline{\mathcal{A}})$ or $[x] \in \underline{\mathcal{B}}$.

Case 1: $[x] \in \underline{\mathcal{B}}$ implies $[x] \subseteq B \subseteq C$. Therefore, $[x] \in \underline{\mathcal{C}}$.

Case 2: $[x] \in (\underline{\mathcal{U}} \backslash \underline{\mathcal{A}})$ implies $[x] \in \underline{\mathcal{U}}$ and $[x] \notin \underline{\mathcal{A}}$. Since $\underline{\mathcal{A}} = \underline{\mathcal{U}} \cap \overline{A}$ and $[x] \in \underline{\mathcal{U}}$, we have $[x] \notin \overline{A}$, implying $[x] \subseteq U \backslash \overline{A}$ and $[x] \in \underline{\mathcal{C}}$.

Remark 16. The closure of the operation $\sim$ in the set $\mathcal{M}(U)$ further substantiates the observation that it captures the property of negation of a rough set more appropriately than $\neg$.

Let us explore the properties that $\mathcal{M}(U)$ satisfies.

Proposition 57. *The algebra* $\mathcal{D}(U) := (\mathcal{M}(U), (\underline{\mathcal{U}}, \overline{\mathcal{U}}), (\emptyset, \emptyset), \cap, \cup, \rightarrow, \sim)$ *over* U, *where* $\cap$, $\cup$, $\sim$, $\rightarrow$ *are as in Definition 105, satisfies the following properties.*

(1). $\sim(\underline{\mathcal{U}}, \overline{\mathcal{U}}) = (\underline{\mathcal{U}} \backslash \overline{\mathcal{U}}, \overline{\mathcal{U}} \backslash \underline{\mathcal{U}}) = (\emptyset, \overline{\mathcal{U}} \backslash \underline{\mathcal{U}})$.

(2). $\sim(\emptyset, \emptyset) = (\underline{\mathcal{U}}, \overline{\mathcal{U}})$.

*(3). $\sim\sim(A_1, A_2) = (A_1, A_2 \cup (\overline{\mathcal{U}} \backslash \underline{\mathcal{U}}))$, which is not the same as (A_1, A_2).

(4). $(A_1, A_2) \subseteq \sim\sim(A_1, A_2)$.

(5). $\sim\sim\sim(A_1, A_2) = \sim(A_1, A_2)$.

(6). $(A_1, A_2) \cup \sim(A_1, A_2) = (\underline{\mathcal{U}}, \overline{\mathcal{U}})$.

*(7). $(A_1, A_2) \cap \sim(A_1, A_2) = (\emptyset, A_2 \backslash A_1)$, which may not be equal to $(\emptyset, \emptyset)$.

(8). $\sim((A_1, A_2) \cup (B_1, B_2)) = \sim(A_1, A_2) \cap \sim(B_1, B_2)$.

(9). $\sim((A_1, A_2) \cap (B_1, B_2)) = \sim(A_1, A_2) \cup \sim(B_1, B_2)$.

(10). $\sim\sim((A_1, A_2) \cup \sim(A_1, A_2)) = (\underline{\mathcal{U}}, \overline{\mathcal{U}})$.

(11). $(A_1, A_2) \rightarrow (B_1, B_2) \subseteq \sim(B_1, B_2) \rightarrow \sim(A_1, A_2)$.

(12). $\sim(A_1, A_2) = (A_1, A_2) \rightarrow \sim(\underline{\mathcal{U}}, \overline{\mathcal{U}})$.

Proof. (1) and (2) are direct using the definition of $\sim$.

(3) and (4) : $\sim\sim(A_1, A_2) = \sim(\underline{\mathcal{U}} \backslash A_2, \overline{\mathcal{U}} \backslash A_1) = ((\underline{\mathcal{U}} \backslash (\overline{\mathcal{U}} \backslash A_1)), (\overline{\mathcal{U}} \backslash (\underline{\mathcal{U}} \backslash A_2))) = (A_1, A_2 \cup (\overline{\mathcal{U}} \backslash \underline{\mathcal{U}}))$.

(5) : $\sim\sim\sim(A_1, A_2) = \sim(A_1, A_2 \cup (\overline{\mathcal{U}} \backslash \underline{\mathcal{U}})) = (\underline{\mathcal{U}} \backslash (A_2 \cup (\overline{\mathcal{U}} \backslash \underline{\mathcal{U}})), \overline{\mathcal{U}} \backslash A_1) = (\underline{\mathcal{U}} \backslash A_2, \overline{\mathcal{U}} \backslash A_1) = \sim(A_1, A_2)$.

(6) : $(A_1, A_2) \cup \sim(A_1, A_2) = (A_1, A_2) \cup (\underline{\mathcal{U}} \backslash A_1, \overline{\mathcal{U}} \backslash A_1) = (A_1 \cup (\underline{\mathcal{U}} \backslash A_1), A_2 \cup (\overline{\mathcal{U}} \backslash A_1)) = (\underline{\mathcal{U}}, \overline{\mathcal{U}}) = 1$.

(7) : $(A_1, A_2) \cap \sim(A_1, A_2) = (A_1, A_2) \cap (\underline{\mathcal{U}} \backslash A_1, \overline{\mathcal{U}} \backslash A_1) = (A_1 \cap (\underline{\mathcal{U}} \backslash A_1), A_2 \cap (\overline{\mathcal{U}} \backslash A_1)) = (\emptyset, A_2 \backslash A_1)$.

(8) and (9) can be directly obtained using deMorgan's law of sets. Let us see (9).
$\sim((A_1, A_2) \cap (B_1, B_2)) = (\underline{\mathcal{U}} \backslash (A_1 \cap B_1), \overline{\mathcal{U}} \backslash (A_1 \cap B_1)) = (\underline{\mathcal{U}} \backslash A_1 \cup \underline{\mathcal{U}} \backslash B_1, \overline{\mathcal{U}} \backslash A_1 \cup \underline{\mathcal{U}} \backslash B_1) = (\underline{\mathcal{U}} \backslash A_1, \overline{\mathcal{U}} \backslash A_1) \cup (\underline{\mathcal{U}} \backslash B_1, \overline{\mathcal{U}} \backslash B_1) = \sim(A_1, A_2) \cup \sim(B_1, B_2)$.

(10) : $\sim\sim((A_1, A_2) \cup \sim(A_1, A_2)) = \sim\sim(\underline{\mathcal{U}}, \overline{\mathcal{U}}) = (\underline{\mathcal{U}}, \overline{\mathcal{U}})$, using (6) and (3).

(11) : $(A_1, A_2) \rightarrow (B_1, B_2) = (B_1 \cup (\underline{\mathcal{U}} \backslash A_1), B_2 \cup (\overline{\mathcal{U}} \backslash A_2))$, $\sim(B_1, B_2) \rightarrow \sim(A_1, A_2) = (B_1 \cup (\underline{\mathcal{U}} \backslash A_1), B_1 \cup (\overline{\mathcal{U}} \backslash A_1))$.
We need to verify $B_2 \cup (\overline{\mathcal{U}} \backslash A_2) \subseteq B_1 \cup (\overline{\mathcal{U}} \backslash A_1)$.
Let $[x] \in B_2 \cup (\overline{\mathcal{U}} \backslash A_2)$. If $[x] \in (\overline{\mathcal{U}} \backslash A_2)$, then we have $[x] \in (\overline{\mathcal{U}} \backslash A_1)$. Suppose

$[x] \in B_2$ such that $[x] \notin B_1$. Then we have, $[x] \notin B_2 \cap \mathcal{U}$, that is, $[x] \notin \mathcal{U}$, that is, $[x] \notin A_1$. Therefore, $[x] \in \overline{\mathcal{U}} \backslash A_1$.

$(12) : \sim(A_1, A_2) = (\mathcal{U} \backslash A_2, \overline{\mathcal{U}} \backslash A_1) = (\mathcal{U} \backslash A_1, \overline{\mathcal{U}} \backslash (A_2 \cap \mathcal{U})) = (\mathcal{U} \backslash A_1, \overline{\mathcal{U}} \backslash A_2) \cup (\emptyset, \overline{\mathcal{U}} \backslash \mathcal{U}) = (A_1, A_2) \rightarrow (\emptyset, \overline{\mathcal{U}} \backslash \mathcal{U})$.

Note that since $(\mathcal{M}(U), (\mathcal{U}, \overline{\mathcal{U}}), (\emptyset, \emptyset), \cap, \cup, \rightarrow, \neg)$ forms a Boolean algebra, the reduct $(\mathcal{M}(U), (\mathcal{U}, \overline{\mathcal{U}}), \cap, \cup, \rightarrow)$ is an *rpc* lattice. Further, using properties (6) and (12) in the above Proposition, and the Definition 6 of $c\vee c$ lattices, we have the following result.

Theorem 18. *For every object $(\mathcal{U}, \overline{\mathcal{U}})$ of RSC, the algebra of strong subobjects of $(\mathcal{U}, \overline{\mathcal{U}})$, that is, $\mathcal{D}(U) := (\mathcal{M}(U), (\mathcal{U}, \overline{\mathcal{U}}), (\emptyset, \emptyset), \cap, \cup, \rightarrow, \sim)$, is a $c\vee c$ lattice with the least element $(\emptyset, \emptyset)$.*
Thus, we obtain a class of $c\vee c$ lattices, *from the objects of RSC.*

We note here that the above algebra is not a Boolean algebra because of the properties (3) and (7) in Proposition 57.

Proposition 58.

(A). *In the algebra* $\mathcal{D}(U) := (\mathcal{M}(U), (\mathcal{U}, \overline{\mathcal{U}}), (\emptyset, \emptyset), \cap, \cup, \rightarrow, \sim)$, $(A_1, A_2) \cap \sim(A_1, A_2) = (\emptyset, \emptyset)$ *for all* $(A_1, A_2) \in \mathcal{M}(U)$, *if and only if* $\mathcal{U} = \overline{\mathcal{U}}$.
(B). *If* $\mathcal{U} = \overline{\mathcal{U}}$, $\mathcal{D}(U)$ *is a Boolean algebra and* $\mathcal{D}(U) \cong \mathcal{P}(\overline{\mathcal{U}})$.

Remark 17. In the algebra $\mathcal{D}(U)$, the operators $\rightarrow$ and $\sim$ deviate from the properties of Boolean algebra (Proposition 2), as there are $(A_1, A_2), (B_1, B_2) \in \mathcal{M}(U)$ such that

(1) $\sim(A_1, A_2) \cup (B_1, B_2) \nsubseteq (A_1, A_2) \rightarrow (B_1, B_2)$,
(2) $\sim(A_1, A_2) \nsubseteq (A_1, A_2) \rightarrow (B_1, B_2)$,
(3) $((A_1, A_2) \rightarrow (B_1, B_2)) \cap ((A_1, A_2) \rightarrow \sim(B_1, B_2)) \neq \sim(A_1, A_2)$,
(4) $\sim(B_1, B_2) \rightarrow \sim(A_1, A_2) \nsubseteq (A_1, A_2) \rightarrow (B_1, B_2)$.
(5) $\sim((B_1, B_2) \rightarrow (B_1, B_2)) \rightarrow (A_1, A_2) \neq (\mathcal{U}, \overline{\mathcal{U}})$.
(6) $\sim\sim(B_1, B_2) \nsubseteq (B_1, B_2)$.

These can be checked through an example. For (1) above, consider $(\mathcal{U}, \overline{\mathcal{U}}) := (\emptyset, \{[x], [y]\})$ for an appropriate U, where x and y are two elements of $\mathbb{U}$ in distinct equivalence classes. Take $(A_1, A_2) := (\emptyset, \{[x]\})$ and $(B_1, B_2) := (\emptyset, \{[y]\})$. Here, $\sim(A_1, A_2) = (\emptyset, \{[x], [y]\}) = (\emptyset, \overline{\mathcal{U}})$ and $\sim(B_1, B_2) = (\emptyset, \{[x], [y]\}) = (\emptyset, \overline{\mathcal{U}})$. Now, $(A_1, A_2) \rightarrow (B_1, B_2) = (\emptyset, \{[y]\}) = (B_1, B_2)$, and $\sim(A_1, A_2) \cup (B_1, B_2) = (\emptyset, \overline{\mathcal{U}}) \cup (B_1, B_2) = (B_1, \overline{\mathcal{U}})$. Since $B_2 \neq \overline{\mathcal{U}}$, we have the inequality (1). The same example will work for the other inequalities.

Note: In Proposition 64, we shall see that the construction of the algebra $\mathcal{D}(U)$ can be directly generalized to generate examples of classes of $c\vee c$ lattices obtained from Boolean algebras.

Let us now return to the quasitopos $RSC(\mathscr{C})$. Bringing Iwiński's *rough difference* operator into the scenario of $RSC(\mathscr{C})$ as well, we define a new negation $\sim$ on $\mathcal{M}((U_1, U_2, m))$ as follows:

$$\sim \, : \, \sim(f', f) := (\neg f', \neg(m \circ f')).$$

where $(f', f) : (X_1, X_2, m_1) \to (U_1, U_2, m)$ is a strong monic with codomain (U_1, U_2, m).

Is $\sim$ well-defined? We show that $(\neg f', \neg(m \circ f'))$ (i) forms an $RSC(\mathscr{C})$-arrow, and (ii) is a strong monic. For (i), we first have to find a suitable $RSC(\mathscr{C})$-object that will be the domain of $(\neg f', \neg(m \circ f'))$.

Since (f', f) is a strong monic in $\mathscr{C}$, we have the following pullback diagram in $\mathscr{C}$ (Proposition 35).

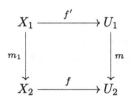

Both f' and f are monics (Proposition 33). Using Proposition 12, $m \circ f' = f \circ m_1 = f \cap m$ in $Sub(U_2)$. Note that $(m \circ f') \in Sub(U_2)$, $\neg f' \in Sub(U_1)$ and the domains of the $\mathscr{C}$-arrows f' and $m \circ f'$ are same, i.e. X_1. Consider $\neg X_1$ and $\neg \tilde{X}_1$ as the domains of $\neg(m \circ f')$ and $\neg f'$ in $Sub(U_2)$ and $Sub(U_1)$ respectively. Consider the following diagram:

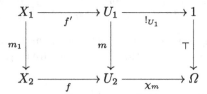

The left hand side square is a pullback because (f', f) is a strong monic. Right hand side is a pullback using the subobject classifier property in $\mathscr{C}$. Proposition 4 implies that the outer square is a pullback. Using the property of the subobject classifier, $\chi_{m_1} = \chi_m \circ f$. Now, consider the following diagram:

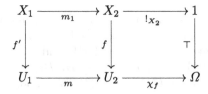

Again the right hand side square is a pullback and using Proposition 4, we have the outer square as pullback. Using the property of the subobject classifier, $\chi_{f'} = \chi_f \circ m$. We also have the following pullback in $\mathscr{C}$, and $\top \circ !_{U_1} = \chi_{1_{U_1}}$.

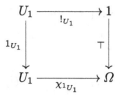

Now, consider the diagram in Fig. 45a. The inner square is a pullback using the definition of $\neg$ in the topos $\mathscr{C}$ (Fig. 18). Here, $\chi_{(m \circ f')} \circ m = \chi_{f \cap m} \circ m = \cap \circ \langle \chi_f, \chi_m \rangle \circ m = \cap \circ \langle \chi_f \circ m, \chi_m \circ m \rangle = \cap \circ \langle \chi_{f'}, \chi_m \circ m \rangle = \cap \circ \langle \chi_{f'}, \top \circ !_{U_1} \rangle = \cap \circ \langle \chi_{f'}, \chi_{1_{U_1}} \rangle = \chi_{(f' \cap 1_{U_1})}$. Moreover, in $Sub(U_1)$, $(f' \cap 1_{U_1}) = f'$.

$$\therefore \ \chi_{(m \circ f')} \circ m = \chi_{f'}. \tag{5.1}$$

We also have the pullback diagram in Fig. 45b. This gives $\chi_{(m \circ f')} \circ m \circ \neg f' = \chi_{f'} \circ \neg f' = \bot \circ !_{\neg \tilde{X}_1}$, and thus the outer diagram in Fig. 45a commutes. Using pullback property of the inner square, there exists a unique arrow $\alpha : \neg \tilde{X}_1 \to \neg X_1$ such that $\neg(m \circ f') \circ \alpha = m \circ \neg f'$. Therefore, the diagram in Fig. 46 commutes.

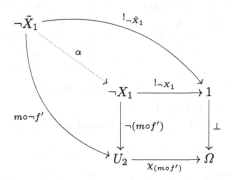

(a) Negation of $(m \circ f')$

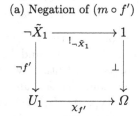

(b) Negation of f'

Fig. 45. Negations in $\mathscr{C}$

Claim: α is a monic. Let $h, k : D \to \neg \tilde{X}_1$ such that $\alpha \circ k = \alpha \circ h$. Then $\neg(m \circ f') \circ \alpha \circ k = \neg(m \circ f') \circ \alpha \circ h \Rightarrow m \circ \neg f' \circ k = m \circ \neg f' \circ h \Rightarrow k = h$, as m and $\neg f'$ are monics in $\mathscr{C}$. So, $(\neg \tilde{X}_1, \neg X_1, \alpha)$ is an $RSC(\mathscr{C})$-object and $(\neg f', \neg(m \circ f')) : (\neg \tilde{X}_1, \neg X_1, \alpha) \to (U_1, U_2, m)$ is an $RSC(\mathscr{C})$-arrow.

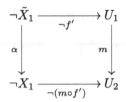

Fig. 46. $\sim(f', f)$ in $RSC(\mathscr{C})$

For (ii), we have the following claim: the diagram in Fig. 46 is a pullback. Consider the following diagram.

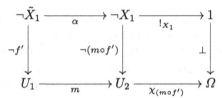

Using the negation property in the topos $\mathscr{C}$ (Fig. 18), the right hand side diagram is a pullback. From Eq. 5.1, $\chi_{(m \circ f')} \circ m = \chi_{f'}$. Using the uniqueness of $\chi_{f'}$, the outer diagram is also a pullback. Proposition 4 implies that the left hand side square is also a pullback, and thus the diagram in Fig. 46 is a pullback. Therefore, $(\neg f', \neg(m \circ f')) : (\neg \tilde{X}_1, \neg X_1, \alpha) \to (U_1, U_2, m)$ is a strong monic in $RSC(\mathscr{C})$. We thus formally have the definition of 'rough' negation.

Definition 106 (Rough negation).
On the set $\mathcal{M}((U_1, U_2, m))$, *define* rough negation *as:*

$$\sim \; : \; \sim(f', f) := (\neg f', \neg(m \circ f')) : (\neg \tilde{X}_1, \neg X_1, \alpha) \to (U_1, U_2, m)$$

where (f', f) *is a strong monic with codomain* (U_1, U_2, m) *and domain* (X_1, X_2, m_1); $\neg X_1$ *and* $\neg \tilde{X}_1$ *are the domains of* $\neg(m \circ f')$ *and* $\neg f'$ *respectively; the domain of* $\sim(f', f)$ *is* $(\neg \tilde{X}_1, \neg X_1, \alpha)$ *(defined as above).*

Observation 33. In RSC, we know that the set $\mathcal{M}((U_1, U_2))$ consists of all RSC-objects (X_1, X_2) such that $X_1 \subseteq X_2 \subseteq U_2$ and $X_1 = X_2 \cap U_1$ (Proposition 36). Note that here a strong subobject is identified with its domain. Consider a strong monic RSC-arrow f with domain (X_1, X_2) and codomain (U_1, U_2). Here, the domains of $\neg f|_{X_1}$ and $\neg(i_{U_1} \circ f|_{X_1})$ are $(U_1 \backslash X_1)$ and $(U_2 \backslash X_1)$ respectively, where $i_{U_1} : U_1 \to U_2$ is the inclusion function. Therefore, in RSC, for any $(X_1, X_2) \in \mathcal{M}((U_1, U_2))$, $\sim(X_1, X_2) = (U_1 \backslash X_2, U_2 \backslash X_1)$. This coincides with the definition of $\sim$ given in Definition 105.

Consider the set $\mathcal{M}((U_1, U_2, m))$ for any $RSC(\mathscr{C})$-object (U_1, U_2, m) and the algebra

$$\mathcal{A} := (\mathcal{M}((U_1, U_2, m)), (1_{U_1}, 1_{U_2}), (i_{U_1}, i_{U_2}), \cap, \cup, \to, \neg, \sim).$$

We have the following proposition.

Proposition 59. *In* $\mathcal{M}((U_1, U_2, m))$,

(1) $\sim(1_{U_1}, 1_{U_2}) = \neg\neg\sim(1_{U_1}, 1_{U_2})$, *and*
(2) $\sim(f', f) = (f', f) \rightarrow (\neg\neg\sim(1_{U_1}, 1_{U_2}))$, *for any* $(f', f) \in \mathcal{M}((U_1, U_2, m))$.

Proof.
(1). $\sim(1_{U_1}, 1_{U_2}) = (\neg 1_{U_1}, 1_{U_2} \rightarrow \neg m) = (i_{U_1}, \neg m)$.
So $\neg\neg\sim(1_{U_1}, 1_{U_2}) = \neg\neg(i_{U_1}, \neg m) = (\neg\neg i_{U_1}, \neg\neg\neg m) = (i_{U_1}, \neg m) = \sim(1_{U_1}, 1_{U_2})$.
(2). Let us take any $(f', f) \in \mathcal{M}((U_1, U_2, m))$ such that $(f', f) : (X_1, X_2, m_1) \rightarrow (U_1, U_2, m)$. In the definition of $\sim$ above, m and f' both are monics with codomain U_2.
Using Observation 2 and Proposition 35, we have $m \circ f' = f \cap m$. Simplifying $\sim(f', f)$, we have $\sim(f', f) = (\neg f', \neg(f \cap m)) =$
$(\neg f', f \rightarrow (\neg m)) = (f' \rightarrow i_{U_1}, f \rightarrow (\neg m)) = (f', f) \rightarrow (i_{U_1}, \neg m)$.
Thus $\sim(f', f) = (f', f) \rightarrow (\neg\neg\sim(1_{U_1}, 1_{U_2}))$.

Using Proposition 59 and Definition 6 of cc lattices, we have the following result.

Proposition 60. *Consider* $\mathcal{M}((U_1, U_2, m))$ *for any* $RSC(\mathscr{C})$*-object* (U_1, U_2, m).
Then $(\mathcal{M}((U_1, U_2, m)), (1_{U_1}, 1_{U_2}), (i_{U_1}, i_{U_2}), \cap, \cup, \rightarrow, \sim)$ *is a cc lattice with the least element* (i_{U_1}, i_{U_2}).

Observation 34. Using Observation 33 and Remark 17, we have the following.

(1) $\sim$ does not satisfy the semi-negation property:

$$\sim((f', f) \rightarrow (f', f)) \rightarrow (g', g) \neq (1_{U_1}, 1_{U_2}),$$

for some $(f', f), (g', g) \in \mathcal{M}((U_1, U_2, m))$.
Note that the semi-negation property holds in a pBa.
(2) $\sim$ does not satisfy the involution property:

$$\sim\sim(f', f) \neq (f', f)$$

for some $(f', f) \in \mathcal{M}((U_1, U_2, m))$.

Note that involution holds in a quasi-Boolean algebra.

Observation 34 suggests that the lattice

$$(\mathcal{M}((U_1, U_2, m)), (1_{U_1}, 1_{U_2}), (i_{U_1}, i_{U_2}), \cap, \cup, \rightarrow, \sim)$$

is neither pseudo-Boolean nor quasi-Boolean. Proposition 53 already gave us that $(\mathcal{M}((U_1, U_2, m)), (1_{U_1}, 1_{U_2}), (i_{U_1}, i_{U_2}), \cap, \cup, \rightarrow, \neg)$ is a pBa. Moreover, the two negations $\sim$ and $\neg$ are connected by the properties given in Proposition 59. Therefore we have obtained an instance of a new algebraic structure involving two distinct negations $\sim$ and $\neg$. We define the new structure as follows.

Definition 107. [64]

An abstract algebra $\mathcal{A} := (A, 1, 0, \vee, \wedge, \rightarrow, \neg, \sim)$ is called a contrapositionally complemented pseudo-Boolean algebra (ccpBa), if $(A, 1, 0, \wedge, \vee, \rightarrow, \neg)$ forms a pseudo-Boolean (Heyting) algebra, and for all $a \in A$, the following condition holds:

$$\sim a = a \rightarrow (\neg\neg\sim 1).$$

If, in addition, for all $a \in A$, $a \vee \sim a = 1$, we call $\mathcal{A}$ a contrapositionally $\vee$ complemented pseudo-Boolean algebra (c$\vee$cpBa).

Proposition 61.

(1) *For each $RSC(\mathscr{C})$-object (U_1, U_2, m), $(\mathcal{M}((U_1, U_2, m)), (1_{U_1}, 1_{U_2}), (i_{U_1}, i_{U_2}), \cap, \cup, \rightarrow, \neg, \sim)$ is a ccpBa.*
(2) *For each RSC-object $(\underline{\mathcal{X}}, \overline{\mathcal{X}})$, $(\mathcal{M}(X), (\underline{\mathcal{X}}, \overline{\mathcal{X}}), \cap, \cup, \rightarrow, \neg, \sim)$ is a c$\vee$cpBa.*

Proof. (1) is using Propositions 53 and 59(1).
(2) is using (1) where $\mathscr{C}$ is RSC, and Proposition 57(6). ∎

5.2   The Algebras $ccpBa$ and $c\vee cpBa$

Let us compare the algebras defined above with some existing lattices having two different negations. One of the most familiar such structure is the *quasi-pseudo Boolean algebra* or *Nelson algebra* [44, 75], in which one of the negations satisfies the involution property. Another example is from fuzzy logics, where the *strict basic logic (SBL)* with an additional negation and the corresponding algebras $SBL_\sim$ are defined [36]. Again, the additional negation is always taken to be involutive in $SBL_\sim$-algebras. Note that both the negations in $ccpBa$ and $c\vee cpBa$ are non-involutive in general.

Observation 35. In a $ccpBa$ $\mathcal{A} := (A, 1, 0, \vee, \wedge, \rightarrow, \neg, \sim)$, the condition $\sim a = a \rightarrow (\neg\neg\sim 1)$ can be equivalently expressed as '(1). $\sim a = a \rightarrow \sim 1$ and (2). $\sim 1 = (\neg\neg\sim 1)$'. Condition (1) is the required condition for negation in cc lattices, and thus the reduct $(A, 1, 0, \vee, \wedge, \rightarrow, \sim)$ involving only one negation $\sim$, forms a cc lattice with 0. This is the primary reason for naming the algebraic structure as 'contrapositionally-complemented' pseudo-Boolean algebra.

In any cc lattice, the negation $\sim$ is completely determined by $\rightarrow$ and the element ~ 1 (Definition 6(1)). Moreover, the element ~ 1 need not be the bottom element of the lattice. Note that in a $ccpBa$, there is a bottom element 0, which defines the negation $\neg$. The property $\sim 1 = (\neg\neg\sim 1)$ then gives the involutive property for the element ~ 1 of a $ccpBa$, with respect to the negation $\neg$. Examples in Sect. 5.3 will show that a $ccHa$ (a) need not have the involutive property for *all* elements, and (b) ~ 1 need not be the bottom element 0. We further note that the property $\sim 1 = (\neg\neg\sim 1)$ is a distinctive property of $ccpBa$s. If we define $\neg a := a \rightarrow 0$ in any bounded cc lattice $(A, 1, 0, \vee, \wedge, \rightarrow, \sim)$ then this property need not hold – see Example 18.

Let us now see some properties of the algebra involving the negation operators.

Proposition 62. *In any ccpBa* $\mathcal{A} := (A, 1, 0, \vee, \wedge, \rightarrow, \neg, \sim)$, *the following hold for all* $a, b \in A$.

(1) $a \rightarrow \sim b = b \rightarrow \sim a$ *(contraposition law for* $\sim$*)*.
(2) $a \leq \sim\sim a$.
(3) $\neg\neg\sim 1 = \sim 1$.
(4) $\sim\sim(\sim 1 \rightarrow a) = 1$.
(5) $\sim a = \neg(a \wedge \neg\sim 1)$.

Proof.
(1) and (2). From Observation 35, $(A, 1, \vee, \wedge, \rightarrow, \sim)$ is a *cc* lattice, thus (1) and (2) hold. Alternatively,

$$\begin{aligned} a \rightarrow \sim b &= a \rightarrow (b \rightarrow \neg\neg\sim 1) & \text{(by Definition 107)} \\ &= b \rightarrow (a \rightarrow \neg\neg\sim 1) & \text{((by Proposition 1(B)(2)))} \\ &= b \rightarrow \sim a \end{aligned}$$

Taking b as $\sim a$ in (1), and using $\sim a \rightarrow \sim a = 1$, we obtain (2).
(3). It is again direct from Observation 35. Alternatively, by Definition 107, $\sim 1 = 1 \rightarrow (\neg\neg\sim 1) = (\neg\neg\sim 1)$ (by Proposition 1(B)(4)).
(4). By Proposition 1(B)(5), $(\sim 1 \rightarrow 0) \wedge (\sim 1 \rightarrow a) = \sim 1 \rightarrow (0 \wedge a) = \sim 1 \rightarrow 0 = \neg\sim 1$. Now,

$$\begin{aligned} \sim(\sim 1 \rightarrow a) &= (\sim 1 \rightarrow a) \rightarrow \neg\neg\sim 1 \\ &= \neg\sim 1 \rightarrow \neg(\sim 1 \rightarrow a) & \text{(Using contraposition law for } \neg) \\ &= \neg\sim 1 \rightarrow ((\sim 1 \rightarrow a) \rightarrow 0) \\ &= (\neg\sim 1 \wedge (\sim 1 \rightarrow a)) \rightarrow 0 & \text{(By Proposition 1(B)(2))} \\ &= ((\sim 1 \rightarrow 0) \wedge (\sim 1 \rightarrow a)) \rightarrow 0 \\ &= (\sim 1 \rightarrow (0 \wedge a)) \rightarrow 0 & \text{(By Proposition 1(B)(5))} \\ &= (\sim 1 \rightarrow 0) \rightarrow 0 = \neg\neg\sim 1. \end{aligned}$$

In particular, $\sim(\sim 1 \rightarrow a) \leq \neg\neg\sim 1 \Rightarrow \sim(\sim 1 \rightarrow a) \rightarrow \neg\neg\sim 1 = 1$ (Proposition 1(B)(7)). Therefore, $\sim\sim(\sim 1 \rightarrow a) = 1$.
(5). $\sim a = a \rightarrow (\neg\neg\sim 1) = a \rightarrow (\neg\sim 1 \rightarrow 0) = (a \wedge \neg\sim 1) \rightarrow 0$ (by Proposition 1(B)(2)). Therefore, $\sim a = \neg(a \wedge \neg\sim 1)$.

Observation 36. The property $\sim\sim(\sim 1 \rightarrow a) = 1$ can also be expressed as follows:

$$((\sim 1 \rightarrow a) \rightarrow \sim 1) \rightarrow \sim 1 = 1$$

This is a special case of the Peirce's law $((b \rightarrow a) \rightarrow b) \rightarrow b = 1$ when $b = \sim 1$ [78].

Some more properties concerning the negation operators are as follows.

Proposition 63. *In a ccpBa* $\mathcal{A} := (A, 1, 0, \vee, \wedge, \rightarrow, \neg, \sim)$, *the following hold for any* $a \in A$.

(1) $\neg a \leq \sim a$.
(2) $a \leq \sim\neg a$.
(3) $\neg\sim a \leq \sim\neg a$
(4) $\sim a = \neg\neg\sim a$.
(5) $\neg\sim\neg a \leq \neg a$.

Proof. (1). We have $0 \leq \sim 1 \Rightarrow a \to 0 \leq a \to \sim 1$ (By Proposition 1(B)(8)). Therefore, $\neg a \leq \sim a$.

(2). From Proposition 2(2) and (1) above, we have $a \leq \neg(\neg a) \leq \sim(\neg a)$. Thus, $a \leq \sim\neg a$.

(3). From (1), $\neg a \leq \sim a$. Therefore $\neg a \wedge \sim a = \neg a$. From Proposition 2(6), $(\neg\sim a \wedge \sim a) = 0$, i.e. $(\neg\sim a \wedge \sim a) \wedge \neg a = 0 \wedge \neg a = 0 \leq \sim 1$. So, $\neg\sim a \wedge (\sim a \wedge \neg a) \leq \sim 1$, implying

$$\neg\sim a \leq (\sim a \wedge \neg a) \to \sim 1 \qquad \text{(By Proposition 1(B)(3))}$$
$$= \sim(\sim a \wedge \neg a) = \sim\neg a.$$

(4). Proposition 62(5) and Proposition 2(3) imply $\neg\neg\sim a = \neg\neg\neg(a \wedge \neg\sim 1) = \neg(a \wedge \neg\sim 1) = \sim a$.

(5). From (2) and Proposition 2(2), we have $a \leq \sim\neg a \leq \neg\neg(\sim\neg a)$. Using Proposition 1(B)(7), we have $a \to \neg\neg\sim\neg a = 1$. Using contraposition law for $\neg$, we have $\neg\sim\neg a \to \neg a = 1$. Using Proposition 1(B)(7) again, we have $\neg\sim\neg a \leq \neg a$.

5.2.1 Examples

Let us first identify the 3-element *ccpBa*.

Example 17. Consider the only 3-element bounded *rpc* lattice $(A, 1, 0, \wedge, \vee, \to_1)$ (Table 3a), upto isomorphism.

Then $(A, 1, 0, \wedge, \vee, \to_1, \neg_1)$ is the only 3-element *pBa*, upto isomorphism (Table 3b).

Now there are exactly two choices for $\sim$ that would make $(A, 1, 0, \wedge, \vee, \to_1, \neg_1, \sim)$ a *ccpBa*, namely $\sim := \neg_1$ (Table 3b) and $\sim := \sim_1$ (Table 3c). Thus, we get exactly two 3-element *ccpBas* $\mathcal{A}' := (A, 1, 0, \vee, \wedge, \to_1, \neg_1, \sim_1)$ and $\mathcal{B}' := (A, 1, 0, \vee, \wedge, \to_1, \neg_1, \neg_1)$, upto isomorphism. Note that $\mathcal{A}'$ is a $c \vee cpBa$, while $\mathcal{B}'$ is not.

Table 3. 3-element *ccpBa*

(a) Implication $\to_1$

$\to_1$	0	a	1
0	1	1	1
a	0	1	1
1	0	a	1

(b) Negation $\neg_1$

x	$\neg_1 x$
0	1
a	0
1	0

(c) Negation $\sim_1$

x	$\sim_1 x$
0	1
a	1
1	1

Using Observation 35 and Definition 107, examples of *ccpBa* can be obtained by starting from an arbitrary *pBa*, choosing an element '~ 1' in the *pBa* such that $\sim 1 = \neg\neg\sim 1$, and defining $\sim a$ as $\sim a := a \to \sim 1$. Let us see an example of a pBa where we get choices for the element ~ 1 apart from the elements 0 and 1.

Example 18. Consider the 6-element pBa $\mathcal{H}_6 := (H_6, 1, 0, \wedge, \vee, \to, \neg)$ given in Example 1. For convenience, we give the Hasse diagram (Fig. 47) and the table for $\to$ (Table 4) again.

As mentioned above, it is sufficient to fix ~ 1 such that $\neg\neg\sim 1 = \sim 1$ and define $\sim a := a \to \sim 1$. Thus the available choices for ~ 1 here are $0, 1, x,$ and y (Table 5).

(1) $\sim 1 := 0$, i.e. $\sim a = \neg a$. The resulting algebra $(H_6, 1, 0, \wedge, \vee, \to, \neg, \neg)$ is a $ccpBa$, but not a $cVcpBa$ because $z \vee \sim z \neq 1$.
(2) $\sim 1 := y$. The resulting algebra $(H_6, 1, 0, \wedge, \vee, \to, \neg, \sim)$ is a $ccpBa$, but again not a $cVcpBa$ because $z \vee \sim z \neq 1$.
(3) $\sim 1 := x$. The resulting algebra $(H_6, 1, 0, \wedge, \vee, \to, \neg, \sim)$ is a $cVcpBa$, and thus a $ccpBa$.
(4) $\sim 1 := 1$. The resulting algebra $(H_6, 1, 0, \wedge, \vee, \to, \neg, \sim)$ is again a $ccpBa$ and a $cVcpBa$.

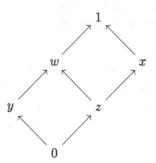

Fig. 47. $\mathcal{H}_6$ - a 6-element rpc lattice

Table 4. Implication in $\mathcal{H}_6$

$\to$	0	y	z	w	x	1
0	1	1	1	1	1	1
y	x	1	x	1	x	1
z	y	y	1	1	1	1
w	0	y	x	1	x	1
x	y	y	w	w	1	1
1	0	y	z	w	x	1

What if we take $\sim 1 := w$ or z? The resulting algebras are not $ccpBas$ because $\neg\neg w = 1 \neq w$ and $\neg\neg z = x \neq z$. In both of these cases, $(A, 1, 0, \wedge, \vee, \to, \sim)$ is a bounded cc lattice where $\sim a := a \to \sim 1$, but $\sim 1 \neq \neg\neg\sim 1$.

Table 5. Possibilities for the negation '$\sim$' in $\mathcal{H}_6$

a	$\sim a := \neg a$	$\sim a := a \to y$	$\sim a := a \to x$	$\sim a := a \to 1$	$\sim a := a \to w$
0	1	1	1	1	1
y	x	1	x	1	1
z	y	y	1	1	1
w	0	y	x	1	1
x	y	y	1	1	w
1	0	y	x	1	w
	$ccpBa$	$ccpBa$	$ccpBa$	$ccpBa$	$\cancel{ccpBa}$
	$\cancel{c\vee cpBa}$	$\cancel{c\vee cpBa}$	$c\vee cpBa$	$c\vee cpBa$	$\cancel{c\vee cpBa}$

We have already observed in Proposition 61 that by defining the operators appropriately on the class of the strong subobjects of an object in the (generalized) category of rough sets, we obtain a $ccpBa$ and $c\vee cpBa$. We now show that entire classes of examples of the algebra can be obtained by starting from an arbitrary pBa or Boolean alegbra and abstracting the category-theoretic constructions [64], as follows.

Consider a pBa $\mathcal{H} := (H, 1, 0, \vee, \wedge, \to, \neg)$, and the set $\mathcal{H}^{[2]} := \{(a, b) : a \leq b, a, b \in H\}$. $\mathcal{H}^{[2]}$ has been studied in various algebraic contexts for the special case when $\mathcal{H}$ is a Boolean algebra, for example by Moisil (cf. [19]).

Proposition 64. *For a pBa $\mathcal{H} := (H, 1, 0, \vee, \wedge, \to, \neg)$ and $u := (u_1, u_2) \in \mathcal{H}^{[2]}$, consider the set $A_u := \{(a_1, a_2) \in \mathcal{H}^{[2]} : a_2 \leq u_2 \text{ and } a_1 = a_2 \wedge u_1\}$. Define the following operators on A_u:*

$$\sqcup : (a_1, a_2) \sqcup (b_1, b_2) := (a_1 \vee b_1, a_2 \vee b_2),$$
$$\sqcap : (a_1, a_2) \sqcap (b_1, b_2) := (a_1 \wedge b_1, a_2 \wedge b_2),$$
$$\to : (a_1, a_2) \to (b_1, b_2) := ((a_1 \to b_1) \wedge u_1, (a_2 \to b_2) \wedge u_2),$$
$$\sim : \sim(a_1, a_2) := (u_1 \wedge \neg a_1, u_2 \wedge \neg a_1), \text{ and}$$
$$\neg : \neg(a_1, a_2) := (a_1, a_2) \to (0, 0).$$

Then $\mathcal{A}_u := (A_u, (u_1, u_2), (0, 0), \sqcap, \sqcup, \to, \neg, \sim)$ is a $ccpBa$.
Moreover, if $\mathcal{H}$ is a Boolean algebra, then $\mathcal{A}_u$ forms a $c\vee cpBa$. Thus we obtain a class of $ccpBas$ and $c\vee cpBas$, from any given pBa and Boolean algebra respectively.

Proof. Let us first verify that $(A_u, (u_1, u_2), \sqcup, \sqcap, \to, \neg)$ forms an *rpc* lattice. Since $(A, \vee, \wedge)$ is a distributive lattice; and $\sqcup$ and $\sqcap$ are defined component-wise over $\vee$ and $\wedge$ respectively, therefore $(A_u, \sqcup, \sqcap)$ forms a distributive lattice. We now have to check that the following property holds: For $(a_1, a_2), (b_1, b_2)$ and (x_1, x_2) in A_u,

$$(a_1, a_2) \sqcap (x_1, x_2) \leq (b_1, b_2) \Leftrightarrow (x_1, x_2) \leq (a_1, a_2) \to (b_1, b_2)$$

i.e., $(a_1 \wedge x_1, a_2 \wedge x_2) \leq (b_1, b_2) \Leftrightarrow (x_1, x_2) \leq ((a_1 \to b_1) \wedge u_1, (a_2 \to b_2) \wedge u_2)$.

This is equivalent to having the following property in $\mathcal{H}$:

$$\left\{ \begin{array}{c} a_1 \wedge x_1 \leq b_1 \\ \text{and } a_2 \wedge x_2 \leq b_2 \end{array} \right\} \Leftrightarrow \left\{ \begin{array}{c} x_1 \leq (a_1 \rightarrow b_1) \wedge u_1 \\ \text{and } x_2 \leq (a_2 \rightarrow b_2) \wedge u_2 \end{array} \right\}$$

$$i.e., \left\{ \begin{array}{c} x_1 \leq a_1 \rightarrow b_1 \\ \text{and } x_2 \leq a_2 \rightarrow b_2 \end{array} \right\} \Leftrightarrow \left\{ \begin{array}{c} x_1 \leq (a_1 \rightarrow b_1) \wedge u_1 \\ \text{and } x_2 \leq (a_2 \rightarrow b_2) \wedge u_2 \end{array} \right\}$$

The right to left implication is direct.

($\Rightarrow$): Using $x_1 \leq u_1$, we have $x_1 \leq (a_1 \rightarrow b_1)$ implies $x_1 \leq (a_1 \rightarrow b_1) \wedge u_1$. Similarly, $x_2 \leq u_2$ and $x_2 \leq (a_2 \rightarrow b_2)$ imply $x_2 \leq (a_2 \rightarrow b_2) \wedge u_2$.

Clearly $(0,0)$ and (u_1, u_2) are the bottom and top elements in A_u respectively, and A_u forms a *pBa*. For A_u to be a *ccpBa*, we have to show that $\sim(a_1, a_2) = (a_1, a_2) \rightarrow (\neg\neg\sim(u_1, u_2))$ for all $(a_1, a_2) \in A_u$. Let us make the following observations:

(1). In A_u, $\neg(a_1, a_2) = (a_1, a_2) \rightarrow (0,0) = ((a_1 \rightarrow 0) \wedge u_1, (a_2 \rightarrow 0) \wedge u_2) = (\neg a_1 \wedge u_1, \neg a_2 \wedge u_2)$.

(2). In A_u, $\sim(u_1, u_2) = (u_1 \wedge \neg u_1, u_2 \wedge \neg u_1) = \neg(u_1, u_1)$.

(3). In $\mathcal{H}$, $a_2 \rightarrow (u_2 \wedge \neg u_1) = (a_2 \rightarrow u_2) \wedge (a_2 \rightarrow \neg u_1)$ (Using Proposition 1(B)(5))

$= (a_2 \rightarrow \neg u_1)$ (Using $a_2 \rightarrow u_2 = 1$ and $1 \wedge x = x$)

$= (a_2 \rightarrow (u_1 \rightarrow 0)) = (a_2 \wedge u_1) \rightarrow 0 = a_1 \rightarrow 0 = \neg a_1$.

Now, using (2), we have $(a_1, a_2) \rightarrow (\neg\neg\sim(u_1, u_2)) = (a_1, a_2) \rightarrow (\neg\neg\neg(u_1, u_1))$. As A_u is a *pBa*, using Proposition 2(3), we have

$$\begin{aligned}
(a_1, a_2) \rightarrow (\neg\neg\sim(u_1, u_2)) &= (a_1, a_2) \rightarrow (\neg(u_1, u_1)) \\
&= (a_1, a_2) \rightarrow (0, u_2 \wedge \neg u_1) && \text{(Using (1))} \\
&= (\neg a_1 \wedge u_1, (a_2 \rightarrow (u_2 \wedge \neg u_1)) \wedge u_2) \\
&= (\neg a_1 \wedge u_1, \neg a_1 \wedge u_2) && \text{(Using (3))} \\
&= \sim(a_1, a_2)
\end{aligned}$$

If $\mathcal{H}$ is a Boolean algebra, then for A_u to form a *c$\vee$cpBa*, we have to show that for all $(a_1, a_2) \in A_u$, $(a_1, a_2) \vee \sim(a_1, a_2) = (u_1, u_2)$.

$$\begin{aligned}
(a_1, a_2) \vee \sim(a_1, a_2) &= (a_1, a_2) \vee (u_1 \wedge \neg a_1, u_2 \wedge \neg a_1) \\
&= (a_1 \vee (u_1 \wedge \neg a_1), a_2 \vee (u_2 \wedge \neg a_1)) \\
&= ((a_1 \vee u_1) \wedge (a_1 \vee \neg a_1), (a_2 \vee u_2) \wedge (a_2 \vee \neg a_1)) \\
&= (u_1 \wedge 1, u_2 \wedge (a_2 \vee \neg a_1))
\end{aligned}$$

We have $(a_2 \vee \neg a_1) = (a_2 \vee \neg(a_2 \wedge u_1)) = (a_2 \vee (\neg a_2 \vee \neg u_1)) = 1$, using Proposition 2 (13)–(14). Therefore, $(a_1, a_2) \vee \sim(a_1, a_2) = (u_1, u_2)$.

Example 19. Consider the 6-element *pBa* $\mathcal{H}_6$ (Fig. 47) from Example 18 and $(u_1, u_2) := (z, w) \in \mathcal{H}_6^{[2]}$. Using Proposition 64, $A_u := (A_u, (u_1, u_2), (0,0), \sqcap, \sqcup, \rightarrow, \neg, \sim)$ forms a *ccpBa*, where $A_u := \{(z, w), (0,0), (z, z), (0, y)\}$ (Fig. 48).

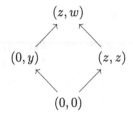

Table 6. Negation $\sim$ in $\mathcal{A}_u$

x	$\sim x$
$(0,0)$	$(0,y)$
$(0,y)$	(z,w)
(z,z)	$(0,y)$
(z,w)	(z,w)

Fig. 48. An example of $\mathcal{A}_u$

Observation 37. Proposition 64 mentions that if $\mathcal{H}$ is a Boolean algebra, then $\mathcal{A}_u$ is a $c\vee cpBa$. Is the converse true, i.e. if $\mathcal{A}_u$ is a $c\vee cpBa$, then is $\mathcal{H}$ a Boolean algebra? Answer is in the negative. It can be observed from Table 6 that for the above example, $\mathcal{A}_u$ is a $c\vee cpBa$. However $\mathcal{H}_6$ is not a Boolean algebra.

As shown in Proposition 58, the $c\vee c$ lattice $\mathcal{D}(U)$ is not a Boolean algebra in general. It is isomorphic to the Boolean algebra $\mathcal{P}(\overline{\mathcal{U}})$ if and only if $\overline{\mathcal{U}} = \mathcal{U}$. One may reiterate this point here as well: the $c\vee c$ lattice $\mathcal{A}_u$ obtained in Proposition 64 when $\mathcal{H}$ is a Boolean algebra, is not isomorphic to $\mathcal{H}$ in general. This is because $(u_1, u_2) \wedge \sim(u_1, u_2) = (0, u_2 \wedge \neg u_1)$, and $u_2 \wedge \neg u_1$ is not necessarily equal to 0 in $\mathcal{H}$.

5.3   Comparison with Other Algebras

Based on Observation 35 and Proposition 62, the following is straight-forward.

Proposition 65. *Consider an algebraic structure $\mathcal{A} := (A, 1, 0, \vee, \wedge, \rightarrow, \neg, \sim)$ such that the reduct $(A, 1, 0, \vee, \wedge, \rightarrow, \neg)$ is a pBa. Then the following are equivalent.*

1. $\mathcal{A}$ is a ccpBa.
2. $(A, 1, \vee, \wedge, \rightarrow, \sim)$ is a cc lattice and $\neg\neg\sim 1 = \sim 1$.
3. $(A, 1, \vee, \wedge, \rightarrow, \sim)$ is a cc lattice and $\sim\sim(\sim 1 \rightarrow a) = 1$ for any $a \in A$.

Proof. $(1) \Rightarrow (3)$ is directly using Observation 35 and Proposition 62.
$(3) \Rightarrow (2)$. $\sim\sim(\sim 1 \rightarrow a) = 1 \Rightarrow \sim(\sim 1 \rightarrow a) \rightarrow \sim 1 = 1$ because in a cc lattice, $\sim x = x \rightarrow \sim 1$. Therefore, we have $\sim(\sim 1 \rightarrow a) \leq \sim 1$ (by Proposition 1(B)(7)). In particular $\sim(\sim 1 \rightarrow 0) \leq \sim 1$. Now,

$$0 \leq \sim 1$$
$$\Rightarrow \neg\neg\sim 1 \wedge \neg\neg\sim 1 \leq \sim 1 \qquad \text{(by Proposition 2(6))}$$
$$\Rightarrow \neg\neg\sim 1 \leq \neg\neg 1 \rightarrow \sim 1 \qquad \text{(by Proposition 1(B)(3))}$$
$$= (\sim 1 \rightarrow 0) \rightarrow \sim 1 = \sim(\sim 1 \rightarrow 0) \leq \sim 1$$

$(2) \Rightarrow (1)$. By Definition 6, $\sim a = a \rightarrow \sim 1 = a \rightarrow (\neg\neg\sim 1)$.

Observe that Condition (3) above does not have any occurrence of $\neg$, but the algebra $\mathcal{A} := (A, 1, 0, \vee, \wedge, \rightarrow, \neg, \sim)$ has an underlying pBa $(A, 1, 0, \vee, \wedge, \rightarrow, \neg)$.

The following example shows that there exists an algebra $\mathcal{A} := (A, 1, \vee, \wedge, \rightarrow, \sim)$ satisfying Condition (3), such that there is no least element $0 \in A$ and no negation $\neg$, with the help of which $\mathcal{A}$ can be extended to a $ccpBa$ $(A, 1, 0, \vee, \wedge, \rightarrow, \neg, \sim)$.

Example 20. Consider the linear lattice $(L, 0, \vee, \wedge)$, where L consists of all the negative integers including 0. Note that L has 0 as its top element, and has no bottom element. Define an implication operator (cf. [53]) as

$$a \rightarrow b := b \quad \text{if } b < a,$$
$$:= 0 \quad \text{otherwise.}$$

We claim that $(L, 0, \vee, \wedge, \rightarrow)$ forms an *rpc* lattice. For this, we have to show that

$$a \wedge x \leq b \Leftrightarrow x \leq a \rightarrow b \text{ for all } x, a, b \in L.$$

Let $x \wedge a \leq b$. Since L is a linear lattice, we have two cases:
(1). $a \leq b$, i.e. $a \rightarrow b = 0$. This implies that $x \leq 0$ for any $x \in L$ as 0 is an upper bound in L.
(2). $b < a$, i.e. $a \rightarrow b = b$. Now, if $x \not\leq b$ then $b < x$. This along with $b < a$ implies $b < x \wedge a$ - a contradiction. Therefore, we have $x \leq b$.
For the converse, assume $x \leq a \rightarrow b$. Now if $a \rightarrow b = 0$, i.e. $a \leq b$, then $x \wedge a \leq x \wedge b \leq b$; else if $a \rightarrow b = b$, i.e. $b < a$, then $x < a$, i.e. $x \wedge a = x \leq b$.
Thus, $(L, 0, \vee, \wedge, \rightarrow)$ is an *rpc* lattice.
Now define $\sim$ as $\sim a := a \rightarrow 0$. In other words, for all $a \in L$, $\sim a = 0$, the top element of L. Thus, trivially, $(L, 0, \vee, \wedge, \rightarrow, \sim)$ becomes a *cc* lattice satisfying Condition (3) of Proposition 65. However, $(L, \leq)$ does not have a lower bound. So, $(L, 0, \vee, \wedge, \rightarrow)$ cannot be extended to a pBa, and hence the *cc* lattice $(L, 0, \vee, \wedge, \rightarrow, \sim)$ cannot be extended to a $ccpBa$.

As mentioned earlier, one of the most well-known algebraic structures with two negations and having the same algebraic type as $ccpBa$ is the *Nelson algebra* [75]. Let us compare the two structures. Recall the definition of Nelson algebra (called $\mathcal{N}$-lattice in [84]).

Definition 108 (Nelson algebras [75,84]).
Consider an abstract algebra $\mathcal{A} := (A, 1, 0, \vee, \wedge, \rightarrow, \neg, \sim)$. Here, $\leq$ is the partial order in the lattice $\mathcal{A}$. Define another relation $<$ as follows: $a < b$ if and only if $a \rightarrow b = 1$. Then the relation $<$ is reflexive and transitive. Let $\mathcal{A}$ satisfy the following properties for all $a, b, c \in A$.

(1). $(A, 1, 0, \vee, \wedge)$ *is a bounded distributive lattice.*
(2). $\sim(a \wedge b) = \sim a \vee \sim b.$
(3). $\sim(a \vee b) = \sim a \wedge \sim b.$
(4). $\sim \sim a = a.$
(5). $a \wedge b < c \Leftrightarrow a < b \rightarrow c.$

(6). $a \leq b \Leftrightarrow (a < b)$ & $(\sim b < \sim a)$.
(7). $(a < c)$ & $(b < c) \Rightarrow a \vee b < c$.
(8). $(a < b)$ & $(a < c) \Rightarrow a < b \wedge c$.
(9). $a \wedge \sim b < (\sim(a \to b))$.
(10). $\sim(a \to b) < (a \wedge \sim b)$.
(11). $\sim\neg a < a$.
(12). $a \wedge \sim a < b$.
(13). $\neg a = a \to \sim 1$.

Then $\mathcal{A}$ is called a Nelson algebra *(or $\mathcal{N}$-lattice).*

Remark 18. Note the following differences between Nelson algebra and *ccpBa*. These will be verified through an example later.

(1) The implication $\to$ is a pseudo-complement operator in a *ccpBa*, which may not be true for $\to$ in a Nelson algebra.
(2) The negation $\sim$ satisfies the property $\sim\sim a = a$ for all $a \in \mathcal{A}$ in a Nelson algebra $\mathcal{A}$. This may not be true in an arbitrary *ccpBa*, i.e. neither of the two negations in a *ccpBa* may satisfy this property.
(3) The negation $\neg$ satisfies the property $\sim a = \neg\neg\sim a$ for all $a \in \mathcal{A}$ in a *ccpBa* $\mathcal{A}$ (Proposition 63(4)). This property may not hold true for an arbitrary Nelson algebra.

Let us validate the above remark using an example of a 3-valued Nelson algebra.

Example 21. Consider the 3-element Nelson algebra $\mathcal{A} := (A, 1, 0, \vee, \wedge, \to, \neg, \sim)$ from [75], where $A := \{0, a, 1\}$ with the ordering as $0 \leq a \leq 1$. The operators $\to$, $\neg$ and $\sim$ are defined in Table 7.

Table 7. 3-element Nelson algebra

$\to$	0	a	1		x	$\neg x$		x	$\sim x$
0	1	1	1		0	1		0	1
a	1	1	1		a	1		a	a
1	0	a	1		1	0		1	0

In [75], it has been observed that every 3-element Nelson algebra is isomorphic to $\mathcal{A}$. As shown in Example 17, $\mathcal{A}' := (A, 1, 0, \vee, \wedge, \to_1, \neg_1, \sim_1)$ and $\mathcal{B}' := (A, 1, 0, \vee, \wedge, \to_1, \neg_1, \neg_1)$ are the only 3-element *ccpBas*, upto isomorphism, on A.

(1). In $\mathcal{A}$, $a \leq a \to 0 = 1$ holds, but $a = a \wedge a \not\leq 0$. Thus the operator $\to$ is not a pseudo-complement in $\mathcal{A}$, and $\mathcal{A}$ is not a *ccpBa*.

(2). In $\mathcal{A}'$, $\sim_1\sim_1 a = 1 \neq a$, and in $\mathcal{B}'$, $\neg_1\neg_1 a = 1 \neq a$. So, neither negation is involutive and $\mathcal{A}'$, $\mathcal{B}'$ are thus not Nelson algebras.

Since, these are the only examples of 3-element Nelson algebras and *ccpBas* upto isomorphism, we can conclude that neither of the two algebraic classes is a sub-class of the other.

5.4   Representation Theorems

The class of pBa's has a representation theorem (Theorem 9): for every pBa $\mathcal{A}$, there exists a monomorphism h from $\mathcal{A}$ into the pseudo-field of all open subsets of a topological space. On similar lines, we obtain the representation theorem for $ccpBas$.

Definition 109 (Contrapositionally complemented pseudo-fields [64]).
Let $\mathscr{G}(X) := (\mathcal{G}(X), X, \emptyset, \cap, \cup, \rightarrow, \neg)$ be a pseudo-field of open subsets of a topological space X. Define

$$\sim X := \neg\neg Y_0, \quad \text{for some } Y_0 \text{ belonging to } \mathcal{G}(X),$$
$$\sim Z := Z \rightarrow (\neg\neg\sim X), \quad \text{for each } Z \in \mathcal{G}(X).$$

The algebra $(\mathcal{G}(X), X, \emptyset, \cap, \cup, \rightarrow, \neg, \sim)$ is called contrapositionally complemented pseudo-field (cc pseudo-field) of open subsets of X.

Observation 38. Any pseudo-field is a pBa (Definition 9). In a cc pseudo-field $(\mathcal{G}(X), X, \emptyset, \cap, \cup, \rightarrow, \neg, \sim)$, moreover, we have $\neg\neg\sim X = \neg\neg\neg\neg Y_0 = \neg\neg Y_0 = \sim X$, and $\sim Z = Z \rightarrow \sim X$, for each $Z \in \mathcal{G}(X)$. Therefore, by Proposition 65, we have the following result.

Proposition 66. *Any cc pseudo-field of open subsets of a topological space X forms a ccpBa.*

Proposition 67. *For every ccpBa $\mathcal{A} := (A, 1, 0, \cap, \cup, \rightarrow, \neg, \sim)$, there exists a monomorphism h from $\mathcal{A}$ into a cc pseudo-field of all open subsets of a topological space X.*

Proof. By the representation theorem for pBa (Theorem 3), there exists a monomorphism h from $\mathcal{H} := (A, 1, 0, \cap, \cup, \rightarrow, \neg)$ into the pseudo-field

$$\mathscr{G}(X) := (\mathcal{G}(X), X, \emptyset, \cap, \cup, \rightarrow, \neg)$$

of all open subsets of a topological space X. Fixing $Y_0 := h(\sim 1)$ and defining $\sim$ on $\mathcal{G}(X)$ as in Definition 109, $(\mathcal{G}(X), X, \emptyset, \cap, \cup, \rightarrow, \neg, \sim)$ forms a cc pseudo-field of all open subsets of X. We now have

$$\sim X = \neg\neg Y_0 = \neg\neg h(\sim 1) = h(\neg\neg\sim 1) = h(\sim 1), \tag{5.2}$$

because $\neg h(a) = h(\neg a)$ for all $a \in A$, and $\neg\neg\sim 1 = \sim 1$. For h to be a monomorphism from $\mathcal{A}$ into a cc pseudo-field, we must have $h(\sim a) = \sim h(a)$. Indeed, using Condition 5.2 and Observation 38, we get the following.

$$h(\sim a) = h(a \rightarrow \sim 1) = h(a) \rightarrow h(\sim 1) = h(a) \rightarrow \sim X = \sim h(a).$$

Another representation for the $ccpBa$ can be obtained using the duality results for pBa (Theorem 1). This duality result can be extended to $ccpBas$ and $c\vee cpBas$. In an Esakia space $(X, \tau, \leq)$, for clopen upsets to form a $ccpBa$, we have to define $\sim X$ such that $\neg\neg\sim X = \sim X$ and $\sim U := U \to \sim X$ for all $U \in CpUp(X)$. For this, we have to choose some $Y_0 \in CpUp(X)$ such that $\neg\neg Y_0 = Y_0$. Since $CpUp(X)$ is a pBa, $Y_0 \subseteq \neg\neg Y_0$ is always true (Proposition 2(2)). The condition $\neg\neg Y_0 \subseteq Y_0$ is equivalent to $X\backslash \downarrow (X\backslash \downarrow Y_0) \subseteq Y_0$ (Condition 2.1), which can be written in expanded form as follows:

$$\forall x \in X(\forall y \in X(x \leq y \Rightarrow \exists z \in X(y \leq z \ \& \ z \in Y_0)) \Rightarrow x \in Y_0).$$

This leads us to the duality result for $ccpBa$. A similar approach gives us the duality result for $c\lor cpBa$, where the condition $U\cup\sim U = X$ for all $U \in CpUp(X)$ is expanded.

Theorem 19 (Duality for $ccpBa$ and $c\lor cpBa$). *The class of $ccpBas$ is dual to that of ordered topological spaces of the form $(X, \tau, \leq, Y_0)$, where $(X, \tau, \leq)$ is an Esakia space and Y_0 is a clopen set satisfying the following.*

$$\forall x \in X(\forall y \in X(x \leq y \Rightarrow \exists z \in X(y \leq z \ \& \ z \in Y_0)) \Rightarrow x \in Y_0). \tag{5.3}$$

Moreover, the class of $c\lor cpBas$ is dual to that of ordered topological spaces of the above kind, satisfying the following.

$$\forall x, y \notin Y_0 \Rightarrow (x \leq y \Rightarrow y \leq x). \tag{5.4}$$

Proof.
(For $ccpBa$) Given a $ccpBa$ $\mathcal{A} := (A, 1, 0, \land, \lor, \to, \neg, \sim)$, we have an underlying pBa $(A, 1, 0, \land, \lor, \to, \neg)$. Using the duality result for pBa (Theorem 1), we obtain an Esakia space $(X_A, \tau_A, \subseteq)$ such that $(A, 1, 0, \land, \lor, \to, \neg)$ is isomorphic to

$$\mathcal{D}(X_A) := (CpUp(X_A), X_A, \emptyset, \cap, \cup, \to, \neg),$$

where the isomorphism is given by $a \mapsto \sigma(a)$. Define $Y_0 := \sigma(\sim 1)$. Y_0 is an upset. Since $\{\sigma(a) \mid a \in A\} \cup \{\sigma(a)^c \mid a \in A\}$ forms the sub-basis for τ_A, $\sigma(\sim 1)$ is clopen. In any $ccpBa$ we have $\sim 1 = \neg\neg\sim 1$, that is, $\sigma(\sim 1) = \sigma(\neg\neg\sim 1)$. Using the facts that σ is a homomorphism, and $\neg a = a \to 0$, we have

$$\sigma(\sim 1) = \sigma(\neg\neg\sim 1) = \neg\neg\sigma(\sim 1) = (\sigma(\sim 1) \to \emptyset) \to \emptyset$$
$$\text{i.e., } Y_0 = (Y_0 \to \emptyset) \to \emptyset$$
$$\text{i.e., } Y_0 = X_A\backslash \downarrow (X_A\backslash \downarrow Y_0) \quad \text{(Using Condition 2.1)}$$

In particular, we have $X_A\backslash \downarrow (X_A\backslash \downarrow Y_0) \subseteq Y_0$. Expanding this, we obtain Condition 5.3, as observed before the theorem.

Conversely, consider an ordered topological space $(X, \tau, \leq, Y_0)$ such that the tuple $(X, \tau, \leq)$ forms an Esakia space and Y_0 is a clopen set satisfying Condition 5.3. We already know that $\mathcal{D}(X)$ forms a pBa. Define $\sim X := Y_0$ and $\sim U := U \to \sim X$ for all $U \in CpUp(X)$. Then using Conditions 5.3 and 2.1, we have $\neg\neg\sim X \subseteq \sim X$. Any pBa has the property: $a \leq \neg\neg a$ for all $a \in A$. In particular in $\mathcal{D}(X)$, we have $\sim X \subseteq \neg\neg\sim X$. Therefore, $\sim X = \neg\neg\sim X$.

(**For** $c\lor cpBa$) If A is a $c\lor cpBa$, we have $\sigma(a) \cup (\sigma(a) \to Y_0) = X_A$ for all $a \in A$, i.e.

$$\sigma(a) \cup X_A \setminus \downarrow (\sigma(a) \setminus Y_0) = X_A \text{ for all } a \in A. \tag{5.5}$$

For the tuple $(X_A, \tau_A, \subseteq, Y_0)$ to satisfy Condition 5.4, we have to show that for any two prime filters $F, G \in X_A$, if $F, G \notin Y_0$ and $F \subseteq G$ then $G \subseteq F$. Suppose not, i.e. $F, G \notin Y_0$ and $F \subseteq G$ and there exists $x \in G$ such that $x \notin F$. Thus $F \notin \sigma(x)$ and $G \in \sigma(x)$. In Condition 5.5 for $a = x$, since $F \in X_A$, we must have $F \in \sigma(x) \cup X_A \setminus \downarrow (\sigma(x) \setminus Y_0)$. We already have $F \notin \sigma(x)$. Therefore $F \in X_A \setminus \downarrow (\sigma(x) \setminus Y_0)$, i.e. $F \notin \downarrow (\sigma(x) \setminus Y_0)$. Since $F \subseteq G$, using the definition of $\downarrow (\sigma(x) \setminus Y_0)$, we have $G \notin \sigma(x) \setminus Y_0$, i.e. $G \notin \sigma(x)$ (because $G \notin Y_0$), a contradiction.

Conversely, consider an ordered topological space $(X, \tau, \leq, Y_0)$ such that $(X, \tau, \leq)$ forms an Esakia space and Y_0 satisfies Conditions 5.3 and 5.4. We have already shown that $\sim X = \neg\neg\sim X$.

Y_0 satisfies Condition 5.4. We have to show that for any $V \in CpUp(X)$, $V \cup (V \to \sim X) = X$, i.e. $V \cup (X \setminus \downarrow (V \setminus Y_0)) = X$. Suppose not, i.e. there exists $x \in X$ such that $x \notin V$ and $x \in \downarrow (V \setminus Y_0)$. Then there exists $y \in X$ such that $x \leq y$ and $y \in V \setminus Y_0$. We have $x \leq y$, $y \in V$ and $y \notin Y_0$, i.e. $x \notin Y_0$ (because Y_0 is an upset). Using $x \leq y$ and Condition 5.4, we have $y \leq x$. Since $y \in V$ and V is an upset, we have $x \in V$, a contradiction.

5.5 Conclusions

In this section, we discussed algebras obtained from the strong subobjects of an object in the quasitopos $RSC(\mathscr{C})$. We incorporate the *relative rough complementation* given by Iwiński [48] as the relative rough negation, and obtain two new classes of algebras $ccpBa$ and $c\lor cpBa$.

In Sect. 3, other categories of rough sets, like RSC^*, RSC_1, RSC_2, and their generalizations (Sects. 3.4 and 3.5) are given. Some of them are observed to form a topos or a quasitopos. For these, one may study the internal algebra of subobjects in a similar way as done in this section.

In case of the quasitopos RSC and topos $\xi\text{-}RSC$, we saw that both have the same internal algebra of subobjects. We could make a comparison of the algebras obtained from these other categories of rough sets that are non-equivalent to RSC or $\xi\text{-}RSC$.

We defined a new negation $\sim$ in the internal algebra of $RSC(\mathscr{C})$. This operator is defined in purely algebraic terms. However, the other connectives, namely $\cap$, $\cup$, $\neg$ and $\to$, are all obtained first as $\mathscr{C}$-arrows, where $\mathscr{C}$ is a topos or a quasitopos. A question in this direction is - can we define $\sim$ as a $\mathscr{C}$-arrow in a topos or quasitopos $\mathscr{C}$?

We obtained a representation of $ccpBas$ as cc pseudo-fields of all open subsets of a topological space (Proposition 67). This result, however, is not extended to $c\lor cpBas$, and stands as an open question. Our hunch is that one should be able to get a condition on $X \setminus Y_0$ in a cc pseudo-field $\mathscr{G}(X) := (\mathcal{G}(X), X, \emptyset, \cap, \cup, \to, \neg)$ (Definition 109) that would make such a representation possible.

6   The Logic ILM

We saw two new algebraic structures called
(1) contrapositionally complemented pseudo-Boolean algebras (*ccpBa*) and
(2) contrapositionally $\vee$ complemented pseudo-Boolean algebras (*c$\vee$cpBa*)
in the previous section. In this section, we shall study the logics corresponding
to these algebraic structures. We shall use the terminology of [75].

In the next section, two logics - ILM and ILM-$\vee$ are defined, and studied in
details. Two Kripke-style relational semantics for both these logics are obtained
in Sect. 6.2. We also study the inter-translation between these two semantics in
this section. Connections between algebraic semantics and relational semantics
are discussed in Sect. 6.3. We end this section by defining two more relational
semantics for ILM and ILM-$\vee$ in Sect. 6.4.

6.1   Intuitionistic Logic with Minimal Negation

The logic corresponding to the class of contrapositionally complemented pseudo
Boolean algebras is named 'Intuitionistic logic with minimal negation' [64].

Definition 110 (Intuitionistic logic with minimal negation (ILM)).
The alphabet of the language $\mathcal{L}$ of ILM *and* ILM-$\vee$ *is that of* IL *along with a
unary connective* $\sim$, *and the formulas are given by the scheme:*

$$\top \mid \bot \mid p \mid \alpha \wedge \beta \mid \alpha \vee \beta \mid \alpha \rightarrow \beta \mid \neg\alpha \mid \sim\alpha$$

Axioms:

>   (A1). $\alpha \rightarrow (\beta \rightarrow \alpha)$
>   (A2). $(\alpha \rightarrow (\beta \rightarrow \gamma)) \rightarrow ((\alpha \rightarrow \beta) \rightarrow (\alpha \rightarrow \gamma))$
>   (A3). $\alpha \rightarrow (\alpha \vee \beta)$
>   (A4). $\beta \rightarrow (\alpha \vee \beta)$
>   (A5). $(\alpha \rightarrow \gamma) \rightarrow ((\beta \rightarrow \gamma) \rightarrow ((\alpha \vee \beta) \rightarrow \gamma))$
>   (A6). $(\alpha \wedge \beta) \rightarrow \alpha$
>   (A7). $(\alpha \wedge \beta) \rightarrow \beta$
>   (A8). $(\alpha \rightarrow \beta) \rightarrow ((\alpha \rightarrow \gamma) \rightarrow (\alpha \rightarrow (\beta \wedge \gamma)))$
>   (A9). $\alpha \rightarrow \top$
>   (A10). $\bot \rightarrow \alpha$
>   (A11). $(\alpha \rightarrow \beta) \rightarrow ((\alpha \rightarrow \neg\beta) \rightarrow \neg\alpha)$
>   (A12). $\neg\alpha \rightarrow (\alpha \rightarrow \beta)$
>   (A13). $\sim\alpha \leftrightarrow (\alpha \rightarrow \neg\neg\sim\top)$

Modus ponens (MP) *is the only rule of inference in* ILM.
Addition of the following axiom gives the axiomatic system for ILM-$\vee$.

>   (A14). $(\alpha \vee \sim\alpha)$

Definition 111 (Extension of ILM).
A logic S *is called an extension of* ILM *if the language of* S *is same as that of*
ILM *(i.e. $\mathcal{L}$),* S *is deductively closed under* MP, *and all the theorems of* ILM *are
theorems of* S.

Consider F to be the set of well-formed formulas in the language $\mathcal{L}$. For any extension S of ILM, we can define $\Gamma \vdash_S \alpha$ using Definition 49, where $\Gamma \cup \{\alpha\} \subseteq F$. The following extends directly from Proposition 18.

Proposition 68. *The following hold for any set $\Delta \cup \Delta' \cup \{\alpha, \beta\} \subseteq F$ of formulas, in any extension S of* ILM.

(a) $\alpha \in \Delta \Rightarrow \Delta \vdash_S \alpha$.
(b) $\Delta \vdash_S \alpha \Rightarrow \Delta \cup \Delta' \vdash_S \alpha$.
(c) $\Delta \vdash_S \alpha$ and $\Delta' \cup \{\alpha\} \vdash_S \beta \Rightarrow \Delta \cup \Delta' \vdash_S \beta$.

Proposition 69 (Deduction theorem (DT)).
In any extension S of ILM*, for a set $\Delta \cup \{\alpha, \beta\} \subseteq F$ of formulas, we have the following.*

$$\Delta \cup \{\alpha\} \vdash_S \beta \text{ implies } \Delta \vdash_S (\alpha \rightarrow \beta).$$

The converse also holds in S: for all $\Delta \cup \{\alpha, \beta\} \subseteq F$, $\Delta \vdash_S (\alpha \rightarrow \beta)$ implies $\Delta \cup \{\alpha\} \vdash_S \beta$.

Proof. The proof is standard, by induction on the number 'n' of steps in the proof of $\Delta \cup \{\alpha\} \vdash_S \beta$. For the sake of completeness of the study, we include it. For $n = 1$, either $\beta \in \Delta \cup \{\alpha\}$ or β is an axiom in S. Let $\beta \in \Delta \cup \{\alpha\}$ where $\beta = \alpha$. Then $\Delta \vdash_S (\alpha \rightarrow \alpha)$ gives the result. Now, if $\beta \in \Delta$ or β is an axiom in S then $\Delta \vdash_S \beta$. Using (A1): $\Delta \vdash_S \beta \rightarrow (\alpha \rightarrow \beta)$ and MP imply $\Delta \vdash_S (\alpha \rightarrow \beta)$. Let us assume that the theorem holds for all $n < k$, where $1 \neq k \in \mathbb{N}$, i.e. the proof of $\Delta \cup \{\alpha\} \vdash_S \beta$ has less than k steps. We have three cases: (1). $\beta \in \Delta \cup \{\alpha\}$, (2). β is an axiom in S, or (3). β is derived using MP. For cases (1) and (2), we have $\Delta \vdash_S (\alpha \rightarrow \beta)$ using the argument of the basis step.
For case (3), there exists $\gamma \in F$ such that $\Delta \cup \{\alpha\} \vdash_S \gamma$ and $\Delta \cup \{\alpha\} \vdash_S \gamma \rightarrow \beta$. Since the proof of both of these has less than k steps, using induction hypothesis,

$$\Delta \vdash_S \alpha \rightarrow \gamma$$

$$\Delta \vdash_S \alpha \rightarrow (\gamma \rightarrow \beta)$$

Then using (A2): $\Delta \vdash_S (\alpha \rightarrow (\gamma \rightarrow \beta)) \rightarrow ((\alpha \rightarrow \gamma) \rightarrow (\alpha \rightarrow \beta))$, and MP twice, we obtain $\Delta \vdash_S (\alpha \rightarrow \beta)$.
The converse: using Proposition 68 and MP, $\Delta \vdash_S (\alpha \rightarrow \beta)$ implies $\Delta \cup \{\alpha\} \vdash_S (\alpha \rightarrow \beta)$. Now, using $\Delta \cup \{\alpha\} \vdash \alpha$ and MP, we have $\Delta \cup \{\alpha\} \vdash_S \beta$.

Proposition 70. *In any extension S of* ILM*, for any α, β and γ in F,*

(1) If $\vdash_S \alpha \rightarrow \beta$ and $\vdash_S \beta \rightarrow \gamma$, then $\vdash_S \alpha \rightarrow \gamma$.
(2) $\vdash_S (\alpha \rightarrow \beta) \rightarrow (\sim\beta \rightarrow \sim\alpha)$.

Proof. (1) is easy using deduction theorem and MP. We prove (2).

$$\{\alpha \to \beta, \sim\beta\} \vdash_S \sim\beta \to (\beta \to \neg\neg\sim\top) \qquad \text{(by (A13))}$$
$$\{\alpha \to \beta, \sim\beta\} \vdash_S \sim\beta \qquad \text{(Proposition 68(1))}$$
$$\{\alpha \to \beta, \sim\beta\} \vdash_S \beta \to \neg\neg\sim\top \qquad \text{(by MP)}$$
$$\{\alpha \to \beta, \sim\beta\} \vdash_S \alpha \to \beta \qquad \text{(Proposition 68(1))}$$
$$\{\alpha \to \beta, \sim\beta\} \vdash_S \alpha \to \neg\neg\sim\top \qquad \text{(Using (1) above)}$$
$$\{\alpha \to \beta, \sim\beta\} \vdash_S \sim\alpha \qquad \text{(By (A13) and MP)}$$
$$\vdash_S (\alpha \to \beta) \to (\sim\beta \to \sim\alpha) \qquad \text{(By DT)}$$

Proposition 71 (Equivalence theorem). *Let* S *be an extension of* ILM, *and* $\alpha, \beta, \gamma \in F$ *such that* β *is a subformula of* α. *Suppose* α' *is a formula obtained by replacing some (or all) occurrences of* β *in* α *by* γ. *Then we have*

$$\vdash_S \beta \leftrightarrow \gamma \text{ implies } \vdash_S \alpha \leftrightarrow \alpha'.$$

Proof. The proof is by induction on the number of connectives in α. If α is a propositional variable p, then $sub(\alpha) = \{p\}$, i.e. $\beta = p$. Then α' is either p or γ, and the result is trivially obtained.

Similarly, if α is a propositional constant, then $sub(\alpha) = \{\alpha\}$, $\beta = \alpha$, and $\alpha' = \alpha$ or γ.

Now, assume the induction hypothesis, i.e. if α has less than k connectives then $\vdash_S \beta \leftrightarrow \gamma$ implies $\vdash_S \alpha \leftrightarrow \alpha'$.

We prove the case when $\alpha := \sim\delta$ for some $\delta \in F$. Then $sub(\sim\delta) = sub(\delta) \cup \{\sim\delta\}$.
Case 1: $\beta \in sub(\delta)$. Since δ has less than k connectives, we can use induction hypothesis on δ. Let δ' be obtained by replacing some (or all) occurrences of β in δ by γ. Note that in this case $\alpha' = \sim\delta'$. Now, $\vdash_S \beta \leftrightarrow \gamma$ implies $\vdash_S \delta \leftrightarrow \delta'$ by induction hypothesis. Using Proposition 70, we have $\vdash_{ILM} \sim\delta' \leftrightarrow \sim\delta$, i.e. $\vdash \alpha' \leftrightarrow \alpha$.
Case 2: $\beta = \sim\delta = \alpha$. In this case, α' is either γ or α, and we get the result trivially as before.
For other connectives, the proof follows from Proposition 20.

Since ILM-∨ is an extension of ILM, the above propositions also hold for ILM-∨. Let us now see the correspondence of the logic and algebra for S using Definitions 50 and 51.

Definition 112 (Valuation and validity).
Let $\mathcal{A} := (A, 1, 0, \vee, \wedge, \to, \neg, \sim)$ *be an algebra associated with the language* $\mathcal{L}$, *and consider any map* $v_0 : PV \to A$, *where* PV *denotes the set of propositional variables in the language* $\mathcal{L}$. *The map* v_0 *can be extended recursively to get a map* $v : F \to A$, *in the following way.*

$$v(p) := v_0(p) \text{ for all } p \in PV$$
$$v(\alpha \vee \beta) := v(\alpha) \vee v(\beta)$$
$$v(\alpha \wedge \beta) := v(\alpha) \wedge v(\beta)$$
$$v(\alpha \rightarrow \beta) := v(\alpha) \rightarrow v(\beta)$$
$$v(\neg\alpha) := \neg v(\alpha)$$
$$v(\sim\alpha) := \sim v(\alpha)$$
$$v(\top) := 1$$
$$v(\bot) := 0$$

v is called a valuation *on the algebra* $\mathcal{A}$, *and a formula* α *is said to be* true *for the valuation* v *on* $\mathcal{A}$ *if* $v(\alpha) = 1$. *This is denoted by* $v \vDash_\mathcal{A} \alpha$.

If $v \vDash_\mathcal{A} \alpha$ *for any valuation* v *on* $\mathcal{A}$, *then we say that* α *is* valid in $\mathcal{A}$, *denoted by* $\vDash_\mathcal{A} \alpha$.

Let C *denote a sub-class of the class of algebras of the type* $(A, 1, 0, \vee, \wedge, \rightarrow, \neg, \sim)$. *We say* α *is* valid in C *(denoted by* $\vDash_C \alpha$*), if* $\vDash_\mathcal{A} \alpha$ *for all* $\mathcal{A} \in C$.

We now proceed to show that ILM (ILM-$\vee$) is sound and complete with respect to the class of *ccpBas* (*c$\vee$cpBas*). Recall the logic-algebra connections presented in Definitions 52 and 54 and the brief discussion there on soundness-completeness results of logics with respect to classes of algebras.

Proposition 72. *Any ccpBa (c$\vee$cpBa) is an* ILM*-algebra (*ILM-$\vee$*-algebra).*

Proof. Consider a *ccpBa* $\mathcal{A}$ and a valuation v on $\mathcal{A}$. We require to verify that all ILM-axioms are valid, and MP preserves validity.

We just consider (A13) and establish its validity. So let α be of the form $\sim\beta \leftrightarrow (\beta \rightarrow \neg\neg\sim\top)$. Then we have to show that $v(\sim\beta \leftrightarrow (\beta \rightarrow \neg\neg\sim\top)) = 1$, i.e. $\sim v(\beta) \leftrightarrow (v(\beta) \rightarrow \neg\neg\sim 1) = 1$, i.e. $\sim v(\beta) = v(\beta) \rightarrow \neg\neg\sim 1$, which is true by Definition 107 of *ccpBa*. MP clearly preserves validity.

For ILM-$\vee$, it is sufficient to show that for any valuation v on a *c$\vee$cpBa* $\mathcal{A}$, (A14) is valid, i.e. $v(\alpha \vee \sim\alpha) = 1$, i.e. $v(\alpha) \vee \sim v(\alpha) = 1$. This is true by Definition 107 of *c$\vee$cpBa*.

Proposition 73. [64] *The Lindenbaum-Tarski algebra* $\mathcal{U}$(ILM) *(*$\mathcal{U}$(ILM $-\vee$)) *is a ccpBa (c$\vee$cpBa).*

Proof. Consider the Lindenbaum-Tarski algebra for ILM:

$$\mathcal{U}(\text{ILM}) := (F/\simeq, [\top], [\bot], \vee, \wedge, \rightarrow, \neg, \sim).$$

The operations on $F/\simeq$ are as follows: for any $[\alpha], [\beta] \in F/\simeq$,

$$[\alpha] \vee [\beta] := [\alpha \vee \beta],$$
$$[\alpha] \wedge [\beta] := [\alpha \wedge \beta],$$
$$[\alpha] \rightarrow [\beta] := [\alpha \rightarrow \beta],$$
$$\neg[\alpha] := [\neg\alpha],$$
$$\sim[\alpha] := [\sim\alpha], \text{ and}$$
$$(\text{Relation}) \ [\alpha] \leq [\beta] \Leftrightarrow \vdash_{\text{ILM}} \alpha \rightarrow \beta.$$

We have to check that all the above operators are well-defined. We only show this for the operation $\sim$. Let $[\alpha] = [\beta]$, i.e. $\vdash_{\text{ILM}} \alpha \leftrightarrow \beta$. We have to show that $\sim[\alpha] = \sim[\beta]$, i.e. $\vdash_{\text{ILM}} \sim\beta \leftrightarrow \sim\alpha$, which holds using Proposition 70 and MP. For any $[\alpha] \in F/\simeq$, we can easily observe the following.

(1) $\alpha \in [\top]$ if and only if $\vdash_{\text{ILM}} \alpha$,
(2) $[\alpha] \leq [\beta]$ and $[\beta] \leq [\alpha]$ imply $[\alpha] = [\beta]$,
(3) $[\top]$ is the top element of $F/\simeq$ with respect to $\leq$, i.e. $[\alpha] \leq [\top]$, and
(4) $[\bot]$ is the bottom element of $F/\simeq$ with respect to $\leq$, i.e. $[\bot] \leq [\alpha]$.

We have to show that the algebra $(F/\simeq, [\top], [\bot], \vee, \wedge, \rightarrow, \neg, \sim)$ forms a $ccpBa$. Using (A13) and the definition of $\leq$, we have $[\sim\alpha] = [\alpha \rightarrow \neg\neg\sim\top]$. Thus, $\sim[\alpha] = [\alpha] \rightarrow \neg\neg\sim[\top]$. Similarly, using the other axioms, $F/\simeq$ can be shown to form a pBa.
For ILM-$\vee$, we have to only show that $\sim[\alpha] \vee [\alpha] = [\top]$ holds for any $[\alpha], [\beta] \in F/\simeq$. $\sim[\alpha] \vee [\alpha] = [\sim\alpha \vee \alpha] = [\top]$ (using (A14)).

The above two propositions establish the soundness and completeness of ILM (ILM-$\vee$) with respect to the class of $ccpBas$ ($c\vee cpBas$).

Corollary 7. *For any $\alpha \in F$, $\vdash_{\text{ILM}} \alpha$ ($\vdash_{\text{ILM-}\vee} \alpha$) if and only if $\vDash_{\mathcal{A}} \alpha$ for every $ccpBa$ ($c\vee cpBa$) $\mathcal{A}$.*

Using the above result, we can easily obtain theorems of ILM, which also hold in any extension S of ILM. We can also show that $\sim\top$ is not logically equivalent to $\bot$.

Proposition 74. *(a)* $\vdash_{\text{ILM}} (\alpha \rightarrow \sim\beta) \rightarrow (\beta \rightarrow \sim\alpha)$ *($\sim$ contraposition)*
(b) $\vdash_{\text{ILM}} \alpha \rightarrow \sim\sim\alpha$
(c) $\vdash_{\text{ILM}} \neg\neg\sim\top \leftrightarrow \sim\top$, *i.e.* $\vdash_{\text{ILM}} ((\sim\top \rightarrow \bot) \rightarrow \bot) \leftrightarrow \sim\top$
(d) $\vdash_{\text{ILM}} \sim\sim(\sim\top \rightarrow \alpha)$
(e) $\vdash_{\text{ILM}} \sim\alpha \leftrightarrow \neg(\alpha \wedge \neg\sim\top)$
(f) $\vdash_{\text{ILM}} \neg\alpha \rightarrow \sim\alpha$
(g) $\vdash_{\text{ILM}} \alpha \rightarrow \sim\neg\alpha$
(h) $\vdash_{\text{ILM}} \neg\sim\alpha \rightarrow \sim\neg\alpha$
(i) $\vdash_{\text{ILM}} \sim\alpha \leftrightarrow \neg\neg\sim\alpha$
(j) $\vdash_{\text{ILM}} \neg\sim\neg\alpha \rightarrow \neg\alpha$
(k) $\vdash_{\text{ILM}} \sim\alpha \leftrightarrow (\alpha \rightarrow \sim\top)$
(l) $\vdash_{\text{ILM}} \neg\alpha \leftrightarrow (\alpha \rightarrow \bot)$
(m) $\vdash_{\text{ILM}} \neg\bot \leftrightarrow \top$
(n) $\vdash_{\text{ILM}} \bot \leftrightarrow \neg\top$
(o) $\vdash_{\text{ILM}} \sim\bot$
(p) $\vdash_{\text{ILM}} (\top \rightarrow \alpha) \leftrightarrow \alpha$
(q) $\nvdash_{\text{ILM}} \sim\top \leftrightarrow \bot$

Proof. Using Proposition 62 and Corollary 7, (a)–(e) are direct.
(f)-(j) are using Proposition 63 and Corollary 7.
(k) is using (A13) and (c).

(1), (m) and (n) are using Corollary 7 and the properties $\neg a = a \to 0$, $\neg 0 = 1$ and $0 = \neg 1$ for all $a \in A$ respectively (Proposition 2), where $\mathcal{A} := (A, 1, 0, \vee, \wedge, \to, \neg, \sim)$ is any $ccpBa$, and thus a pBa.
(o) is using (k) and (A10).
(p) is using Proposition 1(B)(4) and Corollary 7.
For (q) we only need to show $\nvdash_{\mathrm{ILM}} \sim\top \to \bot$. This is obtained using soundness and one of the $ccpBas$ mentioned in Example 18 where $0 \neq \sim 1$.

Let us now look at the connection of ILM with IL and ML. We have discussed this connection in [64]. ILM can be interpreted in IL, where 'interpretation' of one logic to another is as in Definition 56. The proof is similar to the one that is used to show the connections between 'constructive logic with strong negation' and IL [75, Chapter XII]. Let us first see the definition of substitution from one logic to another.

Definition 113 (Substitution [75]).
Let L_1 *and* L_2 *be two logics on the same language* $\mathcal{L}$, *and* F *be the set of formulas in* $\mathcal{L}$. *A substitution from* L_1 *to* L_2 *is a map*

$$T : PV \to F$$

extended uniquely to the set F *as follows.*

(1) *For every constant* c *in* $\mathcal{L}$, $T(c) := c$.
(2) *For every unary connective* $\circ$ *in* $\mathcal{L}$, $T(\circ \alpha) := \circ T(\alpha)$.
(3) *For every binary connective* $*$ *in* $\mathcal{L}$, $T(\alpha * \beta) := T(\alpha) * T(\beta)$.

Using (A13) for ILM, we obtain the following proposition.

Proposition 75. *Let* $\alpha \in F$ *such that* $p_1, p_2, \ldots, p_n$ *are all the distinct propositional variables in* α. *Then there exists a formula* $\alpha^* \in F$ *such that (i) there is no occurrence of* $\sim$ *sign in* α^*, *(ii)* α^* *contains* $p_1, p_2, \ldots, p_n$ *and a propositional variable* q *distinct from the* p_i's, *and (iii) the following condition is satisfied.*

(K) *For any substitution* T *such that* $T(p_i) := p_i$ *and* $T(q) := \sim\top$ *for all* $i = 1, \ldots, n$, *we have* $\vdash_{\mathrm{ILM}} \alpha \leftrightarrow T(\alpha^*)$.

Proof. For the formula α, define α^* by (i) replacing all the occurrences of $\sim\gamma$ by $\gamma \to \neg\neg\sim\top$, where $\sim\gamma$ is any subformula of α other than $\sim\top$, and (ii) replacing $\sim\top$ by a propositional variable q, where q is distinct from $p_1, p_2, \ldots, p_n$. By the Equivalence Theorem 71, in the language $\mathcal{L}$, for any $\delta \in F$ and a subformula β of δ, if $\vdash_{\mathrm{ILM}} \beta \leftrightarrow \gamma$ for some $\gamma \in F$, then $\vdash_{\mathrm{ILM}} \delta \leftrightarrow \overline{\delta}$, where $\overline{\delta}$ is obtained by replacing all the occurrences of β by γ in the formula δ. Since $\vdash_{\mathrm{ILM}} \sim\gamma \leftrightarrow (\gamma \to \neg\neg\sim\top)$ using (A13), we have $\vdash_{\mathrm{ILM}} \alpha \leftrightarrow T(\alpha^*)$ for any substitution T as defined in the Condition (K).

Using Proposition 75 and Corollary 7, we have the following result.

Theorem 20. *For any formula* $\alpha \in F$, *consider* α^* *and a propositional variable* q *as in Proposition 75. Let* $\beta := \neg\neg q \to q$. *Then* $\vdash_{\mathrm{ILM}} \alpha$ *if and only if* $\{\beta\} \vdash_{\mathrm{IL}} \alpha^*$.

Proof. Let $\nvdash_{\mathrm{ILM}} \alpha$. By Proposition 75, α^* is such that for any substitution T, where $T(p_i) := p_i$ for all propositional variables p_i, $i = 1, \ldots, n$, occurring in α and $T(q) := \sim\top$, we have $\nvdash_{\mathrm{ILM}} T(\alpha^*)$. Using the completeness of ILM (Corollary 7), there exist a *ccpBa* $\mathcal{A}$ and a valuation v on $\mathcal{A}$ such that $v(T(\alpha^*)) \neq 1$. Define a valuation Tv on $\mathcal{A}$ such that $(Tv)(p) := v(T(p))$ for all propositional variables p. We can show by induction, for any formula $\gamma \in F$, $(Tv)(\gamma) = v(T(\gamma))$. Therefore $(Tv)(\alpha^*) \neq 1$. For β,

$$
\begin{aligned}
(Tv)(\beta) = (Tv)(\neg\neg q \to q) &= \neg\neg(Tv)(q) \to (Tv)(q) \\
&= \neg\neg v(Tq) \to v(Tq) = \neg\neg v(\sim\top) \to v(\sim\top) \\
&= \neg\neg\sim 1 \to \sim 1 = 1 \qquad \text{(Proposition 62(3))}
\end{aligned}
$$

Now, Tv is a valuation on the *ccpBa* $\mathcal{A}$. As $\mathcal{A}$ is a *pBa*, $(Tv)(\beta) = 1$ and $(Tv)(\alpha^*) \neq 1$ imply $\{\beta\} \nvDash_{\mathrm{IL}} \alpha^*$, and by Theorem 6, $\{\beta\} \nvdash_{\mathrm{IL}} \alpha^*$.

Let $\{\beta\} \nvdash_{\mathrm{IL}} \alpha^*$. Using completeness of IL with respect to the class of *pBa*'s (Theorem 6), we have $\{\beta\} \nvDash_{\mathrm{IL}} \alpha^*$. Thus, there exist a *pBa* $\mathcal{H} := (H, 1, 0, \cap, \cup, \to, \neg)$ and a valuation v on $\mathcal{H}$ such that $v(\alpha^*) \neq 1$ and $v(\beta) = 1$. Define a *ccpBa* $\mathcal{B} := (H, 1, 0, \cap, \cup, \to, \neg, \sim)$ such that $\sim 1 := v(q)$ and $\sim a := a \to (\neg\neg\sim 1)$, for any $a \in H$. Note that we have $\neg\neg\sim 1 = v(\neg\neg q) = v(q)$ as $v(\beta) = v(\neg\neg q \to q) = 1$. Therefore, the operator $\sim$ is well defined on $1 \in H$. The map $v : PV \to H$ can also be considered as a valuation on $\mathcal{B}$. So, by soundness (Corollary 7), $\nvdash_{\mathrm{ILM}} \alpha^*$. Finally using Proposition 75, we have $\nvdash_{\mathrm{ILM}} \alpha$.

Therefore, we have an interpretation of ILM in IL. What about a connection between ILM and ML? The interpretation of ILM in ML (according to Definition 56) can be obtained by a composition of interpretations. For instance, one can take the mapping r between F^* and $\bar{F}$, the set of formulas of ML, given in [74, Theorem B]: for any $\alpha \in F^*$, $r(\alpha)$ is obtained by induction, by replacing every subformula β of α with $\beta \vee \neg\top$. We then have $\vdash_{\mathrm{IL}} \alpha$ if and only if $\vdash_{\mathrm{ML}} r(\alpha)$. Composing r with the interpretation of ILM in IL (cf. Theorem 20), we have the following.

Corollary 8. *There exists an interpretation* $t : F \to \bar{F}$ *of* ILM *in* ML.

The following connection of ML and IL with ILM is clear.

Theorem 21. ML *and* IL *are both embedded in* ILM.

Proof. Let α be a theorem in ML, where negation in the language is denoted by '$\neg$'. Since, all the axioms of ML are axioms in ILM (Definition 60), and the rule of inference MP is also a rule in ILM, α can be proved in ILM using the same steps as in the proof of α in ML. In other words, for any formula α of ML, $\vdash_{\mathrm{ML}} \alpha \Leftrightarrow \vdash_{\mathrm{ILM}} \alpha$.

Similarly, for any formula α in IL (Definition 61), α is a theorem in IL (with negation denoted by the symbol $\neg$) if and only if α is a theorem in ILM. So, by Definition 57, we get the result.

As observed in the case of IL (Observation 10), we can express $\neg$ in terms of $\perp$ and $\rightarrow$. We can get a variant of ILM following this. Remove the connective $\neg$ from the alphabet of the language $\mathcal{L}$ and the Axioms (A11), (A12) and (A13) (involving $\neg$) from the axiomatic system of ILM. Then define $\neg$ as $\neg\alpha := \alpha \rightarrow \perp$. (A11)-(A13) are obtained as theorems in the new logic. Formally, we have the following.

Definition 114 (ILM$_1$).
The formulas of the language $\mathcal{L}_1$ are given by the scheme:

$$\top \mid \perp \mid p \mid \alpha \wedge \beta \mid \alpha \vee \beta \mid \alpha \rightarrow \beta \mid \sim\alpha$$

Define a logic ILM$_1$ *with the axiomatic system given by the Axioms* (A1)–(A10), *and*

(A15). $(\alpha \rightarrow \beta) \rightarrow ((\alpha \rightarrow \sim\beta) \rightarrow \sim\alpha)$, *and*
(A16). $((\sim\top \rightarrow \perp) \rightarrow \perp) \leftrightarrow \sim\top$;

and MP *as the rule of inference.*

Observation 39. We can obtain the Deduction theorem for ILM$_1$ as obtained in Proposition 69. Do (1) and (2) of Proposition 70 hold for ILM$_1$, where S is replaced by the logic ILM$_1$? Yes. (1) holds because the proof of (1) in Proposition 70 does not involve (A11)-(A13) (removed axioms). Let us see the proof of (2).

$$\{\alpha \rightarrow \beta, \sim\beta\} \vdash_{\text{ILM}_1} \sim\beta \qquad \text{(Proposition 68)}$$
$$\{\alpha \rightarrow \beta, \sim\beta\} \vdash_{\text{ILM}_1} \sim\beta \rightarrow (\alpha \rightarrow \sim\beta) \qquad \text{((A1))}$$
$$\{\alpha \rightarrow \beta, \sim\beta\} \vdash_{\text{ILM}_1} \alpha \rightarrow \sim\beta \qquad \text{(MP)}$$
$$\{\alpha \rightarrow \beta, \sim\beta\} \vdash_{\text{ILM}_1} \alpha \rightarrow \beta \qquad \text{(Proposition 68)}$$
$$\{\alpha \rightarrow \beta, \sim\beta\} \vdash_{\text{ILM}_1} \sim\alpha \qquad \text{((A15) and MP)}$$
$$\vdash_{\text{ILM}_1} (\alpha \rightarrow \beta) \rightarrow (\sim\beta \rightarrow \sim\alpha) \qquad \text{(DT)}$$

Therefore, we can obtain Proposition 70 where the logic is ILM$_1$, and using this obtain Proposition 71 for ILM$_1$.

The next question is the relationship between the logics ILM and ILM$_1$. It is expected that the logics are equivalent.

Theorem 22. ILM $\cong$ ILM$_1$.

Proof. Define a map θ from the set of ILM-formulas to that of ILM$_1$-formulas such that for any ILM-formula α, $\theta(\alpha)$ is obtained from α by replacing every subformula of the form $\neg\beta$ by the formula $\beta \rightarrow \perp$. We claim that, using Definition 55, θ and the inclusion map from the set of ILM$_1$-formulas to that of ILM-formulas give the desired equivalence.
Observe that (A16) is just

$$\theta(\neg\neg\sim\top) \leftrightarrow \sim\top. \qquad (6.1)$$

Axioms (A1)–(A10) are common in both. (A11) and (A12) follow from Observation 10(2). What about Axiom (A13)? If we prove $\vdash_{(ILM_1)} \sim\alpha \leftrightarrow (\alpha \to \sim\top)$ then using Formula 6.1 above, we shall obtain θ-image of (A13) as an ILM_1-theorem. $\vdash_{(ILM_1)} (\alpha \to \sim\top) \to \sim\alpha$ is direct using (A15) by taking β as $\top$, using (A9) and MP. For $\vdash_{(ILM_1)} \sim\alpha \to (\alpha \to \sim\top)$:

$$\{\sim\alpha, \alpha\} \vdash_{ILM_1} \top \to \alpha \qquad \text{((A1) and MP)}$$
$$\{\sim\alpha, \alpha\} \vdash_{ILM_1} \top \to \sim\alpha \qquad \text{((A1) and MP)}$$
$$\{\sim\alpha, \alpha\} \vdash_{ILM_1} \sim\top \qquad \text{((A15) and MP)}$$
$$\vdash_{ILM_1} \sim\alpha \to (\alpha \to \sim\top) \qquad \text{(Deduction theorem)}$$

We also have the converse, i.e. the axioms of (ILM_1) are provable in ILM. Axioms (A1)-(A10) are common in both. (A16) is obtained in Proposition 74. Let us see the proof of (A15) in ILM:

$$\{\alpha \to \beta, (\alpha \to (\beta \to \neg\neg\sim\top)), \alpha\} \vdash_{ILM} \neg\neg\sim\top \qquad \text{(Using MP)}$$
$$\vdash_{ILM} (\alpha \to \beta) \to ((\alpha \to (\beta \to \neg\neg\sim\top)) \to (\alpha \to \neg\neg\sim\top))$$
$$\text{(Using Deduction theorem)}$$

Finally using Proposition 71 and (A13), we have (A15) as a theorem in ILM.

Observe that (A15) is the defining condition for *minimal negation*. Historically, various extensions of minimal negation and minimal logic have been studied [71, 78]. One such extension becomes directly relevant to ILM.

6.1.1   An Extension of ML

Let us recall the language and the axiomatic system of ML, with the change that the negation is represented by $\sim$, and not by $\neg$. Consider the language $\mathcal{L}'$ with formulas given by the scheme:

$$\top \mid p \mid \alpha \vee \beta \mid \alpha \wedge \beta \mid \alpha \to \beta \mid \sim\alpha$$

The axioms of ML are given as (A1)-(A9) and (A15),

$$\text{(A15). } (\alpha \to \beta) \to ((\alpha \to \sim\beta) \to \sim\alpha), \qquad (\sim \text{ reductio ad absurdum})$$

and MP as the rule of inference.
In [78], Segerberg has defined the system JP, which is obtained by adding Peirce's law (P) as an axiom to ML.

$$\text{(P)} \quad ((\alpha \to \beta) \to \alpha) \to \alpha$$

Let us consider the logic JP′:

Definition 115 (JP′).
The language of JP′ is $\mathcal{L}'$. The axioms are (A1) − (A9), (A15) and

$$\text{(P′)} \quad \sim\sim(\sim\top \to \beta).$$

MP *is the only rule of inference.*

Using $\vdash_{\mathrm{ML}} \sim\alpha \leftrightarrow (\alpha \to \sim\top)$ (Proposition 21(5)), we observe that (P)$'$ is logically equivalent to the formula

$$((\sim\top \to \beta) \to \sim\top) \to \sim\top. \tag{6.2}$$

Thus, P$'$ is a special case of P, where α is $\sim\top$. Let us now compare the logics JP$'$ and ILM.

Theorem 23. JP$'$ *is embedded in* ILM.

Proof. (P$'$) is a theorem in ILM (Proposition 74(d)). Axioms (A1)–(A9) are common to JP$'$ and ILM. We have already observed in the proof of Theorem 22 that (A15) is a theorem in ILM. Thus, all the axioms of JP$'$ are theorems in ILM. Thus, the inclusion map from JP$'$ to ILM is the required map r to obtain the embedding (Definition 57).

Is JP$'$ equivalent to ILM?

Theorem 24. JP$'$ *is not equivalent to* ILM.

Proof. Suppose we had the equivalence. By Definition 55, we have the inclusion map θ from the set of JP$'$-formulas to that of ILM-formulas, as all n-ary connectives in $\mathcal{L}'$ are present in $\mathcal{L}$. We also have a map ρ from the set of ILM-formulas to that of JP$'$-formulas such that $\rho(\bot)$ does not contain any propositional variable. Thus, if $\mathcal{A} := (A, 1, \vee, \wedge, \to, \sim)$ is any JP$'$-algebra, there must exist $a_0 \in A$ such that for any valuation v on $\mathcal{A}$, $v(\rho(\bot)) = a_0$.
Claim: a_0 is the bottom element of $\mathcal{A}$, that is, $a_0 \leq x$ for all $x \in A$.
Proof of Claim: Consider any $x \in A$ and $p \in PV$. Define $v : PV \to A$ such that $v(p) := x$. Since $\vdash_{\mathrm{ILM}} \bot \to p$ (Axiom (A10)), we have $\vdash_{\mathrm{JP}'} \rho(\bot) \to p$, whence in all JP$'$-algebras, $\rho(\bot) \to p$ is a valid formula. Thus, for the valuation v on $\mathcal{A}$ in particular, $a_0 = v(\rho(\bot)) \leq x$.
However, in Example 20, we have encountered a JP$'$-algebra $(L, 0, \vee, \wedge, \to, \sim)$ that does not have a bottom element. This yields a contradiction.

It is then expected that if we add a new propositional constant $\bot$ to the alphabet of $\mathcal{L}'$ and consider JP$'$ enhanced with Axiom (A10), the resulting system will be equivalent to ILM.

Definition 116 (ILM$_2$).
The formulas of $\mathcal{L}_1$ are given by the scheme:

$$\top \mid \bot \mid p \mid \alpha \wedge \beta \mid \alpha \vee \beta \mid \alpha \to \beta \mid \sim\alpha$$

Define a logic ILM$_2$ *with the axiomatic system given by the Axioms* (A1)–(A10), (A15) *and* (P$'$). MP *is the only rule of inference.*

Theorem 25. ILM $\cong$ ILM$_2$.

Proof. As shown for Theorem 22, (A15) and (P′) are ILM-theorems.

Define a map θ from the set of ILM-formulas to that of ILM$_2$-formulas such that for any ILM-formula α, $\theta(\alpha)$ is obtained from α by replacing every subformula of the form $\neg\beta$ by the formula $\beta \to \bot$.

One can show that the θ-images of ILM- axioms (A9), (A10) and (A11) are ILM$_2$-theorems, using the following results.

(a) $\vdash_{\text{ILM}_2} \beta \leftrightarrow (\top \to \beta)$.   (b) $\vdash_{\text{ILM}_2} \sim\alpha \leftrightarrow (\alpha \to \sim\top)$.

(c) $\{\alpha, \neg\alpha\} \vdash_{\text{ILM}_2} \bot$.   (d) $\vdash_{\text{ILM}_2} ((\sim\top \to \bot) \to \sim\top) \to \sim\top$.

Thus, θ and the inclusion map from the set of ILM$_2$-formulas to that of ILM-formulas give the desired equivalence.

Let us see one final version of a logic equivalent to ILM based on the following observation.

Observation 40. Consider the logic ILM. Let us define a new unary connective $\bot'$ as: $\bot' := \sim\top$. Taking $\alpha := \top$ in (A13): $\sim\alpha \leftrightarrow (\alpha \to \neg\neg\sim\top)$, we have $\vdash_{\text{ILM}} \bot' \leftrightarrow (\top \to \neg\neg\bot')$. Using Proposition 74(p), $\vdash_{\text{ILM}} (\top \to \neg\neg\bot') \leftrightarrow \neg\neg\bot'$. Thus,

$$\vdash_{\text{ILM}} \bot' \leftrightarrow \neg\neg\bot'. \tag{6.3}$$

Using Proposition 71 on (A13) we get

$$\vdash_{\text{ILM}} \sim\alpha \leftrightarrow (\alpha \to \bot').$$

This tells us that in any formula β, any subformula of the type $\sim\alpha$ replaced by $\alpha \to \bot'$ will give a logically equivalent formula, using Proposition 71. This is similar to Observation 9, where in ML, $\neg\alpha$ was replaced by $\alpha \to \bot$, and $\vdash_{\text{ML}} (\alpha \to \beta) \to ((\alpha \to \neg\beta) \to \neg\alpha)$ is obtained.

Based on the above observation, let us define a new logic without the negations $\sim$ and $\neg$. Consider the language $\mathcal{L}_2$ with alphabet:

$$\top \mid \bot \mid \bot' \mid p \mid \alpha \wedge \beta \mid \alpha \vee \beta \mid \alpha \to \beta$$

Definition 117 (ILM$_3$).

Define a logic ILM$_3$ *with Axioms as* (A1)-(A10), *and*

(A17). $((\bot' \to \bot) \to \bot) \leftrightarrow \bot'$,

and MP *as the rule of inference.*

Theorem 26. ILM $\cong$ ILM$_3$.

Proof. Define a map θ from the set of ILM-formulas to that of ILM$_3$-formulas such that for any ILM-formula α, $\theta(\alpha)$ is obtained from α by replacing (a) every subformula of the form $\neg\beta$ by the formula $\beta \to \bot$, (b)every subformula of the form $\sim\beta$ by the formula $\beta \to \bot'$, and (c) constant $\sim\top$ by $\bot'$. Define another map ρ from the set of ILM$_3$-formulas to that of ILM-formulas such that the constant $\bot'$ is mapped to $\sim\top$. One can then show that the θ-images of ILM-axioms are ILM$_3$-theorems and ρ-images of ILM$_3$-axioms are ILM-theorems.

Theorem 27. *Summarizing Theorems 22, 25 and 26,*

$$\text{ILM} \cong \text{ILM}_1 \cong \text{ILM}_2 \cong \text{ILM}_3.$$

Till now in this section, we have restricted our focus to ILM. The results can also be extended to ILM-$\vee$ (Definition 110). Let us define the logics ILM_1-$\vee$, ILM_2-$\vee$ and ILM_3-$\vee$ in the languages $\mathcal{L}_1$, $\mathcal{L}_1$ and $\mathcal{L}_2$ respectively, as follows: the axioms and rules of ILM_i-$\vee$, $i = 1, 2, 3$, are those of ILM_i, along with the axiom (A14). We have the following equivalence result.

Corollary 9.

$$\text{ILM-}\vee \cong \text{ILM}_1\text{-}\vee \cong \text{ILM}_2\text{-}\vee \cong \text{ILM}_3\text{-}\vee.$$

We have already studied the algebraic semantics for ILM in Sect. 1. Next we shall see various relational semantics for ILM and ILM-$\vee$.

6.2   Kripke-Style Relational Semantics for ILM

We have discussed the relational semantics for ML, IL and CL in Sects. 2.5.2.1 and 2.5.2.2. Let us now check for a relational semantics for ILM. In the following subsection we shall study Došen's semantics. In Sect. 6.2.2, Segerberg's semantics for ILM is defined and the connection between these two semantics are investigated. We shall also investigate the connection between Segerberg's semantics and the algebraic semantics for ILM.

6.2.1   Došen's semantics for ILM

In Došen's relational semantics for the language with negation (Sect. 2.5.2.2), there is a binary relation R_N which defines the semantics for the negation. Summarizing Definitions 70 and 74, a strictly condensed J-frame is a triple $\mathcal{F} := (X, R_I, R_N)$ satisfying the following properties:

(1) X is a non-empty set,
(2) $R_I \subseteq X^2$ such that R_I is reflexive and transitive,
(3) $R_N \subseteq X^2$ satisfying $R_I R_N \subseteq R_N R_I^{-1}$,
(4) $R_I R_N \subseteq R_N$ and $R_N R_I^{-1} \subseteq R_N$,
(5) $R_N R_I^{-1}$ is symmetric, and
(6) $\forall x, y \in X(x R_N y \Rightarrow \exists z \in X(x R_I z \ \& \ y R_I z \ \& \ x R_N z))$.

A J-frame $\mathcal{F} := (X, R_I, R_N)$ with $R_N R_I^{-1}$ reflexive, is called an H-frame.

Recall the characterization result (Proposition 25) that the class of strictly condensed J-frames and the class of strictly condensed H-frames, where R_I is a partial order, characterize ML and IL respectively. As R_I is a partial order, let us use the standard notation '$\leq$' for it.

Which class of frames could characterize ILM? In the previous section, we saw various ways of defining the logic ILM. Let us make some observations about such frames, based on the properties of negations $\neg$ and $\sim$ in ILM.

Observation 41.

1. There are two negations in ILM. Therefore, to define the semantics for each negation, the frame should have two relations, say R_{N_1} corresponding to the negation $\neg$, and R_{N_2} corresponding to the negation $\sim$.
2. $\vdash_{\text{ILM}} \neg\alpha \leftrightarrow \alpha \to \bot$ makes $\neg$ an intuitionistic negation. Therefore, R_{N_1} must satisfy the properties corresponding to the relation R_N in an H-frame $\mathcal{F} := (X, R_I, R_N)$.
3. $\vdash_{\text{ILM}} \sim\alpha \leftrightarrow \alpha \to \sim\top$ makes $\sim$ a minimal negation. Therefore, R_{N_2} must satisfy the properties corresponding to the relation R_N in a J-frame $\mathcal{F} := (X, R_I, R_N)$.
4. As $\vdash_{\text{ILM}} \neg\neg(\sim\top) \leftrightarrow \sim\top$, in an ILM-frame, there has to be a property connecting R_{N_1} and R_{N_2} corresponding to the above formula.
5. The logic ILM-$\vee$ has the axiom (A14) for the negation $\sim$. Thus, there must be an extra condition on R_{N_2} corresponding to this axiom, in an ILM-$\vee$ frame.

Based on the above observations, let us define a class of frames.

Definition 118. ($\hat{N}$-frames and $\hat{N}'$-frames).
Consider the quadruple $\mathcal{F} := (X, \leq, R_{N_1}, R_{N_2})$, where X is a non-empty set and $\leq$ is a partial order, satisfying the following conditions.

1. $(X, \leq, R_{N_1})$ *is a strictly condensed H-frame, i.e.*
   (a). $(\leq R_{N_1}) \subseteq (R_{N_1} \leq^{-1})$,
   (b). $(\leq R_{N_1}) \subseteq R_{N_1}$,
   (c). $(R_{N_1} \leq^{-1}) \subseteq R_{N_1}$,
   (d). $R_{N_1} \leq^{-1}$ *is symmetric,*
   (e). $\forall x, y \in X (x R_{N_1} y \Rightarrow \exists z \in X (x \leq z \ \& \ y \leq z \ \& \ x R_{N_1} z))$, *and*
   (f). $R_{N_1} \leq^{-1}$ *is reflexive.*
2. $(X, \leq, R_{N_2})$ *is a strictly condensed J-frame, i.e.*
   (a). $(\leq R_{N_2}) \subseteq (R_{N_2} \leq^{-1})$,
   (b). $(\leq R_{N_2}) \subseteq R_{N_2}$,
   (c). $(R_{N_2} \leq^{-1}) \subseteq R_{N_2}$,
   (d). $R_{N_2} \leq^{-1}$ *is symmetric, and*
   (e). $\forall x, y \in X (x R_{N_2} y \Rightarrow \exists z \in X (x \leq z \ \& \ y \leq z \ \& \ x R_{N_2} z))$.
3. $\forall x \in X \left(\forall y \in X \left(x R_{N_1} y \Rightarrow \exists z \in X (y R_{N_1} z \ \& \ \forall z' \in X \ (z \not R_{N_2} z')) \right) \Rightarrow \forall z'' \in X \ (x \not R_{N_2} z'') \right)$.   (DNE($\sim\top$))

$\mathcal{F}$ *is called an $\hat{N}$-frame. The class of all $\hat{N}$-frames is denoted by $\mathfrak{F}_1$.*
An $\hat{N}$-frame $\mathcal{F}$ satisfying $R_{N_2} \subseteq (\leq^{-1})$ is called an $\hat{N}'$-frame, and the class of all $\hat{N}'$-frames is denoted by $\mathfrak{F}'_1$.

Note:
(1). Hereafter, $\leq$ always denotes a partial order.
(2). Condition (3) above is denoted as 'DNE($\sim\top$)' which stands for 'Double Negation Elimination of $\sim\top$'. After defining the valuation on $\hat{N}$-frames for ILM, we shall see the reason behind naming it so (cf. Observation 44).
Some of the conditions on strictly condensed J-frames and H-frames follow from the others. Let us observe the following for strictly condensed N-frames.

Observation 42.

1. A triple $(X, \leq, R_N)$, where X is non-empty and $\leq$ is a partial order, is a strictly condensed N-frame if and only if $(\leq R_N \leq^{-1}) \subseteq R_N$ [27].
2. In a strictly condensed N-frame $(X, \leq, R_N)$, $(R_N \leq^{-1}) = R_N$ always holds. Thus, the conditions '$R_N \leq^{-1}$ is symmetric' and '$R_N \leq^{-1}$ is reflexive' reduces just to 'R_N is symmetric' and 'R_N is reflexive' respectively.
3. In a strictly condensed N-frame $(X, \leq, R_N)$, 'R_N is reflexive' and '$\forall x, y \in X(xR_N y \Rightarrow \exists z \in X(x \leq z \ \& \ y \leq z \ \& \ xR_N z))$' together imply that '$R_N$ is symmetric'. This is because for all $x, y \in X$, $xR_N y$ implies there exists $z \in X$ such that $x \leq z$, $y \leq z$ and $xR_N z$. Now, $y \leq z$, $zR_N z$ (R_N is reflexive), and $x \leq z$ imply $y(\leq R_N \leq^{-1})x$. Using (1), $yR_N x$.

Therefore, we can simplify the definition of $\hat{N}$-frames.

Proposition 76. *Consider the quadruple $\mathcal{F} := (X, \leq, R_{N_1}, R_{N_2})$, where X is a non-empty set and $\leq$ is a partial order. The following are equivalent.*

a. $\mathcal{F}$ *is an $\hat{N}$-frame.*
b. $\mathcal{F}$ *satisfies the following conditions.*
   1. $(\leq R_{N_1} \leq^{-1}) \subseteq R_{N_1}$,
   2. R_{N_1} *is reflexive,*
   3. $\forall x, y \in X(xR_{N_1} y \Rightarrow \exists z \in X(x \leq z \ \& \ y \leq z \ \& \ xR_{N_1} z))$,
   4. $(\leq R_{N_2} \leq^{-1}) \subseteq R_{N_2}$,
   5. R_{N_2} *is symmetric,*
   6. $\forall x, y \in X(xR_{N_2} y \Rightarrow \exists z \in X(x \leq z \ \& \ y \leq z \ \& \ xR_{N_2} z))$, *and*
   7. $\forall x \in X \ \big(\forall y \in X(xR_{N_1} y \Rightarrow \exists z \in X(yR_{N_1} z \ \& \ \forall z' \in X \ (zR_{N_2} z')))\big) \Rightarrow$ $\forall z'' \in X \ (xR_{N_2} z'')\big).$ $\hspace{2cm}$ (DNE($\sim\top$))

Definition 119 (Valuation on an $\hat{N}$-frame).
A valuation v of $\mathcal{L}$ (the language of ILM) on an $\hat{N}$-frame $\mathcal{F} := (X, \leq, R_{N_1}, R_{N_2})$ is a mapping from the set of propositional variables in $\mathcal{L}$ to the power set $\mathcal{P}(X)$ of X such that for any propositional variable p, $v(p)$ is an upset, i.e. it satisfies the following:
$$\forall x, y \in X(x \leq y \Rightarrow (x \in v(p) \Rightarrow y \in v(p))).$$

Definition 120 (Model on an $\hat{N}$-frame).
Consider a valuation v of $\mathcal{L}$ on an $\hat{N}$-frame $\mathcal{F} := (X, \leq, R_{N_1}, R_{N_2})$. The pair $\mathcal{M} := (\mathcal{F}, v)$ is called an $\hat{N}$ − model on the $\hat{N}$-frame $\mathcal{F} := (X, \leq, R_{N_1}, R_{N_2})$.

Definition 121 (Truth of a formula).
The truth of a formula $\alpha \in F$ at a world $x \in X$ in the model $\mathcal{M} := (\mathcal{F}, v) = ((X, \leq, R_{N_1}, R_{N_2}), v)$ (notation - $\mathcal{M}, x \vDash \alpha$) is defined by extending the valuation map $v : PV \to \mathcal{P}(X)$ to the set F of formulas as follows:

1. $\mathcal{M}, x \vDash p \Leftrightarrow x \in v(p)$ *for all propositional variables p.*
2. $\mathcal{M}, x \vDash \phi \wedge \psi \Leftrightarrow \mathcal{M}, x \vDash \phi$ *and* $\mathcal{M}, x \vDash \psi$.
3. $\mathcal{M}, x \vDash \phi \vee \psi \Leftrightarrow \mathcal{M}, x \vDash \phi$ *or* $\mathcal{M}, x \vDash \psi$.
4. $\mathcal{M}, x \vDash \phi \to \psi \Leftrightarrow$ *for all $y \in X$, if $x \leq y$ and $\mathcal{M}, y \vDash \phi$ then $\mathcal{M}, y \vDash \psi$.*

5. $\mathcal{M}, x \vDash \top$.
6. $\mathcal{M}, x \nvDash \bot$.
7. $\mathcal{M}, x \vDash \neg\phi \Leftrightarrow$ *for all* $y \in X(x R_{N_1} y \Rightarrow \mathcal{M}, y \nvDash \phi)$.
8. $\mathcal{M}, x \vDash \sim\phi \Leftrightarrow$ *for all* $y \in X(x R_{N_2} y \Rightarrow \mathcal{M}, y \nvDash \phi)$.

Note:
(1). The notation '$\mathcal{M}, x \nvDash \phi$' denotes that ϕ is not true at the world $x \in X$ in the model $\mathcal{M}$.
(2). We shall use the notation $x \vDash \phi$ whenever the model $\mathcal{M}$ is clear from the context.

Definition 122 (Validity of a formula).
A formula α *is* true *in a model* $\mathcal{M} := (\mathcal{F}, v)$ *(notation -* $\mathcal{M} \vDash \alpha$*) if* $\mathcal{M}, x \vDash \alpha$ *for all* $x \in X$.
A formula $\alpha \in F$ *is* valid *in the* $\hat{N}$-*frame* $\mathcal{F} := (X, \leq, R_{N_1}, R_{N_2})$ *(notation -* $\mathcal{F} \vDash \alpha$*) if* $\mathcal{M} \vDash \alpha$ *for every model* $\mathcal{M}$ *on the* $\hat{N}$-*frame* $\mathcal{F}$.
A formula $\alpha \in F$ *is* valid *in a class* $\mathcal{C}$ *of* $\hat{N}$-*frames (notation -* $\mathcal{C} \vDash \alpha$*) if for any* $\hat{N}$-*frame* $\mathcal{F} \in \mathcal{C}$ *we have* $\mathcal{F} \vDash \alpha$.

Observation 43.

1. In Definition 121 (7) and (8), take $\phi := \top$, then using (5), we have the following.

$$\mathcal{M}, x \vDash \neg\top \Leftrightarrow \forall y \in X \ (x \not{R}_{N_1} y)$$
$$\mathcal{M}, x \vDash \sim\top \Leftrightarrow \forall y \in X \ (x \not{R}_{N_2} y)$$

2. Definition 121(4) implies that

$$\mathcal{M}, x \vDash \phi \to \psi \text{ and } \mathcal{M}, x \vDash \phi \text{ imply } \mathcal{M}, x \vDash \psi$$

Using this, if $x \vDash \phi \leftrightarrow \psi$, then $x \vDash \phi$ if and only if $x \vDash \psi$. We have the following for any $\hat{N}$-model $\mathcal{M}$.

$$\mathcal{M} \vDash \phi \leftrightarrow \psi \text{ if and only if } \forall x \in W(x \vDash \phi \Leftrightarrow x \vDash \psi)$$

3. Given $x \in X$, let us see the contraposition of the reverse implication in the Condition DNE($\sim\top$):

$$\exists y \in X\big(x R_{N_1} y \ \& \ \forall z \in X(y R_{N_1} z \Rightarrow \exists z' \in X \ (z R_{N_2} z'))\big) \Rightarrow$$

$$\exists z'' \in X \ (x R_{N_2} z'').$$

This is always true: let $y \in X$ such that $x R_{N_1} y$ and $\forall z \in X(y R_{N_1} z \Rightarrow \exists z' \in X \ z R_{N_2} z')$. $x R_{N_1} y$ implies $y R_{N_1} x$. Therefore, there exists $z' \in X$ such that $x R_{N_2} z'$. Take $z'' = z'$.

Observation 44. What exactly is the Condition DNE($\sim\top$)? Observe that this is the only condition in the definition of $\hat{N}$-frames involving both the relations R_{N_1} and R_{N_2}. Further, recall $\vdash_{\text{ILM}} \neg\neg\sim\top \leftrightarrow \sim\top$ involving both the negations.

Let us expand $x \vDash \neg\neg\sim\top \leftrightarrow \sim\top$ in the model $\mathcal{M} := (\mathcal{F}, v)$, where $\mathcal{F} := (W, \leq, R_{N_1}, R_{N_2})$.

$$\forall x \in X (x \vDash \neg\neg\sim\top \leftrightarrow \sim\top)$$
$$\Leftrightarrow \forall x \in X (x \vDash \neg\neg\sim\top \Leftrightarrow x \vDash \sim\top) \qquad \text{(Using Observation 43(2))}$$
$$\Leftrightarrow \forall x \in X (\forall y \in X (x R_{N_1} y \Rightarrow y \nvDash \neg\sim\top) \Leftrightarrow x \vDash \sim\top)$$
$$\Leftrightarrow \forall x \in X (\forall y \in X (x R_{N_1} y \Rightarrow \exists z \in X (y R_{N_1} z \;\&\; z \vDash \sim\top)) \Leftrightarrow x \vDash \sim\top)$$
$$\Leftrightarrow \forall x \in X (\forall y \in X (x R_{N_1} y \Rightarrow \exists z \in X (y R_{N_1} z \;\&\; \forall z' \in X (z \overset{\prime}{R}_{N_2} z'))) \Leftrightarrow$$
$$\forall z'' \in X (x \overset{\prime}{R}_{N_2} z'')) \qquad \text{(Observation 43(1))}$$

One direction of the last bi-implication is always true using Observation 43(3). The other direction is exactly the Condition DNE($\sim\top$). So, we may expect that ILM will be complete with respect to the class $\mathfrak{F}_1$ of $\check{N}$-frames.

Proposition 77. *Consider an extension* S *of ILM and a model* $\mathcal{M} := (\mathcal{F}, v)$ *where* $\mathcal{F} \in \mathfrak{F}_1$. *For any formula* $\alpha \in F$ *and* $x \in X$, *we have the following:*

1. $\forall y \in X ((\mathcal{M}, x \vDash \alpha \;\&\; x \leq y) \Rightarrow \mathcal{M}, y \vDash \alpha)$.
2. $\mathcal{M}, x \vDash \neg\alpha \Leftrightarrow \forall y \in X (\exists z \in X (x \leq z \;\&\; y \leq z) \Rightarrow y \nvDash \alpha)$.
3. $\mathcal{M}, x \vDash \sim\alpha \Leftrightarrow \forall y \in X (x R_{N_1} y \Rightarrow (y \vDash \alpha \Rightarrow \exists z \in X (y R_{N_1} z \;\&\; \forall z' \in X (z \overset{\prime}{R}_{N_2} z'))))$.

Proof.
(1). This can be obtained through induction on the number of connectives in α. For $\top$ and $\bot$, we have $x \vDash \top$ and $x \nvDash \bot$ for all $x \in W$. If α is a propositional variable p, then by definition of valuation v, $x \vDash p$ if and only if $x \in v(p)$. Since, $v(p)$ is an upset by the definition of the valuation v, we have for $x \leq y$, $y \in v(p)$, i.e. $y \vDash \alpha$. Let us assume that the statement holds whenever α has less than n connectives.
Now, suppose $x \vDash \alpha$ and $x \leq y$, where α has n (≥ 1) connectives. Let $\alpha := \sim\beta$ such that β has $(n-1)$ connectives. Let $x \leq y$ and $x \vDash \sim\beta$. We have to show $y \vDash \sim\beta$, i.e. for $z \in X$ such that $y R_{N_2} z$, $\mathcal{M}, z \nvDash \beta$. Let $z \in X$ such that $y R_{N_2} z$. We have $x \leq y$, $z \leq z$ and $y R_{N_2} z$. Using the condition $(\leq R_{N_2} \leq^{-1}) \subseteq R_{N_2}$, we have $x R_{N_2} z$. Using $x \vDash \sim\beta$ and $x R_{N_2} z$, we have $z \nvDash \beta$. If α is of the form of other connectives, we can easily show (1).
(2). ($\Rightarrow$). Let $x \vDash \neg\alpha$ and $y \in X$ such that there exists $z \in X$ satisfying $x \leq z$ and $y \leq z$. We have $z R_{N_1} z$, as R_{N_1} is reflexive. Using the condition $(\leq R_{N_1} \leq^{-1}) \subseteq R_{N_1}$, we have $x R_{N_1} y$. Finally, $x \vDash \neg\alpha$ and $x R_{N_1} y$ imply $y \nvDash \alpha$.
($\Leftarrow$). We have to show that $x \vDash \neg\alpha$. Let $y \in X$ such that $x R_{N_1} y$. To show that $y \nvDash \alpha$. $x R_{N_1} y$ implies that there exists $z \in X$ such that $x \leq z$, $y \leq z$ and $x R_{N_1} z$. Therefore, existence of such $z \in X$ implies $y \nvDash \alpha$.
(3). ($\Rightarrow$). Let $y \in X$ such that $x R_{N_1} y$ and $y \vDash \alpha$. We have to show that there exists $z \in X$ such that $y R_{N_1} z$ and for all $z' \in X$, $z \overset{\prime}{R}_{N_2} z'$. Using Observation 42, R_{N_1} is symmetric. Therefore, $x R_{N_1} y$ implies $y R_{N_1} x$. This implies there exists $z \in X$ such that $x \leq z$, $y \leq z$ and $y R_{N_1} z$. We claim that z satisfies $z \overset{\prime}{R}_{N_2} z'$ for all $z' \in X$. Suppose not, i.e. there exists $z' \in X$ such that $z R_{N_2} z'$. $z R_{N_2} z'$ implies there exists $\delta \in X$ such that $z \leq \delta$, $z' \leq \delta$ and $z R_{N_2} \delta$. Now, using

$(\leq R_{N_2} \leq^{-1}) \subseteq R_{N_2}$, $x \leq z$, $zR_{N_2}\delta$ and $z \leq \delta$ imply $xR_{N_2}z$. Since $x \vDash \sim\alpha$, $xR_{N_2}z$ implies $z \nvDash \alpha$. Finally, using (1), $y \leq z$ and $z \nvDash \alpha$ imply $y \nvDash \alpha$ - a contradiction.

($\Leftarrow$). We shall prove this by contradiction. Let $y' \in X$ such that $xR_{N_2}y'$ and $y' \vDash \alpha$. $xR_{N_2}y'$ implies there exists $y'' \in X$ such that $x \leq y''$, $y' \leq y''$ and $xR_{N_2}y''$. The contraposition of DNE($\sim\top$) is

$$\forall x \in X\big(\exists z'' \in X \ (xR_{N_2}z'') \Rightarrow \exists y \in X\big(xR_{N_1}y \ \& \ \forall z \in X(yR_{N_1}z \Rightarrow$$

$$\exists z' \in X \ (zR_{N_2}z'))\big)\big).$$

Since R_{N_2} is symmetric, we have $y''R_{N_2}x$. Therefore,

$$\exists \delta \in X\big(y''R_{N_1}\delta \ \& \ \forall z \in X(\delta R_{N_1}z \Rightarrow \exists z' \in X \ (zR_{N_2}z'))\big). \qquad (6.4)$$

Now, $y''R_{N_1}\delta$ implies there exists $\delta' \in X$ such that $y'' \leq \delta'$, $\delta \leq \delta'$ and $y''R_{N_1}\delta'$. We prove that for $\delta' \in X$, we have

$$(xR_{N_1}\delta') \ \& \ (\delta' \vDash \alpha) \ \& \ (\forall z \in X(\delta'R_{N_1}z \Rightarrow \exists z' \in X \ (zR_{N_2}z'))).$$

This would give us the result. Indeed, $x \leq y''$, $y''R_{N_1}\delta'$, and $\delta' \leq \delta'$ imply $xR_{N_1}\delta'$, using $(\leq R_{N_1} \leq^{-1}) \subseteq R_{N_1}$. From (1), $y' \vDash \alpha$ and $y' \leq y'' \leq \delta'$ imply $\delta' \vDash \alpha$. Let $z \in X$ be such that $\delta'R_{N_1}z$. $\delta \leq \delta'$, $\delta'R_{N_1}z$, and $z \leq z$ imply $\delta R_{N_1}z$, using $(\leq R_{N_1} \leq^{-1}) \subseteq R_{N_1}$. So, using Condition 6.4, there exists $z' \in X$ such that $zR_{N_2}z'$.

Therefore, we have the claim, which contradicts the given hypothesis.

Our aim now is to show that the logic ILM (ILM-$\vee$) is determined by the class $\mathfrak{F}_1$ ($\mathfrak{F}_1'$) of $\hat{N}$-frames ($\hat{N}'$-frames). In other words, we prove that for any formula $\alpha \in F$, $\vdash_{\text{ILM}} \alpha$ ($\vdash_{\text{ILM-}\vee} \alpha$) if and only if α is valid in $\mathfrak{F}_1$ ($\mathfrak{F}_1'$). This is presented in Theorem 28 and Theorem 29.

Theorem 28 (Soundness). *For any formula $\alpha \in F$,*

(i) $\vdash_{\text{ILM}} \alpha \Rightarrow \mathfrak{F}_1 \vDash \alpha$.
(ii) $\vdash_{\text{ILM-}\vee} \alpha \Rightarrow \mathfrak{F}_1' \vDash \alpha$.

Proof. (i). The proof is using induction on the number of steps in the proof of α. We shall only show the validity of (A13), i.e. for any $\mathcal{F} := (X, \leq, R_{N_1}, R_{N_2}) \in \mathfrak{F}_1$, $\mathcal{F} \vDash \sim\alpha \leftrightarrow (\alpha \rightarrow \neg\neg\sim\top)$. Let $x \in X$.

(A). $x \vDash \sim\alpha \rightarrow (\alpha \rightarrow \neg\neg\sim\top)$: let $x \vDash \sim\alpha$. We have to show that $x \vDash \alpha \rightarrow \neg\neg\sim\top$, i.e. for any $y \in X$, $x \leq y$ and $y \vDash \alpha$ imply $y \vDash \neg\neg\sim\top$. Suppose this is not true, that is, there exists $y \in X$ such that $x \leq y$, $y \vDash \alpha$, and $y \nvDash \neg\neg\sim\top$. Then

$$\exists z \in X\big(yR_{N_1}z \ \& \ \forall z' \in X(zR_{N_1}z' \Rightarrow \exists z'' \in X(z'R_{N_2}z''))\big).$$

Using Observation 43(3), we obtain $z_0 \in X$ such that $yR_{N_2}z_0$. Now, $yR_{N_2}z_0$ implies there exists $z_0' \in X$ such that $y \leq z_0'$, $z_0 \leq z_0'$ and $yR_{N_2}z_0'$. $y \vDash \alpha$ and $y \leq z_0'$ imply $z_0' \vDash \alpha$. Moreover, using $(\leq R_{N_2} \leq^{-1}) \subseteq R_{N_2}$, $x \leq y$, $yR_{N_2}z_0'$ and

$z_0' \leq z_0'$ imply $xR_{N_2}z_0'$. Therefore, we have obtained a $z_0' \in X$ such that $xR_{N_2}z_0'$ and $z_0' \vDash \alpha$. So, $x \nvDash \sim\alpha$, contradicting our assumption.

(B). $x \vDash (\alpha \rightarrow \neg\neg\sim\top) \rightarrow \sim\alpha$: let $x \vDash (\alpha \rightarrow \neg\neg\sim\top)$, i.e. for any $y \in X$, if $x \leq y$ and $y \vDash \alpha$, we obtain $y \vDash \neg\neg\sim\top$. Expanding $y \vDash \neg\neg\sim\top$,

$$\forall z \in X\big(yR_{N_1}z \Rightarrow \exists z' \in X(zR_{N_1}z' \ \& \ \forall z'' \in X(z'R_{N_2}z''))\big)$$

Using DNE($\sim\top$), we get for all $z_0 \in X$, $yR_{N_1}z_0$. Now, we have to show that $x \vDash \sim\alpha$. Suppose this is not true, i.e. there is $y_0 \in X$ such that $xR_{N_2}y_0$ and $y_0 \vDash \alpha$. $xR_{N_2}y_0$ implies there exists $\delta \in X$ such that $x \leq \delta$, $y_0 \leq \delta$ and $xR_{N_2}\delta$. $y_0 \leq \delta$ and $y_0 \vDash \alpha$ imply $\delta \vDash \alpha$. Moreover, $x \leq \delta$ and $\delta \vDash \alpha$ imply $\delta \vDash \neg\neg\sim\top$, i.e. for any $z_0 \in X$, $\delta R_{N_2}z_0$. However, R_{N_2} being symmetric, $xR_{N_2}\delta$ gives $\delta R_{N_2}x$, a contradiction.

(ii). Consider a frame $\mathcal{F} := (X, \leq, R_{N_1}, R_{N_2}) \in \mathfrak{F}_1'$. We have to show that $\mathcal{F} \vDash \alpha \vee \sim\alpha$. Let v be a valuation on $\mathcal{F}$ and $x \in X$. Suppose $x \nvDash \alpha \vee \sim\alpha$ for some $\alpha \in F$. Then, $x \nvDash \alpha$ and $x \nvDash \sim\alpha$. This implies that there exists $y \in X$ such that $xR_{N_2}y$ and $y \vDash \alpha$. $xR_{N_2}y$ implies $y \leq x$, as $\mathcal{F}$ is an $\hat{N}'$-frame. Therefore, we have $x \vDash \alpha$, a contradiction.

The completeness of ILM (ILM-$\vee$) with respect to $\hat{N}$-frames ($\hat{N}'$-frames) is given by

Theorem 29 (Completeness). *For any formula* $\alpha \in F$,

(i) $\mathfrak{F}_1 \vDash \alpha \Rightarrow \ \vdash_{\text{ILM}} \alpha$.
(ii) $\mathfrak{F}_1' \vDash \alpha \Rightarrow \ \vdash_{\text{ILM-}\vee} \alpha$.

To prove the theorem, we shall require the concept of a theory [27,78], that we extend to the context of ILM.

Definition 123 (Theory).
A theory $T \subseteq F$ *(with respect to an extension* S *of ILM) is a non-empty set of formulas in* $\mathcal{L}$ *such that, for formula* $\alpha, \beta \in F$,

1. if $\alpha \in T$ *and* $\alpha \rightarrow \beta \in T$, *then* $\beta \in T$ *(closed under deduction),*
2. $\alpha \in T$, *where* $\vdash_S \alpha$, *and*
3. if $\alpha, \beta \in T$ *then* $\alpha \wedge \beta \in T$ *(closed under* $\wedge$*).*

A theory is consistent *if* $\perp \notin T$, *otherwise* inconsistent. *A* prime *theory is a consistent theory such that for any formula* $\alpha, \beta \in F$, *if* $\alpha \vee \beta \in T$ *then either* $\alpha \in T$ *or* $\beta \in T$.

Using the Axiom (A10): $\vdash_S \perp \rightarrow \alpha$, a theory T is consistent if and only if there exists a formula α such that $\alpha \notin T$.

Now, for an arbitrary set Δ of formulas, the intersection of all the theories containing Δ is also a theory; it is called the *theory generated by* Δ and denoted by $Th(\Delta)$. One can also show

Lemma 3.

$Th(\Delta) = \{\alpha \in F \mid \exists \phi_1, \phi_2 \ldots, \phi_n \in \Delta \text{ such that } \vdash_S (\phi_1 \wedge \phi_2 \wedge \cdots \wedge \phi_n) \to \alpha\}.$

As a consequence, $Th(\{\top\}) = \{\alpha \in F \mid \vdash_S \alpha\}.$

The following lemmas and notions will be useful in proving Theorem 29. The proofs of the results are in the same lines as in [31,33].

Lemma 4. *For theories P and Q, $\alpha \in Th(P \cup Q)$ if and only if there exist $\psi \in P$ and $\phi \in Q$ such that $\vdash_S \psi \to (\phi \to \alpha)$.*

Proof. ($\Leftarrow$). If there exist $\psi \in P$ and $\phi \in Q$ such that $\vdash_S \psi \to (\phi \to \alpha)$, then $\phi, \psi \in Th(P \cup Q)$ implies $\alpha \in Th(T \cup Q)$.
($\Rightarrow$). Using Lemma 3, there exist $\phi_1, \phi_2 \ldots, \phi_n \in P \cup Q$ such that $\vdash_S (\phi_1 \wedge \phi_2 \wedge \cdots \wedge \phi_n) \to \alpha$. We can have three cases here.

Case 1: $\phi_i \in P$ for all $1 \leq i \leq n$.
Take $\psi := \phi_1 \wedge \phi_2 \wedge \cdots \wedge \phi_n$ and $\phi := \top$, then $\{\psi, \phi\} \vdash_S \alpha$. Using Deduction Theorem (Proposition 69), we have $\vdash_S \psi \to (\phi \to \alpha)$

Case 2: $\phi_i \in Q$ for all $1 \leq i \leq n$.
Take $\phi := \phi_1 \wedge \phi_2 \wedge \cdots \wedge \phi_n$ and $\psi := \top$.

Case 3: Suppose $\phi_{i_1}, \phi_{i_2}, \ldots, \phi_{i_k} \in P$, then $\phi_j \in Q$ for all $j \neq i_l$ $(1 \leq l \leq k)$ and $1 \leq j \leq n$. Take $\phi := \wedge_s(\phi_s)$ where $s = i_t$ $(1 \leq t \leq k)$ and $\psi := \wedge_l(\phi_l)$, where $l \neq i_t$ $(1 \leq t \leq k)$ and $1 \leq l \leq n$. We have $\{\psi, \phi\} \vdash_S \alpha$. Using Deduction Theorem, we have $\vdash_S \psi \to (\phi \to \alpha)$.

A direct corollary to this is the following.

Lemma 5. *For a theory T, and formulas $\phi, \alpha \in F$, $\alpha \in Th(T \cup \{\phi\})$ if and only if there exists $\psi \in T$ such that $\vdash_S \psi \to (\phi \to \alpha)$.*

Definition 124 ($\vee$-closed).
A set $F' \subseteq F$ of formulas is closed under $\vee$, if for any $\alpha, \beta \in F'$, $\alpha \vee \beta \in F'$. The set F' is called disjunctive closed *(or $\vee$-closed)*.

In fact, we can extend any arbitrary $\Delta \subseteq F$ to a $\vee$-closed set using the following.

Definition 125 (Disjunctive closure).
For $\Delta \subseteq F$,

$$\mathrm{dc}(\Delta) := \bigcap \{\Delta' \subseteq F \mid \Delta \subseteq \Delta' \text{ and } \forall \alpha, \beta \in F(\alpha, \beta \in \Delta' \Rightarrow \alpha \vee \beta \in \Delta')\}.$$

Observation 45.

(1) For $\Delta \subseteq F$, $\mathrm{dc}(\Delta)$ is $\vee$-closed.
(2) For $\alpha \in F$, if $\beta \in \mathrm{dc}(\{\alpha\})$ then $\vdash_S \alpha \leftrightarrow \beta$.

Lemma 6. (Extension lemma). *Let Δ be a consistent theory and $\Gamma \subseteq F$ be a $\vee$-closed set. If $\Delta \cap \Gamma = \emptyset$ then there exists a prime theory P such that $\Delta \subseteq P$ and $P \cap \Gamma = \emptyset$.*

Proof. Define a set

$$Z := \{\Delta' \subseteq F \mid \Delta \subseteq \Delta', \Delta' \cap \Gamma = \emptyset, \text{ and } \Delta' \text{ is a consistent theory}\}.$$

Note that Z is non-empty, because $\Delta \in Z$. The partially ordered set $(Z, \subseteq)$ satisfies the hypothesis of Zorn's lemma (Lemma 2).
Therefore, there exists $P \in Z$ such that P is maximal. So P is a consistent theory.
P is prime: suppose there exist $\alpha, \beta \in F$ such that $\alpha \vee \beta \in P$, $\alpha \notin P$ and $\beta \notin P$.
Since P is a maximal element in Z, $Th(P \cup \{\alpha\}) \notin Z$ and $Th(P \cup \{\beta\}) \notin Z$.
Therefore, we have 4 cases.

Case 1: $Th(P \cup \{\alpha\})$ and $Th(P \cup \{\beta\})$ are inconsistent, i.e. $\perp \in Th(P \cup \{\alpha\})$ and $\perp \in Th(P \cup \{\beta\})$. Using Lemma 5, there exist $\gamma, \delta \in P$ such that $\vdash_S \gamma \to (\alpha \to \perp)$ and $\vdash_S \delta \to (\beta \to \perp)$, i.e.

$$\vdash_S (\gamma \wedge \alpha) \to \perp \text{ and } \vdash_S (\delta \wedge \beta) \to \perp$$

Now, since P is a theory, $(\gamma \wedge \delta) \wedge (\alpha \vee \beta) \in P$. This implies $(\gamma \wedge \delta \wedge \alpha) \vee (\gamma \wedge \delta \wedge \beta) \in P$. We have

$$\vdash_S ((\gamma \wedge \delta \wedge \alpha) \vee (\gamma \wedge \delta \wedge \beta)) \to ((\gamma \wedge \alpha) \vee (\delta \wedge \beta))$$
$$\text{i.e. } \vdash_S ((\gamma \wedge \delta \wedge \alpha) \vee (\gamma \wedge \delta \wedge \beta)) \to \perp$$

Therefore, $\perp \in P$, implying P is inconsistent, a contradiction.

Case 2: $Th(P \cup \{\alpha\})$ is inconsistent and $Th(P \cup \{\beta\}) \cap \Gamma \neq \emptyset$, i.e. we have $\delta \in \Gamma$ and $\gamma, \delta' \in P$ such that

$$\vdash_S (\gamma \wedge \alpha) \to \perp \text{ and } \vdash_S (\delta' \wedge \beta) \to \delta.$$

Since P is a theory, $(\gamma \wedge \delta') \wedge (\alpha \vee \beta) \in P$. This implies that $(\gamma \wedge \delta' \wedge \alpha) \vee (\gamma \wedge \delta' \wedge \beta) \in P$. Thus, $(\perp \vee \delta) \in P$, implying $\delta \in P$, but this means that $P \cap \Gamma \neq \emptyset$, a contradiction.

Case 3: $Th(P \cup \{\beta\})$ is inconsistent and $Th(P \cup \{\alpha\}) \cap \Gamma \neq \emptyset$: similar as case 2 above.

Case 4: $Th(P \cup \{\alpha\}) \cap \Gamma \neq \emptyset$ and $Th(P \cup \{\beta\}) \cap \Gamma \neq \emptyset$, i.e. there exist $\gamma, \delta \in \Gamma$ and $\gamma', \delta' \in P$ such that

$$\vdash_S (\gamma' \wedge \alpha) \to \gamma \text{ and } \vdash_S (\delta' \wedge \beta) \to \delta$$

Since P is a theory, $(\gamma' \wedge \delta') \wedge (\alpha \vee \beta) \in P$. This implies that $(\gamma' \wedge \delta' \wedge \alpha) \vee (\gamma' \wedge \delta' \wedge \beta) \in P$. Thus, $(\gamma \vee \delta) \in P$. Now, $\gamma, \delta \in \Gamma$ implies $(\gamma \vee \delta) \in \Gamma$ (because Γ is $\vee$-closed). This means that $P \cap \Gamma \neq \emptyset$, a contradiction.
In each case, we have a contradiction. This tells us that no such α, β exist, implying that P is prime.

Corollary 10. *Let Δ be a consistent theory and $\alpha \in F$ be such that $\alpha \notin \Delta$. Then there is a prime theory P such that $\Delta \subseteq P$ and $\alpha \notin P$.*

Proof. Consider $\Gamma := \mathrm{dc}\{\alpha\}$.

Let us now define the 'canonical' frame $\mathcal{F}^c := (X^c, \subseteq, R^c_{N_1}, R^c_{N_2})$ in the standard way.

Definition 126 (Canonical frame).
The canonical frame *for any extension* S *of* ILM *is the quadruple* $\mathcal{F}^c := (X^c, \subseteq , R^c_{N_1}, R^c_{N_2})$, *where*
$X^c := \{P \subseteq F \mid P$ *is a prime theory*$\}$,
$P R^c_{N_1} Q$ *if and only if (for all* $\alpha \in F$, $\neg\alpha \in P \Rightarrow \alpha \notin Q$), *and*
$P R^c_{N_2} Q$ *if and only if (for all* $\alpha \in F$, $\sim\alpha \in P \Rightarrow \alpha \notin Q$).

We shall prove that $\mathcal{F}^c$ is an $\hat{N}$-frame. For that, we shall require the following two results.

Lemma 7. *For any* $\alpha \in F$ *and any* $P \in X^c$,

1. $\neg\alpha \in P$ *if and only if for all* $Q \in X^c$, $P R^c_{N_1} Q \Rightarrow \alpha \notin Q$, *and*
2. $\sim\alpha \in P$ *if and only if for all* $Q \in X^c$, $P R^c_{N_2} Q \Rightarrow \alpha \notin Q$.

Proof. Let us prove (2).
($\Rightarrow$) By the definition of $R^c_{N_2}$.
($\Leftarrow$). Let $\sim\alpha \notin P$. We need to find a prime theory Q such that $P R^c_{N_2} Q$ and $\alpha \in Q$. Consider the theory $\Delta := Th(\{\alpha\})$ generated by $\{\alpha\}$, and $\Gamma := \{\beta \in F \mid \sim\beta \in P\}$. Observe that α is not $\bot$, as $\sim\bot \in P$ by Proposition 74(o).
Claim 1: Δ is consistent. Suppose not, then $\bot \in \Delta$. Using Lemma 3, $\vdash_S \alpha \to \bot$. Using Proposition 70, we have $\vdash_S \sim\bot \to \sim\alpha$. Using Proposition 74(o), $\vdash_S \sim\alpha$ and $\sim\alpha \in P$, which contradicts the assumption.
Claim 2: $\Delta \cap \Gamma = \emptyset$. Suppose not, then there exists $\beta \in \Gamma \cap Th(\{\alpha\})$. Therefore, using Lemma 3, $\vdash_S \alpha \to \beta$. Using Proposition 70, we have $\vdash_S \sim\beta \to \sim\alpha$. $\beta \in \Gamma$ implies $\sim\beta \in P$. Therefore, $\sim\alpha \in P$, a contradiction.
Claim 3: Γ is $\vee$-closed. Let $\gamma, \beta \in \Gamma$, then $\sim\gamma, \sim\beta \in P \Rightarrow \sim\gamma \wedge \sim\beta \in P$, i.e. $(\gamma \to \neg\neg\sim\top) \wedge (\beta \to \neg\neg\sim\top) \in P$ (using (A13)). Therefore, by (A5), $((\gamma \vee \beta) \to \neg\neg\sim\top) \in P \Rightarrow \sim(\gamma \vee \beta) \in P \Rightarrow \gamma \vee \beta \in \Gamma$.
Therefore, using Lemma 6, there exists a prime theory Q such that $\Gamma \cap Q = \emptyset$ and $Th(\{\alpha\}) = \Delta \subseteq Q$. So $\alpha \in Q$.
$P R^c_{N_2} Q$: let $\beta \in F$ be such that $\sim\beta \in P$. This implies that $\beta \in \Gamma \Rightarrow \beta \notin Q$.
(1) is obtained by using similar arguments as in (2) above, where $\sim$ is replaced by $\neg$.

An immediate corollary to the above lemma is obtained for $\alpha := \top$, using the fact that $\top \in Q$ for any prime theory Q.

Lemma 8. *1.* $\neg\top \in P$ *if and only if for all* $Q \in X^c$, $P \not{R}^c_{N_1} Q$, *and*
2. $\sim\top \in P$ *if and only if for all* $Q \in X^c$, $P \not{R}^c_{N_2} Q$.

Let us now see that the canonical frame $\mathcal{F}^c$ indeed satisfies all the properties of an $\hat{N}$-frame.

Proposition 78. $\mathcal{F}^c := (X^c, \subseteq, R^c_{N_1}, R^c_{N_2})$ *is an $\hat{N}$-frame.*

Proof. We shall use Proposition 76 to show that $\mathcal{F}^c$ is an $\hat{N}$-frame.

(1). Let $P, Q \in X^c$ such that $P(\subseteq R^c_{N_1} \subseteq^{-1})Q$. This means there exist $P', Q' \in X^c$ such that $P \subseteq P'$ and $Q \subseteq Q'$ and $P' R^c_{N_1} Q'$. We have to show that $P R^c_{N_1} Q$. Consider $\alpha \in F$ such that $\neg\alpha \in P$. Then $\neg\alpha \in P' \Rightarrow \alpha \notin Q' \Rightarrow \alpha \notin Q$. Therefore, $P R^c_{N_1} Q$.

(2). $P R^c_{N_1} P$: consider $\alpha \in F$ such that $\neg\alpha \in P$. Now, if $\alpha \in P$, then $\alpha \wedge \neg\alpha \in P \Rightarrow \perp \in P$ (Proposition 74(1)), a contradiction to P being consistent. Therefore, $\alpha \notin P$.

(3). Consider $P, Q \in X^c$ such that $P R^c_{N_1} Q$. We have to find $Q' \in X^c$ such that $P \subseteq Q'$, $Q \subseteq Q'$ and $P R^c_{N_1} Q'$.

Consider the theory $\Delta := Th(P \cup Q)$. Δ is consistent, because if not, then using Lemma 4, there exist $\alpha \in P$ and $\beta \in Q$ such that $\vdash_S \alpha \to (\beta \to \perp)$. This implies $\vdash_S \alpha \to \neg\beta \Rightarrow \neg\beta \in P \Rightarrow \beta \notin Q$ (as $P R^c_{N_1} Q$), a contradiction.

Define $\Gamma := \{\alpha \in F \mid \neg\alpha \in P\}$. Observe that Γ is $\vee$-closed, because for $\alpha, \beta \in \Gamma$, we have $\neg\alpha, \neg\beta \in P$ implying $\neg\alpha \wedge \neg\beta \in P \Rightarrow (\alpha \to \perp) \wedge (\beta \to \perp) \in P \Rightarrow (\alpha \vee \beta) \to \perp \in P \Rightarrow \neg(\alpha \vee \beta) \in P \Rightarrow (\alpha \vee \beta) \in \Gamma$.

We claim that $\Gamma \cap \Delta = \emptyset$. Suppose not, i.e. there exists $\alpha \in \Gamma$ such that $\alpha \in Th(P \cup Q)$. This means $\neg\alpha \in P$, and there exist $\beta \in P$, $\gamma \in Q$ such that $\vdash_S \beta \to (\gamma \to \alpha)$, i.e. $\{\beta\} \vdash_S \gamma \to \alpha$. Using contraposition law for $\neg$ (Proposition 21(1)), we have $\{\beta\} \vdash_S \neg\alpha \to \neg\gamma$, i.e. $\vdash_S \beta \to (\neg\alpha \to \neg\gamma)$. Since, $\beta, \neg\alpha \in P$, we have $\neg\gamma \in P$. Using $P R^c_{N_1} Q$, $\gamma \notin Q$, a contradiction.

Thus using Proposition 6, there exists a prime theory $Q' \in X^c$ such that $Th(P \cup Q) \subseteq Q'$ and $\Gamma \cap Q' = \emptyset$. Our final claim is $P R^c_{N_1} Q'$. Let $\alpha \in F$ such that $\neg\alpha \in P$. Then $\alpha \in \Gamma \Rightarrow \alpha \notin Q'$. Therefore, $P R^c_{N_1} Q'$.

(4). Identical argument as in (1), where $\neg$ is replaced by $\sim$.

(5). Consider $P, Q \in X^c$ such that $P R^c_{N_2} Q$. We show $Q R^c_{N_2} P$. Let $\alpha \in F$ such that $\sim\alpha \in Q$ and $\alpha \in P$. Using Proposition 74(b), $\sim\sim\alpha \in P$. Therefore, $\sim\alpha \notin Q$, a contradiction. Therefore, $Q R^c_{N_2} P$.

(6). Similar argument as in (3), where $\neg$ is replaced by $\sim$. The only change in the proof is to show that Δ is consistent. Suppose not, then using Lemma 4, there exist $\alpha \in P$ and $\beta \in Q$ such that $\vdash_S \alpha \to (\beta \to \perp)$. This implies $\vdash_S \alpha \to \neg\beta \Rightarrow \neg\beta \in P$. Now, using Proposition 74(f), $\sim\beta \in P$. Therefore, $\beta \notin Q$, a contradiction.

(7). We have to show the following. For all $P \in X^c$,
$$\left(\forall Q \in X^c \left(P R^c_{N_1} Q \Rightarrow \exists R \in X^c (Q R^c_{N_1} R \ \& \ \forall R' \in X^c \ (R R^c_{N_2} R'))\right)\right) \Rightarrow$$
$$\forall R'' \in X^c \ (P R^c_{N_2} R'').$$
Suppose $\forall Q \in X^c \left(P R^c_{N_1} Q \Rightarrow \exists R \in X^c (Q R^c_{N_1} R \ \& \ \forall R' \in X^c \ (R R^c_{N_2} R'))\right)$. This is equivalent to assuming

$$\forall Q \in X^c \big(PR_{N_1}^c Q \Rightarrow \exists R \in X^c(QR_{N_1}^c R \ \& \ \sim\top \in R)\big) \qquad \text{(Lemma 8)}$$
$$\Leftrightarrow \ \forall Q \in X^c(PR_{N_1}^c Q \Rightarrow \neg\sim\top \notin Q) \qquad \text{(Lemma 7)}$$
$$\Leftrightarrow \ \neg\neg\sim\top \in P \qquad \text{(Lemma 7)}$$
$$\Leftrightarrow \ \sim\top \in P \qquad \text{(Proposition 74(c))}$$
$$\Leftrightarrow \ \forall R'' \in X^c \ (P R_{N_2}^c R'') \qquad \text{(Lemma 8)}$$

The canonical $\hat{N}$-model on the canonical $\hat{N}$-frame is also defined in the standard way.

Define a valuation v^c on the canonical frame $\mathcal{F}^c := (X^c, \subseteq, R_{N_1}^c, R_{N_2}^c)$ as follows: for all $p \in PV$, $v^c(p) := \{P \in X^c \mid p \in P\}$. The pair $\mathcal{M}^c := (\mathcal{F}^c, v^c)$ is called the *canonical $\hat{N}$-model*.

Lemma 9 (Truth Lemma). *For any $\alpha \in F$ and $P \in X^c$, $\mathcal{M}^c, P \vDash \alpha$ if and only if $\alpha \in P$.*

Proof. The proof is using induction on the number of connectives in α.

Basis Step: If α is $\top$, then we know that $\top \in P$, and for any valuation v on the $\hat{N}$-frame $\mathcal{F}^c$ and $P \in X^c$, $(\mathcal{F}^c, v), P \vDash \top$.

If α is $\bot$, then we know that $\bot \notin P$ (as P is consistent), and for any valuation v on a $\hat{N}$-frame $\mathcal{F}^c$ and $P \in X^c$, $(\mathcal{F}^c, v), P \nvDash \bot$.

If α is a propositional variable p, then by definition of v^c, we have $\mathcal{M}^c, P \vDash p$ if and only if $p \in P$.

Let the result hold when α has less than n connectives.

Induction Step: Let α have n connectives. The cases when α is of the form $\sim\beta$ or $\neg\beta$, where $\beta \in F$, are obtained directly from the definition of valuation (Definition 121) and Lemma 7. Let us check the case when $\alpha := \beta \to \gamma$.

($\Leftarrow$): Let $\beta \to \gamma \in P$, and let $Q \in X^c$ be such that $P \subseteq Q$ and $\mathcal{M}^c, Q \vDash \beta$. By induction hypothesis, $\beta \in Q$. As $\beta \to \gamma \in Q$, we have $\gamma \in Q$. By induction hypothesis, $\mathcal{M}^c, Q \vDash \gamma$.

($\Rightarrow$): Let $\beta \to \gamma \notin P$. We need to show that $\mathcal{M}^c, P \nvDash \beta \to \gamma$, i.e. there exists $\Omega \in X^c$ such that $P \subseteq \Omega$, $\beta \in \Omega$ and $\gamma \notin \Omega$. We use Zorn's lemma (Lemma 2) on the set Δ defined as follows:

$$\Delta := \{P' \in X^c \mid P \subseteq P' \ \& \ (\beta \to \gamma) \notin P'\}.$$

The partially ordered set $(\Delta, \subseteq)$ satisfies the hypothesis of Zorn's lemma. Therefore, there exists $\Omega \in \Delta$ such that Ω is maximal in Δ. $\Omega \in \Delta$ implies that Ω is a prime theory such that $P \subseteq \Omega$ and $\beta \to \gamma \notin \Omega$.

Claim: $\gamma \notin \Omega$.

Suppose not, i.e. $\gamma \in \Omega$. We have $\gamma \to (\beta \to \gamma) \in \Omega$, as any prime theory contains all the axioms of ILM. So $\beta \to \gamma \in \Omega$, a contradiction.

Claim: $\beta \in \Omega$. If possible, let $\beta \notin \Omega$. Consider the theory $T := Th(\Omega \cup \{\beta\})$.

Subclaim: $T := Th(\Omega \cup \{\beta\})$ is consistent. Suppose not, i.e. there exists $\delta \in \Omega$ such that $\vdash_{\text{ILM}} \delta \to (\beta \to \bot)$ (by Lemma 5). This implies $\vdash_{\text{ILM}} \delta \to \neg\beta$ (by

Proposition 74 and Equivalence theorem). $\delta \in \Omega$ implies $\neg\beta \in \Omega$. Now, axiom (A12): $\neg\beta \rightarrow (\beta \rightarrow \gamma)$ and $\neg\beta \in \Omega$ imply $\beta \rightarrow \gamma \in \Omega$ - a contradiction. So T is a consistent theory.

Subclaim: $\beta \rightarrow \gamma \notin T$. Suppose this is not true, i.e. $\beta \rightarrow \gamma \in T$. Then using Lemma 5, there exists $\delta \in \Omega$ such that $\vdash_{ILM} \delta \rightarrow (\beta \rightarrow (\beta \rightarrow \gamma))$. So $(\beta \rightarrow (\beta \rightarrow \gamma)) \in \Omega$. Since, $\vdash_{ILM} (\beta \rightarrow (\beta \rightarrow \gamma)) \rightarrow (\beta \rightarrow \gamma)$, we have $(\beta \rightarrow \gamma) \in \Omega$, a contradiction.

Using Corollary 10, there exists a prime theory Q such that $T \subseteq Q$ and $\beta \rightarrow \gamma \notin Q$.

Note that $P \subseteq \Omega \subseteq T \subseteq Q$. Therefore, $Q \in \Delta$. But Ω is a proper subset of Q, as by assumption, $\beta \in Q \backslash \Omega$. This is a contradiction to the fact that Ω is a maximal element in Δ. So, $\beta \in \Omega$.

Let us now see the proof of the completeness result for the logics ILM and ILM-∨.

Proof. (**of Theorem** 29):

(i). Let $\nvdash_{ILM} \alpha$. If $\alpha = \bot$, then by definition of valuation, $\mathcal{F} \nvDash \bot$ for any $\hat{N}$-frame $\mathcal{F}$. Therefore, suppose α is not $\bot$. Then define $\Delta := Th(\top)$ and $\Gamma := dc(\{\alpha\})$. Δ is a consistent theory, because $\bot \notin \Delta$ using Soundness Theorem 28. $\nvdash_{ILM} \alpha \Rightarrow \alpha \notin \Delta$, which implies, by Corollary 10, there is $P \in X^c$ such that $\Delta \subseteq P$ and $\alpha \notin P$. Using the Truth Lemma 9, $\mathcal{M}^c, P \nvDash \alpha$.

(ii). For ILM-∨, we have to show that the canonical frame for ILM-∨ belongs to $\mathfrak{F}_1'$, and for all frames $\mathcal{F} \in \mathfrak{F}_1'$ and $\alpha \in F$, $\mathcal{F} \vDash \alpha \vee \sim\alpha$. Consider the canonical frame $\mathcal{F}^c := (X^c, \subseteq, R_{N_1}^c, R_{N_2}^c)$. We have already shown that it is an $\hat{N}$-frame. We have to show that for all $P, Q, \in X^c$, if $PR_{N_2}^c Q$ then $Q \subseteq P$. Let P and Q be such that $PR_{N_2}^c Q$ and $Q \nsubseteq P$. Then there exists $\alpha \in Q$ such that $\alpha \notin P$. We have $\alpha \vee \sim\alpha \in P$, as $(\alpha \vee \sim\alpha)$ is an axiom in ILM-∨. Therefore, $\sim\alpha \in P$. Using the definition of $R_{N_2}^c$, we have $\alpha \notin Q$, a contradiction.

Next, we shall see another semantics for ILM.

6.2.2 Kripke-Style Semantics for ILM

We now recall the Kripke-style semantics for ML and IL, as mentioned in Sect. 2.5.2.

What would be the class of Kripke-style frames that can characterize ILM? We saw in the last section, in the case of Došen's relational semantics for ILM, there are two binary relations R_{N_1} and R_{N_2} which define the semantics for the negations $\neg$ and $\sim$ respectively. Let us make a few observations before defining the Kripke-style semantics for ILM in this section.

Observation 46.

(1) Consider a partially ordered set $(W, \leq)$. Since negations are defined using upsets in j-frames, we shall require two upsets $Y_{0\neg}$ and $Y_{0\sim}$, one for each negation $\neg$ and $\sim$ respectively.

(2) As noted in Observation 41(2), the negation $\neg$ is intuitionistic. For intuitionistic negation, $Y_{0\neg}$ consists of all the worlds where $\bot$ holds. Therefore, $Y_{0\neg}$ must be the empty set, as is the case with frames corresponding to IL.

(3) As noted in Observation 41(3), the negation $\sim$ is minimal. In this case, we do not deal with $Y_{0\sim}$ as done above for intuitionistic negation. We characterize $Y_{0\sim}$ by those worlds where $\sim\top$ holds, as we shall see in 79(1). Note that, in ILM, $\bot$ is not logically equivalent to $\sim\top$ (Proposition 74(q)). So, $Y_{0\sim}$ may not be empty.

Consider the following class of frames first given by Woodruff [87] to characterize the logic JP$'$ (Definition 115), an extension of ML.

Definition 127 (Sub-normal frames [87]).

Consider the j-frame $\mathcal{F} := (X, \leq, Y_0)$ satisfying the following condition:

$$\forall x \in X (\forall y \in X (x \leq y \Rightarrow \exists z \in X (y \leq z \ \& \ z \in Y_0)) \Rightarrow x \in Y_0). \tag{D}$$

$\mathcal{F}$ is called a sub-normal frame. *The class of all sub-normal frames is denoted by $\mathfrak{F}_2$.*
The sub-class of all sub-normal frames $\mathcal{F} := (X, \leq, Y_0)$ satisfying the following property shall be denoted by $\mathfrak{F}_2'$.

$$\forall x, y \in X \backslash Y_0 \Rightarrow (x \leq y \Rightarrow y \leq x).$$

Theorem 30. [87] *JP$'$ is determined by the class of all sub-normal frames.*

Following is a relationship between sub-normal frames and normal frames. Consider a sub-normal frame $(W, \leq, Y_0)$. Trivially $(W, \leq)$ is a normal frame. However, we have more. For every $w \notin Y_0$, one can generate a sub-normal frame $(W_{v_w}, \leq)$: by the contraposition of (D), we have $v_w \in W$ such that $w \leq v_w$ and $\forall v' \in W (v_w \leq v' \Rightarrow v' \notin Y_0)$. Consider the subframe 'generated by v_w' - $(W_{v_w}, \leq, Y_{0_{v_w}})$, where $W_{v_w} := \{z \in W \mid v_w \leq z\}$ and $Y_{0_{v_w}} := W_{v_w} \cap Y_0$. Then $Y_{0_{v_w}}$ empty. Therefore $(W_{v_w}, \leq)$ is a normal frame.

As observed in Theorem 25, ILM is an extension of JP$'$. So it is expected that Kripke-style models of ILM and ILM-$\vee$ would be sub-normal models.

Definition 128 (Valuation on a sub-normal frame).

A valuation of $\mathcal{L}$ on a sub-normal frame $\mathcal{F} := (W, \leq, Y_0)$ is a mapping $v : PV \to \mathcal{P}(W)$ such that $v_0(p)$ is an upset for each $p \in PV$.

Definition 129 (Sub-normal model).

For a sub-normal frame $\mathcal{F} := (W, \leq, Y_0)$, and a valuation v_0 on $\mathcal{F}$, the pair $\mathcal{M} := (\mathcal{F}, v_0)$ is called a sub-normal model *on the sub-normal frame.*

Definition 130 (Truth of a formula).

The truth *of a formula $\alpha \in F$ at a world $w \in W$ in the model $\mathcal{M} := (\mathcal{F}, v_0) = ((W, \leq, Y_0), v_0)$ (notation - $\mathcal{M}, w \vDash \alpha$) is defined by extending the valuation map $v_0 : PV \to \mathcal{P}(W)$ to the set F of formulas as follows:*

1. $\mathcal{M}, w \vDash p \Leftrightarrow w \in v_0(p)$ *for all propositional variables p.*
2. $\mathcal{M}, w \vDash \phi \wedge \psi \Leftrightarrow \mathcal{M}, w \vDash \phi$ *and* $\mathcal{M}, w \vDash \psi$.
3. $\mathcal{M}, w \vDash \phi \vee \psi \Leftrightarrow \mathcal{M}, w \vDash \phi$ *or* $\mathcal{M}, w \vDash \psi$.

4. $\mathcal{M}, w \vDash \phi \rightarrow \psi \Leftrightarrow$ for all $w' \in W$, if $w \leq w'$ and $\mathcal{M}, w' \vDash \phi$ then $\mathcal{M}, w' \vDash \psi$.

5. $\mathcal{M}, w \vDash \sim\phi \Leftrightarrow$ for all $w' \in W$, if $w \leq w'$ and $\mathcal{M}, w' \vDash \phi$ then $w' \in Y_0$.

6. $\mathcal{M}, w \vDash \neg\phi \Leftrightarrow$ for all $w' \in X(w \leq w' \Rightarrow \mathcal{M}, w' \nvDash \phi)$.

7. $\mathcal{M}, w \vDash \top$.

8. $\mathcal{M}, w \nvDash \bot$.

When the context is clear, we may simply write $w \vDash \beta$.

The validity of a formula $\alpha \in F$ in a sub-normal frame, and in a class of sub-normal frames is defined in the standard way (Definition 68). Let us now look at some properties of the valuation function and Condition (D) of the sub-normal frame.

Observation 47. In Definition 130(4), if we take $w' = w$, then we have

$$w \vDash \phi \rightarrow \psi \text{ and } w \vDash \phi \text{ imply } w \vDash \psi$$

Using this, if $w \vDash \phi \leftrightarrow \psi$, then $w \vDash \phi$ if and only if $w \vDash \psi$. In fact for any sub-normal model $\mathcal{M}$, we have the following.

$$\mathcal{M} \vDash \phi \leftrightarrow \psi \text{ if and only if } \forall w \in W(w \vDash \phi \Leftrightarrow w \vDash \psi)$$

Proposition 79. For $w \in W$,

1. $\mathcal{M}, w \vDash \sim\top \Leftrightarrow w \in Y_0$.
2. $\mathcal{M}, w \vDash \sim\phi \leftrightarrow (\phi \rightarrow \sim\top)$

Proof.
(1). Definition 130(5) gives the semantics of $\sim\top$ as: $w \vDash \sim\top \Leftrightarrow$ for all $w' \in W$, if $w \leq w'$ then $w' \in Y_0$. Taking $w' = w$ gives $w \in Y_0$. Since, Y_0 is an upset, for any $w' \in Y_0$, $w \in Y_0$ and $w \leq w'$ imply $w' \in Y_0$. Therefore, we have $\mathcal{M}, w \vDash \sim\top \Leftrightarrow w \in Y_0$.

(2). Using (1) and Definition 130(4), we obtain for all $w \in W$, $w \vDash \sim\phi$ if and only if $w \vDash \phi \rightarrow \sim\top$. Therefore, by Observation 47, we have $\mathcal{M}, w \vDash \sim\phi \leftrightarrow (\phi \rightarrow \sim\top)$.

This proposition suggests that the semantics of the connective $\sim$ can be obtained using the semantics of the connective $\rightarrow$ and the set Y_0.

What exactly is the Condition (D) in sub-normal frames (Definition 127)?

Observation 48.

(1) For any $w \in W$, first let us note that the contraposition of the reverse implication in Condition (D) is always true, i.e.

$$\exists v \in W(w \leq v \text{ and } \forall v' \in W(v \leq v' \Rightarrow v' \notin Y_0)) \Rightarrow w \notin Y_0.$$

Indeed, let $v \in W$ be such that $w \leq v$ and $\forall v' \in W(v \leq v' \Rightarrow v' \notin Y_0))$. If $w \in Y_0$ then $v \in Y_0$ (as Y_0 is an upset). However, $v \leq v$ implies $v \notin Y_0$, a contradiction.

(2) Now let us proceed as in Observation 44 for $\hat{N}$-frames to study the Condition (D): expand $\mathcal{M} \models \neg\neg\sim\top \leftrightarrow \sim\top$, where $\mathcal{M} := (\mathcal{F}, v)$ and $\mathcal{F} := (W, \leq, Y_0)$ is a sub-normal frame.

$$\forall w \in W(w \models \neg\neg\sim\top \leftrightarrow \sim\top)$$
$$\Leftrightarrow \forall w \in W(w \models \neg\neg\sim\top \Leftrightarrow w \models \sim\top) \qquad \text{(Using Observation 47)}$$
$$\Leftrightarrow \forall w \in W(\forall w' \in W(w \leq w' \Rightarrow w' \nVdash \neg\sim\top) \Leftrightarrow w \models \sim\top)$$
$$\Leftrightarrow \forall w \in W(\forall w' \in W(w \leq w' \Rightarrow \exists w'' \in W(w' \leq w'' \ \& \ w'' \in Y_0)) \Leftrightarrow w \in Y_0).$$
$$\text{(Proposition 79(1))}$$

The reverse of the last bi-implication is always true using Proposition 79(1). The forward direction is exactly Condition (D). Therefore, we expect that ILM will be complete with respect to the class $\mathfrak{F}_2$ of sub-normal frames.

Proposition 80. *For any formula $\alpha \in F$ and model $\mathcal{M} := (\mathcal{F}, v)$ on a sub-normal frame $\mathcal{F}$, we have the following for any $w \in W$.*

1. *For all $w' \in W$, if $\mathcal{M}, w \models \alpha$ and $w \leq w'$, then $\mathcal{M}, w' \models \alpha$.*
2. *$\mathcal{M}, w \models \neg\alpha$ if and only if $\forall w' \in W(\exists w'' \in W(w \leq w'' \ \& \ w' \leq w'') \Rightarrow w' \nVdash \alpha)$.*
3. *$\mathcal{M}, w \models \sim\alpha$ if and only if*

$$\forall w' \in W(w \leq w' \Rightarrow (w' \models \alpha \Rightarrow \exists w'' \in W(w' \leq w'' \ \& \ w'' \in Y_0))).$$

Proof.
(1). This can be obtained through induction on the number of connectives in α. For $\top$ and $\bot$, $\mathcal{M}, w \models \top$ and $\mathcal{M}, w \nVdash \bot$ for all $w \in W$. If α is a propositional variable p, the result follows from the definitions of truth and valuation. Assume that the statement holds whenever α has less than n connectives.
Suppose $\mathcal{M}, w \models \alpha$ and $w \leq w'$, where α has n (≥ 1) connectives. Let $\alpha := \sim\beta$. We have to show $\mathcal{M}, w' \models \sim\beta$, i.e. for any $w'' \in W$ such that $w' \leq w''$ and $\mathcal{M}, w'' \models \beta$, we have to show that $w'' \in Y_0$. Using $\mathcal{M}, w \models \sim\beta$, $w \leq w' \leq w''$ and $\mathcal{M}, w'' \models \beta$, we have $w'' \in Y_0$.
A similar approach will work when α is of the type $\neg\beta$ or $\beta \to \gamma$ where β and γ have less than n connectives. Note that in these cases, only transitivity of $\leq$ is used. However, when α is of the type $\beta \vee \gamma$ or $\beta \wedge \gamma$ such that β and γ has less than n connectives, the induction hypothesis is explicitly used. These are straightforward.
(2). ($\Rightarrow$). Proof by contradiction. Let $w \models \neg\alpha$ and let $w' \in W$ such that there exists $w'' \in W$ satisfying $w \leq w''$ and $w' \leq w''$ and $w' \models \alpha$. Since $w \models \neg\alpha$ and $w \leq w''$, we have $w'' \nVdash \alpha$ (by definition of truth of $\neg$). Using $w' \models \alpha$, $w' \leq w''$ and using (1), we have $w'' \models \alpha$ - a contradiction.
($\Leftarrow$). Suppose for all $w' \in W$, if there exists $w'' \in W$ such that $w \leq w''$ and $w' \leq w''$, then we have $w' \nVdash \alpha$. We claim that $w \models \neg\alpha$, i.e. for any $w \leq w_0$, we show $w_0 \nVdash \alpha$. Indeed, for $w' = w'' = w_0$, we have $w \leq w''$ and $w' \leq w''$, and thus the claim.
(3). ($\Rightarrow$). Let $w \models \sim\alpha$ and let $w' \in W$ be such that $w \leq w'$ and $w' \models \alpha$. This implies $w' \in Y_0$. Take $w'' := w'$.

($\Leftarrow$). Suppose for any $w' \in W$, if $w \leq w'$ and $w' \vDash \alpha$, then there exists a $w'' \in W$ such that $w' \leq w''$ and $w'' \in Y_0$. We have to show that $w \vDash \sim\alpha$. So consider $w_0 \in W$ such that $w \leq w_0$ and $w_0 \vDash \alpha$. To show $w_0 \in Y_0$, we will use Condition (D). Let $w'_0 \in W$ such that $w_0 \leq w'_0$. Then we have $w \leq w_0 \leq w'_0$. Moreover, using $w_0 \vDash \alpha$ and (1), we have $w'_0 \vDash \alpha$. Using hypothesis, there exists $w''_0 \in W$ such that $w'_0 \leq w''_0$ and $w''_0 \in Y_0$, which is the requirement in (D) to get $w_0 \in Y_0$.

In the next section, we shall prove that ILM and ILM-$\vee$ are determined by the classes $\mathfrak{F}_2$ and $\mathfrak{F}'_2$ of sub-normal frames respectively. We shall prove this by obtaining a relationship between the classes of sub-normal frames and $\hat{N}$-frames.

6.2.3   Intertranslation Between $\hat{N}$-frames and Sub-normal Frames

In Sect. 6.2.1, we defined a relational semantics for ILM based on Došen's N-frames [27]. Došen had shown that models on strictly condensed J-frames are 'inter-translatable' with the Kripke-style frames for minimal logic, preserving the truth of any formula α at any world w of the frame (Theorems 8 and 7). The same can be achieved for ILM. Starting from a sub-normal frame, we can obtain an $\hat{N}$-frame; from an $\hat{N}$-frame, we can obtain a sub-normal frame.

Theorem 31. *Let $\mathcal{F} := (W, \leq, Y_0)$ be a sub-normal frame, i.e. $\mathcal{F} \in \mathfrak{F}_2$. Define the relations R_{N_1} and R_{N_2} over W for all $x, y \in W$ as:*

(A) $xR_{N_1}y$ if and only if $\exists z \in W(x \leq z \ \& \ y \leq z)$.
(B) $xR_{N_2}y$ if and only if $\exists z \in W(x \leq z \ \& \ y \leq z \ \& \ z \notin Y_0)$.

Then we have the following.

1. *$Y_0 = \{z \in W \mid \forall x \in W(z\not{R}_{N_2}x)\}$.*
2. *$(W, \leq, R_{N_1}, R_{N_2})$ is an $\hat{N}$-frame. Denote this frame as $\Phi(\mathcal{F})$.*
3. *If v is a valuation on $\mathcal{F}$, then v is a valuation on $\Phi(\mathcal{F})$ such that for all $\phi \in F$ and $x \in W$,*
   (a) $(\mathcal{F}, v), x \vDash \phi \Leftrightarrow (\Phi(\mathcal{F}), v), x \vDash \phi$, and
   (b) $(\mathcal{F}, v) \vDash \phi \Leftrightarrow (\Phi(\mathcal{F}), v) \vDash \phi$.
   (c) $\mathcal{F} \vDash \phi \Leftrightarrow \Phi(\mathcal{F}) \vDash \phi$.

Proof.
(1). ($\supseteq$). Let $x \notin Y_0$. Since $x \leq x$, using (B), we have $xR_{N_2}x$, i.e.

$$x \notin \{z \in W \mid \forall y \in W(z\not{R}_{N_2}y)\}.$$

($\subseteq$). Let $x \notin \{z \in W \mid \forall y \in W(z\not{R}_{N_2}y)\}$, i.e. there exists $y \in W$ such that $xR_{N_2}y$. Using (B), there exists $z' \in W$ such that $x \leq z'$, $y \leq z'$ and $z' \notin Y_0$. Since Y_0 is an upset, $x \notin Y_0$.
(2). We use Proposition 76.

(a) $(\leq R_{N_1} \leq^{-1}) \subseteq R_{N_1}$: let $x, y, z, z' \in W$ be such that $x \leq y$, $yR_{N_1}z$ and $z' \leq z$. Using (A), there exists $z'' \in W$ such that $y \leq z''$ and $z \leq z''$. Then, $x \leq y \leq z''$ and $z' \leq z \leq z''$ imply $xR_{N_1}z'$.

(b) $(\leq R_{N_2} \leq^{-1}) \subseteq R_{N_2}$: let $x, y, z, z' \in W$ be such that $x \leq y$, $yR_{N_2}z$ and $z' \leq z$. Using (B), there exists $z'' \in W$ such that $y \leq z''$, $z \leq z''$ and $z'' \notin Y_0$. Then $x \leq y \leq z''$, $z' \leq z \leq z''$ and $z'' \notin Y_0$ imply $xR_{N_2}z'$.

(c) R_{N_1} is reflexive: take $z = x$ in (A).

(d) $\forall x, y \in W(xR_{N_1}y \Rightarrow \exists z \in W(x \leq z$ and $y \leq z$ and $xR_{N_1}z))$: using (A), $xR_{N_1}y$ implies that there exists $z \in W$ such that $x \leq z$ and $y \leq z$. Again, using (A), $x \leq z$ and $z \leq z$ imply $xR_{N_1}z$.

(e) R_{N_2} is symmetric: the Condition (B) is symmetric with respect to x and y.

(f) $\forall x, y \in W(xR_{N_2}y \Rightarrow \exists z \in W(x \leq z$ and $y \leq z$ and $xR_{N_2}z))$: using (B), $xR_{N_2}y$ implies that there exists $z \in W$ such that $x \leq z$, $y \leq z$ and $z \notin Y_0$. Then $x \leq z$, $z \leq z$ and $z \notin Y_0$ imply $xR_{N_2}z$, by (B).

(g) Condition DNE($\sim\top$) $\forall x \in W \left(\forall y \in W\left(xR_{N_1}y \Rightarrow \exists z \in W(yR_{N_2}z \ \& \ \forall z' \in W \ (z\cancel{R}_{N_2}z')\right)\right) \Rightarrow \forall z'' \in W \ (x\cancel{R}_{N_2}z''))$: using the expression for Y_0 obtained in (1), this condition is equivalent to the following:
$\forall x \in W \left(\forall y \in W\left(xR_{N_1}y \Rightarrow \exists z \in W(yR_{N_1}z \ \& \ z \in Y_0)\right) \Rightarrow x \in Y_0\right)$.
Let $x \in W$ and let
$$\forall y \in W(xR_{N_1}y \Rightarrow \exists z \in W(yR_{N_1}z \ \& \ z \in Y_0)). \tag{*}$$
To show $x \in Y_0$, we shall use Condition (D) of the sub-normal frame $(W, \leq, Y_0)$, i.e.

$$\forall x \in W \left(\forall y \in W\left(x \leq y \Rightarrow \exists z \in W(y \leq z \ \& \ z \in Y_0)\right) \Rightarrow x \in Y_0\right).$$

Claim: $\forall y \in W(x \leq y \Rightarrow \exists z \in W(y \leq z$ and $z \in Y_0))$. Indeed, let $x \leq y$, for $y \in W$. Using (A) and $y \leq y$, we have $xR_{N_1}y$. Then using (*), we have $z \in W$ such that $yR_{N_1}z$ and $z \in Y_0$. Using (A) on $yR_{N_1}z$, there exists $z' \in W$ such that $y \leq z'$ and $z \leq z'$. Since, Y_0 is an upset and $z \in Y_0$, $z' \in Y_0$. Finally, $y \leq z'$ and $z' \in Y_0$ gives us the claim. Therefore, using Condition (D), we have $x \in Y_0$.

(3). Since v is a valuation on $\mathcal{F}$, $v(p)$ is an upset for all propositional variables $p \in F$. This is enough for v to be a valuation on $\Phi(\mathcal{F})$ (Definition 119).

(3)(a). We shall show this by induction on the number of connectives in ϕ.

Basis case: If ϕ is a propositional variable p or constant $\top$ or $\bot$, the statement is true by definition (Definitions 121 and 130).

Induction hypothesis. For all formulas ϕ with less than n connectives and for all $x \in W$, $(\mathcal{F}, v), x \vDash \phi \Leftrightarrow (\Phi(\mathcal{F}), v), x \vDash \phi$.

Induction step. Suppose ϕ has n connectives and $x \in W$ such that $(\mathcal{F}, v), x \vDash \phi$. In this case, ϕ has one of these forms: $\alpha \wedge \beta$ or $\alpha \vee \beta$ or $\alpha \to \beta$ or $\neg\beta$ or $\sim\beta$, where α and β have less than n connectives. Let $\phi := \sim\alpha$. We have to show the following:

$$(\mathcal{F}, v), x \vDash \sim\alpha \Leftrightarrow (\Phi(\mathcal{F}), v), x \vDash \sim\alpha$$

($\Rightarrow$) $(\mathcal{F}, v), x \vDash \sim\beta$, i.e. $\forall y \in W$, if $x \leq y$ and $(\mathcal{F}, v), y \vDash \beta$ then $y \in Y_0$.   (**).
Claim: $(\Phi(\mathcal{F}), v), x \vDash \sim\beta$, i.e. we have to show that for all $y \in W$, if $xR_{N_2}y$ then $(\Phi(\mathcal{F}), v), y \nvDash \beta$. Let $y \in W$ be such that $xR_{N_2}y$. Using (B), there exists $z \in W$ such that $x \leq z$, $y \leq z$ and $z \notin Y_0$. Therefore, using $x \leq z$ and (**), we have $(\mathcal{F}, v), z \nvDash \beta$. Using induction hypothesis, we have $(\Phi(\mathcal{F}), v), z \nvDash \beta$. Finally, using $y \leq z$ and Proposition 77(1), we have $(\Phi(\mathcal{F}), v), y \nvDash \beta$.

($\Leftarrow$). Given that $(\mathcal{F}, v), x \nvDash \sim\beta$, i.e. there exists $y \in W$ such that $x \leq y$, $(\mathcal{F}, v), y \vDash \beta$ and $y \notin Y_0$.

Claim: $(\Phi(\mathcal{F}), v), x \nvDash \sim\beta$, i.e. we have to show that there exists $y \in W$ such that $x R_{N_2} y$ and $(\Phi(\mathcal{F}), v), y \vDash \beta$. Using (B) on $x \leq y$, $y \leq y$ and $y \notin Y_0$, we have $x R_{N_2} y$. Using induction hypothesis on β, we have $(\Phi(\mathcal{F}), v), y \vDash \beta$.

Now, if ϕ is of the form $\alpha \vee \beta$, $\alpha \wedge \beta$, $\alpha \rightarrow \beta$, as the definitions of truth for the connectives $\vee$, $\wedge$ and $\rightarrow$ are identical in Definition 121 and 130, the induction step in each of the cases is obtained.

$\phi := \neg\alpha$: $\vdash_{\text{ILM}} \neg\alpha \leftrightarrow (\alpha \rightarrow \perp)$, and from the above, for any $(\mathcal{F}, v)$, $x \in W$, $(\mathcal{F}, v), x \vDash \alpha \rightarrow \perp$ if and only if $(\Phi(\mathcal{F}), v), x \vDash \alpha \rightarrow \perp$. So we get the result in this case as well (Observation 47 and 43).

(3)(b). This is direct using (3)(a). Let $(\mathcal{F}, v) \vDash \phi$, i.e. for any $x \in W$, we have $(\mathcal{F}, v), x \vDash \phi$. Using (3)(a), $(\Phi(\mathcal{F}), v), x \vDash \phi$ also. Therefore, $(\Phi(\mathcal{F}), v) \vDash \phi$. Similarly, we get the converse.

(3)(c). Given that $\mathcal{F} \vDash \phi$. Let v be a valuation on $\Phi(\mathcal{F})$; we show that $(\Phi(\mathcal{F}), v) \vDash \phi$. Since v is a valuation on $\Phi(\mathcal{F})$, it is also a valuation on $\mathcal{F}$. Therefore, $(\mathcal{F}, v) \vDash \phi$. Using (3)(b), we have $(\Phi(\mathcal{F}), v) \vDash \phi$.

Theorem 32. *Let $\mathcal{G} := (W, \leq, R_{N_1}, R_{N_2})$ be an $\hat{N}$-frame. Define the subset $Y_0 \subseteq W$ as $Y_0 := \{z \in W \mid \forall x \in W(z \acute{R}_{N_2} x)\}$. Then we have the following.*

1. (a) $x R_{N_2} y$ if and only if $\exists z \in W(x \leq z \,\&\, y \leq z \,\&\, z \notin Y_0)$, and
   (b) $x R_{N_1} y$ if and only if $\exists z \in W(x \leq z \,\&\, y \leq z)$.

2. $(W, \leq, Y_0)$ is a sub-normal frame. Denote this frame by $\Psi(\mathcal{G})$.

3. If v is a valuation on $\mathcal{G}$, then v is a valuation on $\Psi(\mathcal{G})$ such that for all $\phi \in F$ and $x \in W$,

   (a) $(\mathcal{G}, v), x \vDash \phi \Leftrightarrow (\Psi(\mathcal{G}), v), x \vDash \phi$, and
   (b) $(\mathcal{G}, v) \vDash \phi \Leftrightarrow (\Psi(\mathcal{G}), v) \vDash \phi$.
   (c) $\mathcal{G} \vDash \phi \Leftrightarrow \Psi(\mathcal{G}) \vDash \phi$.

Proof. (1)(a). ($\Rightarrow$) Let $x R_{N_2} y$. Using the definition of an $\hat{N}$-frame, there exists $z \in W$ such that $x \leq z$, $y \leq z$ and $x R_{N_2} z$. Since R_{N_2} is symmetric, we have $z R_{N_2} x$. Using definition of Y_0, we have $z \notin Y_0$.

($\Leftarrow$) Let there exist $z \in W$ such that $x \leq z$, $y \leq z$ and $z \notin Y_0$. By definition of Y_0, we have $z' \in W$ such that $z R_{N_2} z'$. Using the definition of an $\hat{N}$-frame, there exists $w \in W$ such that $z \leq w$, $z' \leq w$ and $z R_{N_2} w$. We have $x \leq z$, $z R_{N_2} w$ and $y \leq z \leq w$. Using $(\leq R_{N_2} \leq^{-1}) \subseteq R_{N_2}$, we have $x R_{N_2} y$.

(1)(b). ($\Rightarrow$) Let $x R_{N_1} y$. By the definition of an $\hat{N}$-frames, there exists $z \in W$ such that $x \leq z$ and $y \leq z$.

($\Leftarrow$) Let there exist $z \in W$ such that $x \leq z$ and $y \leq z$. Since R_{N_1} is reflexive, we have $z R_{N_1} z$. Using $(\leq R_{N_1}) \subseteq R_{N_1}$, we have $x R_{N_1} z$. Further, using $(R_{N_1} \leq^{-1}) \subseteq R_{N_1}$, we have $x R_{N_1} y$.

(2) To show that $(W, \leq, Y_0)$ is a sub-normal frame, we have to show that Y_0 is an upset, and it satisfies Condition (D), i.e. $\forall x \in W(\forall y \in W(x \leq y \Rightarrow \exists z \in W(y \leq z \text{ and } z \in Y_0)) \Rightarrow x \in Y_0)$.

Y_0 is an upset: let $x, y \in W$ such that $x \leq y$ and $y \notin Y_0$. $y \notin Y_0$ implies there exists $z' \in W$ such that $y R_{N_2} z'$. Using $(\leq R_{N_2} \leq^{-1}) \subseteq R_{N_2}$, $x \leq y$, $y R_{N_2} z'$ and

$z' \leq z'$ imply $xR_{N_2}z'$. Therefore, $x \notin Y_0$ and Y_0 is upset.

Condition (D): let $x \in W$ be such that

$\forall y \in W(x \leq y \Rightarrow \exists z \in W(y \leq z \text{ and } z \in Y_0))$. $\hspace{2cm}$ (***)

We have to show that $x \in Y_0$. Using the Condition DNE($\sim\top$) of $\hat{N}$-frames, we show:

$$\forall y \in W(xR_{N_1}y \Rightarrow \exists z \in W(yR_{N_1}z \text{ and } \forall z' \in W \ (z\mathcal{R}_{N_2}z')))$$

$$\text{i.e. } \forall y \in W(xR_{N_1}y \Rightarrow \exists z \in W(yR_{N_1}z \text{ and } z \in Y_0)).$$

Let $xR_{N_1}y$. Using (1)(b), there exists $z \in W$ such that $x \leq z$ and $y \leq z$. Using $x \leq z$ and (***), we have $z' \in W$ such that $z \leq z'$ and $z' \in Y_0$. Since R_{N_1} is reflexive, we have $z'R_{N_1}z'$. Using $y \leq z \leq z'$ and ($\leq R_{N_1} \leq^{-1}) \subseteq R_{N_1}$, we have $yR_{N_1}z'$.

Thus, DNE($\sim\top$) gives $\forall z'' \in W \ (x\mathcal{R}_{N_2}z'')$, i.e. $x \in Y_0$.

(3). The proof is similar to part (3) of Theorem 31.

Remark 19.

(1) In Theorem 7, for any j-frame $(W, \leq, R_N)$, Y_0 is defined as follows.

$$z \in Y_0 \Leftrightarrow \exists x, y \in W(x \leq z \ \& \ y \leq z \ \& \ x\mathcal{R}_N y).$$

However, in Theorem 32, we have defined $Y_0 := \{x \in W \mid \forall z(x\mathcal{R}_N z)\}$, in order to give a simpler expression for Y_0.

The set Y_0 obtained from either is, in fact, the same. Indeed, one can easily show that in any strictly condensed J-frame $(W, \leq, R_N)$, for any $x \in W$, we have

$$\forall z \in W(x\mathcal{R}_N z) \text{ if and only if } \exists y, z \in W(y \leq x \text{ and } z \leq x \text{ and } y\mathcal{R}_N z).$$

($\Leftarrow$). Let $z' \in W$ be such that $xR_N z'$. We have to show that, given $y, z \in W$, $y \leq x$ and $z \leq x$ imply $yR_N z$. Since $(W, \leq, R_N)$ is a J-frame, $xR_N z'$ implies that there exists $z'' \in W$ such that $x \leq z''$, $z' \leq z''$ and $xR_N z''$. In the strictly condensed frames, we have ($\leq R_N \leq^{-1}) \subseteq R_N$. Therefore, $y \leq x$, $xR_N z''$ and $z \leq x \leq z''$ imply that $yR_N z$.

($\Rightarrow$). Suppose for all $y, z \in W$, if $y \leq x$ and $z \leq x$, then $yR_N z$. We have to find a $z' \in W$ such that $xR_N z'$. Take $y = z = x$.

(2) We have given an intuitive reading of the relation R_N for any N-frame (X, R_I, R_N) after Proposition 25. Here, R_{N_1} has the same intuitive reading: x is R_{N_1}-*compatible* with y if and only if there is a common extension of x and y. From the above inter-translation, we can characterize R_{N_2} as well: x is R_{N_2}-*compatible* with y if and only if there is a common extension of x and y where $\sim\top$ does not hold.

We also have the following.

Theorem 33. *Consider the maps Φ and Ψ obtained in Theorems 31 and 32 above.*

1. $\Phi\Psi$ and $\Psi\Phi$ are identity maps on the classes $\mathfrak{F}_1$ and $\mathfrak{F}_2$ of $\hat{N}$-frames and sub-normal frames respectively.
2. $\mathfrak{F}_1 \vDash \phi \Leftrightarrow \mathfrak{F}_2 \vDash \phi$ for any formula $\phi \in F$.

Proof.
(1). Consider an $\hat{N}$-frame $\mathcal{G} := (W, \leq, R_{N_1}, R_{N_2})$.
Then $\Phi(\Psi(\mathcal{G})) = (W, \leq, R'_{N_1}, R'_{N_2})$, where $xR'_{N_1}y$ if and only if $\exists z \in W(x \leq z$ and $y \leq z)$; $xR'_{N_2}y$ if and only if $\exists z \in W(x \leq z$ and $y \leq z$ and $z \notin Y_0)$; and $Y_0 = \{z \in W \mid \forall x \in W(z R_{N_2} x)\}$. We have to show that $R_{N_i} = R'_{N_i}$ for $i = 1, 2$.
(For $i = 1$). ($\Rightarrow$). Let $xR_{N_1}y$. Using Proposition 76(b)(3), there exists $z \in W$ such that $x \leq z$, $y \leq z$ and $xR_{N_1}z$. This implies that $xR'_{N_1}y$.
($\Leftarrow$). Let $xR'_{N_1}y$. Then there exists $z \in W$ such that $x \leq z$ and $y \leq z$. Using Theorem 32(1)(b), we have $xR_{N_1}y$.
(For $i = 2$). ($\Rightarrow$). Let $xR_{N_2}y$. Using Proposition 76(b)(6), there exists $z \in W$ such that $x \leq z$, $y \leq z$ and $xR_{N_2}z$. Further, since R_{N_2} is symmetric, $xR_{N_2}z$ implies that $zR_{N_2}x$, which further implies that $z \notin Y_0$. Therefore, we have $xR'_{N_2}y$.
($\Leftarrow$). Let $xR'_{N_2}y$. Then there exists $z \in W$ such that $x \leq z$, $y \leq z$ and $z \notin Y_0$. Using Theorem 32(1)(a), we have $xR_{N_2}y$.
Consider a sub-normal frame $\mathcal{F} := (W, \leq, Y_0)$. Then, $\Psi(\Phi(\mathcal{F})) = (W, \leq, Y'_0)$, where $Y'_0 := \{z \in W \mid \forall x \in W(z R_{N_2} x)\}$, and $xR_{N_2}y$ if and only if $\exists z \in W(x \leq z$ and $y \leq z$ and $z \notin Y_0)$. We have to show that $Y_0 = Y'_0$.
($\Rightarrow$). Let $z \notin Y'_0$. Then there exists $x \in W$ such that $zR_{N_2}x$. Further, $zR_{N_2}x$ implies that there exists $z' \in W$ such that $z \leq z'$, $x \leq z'$ and $z' \notin Y_0$. As Y_0 is an upset, $z \notin Y_0$.
($\Leftarrow$). Let $z \notin Y_0$. Then using $z \leq z$, we have $zR_{N_2}z$. This implies that $z \notin Y'_0$.
(2). This follows directly from part (1). Let $\mathfrak{F}_2 \vDash \phi$. Let $\mathcal{G} \in \mathfrak{F}_1$. Then using Theorem 32(2), $\Psi(\mathcal{G}) \in \mathfrak{F}_2$. Therefore, $\Psi(\mathcal{G}) \vDash \phi$. Further, Theorem 31(3)(c) implies $\Phi(\Psi(\mathcal{G})) \vDash \phi$. Using Part (1) above, we have $\Phi(\Psi(\mathcal{G})) = \mathcal{G}$. Therefore, $\mathcal{G} \vDash \phi$. The converse is similar.

Theorems 31, 32 and 33 can be extended to the relational semantics of ILM-$\vee$.

Theorem 34. *Consider the maps Φ and Ψ as defined in Theorems 31 and 32 respectively. We have the following.*

1. $\mathcal{F} \in \mathfrak{F}'_2 \Leftrightarrow \Phi(\mathcal{F}) \in \mathfrak{F}'_1$.
2. $\mathcal{G} \in \mathfrak{F}'_1 \Leftrightarrow \Psi(\mathcal{G}) \in \mathfrak{F}'_2$.
3. $\mathfrak{F}'_1 \vDash \phi \Leftrightarrow \mathfrak{F}'_2 \vDash \phi$, for any formula $\phi \in F$.

Proof. (1) and (2). It is enough to show the following:
(A). In Theorem 31, if $\mathcal{F} := (W, \leq, Y_0) \in \mathfrak{F}'_2$ then R_{N_2} in $\Phi(\mathcal{F})$ satisfies the following: $R_{N_2} \subseteq \leq^{-1}$.
(B). In Theorem 32, if $\mathcal{G} := (W, \leq, R_{N_1}, R_{N_2}) \in \mathfrak{F}'_1$ then Y_0 in $\Psi(\mathcal{F})$ satisfies the following: $\forall x, y \in W\backslash Y_0(x \leq y \Rightarrow y \leq x)$.
(A): Let $x, y \in W$ such that $xR_{N_2}y$. By the definition of R_{N_2} in $\Phi(\mathcal{F})$, there exists $z \in W$ such that $x \leq z$, $y \leq z$ and $z \notin Y_0$. Since Y_0 is an upset, we have

$x, y \in W \backslash Y_0$. Further, $\mathcal{F} \in \mathfrak{F}_2'$ implies $\forall x, y \in W \backslash Y_0 (x \leq y \Rightarrow y \leq x)$. Therefore, we have $z \leq x$ and $z \leq y$. Finally, $y \leq z$ and $z \leq x$ imply $y \leq x$.

(B): Let $x, y \in W \backslash Y_0$ such that $x \leq y$. In $\Psi(\mathcal{G})$, $x R_{N_2} y \Leftrightarrow \exists z \in W (x \leq z \ \& \ y \leq z \ \& \ z \notin Y_0)$. We have $x \leq y$, $y \leq y$ and $y \notin Y_0$. Therefore, $x R_{N_2} y$. Further, $\mathcal{G} \in \mathfrak{F}_1'$ implies $R_{N_2} \subseteq \leq^{-1}$. Thus, $y \leq x$.

(A) and (B) show that $\mathcal{F} \in \mathfrak{F}_2' \Rightarrow \Phi(\mathcal{F}) \in \mathfrak{F}_1'$ and $\mathcal{G} \in \mathfrak{F}_1' \Rightarrow \Psi(\mathcal{G}) \in \mathfrak{F}_2'$. Note that $\Phi\Psi$ and $\Psi\Phi$ are identity maps on classes $\mathfrak{F}_1$ and $\mathfrak{F}_2$ respectively (Theorem 33), and hence on $\mathfrak{F}_1'$ and $\mathfrak{F}_2'$. Thus, we obtain the other direction: for suppose $\Phi(\mathcal{F}) \in \mathfrak{F}_1'$, then $\Psi(\Phi(\mathcal{F})) \in \mathfrak{F}_2'$, implying $\mathcal{F} \in \mathfrak{F}_2'$. Similarly, $\Psi(\mathcal{G}) \in \mathfrak{F}_2' \rightarrow \mathcal{G} \in \mathfrak{F}_1'$.

(3) follows from (1) and (2).

We have already proved that $\mathfrak{F}_1$ ($\mathfrak{F}_1'$) determines ILM (ILM-∨) in Theorem 29. Using Theorems 33(2) and 34(3), we obtain the following corollary.

Corollary 11. *The classes $\mathfrak{F}_2$ and $\mathfrak{F}_2'$ determine* ILM *and* ILM-∨ *respectively.*

Observation 49. It has been shown in [87] that the class of sub-normal frames determines the logic JP′ (Proposition 30). From the above theorem, we have obtained that the same class of frames determines ILM. Thus, we have the same subclass of j-frames - $\mathfrak{F}_2$ - characterizing two non-equivalent logics JP′ and ILM, which, we have seen, have different alphabets, the latter having the extra propositional constant $\perp$, not definable in JP′ using other connectives.

In the next section, we shall define a logic L without the 'implication' connective $\rightarrow$, and show that it is also determined by the class of $\hat{N}$-frames. Moreover, ILM and L will be shown to be non-equivalent and '$\rightarrow$' is not definable in L using other connectives. Therefore, again we shall have two different logics - ILM and L, with two different alphabets, but with the same relational semantics.

The above observations point to limitations of relational frames of the above kind – they are unable to differentiate between non-equivalent logics.

6.3   Connection Between Sub-normal Frames and *ccpBa*

We saw the connection between normal frames and pBa in Theorem 9. Every pBa $\mathcal{A}$ is monomorphic to the complex algebra of the canonical frame of $\mathcal{A}$; every normal frame $\mathcal{F}$ can be embedded into the canonical frame of the complex algebra of $\mathcal{F}$. This construction can be extended to the case of $ccpBas$ ($c \lor cpBas$) and relational frames for ILM (ILM-∨). Let us define embeddings in sub-normal frames.

Definition 131. (Embeddings between sub-normal frames).
Given two sub-normal frames $(W, \leq, Y_0)$ and $(W', \leq', Y_0')$. The mapping $\phi : W \rightarrow W'$ is an embedding *if it satisfies the following conditions, for all $a, b \in W$:*

1. *$a \leq b$ if and only if $\phi(a) \leq' \phi(b)$ (poset embedding), and*
2. *$a \in Y_0$ if and only if $\phi(a) \in Y_0'$.*

Definition 132. (Complex algebra of a sub-normal frame).
Consider a sub-normal frame $\mathcal{F} := (W, \leq, Y_0)$. Let $Up(W)$ denote the set of upsets of W (Definition 10). Define the following operators on $Up(W)$. For any $U, V \in Up(W)$,

1. $U \rightarrow V := \{w \in W \mid \forall v \in W(w \leq v \Rightarrow (v \in U \Rightarrow v \in V))\}$,
2. $\neg U := U \rightarrow \emptyset$, *i.e.* $\neg U := \{w \in W \mid \forall v \in W(w \leq v \Rightarrow v \notin U)\}$, *and*
3. $\sim U := U \rightarrow Y_0$, *i.e.* $\sim U := \{w \in W \mid \forall v \in W((w \leq v \ \& \ v \in U) \Rightarrow v \in Y_0)\}$.

The algebra $Up(\mathcal{F}) := (Up(W), W, \emptyset, \cap, \cup, \rightarrow, \neg, \sim)$, where $\cap, \cup$ are the set intersection, union operations, is called the complex algebra *of the sub-normal frame $\mathcal{F}$.*

Proposition 81. *For a sub-normal frame $\mathcal{F} := (W, \leq, Y_0) \in \mathfrak{F}_2$, the complex algebra $Up(\mathcal{F})$ of $\mathcal{F}$ forms a ccpBa. Further, if $\mathcal{F} \in \mathfrak{F}_2'$, then $Up(\mathcal{F})$ is a cVcpBa.*

Proof. Since $\mathcal{F}$ is a sub-normal frame, $(W, \leq)$ is a normal frame. Note that the set $Up(W)$ and the operators other than $\sim$ are defined only using the set W and $\leq$. Therefore, using Proposition 27, $(Up(W), W, \emptyset, \cap, \cup, \rightarrow, \neg)$ forms a pBa. Using Observation 35 and the definition of $\sim$ in $Up(W)$, for $Up(\mathcal{F})$ to form a $ccpBa$ it is enough to show that (i) $Y_0 = \sim W$ and (ii) $\neg\neg Y_0 = Y_0$.
(i). $W \rightarrow Y_0 = \{w \in W \mid \forall v \in W(w \leq v \Rightarrow v \in Y_0)\}$. Since Y_0 is an upset, $Y_0 \subseteq W \rightarrow Y_0$. Let $w \in W \rightarrow Y_0$. Then $w \leq w \Rightarrow w \in Y_0$.
(ii). Expanding $\neg\neg Y_0 = Y_0$, we get the following.

$$\forall w \in W(\forall w' \in W(w \leq w' \Rightarrow w' \notin (Y_0 \rightarrow \emptyset)) \Leftrightarrow w \in Y_0)$$

i.e., $\forall w \in W(\forall w' \in W(w \leq w' \Rightarrow \exists w'' \in W(w' \leq w'' \ \& \ w'' \in Y_0)) \Leftrightarrow w \in Y_0)$.

The forward direction of the above implication is obtained as Y_0 satisfies Condition (D) of the sub-normal frame $\mathcal{F}$ (Definition 127). The reverse direction is using Observation 79(1).
Now, let $\mathcal{F} \in \mathfrak{F}_2'$, i.e. $\mathcal{F}$ is a sub-normal frame satisfying the following property.

$$\forall x, y \in W \backslash Y_0 \Rightarrow (x \leq y \Rightarrow y \leq x) \qquad (*)$$

We have to show that for any $V \in Up(W)$, $V \cup \sim V = W$. Let $x \in W$ such that $x \notin \sim V = V \rightarrow Y_0$. Then, using definition of $V \rightarrow Y_0$, there exists $v \in W$ such that $x \leq v$, $v \in V$ and $v \notin Y_0$. Y_0 is an upset implies $x \notin Y_0$. Using $(*)$, we have $v \leq x$. Finally, $v \in V$ implies $x \in V$, because V is an upset.

Therefore, given a sub-normal frame we have obtained a $ccpBa$, namely the complex algebra of the frame. The converse is also not difficult to obtain, i.e. given a $ccpBa$, we can obtain a sub-normal frame, called *canonical* sub-normal frame. Recall Definition 75 of canonical normal frames of a pBa. Since any $ccpBa$ is a distributive lattice, the concept of filters (Definition 11) extends to $ccpBa$.

Definition 133 (Canonical frame of a *ccpBa*).
Consider a ccpBa $\mathcal{A} := (A, 1, 0, \wedge, \vee, \rightarrow, \neg, \sim)$. *Let* X_A *denote the set of all prime filters in* $\mathcal{A}$. *Define a set* Y_0 *as follows:*

$$Y_0 := \{P \in X_A \mid \sim 1 \in P\}.$$

The triple $\mathcal{F}_A := (X_A, \subseteq, Y_0)$ *is called the* canonical *frame of* $\mathcal{A}$.

Proposition 82. *For a ccpBa* $\mathcal{A} := (A, 1, 0, \wedge, \vee, \rightarrow, \neg, \sim)$, *the canonical frame* $\mathcal{F}_A := (X_A, \subseteq, Y_0)$ *is a sub-normal frame.*
If $\mathcal{A}$ *is a cVcpBa then* $\mathcal{F}_A \in \mathfrak{F}_2'$.

Proof. For $P \in Y_0$ and $P \subseteq Q$, $\sim 1 \in P$ implies $\sim 1 \in Q$. Therefore $Q \in Y_0$, and Y_0 is an upset. For $\mathcal{F}_A$ to be a sub-normal frame, we have to know that it satisfies Condition (D):

$$\forall P \in X_A (\forall Q \in X_A \ (P \subseteq Q \Rightarrow \exists R \in X_A \ (Q \subseteq R \ \& \ R \in Y_0)) \Rightarrow P \in Y_0)$$
i.e. $\forall P \in X_A (P \notin Y_0 \Rightarrow \exists Q \in X_A \ (P \subseteq Q \ \& \ \forall R \in X_A \ (Q \subseteq R \Rightarrow R \notin Y_0)))$
i.e. $\forall P \in X_A (\sim 1 \notin P \Rightarrow \exists Q \in X_A \ (P \subseteq Q \ \& \ \forall R \in X_A \ (Q \subseteq R \Rightarrow \sim 1 \notin R)))$.

Consider a prime filter P in $\mathcal{A}$ such that $\sim 1 \notin P$. The aim is to find a $Q \in X_A$ such that $P \subseteq Q$ and for all $R \in X_A$, if $Q \subseteq R$ then $\sim 1 \notin R$. Define a set Δ as follows.

$$\Delta := \{P' \subseteq A \mid P' \text{ is a filter in } \mathcal{A}; P \subseteq P'; \sim 1 \notin P'\}.$$

$(\Delta, \subseteq)$ satisfies the hypothesis of Zorn's lemma (Lemma 2). Hence there exists a filter Q of $\mathcal{A}$ that is a maximal element in Δ. We show that Q is, in fact, a maximal filter in $\mathcal{A}$.
Indeed, note that Q is a proper subset of A because $\sim 1 \notin Q$. Further we have the following.
$\neg \sim 1 \in Q$: suppose not. Then the filter generated by $Q \cup \{\neg \sim 1\}$ (Definition 11), $\langle Q \cup \{\neg \sim 1\}\rangle$, is a proper superset of Q and contains P. Since Q is maximal in Δ, $\sim 1 \in \langle Q \cup \{\neg \sim 1\}\rangle$. Using Lemma 1(3), there exists $q \in Q$ such that $q \wedge \neg \sim 1 \leq \sim 1 = \neg \neg \sim 1$.

$$\Rightarrow q \leq \neg \sim 1 \rightarrow \neg \neg \sim 1 = \neg \sim 1 \rightarrow (\neg \sim 1 \rightarrow 0)$$
$$\Rightarrow q \leq (\neg \sim 1 \wedge \neg \sim 1) \rightarrow 0$$
$$\Rightarrow q \leq \neg \sim 1 \rightarrow 0 = \neg \neg \sim 1 = \sim 1$$

Since Q is a filter in $\mathcal{A}$, and any filter is an upset (Lemma 1(1)), $\sim 1 \in Q$. This is a contradiction as $Q \in \Delta$. Thus, $\neg \sim 1 \in Q$.
Q is a maximal filter in $\mathcal{A}$: it is enough to show for any $a \in A$ such that $a \notin Q$, $\langle Q \cup \{a\}\rangle = A$. Since $\langle Q \cup \{a\}\rangle$ contains P and is a proper superset of Q, we have $\langle Q \cup \{a\}\rangle \notin \Delta$, i.e. $\sim 1 \in \langle Q \cup \{a\}\rangle$. Using Lemma 1(3), we have $q \in Q$ such that $q \wedge a \leq \sim 1$, i.e. $q \wedge a \leq \neg \neg \sim 1 = \neg \sim 1 \rightarrow 0$. This gives us $q \wedge a \wedge \neg \sim 1 \leq 0$.

Since $\neg\sim 1 \in Q$, we have $0 \in \langle Q \cup \{a\}\rangle$, and thus, $A = \langle Q \cup \{a\}\rangle$. So Q is prime (Lemma 1), i.e. $Q \in X_A$.

Thus for all $R \in X_A$, $Q \subseteq R$ and Q maximal will imply that $Q = R$, i.e. $\sim 1 \notin R$.

Let $\mathcal{A}$ be a $c \vee c p Ba$. We show that $\mathcal{F}_A \in \mathfrak{F}_2'$. We need to verify that if $P, Q \in X_A \backslash Y_0$ then $P \subseteq Q \Rightarrow Q \subseteq P$. Suppose not, that is, there exist $P, Q \in X_A$ such that $P, Q \notin Y_0$, $P \subseteq Q$ and there exists $x \in Q$ such that $x \notin P$. Then $x \vee \sim x = 1$, $1 \in P$ and P is prime imply $\sim x \in P \subseteq Q$. Since, $\sim x = x \rightarrow \sim 1$ and $x \wedge (x \rightarrow \sim 1) \leq \sim 1$, we have $\sim 1 \in Q$ (Lemma 1(1)). However, $Q \notin Y_0$ implies $\sim 1 \notin Q$ - a contradiction.

We now have the the to and for connection between sub-normal frames $\mathcal{F} := (W, \leq, Y_0) \in \mathfrak{F}_2$ ($\mathfrak{F}_2'$) and $ccpBas$ ($c \vee cpBas$).

Theorem 35. *1. Every $ccpBa$ (or $c \vee cPba$) $\mathcal{A}$ is embeddable into the complex algebra $Up(\mathcal{F}_A)$ of the canonical frame $\mathcal{F}_A$ of $\mathcal{A}$.*

2. Any sub-normal frame $\mathcal{F} \in \mathfrak{F}_2$ (or $\mathfrak{F}_2'$) can be embedded into the canonical frame $\mathcal{F}_{Up(\mathcal{F})}$ of the complex algebra $Up(\mathcal{F})$ of $\mathcal{F}$.

Proof.
(1). Consider the complex algebra $Up(\mathcal{F}_A) := (Up(X_A), X_A, \emptyset, \cap, \cup, \rightarrow, \neg, \sim)$ of the canonical frame $\mathcal{F}_A := (X_A, \subseteq, Y_0)$ of a $ccpBa$ $\mathcal{A} := (A, 1, 0, \wedge, \vee, \rightarrow, \neg, \sim)$. The map $h : A \rightarrow Up(X_A)$ such that for all $a \in A$,

$$h(a) := \{P \in X_A \mid a \in P\},$$

is a monomorphism from the pBa $(A, 1, 0, \wedge, \vee, \rightarrow, \neg)$ to the pBa $(Up(X_A), \emptyset, X_A, \cap, \cup, \rightarrow, \neg)$ (Theorem 9(1)). It remains to show that h preserves $\sim$, i.e.

$$h(\sim 1) = \sim X_A.$$

Now, $\sim X_A = X_A \rightarrow Y_0$, where $Y_0 := \{P \in X_A \mid \sim 1 \in P\}$. As obtained in the proof of Proposition 81, one can show here also that $\sim X_A = Y_0$. Moreover, using the definition of h, $h(\sim 1) = \{P \in X_A \mid \sim 1 \in P\}$. Therefore, $h(\sim 1) = \sim X_A$.

(2). Consider the canonical frame $(X_{Up(W)}, \subseteq, Y_{0_{Up(W)}})$ of the complex algebra $Up(\mathcal{F}) := (Up(W), W, \emptyset, \cap, \cup, \rightarrow, \neg, \sim)$ of the sub-normal frame $\mathcal{F} := (W, \leq, Y_0)$. The map $g : W \rightarrow X_A$ such that for all $w \in W$,

$$g(w) := \{U \in Up(W) \mid w \in U\},$$

is an embedding from $(W, \leq)$ to $(X_{Up(W)}, \subseteq)$ (Theorem 9(2)). Using the definition of embeddings (Definition 131), we have to only show that for all $w \in W$, $w \in Y_0$ if and only if $g(w) \in Y_{0_{Up(W)}}$. Note that, as observed in the proof of Proposition 81, we have here also that $\sim W = Y_0$.

($\Rightarrow$) Let $w \in Y_0$. We need to show that $g(w) \in Y_{0_{Up(W)}}$, i.e. $\sim W \in g(w)$, i.e. $Y_0 \in g(w)$, which is true by our assumption.

($\Leftarrow$) Let $g(w) \in Y_{0_{Up(W)}}$, i.e. $\sim W \in g(w)$. As $\sim W = Y_0$, $Y_0 \in g(w)$, i.e. $w \in Y_0$.

Using Propositions and 81 and 82, we have that if the algebra $\mathcal{A}$ is a $c\vee cpBa$ then the complex algebra $Up(\mathcal{F_A})$ is a $c\vee cpBa$, and if the sub-normal frame $\mathcal{F} \in \mathfrak{F}'_2$ then the canonical frame $\mathcal{F}_{Up(\mathcal{F})} \in \mathfrak{F}'_2$.

We obtain the embeddings in (1) and (2) for $c\vee cBa$'s and sub-normal frames of $\mathfrak{F}'_1$, through the same maps h and g above.

6.4   Conclusions

In the previous sections we defined two different semantics for ILM through the classes $\mathfrak{F}_1$ and $\mathfrak{F}_2$ of frames - $\mathfrak{F}_1$ using Došen's N-frames and $\mathfrak{F}_2$ using Segerberg's j-frames, and gave the characterization results for ILM and ILM-$\vee$. The difference in both lies in the treatment of negation.

Case-1: For $\mathfrak{F}_1$, both negations are considered as unary modal connectives and the semantics is defined using the modal accessibility relations R_{N_1} and R_{N_2}.

Case-2: For $\mathfrak{F}_2$, both negations are treated as unary connectives, their semantics being defined using the relation $\leq$ and Y_0.

This immediately gives us the idea of constructing new frame classes and corresponding semantics, by considering one of the two negations of ILM as a unary connective, and the other negation as a unary 'impossibility' modal connective. What would be an adequate definition of frames in such a treatment of negations for ILM?

Case-3: Consider the quadruple $\mathcal{F} := (X, \leq, R_{N_1}, Y_0)$ satisfying the following:

1. $(X, \leq, R_{N_1})$ is a strictly condensed N-frame, where $\leq$ is a partial order,
2. R_{N_1} is reflexive,
3. $\forall x, y \in X(xR_{N_1}y \Rightarrow \exists z \in X(x \leq z$ and $y \leq z$ and $xR_{N_1}z))$,
4. Y_0 is an upset, and
5. $\forall x \in X \ (\forall y \in X(xR_{N_1}y \Rightarrow \exists z \in X(yR_{N_1}z$ and $z \in Y_0)) \Rightarrow x \in Y_0)$.

Let us denote the class of such frames by $\mathfrak{F}_3$.

Define a subclass $\mathfrak{F}'_3$ of $\mathfrak{F}_3$, such that in any frame $(X, \leq, R_{N_1}, Y_0) \in \mathfrak{F}_3$, $\leq$ is an identity relation on $X \backslash Y_0$, i.e. for any $x, y \in X \backslash Y_0$, if $x \leq y$ then $y \leq x$.

A valuation $v : PV \rightarrow \mathcal{P}(X)$ on a frame $\mathcal{F} := (X, \leq, R_{N_1}, Y_0) \in \mathfrak{F}_3$ is defined in a similar way as defined for $\hat{N}$-frames (Definition 119). The truth of a formula $\alpha \in F$ at a world $x \in X$ can then be obtained inductively. For the connectives $\vee, \wedge, \rightarrow, \top$ and $\perp$, the semantics is the same as that defined for the frame class $\mathfrak{F}_1$ (Definition 121). For the connectives $\sim$ and $\neg$, the definitions are as follows:

1. $\mathcal{M}, x \vDash \neg\phi \Leftrightarrow$ for all $y \in X(xR_{N_1}y \Rightarrow \mathcal{M}, y \nvDash \phi)$.
2. $\mathcal{M}, x \vDash \sim\phi \Leftrightarrow \forall y \in X((x \leq y$ and $\mathcal{M}, y \vDash \phi) \Rightarrow y \in Y_0)$.

The semantics for the negation '$\neg$' is defined using the accessibility relation R_{N_1} as defined in $\hat{N}$-frames; for the negation '$\sim$' is defined using the relation $\leq$ and the subset Y_0 of X as for the Kripke-style frames.

Case-4: Consider the triple $\mathcal{F} := (X, \leq, R_{N_2})$ satisfying the following:

1. $(X, \leq, R_{N_2})$ is a strictly condensed N-frame (with $\leq$ partial ordering),
2. R_{N_2} is symmetric,
3. $\forall x, y \in X(x R_{N_2} y \Rightarrow \exists z \in X(x \leq z$ and $y \leq z$ and $x R_{N_2} z))$, and
4. $\forall x \in X \ (\forall y \in X(x \leq y \Rightarrow \exists z \in X(y \leq z$ and $\forall z' \in X \ (z \acute{R}_{N_2} z'))) \Rightarrow \forall z'' \in X \ (x \acute{R}_{N_2} z''))$.

Let us denote the class of such frames by $\mathfrak{F}_4$.

Define another subclass $\mathfrak{F}_4'$ as the frames in $\mathfrak{F}_4$ such that any frame $(X, \leq, R_{N_2}) \in \mathfrak{F}_4$ satisfies $R_{N_2} \subseteq \leq^{-1}$.

A valuation $v : PV \to \mathcal{P}(X)$ on a frame $\mathcal{F} := (X, \leq, R_{N_2}) \in \mathfrak{F}_4$ is defined in a similar way as for $\hat{N}$-frames (Definition 119). For the connectives $\vee, \wedge, \to, \top$ and $\perp$, the semantics is again defined just as for the frame class $\mathfrak{F}_1$ (Definition 119). For the connectives $\sim$ and $\neg$, we have

1. $\mathcal{M}, x \vDash \neg\phi \Leftrightarrow$ for all $y \in X(x \leq y \Rightarrow \mathcal{M}, y \nvDash \phi)$.
2. $\mathcal{M}, x \vDash \sim\phi \Leftrightarrow$ for all $y \in X(x R_{N_2} y \Rightarrow \mathcal{M}, y \nvDash \phi)$.

In this case the semantics for the negation '$\sim$' is defined using the accessibility relation R_{N_2} as defined in $\hat{N}$-frames; and the semantics for the negation '$\neg$' is defined using the relation $\leq$.

It is naturally expected that both $\mathfrak{F}_3$ ($\mathfrak{F}_3'$) and $\mathfrak{F}_4$ ($\mathfrak{F}_4'$) also determine ILM (ILM-$\vee$). This could be obtained by defining an inter-translation between the classes $\mathfrak{F}_3/\mathfrak{F}_4$ and $\mathfrak{F}_1/\mathfrak{F}_2$ (as done in Theorems 31 and 32 for $\mathfrak{F}_1$ and $\mathfrak{F}_2$). The mapping involved may be given by the diagram below, defined using the following relations.

(1) $x R_{N_1} y \Leftrightarrow \exists z(x \leq z$ and $y \leq z)$,
(2) $x R_{N_2} y \Leftrightarrow \exists z(x \leq z$ and $y \leq z$ and $z \notin Y_0)$, and
(3) $Y_0 := \{x \mid \forall z(x \acute{R}_{N_2} z)\}$.

$$(X, \leq, Y_0) \longleftarrow (X, \leq, R_{N_1}, R_{N_2})$$

$$(X, \leq, R_{N_1}, Y_0) \longleftarrow (X, \leq, R_{N_2})$$

It has been shown that JP' has the finite model property [42], i.e. for any JP'-formula α, α is a theorem in JP' if and only if α is valid in the class of finite sub-normal frames. This helps in proving the decidability of the logic JP' [42]. We defined (an equivalent version of) ILM by extending the alphabet of JP' with the propositional constant $\perp$. However, we observed that both ILM and JP' are characterized by the same class $\mathfrak{F}_2$ of frames. So one expects that the decidability of ILM may be obtained by using that of JP'.

7   The Logic K_{im}

In this section, we shall define and study two logics in Dunn's logical framework, as discussed in Sect. 2.6. We shall see how these particular logics are connected with the intuitionistic logics with minimal negation discussed in the previous section, through their algebraic and relational semantics.

In the next section, we define two logics - K_{im} and $K_{im-\vee}$, and study their properties. We also give algebraic semantics for such logics, and compare them with those of ILM and ILM-$\vee$. A relational semantics using Dunn's compatibility frames for both these logics are obtained in Sect. 7.2. Connections between algebraic semantics and relational semantics are discussed in Sect. 7.3. We end this section in Sect. 7.4.

7.1   The Logics K_{im} and $K_{im-\vee}$

Consider an alphabet that has propositional variables $p, q, r, \ldots$, binary connectives $\vee$ and $\wedge$, unary connectives $\neg$ and $\sim$, and constants $\bot$ and $\top$. Define a language $\mathcal{L}$ with this alphabet and class F of well-formed formulas given by the scheme:

$$p \mid \top \mid \bot \mid \alpha \wedge \beta \mid \alpha \vee \beta \mid \neg\alpha \mid \sim\alpha$$

Definition 134 (The logics K_{im} and $K_{im-\vee}$).
The language of K_{im} is $\mathcal{L}$. The axioms and rules of K_{im} are as follows:

> A1. $\alpha \vdash \alpha$
> A2. $\alpha \vdash \beta, \beta \vdash \gamma / \alpha \vdash \gamma$
> A3. $\alpha \wedge \beta \vdash \alpha; \alpha \wedge \beta \vdash \beta$
> A4. $\alpha \vdash \beta, \alpha \vdash \gamma / \alpha \vdash \beta \wedge \gamma$
> A5. $\alpha \vdash \gamma, \beta \vdash \gamma / \alpha \vee \beta \vdash \gamma$
> A6. $\alpha \vdash \alpha \vee \beta; \beta \vdash \alpha \vee \beta$
> A7. $\alpha \wedge (\beta \vee \gamma) \vdash (\alpha \wedge \beta) \vee (\alpha \wedge \gamma)$
> A8. $\alpha \vdash \top$
> A9. $\bot \vdash \alpha$
> A10. $\alpha \vdash \beta / \neg\beta \vdash \neg\alpha$
> A11. $\neg\alpha \wedge \neg\beta \vdash \neg(\alpha \vee \beta)$
> A12. $\top \vdash \neg\bot$
> A13. $\alpha \vdash \neg\neg\alpha$
> A14. $\alpha \wedge \beta \vdash \gamma / \alpha \wedge \neg\gamma \vdash \neg\beta$
> A15. $\alpha \wedge \neg\alpha \vdash \beta$
> A16. $\sim\alpha \vdash \neg(\alpha \wedge \neg\sim\top)$
> A17. $\neg(\alpha \wedge \neg\sim\top) \vdash \sim\alpha$

The logic $K_{im-\vee}$ is K_{im} enhanced with the following axiom.

> A18. $\top \vdash \alpha \vee \sim\alpha$

Observation 50.

(1) A1–A12 are the axioms and rules present in the logic K_i with preminimal negation, where negation is represented by $\neg$ (Definition 77).
(2) A13, A14 and A15 make the negation $\neg$ intuitionistic. Therefore, A1-A15 give the logic with intuitionistic negation (with negation as $\neg$).
What properties does the negation $\sim$ in K_{im} satisfy?

Proposition 83. *The following can be proved in the system* K_{im}:

$$P1. \ \alpha \vdash \beta, \ \delta \vdash \gamma / \alpha \wedge \delta \vdash \beta \wedge \gamma$$
$$P2. \ \sim\text{-}Contraposition: \alpha \vdash \beta / {\sim}\beta \vdash {\sim}\alpha$$
$$P3. \ \sim\text{-}\vee\text{-}Linearity: {\sim}\alpha \wedge {\sim}\beta \vdash {\sim}(\alpha \vee \beta)$$
$$P4. \ \sim\text{-}Nor: \top \vdash {\sim}\bot$$
$$P5. \ \alpha \vdash {\sim}{\sim}\alpha$$
$$P6. \ \alpha \wedge \beta \vdash \gamma / \alpha \wedge {\sim}\gamma \vdash {\sim}\beta$$
$$P7. \ (\text{DNE}({\sim}\top)) \ \neg\neg{\sim}\top \vdash {\sim}\top$$

Proof.
($P1$): Using $A3$, we have $\alpha \wedge \delta \vdash \alpha$, and then using $A2$, $\alpha \wedge \delta \vdash \beta$. Again using $A3$: $\alpha \wedge \delta \vdash \delta$ and $A4$, we have $\alpha \wedge \delta \vdash \beta \wedge \gamma$.
($P2$): Let $\alpha \vdash \beta$. Using $P1$, we have $\alpha \wedge \neg{\sim}\top \vdash \beta \wedge \neg{\sim}\top$. Now, using $A10$, we have $\neg(\beta \wedge \neg{\sim}\top) \vdash \neg(\alpha \wedge \neg{\sim}\top)$. Finally, using $A16$, $A17$, and $A2$, we have ${\sim}\beta \vdash {\sim}\alpha$.
($P3$):

$$\sim\alpha \wedge {\sim}\beta \vdash \neg(\alpha \wedge \neg{\sim}\top) \wedge \neg(\beta \wedge \neg{\sim}\top) \qquad \text{(Using } A16 \text{ and } P1\text{)}$$
$$\Rightarrow \ \neg(\alpha \wedge \neg{\sim}\top) \wedge \neg(\beta \wedge \neg{\sim}\top) \vdash \neg((\alpha \wedge \neg{\sim}\top) \vee (\beta \wedge \neg{\sim}\top)) \quad \text{(Using } A11\text{)}$$
$$\Rightarrow \ \sim\alpha \wedge {\sim}\beta \vdash \neg((\alpha \wedge \neg{\sim}\top) \vee (\beta \wedge \neg{\sim}\top)) \quad \text{(Using } A2 \text{ and above sequents)}$$

Using $A7$, we have $(\alpha \vee \beta) \wedge \neg{\sim}\top \vdash (\alpha \wedge \neg{\sim}\top) \vee (\beta \wedge \neg{\sim}\top)$. Using $A10$ on this, we obtain $\neg((\alpha \wedge \neg{\sim}\top) \vee (\beta \wedge \neg{\sim}\top)) \vdash \neg((\alpha \vee \beta) \wedge \neg{\sim}\top)$. Using $A2$,

$$\sim\alpha \wedge {\sim}\beta \vdash \neg((\alpha \vee \beta) \wedge \neg{\sim}\top)$$
$$\Rightarrow \ \sim\alpha \wedge {\sim}\beta \vdash {\sim}(\alpha \vee \beta) \qquad \text{(Using } A17 \text{ and } A2\text{)}$$

($P4$): $\bot \wedge \neg{\sim}\top \vdash \bot$ from $A3$. Using $A10$, we have $\neg\bot \vdash \neg(\bot \wedge \neg{\sim}\top)$. Further, $A12$ and $A2$ give $\top \vdash \neg(\bot \wedge \neg{\sim}\top)$. Finally, using $A17$, we have $\top \vdash {\sim}\bot$.
($P5$): Using $A1$ and commutativity of $\wedge$, we have $\neg{\sim}\top \wedge \alpha \vdash (\alpha \wedge \neg{\sim}\top)$. $A14$ implies the following:

$$\neg{\sim}\top \wedge \neg(\alpha \wedge \neg{\sim}\top) \vdash \neg\alpha$$
$$\Rightarrow \ \neg{\sim}\top \wedge {\sim}\alpha \vdash \neg\alpha \qquad \text{(Using } A17, A2 \text{ and } P1\text{)}$$
$$\Rightarrow \ \neg\neg\alpha \vdash \neg(\neg{\sim}\top \wedge {\sim}\alpha) \qquad \text{(Using } A10\text{)}$$
$$\Rightarrow \ \alpha \vdash \neg(\neg{\sim}\top \wedge {\sim}\alpha) \qquad \text{(Using } A13 \text{ and } A2\text{)}$$

Now, applying $A17$ on ${\sim}\alpha$, we get $\neg({\sim}\alpha \wedge \neg{\sim}\top) \vdash {\sim}{\sim}\alpha$. Using $A2$, we have $\alpha \vdash {\sim}{\sim}\alpha$.

(P6): Let $\alpha \wedge \beta \vdash \gamma$. Using $P1$ and $\neg\sim\top \vdash \neg\sim\top$ (A1), we have $\alpha \wedge (\beta \wedge \neg\sim\top) \vdash (\gamma \wedge \neg\sim\top)$. Now using $A14$,

$$\alpha \wedge \neg(\gamma \wedge \neg\sim\top) \vdash \neg(\beta \wedge \neg\sim\top)$$

Again $P1$, $\sim\gamma \vdash \neg(\gamma \wedge \neg\sim\top)$ ($A16$) and $A1$ imply $\alpha \wedge \sim\gamma \vdash \alpha \wedge \neg(\gamma \wedge \neg\sim\top)$. Using A2,

$$\alpha \wedge \sim\gamma \vdash \neg(\beta \wedge \neg\sim\top)$$
$$\Rightarrow \alpha \wedge \sim\gamma \vdash \sim\beta. \qquad\qquad \text{(Using } A17 \text{ and } A2)$$

(P7) (DNE($\sim\top$)): $\top \wedge \neg\sim\top \vdash \neg\sim\top$ (using $A3$). Using $A10$, $\neg\neg\sim\top \vdash \neg(\top \wedge \neg\sim\top)$. Finally using $A17$ and $A2$, we get $\neg\neg\sim\top \vdash \sim\top$.

'DNE' in (P7) stands for 'double negation elimination' as in the previous section.

Observation 51. $P1$, $P2$ and $P3$ make the negation $\sim$ a preminimal negation. Further, $P4$ and $P5$ are the versions of $A13$ and $A14$, where $\neg$ is replaced by $\sim$. Thus, it makes the negation $\sim$ minimal (Remark 7).

We can in fact give another equivalent consequence system for the logic K_{im}.

Proposition 84. *Consider a logic K with the same language as K_{im}. The axioms and rules of K are given by $A1 - A15$ and $P2 - P6$ and $P7$ (DNE($\sim\top$)). Then, K is equivalent to K_{im}, i.e. for any $\alpha, \beta \in F$, $\alpha \vdash_{K_{im}} \beta$ if and only if $\alpha \vdash_K \beta$.*

Proof. First observe that $P1$ holds here, because in the proof of $P1$ in Proposition 83 only $A3$, $A2$ and $A4$ are required, all of which are included in this case. Now let us see the proofs of $A16$ and $A17$.
(A16):

$$\alpha \wedge \top \vdash \alpha \qquad\qquad \text{(Using } A3)$$
$$\Rightarrow \alpha \wedge \sim\alpha \vdash \sim\top \qquad\qquad \text{(Using } P6)$$
$$\Rightarrow \alpha \wedge \neg\sim\top \vdash \neg\sim\alpha \qquad\qquad \text{(Using } A14)$$
$$\Rightarrow \neg\neg\sim\alpha \vdash \neg(\alpha \wedge \neg\sim\top) \qquad\qquad \text{(Using } A10)$$

Finally, using $A13$, $\sim\alpha \vdash \neg\neg\sim\alpha$ and $A2$, we have $A16$.
(A17):

$$\alpha \wedge \neg\sim\top \vdash \alpha \wedge \neg\sim\top \qquad\qquad \text{(Using } A1)$$
$$\Rightarrow \alpha \wedge \neg(\alpha \wedge \neg\sim\top) \vdash \neg\neg\sim\top \qquad\qquad \text{(Using } A14)$$
$$\Rightarrow \alpha \wedge \neg(\alpha \wedge \neg\sim\top) \vdash \sim\top \qquad\qquad \text{(Using } P7 \text{ and } A2)$$
$$\Rightarrow \sim\sim\top \wedge \neg(\alpha \wedge \neg\sim\top) \vdash \sim\alpha \qquad\qquad \text{(Using } P6)$$

We have $\top \vdash \sim\sim\top$ using $P5$. Now, using $P1$ and $A2$, $\top \wedge \neg(\alpha \wedge \neg\sim\top) \vdash \sim\alpha$. $A8$ and $P1$ imply $\neg(\alpha \wedge \neg\sim\top) \vdash \sim\alpha$.

Let us look at the algebraic semantics for K_{im} and $K_{im-\vee}$. We have already defined bounded distributive lattices with various negations in Definition 78. Let us now define 'K_{im}-algebras'. To ensure soundness, the algebra must be of the form $(A, 1, 0, \vee, \wedge, \neg, \sim)$ such that the negation $\neg$ is an intuitionistic negation (Observation 50), and thus the reduct $(A, 1, 0, \vee, \wedge, \neg)$ must be a distributive lattice with intuitionistic negation. Moreover, Observation 51 suggests that $\sim$ is a minimal negation, therefore the reduct $(A, 1, 0, \vee, \wedge, \sim)$ must be a distributive lattice with minimal negation.

Definition 135 (K_{im}-algebras).
A K_{im}-algebra $\mathcal{A}$ is a tuple of the form $(A, 1, 0, \vee, \wedge, \neg, \sim)$ satisfying the following conditions:

(1) the reduct $(A, 1, 0, \vee, \wedge)$ is a bounded distributive lattice,
(2) the negation $\neg$ is an intuitionistic negation,
(3) the negation $\sim$ is a minimal negation, and
(4) $\neg\neg\sim 1 = \sim 1$.

A K_{im}-algebra $\mathcal{A}$ satisfying $a \vee \sim a = 1$ for all $a \in A$ is called a $K_{im-\vee}$-algebra.

Theorem 36. K_{im} ($K_{im-\vee}$) is sound and complete with respect to the class of K_{im}-algebras ($K_{im-\vee}$-algebras).

Proof. Soundness follows, considering the equivalent logic K (Proposition 84). Completeness proof is routine, by construction of the corresponding Lindenbaum-Tarski-style algebra $\mathcal{U}(K)$ (cf. Sect. 2.6).

Is there any relationship between K_{im}-algebras and $ccpBa$? Using Definition 7, for any $ccpBa$ $\mathcal{A} := (A, 1, 0, \vee, \wedge, \rightarrow, \neg, \sim)$, $(A, 1, 0, \vee, \wedge)$ is a bounded distributive lattice. Proposition 2(1), (2), (4), (5) (6) and (7) make $\neg$ an intuitionistic negation. Using Observation 35, $(A, 1, 0, \wedge, \vee, \rightarrow, \sim)$ is a bounded cc lattice. Therefore, Proposition 2(1), (2), (4) and (5) and Proposition 62(5) applied on 0 (giving $\sim 0 = \neg 0 = 1$), make $\sim$ a minimal negation.

Proposition 85. For any $ccpBa$ $(A, 1, 0, \vee, \wedge, \rightarrow, \neg, \sim)$, the reduct $(A, 1, 0, \vee, \wedge, \neg, \sim)$ is a K_{im}-algebra.

Can any K_{im}-algebra $(A, 1, 0, \vee, \wedge, \neg, \sim)$ be extended to a $ccpBa$? Consider the lattice $L := \mathbb{Z} \times \mathbb{Z}$, the set of pairs of integers, with the usual ordering $\leq$: $(m, n) \leq (r, s)$ if and only if $m \leq r$ and $n \leq s$. L is a distributive lattice. Define $L' := L \cup \{\hat{0}, \hat{1}\}$ ($\hat{0} \neq \hat{1}$), such that $\leq$ is extended to L' in the following way: $\hat{0} \leq (m, n) \leq \hat{1}$ for all $(m, n) \in L$. Addition of $\hat{0}$ and $\hat{1}$ makes the lattice bounded, i.e. L' is a bounded distributive lattice. Define two negations $\neg$ and $\sim$ on L' as follows: (1) $\neg(m, n) := \hat{0}$ for all $(m, n) \in L$, $\neg\hat{1} := \hat{0}$, and $\neg\hat{0} := \hat{1}$. (2) $\sim a = \hat{1}$, for all $a \in L'$.

(A): $\neg$ is an intuitionistic negation.

(a) $a \leq b \Rightarrow \neg b \leq \neg a$: let $a \leq b$. Then we have the following cases: (i). $a = (m,n) \in L$ and $b = (r,s) \in L$ such that $a \leq b$, then $\neg a = \neg b = \hat{0}$ and $\hat{0} \leq \hat{0}$. (ii). For $a \in L'$ and $b = \hat{1}$, $\neg b = \hat{0}$ and $\hat{0} \leq a$. (iii). For $a = \hat{0}$ and $b \in L'$, $\neg a = \hat{1}$ and $\neg b \leq \hat{1}$, and (iv). For $a = \hat{1}$, then $b = \hat{1}$ and $\neg b = \neg a = \hat{0}$.

(b). $\neg a \wedge \neg b \leq \neg(a \vee b)$: there are two possible choices for $\neg a \wedge \neg b$, namely $\hat{0}$ and $\hat{1}$. If $\neg a \wedge \neg b = \hat{0}$, then this is trivially true. Let $\neg a \wedge \neg b = \hat{1}$, then we have $\neg a = \neg b = \hat{1}$, i.e. $a = b = \hat{0}$ and $\neg(a \vee b) = \neg \hat{0} = \hat{1}$.

(b). $\hat{1} = \neg \hat{0}$, by definition of $\neg$.

(d). $a \wedge b \leq c \Rightarrow a \wedge \neg c \leq \neg b$: let $a \wedge b \leq c$. There are two possible choices for $\neg c$: $\hat{0}$ and $\hat{1}$. If $\neg c = \hat{0}$, then $a \wedge \neg c = a \wedge \hat{0} = \hat{0} \leq \neg b$. If $\neg c = \hat{1}$, then $c = \hat{0}$ and $a \wedge b = \hat{0}$. This implies $a = b = \hat{0}$ and $a \wedge \neg c = \hat{0} \wedge \hat{1} = \hat{0} \leq \neg b$.

(e). $a \leq \neg \neg a$: if $a = \hat{1}$ or $a = (m,n)$, $\neg \neg a = \neg \hat{0} = \hat{1}$ and $a \leq \neg \neg a$. If $a = \hat{0}$, $\neg \neg a = \neg \hat{1} = \hat{0}$ and $a = \neg \neg a$.

(f). $a \wedge \neg a \leq b$: if $a \wedge \neg a = \hat{0}$, then this is trivially true. If $a \wedge \neg a = \hat{1}$, then $a = \hat{1}$ and $\neg a = \hat{1}$. This implies $a = \hat{0}$, a contradiction, as $\hat{0} \neq \hat{1}$. Therefore, no such case exists. $a \wedge \neg a$ cannot be of the form (m,n), because of the definition of the negation $\neg$.

(B): $\sim$ is a minimal negation. The properties (a) $a \leq b \Rightarrow \sim b \leq \sim a$, (b) $\sim a \wedge \sim b \leq \sim(a \vee b)$, (c) $\hat{1} = \sim \hat{0}$, (d) $a \wedge b \leq c \Rightarrow a \wedge \sim c \leq \sim b$, and (e) $a \leq \sim \sim a$, for $a,b,c \in L'$, are all trivially true because of the definition of $\sim$ on L'. We also have $\neg \neg \sim \hat{1} = \neg \neg \hat{1} = \neg \hat{0} = \hat{1} = \sim \hat{1}$.

Therefore, $\mathcal{L}' := (L', \hat{1}, \hat{0}, \vee, \wedge, \neg, \sim)$ is a K_{im}-algebra. For $\mathcal{L}'$ to be extended to a $ccpBa$, we must be able to define an operator '$\rightarrow$', such that the $(L', \hat{1}, \hat{0}, \vee, \wedge, \rightarrow)$ is an rpc lattice, i.e. the following holds: for all $a, b, x \in L'$, $a \wedge x \leq b \Leftrightarrow x \leq a \rightarrow b$.

Let $a := (1,0)$ and $b := (0,1)$. The possible choices of x for which $a \wedge x \leq b$ are from the set $\{\hat{0}\} \cup \{(m,n) \in \mathbb{Z} \times \mathbb{Z} \mid m \leq 0\}$. Since $x \leq a \rightarrow b$ for all such x, the only possible choice for $a \rightarrow b$ is $\hat{1}$. However, this value would make the converse false, because $(1,1) \leq \hat{1}$, but $(1,0) \wedge (1,1) = (1,0) \not\leq (0,1)$. Thus, we have the following.

Proposition 86. *There exists a K_{im}-algebra $\mathcal{A} := (A, 1, 0, \vee, \wedge, \neg, \sim)$ such that there is no binary operator $\rightarrow$ on A that makes the algebra $(A, 1, 0, \vee, \wedge, \rightarrow, \neg, \sim)$ a $ccpBa$.*

We have made a similar statement in Example 20, where there exists a cc lattice which cannot be extended to a $ccpBa$.

Observation 52. The aim of showing the above two propositions is that, even though there is a K_{im}-algebra that cannot be extended to a $ccpBa$, the negation properties in a K_{im}-algebra are enough to capture the 'non-implicative' version of $ccpBa$, as the reduct of any $ccpBa$ is a K_{im}-algebra (by removing the operator $\rightarrow$ from the algebra).

We have seen in Observations 50 and 51 that $\neg$ is an intuitionistic negation, while $\sim$ is a minimal negation. Note that (i) not every minimal negation, however, satisfies the condition $\neg \neg \sim 1 = \sim 1$, and (ii) the $\sim$ in K_{im}-algebra need not be intuitionistic.

Proposition 87.

1. *There exists a K_{im}-algebra $\mathcal{A} := (A, 1, 0, \vee, \wedge, \neg, \sim)$ such that $\sim$ is not an intuitionistic negation.*
2. *There exists a bounded distributive lattice $\mathcal{A} := (A, 1, 0, \vee, \wedge, \neg, \sim)$ with minimal negation $\sim$ and an intuitionistic negation $\neg$ such that $\neg\neg\sim 1 \neq \sim 1$.*

Proof.
(1). The K_{im}-algebra $\mathcal{L}' := (L', \hat{1}, \hat{0}, \vee, \wedge, \neg, \sim)$ considered to establish Proposition 86 suffices. For $a := (m, n) \in L$, $a \wedge \sim a = a \wedge \hat{1} = a \neq \hat{0}$.
(2). Consider the bounded distributive lattice $\mathcal{H}_6 := (H_6, 1, 0, \vee, \wedge, \neg, \sim)$ from Example 1, where $\neg$ and $\sim$ are given by Table 8.

Table 8. Negations in $\mathcal{H}_6 := (H_6, 1, 0, \vee, \wedge, \neg, \sim)$

a	$\neg a$	$\sim a$
0	1	1
y	x	1
z	y	1
w	0	1
x	y	w
1	0	w

It can be checked that $\neg$ is intuitionistic, $\sim$ is minimal, but

$$\sim 1 = w \neq \neg\neg\sim 1 (= 1).$$

Thus, we can now place the negation $\sim$ in a K_{im}-algebra in Dunn's Kite diagram strictly in between the nodes of minimal and intuitionistic negations (Fig. 49).

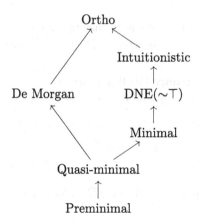

Fig. 49. DNE($\sim\top$) $\neg\neg\sim\top \vdash \sim\top$ in Dunn's Kite

We have studied relational semantics for ILM and ILM-∨. We have also seen the relational semantics for the logic K_i with preminimal negation (Theorem 11). We shall now see a relational semantics for the logics K_{im} and $K_{im-\vee}$.

7.2   Relational Semantics for K_{im}

Recall the definition of Dunn's compatibility frame (Definition 79). Before proceeding any further, let us compare strictly condensed frames and compatibility frames.

Observation 53. Let W be a non-empty set and $\leq$ a partial order. For a triple $(W, \leq, R_N)$ to be a strictly condensed frame, a necessary and sufficient condition is that $(\leq R_N \leq^{-1}) \subseteq R_N$ (Observation 42). Now

$$\forall x, y, x', y' \in W\big((x' \leq x \ \& \ xR_Ny \ \& \ y \leq^{-1} y') \Rightarrow x'(\leq R_N \leq^{-1})y'\big)$$
$$\forall x, y, x', y' \in W\big((x' \leq x \ \& \ xR_Ny \ \& \ y' \leq y) \Rightarrow x'(\leq R_N \leq^{-1})y'\big)$$
$$\therefore \ \forall x, y, x', y' \in W\big((x' \leq x \ \& \ xR_Ny \ \& \ y' \leq y) \Rightarrow x'R_Ny'\big).$$

The last condition is the same as Condition (C) of compatibility frames (Definition 79), with C as the relation R_N. Condition (C) (with C as R_N) clearly implies $(\leq R_N \leq^{-1}) \subseteq R_N$ also. Therefore, the class of strictly condensed frames with partial order is the same as the class of compatibility frames. In fact, the definition of a valuation v on a strictly condensed frame is also the same as that for compatibility frames for the constants $\bot$ and $\top$, and connectives $\vee$, $\wedge$, and $\neg$ (Definitions 80 and 73).

Remark 20. Note that in case of ILM, we had defined four different classes of frames - those of $\hat{N}$-frames (Definition 118) and sub-normal frames (Definition 127), and two other classes in Sect. 6.4. In this section, we shall focus on frames of the type $(X, \leq, R_N)$ presented in Case 4 in Sect. 6.4, because based on the above observation, the frames in this case are the compatibility frames $(W, C, \leq)$ (Definition 79).

We present the definition of the frames in Case 4 (Sect. 6.4) again. Re-write the notation of frames there as $(W, C, \leq)$, and calling them sub-compatibility frames.

Definition 136 (Sub-compatibility frame).
Let $\mathcal{F} := (W, C, \leq)$ *be a compatibility frame such that* C *is symmetric,*

$$\forall x, y \in W(xCy \Rightarrow \exists z \in W(x \leq z \ \& \ y \leq z \ \& \ xCz)), \text{ and} \tag{7.1}$$
$$\forall x \in W\big(\forall y \in W\big(x \leq y \Rightarrow \exists z \in W(y \leq z \ \& \ \forall z' \in W \ (z\cancel{C}z'))\big) \Rightarrow$$
$$\forall z'' \in W \ (x\cancel{C}z'')\big). \tag{7.2}$$

$\mathcal{F}$ *is called a* sub-compatibility frame. *The class of all such frames in denoted by* $\mathfrak{F}_4$.
The sub-class of all the sub-compatibility frames $\mathcal{F} := (W, C, \leq)$ *satisfying the condition* $C \subseteq \leq^{-1}$ *is denoted by* $\mathfrak{F}'_4$.

We now extend definitions of valuation and validity (Definition 80) to the case of K_{im}.

Definition 137 (Valuation on a sub-compatibility frame).

A valuation v_0 of $\mathcal{L}$ (the language of K_{im}) on a sub-compatibility frame $\mathcal{F} := (W, C, \leq)$ is a mapping from the set of propositional variables in $\mathcal{L}$ to the power set $\mathcal{P}(W)$ of W such that for any propositional variable p, $v_0(p)$ is an upset, i.e. $\forall x, y \in W (x \leq y \Rightarrow (x \in v_0(p) \Rightarrow y \in v_0(p)))$.

Definition 138 (Truth of a formula).

The truth of a formula $\alpha \in F$ at a world $x \in W$ in the sub-compatibility frame $\mathcal{F} := (W, C, \leq)$ under a valuation v_0 (notation - $x \vDash \alpha$) is defined as follows:

1. $x \vDash p \Leftrightarrow x \in v_0(p)$ *for all propositional variables p.*
2. $x \vDash \alpha \wedge \beta$ *if and only if $x \vDash \alpha$ and $x \vDash \beta$.*
3. $x \vDash \alpha \vee \beta$ *if and only if $x \vDash \alpha$ or $x \vDash \beta$.*
4. $x \vDash \top$.
5. $x \nvDash \bot$.
6. $x \vDash \neg\alpha$ *if and only if $\forall y \in W$, $(x \leq y \Rightarrow y \nvDash \alpha)$.*
7. $x \vDash \sim\alpha$ *if and only if $\forall y \in W$, $(xCy \Rightarrow y \nvDash \alpha)$.*

Definition 139 (Model on a sub-compatibility frame).

Consider a valuation v_0 of $\mathcal{L}$ on a sub-compatibility frame $\mathcal{F} := (W, C, \leq)$. The pair $\mathcal{M} := (\mathcal{F}, v_0)$ is called a sub-compatibility model *on the sub-compatibility frame $\mathcal{F}$.*

Note:

(1). $x \vDash \alpha$ (Definition 138) will also be denoted as $\mathcal{M}, x \vDash \alpha$.

(2). The notation '$\mathcal{M}, x \nvDash \phi$' denotes that ϕ is not true at the world $w \in W$ in the sub-compatibility model $\mathcal{M}$.

Definition 140 (Validity of a pair of formulas).

We say a pair of formulas (α, β) is valid in a sub-compatibility model $\mathcal{M}$, if for all $x \in W$, $\mathcal{M}, x \vDash \alpha \Rightarrow \mathcal{M}, x \vDash \beta$. This is denoted by $\alpha \vDash_{\mathcal{M}} \beta$.

A rule $\alpha \vdash \beta / \gamma \vdash \delta$ is said to be valid in $\mathcal{M}$, if $\alpha \vDash_{\mathcal{M}} \beta$ implies $\gamma \vDash_{\mathcal{M}} \delta$.

A consequence pair (α, β) is valid in a sub-compatibility frame $\mathcal{F}$, denoted as $\alpha \vDash_{\mathcal{F}} \beta$, if $\alpha \vDash_{\mathcal{M}} \beta$ for every sub-compatibility model $\mathcal{M}$ on $\mathcal{F}$.

For a subclass $\mathfrak{F}$ of sub-compatibility frames, if $\alpha \vDash_{\mathcal{F}} \beta$ for every $\mathcal{F} \in \mathfrak{F}$, we write $\alpha \vDash_{\mathfrak{F}} \beta$.

Observation 54.

1. In Definition 138 of truth, in (7), take $\phi = \top$, then using (4), we have the following: $\mathcal{M}, x \vDash \sim\top \Leftrightarrow \forall y \in W\ (x\cancel{C}y)$.
2. Let us see the contraposition of the reverse implication in Condition 7.2, that is, for any $x \in W$:

$$\exists y \in W \big(x \leq y\ \&\ \forall z \in W(y \leq z \Rightarrow \exists z' \in W\ zCz'))\big) \Rightarrow \exists z'' \in W\ (xCz'').$$

This is always true: let $y \in W$ such that $x \leq y$ and $\forall z \in W(y \leq z \Rightarrow \exists z' \in W\ zCz')$. $y \leq y$ implies there exists $z' \in W$ such that yCz'. Using condition (C), yCz', $x \leq y$ and $z' \leq z'$ imply xCz'.

3. What exactly are the Conditions 7.1 and 7.2? In [33], it has been shown that the Condition 7.1 is 'canonical' to $(P6)$: $\alpha \wedge \beta \vdash \gamma \ / \ \alpha \wedge \sim\gamma \vdash \sim\beta$ from Proposition 83. In other words, $(P6)$ is valid in a compatibility frame $\mathcal{F} := (W, C, \leq)$ if and only if C satisfies Condition 7.1.

Recall the sequent $(\mathrm{DNE}(\sim\top))$: $\neg\neg\sim\top \vdash \sim\top$ from Proposition 83. Validity of $\mathrm{DNE}(\sim\top)$ in a sub-compatibility model $\mathcal{M} := ((W, C, \leq), \vDash)$ means the following.

$$\forall x \in W(x \vDash \neg\neg\sim\top \Rightarrow x \vDash \sim\top))$$
$$\Leftrightarrow \ \forall x \in W(\forall y \in W(x \leq y \Rightarrow y \nvDash \neg\sim\top) \Rightarrow x \vDash \sim\top)$$
$$\Leftrightarrow \ \forall x \in W\big(\forall y \in W(x \leq y \Rightarrow \exists z \in W(y \leq z \ \& \ z \vDash \sim\top)) \Rightarrow x \vDash \sim\top\big)$$
$$\Leftrightarrow \ \forall x \in W\big(\forall y \in W\big(x \leq y \Rightarrow \exists z \in W(y \leq z \ \& \ \forall z' \in W \ (z \mathscr{C} z'))\big) \Rightarrow$$
$$\forall z'' \in W \ (x \mathscr{C} z'')\big),$$

using (1) of this observation. The last is just the Condition 7.2. So we have

Proposition 88. *Condition 7.2 is canonical to* $\mathrm{DNE}(\sim\top)$.

In Proposition 84, we have given an equivalent version K of K_{im}. The axioms/rules involving the negation $\sim$ are $P2 - P7$. Based on the points above, Conditions 7.1 and 7.2 ensure the validity of $P6$ and $P7$. Moreover, any sub-compatibility frame is a compatibility frame, and $P2 - P5$ are valid in any compatibility frame in the language of K_i (Theorem 11). So we get the soundness result.

Theorem 37. *For any formulas* α, β *in* $\mathcal{L}$, $\alpha \vdash_{K_{im}} \beta$ $(\alpha \vdash_{K_{im}-\vee} \beta)$ *implies* $\alpha \vDash_{\mathcal{F}_4} \beta$ $(\alpha \vDash_{\mathcal{F}'_4} \beta)$.

We shall now proceed to prove that K_{im} will be complete with respect to the class of sub-compatibility frames $\mathfrak{F}_4$.

We use similar concepts (theories, canonical frames) and establish similar results as those used for obtaining the completeness result for ILM with respect to $\hat{N}$-frames. Some proofs change, as there is no '$\to$' here - we present only those, and just state the rest of the results.

The following proposition can be obtained immediately. The first result in it is obtained through induction on the number of connectives in α, and the third using Condition 7.2. The second result is immediate, using the first.

Proposition 89. *Consider a sub-compatibility model* $\mathcal{M} := (\mathcal{F}, \vDash)$, *where* $\mathcal{F} := (W, C, \leq) \in \mathfrak{F}_4$. *For any formula* $\alpha \in F$ *and* $x \in W$, *we have the following.*

1. $\forall x, y \in W(x \vDash \alpha \Rightarrow (x \leq y \Rightarrow y \vDash \alpha))$,
2. $x \vDash \neg\alpha \Leftrightarrow \forall y \in W(\exists z \in W(x \leq z \ \& \ y \leq z) \Rightarrow y \nvDash \alpha)$, *and*
3. $x \vDash \sim\alpha \Leftrightarrow \forall y \in W(x \leq y \Rightarrow (y \vDash \alpha \Rightarrow (\exists z \in W(y \leq z \ \& \ \forall z' \in W \ z \mathscr{C} z'))))$.

Definition 141 (Theory [33]).
A theory $T \subseteq F$ with respect to K_{im} is a non-empty set of formulas in $\mathcal{L}$ such that, for formulas $\alpha, \beta \in F$,

1. if $\alpha \in T$ and $\alpha \vdash \beta$, then $\beta \in T$ (closed under deduction),
2. $\top \in T$, and
3. if $\alpha, \beta \in T$ then $\alpha \wedge \beta \in T$ (closed under $\wedge$).

A *theory* is consistent *if* $\perp \notin T$, *otherwise* inconsistent. A prime *theory is a consistent theory such that for any formula* $\alpha, \beta \in F$, *if* $\alpha \vee \beta \in T$ *then either* $\alpha \in T$ or $\beta \in T$.

Using axiom A8: $\perp \vdash \alpha$, a theory T is consistent if and only if there exists a formula α such that $\alpha \notin T$. For an arbitrary set Δ of formulas, $Th(\Delta)$, called *theory generated by* Δ, is defined as the intersection of all the theories containing Δ.

Lemma 10.

$$Th(\Delta) = \{\alpha \in F \mid \exists \phi_1, \phi_2 \ldots, \phi_n \in \Delta \text{ such that } (\phi_1 \wedge \phi_2 \wedge \cdots \wedge \phi_n) \vdash \alpha\}.$$

Lemma 11. *For theories* P *and* Q, $\alpha \in Th(P \cup Q)$ *if and only if there exist* $\psi \in P$ *and* $\phi \in Q$ *such that* $\psi \wedge \phi \vdash \alpha$.

As a corollary, we get

Lemma 12. *For a theory* T, *and formulas* $\phi, \alpha \in F$, $\alpha \in Th(T \cup \{\phi\})$ *if and only if there exists* $\psi \in T$ *such that* $\psi \wedge \phi \vdash \alpha$.

Definition 142 ($\vee$-closed).
A set $F' \subseteq F$ of formulas is $\vee$-closed, if for any $\alpha, \beta \in F'$, $\alpha \vee \beta \in F'$.

An arbitrary set $\Delta \subseteq F$ can be extended to a $\vee$-closed set through the following.

Definition 143.
For $\Delta \subseteq F$,

$$\mathrm{dc}(\Delta) := \bigcap \{\Delta' \subseteq F \mid \Delta \subseteq \Delta' \text{ and } (\forall \alpha, \beta \in F(\alpha, \beta \in \Delta' \Rightarrow \alpha \vee \beta \in \Delta')\}$$

Observation 55.

(1) For a subset $\Delta \subseteq F$, $\mathrm{dc}(\Delta)$ is $\vee$-closed.
(2) For $\alpha \in F$, $\beta \in \mathrm{dc}(\{\alpha\})$ implies $\alpha \vdash \beta$ and $\beta \vdash \alpha$.

The following Lemma is obtained using Zorn's lemma (Lemma 2), and the proof is similar to the Existence lemma proved in [31].

Lemma 13. (Extension lemma [31]). *Let Δ be a consistent theory and $\Gamma \subseteq F$ be a $\vee$-closed set. If $\Delta \cap \Gamma = \emptyset$ then there exists a prime theory P such that $\Delta \subseteq P$ and $P \cap \Gamma = \emptyset$.*

Corollary 12. *Let Δ be a consistent theory and $\alpha \in F$ be such that $\alpha \notin \Delta$. Then there is a prime theory P such that $\Delta \subseteq P$ and $\alpha \notin P$.*

Let us now define the canonical sub-compatibility frame in the following way.

Definition 144 (Canonical frame).
The canonical frame is the tuple $\mathcal{F}^c := (W^c, C^c, \subseteq)$ defined as follows:
$W^c := \{P \subseteq F \mid P \text{ is a prime theory}\}$, *and*
PC^cQ *if and only if (for all $\alpha \in F$, $\sim\!\alpha \in P \Rightarrow \alpha \notin Q$).*

$\mathcal{F}^c$ is a sub-compatibility frame. To show that, we make use of the following.

Lemma 14. *For any $\alpha \in F$ and $P \in W^c$,*

1. *$\neg\alpha \in P$ if and only if for all $Q \in W^c$, $P \subseteq Q \Rightarrow \alpha \notin Q$, and*
2. *$\sim\!\alpha \in P$ if and only if for all $Q \in W^c$, $PC^cQ \Rightarrow \alpha \notin Q$.*

Proof. (1). ($\Rightarrow$). Let $\neg\alpha \in P$ and suppose there exists $Q \in W^c$ such that $P \subseteq Q$, but $\alpha \in Q$. So $\neg\alpha \in Q$. Using the property of prime theories, $\alpha \wedge \neg\alpha \in Q$. Axiom (A15) implies $\beta \in Q$ for all $\beta \in F$. In particular, $\bot \in Q$, implying that Q is inconsistent, a contradiction as $Q \in W^c$.
($\Leftarrow$). Let $\neg\alpha \notin P$. We need to find a prime theory Q such that $P \subseteq Q$ and $\alpha \in Q$.
Consider the theory $\Delta := Th(\{\alpha\})$ generated by the set $\{\alpha\}$, and the set $\Gamma := \{\alpha \in F \mid \neg\alpha \in P\}$. Observe that α is not $\bot$, because if $\alpha = \bot$, then $\neg\bot \notin P$. By A12, $\top \notin P$, a contradiction to the fact that P is a theory.
Claim 1: Δ is consistent. Suppose not, then $\bot \in \Delta$. Using Lemma 10, $\alpha \vdash \bot$. Using A10, $\neg\bot \vdash \neg\alpha$. Using A12, $\neg\bot \in P$, which implies $\neg\alpha \in P$, contradicting the assumption.
Claim 2: $\Delta \cap \Gamma = \emptyset$. Suppose not, i.e. there exists $\beta \in \Gamma \cap \Delta$. Using Lemma 10, $\alpha \vdash \beta$. By A10, we have $\neg\beta \vdash \neg\alpha$. Since $\beta \in \Gamma$, we have $\neg\beta \in P$. Therefore, $\neg\alpha \in P$, a contradiction to our assumption.
Claim 3: Γ is $\vee$-closed. Let $\alpha, \beta \in \Gamma$, then $\neg\alpha, \neg\beta \in P \Rightarrow \neg(\alpha \vee \beta) \in P$ (Using A11). This implies $\alpha \vee \beta \in \Gamma$.
Therefore, using Lemma 13, there exists a prime theory Q such that $\Gamma \cap Q = \emptyset$ and $\Delta \subseteq Q$. Thus $\alpha \in Q$.
We now claim that $\Omega := Th(P \cup Q)$ is consistent. If not, then $\bot \in \Omega$. Using Lemma 11, there exist $\beta \in P$ and $\gamma \in Q$ such that $\beta \wedge \gamma \vdash \bot$. A14 implies $\beta \wedge \neg\bot \vdash \neg\gamma$. Further, by A1, A12 and P1, $\beta \vdash \beta$ and $\top \vdash \neg\bot$ imply $\beta \wedge \top \vdash \beta \wedge \neg\bot$. Therefore, $\beta \wedge \top \vdash \neg\gamma$, by A2. Since $\beta \vdash \beta \wedge \top$ (obtained using A4), we have $\beta \vdash \neg\gamma$ (by A2). Since, $\beta \in P$, we have $\neg\gamma \in P$. So $\gamma \in \Gamma$, a contradiction to the fact that $Q \cap \Gamma = \emptyset$. Therefore Ω is consistent. Finally, again using Lemma

13, there exists a prime theory R such that $\Omega \subseteq R$. This implies that $P \subseteq R$ and $\alpha \in Q \subseteq R$. R is the required prime theory.

(2). The proof follows from [33] directly.

As a corollary, for $\alpha := \top$, we get

Lemma 15. *For any $P \in W^c$,*
(1). $\neg \top \in P$ if and only if for all $Q \in W^c$, $P \not\subseteq Q$, and
(2). $\sim \top \in P$ if and only if for all $Q \in W^c$, $P \not\!\!C^c Q$.

Proposition 90. $\mathcal{F}^c := (W^c, C^c, \subseteq)$ *is a sub-compatibility frame.*

Proof. Let $\Delta = \{\top\}$ and $\Gamma = \emptyset$. Then, there exists a prime theory P such that $\Delta \subseteq P$, implying that W^c is non-empty.

Claim 1: C satisfies Condition (C). Let $P, Q, P', Q' \in W^c$ be such that $P \subseteq P'$, $Q \subseteq Q'$ and $P'C^cQ'$. To show that PC^cQ. Consider $\alpha \in F$ such that $\sim\alpha \in P$. Then $\sim\alpha \in P' \Rightarrow \alpha \notin Q' \Rightarrow \alpha \notin Q$. Therefore, PC^cQ.

Claim 2: C^c is symmetric. Let $P, Q \in W^c$ such that PC^cQ. To show that QC^cP. Suppose not, i.e. there exists $\alpha \in F$ such that $\sim\alpha \in Q$, but $\alpha \in P$. Using Proposition 83(P5), $\alpha \in P$ implies $\sim\sim\alpha \in P$. PC^cQ and $\sim\sim\alpha \in P$ imply $\sim\alpha \notin Q$, a contradiction.

Claim 3: C^c satisfies Condition 7.1, i.e. if $P, Q \in W^c$ such that PC^cQ then there exists $R \in W^c$ such that $P \subseteq R$, $Q \subseteq R$ and PC^cR.

Let $P, Q \in W^c$ such that PC^cQ. Consider the theory $\Delta := Th(P \cup Q)$. Δ is consistent, because if not, then using Lemma 11, there exist $\alpha \in P$ and $\beta \in Q$ such that $\alpha \wedge \beta \vdash \bot$. Proposition 83(P6) implies $\alpha \wedge \sim\bot \vdash \sim\beta$. Proposition 83(P4) implies that $\sim\bot \in P$. This implies $\alpha \wedge \sim\bot \in P \Rightarrow \sim\beta \in P \Rightarrow \beta \notin Q$, a contradiction.

Define $\Gamma := \{\alpha \in F \mid \sim\alpha \in P\}$. Observe that Γ is $\vee$-closed, because for $\alpha, \beta \in \Gamma$, we have $\sim\alpha, \sim\beta \in P$ implying $\sim\alpha \wedge \sim\beta \in P$. Proposition 83(P3) implies $\sim(\alpha \vee \beta) \in P \Rightarrow \alpha \vee \beta \in \Gamma$.

$\Gamma \cap \Delta = \emptyset$: suppose not, i.e. there exists $\alpha \in \Gamma$ such that $\alpha \in Th(P \cup Q)$. This means $\sim\alpha \in P$ and there exist $\beta \in P$ and $\gamma \in Q$ such that $\beta \wedge \gamma \vdash \alpha$. Proposition 83(P6) implies $\beta \wedge \sim\alpha \vdash \sim\gamma$. $\beta \wedge \sim\alpha \in P$ implies $\sim\gamma \in P$. PC^cQ and $\sim\gamma \in P$ gives $\gamma \notin Q$, a contradiction.

Thus using Proposition 13, there exists a prime theory $Q' \in W^c$ such that $Th(P \cup Q) \subseteq Q'$ and $\Gamma \cap Q' = \emptyset$. Observe that $P \subseteq Q'$ and $Q \subseteq Q'$. We have to show that PC^cQ'. Let $\alpha \in F$ such that $\sim\alpha \in P$. Then we have $\alpha \in \Gamma \Rightarrow \alpha \notin Q'$. Therefore, PC^cQ'.

Claim 3: C^c satisfies Condition 7.2, i.e. for all $P \in W^c$,

$$\forall Q \in W^c(P \subseteq Q \Rightarrow \exists R \in W^c(Q \subseteq R \ \& \ \forall R' \in W^c \ (R\not\!\!C^c R'))) \Rightarrow$$

$$\forall R'' \in W^c \ (P\not\!\!C^c R'').$$

Suppose for all $Q \in W^c$ such that $P \subseteq Q$, there exists $R \in W^c$ such that $Q \subseteq R$ and for all $R' \in W^c$, $R \mathcal{Q}^c R'$. This implies

$$\forall Q \in W^c(P \subseteq Q \Rightarrow \exists R \in W^c(Q \subseteq R \ \& \sim\top \in R)) \qquad \text{(Lemma 15)}$$
$$\Rightarrow \forall Q \in W^c(P \subseteq Q \Rightarrow \neg\sim\top \notin Q) \qquad\qquad\qquad \text{(Lemma 14)}$$
$$\Rightarrow \neg\neg\sim\top \in P \qquad\qquad\qquad\qquad\qquad\qquad\qquad \text{(Lemma 14)}$$
$$\Rightarrow \sim\top \in P \qquad\qquad\qquad\qquad\qquad\qquad \text{(Proposition 83(P7))}$$
$$\Rightarrow \forall R'' \in W^c \ (P \mathcal{Q}^c R'') \qquad\qquad\qquad\qquad\qquad \text{(Lemma 15)}$$

Define a *valuation* v^c on the canonical sub-compatibility frame $(W^c, C^c, \subseteq)$ as follows: for all $p \in PV$, $v^c(p) := \{P \in W^c \mid p \in P\}$. The pair $\mathcal{M}^c := (\mathcal{F}^c, \vDash)$ is called the *canonical sub-compatibility model*.

To show the completeness result, we need the 'truth lemma'.

Lemma 16 (Truth Lemma). *For any $\alpha \in F$ and $P \in W^c$, we have $\mathcal{M}^c, P \vDash \alpha$ if and only if $\alpha \in P$.*

Proof. The basis step and induction step for connectives $\vee$ and $\wedge$ are obtained easily.

α is of the form of $\sim\beta$, where $\beta \in F$ such that β has less than n connectives:

($\Rightarrow$): Let $\mathcal{M}^c, P \vDash \sim\beta$, i.e. for all $Q \in W^c$, $PC^cQ \Rightarrow \mathcal{M}^c, Q \nvDash \beta$. Using induction hypothesis, for any $Q \in W^c$, $\mathcal{M}^c, Q \vDash \beta$ if and only if $\beta \in Q$. Therefore, for all $Q \in W^c$, $PC^cQ \Rightarrow \beta \notin Q$. Lemma 14 implies $\sim\beta \in P$.

($\Leftarrow$): Let $\sim\beta \in P$. Lemma 14 implies that for all $Q \in W^c$, $PC^cQ \Rightarrow \beta \notin Q$. Using induction hypothesis, for all $Q \in W^c$, $PC^cQ \Rightarrow \mathcal{M}^c, Q \nvDash \beta$. By Definition 138, $\mathcal{M}^c, P \vDash \sim\beta$.

Suppose α is of the form of $\neg\beta$, where $\beta \in F$ such that β has less than n connectives:

($\Rightarrow$): Let $\mathcal{M}^c, P \vDash \neg\beta$, i.e. for all $Q \in W^c$, $P \subseteq Q \Rightarrow \mathcal{M}^c, Q \nvDash \beta$. Induction hypothesis implies that for all $Q \in W^c$, $P \subseteq Q \Rightarrow \beta \notin Q$, implying $\neg\beta \in P$ (Lemma 14(1)).

($\Leftarrow$): Let $\neg\beta \in P$. Lemma 14 implies that for all $Q \in W^c$, $P \subseteq Q \Rightarrow \beta \notin Q$. Induction hypothesis implies that for all $Q \in W^c$, $P \subseteq Q \Rightarrow \mathcal{M}^c, Q \nvDash \beta$. By Definition 138, $\mathcal{M}^c, P \vDash \neg\beta$.

Let us now see the proof of the characterization (completeness) result for the logics K_{im} and $K_{im\text{-}\vee}$.

Theorem 38. *For any formulas $\alpha, \beta \in F$,*
(i). $\alpha \vDash_{\mathfrak{F}_4} \beta$ implies $\alpha \vdash_{K_{im}} \beta$, and
(ii). $\alpha \vDash_{\mathfrak{F}_4'} \beta$ implies $\alpha \vdash_{K_{im\text{-}\vee}} \beta$.

Proof. We have to show the following for K_{im}: For any $\alpha, \beta \in F$, if for all models $\mathcal{M} := (\mathcal{F}, \vDash)$ and $w \in W$, where $\mathcal{F} := (W, C, \leq) \in \mathfrak{F}_4$, $(\mathcal{M}, w \vDash \alpha \Rightarrow \mathcal{M}, w \vDash \beta)$, then $\alpha \vdash_{K_{im}} \beta$.

Let $\alpha \nvdash_{K_{im}} \beta$, and let $\mathcal{M} := (\mathcal{F}, \vDash)$ be a compatibility model. Let $w \in W$, where $\mathcal{F} := (W, C, \leq) \in \mathfrak{F}_4$. If α is $\perp$, then by Definition 138 of valuation, $\mathcal{M}, w \nvDash \perp$ for all $w \in W$. Thus, $\mathcal{M}, w \vDash \perp \Rightarrow \mathcal{M}, w \vDash \beta$ holds trivially.

Suppose α is not $\perp$. Then define $\Delta := Th(\alpha)$ and $\Gamma := dc(\{\beta\})$. Since α is not $\perp$, Δ is a consistent theory. Moreover, $\Gamma \cap \Delta = \emptyset$, because if not, then there exists a $\gamma \in \Gamma$ such that $\gamma \in Th(\alpha)$, i.e. $\alpha \vdash_{K_{im}} \gamma$. Observation 55 and $\gamma \in \Gamma$ imply $\gamma \vdash_{K_{im}} \beta$. Axiom $A2$ implies $\alpha \vdash_{K_{im}} \beta$, a contradiction. Further, again using Observation 45, Γ is $\vee$-closed.

Using Lemma 13, there exists a prime theory P such that $\Delta \subseteq P$ and $P \cap \Gamma = \emptyset$. $\beta \in \Gamma$ implies $\beta \notin P$. $\Delta \subseteq P$ implies $\alpha \in P$. Using Lemma 16, in the canonical model for K_{im}, we have $\mathcal{M}^c, P \vDash \alpha$ and $\mathcal{M}^c, P \nvDash \beta$.

For $K_{im\text{-}\vee}$, we have to show that the canonical frame for $K_{im\text{-}\vee}$ belongs to $\mathfrak{F}_4'$, i.e. for all $\alpha \in F$, for all models $\mathcal{M} := (\mathcal{F}, \vDash)$ and $w \in W$, where $\mathcal{F} := (W, C, \leq) \in \mathfrak{F}_4'$, $w \vDash \alpha \vee \sim\alpha$.

Consider the canonical frame $\mathcal{F}^c := (W^c, C^c, \subseteq)$ for $K_{im\text{-}\vee}$. We have already shown that it is a sub-compatibility frame. For $\mathcal{F}^c$ to be in $\mathfrak{F}_4'$, we have to show that $C^c \subseteq (\subseteq^{-1})$, i.e. for all $P, Q, \in W^c$, if PC^cQ then $Q \subseteq P$. Suppose not, i.e. there exist $P, Q \in W^c$ such that PC^cQ and $Q \nsubseteq P$. Then there exists $\gamma \in Q$ such that $\gamma \notin P$. Since $\top \in P$, Axiom $A18$ implies that $\gamma \vee \sim\gamma \in P$. Since P is a prime theory and $\gamma \notin P$, we have $\sim\gamma \in P$. PC^cQ implies $\gamma \notin Q$, a contradiction. Consider a frame $\mathcal{F} := (W, C, \leq) \in \mathfrak{F}_4'$ and a model $\mathcal{M} := (\mathcal{F}, \vDash)$. We have to show that for all $\alpha \in F$ and $w \in W$, $w \vDash \alpha \vee \sim\alpha$. Suppose there exists $w \in W$ such that $w \nvDash \alpha \vee \sim\alpha$ for some $\alpha \in F$. Definition 138 implies $w \nvDash \alpha$ and $w \nvDash \sim\alpha$. So, there exists $w' \in W$ such that wCw' and $w' \vDash \alpha$. wCw' gives $w' \leq w$. Therefore, $w' \leq w$ and $w' \vDash \alpha$ imply $w \vDash \alpha$, a contradiction.

7.3   Connection Between Sub-compatibility Frames and K_{im}-algebras

We saw the connection between subnormal frames and $ccpBas$ in Theorem 35. In this section, we shall see the connection between sub-compatibility frames and K_{im}-algebras. Let us first define embeddings between sub-compatibility frames.

Definition 145 (Embeddings between sub-compatibility frames).
Given two sub-compatibility frames $(W, C, \leq)$ and $(W', C', \leq')$, a mapping $\phi : W \to W'$ is an embedding if it satisfies the following conditions: for all $x, y \in W$,

1. $x \leq y$ if and only if $\phi(x) \leq' \phi(y)$, and
2. xCy if and only if $\phi(x)C'\phi(y)$.

Definition 146 (Complex algebra of a sub-compatibility frame).
Consider a sub-compatibility frame $\mathcal{F} := (W, C, Y_0)$. Let $Up(W)$ denote the set of upsets of W. Define the following operators on $Up(W)$.

1. $\neg U := \{w \in W \mid \forall v \in W(w \leq v \Rightarrow v \notin U)\}$, and
2. $\sim U := \{w \in W \mid \forall v \in W(wCv \Rightarrow v \notin U)\}$.

The algebra $Up(\mathcal{F}) := (Up(W), W, \cap, \cup, \neg, \sim)$ is called the complex algebra *of the sub-compatibility frame $\mathcal{F}$.*

Proposition 91. *For a sub-compatibility frame $\mathcal{F} := (W, C, \leq) \in \mathfrak{F}_4$, the complex algebra $Up(\mathcal{F})$ of $\mathcal{F}$ forms a K_{im}-algebra. Further, if $\mathcal{F} \in \mathfrak{F}'_4$, then $Up(\mathcal{F})$ is a $K_{im\text{-}\vee}$-algebra.*

Proof. Since $\mathcal{F}$ is a sub-compatibility frame, $(W, \leq)$ is a normal frame. Proposition 26 implies that $(Up(W), W, \emptyset, \cap, \cup)$ forms a bounded distributive lattice. Let us check the properties of the negation operators. Let $A, B, U \in Up(W)$.

We note that the subset of W that $\neg U$ defines in Definition 132 coincides with that in Definition 146. Moreover, $\neg$ in a pBa satisfies all the properties of an intuitionistic negation here. So we get (2) of Definition 135.

(1). $A \subseteq \sim\sim A$: let $x \in W$ such that $x \notin \sim\sim A$. Then there exists $y \in W$ such that xCy and for all $z \in W$, yCz implies $z \notin A$. Since, C is symmetric, $xCy \Rightarrow yCx$ and $x \notin A$.

(2). $A \cap B \subseteq X \Rightarrow A \cap \sim X \subseteq \sim B$: let $A \cap B \subseteq X$ and $x \in A \cap \sim X$, i.e. $x \in A$ and for all $y \in W$, $xCy \Rightarrow y \notin X$. Suppose to the contrary, $x \notin \sim B$, i.e. there exists $z \in W$ such that xCz and $z \in B$. Using Condition 7.1 of sub-compatibility frames (Definition 136), xCz implies that there exists $z' \in W$ such that $x \leq z'$, $z \leq z'$ and xCz'. xCz' implies $z' \notin X$. $A \cap B \subseteq X$ implies $z' \notin A \cap B$. $z \in B$ and $z \leq z'$ gives $z' \in B$. Therefore, $z' \notin A$ and $x \leq z' \Rightarrow x \notin A$, a contradiction.

(3). $A \subseteq B \Rightarrow \sim B \subseteq \sim A$ follows from (2).

(4). $\sim A \cap \sim B \subseteq \sim(A \cup B)$:

$$x \in \sim(A \cup B)$$
$$\Leftrightarrow \forall y \in W(xCy \Rightarrow y \notin A \cup B)$$
$$\Leftrightarrow \forall y \in W(xCy \Rightarrow (y \notin A \ \& \ y \notin B))$$
$$\Leftrightarrow \forall y \in W(xCy \Rightarrow y \notin A) \ and \ \forall y \in W(xCy \Rightarrow y \notin B)$$
$$\Leftrightarrow x \in \sim A \cap \sim B.$$

(5). $\sim\emptyset = W$, by definition of $\sim$.

(6). $\neg\neg\sim W = \sim W$: As $\neg$ is intuitionistic, $\sim W \subseteq \neg\neg\sim W$. For the converse, let $x \in \neg\neg\sim W$, i.e. for all $y \in W$, $x \leq y$ implies that there exists $z \in W$ such that $y \leq z$ and for all $z' \in W$, $z \mathcal{C} z'$. Condition 7.2 implies that for all $z'' \in W$, $x \mathcal{C} z''$, i.e. $x \in \sim W$, by definition of $\sim W$.

Now, let $C \subseteq \leq^{-1}$, i.e. $\mathcal{F} \subseteq \mathfrak{F}'_4$. We show that $A \cup \sim A = W$, for any $A \subseteq W$. Let $a \in W$ be such that $a \notin A$. We show $a \in \sim A$, i.e. $\forall v \in W(aCv \Rightarrow v \notin A)$. Let $v \in W$ be such that aCv. As $C \subseteq \leq^{-1}$, $v \leq a$. As A is an upset, $v \notin A$.

Therefore, given a sub-compatibility frame we have obtained a K_{im}-algebra. The converse is also not difficult to obtain, i.e. given a K_{im}-algebra $\mathcal{A}$, we can define the 'canonical' sub-compatibility frame. Recall Definition 75 of canonical normal frames of a given pBa. Since any K_{im}-algebra is a distributive lattice, all the concepts of filters (Definition 11, Lemma 1) extend to the case of K_{im}-algebras.

Definition 147 (Canonical frame of a K_{im}-algebra).
Consider a K_{im}-algebra $\mathcal{A} := (A, 1, 0, \wedge, \vee, \neg, \sim)$. Let X_A denote the set of all prime filters in $\mathcal{A}$. Define a relation C on X_A as follows: for $P, Q \in X_A$,

$$PCQ \text{ if and only if (for all } a \in A, \sim a \in P \Rightarrow a \notin Q).$$

The triple $\mathcal{F}_A := (X_A, C, \subseteq)$ is called the canonical frame.

We obtain the following consequence of the definition.

Lemma 17. *For any $P \in X_A$, $\sim 1 \notin P$ if and only if $\exists R \in X_A$ (PCR).*

Proof. ($\Leftarrow$) $\sim 1 \in P \Rightarrow \forall R \in X_A (P\cancel{C}R)$, by definition of C.
($\Rightarrow$) Let $\sim 1 \notin P$. Consider the set $\Gamma := \{a \in A \mid \sim a \in P\}$.
Γ is $\vee$-closed: if $a, b \in \Gamma$ then $\sim a, \sim b \in P \Rightarrow \sim a \wedge \sim b \in P$. As $\sim a \wedge \sim b \leq \sim(a \vee b)$, $\sim(a \vee b) \in P$, implying $a \vee b \in \Gamma$. Now, $\Delta := \{1\}$ is a filter and is proper ($\sim 1 \notin P$ and P is a proper filter). Using Lemma 1, as $\Gamma \cap \Delta = \emptyset$ ($1 \notin \Gamma$), there is $R \in X_A$ such that $\Delta \subseteq R$ and $\Gamma \cap R = \emptyset$. Thus $P \subseteq R$ as for any $a \in A$, $\sim a \in P \Rightarrow a \in \Gamma \Rightarrow a \notin R$.

Proposition 92. *For a K_{im}-algebra $\mathcal{A} := (A, 1, 0, \wedge, \vee, \neg, \sim)$, the canonical frame $\mathcal{F}_A := (X_A, C, \subseteq)$ is a sub-compatibility frame. If $\mathcal{A}$ is a $K_{im\text{-}\vee}$-algebra then $\mathcal{F}_A \in \mathfrak{F}'_4$.*

Proof. Let us first show that $\mathcal{F}_A$ is a compatibility frame. Let $P, Q, P', Q' \in X_A$ such that PCQ, $P' \subseteq P$ and $Q' \subseteq Q$.
$P'CQ'$: let $x \in A$ such that $\sim x \in P'$. PCQ and $\sim x \in P' \subseteq P$ imply $x \notin Q$. $Q' \subseteq Q$ implies $x \notin Q'$.
Let us now show that C is symmetric and that Conditions 7.1 and 7.2 hold.
C is symmetric: let PCQ. Let $x \in A$ such that $\sim x \in Q$. PCQ and $\sim x \in Q$ imply that $\sim\sim x \notin P$, i.e. $x \notin P$, as $x \leq \sim\sim x$ in a K_{im}-algebra.
C satisfies Condition 7.1: let PCQ. We have to find $R \in X_A$ such that $P \subseteq R$, $Q \subseteq R$ and PCR. Consider $\Delta := \langle P \cup Q \rangle$, the filter generated by $P \cup Q$. Δ is a proper filter. If not, i.e. if $0 \in \Delta$, then using Lemma 1, there exist $a \in P$ and $b \in Q$ such that $a \wedge b = 0$. This implies $a \wedge \sim 0 \leq \sim b$, i.e. $a \leq \sim b$. $a \in P \Rightarrow \sim b \in P \Rightarrow b \notin Q$ (as $P \subseteq Q$), a contradiction.
Define $\Gamma := \{a \in A \mid \sim a \in P\}$, as in the proof of the previous lemma. As observed there, Γ is $\vee$-closed. We claim that $\Gamma \cap \Delta = \emptyset$. Suppose not, i.e. there exists $a \in \Gamma \cap \Delta$. This means $\sim a \in P$ and there exist $b \in P$ and $c \in Q$ such that $b \wedge c \leq a$. This implies $b \wedge \sim a \leq \sim c$, i.e. $\sim c \in P$. PCQ implies $c \notin Q$, a contradiction.
Thus using Lemma 1, there exists a prime theory $Q' \in X_A$ such that $\Delta \subseteq Q'$ and $\Gamma \cap Q' = \emptyset$. Our final claim is PCQ'. Let $a \in A$ such that $\sim a \in P$. Then $a \in \Gamma \Rightarrow a \notin Q'$.
C satisfies Condition 7.2: we need to show that

$$\forall P \in X_A \big(\forall Q \in X_A \ (P \subseteq Q \Rightarrow \exists R \in X_A \ (Q \subseteq R \ \& \ \forall R' \in X_A \ (R\cancel{C}R'))) \Rightarrow$$
$$\forall R'' \in X_A \ (P\cancel{C}R''))$$

i.e. $\forall P \in X_A \big(\exists R'' \in X_A \ (PCR'') \Rightarrow \exists Q \in X_A \ (P \subseteq Q \ \& \ \forall R \in X_A \ (Q \subseteq R \Rightarrow$
$$\exists R' \in X_A \ (RCR'))))$$

i.e. $\forall P \in X_A \big(\sim 1 \notin P \Rightarrow \exists Q \in X_A \ (P \subseteq Q \ \& \ \forall R \in X_A \ (Q \subseteq R \Rightarrow \sim 1 \notin R))),$

using Lemma 17.

Consider a prime filter P in $\mathcal{A}$ such that $\sim 1 \notin P$. The aim is to find a $Q \in X_A$ such that $P \subseteq Q$ and for all $R \in X_A$, if $Q \subseteq R$ then $\sim 1 \notin R$. Define a set

$$\Delta := \{P' \subseteq A \mid P' \text{ is a filter in } \mathcal{A}; P \subseteq P'; \sim 1 \notin P'\},$$

as done in the proof of Proposition 82. We have to modify the proof of Proposition 82 in two places, as we do not have '$\rightarrow$' here. Applying Zorn's lemma, we get a maximal element Q in Δ. Indeed, Q is a proper subset of A because $\sim 1 \notin Q$.

$\neg \sim 1 \in Q$: if not, $\langle Q \cup \{\neg \sim 1\}\rangle$ is a proper superset of Q and contains P. Since Q is maximal in Δ, $\sim 1 \in \langle Q \cup \{\neg \sim 1\}\rangle$. Using Lemma 1(3), there exists $q \in Q$ such that $q \wedge \neg \sim 1 \leq \sim 1 \Rightarrow \neg \sim 1 \wedge \neg \sim 1 \leq \neg q \Rightarrow \neg \sim 1 \leq \neg q \Rightarrow \neg \neg q \leq \neg \neg \sim 1 = \sim 1$.

Since $q \leq \neg \neg q$, we have $q \leq \sim 1$. Q is a filter in $\mathcal{A}$, and any filter is an upset (Lemma 1(1)). Therefore, $\sim 1 \in Q$. This is a contradiction as $Q \in \Delta$. Thus, $\neg \sim 1 \in Q$.

Q is a maximal filter: for any $a \in A$ such that $a \notin Q$, if we show that $\langle Q \cup \{a\}\rangle = A$, then Q will be a maximal filter and we are done. Since $\langle Q \cup \{a\}\rangle$ contains P and is a proper superset of Q, we have $\langle Q \cup \{a\}\rangle \notin \Delta$, i.e. $\sim 1 \in \langle Q \cup \{a\}\rangle$. Using Lemma 11, we have $q \in Q$ such that $q \wedge a \leq \sim 1$. $\sim 1 = \neg \neg \sim 1$ implies $q \wedge a \wedge 1 \leq \neg \neg \sim 1 \Rightarrow q \wedge a \wedge \neg \neg \neg \sim 1 \leq \neg 1$.

Now, for any intuitionistic negation '$\neg$', for all $a \in A$, $\neg \neg \neg a = \neg a$ and $\neg 1 = 0$. Therefore, $q \wedge a \wedge \neg \sim 1 \leq 0$. Since $\neg \sim 1 \in Q$, we have $0 \in \langle Q \cup \{a\}\rangle$, and thus, $A = \langle Q \cup \{a\}\rangle$. Thus Q is prime, i.e. $Q \in X_A$.

Let $\mathcal{A}$ be a $K_{im\text{-}\vee}$-algebra. To show that $\mathcal{F}_A \in \mathfrak{F}'_4$, we have to show that if PCQ then $Q \subseteq P$. Let PCQ and $x \in A$ such that $x \notin P$. $x \vee \sim x = 1$ and $1 \in P$ imply $\sim x \in P$, i.e. $x \notin Q$.

Let us now observe the connection between sub-compatibility frames $\mathcal{F} := (W, C, \leq) \in \mathfrak{F}_4$ ($\mathfrak{F}'_4$) and K_{im}-algebras ($K_{im\text{-}\vee}$-algebras).

Theorem 39.

1. *Every K_{im}-algebra (or $K_{im\text{-}\vee}$-algebra) $\mathcal{A}$ can be embedded into the complex algebra $Up(\mathcal{F}_A)$ of the canonical frame $\mathcal{F}_A$ of $\mathcal{A}$.*
2. *Any sub-compatibility frame $\mathcal{F} \in \mathfrak{F}_4$ (or $\mathfrak{F}'_4$) can be embedded into the canonical frame $\mathcal{F}_{Up(\mathcal{F})}$ of the complex algebra $Up(\mathcal{F})$ of $\mathcal{F}$.*

Proof.

(1). Consider the complex algebra $Up(\mathcal{F}_A) := (Up(X_A), X_A, \emptyset, \cap, \cup, \neg, \sim)$ of the canonical frame $\mathcal{F}_A := (X_A, C, \subseteq)$ of $\mathcal{A} := (A, 1, 0, \wedge, \vee, \neg, \sim)$. Consider the map $h : A \rightarrow Up(X_A)$ such that for all $a \in A$, $h(a) := \{P \in X_A \mid a \in P\}$.

The restriction of Theorem 9(1) to distributive lattices implies that h is a monomorphism from the bounded distributive lattice $(A, 1, 0, \wedge, \vee, \neg)$ to the bounded distributive lattice $(Up(X_A), X_A, \emptyset, \cap, \cup, \rightarrow, \neg)$. It remains to show that h preserves $\sim$ and $\neg$, i.e.

$$h(\neg a) = \neg h(a) \text{ and } h(\sim a) = \sim h(a), \text{ for all } a \in A.$$

$h(\neg a) = \{P \in X_A \mid \neg a \in P\}$ and $\neg h(a) = \{P \in X_A \mid \forall Q \in X_A \ (P \subseteq Q \Rightarrow Q \notin h(a))\}$; $h(\sim a) = \{P \in X_A \mid \sim a \in P\}$ and $\sim h(a) = \{P \in X_A \mid \forall Q \in X_A \ (PCQ \Rightarrow Q \notin h(a))\}$.

$h(\neg a) \subseteq \neg h(a)$: let $P \in h(\neg a)$, i.e. $\neg a \in P$. Let $Q \in X_A$ such that $P \subseteq Q$ and $Q \in h(a)$. So, $\neg a, a \in Q$. Since $a \wedge \neg a = 0$, we have $0 \in Q$, a contradiction.

$\neg h(a) \subseteq h(\neg a)$: let $P \in \neg h(a)$. So, $P \notin h(a)$, i.e. $a \notin P$. Since $a \vee \neg a = 1$, $1 \in P$ and P is a prime filter, $\neg a \in P$, i.e. $P \in h(\neg a)$.

$h(\sim a) \subseteq \sim h(a)$: let $P \in h(\sim a)$, i.e. $\sim a \in P$. Let $Q \in X_A$ such that PCQ. PCQ and $\sim a \in P$ imply $a \notin Q$. So $Q \notin h(a)$.

$\sim h(a) \subseteq h(\sim a)$: let $P \notin h(\sim a)$, i.e. $\sim a \notin P$. Note that $a \neq 0$, because $\sim 0 = 1 \in P$. Consider the set $\Delta := \langle \{a\} \rangle$. Here, Δ is a proper filter, because $0 \notin \Delta$. Define $\Gamma := \{b \in A \mid \sim b \in P\}$, as in the proofs of previous results. Γ is $\vee$-closed. We claim that $\Gamma \cap \Delta = \emptyset$. Suppose not, i.e. there exists $b \in \Gamma \cap \Delta$. This means $a \leq b$ and $\sim b \in P$. $a \leq b$ implies $\sim b \leq \sim a$, i.e. $\sim a \in P$, a contradiction. Thus using Lemma 1, there exists a prime filter $Q' \in X_A$ such that $\Delta \subseteq Q'$ and $\Gamma \cap Q' = \emptyset$. Therefore, $a \in Q'$. Our final claim is PCQ'. Let $b \in A$ such that $\sim b \in P$. Then $b \in \Gamma \Rightarrow b \notin Q'$.

(2). Consider the canonical frame $(X_{Up(W)}, C_{Up(W)}, \subseteq)$ of the complex algebra $Up(\mathcal{F}) := (Up(W), W, \emptyset, \cap, \cup, \neg, \sim)$ of the sub-compatibility frame $\mathcal{F} := (W, C, \leq)$. The map $g : W \to X_{Up(W)}$ considered in Theorem 9(2), namely, such that for all $w \in W$, $g(w) := \{U \in Up(W) \mid w \in U\}$, works here as well.

g is already an embedding from $(W, \leq)$ to $(X_{Up(W)}, \subseteq)$. We have to only show that for all $w, w' \in W$, wCw' if and only if $g(w)C_{Up(W)}g(w')$ (cf. Definition 145).

$(\Rightarrow)$ Let wCw'. Let $P \in Up(W)$ such that $\sim P \in g(w)$ i.e. $w \in \sim P$. This implies that for all $v \in W$, $wCv \Rightarrow v \notin P$. wCw' implies $w' \notin P$, i.e. $P \notin g(w')$.

$(\Leftarrow)$ Let $w\cancel{C}w'$. To show that $g(w)\cancel{C}_{Up(W)}g(w')$, i.e. we have to find a $P \in Up(W)$ such that $w \in \sim P$ and $w' \in P$. Define $P := \{v \in W \mid w\cancel{C}v\}$.

Claim: P is the required upset. Let $v \in P$ such that $v \leq v'$. If $v' \notin P$ then wCv'. Using Condition (C), wCv', $w \leq w$ and $v \leq v'$ imply wCv, a contradiction, as $v \in P$. Therefore, $v' \in P$, and P is an upset.

Note that $w' \in P$, because $w\cancel{C}w'$. Further, for any $v'' \in W$, wCv'' implies $v'' \notin P$. So, $w \in \sim P$.

Using Propositions 91 and 92, we have that if the algebra $\mathcal{A}$ is a $K_{im\text{-}\vee}$-algebra then the complex algebra $Up(\mathcal{F}_A)$ is a $K_{im\text{-}\vee}$-algebra, and if the sub-compatibility frame $\mathcal{F} \in \mathfrak{F}_4'$ then the canonical frame $\mathcal{F}_{Up(\mathcal{F})} \in \mathfrak{F}_4'$. The same functions h and g will give the embeddings here as well.

7.4 Conclusions

In case of ILM and ILM-$\vee$, relational semantics have been defined through four classes of frames, namely $\mathfrak{F}_i$ and $\mathfrak{F}_i'$ ($i = 1, 2, 3, 4$) respectively (Sects. 6.2.1, 6.2.2 and 6.4). For K_{im} and $K_{im\text{-}\vee}$, the relational semantics has been defined with respect to the classes $\mathfrak{F}_4$ and $\mathfrak{F}_4'$ of sub-compatibility frames respectively. We can also define the semantics with respect to the other three classes of frames. For instance, let us consider $\mathfrak{F}_3$ and $\mathfrak{F}_3'$. In parity with the notation used in this section, we denote any frame of $\mathfrak{F}_3$ as $\mathcal{F} := (X, C, Y_0, \leq)$. In other words, we have changed the order of the triple and the relation R_{N_1} is replaced by C here. As expected for ILM and ILM-$\vee$ (cf. Sect. 6.4), $\mathfrak{F}_i$ and $\mathfrak{F}'_i$ should determine K_{im} and $K_{im-\vee}$ respectively, for $i = 1, 2, 3$. This fact, and the inter-translations given in Sect. 6.2.3 demonstrate the close connections between classes of frames and corresponding semantics that have been defined in different contexts. To get further insight into the connections, one may investigate categories formed by the frames along with appropriate morphisms.

Investigations of properties of negations have resulted in various schemes of logical systems. Some of the work in this direction may be found in [28,49,76,78] and more recently, in [23,33,70,71,86]. A common approach adopted is that a 'base logic with negation' is first defined, and new logics are obtained by adding axioms over the existing ones. This is followed by defining relational semantics for the base logic, and obtaining canonical properties for various properties of negation. Odintsov [71] studied the class of extensions of ML, presenting a diagram similar to Dunn's kite. In the direction of logics with two negations, Nelson's logic has been extensively studied - its extensions are discussed by Odintsov [69,71] and its subsystems by Vakarelov [86]. Another relevant and independent work on logics with two negations is done in [33]. Dunn proposed the logic K_-, over the same language as that of K_{im}, which is a BDLL having two negations $\sim, \neg$. The negation $\sim$ is preminimal, and $\neg$ satisfies certain axioms [33]. Various extensions of K_- are then defined and represented by the *united* kite of negations, obtained by combining the lopsided kite of negations with its dual version. In the united kite, the base logic is K_-, where the two negations are preminimal and dual preminimal. This opens up a different direction of study for schemes of logical systems and their semantics, where the logic at the base has two negations. Schemes may be developed for logics with and without implication, in which the base logics are extendable to existing systems with two negations, such as Vakarelov's SUBMIN [86], Nelson's logic, ILM, ILM-$\vee$, Dunn's K_-, K_{im} and $K_{im-\vee}$. For systems with implication, a possibility is to take both the negations in the base to be *basic subminimal*, as in the logic N of [23,67].

8   Conclusions and Future Work

Our work in this thesis can be divided into the following major parts.

1. Categories of rough sets, with focus on the topos-theoretic structures (Sect. 3).
2. A generalization $RSC(\mathscr{C})$ of the category RSC of rough sets, and its properties (Sect. 3).
3. Theory of transformation semigroups for rough sets and some decomposition results (Sect. 4).
4. Study of the internal algebra obtained through the quasitopos $RSC(\mathscr{C})$ (Sect. 5).
5. New algebraic structures called contrapositionally complemented pseudo-Boolean algebras - $ccpBa$ and $c\vee pBa$ (Sect. 5).
6. The logics ILM and ILM-$\vee$, corresponding to the above algebraic structures (Sect. 6).
7. The logics K_{im} and $K_{im-\vee}$, based on Dunn's logical framework (Sect. 7).

The work starts from known results on categories of rough sets mentioned in Sect. 2.3. In [4], two categories $ROUGH$ and ξ-$ROUGH$ of rough sets are defined, and it was shown that $ROUGH$ does not form a topos. Another category RSC of rough sets is defined in [57] and shown to not form a topos. Let us present our main results, sectionwise.

Section 3: This section focusses on categories of rough sets, and their topos-theoretic properties.

(a) RSC is equivalent to $ROUGH$, and thus, they share all category-theoretic properties between them.

(b) ξ-RSC is defined on similar lines as ξ-$ROUGH$, and is observed to be equivalent to ξ-$ROUGH$ and SET^2. Thus, ξ-$ROUGH$ forms a topos.

(c) A topos-theoretic generalization $RSC(\mathscr{C})$, where $\mathscr{C}$ is a non-degenerate (elementary) topos, is defined. In the special case when $\mathscr{C}$ is the category SET of sets, RSC is equivalent to $RSC(SET)$.

(d) $RSC(\mathscr{C})$ forms a quasitopos. All the constructions for finite limits, finite colimits, cartesian closedness, subobject classifier, and representation of partial morphism are explicitly obtained.

(e) A special case of $RSC(\mathscr{C})$ is when $\mathscr{C}$ is the topos **M-Set**. This results in the definition of *monoid actions on rough sets*.

(f) A generalization of the category ξ-$ROUGH$, namely ξ-$RSC(\mathscr{C})$, is defined. ξ-$RSC(\mathscr{C})$ is a topos if and only if $\mathscr{C}$ is a Boolean topos.

(g) RSC and $ROUGH$ also form quasitoposes.

(h) Some more categories of rough sets - RSC_1 and RSC_2 - are defined based on (i) other definitions of rough sets, and (ii) mappings between two rough sets. Their generalizations $RSC_1(\mathscr{C})$ and $RSC_2(\mathscr{C})$ are also defined.

Section 4: In this Section, we study *transformation semigroups (ts) for rough sets*, their algebraic properties, and some decomposition results.

(a) Taking motivation from the category $RSC($**M-Set**$)$ of monoid actions on rough sets, *ts* for rough sets is defined.

(b) The category **RTS** of *ts* for rough sets is observed to be a full subcategory of the category **TS** of *ts* for sets.

(c) Basic algebraic constructions in *ts* theory like resets, direct and wreath products, coverings, admissible partitions and quotients are obtained for *ts* for rough sets.

(d) Decomposition result for the reset *ts* for rough sets is presented. It is observed that it differs from that for reset *ts*. The basic entities in the decomposition are reset ts for rough sets of type $\overline{(1, \mathbf{n})}$.

(e) Decomposition results are also obtained for some special cases of *ts* for rough sets. For example, complete decomposition in case of transformation groups for rough sets is given.

(f) A semiautomaton for rough sets is defined, and we relate it with *ts* for rough sets, just as *ts* and algebraic automata are related.

(g) The structures - *ts* for rough sets and semiautomata for rough sets - differ from rough semi-automata and rough transformation semigroups defined in [9,79].

Section 5: This section is in continuation with Sect. 3. Since $RSC(\mathscr{C})$ forms a quasitopos, the set of strong subobjects of an $RSC(\mathscr{C})$-object forms a pseudo-Boolean algebra (*pBa*) with appropriate operations.

(a) The operations $\cap$, $\cup$, $\rightarrow$ and $\neg$ on strong subobjects of an $RSC(\mathscr{C})$-object are obtained.
(b) With the above operations, the set of strong subobjects of an $RSC(\mathscr{C})$-object forms a pseudo-Boolean algebra. In case of *RSC*, it forms a Boolean algebra.
(c) We observe that the notion of negation $\neg$ is actually *relative negation*, and it differs from the known notion of *relative rough complementation* $\sim$ given by Iwiński [48].
(d) Both the negations on strong subobjects are compared by giving an example of a medical information system.
(e) We incorporate Iwiński's negation in the algebraic structure obtained above, and obtain two new classes of algebras, namely *contrapositionally complemented pseudo Boolean algebras* (*ccpBa*) and *contrapositionally* $\vee$ *complemented pseudo Boolean algebras* (*c$\vee$cpBa*).
(f) Various properties and examples of *ccpBa* and *c$\vee$cpBa* are studied, with focus on the two negations $\neg$ and $\sim$.
(g) A comparison of these algebras with *pBa*'s and Nelson algebras is made.
(h) Representation theorems for *ccpBa* and *c$\vee$cpBa* are obtained with respect to certain restrictions of Esakia spaces.

Section 6: This section presents logics corresponding to the algebras *ccpBa* and *c$\vee$cpBa*, obtained in the previous section.

(a) The logics ILM and ILM-$\vee$ are defined, and shown to correspond to the classes of *ccpBa* and *c$\vee$cpBa* respectively.
(b) 'Interpretations' of ILM into IL and ML are given. Moreover, ML and IL both are observed to be *embedded* in ILM.
(c) Various equivalent logical systems are defined for ILM. One of them is found to be an extension of JP', Glivenko's logic.
(d) Next, we give two relational 'Kripke-style' semantics for ILM and ILM-$\vee$, where frames are of the type $(W, \leq, R_{N_1}, R_{N_2})$ (Došen's semantics) and $(W, \leq, Y_0)$ (Segerberg's semantics). Characterization results for ILM and ILM-$\vee$ with respect to Došen's semantics are obtained using the concept of prime theories.
(e) For Segerberg's semantics, sub-normal frames are used. The characterization results for ILM and ILM-$\vee$ with respect to sub-normal frames are obtained using inter-translatability of frames.

(f) Any $ccpBa$ $(c\lor cpBa)$ $\mathcal{A}$ is shown to be embedded into the $ccpBa$ $(c\lor cpBa)$ formed by the 'canonical' frame $\mathcal{F}$ of $\mathcal{A}$ (the 'complex algebra' of $\mathcal{F}$). Similarly, a sub-normal frame $\mathcal{F}$, satisfying certain conditions, is shown to be embeddable into the canonical frame of the complex algebra of $\mathcal{F}$.

Section 7: This section studies logics with negation - K_{im} and $K_{im-\lor}$ - following Dunn's logical frameworks.

(a) The logics K_{im} and $K_{im-\lor}$ are defined, and shown to correspond to certain classes of bounded distributive lattices with two negations. These algebras are called K_{im}-algebras and $K_{im-\lor}$-algebras respectively.
(b) The reduct of any $ccpBa$ obtained by removing implication $\to$, is a K_{im}-algebra. However, not every K_{im}-algebra can be extended to a $ccpBa$. It is concluded that even though they are not the same algebraic classes, the negation properties of $ccpBa$s are sufficiently captured in K_{im}-algebras.
(c) Relational semantics for K_{im} and $K_{im-\lor}$ are obtained, using Dunn's compatibility frames. The characterizing conditions on compatibility frames for these logics are again obtained using the concept of prime theories.
(d) Just as obtained in the previous section for ILM and ILM-$\lor$, a natural connection between the relational and algebraic semantics of these logics is given.

8.1 Future Work

We have raised questions and indicated directions of future work at the end of every section, giving some details thereof. We end the thesis with some broad questions.

We started with some known categories of rough sets, and studied their properties. Some new categories, along with their generalizations, have also been defined. All these categories have objects as pairs of the form (X_1, X_2) or the triple (U, X_1, X_2). However, in the first category-theoretic work on rough sets by Banerjee and Chakraborty [4], an object in the categories $ROUGH$ and ξ-$ROUGH$ was defined as a triple $\mathcal{U} := (U, R, X)$, including the equivalence relation R in an approximation space (U, R) on which the rough set $\mathcal{U}$ is based. There are concepts of generalizations of (equivalence) relations in category theory. So we have the following.

Open Question 1. *Can generalizations of ROUGH and ξ-ROUGH be given in category-theoretic terminology?*

Of the various categories defined in Sect. 3, not all are studied in details. For example, we have the following question.

Open Question 2. *Does $RSC_2(\mathscr{C})$ form a topos or quasitopos?*

In *ts* for rough sets, we have already mentioned some open questions related to the decomposition results. We know that for semigroups, groups and *ts*, there are decomposition results, which help in identifying their basic algebraic structures.

Open Question 3. *Can a complete decomposition of ts for rough sets be obtained?*

A more application-based direction of work that emerges from *ts* theory for rough sets, is through the notion of semiautomata for rough sets. Applications of semiautomaton or automata theory are far and varied in computer science and biological systems. Some of these areas have applications in rough sets too.

Open Question 4. *Can semiautomata for rough sets be used in applications, such as in the field of cellular automata?*

The relational semantics studied for ILM are based on either Došen's N-frames or Segerberg's j-frames. For constructive logic with strong negation, the logic corresponding to the class of Nelson algebras, Kripke-style models are defined by [44]. Since, Nelson algebra involves two negations just as *ccpBas* or K_{im}-algebras do, one may ask the following question.

Open Question 5. *What other relational semantics are possible for* ILM, ILM-$\vee$, K_{im} *and* $K_{im-\vee}$?

Developing various schemes of logical systems, based on the properties of negation, is an active area of research. Some recent examples can be found in [23, 28, 33, 71, 86]. A common approach adopted is that a 'base logic with negation' is first defined, and new logics are obtained by adding axioms over the existing ones. This is followed by defining relational semantics for the base logic, and obtaining canonical properties for various properties of negation.

Open Question 6. *Can a scheme for logics with two negations, with or without implication, be given such that the base logics are extendable to existing systems with two negations?*

Acknowledgements. The paper is the full version of my phd dissertation supervised by Prof. Mohua Banerjee, approved in June 2019 by the Department of Mathematics and Statistics, Indian Institute of Technology Kanpur, India.

I express my deepest gratitude to Prof. Mohua Banerjee for her invaluable time and effort. Her insights in logic and rough sets were instrumental in assembling my research work in the form of this thesis. She taught me how to approach a research problem and meaningfully express the results. She always encouraged me to attend conferences and exchange ideas with other researchers. I look forward to her guidance and inspiration throughout my life.

I am indebted to all the teachers at IIT Kanpur both for their teaching and discussions I have had with them. I am heartily thankful to Prof. Mihir Chakraborty for the discussion sessions and inputs during the various stages during the PhD. I am grateful to all the members of the Indian Logic group, especially Prof. Sujata Ghosh, for inviting me to the annual logic meets and conferences in India. I have highly benefited from the interactions with Prof. Amit Kuber in the last two years, both as a friend and logician.

I would like to thank the Dean of Research and Development, IIT Kanpur and all the faculty and staff members of Department of Mathematics and Statistics for their unconditional co-operation and support, including the Heads - Prof. Debasis Kundu, Prof. Sobha Madan and Prof. Arbind Lal, during my PhD period.

I acknowledge the Council of Scientific and Industrial Research (CSIR), India and IIT Kanpur for providing me scholarship during my tenure as a Ph.D. student. I am also indebted to Indo-European Research Training Network in Logic (IERTNiL) and Association of Symbolic Logic (ASL) for providing research travel grants to Europe in 2015 and 2017. In addition, I would also like to thank Prof. Dominik Ślęzak, Prof. Andrzej Skowron, and Prof. Lech Polkowski for hosting me during these two European visits. I also acknowledge the assistance I got from Dr. Jose Luis Castiglioni and Dr. Zhen Lin through discussions.

This acknowledgement would be incomplete without acknowledging my family for the support that they have provided me throughout. My special thanks is to my mother who is a perennial source of encouragement. They have been such an immovable shield for me especially at times of difficulty. This thesis becomes a reality with the kind support and help of many individuals. I would like to extend my sincere thanks to all of them.

Appendix A   List of abbreviations

$ccpBa$ contrapositionally complemented pseudo-Boolean algebra
$c\lor cpBa$ contrapositionally $\lor$ complemented pseudo-Boolean algebra
cc contrapositionally complemented
$c\lor c$ contrapositionally $\lor$ complemented
CL classical logic
DLL distributive lattice logic
$DNE(\sim\top)$ double negation elimination for $\sim\top$
DT deduction theorem
IL intuitionistic logic
ILM intuitionistic logic with minimal negation
ML minimal logic
MP modus ponens
pBa pseudo-Boolean algebra
PF(Q) set of partial functions on Q
PL positive logic
$poset$ partially ordered set
PV set of propositional variables
rpc relatively pseudo-complemented
tg transformation group
tm transformation monoid
ts transformation semigroup

Appendix B   List of symbols

i or ! : $0 \to 1$ unique arrow from initial object 0 to terminal object 1
$!_A : A \to 1$ unique arrow from $\mathscr{C}$-object A to terminal object 1
$i_A : 0 \to A$ unique arrow from initial object 0 to $\mathscr{C}$-object A

$\vdash_L \alpha$ α is a theorem in logic L

$\circ_{\mathcal{A}}$ operator on an algebra $\mathcal{A}$

$\cap : \Omega \times \Omega \to \Omega$ characteristic morphism of $\langle \top, \top \rangle$

$\cup : \Omega \times \Omega \to \Omega$ characteristic morphism of $[\langle \top_\Omega, 1_\Omega \rangle, \langle 1_\Omega \rangle, \top_\Omega \rangle]$

$\to : \Omega \times \Omega \to \Omega$ characteristic morphism of equalizer $e : E \to \Omega \times \Omega$

$\neg : \Omega \to \Omega$ characteristic morphism of $\bot : 1 \to \Omega$

1_A identity arrow on a $\mathscr{C}$-object A

$\alpha \vdash \beta$ consequence pair or sequent (α, β)

$|A|$ number of elements in a finite set A

$\mathcal{A}^\bullet$ (a). $ts\ (Q, S^\bullet)$, where $\mathcal{A} := (Q, S)$

  (b). $ts\ (\mathcal{U}, S^\bullet)$ for rough set, where $\mathcal{A} := (\mathcal{U}, S)$

$\overline{\mathcal{A}}$ (a). $(Q, \langle S \cup \{\overline{q} \mid q \in Q\} \rangle)$, closure of a $ts\ \mathcal{A} := (Q, S)$

  (b). $(\mathcal{U}, S')$, where $S' := \langle S \cup \{\overline{q} \mid q \in \underline{\mathcal{X}}\} \cup \{\widetilde{q} \mid q \in \overline{\mathcal{X}} \backslash \underline{\mathcal{X}}\} \rangle$

$\mathcal{A}_P$ restriction $ts\ (P, T/\simeq)$ (Observation 6)

$\mathcal{A}|P$ restriction $ts\ (P, S')$ (Observation 6)

$\mathcal{A}_u$ $\{(a_1, a_2) \in \mathcal{H}^{[2]} : a_2 \leq u_2\ \text{and}\ a_1 = a_2 \wedge u_1\}$

$\mathcal{A} \simeq \mathcal{B}$ ts homomorphism (for rough sets)

$\mathcal{A} \preccurlyeq \mathcal{B}$ covering of $ts\ \mathcal{A}$ by $\mathcal{A}$

$\mathcal{A} \times \mathcal{B}$ direct product of two ts $\mathcal{A}$ and $\mathcal{B}$ (for rough sets)

$\mathcal{A} \circ \mathcal{B}$ wreath product of two ts $\mathcal{A}$ and $\mathcal{B}$ (for rough sets)

$\mathcal{A}/\langle \pi \rangle$ quotient ts (for a rough set)

$A \simeq B$ A is isomorphic to B, where A and B are $\mathscr{C}$-objects

$(A \times B, p_1, p_2)$ product of two $\mathscr{C}$-objects A and B

$\mathcal{A}, v \vDash \alpha$ α is true for the valuation v on algebra $\mathcal{A}$

$\mathcal{A} \vDash \alpha$ $\mathcal{A}, v \vDash \alpha$ for each valuation v on algebra $\mathcal{A}$

$(A + B, i_1, i_2)$ coproduct of two $\mathscr{C}$-objects A and B

  In SET, $A + B = \{(a, 0) \mid a \in A\} \cup \{(b, 1) \mid b \in B\}$.

$A \cap B$ domain of $f \cap g$ where $dom(f) = A$ and $dom(g) = B$

$A \cup B$ domain of $f \cup g$ where $dom(f) = A$ and $dom(g) = B$

$A \to B$ domain of $f \to g$ where $dom(f) = A$ and $dom(g) = B$

$\neg A$ domain of $\neg f$ where $dom(f) = A$

$Arr(\mathscr{C})$ the class of arrows in category $\mathscr{C}$

$\mathcal{B}_4$ 4-element Boolean algebra

B^A exponent object in $\mathscr{C}$. In SET, $B^A := \{f \mid f : A \to B\}$

$\bot : 1 \to \Omega$ characteristic morphism of $! : 0 \to 1$

(C) If $x' \leq x$, $y' \leq y$ and xCy then $x'Cy'$

$\mathscr{C}$ a category

$\mathscr{C}^\to$ the arrow category of $\mathscr{C}$

$\mathscr{C}^2$ product category $\mathscr{C} \times \mathscr{C}$

$\mathscr{C}$-arrow an arrow in the category $\mathscr{C}$

$\mathscr{C}$-object an object in the category $\mathscr{C}$

$\mathscr{C} \times \mathscr{D}$ product of two categories $\mathscr{C}$ and $\mathscr{D}$

$\mathscr{C} \simeq \mathscr{D}$ equivalence between two categories $\mathscr{C}$ and $\mathscr{D}$

$cod(f)$ codomain of an arrow $f : A \to B$ in $\mathscr{C}$

$CpUp(X)$ set of clopen upsets in an ordered topological space $(X, \tau, \leq)$ of X

$dc(\Delta)$ disjunctive closure of set Δ of formulas

$\mathcal{D}(X)$ algebra on the set $CpUp(X)$ of clopen upsets

$dom(f)$ domain of an arrow $f : A \to B$ in $\mathscr{C}$

$ev : B^A \times A \to B$ evaluation map for the exponent object B^A in $\mathscr{C}$

$(\mathcal{F}, v_0)$ a model on the frame $\mathcal{F}$

$\mathfrak{F}_1$ class of $\hat{N}$-frames

$\mathfrak{F}_2$ class of sub-normal frames

$\mathfrak{F}_4$ class of sub-compatibility frames

$\langle F \rangle$ filter generated by a set F

F set of well-formed formulas

$F : \mathscr{C} \to \mathscr{D}$ functor between two categories $\mathscr{C}$ and $\mathscr{D}$

$f \times g$ $\langle f \circ p_1, g \circ p_2 \rangle : A \times B \to C \times D$

$f + g$ $[i_3 \circ f, i_4 \circ g] : A + B \to C + D$

$f : A \to B$ an arrow in $\mathscr{C}$ or a set (partial) function from A to B

$f \cap g$ monic for which characteristic morphism is $\cap \circ \langle \chi_f, \chi_g \rangle$

$f \cup g$ monic for which characteristic morphism is $\cup \circ \langle \chi_f, \chi_g \rangle$

$f \to g$ monic for which characteristic morphism is $\to \circ \langle \chi_f, \chi_g \rangle$

$\neg f$ monic for which characteristic morphism is $\neg \circ \chi_f$

$\Gamma \vdash_{\mathrm{L}} \alpha$ α is provable from Γ in L

$g \circ f$ composition of $\mathscr{C}$-arrows $f : A \to B$ and $g : B \to C$

$\mathcal{G}(X)$ algebra on open subsets of a topological space X

$\mathcal{H}_6$ 6-element pseudo-Boolean algebra

$\mathcal{H}^{[2]}$ $\{(a, b) : a \leq b, a, b \in H\}$

$Hom_{\mathscr{C}}(A, B)$ collection of all arrows between two $\mathscr{C}$-objects A and B

$Im(f)$ image of $\mathscr{C}$-arrow $f : A \to B$

K_i bounded distributive lattice logic with preminimal negation

K_i^m bounded distributive lattice logic with minimal negation

$\mathcal{L}$ the language of the logic L

$\mathrm{L} \cong \mathrm{L}'$ logic L is equivalent to logic L'

$(\mathbf{m}, \mathbf{n})$ reset ts for a rough set

$\mathcal{M}, w \vDash \phi$ ϕ is true at the world $w \in W$ in the model $\mathcal{M}$

$\mathcal{M}, w \nvDash \phi$ ϕ is not true at the world $w \in W$ in the model $\mathcal{M}$

$\mathbf{M\text{-}Set}$ category of monoid actions

$\mathbf{n}$ transformation semigroup $(Q, \emptyset)$, where $|Q| = n$

$\mathbb{N}$ set of natural numbers

$\nu_B : B \to \tilde{B}$ $\mathscr{C}$-arrow that represents partial morphisms with codomain B

$Obj(\mathscr{C})$ the class of objects in a category $\mathscr{C}$

Ω subobject in a topos/quasitopos

(P) $((\alpha \to \beta) \to \alpha) \to \alpha$

(P') $\sim\sim(\sim\top \to \beta)$

π partition of a set in a ts $\mathcal{A}$

(π_1, π_2) admissible partition on a rough set $\mathcal{A}$

$\pi \cap \tau = 1_Q$ $|H_i \cap K_j| \leq 1 \ \forall \ i \in I, j \in J$, for $\pi := \{H_i\}_{i \in I}, \tau := \{K_j\}_{j \in J}$

$\mathcal{P}(A)$ power set of A

Ps $\{qs \mid q \in P$ and qs is defined$\}$, where $P \subseteq Q$

(Q, G) transformation group

(Q, M) transformation monoid

(Q, S) transformation semigroup

(Q, Σ, Δ) semiautomaton

$\overline{q}$ constant function from Q to Q, mapping elements of Q to q

$\widetilde{q}$ $\widetilde{q}(x) := q$ if $x \in \overline{\mathcal{X}} \backslash \underline{\mathcal{X}}$, and undefined otherwise, where $q \in \overline{\mathcal{X}}$

$[r, s]$ Fig. 7b

RTS category of ts for rough sets

$\langle S \rangle$ semigroup generated by set S

qs $\delta(q, s)$, where δ is the partial semigroup action

$S^{\cdot}$ $S \cup \{1_S\}$, where 1_S is the identity function on S

SET category of sets

$\sigma(a)$ the set of prime filters containing $a \in A$

$Sub(D)$ $\{[f] \mid f$ is a monic (strong monic) with codomain $D\}$

$sub(\alpha)$ the set of subformulas of α

$S/{\simeq}$ quotient semigroup of S, induced by the equivalence relation $\simeq$

$Th(\Delta)$ theory generated by a set Δ of formulas

(T, p_A, p_B) pullback of $\mathscr{C}$-arrows $f : A \to C$ and $g : B \to C$

$\top : 1 \to \Omega$ $\mathscr{C}$-arrow 'true' associated with the subobject classifier Ω

$\top_\Omega$ $\top \circ !_\Omega : \Omega \to \Omega$

TS category of transformation semigroups

t_s cover of $s \in S$

U/R the quotient set of U with respect to equivalence relation R

$(\mathcal{U}, S)$ transformation semigroup for a rough set

$(\mathcal{U}, \Sigma, \delta)$ semiautomata for a rough set

$\mathcal{U}(\text{L})$ Lindenbaum-Tarski algebra for the logic L

(U, R) an approximation space

(U, R, X) a rough set

$\mathcal{U}_\pi$ (U, R_π, X_π) - the quotient ts for rough set

$\langle u, v \rangle$ Fig. 7a

$Up(W)$ set of upsets of W

$(W, \leq, Y_0)$ a $j\text{-}frame$ or a sub-normal frame

$(W, \leq)$ a normal frame

$(W, C, \leq)$ compatibility frame

(W, R_I, R_N) N-frame

$[x]$ equivalence class containing x

$[x]_R$ equivalence class containing x for equivalence relation R

$\overline{\mathcal{X}}$ $\{[x] \mid [x] \cap X \neq \emptyset\} \subseteq U/R$

$\underline{\mathcal{X}}$ $\{[x] \mid [x] \subseteq X\} \subseteq U/R$

$\overline{X}$ $\bigcup \overline{\mathcal{X}}$ - upper approximation of set X

$\underline{X}$ $\bigcup \underline{\mathcal{X}}$ - lower approximation of set X

$(X, \leq, R_{N_1}, R_{N_2})$ $\hat{N}$-frame

$(X^c, \subseteq, R^c_{N_1}, R^c_{N_2})$ canonical $\hat{N}$-frame

$(\underline{X}, \overline{X})$ a rough set

X_A the set of prime filters in a distributive lattice $\mathcal{A} := (A, \vee, \wedge)$

$(X_A, \subseteq, Y_0)$ canonical j-frame

$(X, \tau, \leq)$ ordered topological space τ on X

$\chi_m : D \to \Omega$ characteristic arrow of monic arrow $m : A \to D$

$\quad \mathbb{Z}$ set of integers

$\quad \uparrow Z$ upset generated by Z

$\quad \downarrow Z$ downset generated by Z

References

1. Arbib, M.A.: Theories of Abstract Automata. Prentice-Hall, Hoboken (1969)
2. Arbib, M.A., Krohn, K., Rhodes, J.L.: Algebraic Theory of Machines, Languages, and Semigroups. Academic Press, Cambridge (1968)
3. Awodey, S.: Category Theory. Oxford University Press, Oxford (2010)
4. Banerjee, M., Chakraborty, M.K.: A category for rough sets. Found. Comput. Decis. Sci. **18**(3–4), 167–180 (1993)
5. Banerjee, M., Chakraborty, M.K.: Rough sets through algebraic logic. Fund. Inform. **28**(3–4), 211–221 (1996)
6. Banerjee, M., Chakraborty, M.K.: Foundations of vagueness: a category-theoretic approach. Electron. Notes Theor. Comput. Sci. **82**(4), 10–19 (2003)
7. Banerjee, M., Chakraborty, M.K.: Algebras from rough sets. In: Pal, S.K., Polkowski, L., Skowron, A. (eds.) Rough-Neural Computing: Techniques for Computing with Words, pp. 157–184. Springer, Heidelberg (2004). https://doi.org/10.1007/978-3-642-18859-6_7
8. Banerjee, M., Yao, Y.: A categorial basis for granular computing. In: An, A., Stefanowski, J., Ramanna, S., Butz, C.J., Pedrycz, W., Wang, G. (eds.) RSFDGrC 2007. LNCS (LNAI), vol. 4482, pp. 427–434. Springer, Heidelberg (2007). https://doi.org/10.1007/978-3-540-72530-5_51
9. Basu, S.: Rough finite-state automata. Cybern. Syst. **36**(2), 107–124 (2005)
10. Bavel, Z.: The source as a tool in automata. Inf. Control **18**(2), 140–155 (1971)
11. Bezhanishvili, G., Holliday, W.H.: A semantic hierarchy for intuitionistic logic. Indagationes Mathematicae **30**(3), 403–469 (2019)
12. Bezhanishvili, N.: Lattices of intermediate and cylindric modal logics. Ph.D. thesis, Institute for Logic, Language and Computation, University of Amsterdam (2006)
13. Borzooei, R.A., Estaji, A.A., Mobini, M.: On the category of rough sets. Soft. Comput. **21**(9), 2201–2214 (2017)
14. Carnielli, W.A., D'Ottaviano, I.M.L.: Translations between logical systems: a manifesto. Logique et Anal. **40**(157), 67–81 (1997)
15. Cattaneo, G., Giuntini, R., Pilla, R.: BZMV^{dM} algebras and Stonian MV-algebras (applications to fuzzy sets and rough approximations). Fuzzy Sets Syst. **108**(2), 201–222 (1999)
16. Celani, S.A., Jansana, R.: Easkia duality and its extensions. In: Bezhanishvili, G. (ed.) Leo Esakia on Duality in Modal and Intuitionistic Logics. OCL, vol. 4, pp. 63–98. Springer, Dordrecht (2014). https://doi.org/10.1007/978-94-017-8860-1_4
17. Chagrov, A., Zakharyaschev, M.: Modal Logic. Clarendon Press, Oxford (1997)
18. Chomsky, N.: Three models for the description of language. IRE Trans. Inf. Theory **2**(3), 113–124 (1956)
19. Cignoli, R.: The algebras of Łukasiewicz many-valued logic: a historical overview. In: Aguzzoli, S., Ciabattoni, A., Gerla, B., Manara, C., Marra, V. (eds.) Algebraic and Proof-theoretic Aspects of Non-classical Logics. LNCS (LNAI), vol. 4460, pp. 69–83. Springer, Heidelberg (2007). https://doi.org/10.1007/978-3-540-75939-3_5

20. Clark, D.M., Davey, B.A.: Natural Dualities for the Working Algebraist. Cambridge Studies in Advanced Mathematics, vol. 57. Cambridge University Press, Cambridge (1998)
21. Clifford, A.H., Preston, G.B.: The Algebraic Theory of Semigroups, vol. 1. American Mathematical Society, Providence (1961)
22. Clifford, A.H., Preston, G.B.: The Algebraic Theory of Semigroups, vol. 2. American Mathematical Society, Providence (1961)
23. Colacito, A., de Jongh, D., Vargas, A.L.: Subminimal negation. Soft Comput. **21**(1), 165–174 (2017)
24. Davey, B.A., Priestley, H.A.: Introduction to Lattices and Order. Cambridge University Press, Cambridge (2002)
25. Diker, M.: Categories of rough sets and textures. Theoret. Comput. Sci. **488**, 46–65 (2013)
26. Diker, M.: A category approach to relation preserving functions in rough set theory. Int. J. Approx. Reason. **56**, 71–86 (2015)
27. Došen, K.: Negation as a modal operator. Rep. Math. Logic **20**, 15–27 (1986)
28. Došen, K.: Negation in the light of modal logic. In: Gabbay, D.M., Wansing, H. (eds.) What is Negation? Applied Logic Series, vol. 13, pp. 77–86. Springer, Dordrecht (1999). https://doi.org/10.1007/978-94-015-9309-0_4
29. Dunn, J.M.: Gaggle theory: an abstraction of Galois connections and residuation, with applications to negation, implication, and various logical operators. In: van Eijck, J. (ed.) JELIA 1990. LNCS, vol. 478, pp. 31–51. Springer, Heidelberg (1991). https://doi.org/10.1007/BFb0018431
30. Dunn, J.M.: Star and perp: two treatments of negation. Philos. Perspect. **7**, 331–357 (1993)
31. Dunn, J.M.: Positive modal logic. Stud. Logica. **55**(2), 301–317 (1995)
32. Dunn, J.M.: Generalized ortho negation. In: Wansing, H. (ed.) Negation: A Notion in Focus, pp. 3–26. W. De Gruyter, Berlin (1996)
33. Dunn, J.M., Zhou, C.: Negation in the context of gaggle theory. Stud. Logica. **80**(2–3), 235–264 (2005)
34. Eilenberg, S., Tilson, B.: Automata, Languages, and Machines. Volume B. Pure & Applied Mathematics. Academic Press, Cambridge (1976)
35. Eklund, P., Galán, M.A.: Monads can be rough. In: Greco, S., et al. (eds.) RSCTC 2006. LNCS (LNAI), vol. 4259, pp. 77–84. Springer, Heidelberg (2006). https://doi.org/10.1007/11908029_9
36. Esteva, F., Godo, L., Hájek, P., Navara, M.: Residuated fuzzy logics with an involutive negation. Arch. Math. Logic **39**(2), 103–124 (2000)
37. Ferreira, G., Oliva, P.: On the relation between various negative translations. In: Logic, Construction, Computation. Ontos Mathematical Logic, vol. 3, pp. 227–258. Ontos Verlag, Heusenstamm (2012)
38. Fu, T.K., Kutz, O.: The analysis and synthesis of logic translation. In: FLAIRS Conference (2012)
39. Geisler, J., Nowak, M.: Conditional negation on the positive logic. Bull. Sect. Logic **23**(3), 130–136 (1994)
40. Ginsburg, S.: Some remarks on abstract machines. Trans. Am. Math. Soc. **96**(3), 400–444 (1960)
41. Goguen, J.A.: Concept representation in natural and artificial languages: axioms, extensions and applications for fuzzy sets. Int. J. Man Mach. Stud. **6**(5), 513–561 (1974)
42. Goldblatt, R.I.: Decidability of some extensions of. J. Z. Math. Logik Grundlagen Math. **20**, 203–205 (1974)

43. Goldblatt, R.I.: Topoi: The Categorial Analysis of Logic. Dover Books on Mathematics. Dover Publications, Mineola (2006)
44. Gurevich, Y.: Intuitionistic logic with strong negation. Stud. Logica. **36**(1), 49–59 (1977)
45. Höhle, U., Stout, L.N.: Foundations of fuzzy sets. Fuzzy Sets Syst. **40**(2), 257–296 (1991)
46. Holcombe, W.M.L.: Algebraic Automata Theory. Cambridge University Press, Cambridge (1982)
47. Hopcroft, J.E., Motwani, R., Ullman, J.D.: Introduction to Automata Theory, Languages, and Computation. Pearson/Addison Wesley, Reading (2007)
48. Iwiński, T.B.: Algebraic approach to rough sets. Bull. Polish Acad. Sci. Math. **35**, 673–683 (1987)
49. Johansson, I.: Der Minimalkalkül, ein reduzierter intuitionistischer Formalismus. Compositio Math. **4**, 119–136 (1937)
50. Johnstone, P.T.: Stone Spaces, vol. 3. Cambridge University Press, Cambridge (1986)
51. Johnstone, P.T.: Sketches of an Elephant: A Topos Theory Compendium, vol. 2. Oxford University Press, Oxford (2002)
52. Johnstone, P.T., Lack, S., Sobociński, P.: Quasitoposes, quasiadhesive categories and artin glueing. In: Mossakowski, T., Montanari, U., Haveraaen, M. (eds.) CALCO 2007. LNCS, vol. 4624, pp. 312–326. Springer, Heidelberg (2007). https://doi.org/10.1007/978-3-540-73859-6_21
53. Kiszka, J.B., Gupta, M.M., Trojan, G.M.: Multivariable fuzzy controller under Gödel's implication. Fuzzy Sets Syst. **34**(3), 301–321 (1990)
54. Kripke, S.A.: Semantical analysis of modal logic. I. Normal modal propositional calculi. Z. Math. Logik Grundlagen Math. **9**, 67–96 (1963)
55. Kripke, S.A.: Semantical analysis of intuitionistic logic. I. In: Formal Systems and Recursive Functions (Proceedings of the Eighth Logic Colloquium, Oxford, 1963), pp. 92–130. North-Holland, Amsterdam (1965)
56. Kumar, A., Banerjee, M.: Kleene algebras and logic: boolean and rough set representations, 3-valued, rough set and perp semantics. Stud. Logica. **105**(3), 439–469 (2017)
57. Li, X.S., Yuan, X.H.: The category *RSC* of *I*-rough sets. In: Fifth International Conference on Fuzzy Systems and Knowledge Discovery, vol. 1, pp. 448–452 (2008)
58. Lindenmayer, A.: Mathematical models for cellular interactions in development I. Filaments with one-sided inputs. J. Theor. Biol. **18**(3), 280–299 (1968)
59. Linton, S.A., Pfeiffer, G., Robertson, E.F., Ruškuc, N.: Groups and actions in transformation semigroups. Math. Z. **228**(3), 435–450 (1998)
60. Lu, J., Li, S.-G., Yang, X.-F., Fu, W.-Q.: Categorical properties of M-indiscernibility spaces. Theoret. Comput. Sci. **412**(42), 5902–5908 (2011)
61. Mikolajczak, B.: Algebraic and Structural Automata Theory, vol. 44. Elsevier, Amsterdam (1991)
62. Monro, G.: Quasitopoi, logic and Heyting-valued models. J. Pure Appl. Algebra **42**(2), 141–164 (1986)
63. More, A.K., Banerjee, M.: Categories and algebras from rough sets: new facets. Fund. Inform. **148**(1–2), 173–190 (2016)
64. More, A.K., Banerjee, M.: New algebras and logic from a category of rough sets. In: Polkowski, L., et al. (eds.) IJCRS 2017. LNCS (LNAI), vol. 10313, pp. 95–108. Springer, Cham (2017). https://doi.org/10.1007/978-3-319-60837-2_8

65. More, A.K., Banerjee, M.: Transformation semigroups for rough sets. In: Nguyen, H.S., Ha, Q.-T., Li, T., Przybyła-Kasperek, M. (eds.) IJCRS 2018. LNCS (LNAI), vol. 11103, pp. 584–598. Springer, Cham (2018). https://doi.org/10.1007/978-3-319-99368-3_46

66. Munkres, J.R.: Topology. Prentice Hall, Hoboken (2000)

67. Niki, S.: Subminimal logics in light of Vakarelov's logic. Stud. Logica. **108**(5), 967–987 (2020)

68. Nowak, M.: The weakest logic of conditional negation. Bull. Sect. Logic **24**(4), 201–205 (1995)

69. Odintsov, S.P.: The class of extensions of Nelson's paraconsistent logic. Studia Logica **80**(2–3), 291–320 (2005)

70. Odintsov, S.P.: The lattice of extensions of the minimal logic. Siberian Adv. Math. **17**(2), 112–143 (2007)

71. Odintsov, S.P.: Constructive Negations and Paraconsistency. Trends in Logic-Studia Logica Library, vol. 26. Springer, New York (2008). https://doi.org/10.1007/978-1-4020-6867-6

72. Pagliani, P.: Rough set theory and logic-algebraic structures. In: Orłowska, E. (ed.) Incomplete Information: Rough Set Analysis, pp. 109–190. Physica-Verlag, Heidelberg (1998). https://doi.org/10.1007/978-3-7908-1888-8_6

73. Polkowski, L.: Rough Sets: Mathematical Foundations. Advances in Intelligent and Soft Computing, Physica-Verlag HD, Heidelberg (2002). https://doi.org/10.1007/978-3-7908-1776-8

74. Prawitz, D., Malmnäs, P.E.: A survey of some connections between classical, intuitionistic and minimal logic. Stud. Logic Found. Math. **50**, 215–229 (1968)

75. Rasiowa, H.: An Algebraic Approach to Non-classical Logics. Studies in Logic and the Foundations of Mathematics, North-Holland Publishing Company, Amsterdam (1974)

76. Rasiowa, H., Sikorski, R.: Algebraic treatment of the notion of satisfiability. Fundam. Math. **40**, 62–95 (1953)

77. Restall, G.: Defining double negation elimination. Log. J. IGPL **8**(6), 853–860 (2000)

78. Segerberg, K.: Propositional logics related to Heyting's and Johansson's. Theoria **34**, 26–61 (1968)

79. Sharan, S., Srivastava, A.K., Tiwari, S.P.: Characterizations of rough finite state automata. Int. J. Mach. Learn. Cybern. **8**(3), 721–730 (2017)

80. Shramko, Y.: Dual intuitionistic logic and a variety of negations: the logic of scientific research. Stud. Logica. **80**(2–3), 347–367 (2005)

81. Tiwari, S.P., Sharan, S.: On coverings of rough transformation semigroups. In: Kuznetsov, S.O., Ślęzak, D., Hepting, D.H., Mirkin, B.G. (eds.) RSFDGrC 2011. LNCS (LNAI), vol. 6743, pp. 79–86. Springer, Heidelberg (2011). https://doi.org/10.1007/978-3-642-21881-1_14

82. Tiwari, S.P., Sharan, S., Singh, A.K.: On coverings of products of rough transformation semigroups. Int. J. Found. Comput. Sci. **24**(03), 375–391 (2013)

83. Tripathy, B.K., Acharjya, D.P., Cynthya, V.: A framework for intelligent medical diagnosis using rough set with formal concept analysis. Int. J. Artif. Intell. Appl. **2**(2), 45–66 (2011)

84. Vakarelov, D.: Notes on $\mathcal{N}$-lattices and constructive logic with strong negation. Studia Logica **36**(1–2), 109–125 (1977)

85. Vakarelov, D.: Consistency, completeness and negation. In: Priest, G., Routley, R., Norman, J. (eds.) Paraconsistent Logic: Essays on the Inconsistent, pp. 328–369. Philosophia Verlag, Munich (1989)

86. Vakarelov, D.: Nelson's negation on the base of weaker versions of intuitionistic negation. Stud. Logica. **80**(2–3), 393–430 (2005)
87. Woodruff, P.W.: A note on JP′. Theoria **36**(2), 183–184 (1970)
88. Wyler, O.: Lecture Notes on Topoi and Quasitopoi. World Scientific, Singapore (1991)

Author Index